Berichtigung

Auf der Seite **S 1** des Titelblattes des Symposiums "Genetics of Seed Proteins" muß es

statt **Vol. 32**
richtig **Vol. 32/I**

heißen.

2052/32 Kulturpflanze

Die Kulturpflanze

Mitteilungen aus dem Zentralinstitut

für Genetik und Kulturpflanzenforschung

Gatersleben

der Akademie der Wissenschaften der DDR

Band 32

Herausgegeben von

H. Böhme, D. Mettin, W. R. Müller-Stoll,

K. Müntz, R. Rieger, A. Rieth, F. Scholz,

H. Stubbe

Schriftleitung: H. Böhme

Redaktion: I. Neumann

Mit 74 Abbildungen, 16 Tabellen und 5 Tafeln
und 87 Abbildungen und 34 Tabellen des Symposiums

AKADEMIE-VERLAG · BERLIN

1984

ISSN 0075-7209

Erschienen im Akademie-Verlag, DDR - 1086 Berlin, Leipziger Straße 3–4

Lizenznummer: 202 · 100/471/84
Printed in the German Democratic Republic
Gesamtherstellung: IV/2/14 VEB Druckerei „Gottfried Wilhelm Leibniz",
DDR - 4450 Gräfenhainichen · 6395
Umschlaggestaltung: Annemarie Wagner
LSV 1355
Bestellnummer: 763 247 8 (2052/32)

09800

Inhalt

IV. Das Institut im Jahre 1983

[1] Arbeiten in englischer Sprache

Content

Содержание

Die Kulturpflanze

Vol. 32

Genetics of Seed Proteins

Edited by

KLAUS MÜNTZ and CHRISTIAN HORSTMANN

Proceedings of the 3rd Seed Protein Symposium

Gatersleben
August 31 – September 2, 1983

AKADEMIE-VERLAG · BERLIN
1984

Content

Poster Abstracts

Inhalt

Poster

Содержание

Доклады и краткие сообщения

Постерные сообщения

Preface

The Proceedings of the 3rd Seed Protein Symposium 'Genetics of Seed Proteins' are edited as part of the Annual Publication of the Zentralinstitut für Genetik und Kulturpflanzenforschung, Gatersleben, of the Academy of Sciences of the GDR, whereas the proceedings of the preceding two symposia [1] of this series appeared in 'Abhandlungen der Akademie der Wissenschaften der DDR, Abteilung Mathematik, Naturwissenschaften, Technik'. The Annual Publication 'Die Kulturpflanze' offers a fixed rhythm of edition and a high standard of printing quality since more than 30 years. Therefore, the editors of the present Proceedings preferred to include the full-length papers of invited lectures as well as the abstracts of the short oral contributions and of the posters presented at the 3rd Seed Protein Symposium into 'Die Kulturpflanze'. The publishing house, Akademie-Verlag, Berlin, kindly agreed to distribute approximately fifty percent of the number of copies as a volume separated from the institute's report, thus offering specialists in the field of seed protein research the possibility to buy the Proceedings of the 3rd Seed Protein Symposium as a special issue. The editors of the Proceedings gratefully acknowledge the concessions and co-operation granted by the editors of 'Die Kulturpflanze' and by the 'Akademie-Verlag, Berlin'.

No special date of reception is given on the papers since lectures and posters have been presented at the symposium and the majority of manuscripts was already handed over to the editors during the symposium. All contributions are written in English, and no corrections in the language of the manuscripts have been made, but the editors take the responsibility of translating the English summaries of the plenary lectures into German and Russian, which is in accordance with the author instructions of 'Die Kulturpflanze'.

Last not least, the editors want to express their gratitude to Mrs. Eva Bielig and Ms. Ilse Neumann for their skilful technical assistance to prepare the manuscripts of these Proceedings for edition.

Gatersleben, January, 1984

The editors

1 Seed Proteins of Dicotyledonous Plants. Proc. Symp. 1977. – In: Abhdlg. Akad. Wiss. DDR, Abt. Math., Naturwiss., Techn. **N4** (1978).
Regulation of Protein Biosynthesis and Degradation During Embryogenesis and Germination of Plant Seeds. Proc. 2nd Symp. Seed Proteins, 1980. – In: Abhdlg. Akad. Wiss. DDR, Abt. Math., Naturwiss., Techn. **N5** (1981).

Opening Address

Ladies and gentlemen, dear colleagues,
It is a privilege and a great pleasure for me to welcome you, the participants of the Third Symposium on Seed Proteins in Gatersleben. You have gathered as a group of 60 scientists coming from 13 different countries and I hope you'll find everything you need both in the hotels at Quedlinburg and in our institute. Provided weather will continue to be as bright as during the last weeks I am sure you will enjoy to be accomodated in Quedlinburg, since the centre of this town represents one of the few architectural monuments of medieval half-timbered house construction and town formation that remained undestroyed in Germany after the end of World War II. But Quedlinburg not only played an important role in German history and development of architecture, this town and the surrounding region is also one of the most traditional German places of plant breeding, horticultural and agronomic research, as well as seed production. This tradition was continued and even amplified after the foundation of the GDR as exemplified by the large Institute of Breeding Research of the Academy of Agricultural Sciences of the GDR which was established at Quedlinburg after the end of the World War II.

When in 1945 Prof. Stubbe, the founder and first director of our institute, looked for a place to re-establish the former Kaiser Wilhelm Institute of Crop Plant Research he was preferentially attracted by the scientific and plant breeding traditions of the Quedlinburg region. Based on the efficient assistance by the Soviet Military Administration at Halle granted to Dr. Stubbe during this time, our institute started to work at Gatersleben in the end of 1945. During the first years the laboratories were accomodated in the mansion of the former Gatersleben farm, but generously supported by the government of the GDR and by the Academy of Sciences, that was founded in 1948, Dr. Stubbe could start to put into action his ambicious projects concerning the institute's development.

The period of continuous construction activities was initiated in 1948/49, and it only ended at October 7th, 1964, the day when in 1949 the GDR had been founded, with the inauguration of the lecture hall in which we now have gathered together to open our 3rd Seed Protein Symposium.

The institute was headed by Dr. Stubbe for 25 years, i.e., from 1943–1968. In this time period it developed into the main site of basic and applied genetic research in our country. Therefore, when Dr. Böhme, a student of Dr. Stubbe, became director in 1969 the name of our institute was changed into 'Central

Institute of Genetics and Research in Cultivated Plants' of the Academy of Sciences, but we continued to follow three basic ideas dominating in the institute's development since the time of its foundation:

1. Cooperation of different biological disciplines to resolve problems of crop plant research;
2. Combination of basic and applied research and
3. Close contact to practical plant breeding and crop plant cultivation.

When in 1970 the group working on plant protein metabolism was established in our institute, these principles also determined its research programme. This group, in cooperation with the Plant Germ Plasm Bank of our institute (then called 'World Collection of Cultivated Plants'), with the research group working in applied genetics and with our computer station screened from 1970 to 1980 the cereal and some grain legume collections with respect to the genetic variability for the quantity of seed proteins and with respect to the relative amount of limiting essential amino acids in seed proteins. The screening results are used in practical plant breeding. The major part of the research capacity was involved in basic research on proteins and protein formation in grain legumes. *Vicia faba*, which is one of the most important legume crops in our country, was chosen to be the main research subject. It is not necessary to explain the development of the group working on plant protein metabolism in any detail since you are specialists in the field of plant seed proteins and know about the results of these investigations in our institute.

In 1975, the idea was born to arrange a plant seed protein symposium at Gatersleben in 1977 which was to deal with 'Structure, Accumulation, and Genetic Variability of Seed Proteins from Dicotyledonous Plants'. The success of this first Seed Protein Symposium induced the following ones, on the 'Regulation of Biosynthesis and Degradation of Proteins in Developing and Germinating Seeds', held in 1980, and third one, which is now opened and concerns the 'Genetics of Plant Seed Proteins'. The topics of the three subsequent symposia seem to reflect the development of your field of research during the last ten years: Description of characters, structure, biosynthesis, accumulation and compartmentation of seed reserve proteins. predominating up to 1980, and then the entrance into molecular genetic investigations with the introduction of genetic engineering methods since 1980.

Beyond the international scientific contacts and cooperation which are indispensable for fruitful high-standard research, symposia of this kind promote personal contacts and the mutual knowledge on history, culture, social and political situation in the world and in the countries where participants come from. In the dangerous political situation we face today in the world these contacts are from my point of view even more imperative than at any time before. Therefore, the social events of our symposium include an excursion which is intended to direct our attention to the responsibility that scientists have in the present world to prevent war and inhumanity, as well as to protect human culture against being exterminated. Science and scientific cooperation are part of this culture. They only can prosper in peace. Opening the 3rd Seed Protein Symposium I very much hope that the meeting not only will be a successful scientific event but also a small contribution in maintaining peaceful condition for the promotion of science by being in international contact.

As hosts of the Symposium, the scientific and technical staff of our institute will do their best to provide you with everything that is necessary. Just now I want to express my gratitude to all those coworkers of our institute and of the Academy at Berlin who actively contributed to the preparation of this Symposium. I hope that each participant will find fair partners for discussion of her or his scientific problems in order to keep up with the highest standards of research. My wish is that the Symposium will be a successful one, that everybody enjoys the following three days at Gatersleben and Quedlinburg and will finally return to his home country with new informations and stimulation for his own research.

R. RIEGER

Lectures

Kulturpflanze 32 · 1984 · S 21–S 32

Chromosomal location and expression of genes encoding low molecular weight proteins in wheat and related species

F. Garcia-Olmedo, P. Carbonero, G. Salcedo, C. Aragoncillo,
C. Hernandez-Lucas, J. Paz-Ares and F. Ponz
(Madrid, Spain)

Summary

A review of recent work on the chromosomal location and expression of genes encoding moderately abundant, presumably non-storage endosperm proteins from wheat, barley and rye is presented. Related aspects, such as regulatory genetic effects, *in vivo* and *in vitro* synthesis, types of processing, deposition sites, and molecular cloning are also discussed. The relevance of these studies in connection with basic endosperm biology, genetic manipulation of quality and agronomic traits, and the evolution of these important crops is briefly emphasized.

Introduction

The purpose of this paper is to review some recent findings, together with previously published work, concerning the genetics and molecular biology of different types of low molecular weight proteins, which are moderately abundant in the cereal endosperm, but are not typical storage proteins by a number of criteria. The study of these proteins is of interest in connection with the achievement of a more complete understanding of endosperm biology, which will be required in future genetic manipulations of the composition of this tissue. Knowledge of the chromosomal location of genes encoding these proteins has also allowed their use as biochemical chromosome markers in the interspecific transfer ot genes controlling agronomic traits, such as resistance to certain diseases (see Rodriguez-Loperena et al. 1975b; Hart et al. 1976; Delibes et al. 1981; Doussinault et al. 1983). Certain of these proteins have been of diagnostic value in relation to the study of the origin and evolution of cereal species (Johnson and Hall 1965; Hall et al. 1966; Johnson 1972, 1975; Salcedo et al. 1984). The general features of endosperm protein composition have been found to be quite similar in many of the cereal species investigated. Therefore, this review will not be restricted to wheat, but will integrate information concerning other Gramineae, specially barley and rye.

Types of proteins investigated

The albumin and globulin fractions of cereals are made up of over 20 major and many minor components, most of which are under 25,000 molecular weight and are included in the main peak of a typical gel filtration profile (Caldwell and Kasarda 1978; Salcedo et al. 1980b; Rahman et al. 1982).

The complexity of these fractions has made difficult to study the genetic control of many of their individual components by one-dimensional electrophoresis or electrofocusing. This has led to the development of different procedures to partially purify or selectively extract subsets of proteins and to the application of a variety of two-dimensional separation methods (Fig. 1). Thus, a number of components of the albumin and globulin fractions are extracted by aqueous alcohols, and a subset of those soluble in aqueous alcohols, designated CM-proteins, are soluble in chloroform: methanol (2 : 1, v/v) (Garcia-Olmedo and Garcia-Faure 1969; Rodriguez-Loperena et al. 1975a). The CM-proteins also appear residually in prolamin preparations, where they are among the most abundant low molecular weight components (see Salcedo et al. 1980b). This information is summarized in Fig. 2.

Besides the CM-proteins, the fraction under 25,000 molecular weight in prolamin preparations also includes another group of prolamin-like proteins, designated low-molecular-weight prolamins (LMWG in wheat, LMWH in barley), which are insoluble in H_2O and in 0.5 M NaCl (Salcedo et al. 1979; 1982; Aragoncillo et al. 1981; Prada et al. 1982).

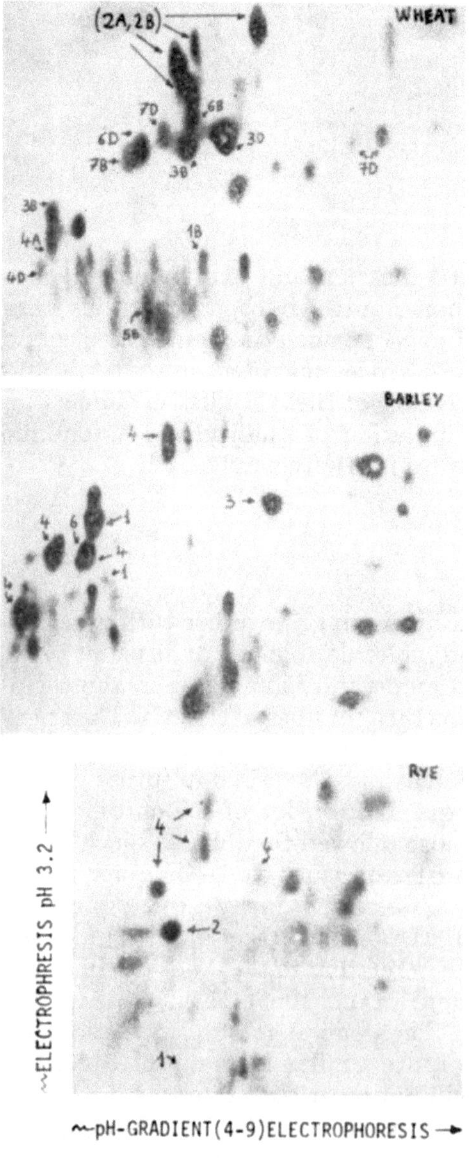

Fig. 1 Two-dimensional fractionation by combined non-equilibrium pH-gradient (4–9) electrophoresis × electrophoresis pH 3.2 of proteins extracted by 0.5 M NaCl from the endosperms of wheat, barley, and rye. Chromosomes where the genes encoding different components of the two-dimensional map are indicated in the figures (Fra-Mon et. al. 1984)

Petroleum ether extracts of cereal endosperms include two types of proteins, the thionins (see GARCIA-OLMEDO et al. 1982) and the lipid binding proteins (LBP), both of which are also extracted with 0.5 M NaCl (PONZ et al. 1984).

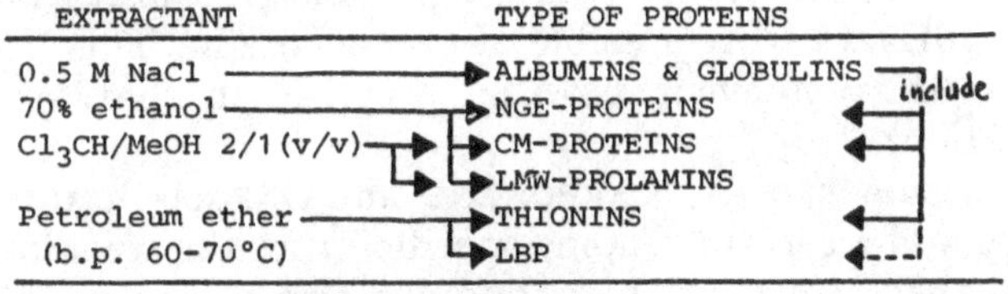

Fig. 2 Relationships among the types of proteins extracted with different solvents from the endosperms of wheat, barley, and rye

Genetic variability and evolutionary implications

In contrast with the considerable intraspecific variability of the main storage prolamins (DOEKES 1968; DOLL and BROWN 1979; SHEWRY et al. 1979), most of the types of proteins described above show little intraspecific and even interspecific variability. For this reason, these proteins have been useful in the elucidation of genome relationships and the analysis of phylogenetic affinities in the Triticineae.

The electrophoretic analysis of the low molecular weight components of the 70 % ethanol extracts of kernel proteins was used by JOHNSON and coworkers (JOHNSON and HALL 1965; HALL et al. 1966; JOHNSON, 1972, 1975) in their extensive studies on the origin of the alloploid Triticineae.

The CM-proteins, which are included in the previous group, have been found to be quite invariant in wheat and barley. In bread-wheat (*Triticum aestivum* L.), protein CM 1 was found to be invariant, while it was absent in tetraploid wheat cultivars (*T. durum* Desf.), which are the adequate ingredient for pasta products, and this protein was used for the detection of the fraudulent use of hexaploid wheat to elaborate these products (GARCIA-OLMEDO and GARCIA-FAURE 1969). Proteins CM 2 and CM 3 have been found to be invariant among tetraploid and hexaploid wheat cultivars, except for two closely related tetraploid ones that presented an allelic variant of CM 3, designated CM 3' (GARCIA-OLMEDO and GARCIA-FAURE 1969; RODRIGUEZ-LOPERENA et al. 1975a; SALCEDO et al.

Table 1
Variability of CM-proteins in *Hordeum vulgare* and *H. spontaneum*

Variant Protein	N° of Samples *H. vulgare*	*H. spontaneum*
CMa-1	38	13
CMa-2	0	4
CMb-1	30	15
CMb-2	7	1
CMb-3	1	1
CMc-1	38	16
CMc-2	0	1
CMd-1	38	17
CMe-1	21	8
CMe-2, 2'	17	7
CMe-3	0	2

1978). In a recent study (SALCEDO et al. 1984),[1] proteins CMa-1, CMc-1, and CMd-1 were found to be invariant among *Hordeum vulgare* cultivars, the latter being invariant among *Hordeum spontaneum* accessions (Fig. 3 and Table 1). Certain variants of CMb-1 and CMe-1 seem to be restricted to *H. vulgare* and *H. spontaneum* samples from Morocco and to cultivars with possible Moroccan origin. This has led to speculate that a domestication event possibly occurred in that area (MOLINA-CANO et al. in preparation).

In a survey of *Aegilops-Triticum* species, CARBONERO and GARCIA-OLMEDO (1969) observed little variability and a certain genome specificity for the thionins. Subsequent sequence studies have confirmed these findings (for a review, see GARCIA-OLMEDO et al. 1982). The amino acid sequence of β-thionin has not changed in the evolution from diploid to hexaploid wheat and, in fact, there is considerable homology between thionins in cereals and similar proteins in very distant *taxa*, such as the viscotoxins of mistletoe (*Viscum album*) and crambin from *Crambe abyssinica*.

Considerable variability was observed for the low molecular weight gliadins among tetraploid and hexaploid wheat cultivars, although they were not as variable as the classical gliadins (SALCEDO et al. 1980a).

The variability of LBPs has not been investigated, but PONZ et al. (1984) have found close homology between the LBPs of wheat and oats.

Chromosomal locations of genes encoding low molecular weight proteins in wheat and related species

A considerable number of genes encoding proteins alluded in this review have been assigned to their corresponding chromosomes by aneuploid genetic analysis. A summary of previously published information is given in Table 2 (for a detailed review, see GARCIA-OLMEDO et al. 1982). More recent work along this line has dealt with the chromosomal location of genes encoding salt-soluble proteins in wheat, rye and barley (FRA-MON et al., 1984) and of genes encoding CM-proteins in barley (SALCEDO et al. 1984).

The 0.5 M NaCl extracts of the nulli-tetrasomic series of Chinese Spring wheat, the Imperial rye/Chinese Spring wheat addition lines, and the Betzes barley/Chinese Spring addition lines were analyzed by two-dimensional pH gradient (4–9) polyacrylamide gel electrophoresis × starch-gel electrophoresis (pH 3.2) as indicated in Fig. 1. Genes for 17 of the most prominent components of the wheat two-dimensional map have been assigned to 12 different chromosomes. In 9 of the cases (2 proteins associated with chromosome 3B, 1 with 3D, 1 with 4A, 1 with 4D, 1 with 6B, 1 with 6D, 1 with 7B, 1 with 7D), the genes had been previously located using more selective protein extraction procedures (GARCIA-OLMEDO and CARBONERO 1970; ARAGONCILLO et al. 1975). The remaining four cases (1 protein in 1B, 1 in 5B, 2 in 7D) correspond to new assignments. Additionally, it has been observed, that group 2 chromosomes affect the expression level of 4 more proteins. Genes encoding a number of minor components have been also tentatively located.

Due to the complexity of the background of wheat proteins, a smaller number of map components of rye and barley, have been assigned through the analysis of

Table 2

Summary of chromosomal locations of genes for non-storage endosperm proteins from wheat and related species

Type of protein+ (Separation method)	Refe-ren-ces++	Chromosome homoeology group N°* 1	2	3	4	5	6	7
Thionins (E)	a	AL (1) BL (1) DL (1) RL (1)						
70 % ethanol (E) (low mol. wt.)	b			D (1)	A (2?) D (1?)			B (1) D (1)
(E×IEF)	c			BS (2) DS (1) AgS (1) R (1)	Aβ (1+1?) D (1)	D (1)	BS (2)	BS (1) DS (2) AgS (2)
CM-proteins (E×IEF)	c				Aβ (2) D (1)			BS (1) DS (2) AgS (2)
	d			H (1)	H (2)			H (2)
	e				R (2?)			
Globulins (E)	f			A (1) B (2) D (4)				
Albumins (imm)	g			D (1)	D (1)			
Buffer sol. (IEF)	h	B (1?)		AS (1?) BS (1+1?) DS (1)	A (2?)		A (2?)	
0.5 M NaCl (E×E)	e	B (1) R (1)	A } B } (4?) R (1)	B (2) D (1) H (1)	A (1) D (1) H (4) R (4)	B (1)	B (1) D (1)	B (1) D (3) H (2)
LMWG (E×E)	i				B (2)			AS (2) DL (2)

\+ Type of protein is indicated by common designation or by extractant used; method of separation is indicated in parenthesis, E = one-dimensional electrophoresis, IEF = isoelectrofocusing, E×E = two-dimensional electrophoresis, imm = immunochemical analysis

++a) GARCIA-OLMEDO et al., 1976; FERNANDEZ DE CALEYA et al., 1976; SANCHEZ-MONGE et al., 1979. b) WAINES, 1973. c) GARCIA-OLMEDO and CARBONERO, 1970; ARAGONCILLO et al., 1975; RODRIGUEZ-LOPERENA et al., 1975. d) SALCEDO et al., 1984. e) FRAMON et al., 1984. f) CUBADDA, 1975. g) BOZZINI et al., 1971. h) NODA and TSUNEWAKI, 1972. i) SALCEDO et al., 1980.

* The genome is indicated first (ABD, wheat; Ag, Agropyron; H, Hordeum; R, rye), followed by the chromosome arm (L, long; S, short; α or β). The number of components assigned is indicated in parenthesis, followed by the question mark when there is overlapping of bands. Chromosome 1 of barley has been placed under homeology group n° 7.

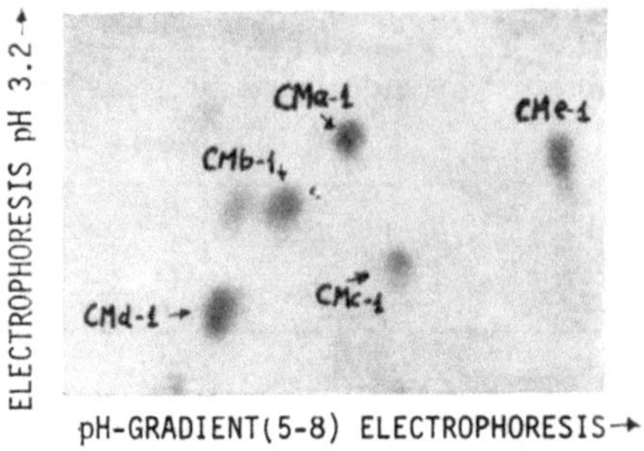

Fig. 3 Two-dimensional fractionation of CM-proteins from barley *(Hordeum vulgare)*

addition lines. In the case of rye, one map component was associated with chromosome 1, one with chromosome 2, and four with chromosome 4. In the case of barley, genes for eight components were located in four chromosomes: two in chromosome 1, one in chromosome 3, four in chromosome 4, and one in chromosome 6.

Among the four salt-soluble proteins associated with chromosome 4 of rye, two of them, and possibly a third one, have been tentatively identified as CM-proteins. In the case of barley, the chromosomal location of genes encoding CM-proteins has been directly investigated in the wheat-barley addition lines (Salcedo et al., 1984). Genes encoding proteins CMa and CMc were located in chromosome 1, genes for proteins CMb and CMd were assigned to chromosome 4, and the newly reported protein CMe was associated with chromosome 3 (Fig. 3). Sequence homology, among proteins CMa, CMb, CMc, and CMd purified from barley and between the barley and the wheat CM-proteins has been inferred on the basis of compositional divergence indexes and of immunological evidence (Salcedo et al. 1982; Paz – Ares et al. 1983a). In particular, CM 3 from wheat and CMd from barley showed the closest interspecific relationship: complete antigenic identity, the only CM-proteins in each species to be extracted with the 7 : 1 (v/v) mixture of chloroform: methanol, and the lowest compositional divergence index. Genes for the chloroform : methanol soluble proteins of wheat have been ascribed to chromosomes of groups 4 and 7, and, in particular, the gene for protein CM 3 was found to be located in chromosome 4A (Garcia-Olmedo and Carbonero 1970; Aragoncillo et al. 1975). The present assignment of genes for the four barley proteins to chromosomes 1 and 4, further supports the previously proposed homoeology between chromosome 1 of barley and group 7 of wheat and between chromosome 4 of barley and group 4 of wheat (Hart et al. 1980; Powling et al. 1981).

Regulatory effects

A quantitative study of gene-dosage responses was conducted by Aragoncillo et al. (1978) with a group of six low molecular weight, 70 % ethanol soluble proteins from wheat endosperm, which were encoded by incomplete (not triplicate) homoeologous gene sets. Approximately linear dosage responses were observed for all the proteins. For two of the proteins, and probably for a third one, the net output of protein for each dose of its structural gene was 30–80 % higher when the chromosome carrying an active homoeogene was absent. These observations indicated that gene-dosage responses for some endosperm proteins can be modified by genetic elements located in a chromosome different from that containing their structural genes. In a related study, SALCEDO et al. (1978) showed differences in gene-dosage responses among alleles at a locus encoding proteins CM 3 and CM 3′ in *T. turgidum*. The net number of protein molecules present in the mature endosperm was measured when each of the alleles was present in one, two, and three doses. Linear gene-dosage responses were again observed, but for a given dosage, about twice as much CM 3 as CM 3′ protein was found. Genetic evidence indicated that the observed quantitative differences either resulted from differences, in the structural genes themselves or were controlled by regulatory or modifier gene(s) linked to them.

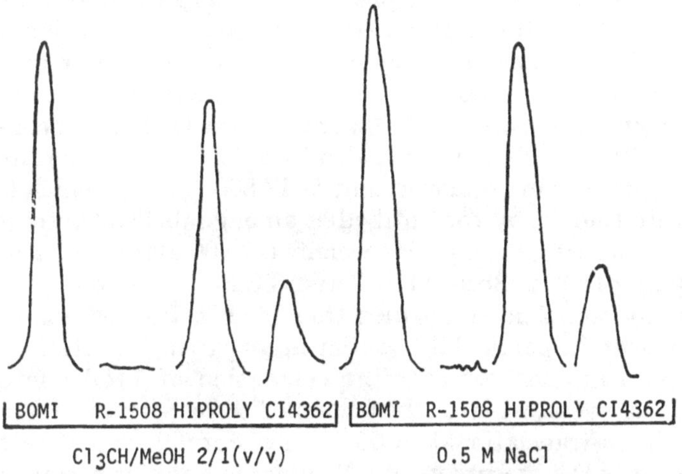

Fig. 4 Densitometric evaluation of protein CMe-1 extracted with chloroform/methanol 2/1 (v/v) or with 0.5 M NaCl after two-dimensional fractionation. Bomi barley and its mutant Ris 1508, HIPROLY barley and its sister line CI4362 were compared in the same gel slab. In the legend of the curves HIPROLY and C 14362 must be exchanged.

In our recent study of the genetic variability and control of CM-proteins in barley (SALCEDO et al. 1984), a different type of genetic effect has been shown to occur: expression of gene(s) encoding protein CMe, located in chromosome 3, is completely blocked by the "high lysine" mutation in Riso 1508, which locus is in chromosome 7. The accumulation of this protein is markedly decreased in the "high lysine" barley Hiproly (Fig. 4). It has been shown that B and C-hordeins are drastically decreased in the Ris 1508 mutant and, to a lesser extent, in Hiproly barley, while the salt-soluble fraction is increased in both mutants (see MIFLIN and SHEWRY 1979). RHODES and GILL (1980) reported that a salt-soluble component was decreased in Ris 1508 and HEJGAARD (1982) found that protein Z was decreased in Ris 1508, but greatly increased in Hiproly. To our knowledge, CMe is the only salt-soluble protein whose accumulation is totally or partially blocked in both Ris 1508 and Hiproly.

Synthesis, processing and deposition of proteins in endosperm

Synthesis, processing and deposition of proteins in the cereal endosperm have been intensively studied in recent times. Storage proteins from maize (BURR et al. 1978; LARKINS and HURKMAN 1978; WIENAND and FEIX 1978; VIOTTI et al. 1979), barley (BRANDT and INGVERSEN 1978; MATTHEWS and MIFLIN 1980), wheat (GREENE 1981; OKITA and GREENE 1982; DONOVAN et al. 1982), rice (YAMAGATA et al. 1982) and oats (LUTHE and PETERSON 1975; MATLASHEWSKI et al. 1982) have atracted most of the interest. In this general context, we have focused our attention on the thionins and CM-proteins of wheat, barley and rye. The initial *in vivo* and *in vitro* studies have been carried out in developing barley endosperm (PAZ-ARES et al. 1983b; PONZ et al. 1983) and some of the observations have been also extended to wheat (unpublished).

PAZ-ARES et al. (1983b) have reported that CM-proteins in barley are synthesized non-synchronously from 10 to 30 d after anthesis by membrane-bound polysomes as precursors of higher apparent molecular weight (13,000–21,000) than the mature proteins (12,000–16,000). These precursors are processed and the mature protein exported into the cytosol. PONZ et al. (1983) detected thionin synthesis from ~8 d to ~30 d after anthesis. They identified two thionin precursors (THP 1 and THP 2) using monospecific antibodies raised against the mature protein. THP 1 is the only polypeptide among the *in vitro* products that is recognized by the monospecific antibodies. THP 1 is encoded by a 7.5 S mRNA and its alkylated derivative has an apparent molecular weight of 17,800. THP 2, which is selected together with mature thionin by the antibodies among labelled proteins *in vivo*, differs from THP 1 in apparent molecular weight (17,400 alkylated) and in electrophoretic mobility at pH 3.2. Both THP 1 and THP 2 were competed out from the antigen-antibody complex by purified thionin. Final deposition of the mature protein takes place in the particulate fraction as an extrinsic protein.

Pulse-chase *in vivo* experiments showed that the conversion of THP 2 into thionin was a post-translational event, whereas the presumed conversion of THP 1 into THP 2 was assumed to be co-translational on the basis of our failure to detect THP 1 *in vivo*. The processing of the precursors of CM-proteins was also assumed to be co-translational by the same criterium. We have now obtained additional evidence in support of these tentative conclusions through *in vitro* translation experiments with initiation-inhibited bound polysomes and with the dog-pancreas *in vitro* processing system.

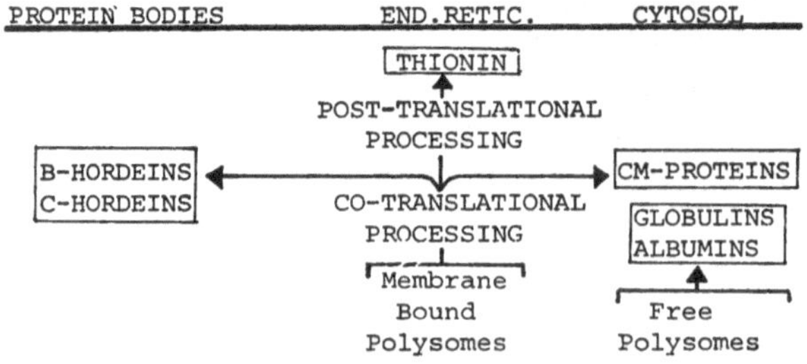

Fig. 5 Pathways of protein synthesis, processing and deposition in barley endosperm

Figure 5 summarizes the reported pathways of protein synthesis, processing and deposition in barley endosperm: i) synthesis of many cytosol albumins and globulins by free polysomes (BRANDT and INGVERSEN 1976); ii) synthesis of CM-proteins by membrane bound polysomes, with co-translational processing, and export into the cytosol; iii) synthesis of B, C, and D-hordein by membrane-bound polysomes, with co-translational excision of signal peptides, and deposition in protein bodies; iv) synthesis of thionins by membrane-bound polysomes, with co-translational and post-translational processing, and deposition as extrinsic membrane proteins.

Molecular cloning

It is evident that the availability of c-DNA and genomic clones of cereal endosperm proteins would greatly enhance our capacity to carry out studies in wide-ranging areas of the basic and applied biology of this tissue, such as genome structure, gene expression, genetic modification of quality-related characters, manipulation of agronomic traits, etc. Our general aim at this level was to obtain c-DNA and genomic clones for the most relevant proteins alluded in this review. Our initial results concern the molecular cloning of c-DNA corresponding to the CM-proteins and thionins from barley. The following standard steps were followed: a) Preparation of total polysomal RNA from barley endosperm collected at about 20 d after anthesis. b) Purification of poly A^+RNA by affinity chromatography (oligo-dT cellulose column). c) Synthesis of double stranded c-DNA, using the poly A^+RNA as template. d) Tailing of the ds c-DNA with poly C. e) Selecting tailed c-DNA molecular greater than $\sim$350 bp long. f) Annealing with PstI restricted, poly G tailed pBR 322. g) Cloning into *E. coli* MC 1601. h) Selection of tetracycline resistant, ampicillin sensitive clones. i) Screening of the selected clones by hybridization with a c-DNA probe prepared from an RNA sucrose gradient fraction of poly A^+RNA from membrane bound polysomes, enriched in mRNAs for thionins and CM-proteins. j) Screening of clones selected in the previous step by hybrid release translation and identification of *in vitro* translation products with monospecific antibodies.

Clones obtained as described are being characterized and used as probes to select genomic clones.

Zusammenfassung

Chromosomale Lokalisierung und Expression von Genen für Proteine mit niedrigem Molekulargewicht in Weizen und verwandten Arten

Es wird ein Überblick über neuere Arbeiten zur chromosomalen Lokalisierung und Expression von Genen für Endospermproteine aus Weizen, Gerste und Roggen gegeben, die wahrscheinlich nicht zu den Reserveproteinen gehören und in mäßiger Menge vorliegen. Verwandte Aspekte, wie regulatorische genetische Effekte, *in vivo*- und *in vitro*-Synthese, Arten des processing, Speicherorte und molekulare Klonierung, werden ebenfalls diskutiert. Die Bedeutung dieser Untersuchungen für die Biologie des Endosperms, die genetische Manipulation von Qualität und agronomischen Eigenschaften und die Evolution dieser wichtigen Getreidearten wird kurz behandelt.

Краткое содержание

Хромосомальная локализация и экспрессия генов, регулирующих протеины с низким молекулярным весом у пшеницы и родственных видов

Дается обзор новых работ о хромосомальной локализации и экспрессии генов, регулирующих протеины эндосперма пшеницы, ячменя и ржи, которые, вероятно не являются запасными белками и встречаются в небольших количествах. Обсуждаются также близкие аспекты, как регулирующие генетические эффекты, сиитез in vivo и in vitro, способ химической обработки, места запасания и молекулярное клонирование. Кратко рассматриваются значение этих исследований для биологии эндосперма, генетичесике манипуляции качества и агрономических особенностей, а также эволюция этих важных видов зерновых злаков.

Literature

ARAGONCILLO, C., M. A. RODRIGUEZ-LOPERENA, P. CARBONERO and F. GARCIA-OLMEDO, 1975: Chromosomal control of non-gliadin proteins from the 70 % ethanol extract of wheat endosperm. – Theor. Appl. Genet. **45**, 322–326.

–, –, G. SALCEDO, P. CARBONERO and F. GARCIA-OLMEDO, 1978: Influence of homoeologous chromosomes on gene-dosage effects in allohexaploid wheat (*Triticum aestivum* L.) – Proc. Natl. Acad. Sci. **75**, 1446–1450.

–, R. SANCHEZ-MONGE and G. SALCEDO, 1981: Two groups of low molecular weight hydrophobic proteins from barley endosperm. – J. Exp. Bot. **32**, 1279–1286.

BOZZINI, A., P. CANTAGALLI and S. E. PIAZZI, 1971: Chromosomal location of the genetic control of two proteins specifically attributed to hexaploid wheat by immunological methods. E.W.A.C. Newsl. 16–17.

BRANDT, A., and J. INGVERSEN, 1976: *In vitro* synthesis of barley endosperm proteins on wild type and mutant templates. Carlsberg Res. Commun. **41**, 311–320.

–, and –, 1978: Isolation and translation of hordein messenger RNA from wild type and mutant endosperms in barley. – Carlsberg Res. Commun. **43**, 451–469.

BURR, B., F. A. BURR, I. RUBENSTEIN and M. N. SIMON, 1978: Purification and translation of zein messenger RNA from maize endosperm protein bodies. – Proc. Natl. Acad. Sci. **75**, 696–700.

CALDWELL, K. A. and D. D. KASARDA, 1978: Assessment of genomic and species relationships in *Triticum* and *Aegilops* by PAGE and by differential staining of seed albumins and globulins. – Theor. Appl. Genet. **52**, 273–280.

CARBONERO, P. and F. GARCIA-OLMEDO, 1969: Purothionins in *Aegilops-Triticum* spp. – Experientia **25**, 1110.

CUBADDA, R., 1975: Chromosomal location of genes controlling the synthesis of some soluble proteins in *T. diurum* and *T. aestivum*. In: Genetics and Breeding of Durum Wheat (ed. G. T. SCARASCIA-MUGUOZZA). Library of the Faculty of Agric., Bari, and Natl. Inst. Nutr., Rome, 653–659.

DELIBES, A., F. DOSBA, C. OTERO and F. GARCIA-OLMEDO, 1981: Biochemical markers associated with two M^v chromosomes from *Aegilops ventricosa* in wheat-*Aegilops* addition lines. – Theor. Appl. Genet. **60**, 5–10.

DOEKES, G. J., 1968: Comparison of wheat varieties by starch gel electrophoresis of their grain proteins. – J. Sci. Food Agric. **19**, 169–176.

DOLL, H. and A. H. D. BROWN, 1979: Hordein variation in wild (*Hordeum spontaneum*) and cultivated (*H. vulgare*) barley. – Can. J. Genet. Cytol. **21**, 391–404.

DONOVAN, G. R., J. W. LEE and T. J. LONGHURST, 1982: Cell-free synthesis of wheat prolamins. – Aust. J. Plant Physiol. **9**, 59–68.

DOUSSINAULT, G., A. DELIBES, R. SANCHEZ-MONGE and F. GARCIA-OLMEDO, 1983: Transfer of a dominant gene for resistance to eyespot disease from a wild grass to hexaploid wheat. – Nature **303**, 698–700.

FERNANDEZ DE CALEYA, R., C. HERNANDEZ-LUCAS, P. CARBONERO and F. GARCIA-OLMEDO, 1976: Gene expression in alloploids: genetic control of lipopurothionins in wheat. – Genetics **83**, 687–699.
FRA-MON, P., G. SALCEDO, G. ARAGONCILLO and F. GARCIA-OLMEDO, 1984: Chromosomal assignment of genes controlling salt-soluble proteins (albumins and globulins) in wheat and related species. – Theor. Appl. Genetic, in press.
GARCIA-OLMEDO, F. and P. CARBONERO, 1970: Homoeologous protein synthesis controlled by homoeologous chromosomes in wheat. – Phytochemistry **9**, 1495–1497.
–, –, C. ARAGONCILLO, R. FERNANDEZ DE CALEYA and J. V. TORRES, 1976: Expression of homoeologous molecular systems in wheat alloploids. In: Heterosis in Plant Breeding (ed. A. JANOSSY and F. G. H. LUPTON, Proc. 5th Eucarpia Congress. Akadémiai Kiadó, Budapest, 51–57.
–, – and B. L. JONES, 1982: Chromosomal locations of genes that control wheat endosperm proteins. – In: Advances in Cereal Science and Technology **5**, (ed. Y. POMERANZ) 1–47.
– and R. GARCIA-FAURE, 1969: A new method for the estimation of common wheat (*T. aestivum* L.) in pasta products. – Lebensm. Wiss. Technol. **2**, 94–96.
GREENE, F. C., 1981: *In vitro* synthesis of wheat (*Triticum aestivum* L.) storage proteins. – Plant Physiol. **68**, 778–783.
HALL, O., B. L. JOHNSON and R. OLERED, 1966: Evaluation of genome relationships in wheat from their protein homologies. – Hereditas, Suppl. **2**, 47–54.
HART, G. E., D. E. MCMILLIN and E. R. SEARS, 1976: Determination of the chromosomal location of a glutamate oxalacetate transaminase structural gene using *Triticum-Agropyron* translocations. – Genetics **83**, 49–61.
–, A. K. H. R. ISLAM and K. M. SHEPHERD, 1980: Use of isozymes as chromosome markers in the isolation and characterization of wheat-barley chromosome addition lines. – Genet. Research **36**, 311–325.
HEJGAARD, J., 1982: Purification and properties of protein Z a major albumin of barley endosperm. – Physiol. Plant. **54**, 174–182.
JOHNSON, B. L., 1972: Protein electrophoretic profiles and the origen of the B genome of wheat. – Proc. Natl. Acad. Sci. **69**, 1398–1402.
–, 1975: Identification of the apparent B genome donor of wheat. – Can. J. Genet. Cytol. **17**, 21–39.
–, and O. HALL, 1965: Analysis of phylogenetic affinities in the Triticinae by protein electrophoresis. – Am. J. Bot. **52**, 506–513.
LARKINS, B. A. and W. J. HURKMAN, 1978: Synthesis and deposition of zein in protein bodies of maize endosperm. – Plant Physiol. **62**, 256–263.
LUTHE, D. S., and D. M. PETERSON, 1975: Cell-free synthesis of globulin by developing oat (*Avena sativa* L.) seeds. – Plant Physiol. **56**, 256–263.
MATLASHEWSKI, G. J., K. ADELI, I. ALTOSAAR, P. R. SHEWRY and B. J. MIFLIN, 1982: *In vitro* synthesis of oat globulin. – FEBS Lett. **145**, 208–212.
MATTHEWS, J. A. and B. J. MIFLIN, 1980: *In vitro* synthesis of barley storage proteins. – Planta **149**, 262–268.
MIFLIN, B. J., and P. R. SHEWRY, 1979: The biology and biochemistry of cereal seed prolamins. – In: Seed protein improvement in cereals and grain legumes. Proc. Symp. IAEA/FAO. IAEA, Vienna, 137–158.
MIFLIN, B. J., and P. R. SHEWRY, 1979: The synthesis of proteins in normal and high lysine barley seeds. In: Recent Advances in the Biochemistry of Cereals (eds. D. LAIDMAN and R. G. WYN JONES). Academic Press, 234–273.
MOLINA-CANO, J. L., P. FRA-MON, G. SALCEDO, C. ARAGONCILLO and F. GARCIA-OLMEDO. – (in preparation).
NODA, K., and K. TSUNEWAKI, 1972: Analysis of seed proteins in ditelosomes of common wheat. Japan J. Genet. **47**, 315–318.
OKITA, T. W., and F. C. GREENE, 1982: Isolation and characterization of the gliadin messenger RNAs. – Plant Physiol. **69**, 834–839.
PAZ-ARES, J., C. HERNANDEZ-LUCAS, G. SALCEDO, C. ARAGONCILLO, F. PONZ and F. GARCIA-OLMEDO, 1983a: The CM-proteins from cereal endosperm: Immunochemical relationships. – J. Exp. Bot. **34**, 388–395.

– PAZ-ARES, J., F. PONZ, C. ARAGONCILLO, C. HERNANDEZ-LUCAS, G. SALCEDO, P. CARBONERO and F. GARCIA-OLMEDO, 1983b: *In vivo* and *in vitro* synthesis of CM-proteins (A-hordeins) from barley (*Hordeum vulgare* L.). – Planta **157**, 74–80.

PONZ, F., C. HERNANDEZ-LUCAS, P. CARBONERO and F. GARCIA-OLMEDO, 1984: Lipid binding proteins from the endosperms of wheat and oat. – Phytochemistry, **23**, 2179–2181.

–, J. PAZ-ARES, C. HERNANDEZ-LUCAS, P. CARBONERO and F. GARCIA-OLMEDO, 1983: Synthesis and processing of thionin precursors in developing endosperm from barley (*Hordeum vulgare* L.). – EMBO J. **2**, 1035–1040.

POWLING, A., A. K. M. R. ISLAM and K. W. SHEPHERD, 1981: Isozymes in wheat-barley hybrid derivative lines. – Biochem. Genet. **19**, 237–254.

PRADA, J., R. SANCHEZ-MONGE, G. SALCEDO and C. ARAGONCILLO, 1982: Isolation of the major low molecular weight gliadins from wheat. – Plant Sci. Lett. **25**, 281–289.

RAHMAN, S., P. R. SHEWRY and B. J. MIFLIN, 1982: Differential protein accumulation during barley grain development. – J. Exp. Bot. **33**, 717–728.

RHODES, A. P., and A. A. GILL, 1980: Fractionation and amino acid analysis of the salt-soluble protein fractions of normal and high-lysine barleys. – J. Sci. Food Agric. **31**, 467–473.

RODRIGUEZ-LOPERENA, M. A., C. ARAGONCILLO, P. CARBONERO and F. GARCIA-OLMEDO, 1975a: Heterogeneity of wheat endosperm proteolipids (CM-proteins). – Phytochemistry 14, 1219–1223.

–, –, J. V. TORRES, P. CARBONERO and F. GARCIA-OLMEDO, 1975b: Biochemical evidence of chromosome homoeology among related plant genera. – Plant Sci. Lett. **5**, 387–393.

SALCEDO, G., C. ARAGONCILLO, M. A. RODRIGUEZ-LOPERENA, P. CARBONERO and F. GARCIA-OLMEDO, 1978: Differencial allelic expression at a locus encoding an endosperm protein in tetraploid wheat (*T. turgidum*). – Genetics **89**, 147–156.

–, P. FRA-MON, J. L. MOLINA-CANO, C. ARAGONCILLO and F. GARCIA-OLMEDO. Genetics of CM-proteins (A-hordeins) in barley. – Theor. Appl. Genet. **68**, 53–59.

–, J. PRADA and C. ARAGONCILLO, 1979: Low molecular weight gliadin like proteins from wheat endosperm. – Phytochemistry **18**, 725–727.

–, –, R. SANCHEZ-MONGE and C. ARAGONCILLO, 1980a: Aneuploid analysis of low molecular weight gliadins from wheat. – Theor. Appl. Genet. **56**, 65–69.

–, R. SANCHEZ-MONGE and C. ARAGONCILLO, 1982: The isolation and characterization of low molecular weight hydrophobic salt soluble proteins from barley. – J. Exp. Bot. **33**, 1325–1331.

–, –, A. ARGAMENTERIA and C. ARAGONCILLO, 1980b: The A-hordeins as a group of salt soluble hydrophobic proteins. – Plant Sci. Lett. **19**, 109–119.

SANCHEZ-MONGE, R., A. DELIBES, C. HERNANDEZ-LUCAS, P. CARBONERO and F. GARCIA-OLMEDO, 1979: Homoeologous chromosomal location of the genes encoding thionins in wheat and rye. – Theor. Appl. Genet. **54**, 61–63.

SHEWRY, P. R., H. M. PRATT, A. J. FAULKS, S. PARMAR and B. J. MIFLIN, 1979: The storage protein (hordein) polypeptide pattern of barley (*Hordeum vulgare* L.) in relation to varietal identification and disease resistance. – J. Nat. Inst. Agric. Bot. **15**, 34–50.

VIOTTI, A., E. SALA, R. MAROTTA, P. ALBERI, C. BALDUCCI and C. SOAVE, 1979: Genes and mRNAs coding for zein polypeptides in *Zea mays*. – Eur. J. Biochem. 102, 211–222.

WAINES, J. G., 1973: Chromosomal location of genes controlling endosperm protein production in *Triticum aestivum* cv. Chinese Spring. Proc. 4th Int. Wheat. Genet. Symp. (eds. E. R. SEARS and L. M. S. SEARS) Agric. Exp. Stn., Univ. of Missouri, Columbia, Mo. 873–877.

WIENAND, U., and G. FEIX, 1978: Electrophoretic fractionation and translation *in vitro* of poly (A)-containing RNA from maize endosperm. – Eur. J. Biochem. **92**, 605–611.

YAMAGATA, H., K. TANAKA and Z. KASAI, 1982: Biosynthesis of storage proteins in developing rice seeds. – Plant Physiol. **70**, 1094–1100.

F. GARCIA-OLMEDO, P. CARBONERO, G. SALCEDO, C. ARAGONCILLO,
C. HERNANDEZ-LUCAS, J. PAZ-ARES and F. PONZ
Departamento de Bioquímica, E.T.S. Ingenieros Agrónomos
Universidad Politécnica, Madrid-3, Spain

Kulturpflanze 32 · 1984 · S 33—S 52

Genetic relationships of wheat gliadin proteins

D. D. KASARDA, D. LAFIANDRA, R. MORRIS and P. R. SHEWRY
(Berkeley, California, U.S.A.)

Summary

Chromosomal assignments for gliadin protein components resolved in 2-dimensional electrophoretic patterns of several different bread wheat (*Triticum aestivum* L. em. Thell.) cultivars have been analyzed. The patterns for A genome-coded proteins included α-gliadins as did the patterns of D genome coded proteins. Proteins controlled by the B genome were notable for the absence of α-gliadins and of slower-moving ω-gliadins. On the basis of N-terminal amino acid sequences, gliadins can be classified into three major types: α-type, γ-type, and ω-type. These classifications apparently correspond to separate subfamilies of the gliadin family insofar as all gliadin genes are apparently related and may be derived from an ancestral repeating sequence. Parts of gliadin polypeptide chains, however, may be the result of independent gene modifications that affected some gliadin genes at later developmental stages. These modified regions of gliadin genes may not be homologous with other gliadin genes. Gliadin genes may be very young genes having evolved within the grass family, which is itself a late evolutionary development.

Introduction

Development of gel electrophoretic methods for resolution of wheat gliadin proteins established the complexity of the traditional gliadin fraction and provided a basis for genetic analysis of the complex mixture of protein components that constitutes gliadin (for review see EWART 1969 and KASARDA et al. 1976). Although some earlier work on the genetic relationships of gliadins had been carried out, the papers by SHEPHERD (1968) and WRIGLEY and SHEPHERD (1973) are worthy of special note.

By means of 1-dimensional starch gel electrophoresis (aluminum lactate buffer, pH 3.1) and by using the aneuploid lines of 'Chinese Spring' produced by E. R. SEARS (1954), SHEPHERD (1968) defined the chromosomal relationships of many of the gliadin components in this cultivar (cv.). The genes coding for gliadins were located on what are now called the short arms of chromosomes of homoeologous groups 1 and 6 (chromosomes 1A, 1B, 1D, 6A, 6B, 6D).

The resolution obtainable by electrophoresis was greatly enhanced by WRIGLEY'S (1970) development of a 2-dimensional method of gliadin electrophoresis in which the 1st dimension was isoelectric focusing (IEF) in an acrylamide gel and the 2nd-dimension was starch gel electrophoresis in aluminum lactate buffer (pH 3.1). WRIGLEY and SHEPHERD (1973) used the 2-dimensional approach to resolve about 46 gliadin protein components in 'Chinese Spring', and by using SEARS' aneuploids, they were able to determine the chromosomal locations of 34 of these components.

It was established by these investigations that gliadins are coded by a complex locus or by several loci on the short arms of chromosomes 1 and 6 of each of the three genomes that are combined in the polyploid (6x) genome of bread wheats. That the locations of structural genes and not merely of controlling sequences were being defined was indicated by the absence of only one chromosome resulting in the absence of a protein in the electrophoretic pattern. If the locations of controlling sequences were being defined rather than the locations of structural genes, and if these controlling sequences were located on different chromosomes from the structural genes, the absence of a protein component should have been associated with the absence of more than one chromosome, which was not the case. Homoeologous controlling sequences might be effective in activating gliadin genes of other genomes, but this would not interfere with the location of gliadin structural genes by aneuploid analysis.

Subsequent investigations from a number of different laboratories supported the findings of SHEPHERD (1968) and WRIGLEY and SHEPHERD (1973) and extended them partially to other cultivars. These investigations have been reviewed by PAYNE et al. (1982a), WRIGLEY (1982), and GARCIA-OLMEDO et al. (1982).

Here we discuss some recent work from our own laboratories on the genetic relationships of gliadin proteins, discuss briefly some work from two other groups, and speculate on the origins and evolution of gliadin genes. We shall not cover the high-molecular-weight (HMW) subunits (subunit MW near 10^5) (see review of PAYNE et al. 1982a), or the low-molecular-weight (LMW) gliadin-like proteins described by SALCEDO et al. (1979), and will touch only in passing on the LMW-glutenin subunits (subunit MW range $40–50 \times 10^3$) (JACKSON et al. 1983). Because the HMW and LMW appelations as applied to glutenin refer to the subunit MW's rather than to the aggregate weights, it might be preferable to refer to glutenin HMW subunits and glutenin LMW subunits, but we shall use the former nomenclature here. LMW-glutenin subunits have also been named "aggregated gliadins" (SHEWRY et al. 1983).

Chromosomal locations of gliadin genes for the cultivars 'Cheyenne' and 'Wichita'

In recent work[1], we have used the 2-dimensional electrophoretic analysis system of MECHAM et al. (1978) (and also a modification of this method[2]) for the analysis of various aneuploids and substitution lines to locate the structural genes of nearly all the monomeric gliadin components of the cvs. 'Cheyenne', 'Wichita', and 'Chinese Spring'. This electrophoretic method combines aluminum lactate

1 D. LAFIANDRA, R. MORRIS and D. D. KASARDA, unpublished results.

electrophoresis (pH 3.2) in the 1st dimension with electrophoresis at pH 9.2 (Tris-glycine buffer) in the second dimension to separate proteins. Separation in the 1st dimension results largely from the sum of the positive charges of the histidine, lysine, and arginine side chains of the proteins (the free carboxyl groups of the proteins are largely uncharged at pH 3.2). Separation in the 2nd dimension results from the net charge on the proteins at pH 9.2 as a consequence of ionization of carboxyl groups and dissociation of protons from histidine side chains to produce the neutral form of this amino acid; lysine side chains may be partially titrated to their uncharged form as well.

The resulting pattern and chromosomal assignments for the cv. 'Cheyenne', a hard red winter wheat of strong mixing character and good quality, are shown in Figure 1. About 35 components were resolved for 'Cheyenne', excluding some

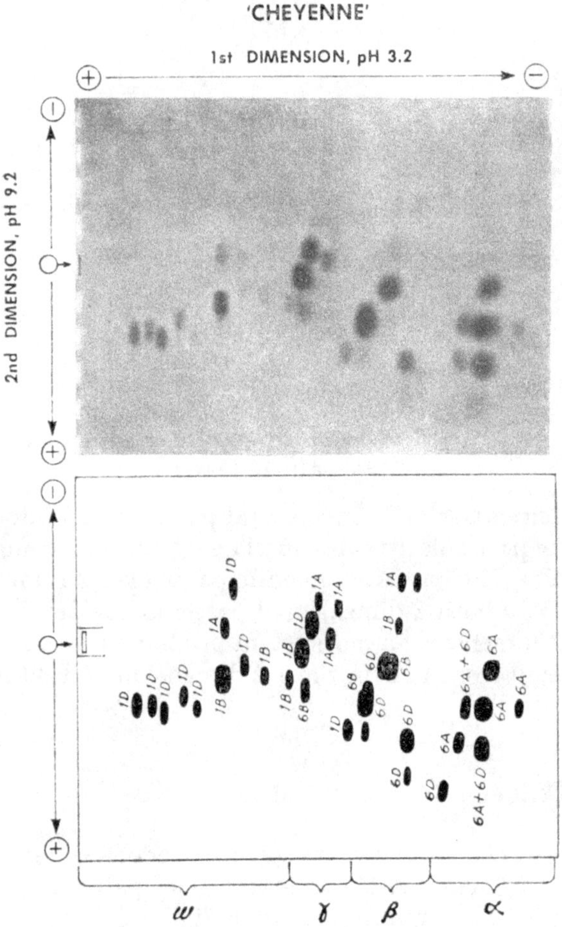

Fig. 1 Two-dimensional (2-pH) electrophoretic pattern of gliadins from 'Cheyenne' with chromosomal assignments shown on diagram below pattern. Procedure of MECHAM et al. (1978)

faint bands that were not counted. The pattern of 'Wichita' (not shown), a hard red winter wheat of medium dough strength, was somewhat more complicated than those of 'Chinese Spring' (not shown), a hard red spring wheat of poor baking quality, and 'Cheyenne'. About 46 bands were readily resolved for 'Wichita'.

Although the improved electrophoretic procedure[2] used for analysis of 'Wichita' gliadins tends to resolve more components, 'Wichita' gave a more complex pattern than the other two cvs. even when the procedure of MECHAM et al. (1978) was used.

The general approach to making assignments is illustrated by the pattern of Figure 2, which is of the gliadins extracted from a 'Cheyenne' nullisomic 1D seed, It may noted that the slow ω-gliadin bands and one of the strong γ-gliadin bands are missing. A careful analysis of the pattern allowed us to assign all missing bands to chromosome 1D. We examined ditelocentrics for only a few chromosomes, but when these were analyzed, the equivalent genes corresponded to locations on the short arms of the respective chromosomes. We shall assume that our chromosomal assignments for gliadins also indicate a short arm location for the locus.

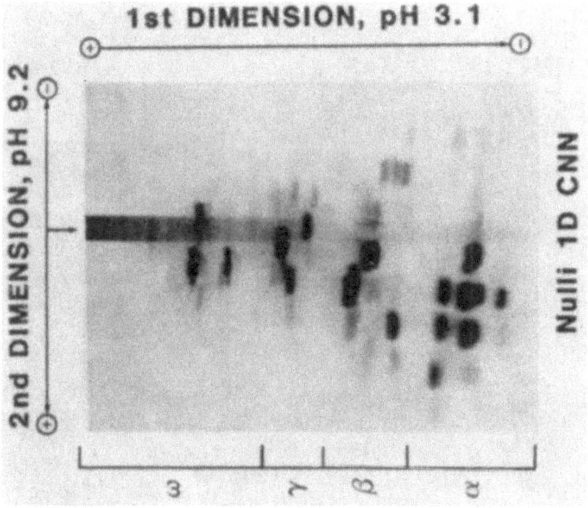

Fig. 2 Two-dimensional (2-pH) electrophoretic pattern of gliadins from a nullisomic 1D seed of 'Cheyenne'. Procedure of MECHAM et al. (1978) as modified by LAFIANDRA and KASARDA (unpublished)

The patterns showed strong similarities to the 2-dimensional patterns of WRIGLEY and SHEPHERD (1973), which were particularly evident when we compared our pattern for 'Chinese Spring' with theirs. The bands assigned to particular chromosomes grouped in much the same way in both 2-dimensional patterns. Evidently, electrophoresis at pH 9.2 for the 2nd dimension in our system produces a separation of gliadins similar to that resulting from IEF in the 1st dimension of WRIGLEY's method.

Extracted genomic patterns of 'Chinese Spring' gliadins

By reproducing separately the patterns corresponding to gliadins coded by genes of a single genome, we can more readily compare these "extracted" patterns for the different genomes. This is illustrated for 'Chinese Spring' in Figure 3. We have indicated the components associated with chromosomes of group 1 by stars and those associated with chromosomes of group 6 by plus signs. No attempt has been made to represent band intensities.

2 D. LAFIANDRA and D. D. KASARDA, unpublished results.

The A genome of 'Chinese Spring' in our patterns is represented by four components coded on the short arm of chromosome 1A and five components coded on the short arm of chromosome 6A. The A-genome is notable in all three cvs. for 6A-coded gliadins that migrate in the 1st dimension with mobilities corresponding to α-gliadins. In all three of the wheat cvs. we examined, 6A coded α-gliadins are always the fastest-moving gliadin components in a lactate gel electrophoretic pattern, and this is probably the general case.

In 'Chinese Spring' there is only one fast-moving ω-gliadin coded by the A genome. (ω-gliadins tend to group into a slower-moving group and a faster-moving group in most wheat cvs.) In general, the pattern is similar to that of some accessions of *Triticum monococcum*[3], but the patterns of the other two cvs. resembled *T. urartu*[4] patterns in some ways. Although the patterns of *T. monococcum*, *T. boeoticum*, and *T. urartu* fit our A genome patterns better than did *Aegilops speltoides*, *Ae. searsii*, *Ae. longissima*, or *Ae. squarrosa*, there was little basis for favoring *T. boeoticum* (or *T. monococcum*) over *T. urartu* as the likely donor of the A genome to bread wheats.

The B genome pattern of 'Chinese Spring' (Figure 3) is notable for the absence of α-gliadins and slow-moving ω-gliadins. This was found also for the B genomes of the other two cvs. as will be discussed.

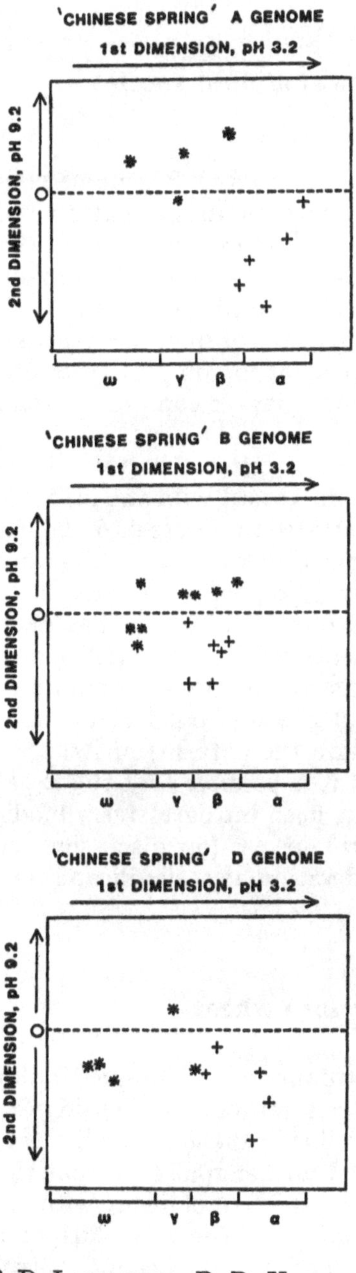

Fig. 3 Diagrams of two-dimensional (2-pH) electrophoretic patterns of gliadins from 'Chinese Spring' corresponding to each of the three genomes of the hexaploid (A, B, and D). Method was the same as in Figure 1. Components associated with chromosomes of homoeologous group 1 are indicated by stars (*) whereas components associated with chromosomes of group 6 are indicated by plus (+) signs

3 D. Lafiandra, D. D. Kasarda and B. L. Johnson, unpublished results.
4 D. K. Mecham and D. D. Kasarda, unpublished results.

The D genome pattern of 'Chinese Spring' is notable for having both slow-moving ω-gliadin components and α-gliadin components. This is characteristic of *Ae. squarrosa* ssp. *strangulata*; it seems virtually certain that this species contributed the D-genome to bread wheats. The electrophoretic patterns of Bushuk and Kerber (1978) also led them to a similar conclusion.

Comparison of the B genome patterns with those of diploid species of *Aegilops*

In Figure 4 we show diagrams of the patterns of the extracted B genome controlled gliadins of our three cvs. and compare them with composite patterns of gliadins from three diploid species of *Aegilops* (*Sitopsis* section) that have been considered possible donors of the B genome to hexaploid wheat. The diploid patterns represent overlays of patterns from several different accessions of each species[4], and are accordingly more complex than the patterns of single accessions[4]. The patterns of single accessions are frequently equivalent in complexity to our extracted patterns although diploid patterns with much greater complexity than those of our extracted patterns have been noted[3,4].

As mentioned in the previous section, our extracted B genome patterns have neither α-gliadins nor slow-moving ω-gliadins. This is in accord with the patterns of *Ae. searsii*, *Ae. longissima*, and *Ae. speltoides* as illustrated in Figure 4. Although it is not possible to exclude any of these species from consideration as a potential donor of the B genome on the basis of our patterns, the composite pattern of *Ae. searsii* seems most similar to our extracted patterns. *Ae. searsii* has been suggested as the donor of the B genome to bread wheats by Feldman (1978) and the SDS-polyacrylamide gel electrophoretic patterns of the HMW subunits of *Ae. searsii* are more comparable to the patterns of B genome-coded HMW subunits in bread wheat cvs. (Payne et al. 1981) than are the patterns of *Ae. speltoides* or *Ae. longissima* (Cole et al. 1981). Although it is possible that the B genome of bread wheats is a composite genome that has been introgressively modified through crosses involving several different diploid species (for discussion see Cole et al. 1981), *Ae. searsii* deserves serious consideration as a significant contributor of all or part of the B genome.

Chromosomal assignments for a tetraploid (*durum*) wheat

Du Cros et al. (1983) have used a 2-dimensional electrophoretic technique similar to that of Wrigley (1970), but with acrylamide gels for both dimensions to analyze the chromosomal assignments of gliadin proteins in the *durum* wheat cv. 'Langdon'. The results were similar to those obtained for hexaploids in that the proteins coded by the B genome were mainly of β- and γ-mobilities with no α-gliadins although there were some slow-moving ω-gliadins. The α-gliadins were coded only by the A genome. Du Cros et al. (1983) found no evidence for control of gliadin bands by chromosomes 2A or 2B, or any chromosomes other than those

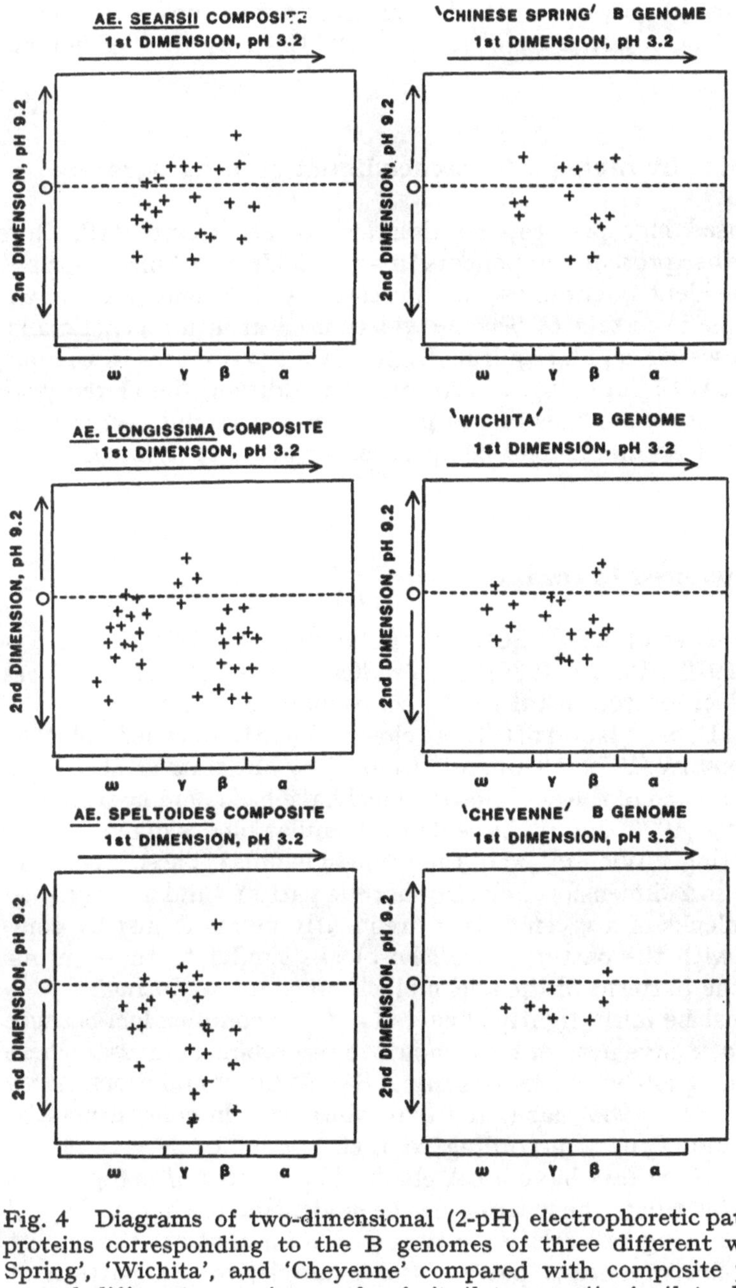

Fig. 4 Diagrams of two-dimensional (2-pH) electrophoretic patterns (as in Figure 3) of proteins corresponding to the B genomes of three different wheat cultivars, 'Chinese Spring', 'Wichita', and 'Cheyenne' compared with composite patterns of gliadins from several different accessions each of *Aegilops searsii*, *Aegilops longissima*, and *Aegilops speltoides*. Data for diploids from D. K. Mecham and D. D. Kasarda, unpublished results

of groups 1 and 6. Possible involvement of group 2 chromosomes in control of gliadin expression had been noted by others (see DU CROS et al. 1983 for discussion).

Analysis of aneuploids by high-performance liquid chromatography

BIETZ et al. (1983) used high-performance liquid chromatography (HPLC) to analyze gliadin and other protein components in aneuploids of 'Chinese Spring'. They were readily able identify chromosomal associations of various peaks in the patterns. This technique is certain to become widely used in future genetic analysis because the analyses are rapid, requiring from only a few minutes to an hour, and can be carried out with submilligram samples. In addition, the eluted peak can be recovered for further analysis of the purified or partly purified protein, such as determination of its amino acid composition or of its N-terminal amino acid sequence.

Gliadin genes are clustered in complex loci

Recent studies (MECHAM et al. 1978; SOZINOV and POPERELYA 1979, 1980; RYBALKA and SOZINOV 1979; PAYNE 1984) have indicated that gliadin genes are clustered in complex loci located distal to the centromere on the short arms of chromosomes 1A, 1B, 1D, 6A, 6B and 6D. These clusters have been called "blocks" by SOZINOV and POPERELYA (1979, 1980) and "isoloci" by MECHAM et al. (1978). Various allelic blocks for each of the chromosomal loci have been demonstrated by SOZINOV and POPERELYA (1980) and we have defined similar blocks in our genetic analysis of 'Chinese Spring', 'Wichita', and 'Cheyenne' gliadins[1]. These blocks are fairly easy to recognize in 2-dimensional electrophoretic patterns and a fairly good analysis of the allelic blocks of a given cv. can frequently be made just by comparison of its pattern with the patterns of allelic blocks similar to those in our Figure 3, which gives the patterns of the allelic blocks of 'Chinese Spring'.

The gliadin genes must be fairly tightly clustered in these complex loci because the components of a block have not yet been shown to recombine in crosses where the parents have different allelic blocks (PAYNE 1984; SOZINOV and POPERELYA 1980, RYBALKA[5]). KASARDA (1980) has noted that wheat cvs. in general may be divided into Type 1 and Type 2 according to their α-gliadin electrophoretic patterns (pH 3.2). Type 1 wheats have an A-gliadin-like cluster of components (coded by chromosome 6A where the location has been identified) whereas Type 2 wheats are notable for having two relatively faint α-gliadins that are the fastest moving bands in the electrophoretic pattern (for example, in the pattern of 'Chinese Spring'). These latter components were coded by chromosome 6A when chromosomal location has been determined. The two types of pattern apparently result from two different allelic blocks of the 6A locus. In our experience, the two fastest bands of the Type 2 pattern appear in every cv. of this type we have

5 A. I. RYBALKA, personal communication.

examined (at least several dozen). Because of the extensive crossing that has taken place in the development of various wheat cvs., it seems reasonable to conclude that the two genes coding for the fastest two 6A components in Type 2 wheats must be extremely closely linked in the corresponding DNA strand. On the other hand, we have noted[1] some minor variations in Type 1 cvs. of the patterns corresponding to the 6A allelic block characteristic of this type. This suggests to us that recombination of the multiple genes (KASARDA 1980) characteristic of this block is possible–at least to a limited extent.

N-terminal amino acid sequences of gliadins

At present, gliadins can be broken down into three major groups or types according to their N-terminal amino acid sequences (KASARDA et al. 1983). These three major groups, the α-type, the γ-type, and the ω-type, are illustrated in Figure 5.

α–type (α-GLIADINS, β-GLIADINS, γ_1-GLIADIN); CODED ON GROUP 6 CHROMOSOMES

NH_2-VAL-ARG-VAL-PRO-VAL-PRO-GLN-LEU-GLN-PRO-GLN-ASN-PRO-SER-GLN-GLN-GLN-PRO-GLN-GLU-
V R V P V P Q L Q P Q N P S Q Q Q P Q E

γ–type (γ_2-, γ_3-GLIADINS); CODED ON GROUP 1 CHROMOSOMES

NH_2-ASN-ILE-GLN-VAL-ASP-PRO-SER-GLY-GLN-VAL-GLN-TRP-LEU-GLN-GLN-GLN-LEU-VAL-PRO-GLN-
N I Q V D P S G Q V Q W L Q Q Q L V P Q

ω–type (ω-GLIADINS); CODED ON GROUP 1 CHROMOSOMES

NH_2-ALA-ARG-GLU-LEU-ASN-PRO-SER-ASN-LYS-GLU-LEU-GLN-SER-PRO-GLN-GLN-SER-PHE-SER-HIS-
A R E L N P S N K E L Q S P Q Q S F S H

NH_2-LYS-GLU-LEU-GLN-SER-PRO-GLN-GLN-SER-PHE-SER-HIS-
K E L Q S P Q Q S F S H

NH_2-SER-ARG-LEU-LEU-SER-PRO-ARG-GLY-LYS-GLU-LEU-HIS-THR-PRO-GLN-GLN-GLN-PHE-PRO-GLN-
S R L L S P R G K E L H T P Q Q Q F P Q

Fig. 5 Gliadin N-terminal amino acid sequences for three major types of gliadin sequences. Three different variant sequences are shown for ω-gliadins. The single-letter code for the amino acids is given below the three-letter code

The α-type is coded on the short arms of chromosomes of group 6. It shows little variation through about 25 amino acid residues, although some substitutions have been found in purified components[6] (see also KASARDA 1980). The α-type includes all α-gliadins and most β-gliadins; there are occasional gliadin components coded by genes on chromosomes of group 6 with mobilities corresponding to γ-gliadins (for example, see Figure 1). We consider it likely that these γ-gliadins have α-type N-terminal amino acid sequences. The γ_1-gliadin sequenced by BIETZ et al. (1977), which had an α-type sequence, probably corresponds to a group 6-coded γ-gliadin.

6 A. I. RYBALKA, M. D. DIETLER, and D. D. KASARDA, unpublished results.

The number of components coded by each of the group 6 gliadin loci is apparently about equal to or greater than the number of components coded by group 1 gliadin loci (for example, see Figure 3), which is somewhat surprising in consideration of the possibility that the group 6 loci have been more recently formed than the group 1 loci. Perhaps the gliadin genes of the group 6 locus undergo gene duplication more readily than do those of the group 1 locus. An equivalent to the group 6 locus does not seem to exist in barley and rye (SHEWRY et al. 1984), which are thought to have diverged before formation of *Triticum* and *Aegilops* species (SMITH and FLAVELL 1974). The group 1 loci may code for LMW-glutenin subunits as well (JACKSON et al. 1983), however, which would not show up in an unreduced gliadin pattern.

The γ-type gliadins have sequences corresponding to the γ_2- and γ_3-gliadin components sequenced by BIETZ et al. (1977). These two variants differ mainly at two positions[7], for example, γ_2-gliadin has isoleucine at position 2 whereas γ_3-gliadin has methionine at that position. The γ-type gliadins that we have sequenced[7] have had mobilities corresponding to γ-gliadins and have been coded by the complex locus on group 1 chromosomes that also codes for ω-gliadins (MECHAM et al. 1978; SOZINOV and POPERELYA 1980; PAYNE 1984). As can be seen in Figures 1 and 3, however, there are some gliadins with mobilities equivalent to β-gliadins that are coded by genes located on chromosomes of group 1. These proteins are usually fairly basic–retaining a net positive charge even at pH 9.2 and migrating toward the negative pole in the 2nd dimension of our electrophoretic method. These basic gliadins have not been purified and characterized as far as we know. They may have γ-type sequence, but this needs to be verified.

The ω-type gliadins have at least five different variants, all of which appear to be homologous with one another on the basis of N-terminal amino acid sequences (KASARDA et al. 1983). Three variants are illustrated in Figure 5 for ω-type gliadins. The divergence between the two 20-residue sequences in Figure 5 is notable for there being only 11 identical residues. This divergence may have resulted to some extent from species divergence. The 1st sequence (A–R–E–. . .) was obtained for a protein controlled by a gene on chromosome 1D whereas the 2nd sequence (S–R–L–. . .) was obtained for a protein apparently coded by a gene on chromosome 1B (KASARDA et al. 1983), and these proteins were likely to have been contributed to hexaploid bread wheats by different diploid species. Even so, the considerable divergence between the genes coding for these proteins probably indicates that gliadin genes have rapid rates of point mutational change (see WILSON et al. 1977). These rates may be obscured to some extent by homogenizing effects, such as may result from concerted evolution (ZIMMER et al. 1980) or gene conversion (BALTIMORE 1981; EGEL 1981).

In general, the N-terminal amino acid sequences of gliadins are more likely to define correct genetic relationships than are electrophoretic mobilities, but in consideration of the arbitrariness of the classification of gliadins by electrophoretic mobility, this latter system agrees rather well with the classification by sequences. The major difference being that α- and β-gliadins fall into a single grouping (α-type) in the sequencing approach.

7 M. D. DIETLER and D. D. KASARDA, unpublished results.

The γ-type gliadins are coded at the same locus as the ω-type gliadins, but these two types differ considerably in amino acid composition. The ω-gliadins have as much as 55 mole percent glutamine, 30 mole percent proline, and 10 % phenylalanine. Together, these three amino acids make up about 80 mole percent of ω-gliadins, but make up only about 65 mole percent of γ-gliadins (MIFLIN et al. 1983). The α-type gliadins coded by genes at the group 6 locus are similar in amino acid composition to γ-type gliadins, but differ in N-terminal amino acid sequence. All three types of N-terminal amino acid sequences differ to an extent that makes homology seem unlikely on the basis of a superficial comparison.

Comparison of the N-terminal sequence of an ω-gliadin from *T. monococcum* (SHEWRY et al. 1980) with that of γ_3-gliadin [7], however, indicates 9 out of 20 identities and an additional 8 residues that might differ as a consequence of single base changes in the corresponding DNA codons (Figure 6). This is suggestive of signi-

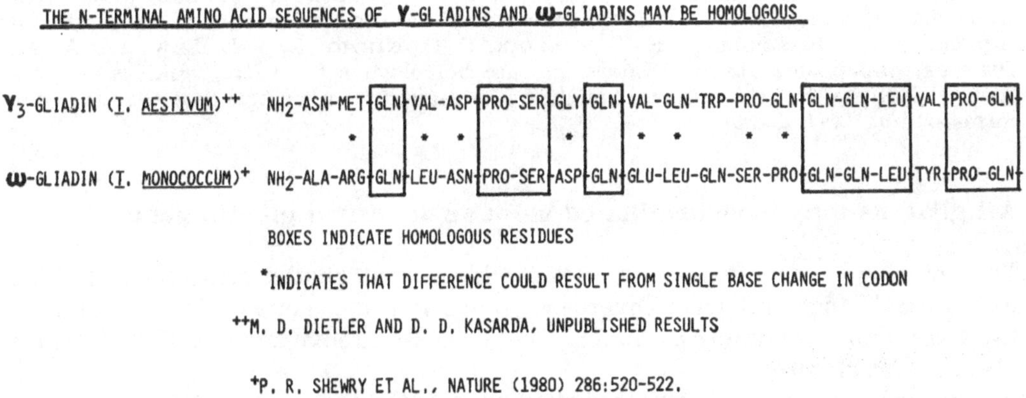

Fig. 6 Comparison of N-terminal amino acid sequences of γ_3-gliadin (data from M. D. DIETLER and D. D. KASARDA, unpublished results) and an ω-gliadin from *Triticum monococcum* (SHEWRY et al. 1980). Boxes indicate homologous (identical) residues. Stars (*) indicate the difference could result from a single base change in the corresponding DNA codon

ficant homology between the N-terminal sequences of ω-gliadins and γ-gliadins, but stronger evidence for such homology is needed particularly since other ω-gliadins we have sequenced [7] do not fit as well with the γ_3-gliadin sequence. The likelihood of some homology between ω-gliadins and γ-gliadins is supported by homologies found in cDNA sequences corresponding to mRNA of 'C' hordeins and 'B' hordeins at the Rothamsted Experimental Station (SHEWRY et al. 1984). The 'C' hordeins are homologous to ω-gliadins whereas 'B' hordeins are homologous to γ-gliadins at least for some parts of the peptide chains (SHEWRY et al. 1984).

The α-gliadins and γ-gliadins also share some regions of significant homology as indicated by the strong homology between a peptide derived from γ_3-gliadin [7] and part of the A-gliadin sequence [8] as illustrated in Figure 7. Eleven out of seventeen residues were identical and an additional four could have resulted from single base changes in the equivalent codons.

8 D. D. KASARDA, J. E. BERNARDIN, C. C. NIMMO, E. J.-L. LEW, and M. D. DIETLER, unpublished results.

A-GLIADIN AND γ_3-GLIADIN ARE HOMOLOGOUS IN AMINO ACID SEQUENCE

COMPARISON OF A-GLIADIN WITH A PEPTIDE DERIVED FROM γ_3-GLIADIN

PEPTIDE FROM γ_3-GLIADIN +:	-SER	LEU	VAL	LEU-GLN-THR-LEU-PRO	THR	MET-CYS-ASN-VAL	TYR	VAL	PRO	ILE-
	*		*		*		*	*		
SEQUENCE FROM A-GLIADIN++:	-ASN	LEU	ALA	LEU-GLN-THR-LEU-PRO	ALA	MET-CYS-ASN-VAL	THR	ILE	PRO	PRO-

BOXES INDICATE HOMOLOGOUS RESIDUES

* INDICATES THAT DIFFERENCE COULD RESULT FROM SINGLE BASE CHANGE IN CODON

\+ M. D. DIETLER AND D. D. KASARDA, UNPUBLISHED RESULTS

++ D. D. KASARDA ET AL., UNPUBLISHED RESULTS

Fig. 7 Comparison of amino acid sequences of a peptide from γ_3-gliadin (data from M. D. DIETLER and D. D. KASARDA, unpublished results) and a part of the A-gliadin sequence (D. D. KASARDA, J. E. BERNARDIN, C. C. NIMMO, E. J.-L. LEW, and M. D. DIETLER, unpublished results). Boxes indicate homologous (identical) residues whereas stars (*) indicate that the difference might have resulted from a single base change in the corresponding DNA codon

All gliadins may have originated from an ancestral gliadin gene

The homologies discussed above make it likely that gliadin genes have derived from a single ancestral gene. Divergence by various mechanisms has given rise to the three main subfamilies of gliadin genes discussed above and to LMW glutenins (JACKSON et al. 1983).

The genetic relationship to other gliadins of LMW glutenin protein components has not been established. There is some indication that they contain γ-type sequences, but other types of N-terminal sequences are present as well (BIETZ and WALL 1980; SHEWRY et al. 1983). The amino acid sequence derived from a cDNA sequence corresponding to a wheat mRNA clone that has been described by BARTELS and THOMPSON (1983) showed homology with 'B' hordein and with our γ_3-gliadin peptide; it could correspond to a LMW glutenin subunit, but there is no evidence for this possibility. Unfortunately, the cDNA sequence did not include the coding sequence for the N-terminal part of the protein. Amino acid or DNA sequences clearly identified with LMW-glutenin subunits are needed to determine the relationship of LMW glutenin subunits to other gliadins.

The likelihood that all gliadin sequences are homologous, at least in part, provides a justification for considering all gliadin genes to belong to a single family. The three different types of gliadin N-terminal sequences may be considered as corresponding to three different subfamilies of the gliadin gene family with the genes of each subfamily grouped at a locus–or perhaps sublocus in the case of the Gli 1 locus on the short arm of chromosomes of homoeologous group 1 (PAYNE et al. 1982b).

Speculations on the mechanism of origin of gliadin genes

SHEWRY et al. (1980) reported the N-terminal amino acid sequence of an ω-gliadin from *T. monococcum* in which the sequence –Pro–Gln–Gln–Pro–Tyr– was repeated. This finding combined with the predominance of glutamine and proline in the compositions of gliadins (especially ω-gliadins where they make up about 75 mole percent) led KASARDA (1980, 1981) to suggest that an ancestral gliadin gene might have resulted from many duplications of a short DNA sequence in which the codons for glutamine and proline predominated. This might have occurred within the context of suitable transcription controls or these may have been added later.

In this respect, gliadins might be considered closely related to the extensive repeating sequences that make up most of the non-coding DNA in higher organisms. They may represent a transition from non-coding to coding DNA as a consequence of selective advantage accruing to certain grasses from development of a new type of storage protein with large amounts of glutamine and proline. Both glutamine and proline are readily transformed to other amino acids, so that proteins with large amounts of these amino acids are likely to provide the developing embryo with an efficient way to store nitrogen.

In addition, if any other amino acids included in an ancestral repeating sequence were also neutral, the resulting proteins would have had low charge (as do gliadins). The low charge of gliadins may be advantageous in permitting extensive mutational change in gliadin genes without drastic effects on the deposition and packing of these storage proteins in protein bodies. Extensive surface charges on storage proteins might require greater ordering of interacting protein molecules in precipitates or deposits (protein bodies) in order that the requirement for essential neutralization of charged side chains be maintained (although neutralization might be effected through bound ions). Enhanced tolerance of mutational change for gliadin genes, on the other hand, might enhance the possibilities for fixation of advantageous mutations in a population.

The multiple gliadin components coded within an allelic block must have resulted from duplications of complete gliadin genes followed by divergence of duplicated genes (at least for those components of the same type, e.g. α-type). Translocation of a diverged gene (or genes) to another chromosome from an original locus (chromosome 1 to chromosome 6?) may have given rise to the present two locus situation known to be characteristic of some species of *Triticum* and *Aegilops* (SHEPHERD 1973) (and probably characteristic of all species). Similarly, translocation of a gene (or part of a gene) to an adjacent region of the chromosome 1 locus followed by divergence and duplications of the translocated gene may have given rise to the complex locus characteristic of the chromosome 1 loci in *Triticum* and *Aegilops*.

Recently, repeating sequences have been found in different types of gliadins (or in related proteins, such as the 'C' hordeins of barley). Some of these repeating sequences are illustrated in Figure 8. Some are based on amino acid sequences; others on cDNA sequences. These evidences of repeating sequences provide additional support for the likelihood that gliadin genes resulted from extensive duplications of short DNA sequences.

REPEATING SEQUENCES FOUND IN VARIOUS WHEAT GLUTEN COMPONENTS AND C-HORDEIN

PROTEIN	REPEATING SEQUENCE	NUMBER OF AA RESIDUES IN REPEAT
A-GLIADIN+	-P-Q-P-Q-P-F-P-P-Q-Q-P-Y-	12 (6)
ω-GLIADIN+	-P-Q-Q-P-Y-	5
LMW-GLUTENIN (?)++	-S-Q-Q-Q-Q-P-V-L-P-Q-Q-P-P-F-	14
C-HORDEIN+++	-P-Q-Q-P-Q-Q-P-F	8

+D. D. KASARDA ET AL. (USDA, WRRC)

++D. BARTELS AND R. THOMPSON (PLANT BREEDING INSTITUTE)

+++P. R. SHEWRY (ROTHAMSTED EXPERIMENTAL STATION)

Fig. 8 Repeating sequences found in various wheat gluten protein components and in 'C' hordein. Data for A-gliadin represents an average of several repeats (D. D. KASARDA, J. E. BERNARDIN, C. C. NIMMO, E. J.-L. LEW, and M. D. DIETLER, unpublished results). Data for a possible LMW-glutenin protein is based on the cDNA sequence of BARTELS and THOMPSON (1983). The 'C' hordein sequence is based on data of P. R. SHEWRY et al. (unpublished results)

The differences that may be noted for the repeating sequences of Figure 8 are likely to have resulted from the same mechanisms that gave rise to the ancestral gene. Although we speak of an ancestral gene, it is possible that early replications of a simple sequence, followed by translocation and divergence of these relatively short sequences may have led to independent, but nearly simultaneous, formation of gliadin genes. The duplications of repeating sequences and of complete genes are almost certain to have taken place over extended periods of evolutionary time–allowing for mutational modification of sequences to occur in some copies.

Although the mechanisms that gave rise to the repeated sequences and multiple copies of gliadin genes, are not known, it is possible to speculate that unequal crossing-over played an important role in the process. This mechanism has been discussed by SMITH (1976) for noncoding DNA, but similar mechanisms may have been involved in formation and diversification of gliadin genes. Figure 9 provides a schematic diagram of the process by which unequal crossing-over might increase or decrease in copy number in a repeating sequence or of complete genes. Also illustrated in Figure 9 is the process of concerted evolution (ZIMMER et al. 1980) by which divergence may be minimized in a gene family through unequal crossing-over followed by selective fixation. Gene conversion (BALTIMORE 1981; EGEL 1981) might also diminish sequence divergence within a gene family.

An additional speculative mechanism for change in the copy number of repeating sequences within gliadin genes is mismatch of repeated sequence DNA followed by excision and repair as illustrated by the schematic diagram of Figure 10. For purposes of speculation, we show repair by DNA synthesis at both loops (Figure 10), but repair of heteroduplex DNA by excision of DNA or by synthesis of DNA may very well be randomly determined. Nonetheless, this randomness could result in fixation of DNA with greater copy number.

Gliadin components differ considerably in MW, even within a type (see Figure 1 of PAYNE et al. 1982a), and both unequal crossing-over (as suggested by LAWRENCE and SHEPHERD 1980) and mismatch and repair might contribute to the differences. Gene conversion events (EGEL 1981) might also contribute to these

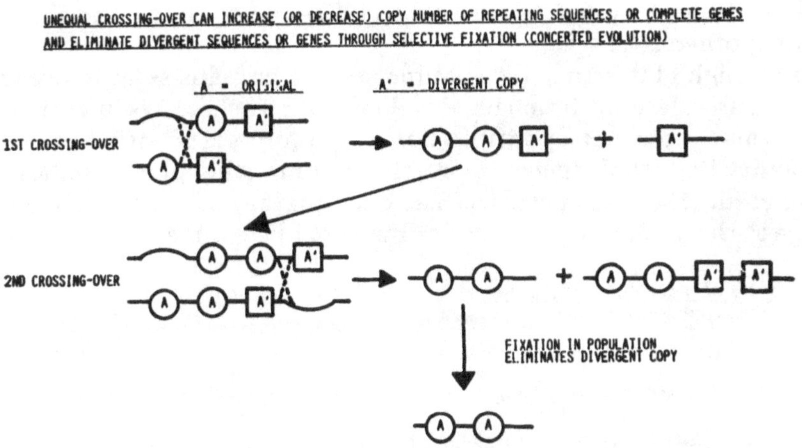

Fig. 9 Schematic diagram of unequal crossing-over between chromosome arms that could lead to an increase or decrease in the copy number of repeated sequences or of complete genes. Sequences from a diverging family can also be homogenized through redistribution of gene copies during several stages of crossing-over followed by selective fixation of homogenized families (concerted evolution—ZIMMER et al. 1980)

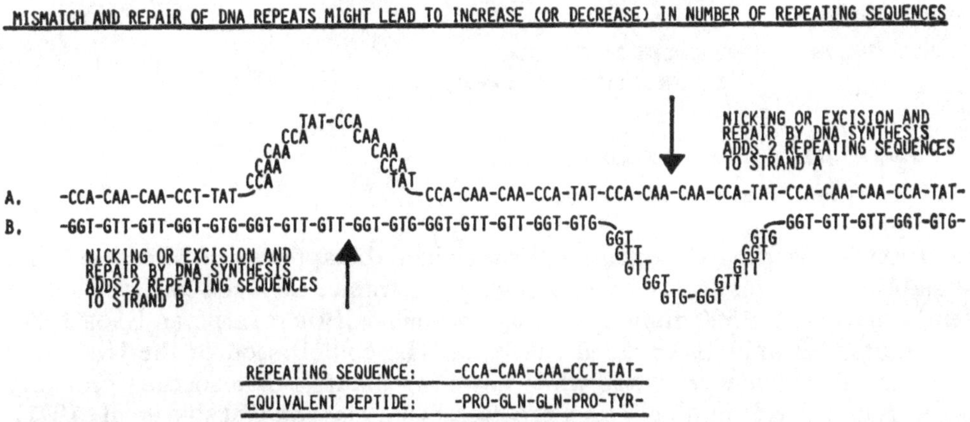

Fig. 10 Schematic diagram of a hypothetical scheme for increase (or decrease) in copy number of mispaired repeating sequences through nicking or excision of DNA from the short strand followed by repair through DNA synthesis following the code of the long strand. The 3rd lines of the legends in the figure should read: add repeating sequence

differences. Extension of repeated sequences or even formation of new ones, may be continuing events in the evolution of gliadin genes. The homologies we have discussed here for different gliadins represent relatively small parts of the gene or protein molecule. It is conceivable that parts of gliadin genes are homologous as a consequence of derivation from an early gliadin gene whereas other parts may be more recent and are not homologous—having resulted from more recently-formed repeating sequences or from extension or deletion of existing repeats followed by divergence. The processes of amplification, deletion, and transposition of repeating

DNA sequences have been discussed by FLAVELL (1982) and by ROSE and DOOLITTLE (1983) among others.

Gliadins and other high-glutamine, high-proline storage proteins seem to occur only in grasses, which are a late evolutionary development. (It should be mentioned, however, that there does not appear to have been any systematic investigation of the possibility that similar genes or proteins occur in other plant families.) This led KASARDA et al. (1983) to speculate that gliadin genes may be relatively young genes–perhaps the youngest gene family. Figure 11 lists some evolutionary

PROLAMIN GENES MAY BE THE MOST RECENTLY EVOLVED GENES

4,000,000,000 YEARS AGO	EARLIEST FORMS OF LIFE APPEAR.
500,000,000 YEARS AGO	COLLAGEN-LIKE GENES APPEAR.
400,000,000 YEARS AGO	PLANTS AND ANIMALS COLONIZE THE LAND.
200,000,000 YEARS AGO	FLOWERING PLANTS (ANGIOSPERMS) DIVERGE.
100,000,000 YEARS AGO	GRASSES (GRAMINEAE) BEGIN TO DEVELOP. GENES CODING FOR HIGH-GLUTAMINE, HIGH-PROLINE STORAGE PROTEINS DEVELOP FROM REPEATED SEQUENCES IN SOME SPECIES OF GRASSES.
10,000 YEARS AGO	BEGINNINGS OF AGRICULTURE. MAN DOMESTICATES DIPLOID WHEAT.
8,000 YEARS AGO	ORIGIN OF HEXAPLOID WHEAT.

Fig. 11 Hypothetical events in evolution

events (very approximately estimated) that place the appearance of grasses and, presumably, prolamins in the evolutionary time frame. Collagen-like genes were evidently in existence 500 million years ago or longer (BORNSTEIN and SAGE 1980) before plants and animals invaded the land. The colonization of the land took place about 400 million years ago while flowering plants (Angiosperms) probably diverged roughly 200 million years ago (see discussion of WILSON et al. 1977). Grasses developed within the flowering plant group at some later time. Fossil evidence for the evolution of grasses is sparse and acceptably defined and dated fossils go back no more than about 65 million years (THOMASSON 1980). Even allowing for a period of development, grasses may not be much older than about 100 million years, which would be relatively young in comparison with older gene families, such as that of collagen.

Acknowledgements

We thank D. K. MECHAM, M. D. DIETLER, A. RYBALKA, B. G. FORDE, B. J. MIFLIN, B. L. JOHNSON, C. C. NIMMO, E. J.-L. LEW and J. E. BERNARDIN for permission to discuss unpublished results.

Zusammenfassung

Genetische Beziehungen zwischen Gliadinproteinen des Weizens

Die chromosomalen Zuordnungen von Komponenten der Gliadinproteine aus einigen unterschiedlichen Brotweizensorten (*Triticum aestivum* L. em. Thell.), die durch zweidimensionale Elektrophorese aufgetrennt wurden, wurden analysiert. Die Muster der Proteine, die sowohl vom A-Genom als auch vom D-Genom kodiert wurden, enthielten α-Gliadine. Die durch das B-Genom kodierten Proteine zeichneten sich durch das Fehlen von α-Gliadinen und von langsamer wandernden ω-Gliadinen aus. Auf der Grundlage der N-terminalen Aminosäuresequenzen können die Gliadine in drei Haupttypen unterteilt werden: α-Typ, γ-Typ und ω-Typ. Diese Einteilung entspricht offensichtlich den gesonderten Unterfamilien der Gliadinfamilie, da wahrscheinlich alle Gliadingene miteinander verwandt sind und von einer sich wiederholenden Vorläufersequenz abgeleitet werden können. Jedoch können Teile der Gliadinpolypeptidketten das Ergebnis unabhängiger Genmodifikationen sein, die einige Gliadingene in späteren Entwicklungsperioden betrafen; diese modifizierten Regionen der Gliadingene müssen nicht mit anderen Gliadingenen homolog sein. Gliadingene sind wahrscheinlich sehr junge Gene, die sich innerhalb der Familie der Gräser entwickelt haben, die selbst eine späte Entwicklungsstufe darstellen.

Краткое содержание

Генетические связи между глиадиновыми протеинами у пшеницы

Анализируется хромосомальная принадлежность компонентов глиадиновых протеинов из различных хлебных пшениц (*Triticum aestivum* L.), выделенных путём двухмерного электрофореза. Структура протеинов, которые кодируются как геномом A, так и геномом D, включает α-глиадины. Протеины, кодируемые геном B, отличались отсутствием α-глиадинов и более медленно перемещающимися ω-глиадинами. По типу N-терминальных аминокислотных последовательностей глиадины могут быть отнесены к трём основным типам: тип α, тип γи тип ω. Эта группировка, очевидно, соответствует отдельным подгруппам глиадинов, так как все гены глиадинов родственны между собой и могут вести начало от одной, повторяющейся предварительной последовательности. Однако, части полипептидной цепи могут быть и результатом независимых генных модификаций, которые произошли у некоторых глиадиновых генов в более поздних периодах развития; эти модифицированные регионы глиадиновых генов не должны быть гомологичны другим глиадиновым генам. Вероятно, глиадиновые гены являются очень молодыми, развившимися в семействе злаков, которое и само является поздней стадией развития.

Literature

BALTIMORE, D., 1981: Gene conversion: some implications for immunoglobulin genes. – Cell **24**, 592–594.

BARTELS, D., and R. D. THOMPSON, 1983: The characterization of cDNA clones coding for wheat storage proteins. – Nucleic Acids Res. **11**, 2961–2977.

BIETZ, J. A., and J. S. WALL, 1980: Identity of high molecular weight gliadin and ethanol-soluble glutenin subunits of wheat: relation to gluten structure. – Cereal Chem. **57**, 415–421.

–, T. BURNOUF, L. COBB, and J. S. WALL, 1984: Wheat varietal identification and genetic analysis by reversed-phase high-performance liquid chromatography. – Cereal Chem. **61**, 124–129.

–, F. R. HUEBNER, J. E. SANDERSON and J. S. WALL, 1977: Wheat gliadin homology revealed through N-terminal amino acid analysis. – Cereal Chem. **54**, 1070–1083.

BORNSTEIN, P., and H. SAGE, 1980: Structurally distinct collagen types. – Annual Rev. Biochem. **49**, 957–1003.

BUSHUK, W., and E. R. KERBER, 1978: The role of *Triticum carthlicum* in the origin of bread wheat based on gliadin electropherograms. – Can. J. Plant Sci. **58**, 1019–1024.

COLE, E. W., J. G. FULLINGTON, and D. D. KASARDA, 1981: Grain protein variability among species of *Triticum* and *Aegilops*: quantitative SDS-PAGE studies. – Theor. Appl. Genet. **60**, 17–30.

DU CROS, D. L., L. R. JOPPA, and C. W. WRIGLEY, 1983: Two-dimensional analysis of gliadin proteins associated with quality in *durum* wheat: chromosomal location of genes for their synthesis. – Theoret. Appl. Genet., **66**, 297–302.

EGEL, R., 1981: Intergenic conversion and reiterated genes. – Nature **290**, 191–192.

EWART, J. A. D., 1969: Comparison of seed proteins of some gramineae. – J. Sci. Food Agric. **20**, 221–224.

FELDMAN, M., 1978: New evidence on the origin of the B genome of wheat. – In: Proc. 5th Wheat Genetics Symp. (Ed. S. RAMANUJAM) pp. 120–132. Indian Society Genet. Plant Breeding, New Delhi.

FLAVELL, R. B., 1982: Sequence amplification, deletion, and rearrangement: major sources of variation during species divergence. – In: Genome Evolution (Ed. G. A. DOVER and R. B. FLAVELL) pp. 301–324. Academic Press, London.

GARCIA-OLMEDO, F., P. CARBONERO, and B. L. JONES, 1982: Chromosomal locations of genes that control wheat endosperm proteins. – In: Advances in Cereal Science and Technology. Vol. V (Ed. Y. POMERANZ) pp. 1–47. American Assn. Cereal Chemists, St. Paul, Minn.

JACKSON, E. A., L. M. HOLT, and P. I. PAYNE, 1983: Characterization of high-molecular-weight gliadin and low-molecular-weight glutenin subunits of wheat endosperm by two-dimensional electrophoresis and the chromosomal localisation of their controlling genes. – Theor. Appl. Genet. **66**, 29–37.

KASARDA, D. D., 1980: Structure and properties of α-gliadins. – Ann. Technol. Agric. **29**, 151–173.

–, 1981: Toxic proteins and peptides in celiac disease: relations to cereal genetics. – In: Food, Nutrition, and Evolution: Food as an Environmental Factor in the Genesis of Human Variability (Eds. D. WALCHER, and N. KRETCHMER) pp. 201–216. Masson Publishing Co., N. Y.

–, J. E. BERNARDIN, and C. C. NIMMO, 1976: Wheat proteins. – In: Advances in Cereal Science and Technology, Vol. I (Ed. Y. POMERANZ) pp. 158–236.

–, J.-C. AUTRAN, C. C. NIMMO, E. J.-L. LEW, and P. R. SHEWRY, 1983: N-Terminal amino acid sequences of ω-gliadins and ω-secalins: implications for the evolution of prolamin genes. Biochim.-Biophys. Acta, in press.

LAWRENCE, G. J., and K. W. SHEPHERD, 1980: Variation in glutenin subunits of wheat. – Aust. J. Biol. Sci. **33**, 221–233.

MECHAM, D. K., D. D. KASARDA, and C. O. QUALSET, 1978: Genetic aspects of wheat gliadin proteins. – Biochem. Genet. **16**, 831–853.

MIFLIN, B. J., J. M. FIELD, and P. R. SHEWRY, 1983: Cereal storage proteins and their effect on technological properties. – In: Plant Proteins (Eds. J. DAUSSANT, J. MOSSE and J. VAUGHN) pp. 255–319. Academic Press, New York.

–, L. M. HOLT, and C. M. LAW, 1981: Structural and genetic studies on the high-molecular-weight subunits of wheat glutenin. I. Allelic variation in subunits amongst varieties of wheat *(Triticum aestivum)*. – Theor. Appl. Genet. **60**, 229–236.

–, –, G. J. LAWRENCE, and C. N. LAW, 1982a: The genetics of gliadin and glutenin, the major storage proteins of the wheat endosperm. – Qualitas Plantarum **31**, 229–241.

–, –, A. J. WORLAND, and C. N. LAW, 1982b: Structural and genetical studies on the high-molecular-weight subunits of wheat glutenin. Part 3. Telocentric mapping of the subunit genes on the long arms of the homoeologous group 1 chromosomes. – Theor. Appl. Genet. **63**, 129–138.

PAYNE, P. I., L. M. HOLT, E. A. JACKSON, and C. N. LAW, 1984: Wheat storage proteins: their genetics and their potential for manipulation by plant breeding. – Phil. Trans. R. Soc. London B **304**, 359–371.

ROSE, M. R., and W. F. DOOLITTLE, 1983: Molecular biological mechanisms of speciation. – Science **220**, 157–162.

RYBALKA, A., and A. A. SOZINOV, 1979: Mapping of the Gld 1B locus, which controls storage protein biosynthesis in bread wheats. (in Russian). – Tsitologiya i Genetika **13**, 276–282.

SALCEDO, G., J. PRADA, and C. ARAGONCILLO, 1979: Low-MW gliadin-like proteins from wheat endosperm. – Phytochemistry **18**, 725–727.

SEARS, E. R., 1954: The aneuploids of common wheat. – Missouri Agricultural Experiment Station Research Bulletin 572, Columbia, Mo.

SHEPHERD, K. W., 1968: Chromosomal control of endosperm proteins in wheat and rye. – In: Proc. 3rd Int. Wheat Genetics Symp. (Eds. K. W. FINLAY, and K. W. SHEPHERD) pp. 86–96. Australian Academy of Sciences, Canberra.

–, 1973: Homoeology of wheat and alien chromosomes controlling endosperm protein phenotypes. – In: Proc. 4th Int. Wheat Genetics Symp. (Eds. E. R. SEARS, and L. M. S. SEARS) pp. 745–760. Agricultural Experiment Station, Univ. of Missouri, Columbia Mo.

SHEWRY, P. R., B. J. MIFLIN, and D. D. KASARDA, 1984: The structural and evolutionary relationships of the prolamin storage proteins of barley, rye, and wheat. – Phil. Trans. R. Soc. London B **304**, 297–308.

–, J.-C. AUTRAN, C. C. NIMMO, E. J.-L. LEW, and D. D. KASARDA, 1980: N-Terminal amino acid sequence homology of storage protein components from barley and a diploid wheat. – Nature **286**, 520–522.

–, B. J. MIFLIN, E. J.-L. LEW, and D. D. KASARDA, 1983: The preparation and characterization of an aggregated gliadin fraction from wheat. – J. Exp. Bot. **34**, 1403–1410.

SMITH, G. P., 1976: Evolution of repeated DNA sequences by unequal crossover. – Science **191**, 528–535.

SMITH, D. B., and R. B. FLAVELL, 1974: The relatedness and evolution of repeated nucleotide sequences in the genomes of some Gramineae species. – Biochem. Genet. **12**, 243–256.

SOZINOV, A. A., and F. A. POPERELYA, 1979: Polymorphism of prolamins and breeding. (In Russian.) – Vestnik Sel'skokhozyaystvennoy Nauki **10**, 21–34.

–, and –, 1980: Genetic classification of prolamines and its use in plant breeding. – Ann. Technol. Agric. **29**, 229–245.

THOMASSON, J. R., 1980: Paleoagrostology: a historical review. – Iowa State Journal of Research **54**, 301–317.

WILSON, A. W., S. S. CARLSON, and T. J. WHITE, 1977: Biochemical evolution. – Ann. Rev. Biochem. **46**, 573–639.

WRIGLEY, C. W., 1970: Protein mapping by combined gel electrofocusing and electrophoresis: application to the study of genotypic variations in wheat gliadins. – Biochem. Genet. **4**, 509–516.

–, 1982: The use of genetics in understanding protein composition and grain quality in wheat. – Qualitas Plantarum **31**, 205–227.

WRIGLEY, C. W. and K. W. SHEPHERD, 1973: Electrofocusing of grain proteins from wheat genotypes. – Ann. N. Y. Acad. Sci. **209**, 154–162.
ZIMMER, E. A., S. L. MARTIN, S. M. BEVERLY, Y. W. KAN, and A. C. WILSON, 1980: Rapid duplication and loss of genes coding for the α chains of hemoglobin. – Proc. Natl. Acad. Sci. USA **77**, 2158–2162.

D. D. KASARDA
Food Proteins Research Unit
Western Regional Research Center
Agricultural Research Service
U. S. Department of Agriculture
Berkeley, California 94710
U.S.A.

D. LAFIANDRA
Germplasm Institute
National Research Council
Via G. Amendola 165/A
70126 Bari
Italy

R. MORRIS
Department of Agronomy
University of Nebraska-Lincoln
Lincoln, Nebraska 68583
U.S.A.

P. R. SHEWRY
Biochemistry Department
Rothamsted Experimental Station
Harpenden, Herts. AL5 2JQ
U.K.

Kulturpflanze 32 · 1984 · S 53–S 62

The structure and expression of barley storage protein genes

P. R. SHEWRY, B. G. FORDE, M. KREIS, S. RAHMAN and B. J. MIFLIN
(Harpenden, Herts., U.K.)

Summary

The hordein storage proteins of barley are classified into three groups called B, C and D hordein. Each is a complex mixture of homologous polypeptides and is encoded by a single, probably polygenic, locus. These have been designated *Hor 1* (C hordein), *Hor 2* (B hordein) and *Hor 3* (D hordein). The B hordeins can be divided into two sub-families which have different cyanogen bromide cleavage patterns and are encoded by separate sub-families of mRNAs. The numbers of genes present at the *Hor* loci are not known, but at least 10–13 are present at *Hor 2*.

The amounts of D hordein, C hordein and the two sub-families of B hordein polypeptides accumulated in the grain vary as a function of stage of development and the relative availabilities of N and S. They are also affected by the presence of "regulatory" high-lysine genes. Analyses of RNA fractions from such grain show a close relationship between the rates of accumulation of the different groups of hordein polypeptides and the populations of RNAs related to them. This suggests that rates of hordein synthesis are determined largely by the abundances of mRNAs. How these populations are controlled is not known. There is also evidence, from the studies of the effects of sulphur starvation, that changes in hordein composition may also result from differences in efficiency of translation of mRNAs.

Introduction

Barley is the fourth most widely grown cereal in the world, with a total yield of almost 160×10^6 metric tonnes in 1981. It is well suited to temperate climates, and is the most important crop in a number of countries in northern Europe, including the U.K. Its main use is for feeding livestock, although significant amounts (up to 20 % in the U.K.) are used for malting and distilling.

Its nutritional value for feeding to monogastric animals (such as pigs, chickens and humans) is limited by the low level of the essential basic amino acid lysine. This is determined by the prolamin storage proteins (hordein), which account for about half of the total grain N, but contain less than 1 mol % lysine. There is therefore considerable interest in improving the nutritional quality of the grain by altering the amount or lysine content of hordein. To this end we have been studying the chemistry of hordein and the chromosomal locations, numbers, and control of expression of the hordein structural genes. The results of these studies, and similar ones carried out in other laboratories, are reviewed here.

The Hordein Proteins

Hordein is a mixture of polypeptides which can be separated by sodium dodecyl-sulphate polyacrylamide gel electrophoresis (SDS-PAGE) into three groups called B, C and D hordein. The general characteristics of these groups are summarized in Fig. 1. In addition at least two groups of minor polypeptides with M_rs below 20,000 are also present (Salcedo et al. 1980, 1982; Aragoncillo et al. 1981). These will not be considered here.

D hordein

This accounts for less than 5 % of the total fraction, and in most European cultivars (including Sundance which is shown in Fig. 1) gives a single band on SDS-PAGE with an M_r of about 105,000 (Shewry and Miflin 1982). D hordein is characterized by a high content of glycine (about 14 mol %) and relatively low proline (11 mol %). The amino acid sequence of D hordein has not been reported, but the N-terminal amino acid sequences of a number of the homologous "high molecular weight" (HMW) prolamin subunits of wheat have been determined (Shewry et al. 1984). Forde et al. (1983) have also determined the nucleotide sequence of a cDNA clone corresponding to the C-terminal end of an HMW subunit. This showed that the central part of the polypeptide chain is based on repeated blocks of six and nine amino acids. Since this cDNA clone hybridized to one for D hordein it is probable that D hordein has a similar structure.

C hordein

C hordein accounts for about 10–20 % of the fraction, and gives a number of bands on SDS-PAGE with M_rs between 55,000 and 75,000. It is extremely rich in glutamine (about 40 mol %), proline (30 mol %) and phenylalanine (9 mol %). Amino acid sequences have been determined for the N-termini of whole proteins (Shewry et al. 1980a, 1981) and peptides produced by digestion with chymotrypsin (Shewry and March, unpublished). Further sequence information has been derived from the nucleotide sequence of a C hordein-related cDNA clone (Forde, Kreis, Fry and Wlliamson unpublished). At the N-terminus a distinct domain of 12 residues is followed by at least three repeated blocks of five residues. The C-terminal domain is shorter, only six residues, and is preceded by repeated blocks of eight rather than five residues. This octapeptide appears to correspond to the N-terminal pentapeptide plus a tripeptide. Similar octapeptide repeats were present in the unordered chymotryptic peptides. The consensus composition of the octapeptide (4 gln : 3 pro : 1 phe) is similar to the ratio of these amino acids in C hordein (40 : 30 : 9). Since gln, pro and phe together account for almost 80 %

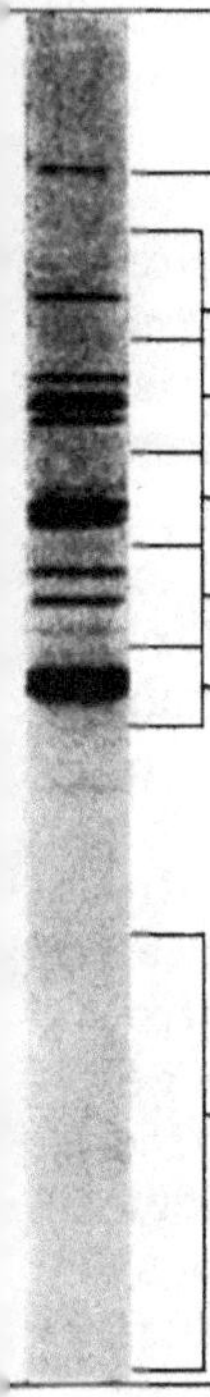

SDS-PAGE (cv. Sundance)	Hordein Group	M_r by SDS-PAGE	SEU[a]	Partial Amino Acid Composition (mol %) Glx[b]	Pro	Gly	Phe	Lys	Cys	Met	% total fraction
	D	105,000	56,000	29.6	11.4	13.6	1.3	0.8	1.7	0.6	2–4
	C2, C1	55–75,000	52,000	41.2	30.6	0.3	8.8	0.2	0	0.2	10–25
	B3	46,000	35,000	35.1	21.6	1.8	4.1	0.2	2.2	1.0	70–85
	B2	–	–	–	–	–	–	–	–	–	
	B1	35,000	32,000	32.0	20.8	2.9	5.1	0.8	2.7	0.9	
	A hordein	12–15,000	–	13.6	10.1	8.7	3.3	3.1	6.4	1.9	1–2
	LMW hordein	10–16,000	–	25.5	17.0	5.5	4.4	0.1	5.3	3.7	?

[a] sedimentation equilibrium ultracentrifugation; [b] about 90% amidated in B and C hordein.

Based on SHEWRY et al. 1980 b; FIELD et al. 1982; ARAGONCILLO et al. 1981; SALCEDO et al. 1980; KIRKMAN et al. 1982; MIFLIN et al. 1983.

Fig. 1 The characteristics of the groups of hordein polypeptides

of the total residues, the octapeptide is probably the basic building block for most of the protein. The pentapeptide found in the N-terminal region may have been derived from this by deletion.

B hordein

This is the quantitatively major group, and is composed of a number of polypeptides with M_rs of 35,000 to 46,000 by SDS-PAGE. They have less glutamine and proline but more cysteine (about 2.5 mol %) than C hordein. The N-terminus is blocked to Edman degradation, but amino acid sequences have been determined for unordered peptides (SCHMITT and SVENDSEN, 1980) and by determining the nucleotide sequence of a cDNA clone corresponding to the C-terminal part of a B hordein polypeptide (FORDE et al., 1981). These showed no evidence of repeated units at the nucleotide or protein levels. However, we have recently determined the nucleotide sequences of two longer cDNA clones encoding B1 and B3 type hordeins (FORDE, KREIS, FRY and WILLIAMSON, unpublished results). Both have long 3′ (carboxy-terminal) domains with no repeated units, and shorter 5′ domains with repeated blocks of eight residues. These octapeptides are related to those present in C hordein, indicating a common origin for C hordein genes and at least part of the B hordein genes.

Chromosomal location of hordein structural loci

The B, C and D groups of hordein polypeptides are encoded by single structural loci designated *Hor 2*, *Hor 1* and *Hor 3* respectively. The mapping of *Hor 1* and *Hor 2* has been reviewed by SHEWRY and MIFLIN (1982) and will only be summarized here. Both are located on the short arm of chromosome 5, with *Hor 1* closest to the centromere. JENSEN (1983) has calculated the map distance between them as 7.6 cM. The position of the centromere on chromosome 5 is not known for certain, but is probably about 50 cM from *Hor 1* (see JENSEN 1983). The *Hor 3* locus is located on the long arm of chromosome 5 (LAWRENCE and SHEPHERD 1981), and is loosely linked to *Hor 1*. SHEWRY et al. (1983a) have reported a map distance from *Hor 1* to *Hor 3* of 65 cM.

The presence of additional structural loci for hordeins has been reported by SOZINOV and coworkers. These workers have designated *Hor 1* and *Hor 2* as *Hrd A* and *Hrd B* respectively. Three additional loci, *Hrd C, D,* and *E,* control minor polypeptides which migrate slightly faster than C hordein on starch gel electrophoresis. These loci are closely linked to each other, with *Hrd C* located about 2.4 cM distally to *Hor 2 (Hrd B)* on the short arm of chromosome 5 (SOZINOV et al. 1978). A further hordein structural locus, *Hrd F,* is located about 0.3 cM distally to *Hor 2 (Hrd B)* (NETSVETAEV and SOZINOV 1982). This appears to encode the B hordein polypeptides with the fastest mobilities on starch gel electrophoresis. It possibly corresponds to one of the two sub-families of genes at the *Hor 2* locus, which have been described by KREIS et al. (1983a) (see below).

Numbers and Sub-families of Structural Genes

FAULKS et al. (1981) separated B hordein fractions from 8 cultivars with different *Hor 2* alleles by 2-D IEF/SDS-PAGE. A total of 47 major polypeptides were present, with 8 to 17 present in any one cultivar. The polypeptides were classified into three groups (called classes I, II and III) on the basis of their cyanogen bromide peptide maps. One class of polypeptide (class III) was present in all 8 cultivars, while 7 of the cultivars also had polypeptides of class I or class II, but not of both classes. In the cultivar Sundance the class III polypeptides correspond to the B2 and B3 hordein bands and the class I polypeptides to the B1 hordein band.

FORDE et al. (1981) reported the construction of a library of cDNA clones, using as a template an mRNA fraction prepared from developing endosperms of the cultivar Sundance. About 50 of these clones were identified as related to mRNAs for B hordein polypeptides. Detailed analysis of these clones by hybrid selected translations using mRNA fractions from a range of cultivars (KREIS et al. 1983a) showed the presence of two sub-families of cDNAs (and hence mRNAs), which encoded either class III or class I and class II polypeptides. It was concluded that the *Hor 2* locus consists of two sub-families of genes, encoding mRNAs for class III and class I/II polypeptides. The latter sub-family has diverged in different cultivars to encode polypeptides with two different CNBr cleavage patterns. This sub-family may also correspond to the *Hrd F* locus of NETSVETAEV and SOZINOV (1982). The relationships of the subfamilies of cDNAs and classes of polypeptides are summarized in Fig. 2.

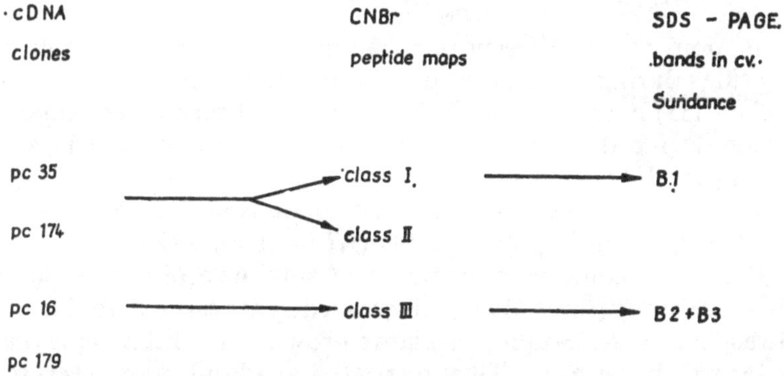

Fig. 2 Summary of subfamilies of B hordein polypeptides and B hordein-related cDNA clones. CNBr, cyanogen bromide

The total number of genes at the *Hor 2* locus is not known for certain. However KREIS et al. (1983b, 1984) have prepared total DNA from barley shoots, digested it with a restriction endonuclease and separated the fragments on an agarose gel. The DNA fragments were then blotted onto nitrocellulose and hybridized to ^{32}P-labelled DNA from the cDNA clone pc179. This clone has an insert of about 500 base pairs, which encodes the C-terminal part of a class III B hordein polypeptide. However, under the conditions of hybridization used this should have hybridized to DNA sequences related to the major classes of B hordein polypeptides. From these experiments it was concluded that at least 10 B hordein genes per haploid genome are present in the cultivar Bomi and 13 in the cultivar Carlsberg II. These numbers may be underestimates, as it has not been proved that this cDNA clone hybridizes to all B hordein genes. There were no differences in the numbers of genes per haploid genome present in endosperm and shoot tissues. The numbers of genes in each sub-family have not been determined.

Nothing is known about the organisation and numbers of genes at the *Hor 1* and *Hor 3* loci, but SHEWRY et al. (1981) have suggested that the C hordein polypeptides present in the cultivar Maris Mink can be classified into two sub-families differing in their M_rs and peptide maps.

Control of Expression of *Hor* loci

Hor 1, *Hor 2* and *Hor 3* are co-ordinately regulated in that they are only expressed in the endosperm and at a certain stage of development. Also, they always appear to be expressed together, no one group of hordein polypeptides being synthesized without the others. Within this broadly synchronized expression there are differences in the relative levels of synthesis and accumulation of the

D hordein, C hordein and sub-families of B hordein polypeptides. These occur as a function of time of endosperm development, in response to the availability of N and S and as a result of the presence of mutant high-lysine genes. We have studied these effects at the protein and mRNA levels.

Effects on Protein Accumulation

RAHMAN et al. (1982) demonstrated differences in the rates of accumulation of B and C hordein polypeptides during endosperm development. In the cultivar Sundance the C and B3 (class III) hordein polypeptides accounted for higher proportions of the total fraction during the early stage of development, with B1 (class I) hordein increasing during the later stages.

High availability of N results in increased total accumulation of hordein, and an increased proportion of C hordein polypeptides (KØIE et al. 1976, KIRKMAN et al. 1982). Because C hordein contains only traces of sulphur amino acids (see Fig. 1), KIRKMAN et al. (1982) suggested that its increased synthesis resulted from low availability of S relative to N. Studies of plants grown with different levels of S provide support for this hypothesis. Thus restricted availability of S results in decreases in the amounts of B and D hordein per seed but little or no effect on the amount of C hordein (SHEWRY et al. 1983d). There is also a decrease in the proportion of B1 (class I) hordein within the B hordein group (RAHMAN et al. 1983b).

Several mutant high lysine genes also affect the proportions of hordein polypeptides (KØIE and DOLL 1979, DOLL 1983). Risø mutants 7 and 13 both affect the proportions of B hordein polypeptides. These were induced in Bomi, which has the same structural allele at the *Hor 2* locus (*Hor 2* Ze) as Sundance. Risø 7 has an increased proportion of B1 hordein and Risø 13 an increased proportion of B3 hordein. Risø mutants 29 and 86 were induced in Carlsberg II and are allelic to Risø 13. They probably have similar effects on the different classes of B hordein polypeptides as mutant 13, but this cannot be definitely concluded as Carlsberg II has a different *Hor 2* allele (*Hor 2* Ca) to Bomi.

The *lys 3a* gene of Risø mutant 1508 has a more dramatic effect on the hordein phenotype (KREIS et al. 1984). D hordein is apparently not affected but C hordein is eliminated, or almost so. The total amount of B hordein is drastically reduced, and the electrophoretic properties of its component polypeptides are altered. We have studied the effects of the mutation on 8 lines with different genetic backgrounds and with two different *Hor 2* structural alleles (*Hor 2* Ze and *Hor 2* Pr). This showed that the mutant gene had a greater effect on class I than on class III B hordein polypeptides.

All the above mutant genes can be called regulatory in that they are not located at the hordein structural loci. A further mutation, present in Risø 56, causes the elimination of specific B hordein polypeptides. This maps at or close to the *Hor 2* locus (DOLL 1980) and appears to be a deletion of some of the B hordein structural genes (KREIS et al. 1983b).

Effects on mRNA populations

The populations of RNAs related to different groups of hordein polypeptides were studied by *in vitro* translation and by hybridization to specific cDNA clones using the dot hybridization procedure of THOMAS (1980). The RNA fractions analyzed were total cellular RNA, polysomal RNA and polysomal poly A^+ RNA (derived from membrane-bound polysomes). These were prepared from endosperms at different stages of development (RAHMAN et al. 1984); from plants grown under different availability of S (RAHMAN et al. 1983) and from Bomi and Risø mutant 1508 (KREIS et al. 1984), the latter two lines differing from each other only in the presence of the *lys 3a* gene.

The *in vitro* translations showed good agreement between the proportions of B1 (class I) and B3 (class III) hordeins accumulated in the grain and the proportions of the *in vitro* products. The proportions of C hordein products were always less than would be expected from the amounts of C hordein accumulated, but showed similar trends during development and in reponse to availability of S. Little or no C hordein products were observed with RNA fractions from Risø 1508. There was also little or no synthesis of D hordein with any of the RNA fractions. The unexpectedly low amounts of C and D hordein products may have been due to less efficient translation of large mRNAs.

A more precise determination of RNA populations is given by the hybrid dot procedure. Specific cDNA probes were used for C hordein, B1 (class I) hordein, B3 (class III) hordein and, in the study of Risø 1508 and Bomi, D hordein. In the study of barley endosperm development and in the comparison of Risø 1508 and Bomi there was good agreement between the populations of RNAs and the relative rates of protein accumulation. These results indicate that the relative rates of protein synthesis are regulated by the abundances of their respective mRNAs. This may be due to control of transcription, but we have not measured this directly and cannot rule out the possibility that it is partly or wholly due to changes in the rates of mRNA degradation.

There was also evidence for translational effects on hordein synthesis in the grain grown under low availability of S. Thus whereas the amount of C hordein per unit weight was increased two-fold compared to that in grain grown with adequate sulphur, the amount of C hordein-related RNA was the same. The population of RNAs related to D hordein was not determined, but those for B1 and B3 hordeins were greater than would be expected from the amounts of polypeptides synthesized. Thus the C hordein mRNA appeared to be translated more efficiently under sulphur stress, probably due to reduced competition for the protein synthetic machinery from mRNAs for B and D hordeins.

Acknowledgements

We would like to thank the following colleagues at Rothamsted for advice, discussion and practical assistance – SAROJ PARMAR, AUDREY FAULKS, JACKIE PYWELL, RICHARD FRY, MARTIN WILLIAMSON, LOURDES BLANCO, SUSAN SMITH and JULIAN FRANKLIN.

We are also grateful to Dr. H. DOLL (Risø National Laboratory, Roskilde, Denmark) and Dr. D. D. KASARDA (USDA, Albany, California, USA) for advice and discussions. We are grateful to the EEC for financial support (Grant No. 470).

Zusammenfassung

Struktur und Expression der Reserveproteingene der Gerste

Die Hordeinreserveproteine werden in drei Gruppen, B-, C-, und D-Hordein, eingeteilt. Jede ist ein komplexes Gemisch homologer Polypeptide und wird von einem einzigen, wahrscheinlich polygenen, Locus kodiert. Diese sind mit *Hor 1* (C-Hordein), *Hor 2* (B-Hordein) und *Hor 3* (D-Hordein) bezeichnet worden. Die B-Hordeine können in zwei Unterfamilien eingruppiert werden, die bei der Bromcyanspaltung unterschiedliche Muster ergeben und die durch verschiedene mRNS-Unterfamilien kodiert werden.

Die Mengen an D-Hordein, C-Hordein und den beiden Unterfamilien von B-Hordein, die im Korn akkumuliert werden, variieren in Abhängigkeit von Entwicklungsstadium und den relativen Verfügbarkeiten von N und S. Sie werden auch durch die Gegenwart „regulatorischer" Hochlysingene beeinflußt. Analysen der RNS-Fraktionen solcher Körner zeigen eine enge Beziehung zwischen den Akkumulationsraten der verschiedenen Gruppen von Hordeinpolypeptiden und den ihnen entsprechenden RNS-Populationen. Das deutet darauf hin, daß die Syntheseraten für Hordein hauptsächlich durch die Menge an mRNS bestimmt werden. Wie diese Populationen kontrolliert werden, ist nicht bekannt. Aus Untersuchungen über die Wirkungen von Schwefelmangel gibt es auch Hinweise, daß Änderungen der Hordeinzusammensetzung aus Unterschieden in der Translationseffizienz von mRNS's resultieren.

Краткое содержание

Структура и экспрессия запасных протеинов ячменя

Хордеиновые запасные белки подразделяются на три группы: хордеины B, C и D. Каждый из них является комплексной смесью гомологичных полипептидов и кодируется только одним, вероятно полигенным локусом. Они обозначаются как Hor 1 (хордеин C), Hor 2 (хордеин B) и Hor 3 (хордеин D). Хордеины B образуют две подгруппы, которые при расщеплении бромцианом дают различные структуры и кодируются различными подгруппами и-РНК.

Количество хордеина D и C, а также обеих подгрупп хордеина B, которое аккумулируется в зерне, варииируют в зависимости от стадии развития и относительной возможности использования азота и серы. Они подвержены также влиянию «регулирующих» высоколизиновых генов. Анализ фракций РНК таких зёрен показывает тесную связь между нормами аккумуляции различных групп хордеиновых полипептидов и соответствующих популяций РНК. Это указывает на то, что нормы синтеза для хордеинов определяются, главным образом, количеством и-РНК. Как эти популяции контролируются — не известно. Исследования о влиянии нехватки серы позволяют предположить, что изменения состава хордеинов являются результатом различий в эффективности трансляции и-РНК.

Literature

ARAGONCILLO, C., R. SANCHEZ-MONGE, and G. SALCEDO, 1981: Two groups of low molecular weight hydrophobic proteins from barley endosperm. – J. Exp. Bot. **32**, 1279–1286.

DOLL, H., 1980: A nearly non-functional mutant allele of the storage protein locus *Hor 2* in barley. – Hereditas **93**, 217–222.

–, 1983: Barley seed proteins and possibilities for their improvement. – In: The Genetics of Seed Proteins of Cereals and Legumes (Eds. W. GOTTSCHALK and H. MÜLLER) pp. 207–223. Martinus Nijhoff, The Netherlands.

FAULKS, A. J., P. R. SHEWRY, and B. J. MIFLIN, 1981: The polymorphism and structural homology of storage polypeptides (hordein) coded by the *Hor-2* locus in barley (*Hordeum vulgare* L.). – Biochem. Genet. **19**, 841–858.

FIELD, J. M., P. R. SHEWRY, B. J. MIFLIN, and J. F. MARCH, 1982: The purification and characterization of homologous high molecular weight storage proteins from grain of wheat, rye and barley. – Theor. Appl. Genet. **62**, 329–336.

FORDE, B. G., M. KREIS, M. B. BAHRAMIAN, J. A. MATTHEWS, B. J. MIFLIN, R. D. THOMPSON, D. BARTELS, and R. B. FLAVELL, 1981: Molecular cloning and analysis of cDNA sequences derived from poly A^+ RNA from barley endosperm: identification of B hordein related clones. – Nucl. Acids Res. **9**, 6689–6707.

FORDE, J., B. G. FORDE, R. FRY, M. KREIS, P. R. SHEWRY, and B. J. MIFLIN, 1983: Identification of barley and wheat cDNA clones related to the high Mr polypeptides of wheat gluten. – FEBS Lett. **162**, 360–366.

JENSEN, J., 1983: Coordinators report: chromosome 5. – Barley Genetics Newsletter 13, in press.

KIRKMAN, M. A., P. R. SHEWRY, and B. J. MIFLIN, 1982: The effect of nitrogen nutrition on the lysine content and protein composition of barley seeds. – J. Sci. Food Agric. **33**, 115–127.

KØIE, B., and H. DOLL, 1979: Protein and carbohydrate components in the Risø high lysine barley mutants. – In: Seed Protein Improvement in Cereals and Grain Legumes, pp. 205–215. IAEA, Vienna.

–, J. INGVERSEN, A. J. ANDERSEN, H. DOLL, and B. O. EGGUM, 1976: Composition and nutritional quality of barley protein. – In: Evaluation of Seed Protein Alterations by Mutation Breeding. pp. 55–61. IAEA, Vienna.

KREIS, M., S. RAHMAN, B. G. FORDE, J. PYWELL, P. R. SHEWRY, and B. J. MIFLIN, 1983a: Sub-families of hordein mRNA encoded at the *Hor 2* locus of barley. – Mol. Gen. Genet. **191**, 194–200.

–, P. R. SHEWRY, B. G. FORDE, S. RAHMAN, M. B. BAHRAMIAN, and B. J. MIFLIN, 1984: Molecular analysis of the effects of the mutant *lys 3a* gene on the expression of *Hor* loci in developing endosperms of barley (*Hordeum vulgare* L.). – Biochem. Genet. **62**, 231–255.

–, –, –, –, and B. J. MIFLIN, 1983b: Molecular analysis of a mutation conferring the high lysine phenotype on the grain of barley *(Hordeum vulgare)*. – Cell **34**, 161–167.

LAWRENCE, G. J., and K. W. SHEPHERD, 1981: Chromosomal location of genes controlling seed proteins in species related to wheat. – Theor. Appl. Genet. **59**, 25–31.

MIFLIN, B. J., J. M. FIELD, and P. R. SHEWRY, 1983: Cereal storage proteins and their effect on technological properties. – In: Seed Proteins (Eds. DAUSSANT, J., J. MOSSE, and J. VAUGHAN) pp. 255–319. Academic Press.

NETSVETAEV, V. P., and A. A. SOZINOV, 1982: Linkage studies of genes *Gle 1* and *Hrd F* in barley chromosome 5. – Barley Genetics Newsletter **12**, 13–17.

RAHMAN, S., P. R. SHEWRY, and B. J. MIFLIN, 1982: Differential protein accumulation during barley grain development. – J. Exp. Bot. **33**, 717–728.

–, M. KREIS, B. G. FORDE, P. R. SHEWRY, and B. J. MIFLIN, 1984: Hordein mRNAs in developing barley grain. – Biochem. J. **233**, 315–322.

–, P. R. SHEWRY, B. G. FORDE, M. KREIS, and B. J. MIFLIN, 1983b: Nutritional control of storage protein synthesis in developing grain of barley (*Hordeum vulgare* L.) – Planta **159**, 366–372.

SALCEDO, G., R. SANCHEZ-MONGE, A. ARGAMENTERIA, and C. ARAGONCILLO, 1980: The A

hordeins as a group of salt-soluble hydrophobic proteins. – Plant Sci. Lett. **19**, 109–119.
–, –, –, –, 1982: Low molecular weight prolamins: purification of a component from barley endosperm. – J. Agr. Food Chem. **30**, 1155–1157.
SCHMITT, J. M., and I. SVENDSEN, 1980: Partial amino acid sequence from hordein polypeptide B1. – Carlsberg Res. Commun. **45**, 549–555.
SHEWRY, P. R., and B. J. MIFLIN, 1982: Genes for the storage proteins of barley. – Qual. Plant. Foods Hum. Nutr. **31**, 251–267.
–, J.-C. AUTRAN, C. C. NIMMO, E. J.-L. LEW, and D. D. KASARDA, 1980a: N-terminal amino acid sequence homology of storage protein components from barley and a diploid wheat. – Nature **286**, 520–522.
–, J. M. FIELD, M. A. KIRKMAN, A. J. FAULKS, and B. J. MIFLIN, 1980b: The extraction, solubility and characterization of two groups of barley storage polypeptides. – J. Exp. Bot. **31**, 393–407.
–, E. J.-L. LEW, and D. D. KASARDA, 1981: Structural homology of storage proteins coded by the *Hor 1* locus of barley (*Hordeum vulgare* L.). – Planta **153**, 246–253.
–, J. M. FIELD, A. J. FAULKS, S. PARMAR, B. J. MIFLIN, M. D. DIETLER, E. J.-L. LEW, and D. D. KASARDA, 1984: Purification and N-terminal amino acid sequence analysis of high molecular weight (HMW) gluten polypeptides of wheat. – Biochim. Biophys. Acta **788**, 23–24.
–, R. FINCH, S. PARMAR, J. FRANKLIN, and B. J. MIFLIN: 1983a: Chromosomal location of *Hor 3*, a new locus governing storage proteins in barley. – Heredity, 50, 179–189.
–, J. FRANKLIN, S. PARMAR, S. J. SMITH, and B. J. MIFLIN, 1983b: The effects of sulphur starvation on the amino acid and protein compositions of barley grain. – J. Cer. Sci. **1**, 21–31.
SOZINOV, A. A., V. P. NETSVETAEV, E. M. GRIGORYAN, and I. S. OBRAZTSOV, 1978: Mapping of *Hrd* locuses in barley (*H. vulgare* L. emed. Vav et Bach). – Genetika, **14**, 1610–1619 (translated into English in Soviet Genetics 1979, **14**, 1137–1147).
THOMAS, P. S., 1980: Hybridization of denatured RNA and small DNA fragments transferred to nitrocellulose. – Proc. Natl. Acad. Sci. USA **77**, 5201–5205.

P. R. SHEWRY
Rothamsted Experimental Station
Biochemistry Department
Harpenden, Herts. AL5 2JQ
U.K.

Kulturpflanze 32 · 1984 · S 63

The genetics of gliadin and glutenin, storage proteins of the wheat endosperm

P. I. Payne, L. M. Holt and E. A. Jackson
(Cambridge, U.K.)

The gliadin proteins, which amount to about 50 % of the storage proteins in the endosperm, consist of a complex mixture of polypeptides which do not have a disulphide-bonded, subunit structure. The other storage protein group is glutenin, a large aggregated complex built up from 3–5 high-molecular-weight (HMW) subunits which amount to 10 % of the total storage protein complement and about 15 low-molecular-weight (LMW) subunits which amount to 40 %. The genes which code for all these proteins occur at nine complex loci on six different chromosomes. *Glu-A1, Glu-B1* and *Glu-D1* contain the genes of HMW glutenin subunits and are situated close to the centromere on the long arms of chromosomes 1 A, 1 B and 1 D respectively. Occurring on the short arms of the same chromosomes, but distant from the centromere, are *Gli-A1, Gli-B1* and *Gli-D1.* Each of these loci carry three major gene families coding for ω-gliadins, γ-gliadins and LMW glutenin subunits. The remaining loci, *Gli-A2, Gli-B2* and *Gli-D2* occur near the ends of the short arms of chromosomes 6 A, 6 B and 6 D respectively and each code for α- and β-gliadins. Recombination of genes within a locus is very rare and has so far only been detected at *Glu-B1,* at the rate of one recombinant in one thousand progeny. Each locus displays allelic variation and this is responsible for differences amongst varieties in protein quality for making bread. The storage protein genes are being mapped to genes determining disease resistance, morphological characters and isoenzymes.

P. I. Payne, L. M. Holt, E. A. Jackson
Plant Breeding Institute
Maris Lane
Cambridge CB2 2LQ
U.K.

Kulturpflanze 32 · 1984 · S 65

Analysis of storage protein genes in the developing endosperm of wheat (*Triticum aestivum*) using cDNA clones

D. BARTELS, N. P. HARBERD, and R. D. THOMPSON
(Cambridge, U.K.)

Gliadins and glutenins, the major storage proteins in wheat (*Triticum aestivum*), can be grouped into a number of families of closely related polypeptides mapping as three complex loci (*Glu-1* and *Gli-1* on group 1 chromosomes (1 A, 1 B, 1 D) and *Gli-2* on group 6 chromosomes (6 A, 6 B, 6 D) (PAYNE et al. 1982).

In order to study the gene families encoding gliadins and glutenins a cDNA clone bank has been constructed using mRNA from the developing endosperm of wheat (var. Chinese Spring) as a template. Within this clone bank cDNA clones have been grouped into a number of sequence-related groups (BARTELS and THOMPSON 1983).

cDNA clones from different groups were used to select complementary RNA which was translated *in vitro* to identify polypeptides encoded. This method has allowed us to identify clones coding for several different, but related, gliadin families as well as high molecular weight (HMW) glutenin subunits.

DNA sequencing of a clone encoding a γ-like gliadin and one encoding HMW glutenin subunits has been carried out. Both sequences show characteristic amino acid compositions for these proteins. The sequences both contain internal tandem repeats in the coding regions. The γ-like gliadin clone shows homology to a B-hordein sequence.

By analysis of the organization of the storage protein sequences in the wheat genome we have identified the chromosome location of some gliadin families and the HMW glutenin family and obtained an estimate of the copy number at each locus. Use of selected cDNA clones to determine complementary RNA levels has indicated that the gliadin and glutenin genes are coordinately transcribed during endosperm development.

Literature

BARTELS, D. and R. D. THOMPSON, 1983: The characterization of cDNA clones coding for wheat storage proteins. – Nucl. Acids. Res. **10**, 2961–2978.

PAYNE, P., L. M. HOLT, G. J. LAWRENCE and C. N. LAW, 1982: The genetics of gliadin and glutenin, the major storage proteins of the wheat endosperm. – Qual. Plant. Foods Hum. Nutr. **31**, 229–241.

D. BARTELS, N. P. HARBERD, R. D. THOMPSON
Cytogenetics Department, Plant Breeding Institute
Cambridge CB2 2LQ
U.K.

Kulturpflanze 32 · 1984 · S 67–S 68

Expression of the zein genes of maize

G. FEIX and P. LANGRIDGE
(Freiburg, Fed. Rep. Germany)

The zein genes are the major proteins of maize endosperm, but they consist of only two predominant size classes. The expression of the genes coding for these proteins has proved to be very complex. The complexity derives firstly from the large number of zein genes in the maize genome, probably in excess of 100 copies, and secondly, from the unusual nature of the zein mRNAs.

Using lambda vectors we have isolated 140 zein specific genomic clones from maize DNA, of which 10 have been analysed to some extent. All clones analysed to date are unique.

Our interests have been primarily with the regions surrounding the zein genes and with the interpretation of the zein gene expression mechanisms. These studies are directed towards an understanding not only of the extremely high expression rate of the zein genes, but also the tissue and developmental specificity and the coordination of zein synthesis.

We observed that the zein mRNAs are present in maize endosperms principally as large precursor mRNAs (LANGRIDGE et al. 1982) even though the zein genes contain no intervening sequences, an unusual feature for eukaryotic genes. The origin of the smallest of these precursor RNAs (twice the size of the final 900 base zein mRNA) has been studied in detail using a genomic clone containing a 21,000 dalton zein gene. Using electron microscopy, *in vitro* transcription in *Hela* cell extracts and *Xenopus* oozyte germinal vesicle extracts, sequencing and S1 mapping with poly rA RNA from endosperm we have shown that the zein mRNAs are transcribed from at least two promoter regions (LANGRIDGE and FEIX 1983). The promoters are separated by about 1,000 base pairs of DNA with no protein coding function and each promoter has double RNA starts. It is presumed that this unusual promoter organisation is related to the high expression rate of the zein genes.

We have recently been able to obtain expression of cloned zein genes in *Acetabularia* and yeast, allowing now the *in vivo* study of zein expression. These systems allow us to overcome to some extent the limitations associated with the *in vitro* systems used to date and will, hopefully, allow a clear picture of the mechanisms of zein gene expression.

Literature

Langridge, P. and G. Feix, 1983: A zein gene of maize is transcribed from two widely separated promoter regions. – Cell **34**, 1015–1022.
–, J. A. Pintor-Toro and G. Feix, 1982: Zein precursor mRNAs from maize endosperm. – Mol. Gen. Genet. **187**, 432–438.

G. Feix and P. Langridge
Institut für Biologie III
Universität Freiburg
Schänzlestr. 1
D - 7800 Freiburg i. Br. (Fed. Rep. Germany)

Kulturpflanze 32 · 1984 · S 69 – S 70

Analysis of the processing steps of phaseolin in the endoplasmic reticulum and in the protein bodies of developing *Phaseolus vulgaris* cotyledons

R. Bollini
(Milano, Italy)

Mature bean seeds contain a major protein, phaseolin, which is synthesized during seed development by membrane-bound polysomes and accumulates in the endoplasmic reticulum (ER) before its transport to the protein bodies (PB) (Bollini et al. 1982). Processing of phaseolin in the ER has been studied in detail and consists in the detachment of a signal peptide and in the addition of one or two oligosaccharide side chains (Bollini et al 1983). *In vitro* translation of polysomes in a run-off system results in the formation of phaseolin polypeptides containing one or no oligosaccharide side chains, indicating that the first glycosylation is a cotranslational event. We could never detect the fully glycosylated polypeptides unless membranes were present in the *in vitro* system. *In vitro* translation, in the presence of unlabeled amino acids, of polysomes isolated from developing cotyledons labeled *in vivo* with [^{3}H] glucosamine results in the formation, among other, of two labeled polypeptides of similar electrophoretic mobility of the fully glycosylated phaseolin, indicating that also the second glycosylation is a cotranslational event.

When phaseolin is transferred to the PB it is further modified. The nature of this processing is not fully understood and we suggested that a proteolytic event is involved (Bollini et al. 1982). When developing cotyledons are incubated in the presence of tunicamycin, to inhibit glycosylation, unglycosylated phaseolin accumulates in the ER. I now show that these unglycosylated polypeptides are transferred to the PB where they have an higher electrophoretic mobility than their counterpart in the ER, confirming the proteolytic nature of the processing. Phaseolin isolated from the ER and PB of developing cotyledons labeled *in vivo* with [^{3}H] tyrosine, which is the C-terminal amino acid of the T type phaseolin (Slightom et al. 1983), was digested with carboxypeptidase A and the digests compared on TLC plates. No differences were detected between the ER and PB forms, suggesting that the proteolytic processing in the PB does not involve the C-terminus of the molecule.

Literature

Bollini, R., W. van der Wilden and M. J. Chrispeels, 1982: A precursor of the reserve protein, phaseolin, is transiently associated with the endoplasmic reticulum of developing *Phaseolus vulgaris* cotyledons. – Physiol. Plant. **55**, 82–92.

—, A. VITALE and M. J. CHRISPEELS, 1983: *In vivo* and *in vitro* processing of seed reserve protein in the endoplasmic reticulum: evidence for two glycosylation steps. — J. Cell. Biol. **96**, 999–1007.

SLIGHTOM, J. L., S. M. SUN and T. C. HALL, 1983: Complete nucleotide sequence of a French bean storage protein gene: Phaseolin. — Proc. Natl. Acad. Sci. USA **80**, 1897–1901.

R. BOLLINI
Istituto Biosintesi Vegetali
CNR
Via Bassini 15
Milano, Italia

Kulturpflanze 32 · 1984 · S 71–S 74

Cotranslational processing of storage globulin polypeptides of *Vicia faba* L. by signal recognition particle and microsomal membranes in a cell-free translation system

R. Bassüner, U. Wobus and T. A. Rapoport*
Gatersleben, GDR, *Berlin-Buch, GDR

Poly(A)$^+$ -RNA from developing field bean cotyledons was used to construct cDNA clones which where further identified and characterized as described (cf. Wobus et al. 1984). Using these clones a hybridization procedure was employed to select individual mRNAs from total poly(A)$^+$ RNA which code for seed proteins. The purified mRNA species were translated in the wheat germ cell-free system using either [^{3}H]leucine or [^{35}S]methionine (Fig. 1). The data support the previous observation made on total poly(A)$^+$-RNA (Bassüner et al. 1983b) that only one of the two main primary legumin translation products can be labeled with [^{35}S] methionine (i.e. preprolegumin A [ppLeA], M_r 67,000), whereas the other cannot (i.e. preprolegumin B [ppLeB], M_r 64,000). In addition, the main vicilin component (previcilin [pVi], M_r 54,000) was not labeled with [^{35}S] methionine either, in contrast to an unidentified product of M_r 30,000. The data are in agreement with the common observation that in most cases the initiator methionine residue is rapidly cleaved off by a specific methionyl-peptide exopeptidase contained in eucaryotic tissue extracts (cf. Housman et al. 1970; Yoshida and Lin 1972). Our data also support findings on the very low Met contents in the vicilin fraction (Derbyshire et al. 1976) and on the lack of Met in one of the two main types of legumin subunits from field beans (Horstmann 1983).

Individual mRNAs coding for storage globulin polypeptides and for the M_r-30,000 unidentified product were also used in cotranslational processing assays for the respective polypeptide products employing a reconstituted heterologous cell-free system. For this purpose, the translocating machinery contained in dog pancreas rough microsomes (RM) was fractionated into signal recognition particle (SRP) and potassium-washed rough microsomes (K-RM), devoid of SRP (Warren and Dobberstein 1978, Dobberstein 1978, Walter and Blobel 1980). The role of these constituents in cotranslational translocation of mammalian secretory proteins was described previously (Walter et al. 1981, Walter and Blobel 1981a, b).

In the presence of SRP the synthesis of storage globulin polypeptides was strongly inhibited. Added K-RM, which by themselves had no effect, restored translation and gave rise to processed products which resisted proteinase K attack and which were smaller than the original precursor by about M_r 1,500 (Bassüner et al. 1984b). These results indicate a cotranslational mode of translocation of the polypeptides across the ER membrane and shortening by removal of the signal peptide. Similar observations were made for an unidentified product of M 30,000 (unpublished results).

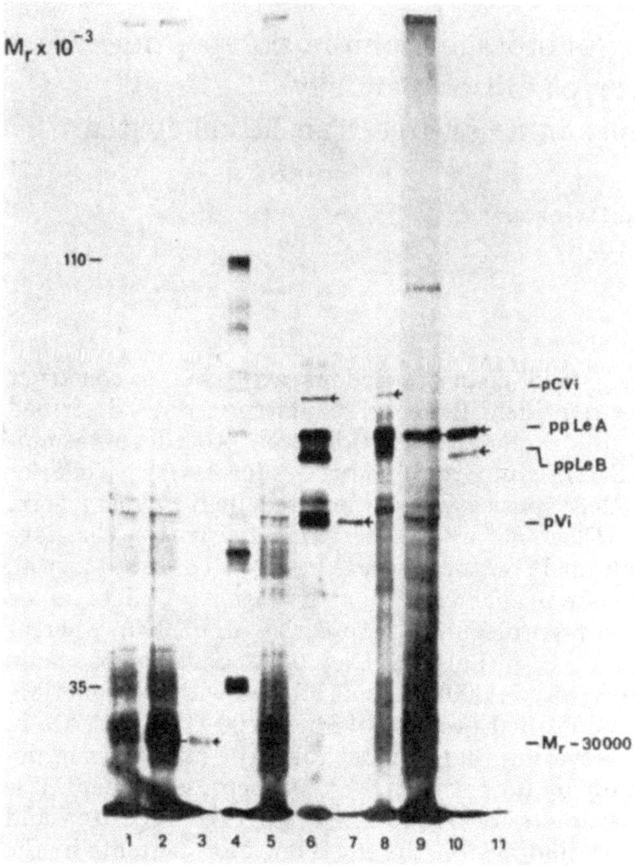

Fig. 1 Cell-free translation with either [^{3}H]leucine or [^{35}S]methionine of hybridization-selected mRNA species from developing field bean cotyledons (*Vicia faba* L.) coding for individual polypeptides.
mRNA species were isolated from cotyledonary poly(A)-containing RNA using plasmids of field bean cDNA clones and translated in the wheat germ cell-free system with either [^{3}H]leucine or [^{35}S]methionine.
The translation products were separated in an 8 % acrylamide SDS-gel and processed for fluorography.
Storage globulin-related translation products are designated as pCVi, preconvicilin; ppLeA, preprolegumin A; ppLeB, preprolegumin B; pVi, previcilin whereas "M_r-30,000" refers to a nonidentified translation product.
The M -scale is given by BMV RNA translation products (track 4).
Tracks represent:

1, 11 – endogenous translation products of the wheat germ system, labeled with [^{35}S] methionine (track 1) or with [^{3}H] leucine (track 11);
2, 3 – translation products of specific mRNA selected with pVfc 25 and labeled with [^{35}S] methionine (track 2) or [^{3}H] leucine (track 3);
5, 7 – translation products of specific mRNA selected with pVfc 28 and labeled with [^{35}S] methionine (track 5) or [^{3}H] leucine (track 7);
6, 8 – translation products of total poly(A)-containing RNA labeled with [^{3}H] leucine (track 6) or with [^{35}S] methionine (track 8);
9, 10 – translation products of mixed specific mRNA selected with either pVfc 67 or PVfc 77 and labeled with [^{35}S] methionine (track 9) or [^{3}H] leucine (track 10)

Storage globulin polypeptides	Primary translation products, shown by – *in vitro* translation of mRNA in a cell-free system (Bassüner et al. 1983b) (reduced)	Signal peptide-processed translation products, shown by – *in vitro* translation of mRNA in oocytes (Bassüner et al. 1983a) – *in vitro* translation of mRNA in the presence of SRP and K-RM (Bassüner et al. 1984b) – *in vivo* translation of *in vivo* initiated polysomes ('read out') (Bassüner et al. 1983b, 1984a) (reduced)	Authentic globulin polypeptides: (Bassüner et al. 1983b, Horstmann 1983) (not reduced)	(reduced)
Convicilin	74,000	72,000	72,000	72,000
Legumin A	67,000	66,000	58,000	
Legumin B	64,000	62,000	61,000	
Vicilin	54,000	52,000	52,000	52,000
Legumin A + B				
– acidic α chains				38,000
– basic β chains				ca. 22,000

From the results presented the following processing scheme of *Vicia faba* storage globulin polypeptides can be derived:

It is concluded, that mammalian SRP and signal peptidase recognize plant signal peptides, thus supporting our previous conclusion (Bassüner et al. 1983a) that storage globulin polypeptides of legumes use a translocation mechanism across the ER membrane homologous to that of animal secretory proteins.

Literature

BASSÜNER, R., A. HUTH, R. MANTEUFFEL, and T. A. RAPOPORT, 1983a: Secretion of plant storage globulin polypeptides by *Xenopus leavis* oocytes. – Eur. J. Biochem. **133**, 321–326.

–, R. MANTEUFFEL, K. MÜNTZ, M. PÜCHEL, P. SCHMIDT, and E. WEBER, 1983b: Analysis of *in vivo* and *in vitro* globulin formation during cotyledon development of field beans (*Vicia faba* L. var. *minor*). – Biochem. Physiol. Pflanzen **178**, 665–684.

–, A. HUTH, R. MANTEUFFEL, and T. A. RAPOPORT, 1984a: Secretion of plant storage globulin polypeptides by *Xenopus* oocytes and legumin processing steps. – Proc. 3rd Seed Protein Symp. 'Genetics of Seed Proteins' (MÜNTZ, K. and C. HORSTMANN, eds.). Kulturpflanze **32**, S 215–S 218.

–, U. WOBUS, and T. A. RAPOPORT, 1984b: Signal recognition particle triggers the translocation of storage globulin polypeptides from field beans (*Vicia faba* L.) across mammalian endoplasmic reticulum membrane. – FEBS Letters, **166**, 314–320.

DERBYSHIRE, E., D. J. WRIGHT, and D. BOULTER, 1976: Legumin and vicilin, storage proteins of legume seeds. – Phytochemistry **15**, 3–24.

DOBBERSTEIN, B., 1978: Assembly of functional rough microsomal membranes. Hoppe-Seyler's Z. Physiol. Chem. **359**, 1469–1470.

HORSTMANN, C., 1983: Specific subunit pairs of legumin from *Vicia faba*. – Phytochemistry **22**, 1861–1866.

HOUSMAN, D., M. JACOBS-LORENA, U. L. RAJBHANDARY, and H. F. LODISH, 1970: Initiation of haemoglobin synthesis by methionyl-RNA. – Nature **227**, 913–918.

WALTER, P., and G. BLOBEL, 1980: Purification of a membrane-associated protein complex required for protein translocation across the endoplasmic reticulum. – Proc. Natl. Acad. Sci. USA **77**, 7112–7116.

–, –, 1981a: Translocation of proteins across the endoplasmic reticulum. II. Signal Recognition Protein (SRP) mediates the selective binding to microsomal membranes of *in-vitro*-assembled polysomes synthesizing secretory protein. – J. Cell Biol. **91**, 551–556.

–, –, 1981b: Translocation of proteins across the endoplasmic reticulum. III. Signal Recognition Protein (SRP) causes signal sequence-dependent and site-specific arrest of chain elongation that is released by microsomal membranes. – J. Cell Biol. **91**, 557–561.

–, I. IBRAHIMI, and G. BLOBEL, 1981: Translocation of protein across the endoplasmic reticulum. I. Signal Recognition Protein (SRP) binds to *in-vitro*-assembled polysomes synthesizing secretory protein. – J. Cell Biol. **91**, 545–550.

WARREN, G., and B. DOBBERSTEIN, 1978: Protein transfer across microsomal membranes reassembled from separated membrane components. – Nature **273**, 569–571.

WOBUS, U., H. BÄUMLEIN, R. BASSÜNER, U. HEIM, R. JUNG, K. MÜNTZ, R. PANITZ, G. SAALBACH, and W. WESCHKE, 1984: Molecular characterization of *Vicia faba* storage protein specific DNA. – Proc. 3rd Seed Protein Symp. 'Genetics of Seed Proteins' (MÜNTZ, K. and C. HORSTMANN, eds.). Die Kulturpflanze **32**, S 117–S 126.

YOSHIDA, A., and M. LIN, 1972: NH_2-terminal formylmethionine- and NH_2-terminal methionine-cleaving enzymes in rabbits. – J. Biol. Chem. **247**, 952–957.

R. BASSÜNER, U. WOBUS
Zentralinstitut für Genetik und Kulturpflanzenforschung
der Akademie der Wissenschaften der DDR
DDR - 4325 Gatersleben
Corrensstraße 3

T. A. RAPOPORT
Zentralinstitut für Molekularbiologie
der Akademie der Wissenschaften der DDR
DDR - 1115 Berlin-Buch
Lindenberger Weg 70

Kulturpflanze 32 · 1984 · S 75–S 78

Molecular cloning of A_3B_4 intermediary subunit precursor cDNA and kinetics of the mRNA levels during the seed development of soybean (*Glycine max* L.)

C. FUKAZAWA and K. UDAKA
(Tsukuba, Japan)

Glycinin which is the most predominant globulin component stored in subcellular particles designated as protein bodies contains at least six acidic subunits (Ala, Alb, A2, A3, A4 and A5) and five basic subunits (Bla, Blb, B2, B3 and B4) (MOREIRA et al. 1979). In the seed development, A3-subunit which has the molecular weight of 40,000 appeared later than the other acidic subunits (FUKAZAWA and UDAKA 1979; MEINKE et al. 1981). This indicated that glycinin was composed of at least two isomers. In *Glycine max* var. Bonminori, we designated that the glycinin component which contained A3-subunit was as α-component and another component which lacked the acidic subunit as β-component. The content of α-glycinin was about 96 % of total glycinin, whereas the β-component was only 4 % of the mature seeds in an immunological sense.

As it is of interest to know how to control the biosynthesis of these glycinin isomers during the development, hitherto little has been known on the molecular mechanism. For the purpose of the understanding of molecular processes controlling protein accumulation during development, it is necessary to isolate the mRNA coding for specific a component. As there is a strong relatedness among the glycinin subunits in amino acid sequence and in immunological analysis (MOREIRA et al. 1979, HIRANO et al. 1983), the isolation of a specific glycinin subunit precursor mRNA has been difficult.

In this communication, we demonstrate the molecular cloning of glycinin A_3B_4 intermediary subunit precursor cDNA and kinetics of the mRNA levels during the seed development of soybean.

Glycinin A_3 and A_4 subunits were purified in homogeneous form, respectively, from dry mature seeds (*Glycine max* var. Bonminori) and the preparation of monospecific antibodies to A_3 and A_4-subunits was achieved by absorption technique, respectively. Figure 1 indicates that the both antibodies absorbed are monospecific in an immunoblot assay system. At first time, positive clones were screened in a soybean cDNA library (prepared to the cotyledonary total poly (A)-RNA at 38 days after flowering) by using a ^{32}p-labeled cDNA probe prepared from a glycinin-enriched fraction which was fractionated through an isokinetic sucrose gradient. About 120 clones in 2,500 clones of the cDNA library were positive. A recombinant plasmid, pGA_3B_472, containing a ds-cDNA insert of about 2.0 Kbp, was selected as a possible candidate for containing the A_3B_4 subunit sequence.

1 Abbreviations: Kbp, kilobase pairs; GAR-HRP, horseradish peroxidase-conjugated goat anti-rabbit immunoglobulin G (Bio-Rad Co.,); SDS, sodium dodecyl sulfate.

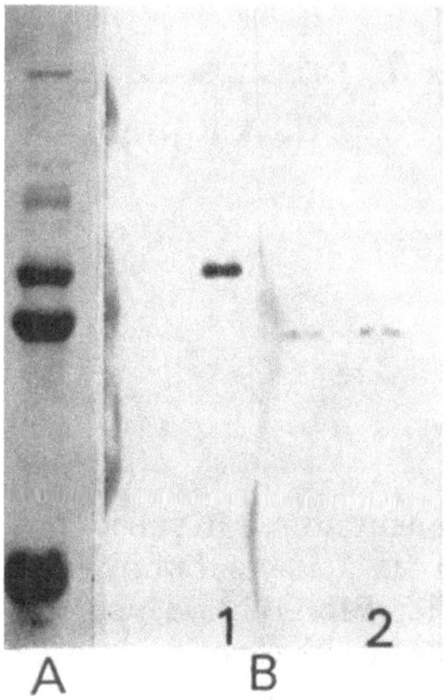

Fig. 1 Immun-blot GAR-HRP/rabbit anti-A_3 subunit first antibody. A. amido black stained gel; B. immunblot after electrophoretic transfer to nitrocellulose. Lane 1, absorbed anti-A_3 subunit serum; lane 2, absorbed anti-A4 subunit serum. Total glycinin was electrophoresed on 12.5 % polyacrylamide gels containing 0.1 % SDS (LAEMMLI 1970)

To establish the identity of the putative A_3B_4 subunit plasmid, this plasmid was digested with Pstl and subcloned into the Pstl site of bacteriophage M13mp7 (McLEAN et al. 1983). The recombinant M13 bacteriophages containing the coding strand (M13GA_3B_4 72m) and the noncoding complementary strand (M13GA_3B_4 72C) corresponding to the 3'-end of A_3B_4 subunit mRNA were then prepared. M13GA_3B_4 72C DNA immobilized on nitrocellulose and hybridized to the total cotyledonary poly(A)-RNA. The specifically bound RNA was eluted and translated in the cell-free, protein-synthesizing system. As shown in Fig. 2, lane 2, the M13GA_3B_4 72C DNA specifically hybridized to a mRNA that directed the synthesis of a single polypeptide of Mr = 63,000. This translation product was immunoprecipitated by specific antibodies to A_3-subunit (Fig. 2, lane 3), thus establishing the identity of the recombinant plasmid.

The size of A_3B_4 intermediary subunit was estimated by hybridizing ^{32}P-labeled cDNA prepared from M13GA_3B_4 72m DNA to nitrocellulose filter blots of poly(A)-RNA. The cDNA probe hybridized to a RNA species of about 2,300 nucleotides

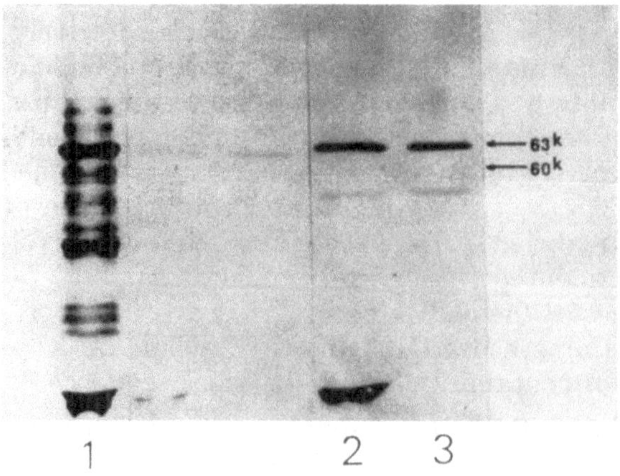

Fig. 2 Positive hybrid selection and cell-free translation of mRNA. Lane 1, total translation products of soybean cotyledonary poly(A)-RNA from membrane-bound polysomal fraction precipitated with a indirect immunoprecipitation procedure using anti-A3 subunit serum as a first antibody; lane 2, total translation products of hybrid-selected mRNA using M13 GA_3B_4 72C DNA; lane 3, immunoprecipitate of the translation products of lane 2 with a specific antibody to A_3 subunit

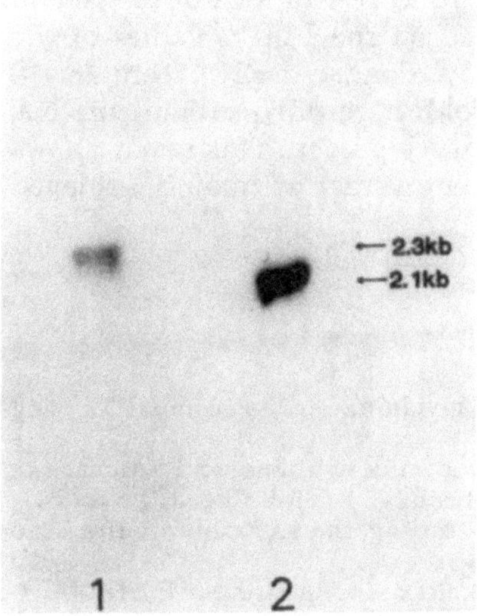

Fig. 3 Size of A_3B_4 subunit mRNA. Poly (A)-RNA(5 μg) from the developing cotylodonary tissue at 38 days after flowering were denatured, electrophoresed in agarose, transferred to nitrocellulose, and hybridized to ^{32}p-labeled pGA_3B_472 (A_3B_4 cDNA), lane 1, and $pGA_5A_4B_3103$ ($A_5A_4B_3$ cDNA), lane 2, respectively. The molecular weight markers used are Hind III restriction endonuclease fragments of λDNA that were labeled with [γ-^{32}P] ATP at the 5′ ends

in length, consistent with the size predicted for mRNA encoding the A_3B_4 subunit as shown in Fig. 3.

To determine A_3B_4 subunit mRNA levels during the development, $M13GA_3B_4$ 72 C DNA was employed as a template for oligo(dT)-primed cDNA synthesis (RICCA et al. 1982). The labeled cDNA probes were then used in RNA-excess DNA hybridization reaction in solution. Fig. 4 demonstrates the hybridization of the A_3B_4 subunit cDNA with total soybean cotyledonary poly(A)-RNA obtained from various stages of the development. The hybridization reaction went to

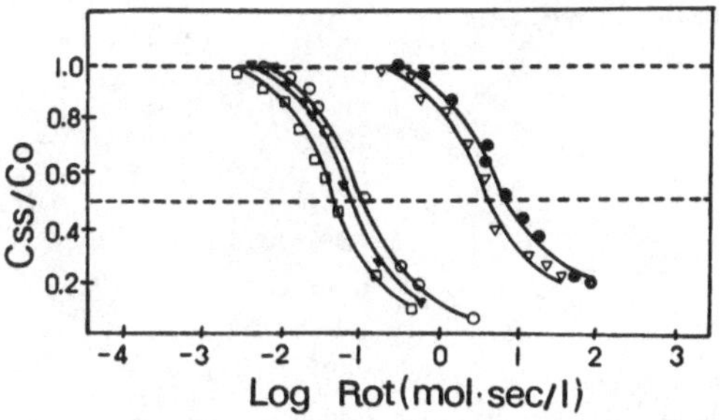

Fig. 4 Quantitation of soybean A_3B_4 subunit mRNA sequences in cotyledonary tissue during the development. Total poly(A)-RNA samples isolated from developing cotyledon at various stages were hybridized in excess to ^{32}P-labeled A_3B_4 subunit cDNA derived from $M13GA_3B_4$ 72m. Hybridization was monitored by assaying the completed reactions for S_1-nuclease-resistant radioactive material. The Rot $^1/_2$ values for the hybridization reactions were obtained by using a computer analysis to analyze the data to solve the equation $C/Co = 1 - \exp [(-\ln z) (Rot/Rot\ ^1/_2)]$, where Co is the initial amount of single-stranded cDNA, and C represents the amount of cDNA in hybrid at time t. RNA for each time point was prepared from pooled cotyledonary tissues. Total poly(A)-RNA from (●) early stage (17 days after flowering), (○) early middle stage (28 days after flowering), (□) middle stage (38 days after flowering), (▼) late stage (48 days after flowering) and (▽) mature stage (58 days after flowering)

greater than 90 % of completion with pseudo-first order kinetics occuring over a log Rot range of approximately two. The S_1-nuclease resistant radioactivity in the absence of added RNA was about 5 %. Based on the Rot $^1/_2$ values of the hybridization curves, it can be calculated that cotyledonary mRNA from 28–48 days after flowering contain approximately 120-fold more A_3B_4 subunit mRNA than those from early and mature stages of the development. This result shows that regulation of the expression of A_3 in developmental stage are mainly achieved in the transcriptional level.

Literature

FUKAZAWA, C. and K. UDAKA, 1979: Heterogeneity of glycinin. – Abst. Annual Meeting of Agricultural Chemistry in Japan p. 82.

HIRANO, H., C. FUKAZAWA and K. HARADA, 1984: The complete amino acid sequence of a subunit of the glycinin seed storage protein of soybean. – J. Biol. Chem., in press.

LAEMMLI, U. K., 1970: Cleavage of structural proteins during the assembly of the head of bacteriophage T4. – Nature, **227**, 680–685.

MCLEAN, J. W., C. FUKAZAWA and J. M. TAYLOR, 1983: Rat Apolipoprotein E mRNA. – J. Biol. Chem. **258**, 8993–9000.

MEINKE, D. W., J. CHEN and R. N. BEACHY, 1981: Expression of storageprotein genes during soybean seed development. – Planta **153**, 130–139.

MOREIRA, M. A., M. A. HARMODSON, B. A. LARKINS, and N. C. NIELSEN, 1979: Partial characterization of the acidic and basic polypeptides of glycinin. – J. Biol. Chem. **254**, 9921–9926.

RICCA, G. A., J. M. TAYLOR and J. E. KALINYAK, 1982: A simple, rapid method for the synthesis of radioactively labeled cDNA hybridization probes utilizing bacteriophage M13mp7. – Proc. Natl. Acad. Sci. U.S.A. **79**, 724–728.

CHIKAFUSA FUKAZAWA
National Food Research Institute
2-1-2 Kannondai, Yatabe-cho, Tsukuba Academic Town,
Ibaraki 305
Japan

Kulturpflanze 32 · 1984 · S 79–S 80

Expression of homoeologous lectin genes in diploid and polyploid cereal species

W. PEUMANS and H. STINISSEN
(Leuven, Belgium)

Plant lectins represent a heterogenous class of proteins sharing in common their unique ability to recognize and bind–in a highly specific way–carbohydrates or carbohydrate-containing molecules. Although they usually have been classified according to their sugar-binding specificity evidence is accumulating now that phytohemagglutinins represent a number of classes of more or less closely related proteins which are characteristic for a particular group of plants. A striking example of such a lectin family is constituted by the so-called cereal lectins, which form a very homogenous class of proteins occurring in embryonic axes of all species belonging to the Triticeae tribe. From a genetic viewpoint cereal lectins represent a favourable system for a study of qualitative and quantitative aspects of the expression of individual genes in complex (allopolyploid) genomes. Indeed, modern cereals and their primitive relatives represent a group of diploid and allopolyploid species with relatively closely related genomes. A study of lectins in both modern and primitive cereals revealed that they all contain one or more lectins, which are very closely related with respect to their chemical, biological, physicochemical and immunological properties (STINISSEN et al. 1983). Ion-exchange chromatography of lectins in extracts from embryos of diploid species (*Secale cereale, Hordeum vulgare, Triticum monococcum* and *Triticum tauschii*) yielded single peaks, which allowed to characterize unequivocally the lectins coded for by these individual genomes. Similar analyses of extracts from allopolyploid species (*Triticum turgidum, Triticum aestivum,* × *Triticosecale*) revealed that in these complex genomes each individual genome directs the synthesis of its own lectin (PEUMANS et al. 1982a). In fact, the situation *in vivo* is somewhat more complex since the different lectin polypeptide chains, which are coded for by the different genomes, combine in a fully random manner with identical and unidentical partners, giving rise to both homodimeric and heterodimeric lectins (PEUMANS et al. 1982b).

Although it can be stated that in general each of the genomes of an allopolyploid species direct the synthesis of a specific lectin there is at least one exception. Analyses of isolectin patterns in a number of hexaploid wheat varieties revealed that most of them do not contain the B-lectin indicating that either the corresponding lectin genes are not expressed or the gene products are not functional.

Isolectin pattern analysis of polyploid cereals not only yielded qualitative data about the expression of lectin genes but also allowed to draw some conclusions about quantitative aspects of the expression of these genes. Indeed, from numerous analyses it became evident that both in tetraploid and hexaploid wheats and in hexaploid × Triticosecales the relative ratios of their different isolectins varied very strongly between different varieties. The observed quantitative differences can be explained either by gene-dosage effects or by some specific gene regulation mechanism. However, since in polyploid cells all homoeologous lectin genes are found in the same nuclear and cytoplasmic environment, the involvement of one or another specific regulatory mechanism seems unlikely. Most probably the quantitative differences between individual isolectins have to be ascribed to gene-dose effects. Some experimental evidence in favour of gene-dosage effects could be obtained recently. The availability of aneuploid lines of 'Chinese Spring' wheat with different doses of both A and D lectin genes allowed to investigate the effect of gene-dosage on the expression of these genes. Thereby, it became evident that the amount of A and D lectins is directly proportional to the dose of their corresponding genes. It is evident, however, that if quantitative differences between individual isolectins are really due to gene-dosage effects, each of the individual genomes does not posses a single lectin gene but rather a multigene family.

Literature

PEUMANS, W. J., H. M. STINISSEN and A. R. CARLIER, 1982a: A genetic basis for the origin of six different isolectins in hexaploid wheat. – Planta **154**, 562–567.

–, – and –, 1982b: Subunit exchange between lectins from different cereal species. – Planta **154**, 568–572.

STINISSEN, H. M., W. J. PEUMANS and A. R. CARLIER, 1983: Occurrence and immunological relationships of lectins in gramineous species. – Planta **159**, 105–111.

W. PEUMANS,
Laboratorium voor Plantenbiochemie
Kardinaal Mercierlaan 92
B - 3030 Leuven (Heverlee)
Belgium

Kulturpflanze 32 · 1984 · S 81—S 98

Synthesis, cloning and sequence analysis of pea (*Pisum sativum* L.) storage protein specific cDNAs

R. R. D. CROY, G. W. LYCETT, J. A. GATEHOUSE and D. BOULTER
(Durham, U.K.)

Summary

Complementary DNAs (cDNAs) have been transcribed from messenger RNA isolated from developing pea seeds actively synthesizing the major storage proteins, vicilin and legumin. Analysis and sequencing of storage protein specific cDNAs and comparison with protein sequences has revealed the origin and nature of the complex polypeptide constituents of both proteins. Each protein type is synthesized as a family of polypeptide precursors which subsequently undergo endoproteolytic cleavage or 'nicking'. Further comparison of protein and cDNA encoded amino acid sequences, implicates specific amino acid sequences as recognition sites in these polypeptide precursors, for proteolytic processing systems.

Computer assisted sequence comparisons, at the amino acid and nucleotide levels, of the pea genes with published sequences for other legume storage protein genes has revealed some interesting differences. More importantly such comparisons confirm or reveal high degrees of sequence homology of the pea proteins to those from other legumes.

Preliminary results on the structure of the legumin genes are presented.

Introduction

Cotyledons of *Pisum sativum* L. and several other legume species contain large amounts of two types of storage proteins–legumin and vicilin, together comprising up to 80 % of the total seed protein. Of the two storage protein types in pea, vicilin is by far the most complex, containing more than ten differently sized polypeptide components in the mature protein, many of which additionally exhibit different charge forms (THOMSON et al. 1980, GATEHOUSE et al. 1981). Estimates of the molecular weights of these polypeptides vary but the SDS gel electrophoresis pattern shown in Fig. 1a) is usually regarded as representative and for the purposes of this paper our molecular weight values have been used (viz Mr 12,500, 13,500, 16,000, 19,000, 30,000, 33,000, 35,000 and several of about 50,000 Mr) (GATEHOUSE et al. 1982). The complexity of the component polypeptides in vicilin has made the elucidation of its molecular structure and biosynthesis during seed development very difficult to study. Recent work involving '*in vitro*' translation of mRNA from developing pea seeds (CROY et al. 1980b, GATEHOUSE et al. 1981) and pulse chase experiments (GATEHOUSE et al. 1981, CHRISPEELS et al. 1982) have shown that vicilin is synthesized initially as 50,000 Mr polypeptides. Trimeric vicilin molecules are assembled from a range of different 50,000 Mr polypeptides, some of which (precursors) subsequently undergo proteolytic processing or 'nicking' to produce the lower molecular weight polypeptides observed in the dissociated vicilin (GATEHOUSE et al. 1981).

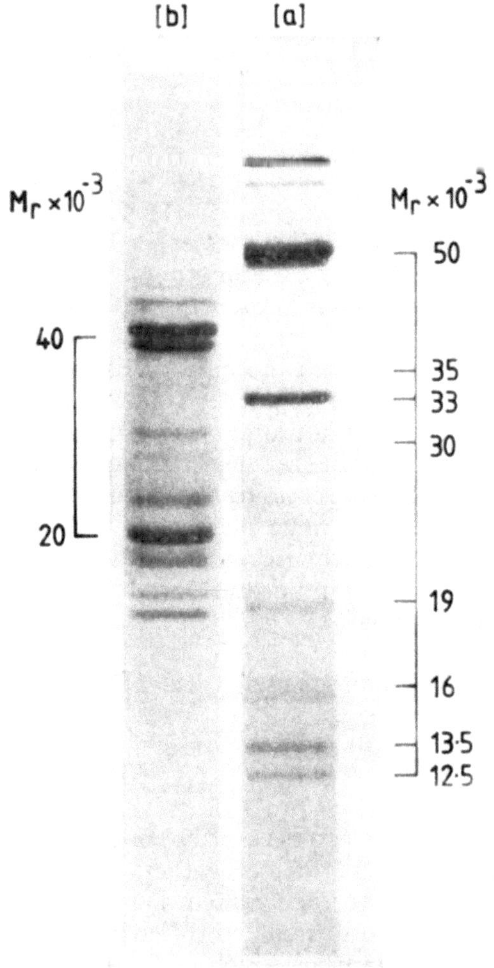

Fig. 1 SDS-polyacrylamide gel electrophoretic analyses of a) pea vicilin and b) pea legumin showing their component polypeptides and molecular weights

Legumin is somewhat less complex than vicilin, consisting essentially of six of a range of different subunit pairs (Mr 60,000), each consisting of an 'acidic' or α subunit (Mr about 40,000) linked by one or more disulphide bonds to a 'basic' or β subunit (Mr about 20,000). Thus on dissociation and reduction the protein displays a range of differently sized α and β subunits (Fig. 1b) (CROY et al. 1979, KRISHNA et al. 1979, GATEHOUSE et al. 1980). Again these subunits display different charge forms (KRISHNA et al. 1979, GATEHOUSE et al. 1980). Using similar techniques to those used to study vicilin biosynthesis, legumin has also been shown to be synthesized initially as a precursor polypeptide (Mr 60,000) (CROY et al.

1980a, SPENCER and HIGGINS 1980). As with vicilin, multimeric legumin is then assembled from a range of different precursors which are subsequently processed to the α and β polypeptides as found in mature legumin (CHRISPEELS et al. 1982, SPENCER and HIGGINS 1980). We describe in this paper the use of cDNAs to elucidate the structure of pea legumin and vicilin precursors and to describe the origins of the complex array of polypeptides which make up these proteins. Furthermore we describe comparisons of cDNA and gene predicted amino acid sequences which reveals or confirms the homologies of legumin and vicilin with other legume storage proteins.

Methodology

cDNA synthesis: The basic method of cDNA synthesis from total or poly(A) rich RNA preparations from developing pea cotyledons has already been reported (EVANS et al. 1980, CROY et al. 1982). More recent improvements to this system include i) addition of human placental RNase inhibitor (RNasin) to all RNA preparations and prior to transcription with AMV reverse transcriptase (MARTYNOFF et al. 1980). This gives appreciably higher yields of long cDNAs. ii) Efficient size fractionation, by gel electrophoresis, of double stranded cDNA before and after ligation to the vectors (pBR 322, pAT 153, pUC 8) so that only full or near full length cDNAs are cloned (DRETZEN et al. 1981). The use of oligonucleotide linkers for the insertion of the cDNAs into the vector has been continued because of the relatively higher efficiency in cDNA cloning and ease of excision of all cloned cDNAs from the vector.

DNA sequencing: The method of SEIF et al. 1980 was used routinely for sequencing excised cDNAs. M13 cloning (MESSING et al. 1981) and sequencing by the 'dideoxy' method of SANGER et al. 1977 was used where subcloning of cDNA restriction fragments was necessary.

Protein isolation and sequencing: Pea vicilin and legumin were isolated and purified by published methods (GATEHOUSE et al. 1980, 1981). The component polypeptides of vicilin were separated by a combination of gel filtration and ion exchange chromatography under dissociating conditions. Legumin subunits were isolated by published procedures (CASEY et al. 1981a, b). Cyanogen bromide fragments were isolated by gel filtration on Sephadex G–100 in 70 % formic acid.

Protein sequencing: Tryptic peptides were prepared and isolated from purified protein subunits by the methods of GATEHOUSE et al. (1982) and LYCETT et al. (1983). Isolated peptides were sequenced by the micro-manual method of CHANG et al. (1978).

Gene cloning and analysis: Pea genomic DNA, isolated from leaves by the method of GRAHAM (1978), was partially restricted with Eco R1 restriction endonuclease, and the fragments size fractionated by agarose gel electrophoresis. DNA fragments of 10–15 kb were isolated from the gel and cloned in the vector λgt WES λ B restricted with Eco RI. Gene banks thus produced were screened with legumin specific cDNAs (pDUB 6) and the hybridising clones isolated.

Southern blotting: DNA restriction fragments separated on agarose gels were blotted onto nitrocellulose paper and hybridised to ^{32}P labelled probes by the methods of JEFFREYS and FLAVELL, 1977.

Vicilin—the origins of the component polypeptides

All the abundant vicilin polypeptides have been purified to homogeneity and tryptic peptides derived from them have been sequenced. Near complete amino acid sequences have been obtained for three polypeptides (Mr 13,500, 12,500 and 16,000) (Hirano et al. 1982, Gatehouse et al. unpublished). Because of the close sequence homology of the 12,500 Mr and 16,000 Mr polypeptides, it is thought that the only difference between the two may be that the latter is glycosylated (Badenoch-Jones et al. 1981, Davey et al. 1981). Several vicilin specific cD NAs have been analysed and completely sequenced (Fig. 2). These include representatives which encode vicilin precursors which are (pDUB4; pDUB7) and which are not (pDUB2; pDUB11) proteolytically processed. Comparison of the predicted amino acid sequences with direct peptide sequence data from the individual vicilin polypeptides has allowed the accurate alignment of these polypeptides relative to a 50,000 Mr precursor. This has confirmed that all the vicilin poly-

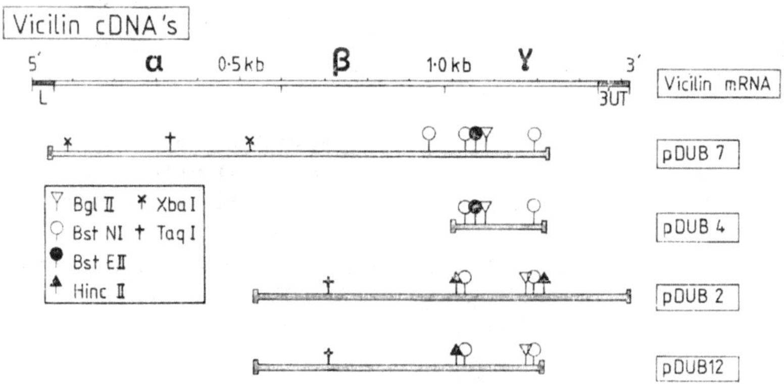

Fig. 2 Restriction maps of sequenced vicilin cDNAs aligned relative to a vicilin precursor mRNA. The mRNA has been divided into regions to show the extent of the α, β and γ polypeptides (see Fig. 3)

peptides of molecular weights less than 50,000, can be derived from a set of 50,000 Mr precursors. Furthermore the nature of the alignment of the polypeptides predicts up to two potential sites for post-translational proteolytic cleavage within any precursor molecule. For convenience if we designate the three fragments produced from the cleavage of a precursor at these sites as α, β and γ, then based on this model, outlined in Fig. 3, we can define all the component vicilin polypeptides from 50,000 Mr down to 12,000 Mr as follows. The 50,000 Mr subunits consist of contiguous α+β+γ polypeptides, representing precursors which have not been cleaved. The 33,000 Mr subunit consists of α+β polypeptides and the 12,500 Mr subunit (also 16,000 Mr) is a γ polypeptide, both produced by cleavage of a precursor at a single site (β : γ site). The 19,000 Mr subunit is an α polypeptide produced by precursor cleavage at the other site (α : β site). The 13,500 Mr subunit is a β polypeptide produced by cleavage of a precursor at both sites (α : β site *and* β : γ site). Thus it is clear that 50,000 Mr precursors (α+β+γ) can give rise to

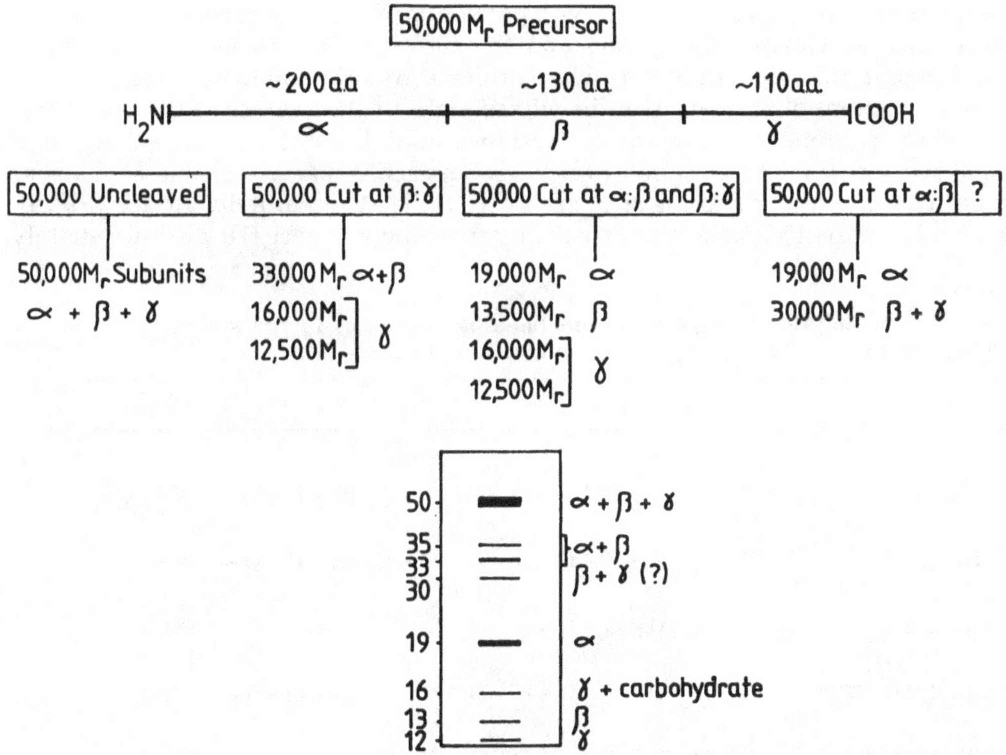

Fig. 3 Proposed model for the origin of the vicilin polypeptides based on alignment of peptide sequences with cDNA predicted amino acid sequences and assignment to bands observed on SDS gels

$\alpha+\beta$ (33,000 Mr) plus γ (12,500 Mr) polypeptides by cleavage at the $\beta:\gamma$ site only; α (19,000 Mr) plus β (13,500 Mr) plus γ (12,500 Mr) polypeptides, by cleavage at both $\alpha:\beta$ and $\beta:\gamma$ sites and α (19,000 Mr) plus $\beta+\gamma$ polypeptides, by cleavage at the $\alpha:\beta$ site only. The $\beta+\gamma$ fragment has not yet been unequivocably identified by protein sequencing although based on tryptic peptide mapping it is thought that the less abundant component of 30,000 Mr is a $\beta+\gamma$ polypeptide. (Figs. 1a and 3). Similar peptide mapping data supports the $\alpha+\beta$ designation of the 35,000 Mr minor vicilin polypeptide (Fig. 3).

Specificity of the vicilin processing sites

While the model shown in Fig. 3 explains the order and origins of the different polypeptides it does not distinguish between partial processing of essentially identical vicilin 50,000 Mr precursors and complete specific processing of unique 50,000 Mr precursors; that is, whether or not different precursors differ in specific amino acid sequences which determine whether processing takes place at none, one or two sites. The observations of complete processing of at least one vicilin precursor and the glycosylation of only one polypeptide class (16,000 Mr) support complete specific processing. Close comparison of N and C terminal sequences of the individual vicilin polypeptides and the corresponding sequences encoded in the different cDNAs or in larger polypeptides allows the following conclusions to be made: potential cleavage sites lie within highly polar regions of the precursors consisting predominantly of acidic residues, and both the β (13,500 Mr) and γ (12,500 Mr, 16,000 Mr) subunits have N terminal aspartate (Table 1). Thus as might be expected processing sites are likely to lie at the protein surface (Fig. 4). At the $\beta:\gamma$ potential processing site the hydrophilic sequence GKEN immediately

Table 1

Potential proteolytic cleavage sites and flanking sequences in the storage protein precursors, encoded by vicilin and legumin cDNAs

Precursor	Site sequences	Cleaved?
Legumin (pDUB6)	–R–R↓–Q–G–D–N↓–G–L–E–E–T–C	Yes
Vicilin (pDUB7)	–N–Q–G–K–E–N↓–D–H–E–E–E–Q–	Yes
Vicilin (pDUB2)	N–Q–Q–G–L–R↓–E–E–D–D–E–E–	No
Vicilin (pDUB7)	H–R–R–S–L–K↓–D–R–R–Q–Q–S–	No
Vicilin (pDUB2)	H–R–R–G–L–R↓–D–K–R–Q–E–I–	No

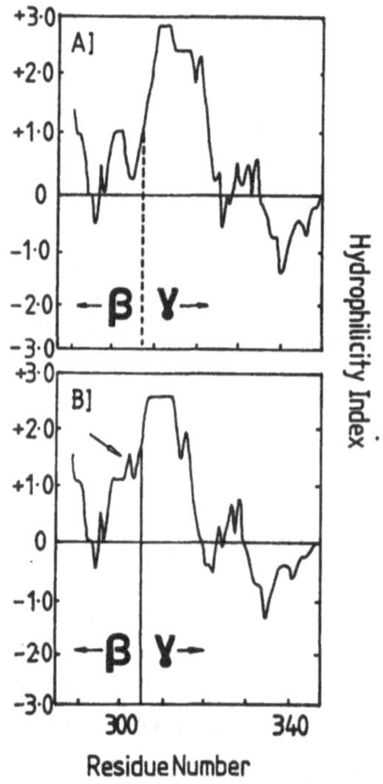

Fig. 4 Hydrophilicity index profiles of cDNA predicted amino acid sequences in the region of the vicilin $\beta:\gamma$ potential processing site in B) cleaved precursor and A) uncleaved precursor, showing the hydrophilic 'spike' in the processed precursor (→). The computer programme uses a hydrophilicity value for each amino acid residue in the sequence; for any particular point plotted the net hydrophilicity index was calculated as the mean value of the residue at that point and the adjacent three 3 residues in both N-terminal and C-terminal directions. The computer programme was adapted directly from Hopp and Woods (1981)

preceding the cleavage point allows proteolysis to take place, whereas the different and less hydrophilic sequence QGLR does not (Table 1). Similarly the sequence SLK immediately preceding the $\alpha:\beta$ potential processing site in uncleaved precursors and polypeptides functionally resembles the GLR sequence at uncleaved $\beta:\gamma$ sites. (Table 1) It is of interest to note that sequences specifying cleavage produce a hydrophilic 'spike' in hydrophilicity index profiles (Fig. 4 B) which is absent in those containing sequences specifying no-cleavage (Fig. 4 A). Thus it seems likely that there is a sequence specificity of cleavage as summarised in Table 1 and that proteolytic processing is complete and of unique precursors. The involvement of asparagine residues at endoproteolytic processing sites in plant proteins may be quite common (cf. DALKIN and BOWLES 1983).

Vicilin – homology with other legume storage proteins

Whereas it has been possible using immunocross reactions to clearly demonstrate vicilin homologues in closely related legume species such as *Vicia faba* and *Lens culinaris* (DUDMAN and MILLERD 1975), this has not been possible in more distantly legumes where immunocross reactions are absent or very poor. With the increasing use of gene cloning and the availability of rapid and complete DNA sequencing, sequence information from many major storage protein genes is accumulating allowing direct comparison of nucleotide or predicted amino acid sequences.

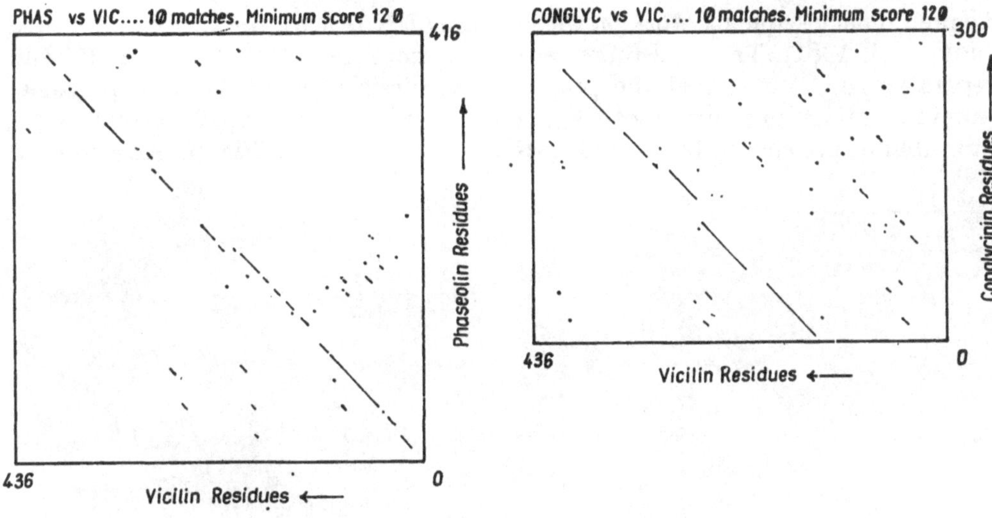

Fig. 5 Dot matrix comparisons of pea vicilin, phaseolin and conglycinin amino acid sequences encoded in cDNA or gene nucleotide sequences. Phaseolin sequences were obtained from SLIGHTOM et al. (1983) and β conglycinin α sequences from SCHULER et al. (1982). The dot matrix programme uses the proportional matching system described by McLACHLAN (1971). A modification for ease of processing is that the scan occurs in the forward direction only, with the 'dot' being printed at the head of the diagonal. The score matrix for the functional values of different residues is that used by STADEN (1982). Comparison of any sequence with itself produces a perfectly straight diagonal line

Thus it has now been possible to compare all or most of the coding sequences of the genes for the major storage proteins from *Phaseolus vulgaris* (phaseolin or glycoprotein II (Sun et al. 1981, Slightom et al. 1983) and *Glycine max* (β conglycinin α) (Schuler et al. 1982). One of the most illustrative ways of comparing such long sequences is by dot matrix comparisons (Fig. 5). As can be seen from Fig. 5 despite the fact that neither phaseolin nor conglycinin react with anti-pea vicilin antibodies (Dudman and Millerd, 1975) they are both homologous to pea vicilin over much of their coding sequence. Both phaseolin and conglycinin have molecular weights similar to pea vicilin (150–200,000 Mr) and also consist of trimeric structures of 50,000 Mr subunits although neither type undergoes proteolytic processing. It is of interest therefore that close examination of nucleotide sequences encoding the regions near to the potential proteolytic processing sites in vicilin show considerable divergence in the phaseolin and conglycinin genes (Lycett et al. 1983). Sequences encoding regions within the small vicilin polypeptides are highly conserved in all three proteins may reflect the importance of these regions in conserving tertiary and quaternary structures. Partial sequence data (Sammour et al. 1983) extends the homology of legume 7 S storage proteins by showing that canavalin, the vicilin of jack bean *(Canavalia ensiformis)*, is also similar in sequence to pea vicilin and phaseolin.

Legumin – the origins of the component polypeptides

The preliminary sequence work on the legumin cDNAs has already been published (Croy et al. 1981). This and more recent sequence data (Lycett et al. 1983 in preparation) has confirmed the proposed synthesis of α (acidic) and β (basic) legumin subunits as a family of single polypeptide precursors (Mr 60,000), since both subunits are encoded in a single mRNA species (cDNA) (Fig. 6). An α subunit

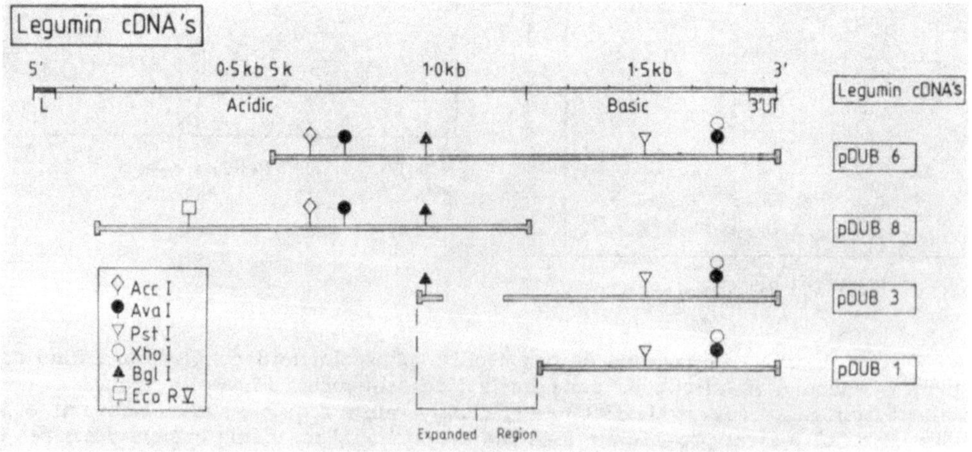

Fig. 6 Restriction maps of sequenced legumin cDNAs aligned relative to a legumin precursor mRNA. The mRNA has been divided into regions to show the extent of the α and β polypeptides

is encoded at the 5′ end (and therefore translated first) followed by a β subunit encoded at the 3′ end in a continuous sequence, with no initiation or termination codons present between the α and β coding sequences. Thus the two subunits are synthesized contiguously and must subsequently undergo endoproteolytic cleavage to form the mature subunits. Two major consequences of this contiguous subunit model are i) that the acidic and basic subunits are 'preselected'–selection of specific separate subunits by post-translational assembly does not take place and ii) subunit pairs i.e. legumin genes must be inherited as single gene loci i.e. there is no independent inheritance of acid and basic subunits within specific subunit pairs (MATTA and GATEHOUSE, 1982).

Specificity of the legumin processing sites

Comparison of the C terminal amino acid sequence of the α (acidic) subunit and the N terminal amino acid sequence of the β (basic) subunit with the cDNA predicted amino acid sequence shows clearly that in the region of the cleavage, a short linking peptide of five amino acids (R-RQGDN-G) could be removed during the proteolytic processing. As with the vicilin precursors, it has also been possible to implicate certain amino acid sequences within the linker peptide in the legumin precursor as potential protease recognition sites. The possible significance of the paired basic residues (R–R), by analogy to the widespread involvement of such sequences in animal polypeptide prohormone processing, has been discussed previously (CROY et al. 1982, GATEHOUSE et al. 1983). It is worthy of further note however that the linker peptide lies in a polar region and like the $\beta:\gamma$ processing site in vicilin, has a functionally similar amino acid sequence (QGDN) immeditely preceding the cleavage point which is also on the C terminal side of an asparagine residue (cf. vicilin–QKEN) Table 1. Thus the possibility exists for two potential processing sites which may function in excising the linker peptide.

Legumin – variations in nucleotide and amino acid sequences

Little variation in the sequences of a number of legumin cDNAs has so far been found, with one important exception. pDUB 3 type cDNAs (Fig. 6) differ from pDUB 6 type cDNAs in that the latter contain an additional 168 nucleotides located in the sequence encoding the C terminal region of the α subunit. This nucleotide sequence is unique in that it encodes three tandem repeats of 18 amino acids each (Fig. 7). The nature of these repeats is vividly illustrated by the hydrophilicity index profile in this region (Fig. 8). Since they consist largely of charged mainly acidic residues the three polar tandem repeats are prominent features of such profiles and are presumably located on the surface of the protein (HOPP and WOOD 1981).

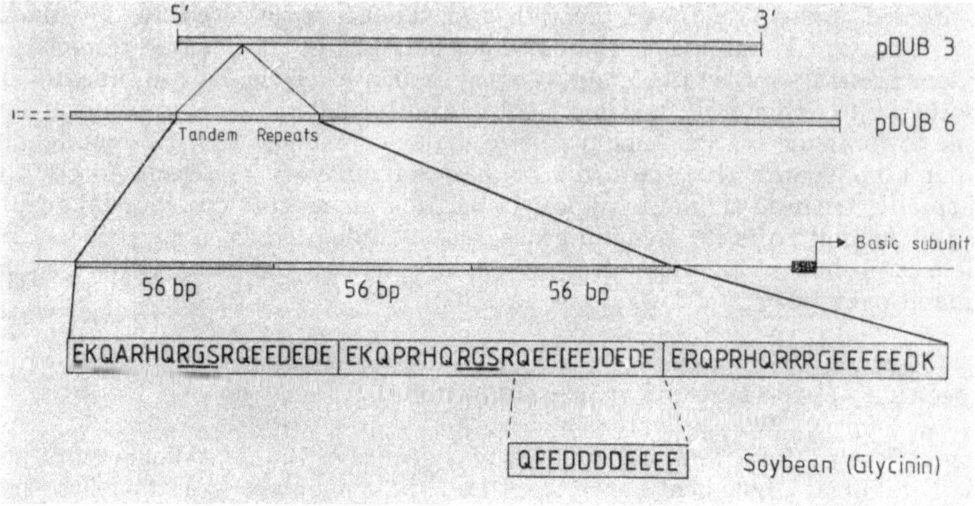

Fig. 7 Position and extent of tandem repeats in pDUB6 type legumin cDNAs. Each repeat unit consists of 18 amino acids. The black box represents the linker sequence and N terminal of β basic sequence

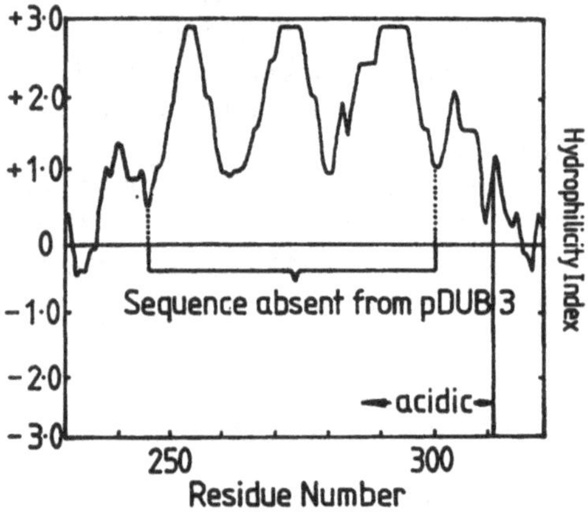

Fig. 8 Hydrophilicity index profile of cDNA predicted amino acid sequences in the region of the legumin tandem repeats. See Fig. 4 legend for details of the hydrophilicity index computer programme

Legumin – homology with other legume storage proteins

It has long been known that proteins homologous to pea legumin occur in other legumes, often despite the absence of immunological cross reactions (Dudman and Millerd 1975). Such early conclusions were based largely on similarities in the physical and chemical properties of the proteins such as solubility, size, disulphide linked subunits etc. (Derbyshire et al. 1976). More recently such homologies have been supported by comparison of N terminal peptide sequence data (Gilroy et al. 1979, Casey et al. 1981a, b). As shown for vicilin, the accumulation of storage protein cDNA and gene nucleotide sequences has made it possible to effect homology comparisons using predicted amino acid sequence data over the whole length of such proteins. So far significant nucleotide sequence data is available for only two legumin-type proteins–pea legumin (GATEHOUSE et al. 1983, LYCETT et al. 1983 in preparation) and soybean glycinin (NIELSEN 1982, NATO FEBS communication) and these have been compared in Fig. 9 as a dot matrix. The proteins are homologous over almost the whole length of the molecule including the linker sequence, except in the region of the tandem repeats in pea legumin. Glycinin shows significant sequence divergence in this region but as shown by the dot matrix in Fig. 9 and by close examination of the sequence data, it does appear to

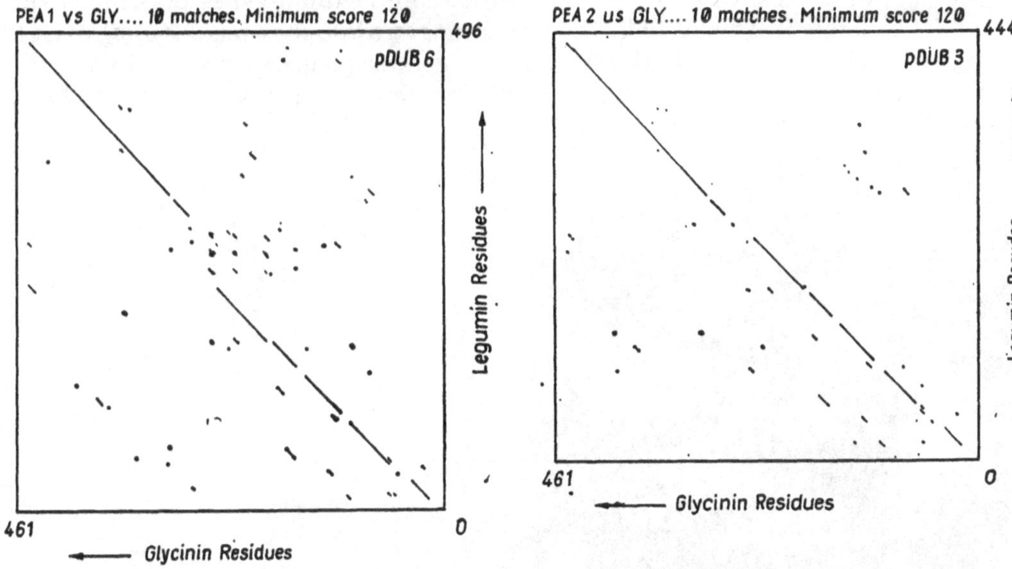

Fig. 9 Dot matrix comparisons of pea legumin and glycinin amino acid sequences encoded in cDNA or gene nucleotide sequences. Pea legumin with (pDUB6) and without (pDUB3) the three tandem repeat sequences are compared with glycinin data (NIELSEN, NATO/FEBS communication 1982). See legend to Fig. 5 for details of the dot matrix computer programme

contain a single copy of a related but degenerative version of one of the pea repeat sequences (Fig. 7). The fact that these repeat sequences are highly polar and may lie on the surface of the protein where they could form antigenic determinants (HOPP and WOOD 1981) may account for the absence or poor immune cross reaction of glycinin with anti-legumin antibodies (DUDMAN and MILLERD 1975).

The widespread occurrence of legumin-type proteins is becoming apparent from the increasing reports of the detection and processing of legumin precursors in a variety of plant species. Thus they have been demonstrated in pea (CROY et al. 1980a), SPENCER and HIGGINS 1980) broad bean (CROY et al. 1980a, WEBER et al. 1981) and soybean (TUMER et al. 1981) as well as in non-legume species such as oats (MATLASHEWSKI et al. 1982, BRINEGAR and PETERSON 1982). Furthermore sequence data is accumulating for the presence of a legumin-type protein in rice (WENMING et al. 1983, FEBS Lett. submitted).

Legumin gene structure

Legumin cDNAs have also been used to isolate cloned legumin gene sequences and in the elucidation of the gene structures. Preliminary work on the sizing of the legumin Eco R1 fragments in total genomic DNA indicated that while some of the genes were located on large R1 fragments (>11 kb) others were on smaller RI fragments (<8 kb) (CROY et al. 1982). By the use of sized, partially Eco RI digested pea DNA it was hoped to isolate the genes on the small RI fragments. To date we have isolated and characterised at least three different genes two of which contain internal Eco RI sites and therefore legumin genes present on the small RI genomic fragments. The three gene clones while showing different overall restriction maps, are very similar within their gene sequences as judged by restriction mapping and southern blots probed with legumin cDNAs. The gene clones have been designated λDUB1, 2 and 3.

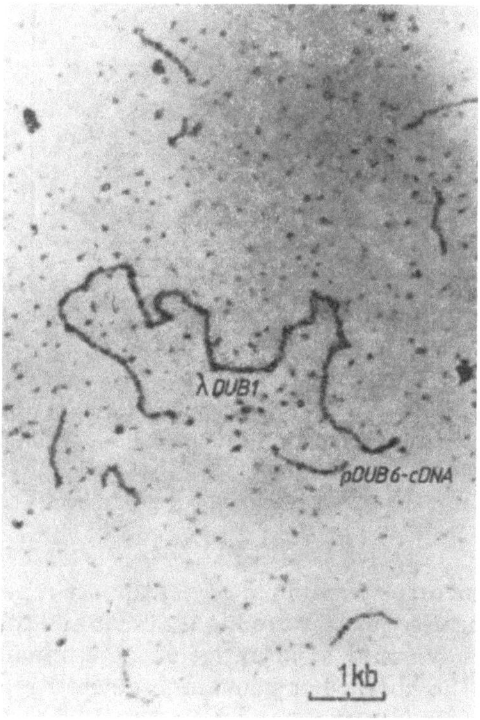

Fig. 10 Heteroduplex mapping of the legumin gene. Insert from λDUB1 (~12.5 kb) was isolated and hybridised, under conditions favouring the formation of intron loops, to isolated legumin cDNA encoding all of the β (basic) subunit. No intron loops were detected indicating the absence of large (>200 bp) introns in this position

Heteroduplex mapping of the first gene (λDUB1) using legumin cDNA (pDUB6), carried out under conditions where large intron loops (>200 bp) are clearly observed, showed no detectable introns (Fig. 10). However, attempted isolation of several known internal cDNA restriction fragments from λDUB1 and λDUB2 (Fig. 11a) showed that the corresponding gene fragments were always larger by more than 100 bp implying the presence of a small intron within the sequence encoding the β (basic) subunit (Fig. 11b).

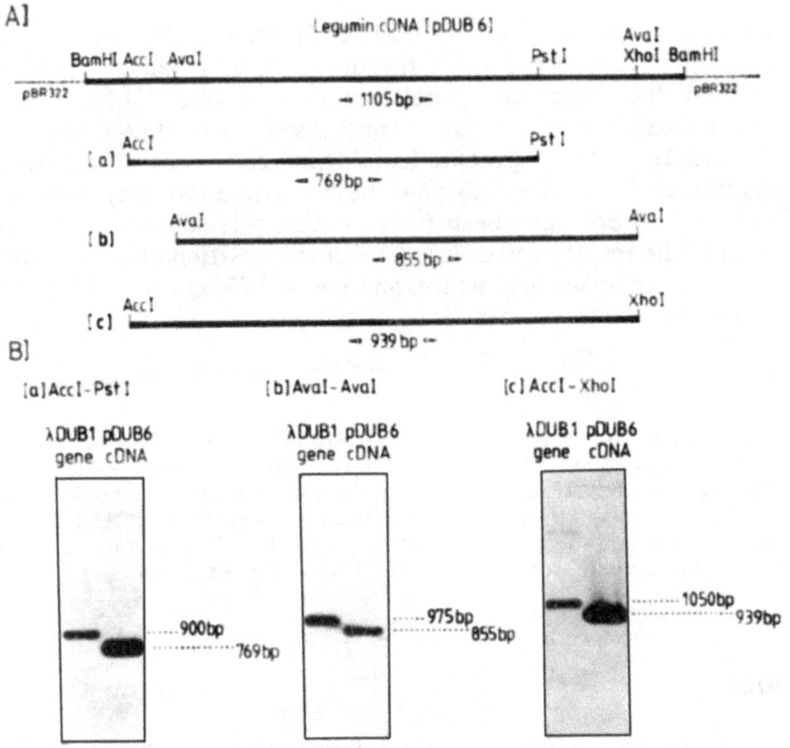

Fig. 11 A) Detailed restriction map of legumin cDNA pDUB6 showing the large internal restriction fragments which can be excised from the cDNA by different enzyme combinations a) b) and c)
B) Southern blots comparing the above cDNA restriction fragments with the corresponding fragments excised from the gene clone λDUB1. The blots were probed with ^{32}P-labelled cDNA

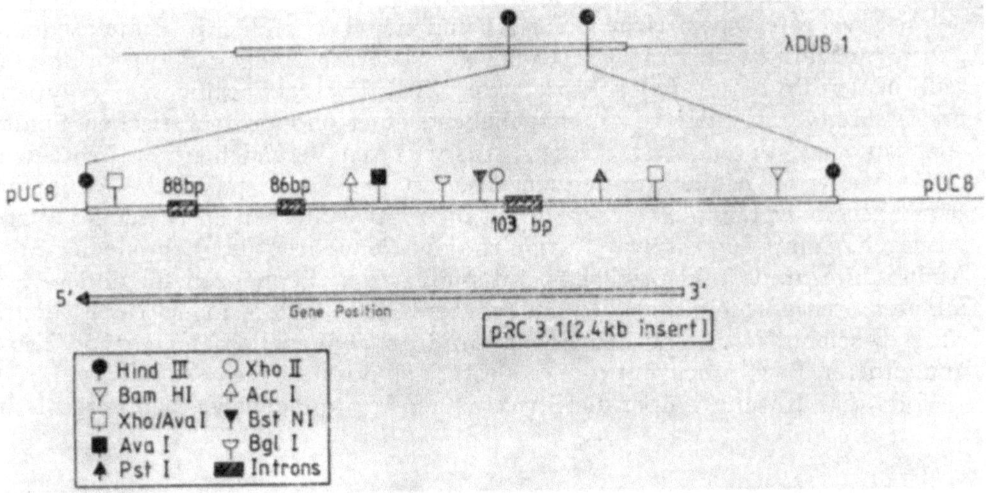

Fig. 12 Restriction map of legumin Hind III gene fragment pDUB21 (pRC 3.1) from λDUB1. Position and size of the introns detected by southern blotting and sequence analysis are shown as well as the polarity of the gene

All the genes have been subcloned into the Hind III site of pUC8 and each lies almost completely within 2.5 kb Hind III fragments. One subclone (pRC 3.1) pDUB21 from λDUB1 has been completely sequenced and contains all the legumin gene minus the leader sequence and 5' control sequences. These have been located on the adjacent Hind III fragment. In addition to the intron mentioned previously, the position and size of which have been verified (103 bp), two other small introns (86 bp and 88 bp) have been found within the sequences encoding the α subunit (Fig. 12). The results agree closely with the position of introns in the glycinin gene except the pea legumin gene introns are of different sizes (NIELSEN, personal communication).

Acknowledgements

The authors would like to acknowledge the excellent technical assistance of PHILLIPA BROWN, MARGARET RICHARDS, TONY PICKARD and DAVE BOWN. They would also like to thank ASHTON DELAUNEY and ANIL SHIRSAT for the use of their unpublished work, JIM COTTRELL who adapted the software for the dot matrix comparison programme and Dr. NICK HARRIS who carried out the heteroduplex mapping. R.R.D.C. acknowledges the financial assistance of the British Council and the University of Durham to present this paper.

Zusammenfassung

Synthese, Klonierung und Sequenzanalyse von reserveproteinspezifischen cDNAs für Reserveproteine der Erbse (*Pisum sativum* L.)

Komplementäre DNAs (cDNAs) wurden von Boten-RNA transkribiert, die aus reifenden Erbsensamen während der aktiven Synthesephase der Hauptreserveproteine, Vicilin und Legumin, isoliert wurde. Analyse und Sequenzaufklärung der reserveproteinspezifischen cDNAs und der Vergleich mit Proteinsequenzen gaben Aufschlüsse über Herkunft und Natur der komplexen Polypeptidbestandteile beider Proteine. Jeder Proteintyp wird als eine Familie von Polypeptidpräkursoren synthetisiert, die anschließend einer endoproteolytischen Spaltung oder „nicking" unterworfen werden. Aus weiteren Vergleichen von Protein- und cDNA-kodierten Aminosäuresequenzen ergibt sich, daß spezifische Aminosäuresequenzen als Erkennungsmerkmale für die proteolytischen „processing"-Systeme vorliegen, Computergestützte Vergleiche der Sequenzen der Erbsengene auf der Aminosäure- und Nukleotidebene mit publizierten Sequenzen für andere Leguminosenreserveproteine ergaben einige interessante Unterschiede. Noch wichtiger ist, daß solche Vergleiche den hohen Grad an Sequenzhomologien von Erbsen- und anderen Leguminosenproteinen bestätigen oder erhellen.

Vorläufige Resultate über die Struktur der Legumingene werden vorgestellt.

Краткое содержание

Синтез, клонирование и анализ последовательностей к-ДНК, специфичных для протеинов гороха

Комплементарные ДНК (к-ДНК) транскрибируются информационной РНК, которая изолировалась из зрелых семян гороха во время активной фазы синтеза главных запасных протеинов, вицилина и легумина. Анализ и выявление последовательностей специфичных для запасных протеинов к-ДНК и сравнение с последовательностями протеинов дают указание о происхождении и природе комплексных полипептидных составных частей обоих протеинов. Каждый тип протеина синтетизируется как одна группа предшественников полипептидов, которые затем подвергаются эндопротеолитическому расщеплению. Дальнейшие сравнения последовательностей аминокислот, кодируемых протеином и к-ДНК показывает, что специфичные последовательности являются как-бы опознавательными признаками для протеолитических систем «процессинга».

Сравнение последовательностей генов гороха с помощью ЭВМ на уровне аминокислот и нуклеотидов с известными последовательностями запасных протеинов других бобовых обнаружило некоторые интересные различия. Ещё важнее, что подобные сравнения подтверждают или разьясняют высокую степень гомологии последовательностей гороха и прочих бобовых.

Сообщается о предварительных результатах выяснения структуры генов легумина.

Literature

Badenoch-Jones, J., D. Spencer, T. J. V. Higgins and A. Millerd, 1981: The role of glycosylation in storage protein synthesis in developing pea seeds. – Planta **153**, 201–209.

Brinegar, A. C. and D. M. Peterson, 1982: Synthesis of oat globulin precursors. Analogy to legumin 11S storage protein synthesis. – Plant Physiol. **70**, 1767–1769.

Casey, R., J. F. March, J. E. Sharman and M. N. Short, 1981a: The purification N-terminal amino acid sequence and some other properties of an α^M -subunit of legumin from pea (*Pisum sativum* L.). – Biochim. Biophys. Acta **670**, 428–432.

–, – and E. Sanger, 1981b: N-terminal amino acid sequence of β-subunits of legumin from *Pisum sativum*. – Phytochemistry **20**, 161–163.

Chang, J. T., D. Braver and B. Wittman-Liebold, 1978: Micro-sequence analysis of peptides and proteins using 4-N-N-Dimethylaminoazobenzene-4-isothiocyanate phenylisothiocyanate double coupling method. – FEBS Letts. **93**, 205–214.

Chrispeels, M. J., T. J. V. Higgins and D. Spencer, 1982: Assembly of storage protein oligomers in the endoplasmic reticulum and processing of the polypeptides in the protein bodies of developing pea cotyledons. – J. Cell Biol. **93**, 306–313.

Croy, R. R. D., E. Derbyshire, T. G. Krishna and D. Boulter, 1979: Legumin of *Pisum sativum* and *Vicia faba*. – New Phytol. **83**, 29–35.

–, J. A. Gatehouse, I. M. Evans and D. Boulter, 1980a: Characterisation of the storage protein subunits synthesised in vitro by polyribosomes and RNA from developing pea (*Pisum sativum* L.) – Legumin. – Planta **148**, 49–56.

–, –, – and –, 1980b: Characterisation of the storage protein subunits synthesised in vitro by polyribosomes and RNA from developing pea (*Pisum sativum* L.) – Vicilin. – Planta **148**, 57–73.

–, –, G. W. Lycett and D. Boulter, 1981: Biochem. Soc. Trans. **9**, 254 P.

Croy, R. R. D., G. W. Lycett, J. A. Gatehouse, J. N. Yarwood and D. Boulter, 1982: Cloning and analysis of cDNAs encoding plant storage protein precursors. – Nature **295**, 76–79.

Dalkin, K. and D. Bowles, 1983: Analysis of inter-relationship of jackbean seed components by two dimensional mapping of iodinated tryptic peptides. – Planta **157**, 536–539

Davey, R. A., T. J. V. Higgins and D. Spencer, 1981: Homologies between two small subunits of vicilin from *Pisum sativum*. – Biochem. Int. **3**, 595–602.

Derbyshire, E., D. J. Wright and D. Boulter, 1976: Legumin and vicilin storage proteins of legume seeds. – Phytochemistry **15**, 3–24.

Dretzen, G., M. Bellard, P. Sassone-Corsi and P. Chambon, 1981: A reliable method for the recovery of DNA fragments from agarose and acrylamide gels. – Anal. Biochem. **112**, 295–298.

Dudman, W. F. and A. Millerd, 1975: Immunochemical behaviour of legumin and vicilin from *Vicia faba*: a survey of related proteins in the Leguminosae subfamily Faboideae. – Biochem. Syst. Ecol. **3**, 25–33.

Evans, I. M., R. R. D. Croy., P. Brown and D. Boulter, 1980: Synthesis of complementary DNAs to partially purified mRNAs coding for the storage proteins of *Pisum sativum* L. – Biochim. Biophys Acta **610**, 81–95.

Gatehouse, J. A., R. R. D. Croy and D. Boulter, 1980: Isoelectric-focusing properties and carbohydrate content of pea *(Pisum sativum)* legumin. – Biochem. J. **185**, 497–503.

–, – and –, 1984: Synthesis and structure of pea seed storage proteins. – CRC Critical Review, in: Plant Sciences **1** Issue 4, 287–314

–, –, H. Morton, M. Tyler and D. Boulter, 1981: Characterisation and subunit structures of the vicilin storage proteins of pea (*Pisum sativum* L.). – Eur. J. Biochem. **118**, 627–633.

–, G. W. Lycett, R. R. D. Croy and D. Boulter, 1982: The post-translational proteolysis of the subunits of vicilin from pea (*Pisum sativum* L.). – Biochem. J. **207**, 629–632.

Gilroy, J., D. J. Wright and D. Boulter, 1979: Homology of basic subunits of legumin from *Glycine max* and *Vicia faba*. – Phytochemistry **18**, 315–316.

Graham, D. E., 1978: Isolation of high molecular weight DNA from whole organisms or large tissue masses. – Anal. Biochem. **85**, 609–613.

Hirano, H., J. A. Gatehouse and D. Boulter, 1982: The complete amino acid sequence of a subunit of the vicilin seed storage protein of pea (*Pisum sativum* L.). – FEBS Letts. **145**, 99–102.

Hopp, T. R. and K. R. Wood, 1981: Prediction of protein antigenic determinants from amino acid sequences. – Proc. Natl. Acad. Sci. USA **78**, 3824–3828.

Jeffreys, A. J., and R. A. Flavell, 1977: A physical map of the DNA regions flanking the rabbit β-globin gene. – Cell **12**, 4291439.

Krishna, T. G., R. R. D. Croy and D. Boulter, 1979: Heterogeneity in subunit composition of the legumin of *Pisum sativum*. – Phytochemistry **18**, 1879–1880.

Lycett, G. W., A. J. Delauney, J. A. Gatehouse, J. Gilroy, R. R. D. Croy and D. Boulter, 1983: The vicilin gene family of pea (*Pisum sativum* L.): a complete cDNA coding sequence for preprovicilin. – Nucl. Acids Res. **11**, 2367–2380.

McLachlan, A. D., 1971: Tests for comparing related amino acid sequences Cytochrome **c** and cytochrome **c** 551. – J. Mol. Biol. **61**, 409–424.

Martynoff, G. de., E. Pays and G. Vassart, 1980: Synthesis of a full length DNA complementary to thyroglobulin 33 S messenger RNA. – Biochem. Biophys. Res. Comm. **93**, 645–653.

Matlashewski, G. J., K. Adeli, I. Altosaar, P. R. Shewry and B. Miflin, 1982: *In vitro* synthesis of oat globulin. – FEBS Letts. **145**, 208–212.

Matta, N. K., and J. A. Gatehouse, 1982: Inheritance and mapping of storage protein genes in *Pisum sativum* L. – Heredity **48**, 383–392.

Messing, J., R. Crea and P. H. Seeburg, 1981: A system for shotgun DNA sequencing. – Nucl. Acids Res. **9**, 309–321.

Sammour, R. H., J. A. Gatehouse, J. Gilroy and D. Boulter, 1984: The homology of the major storage protein of jackbean *(Canavalia ensiformis)* to pea vicilin and its separation from α mannosidase. – Planta **161**, 61–70.

SANGER, F., S. NICKLEN and A. R. COULSON, 1977: DNA sequencing with chain terminating inhibitors. – Proc. Nat. Acad. Sci USA **74**, 5463–5467.
SCHULER, M. A., B. F. LADIN, J. C. POLACCO, G. FREYER and R. N. BEACHY, 1982: Structural sequences are conserved in the genes coding for the α, α' and β subunits of the soybean 7 S seed storage protein. – Nucl. Acid Res. **10**, 8245–8261.
SEIF, I., G. KHOURY and R. DHAR, 1980: A rapid enzymatic sequencing technique: determination of sequence alterations in early simian virus 40 temperature sensitive and deletion mutants. – Nucl. Acids Res. **8**, 2225–2240.
SLIGHTOM, J. L., S. M. SUN and T. C. HALL, 1983: Complete nucleotide sequence of a french bean storage protein gene: Phaseolin. – Proc. Natl. Acad. Sci. USA **80**, 1897–1901.
SPENCER, D. and T. J. V. HIGGINS, 1980: The biosynthesis of legumin in maturing pea seeds. – Biochem. Int. **1**, 502–509.
STADEN, R., 1982: An interactive graphics program for comparing and aligning nucleic acid and amino acid sequences. – Nucleic Acids Research **10**, 2951–2961.
SUN, M., J. L. SLIGHTOM and T. C. HALL, 1981: Intervening sequences in a plant gene–comparison of the partial sequence of cDNA and genomic DNA of french bean phaseolin. – Nature **289**, 37–41.
THOMSON, J. A., H. E. SCHROEDER and A. M. TASSIE, 1980: Cotyledonary storage proteins in *Pisum sativum* V. Further studies on molecular heterogeneity in the vicilin series of holoproteins. – Aust. J. Plant Physiol. **7**, 271–282.
TUMER, N. E., V. E. THANH and N. C. NIELSEN, 1981: Purification and characterisation of mRNA from soybean seeds. Identification of glycinin and β conglycinin precursors. – J. Biol. Chem. **256**, 8756–8760.
WEBER, E., J. INGVERSEN, R. MANTEUFFEL and M. PÜCHEL, 1981: Transfer of *in vitro* synthesised *Vicia faba* globulins and barley prolamins across the endoplasmic reticulum membrane of *Vicia faba*. – Carlsberg Res. Commun. **46**, 383–393.

R. R. D. CROY
University of Durham
Department of Botany
South Road
Durham DH1 3LE
U.K.

Kulturpflanze 32 · 1984 · S 99—S 108

The biochemical genetics of pea seed storage proteins

R. CASEY and C. DOMONEY
(Norwich, England)

Summary

This paper describes the genetics of legumin, vicilin and convicilin, the three major storage proteins from the seeds of the genus *Pisum* (the pea), and considers the genetical behaviour of the proteins in relation to their structure, biochemistry and synthesis.

Introduction

Pisum (pea) seeds contain three major storage proteins called legumin, vicilin and convicilin. Each is a naturally heterogeneous mixture of oligomers and has been the subject of structural, genetical and biosynthetic studies. In this paper we describe the genetics of the three families of proteins in relation to such studies. Although there are some similarities between vicilin and convicilin, we have chosen to discuss the three groups of proteins separately, since they differ appreciably in a number of aspects of their biochemistry and biosynthesis.

Legumin

Structure

Legumin has been isolated from pea seeds or protein bodies by a number of methods (BLAGROVE et al. 1980; CASEY 1979a; CROY et al. 1979; GATEHOUSE et al. 1980; GULDAGER 1978; THOMSON et al. 1978), all of which have resulted in a product of molecular weight of about 4×10^5 (BLAGROVE et al. 1980; CASEY 1979a; CROY et al. 1979; MATTA et al. 1981) containing six acidic (α-) and six basic (β-) subunits which are disulphide-bonded as αβ pairs (DERBYSHIRE et al. 1976) arranged in a trigonal bipyramid (CASEY et al. 1980). In all *Pisum* genotypes examined in any detail, legumin seems to be heterogeneous, as shown, for instance, by the formation of asymmetrical arcs on crossed immunoelectrophoresis (DOMONEY 1981; GULDAGER, 1978) or by band multiplicity on cellulose acetate gel electrophoresis (THOMSON et al. 1978) or on non-denaturing polyacrylamide gel electrophoresis (MATTA et al. 1981). This heterogeneity is probably related to the observed heterogeneity of the α- and β-subunits. Both types of subunit exhibit multiplicity on polyacrylamide gels containing sodium dodecyl sulphate (SDS gels) (CASEY 1979b; THOMSON et al. 1978; MATTA et al. 1981), on polyacrylamide gels containing urea (THOMSON et al. 1978), on two-dimensional isofocusing/SDS gels (CASEY 1979b; KRISHNA et al. 1979) and on two-dimensional non-reducing/reducing SDS gels (DOMONEY 1981; MATTA et al. 1981).

N-terminal sequence analysis of a mixture of legumin β-subunits from a single pea genotype (CASEY et al. 1981a) indicates the presence of several homologous polypeptides, differing at only a few points along their N-terminal sequence. The sequence data suggest the existence of a small gene family, the differences in primary structure between members of the family of polypeptides being responsible for some of the observed multiplicity on gels; heterogeneity may also result from post-translational modification (SPENCER et al. 1980) or deamidation on storage (MATTA and GATEHOUSE 1981). Such sequence studies have also shown the N-terminal amino acid sequence of the β-subunits of pea legumin to be homologous to that of the basic subunits of legumin from *Vicia faba* and of glycinin, the 11 S protein from *Glycine max* (GILROY et al. 1979, MOREIRA et al. 1979).

Two-dimensional gel electrophoresis of pea legumin shows clear multiplicity of the α-subunits and also provides evidence for both major (α^M) and minor (α^m) classes of α-subunit (CASEY 1979b), in agreement with the results of THOMSON et al. (1978). The structural relationship between the α^M- and α^m-subunits is not known, nor is the relationship between individual α^M-subunits. N-terminal sequence analysis of an α^M-subunit of pea legumin (CASEY et al. 1981b) has shown it to be homologous to the acidic subunits of glycinin (MOREIRA et al. 1979) and it seems likely that individual α^M-subunits from pea legumin will show sequence homology, as do the several acidic subunits of glycinin (MOREIRA et al. 1979).

In addition to the legumin species containing the α^M-subunits, there exist other minor types of legumin, which do not conform to the conventional legumin model of DERBYSHIRE et al. (1976). Two-dimensional non-reducing/reducing SDS-gel electrophoresis has pointed to the existence of disulphide-bonded subunit pairs of molecular weight about 40×10^3, in which the acidic subunit is apparently of similar size to the basic subunit (DOMONEY 1981; MATTA et al. 1981). As noted elsewhere (CASEY and DOMONEY 1984; DOMONEY and CASEY 1984) these minor species, which comprise 3–10 % of the total legumin, depending on the genotype (DOMONEY 1981), are significant both to the genetics and to the synthesis of legumin.

Biosynthesis

When cell-free translation products of poly A-containing RNA, purified from developing pea seeds, are immunoprecipitated with antibody to legumin, precursors of molecular weight approximately 60×10^3 are specifically precipitated (CROY et al. 1980a; SPENCER and HIGGINS 1980). A small amount of a polypeptide of molecular weight around 80×10^3 is also precipitated (DOMONEY and CASEY 1984). The $\sim 60 \times 10^3$ mol. wt. precursor molecules contain both an acidic and a basic polypeptide sequence (CROY et al. 1982; DOMONEY 1981) covalently linked by (a) bond(s) other than a disulphide bridge (CROY et al. 1980a; SPENCER and

HIGGINS 1980). This linkage is broken at some stage during the processing of the precursors, three such precursors initially unite to form a 8–9 S intermediate (CHRISPEELS et al. 1982; SPENCER and HIGGINS 1980) and two such trimers, or half-molecules, come together to form the final 12 S species in which the $\alpha\beta$ precursor has been 'nicked' but the α- and β-subunits are held together by an S–S bridge. Since the mRNA for legumin precursor is one which contains contiguous α- and β-subunit sequences, it is to be expected that in any population of legumin precursors a particular α-subunit will be associated with a particular β-subunit (from the same mRNA sequence); that is, there should be non-random association of the α- and β-polypeptides into subunit pairs. Such non-random association has been clearly shown in glycinin (STASWICK et al. 1981), which is synthesized in analogous fashion to pea legumin (BARTON et al. 1982; TUMER et al. 1981). The size of the mRNA for legumin has been estimated at ~1,700 nucleotides (CHANDLER et al. 1983), ~2,200 nucleotides (CROY et al. 1982) and ~2,000–2,100 nucleotides (DOMONEY and CASEY 1984).

The nature of the polypeptide of mol. wt. $\sim 80 \times 10^3$ which is precipitated by anti-legumin is unclear. A legumin-related polypeptide of mol. wt. $\sim 80 \times 10^3$ has been observed in extracts of immature pea cotyledons which have been pulse-labelled with radioactive amino acids (DOMONEY 1981; CHRISPEELS et al. 1982); this labelled polypeptide binds specifically to a column of antilegumin. A cDNA clone, constructed from poly A-containing mRNA from immature pea seeds, has been identified that contains an insert homologous to a mRNA coding for an 80×10^3 molecular weight polypeptide. The latter is immunoprecipitable with antibody to legumin and peptide mapping experiments suggest a structural relationship between it and the conventional (60×10^3 molecular weight) legumin precursor (DOMONEY and CASEY 1984).

If the 40×10^3 minor legumin subunit dimer (referred to above) is made in the same way as 'conventional' legumin, one would expect a precursor of 40×10^3 mol. wt. Such a precursor has not been observed, but genetic evidence (MATTA and GATEHOUSE 1982) suggests that the synthesis of this subunit pair may not follow the standard pattern.

Genetic variation in legumin structure

It is well established that the properties of a given legumin preparation are a function of the genotype from which that legumin was purified. Variation in the net surface charge of native legumin has been reported (CASEY 1979b; MILLERD et al. 1978; THOMSON and SCHROEDER 1978; THOMSON et al. 1978), but such variation is difficult to use for detailed genetic analysis (CASEY 1982). Variation in the net charge and apparent molecular weight of the denatured subunits of legumin, however, does lend itself to such analysis; there have been a number of reports of genetic variation in both the acidic and basic subunits of legumin from which it is clear that charge and size (apparent molecular weight) variants exist for both (CASEY 1979b; CROY et al. 1979; DAVIES 1980; MATTA and GATEHOUSE 1982; PRZYBYLSKA et al. 1979; 1981; THOMSON and SCHROEDER 1978). Extensive variation in the major acidic subunits has been observed on both two-dimensional isofocusing/SDS gels (CASEY 1979b) and on one- and two-dimensional SDS gels (CASEY 1979b;

DOMONEY 1981; MATTA and GATEHOUSE 1982). Variation in the nature of minor acidic subunits of legumin has also been reported (CASEY 1979b; MATTA and GATEHOUSE 1982; THOMSON and SCHROEDER 1978) although the relationship between the α^M-subunits of CASEY (1979b), the Lg-2α-subunits of MATTA and GATEHOUSE (1982) and the products of the LB locus reported by THOMSON and SCHROEDER (1978), is not clear (see below). Only limited variation in the basic subunits of legumin has been observed (MATTA and GATEHOUSE 1982; PRZYBYLSKA et al. 1981).

The minor legumin species, of which the unreduced subunit pair has a molecular weight of approximately 40×10^3 (see above), shows genetic variation in the apparent molecular weight of the unreduced dimer (DOMONEY 1981) and significant variation in both the acidic and basic subunits from the reduced molecule (MATTA and GATEHOUSE 1982).

The genetics of legumin subunits

Each of the three groups to have examined the genetics of *Pisum* legumin has used a different system of nomenclature (CASEY 1979b; MATTA and GATEHOUSE 1982; THOMSON and SCHROEDER 1978). The way in which these systems are related has been described elsewhere (CASEY 1982).

The major acidic (α^M-) subunits of legumin behave as the products of a single Mendelian gene (CASEY 1979b; THOMSON and SCHROEDER 1978) with at least five different possible alleles (MATTA and GATEHOUSE 1982). Present evidence (CASEY 1979b, 1982; MATTA and GATEHOUSE 1982) suggests the existence of a small number of genes for legumin α^M-subunits in a closely linked array (small gene family) which maps near to the *r* locus on chromosome (linkage group) 7 (DAVIES 1980; MATTA and GATEHOUSE 1982).

Further evidence for the existence of a small gene family for legumin comes from specific immunoprecipitation of cell-free translation products and from experiments using recombinant plasmids containing cDNA inserts homologous to legumin mRNAs. When cell-free translation products of poly A-containing RNA, purified from developing pea seeds, are immunoprecipitated with antibody to legumin, several precursors of molecular weight approximately 60×10^3 are specifically precipitated (SPENCER and HIGGINS 1980), although some genotypes appear not to show this type of heterogeneity (CROY et al. 1980a). Hybrid-release translation experiments, using plasmids containing cDNA inserts homologous to legumin, indicate that a number of different mRNA species are selected by such plasmids (CHANDLER et al. 1983; DOMONEY and CASEY 1984) and strongly suggest the existence of several legumin mRNA species in any given preparation of poly A-containing RNA. Direct measurement of the gene copy number for legumin (CROY et al. 1982) suggests the existence of four legumin genes in *Pisum*, although it must be remembered that this value is obtained in one genotype with one clone.

It is clear from biosynthetic studies (CROY et al. 1980a; SPENCER and HIGGINS 1980) and from DNA sequencing (CROY et al. 1982) that legumin polypeptides are synthesized as precursor molecules containing both α- and β-sequences and it is likely that a legumin structural gene codes for both an α- and β-subunit. It follows, then, that the genetic behaviour of β-subunits should be similar to that of α-subunits; so far, however, detailed study of the inheritance of β-subunits has been hindered by a lack of suitable variants.

The inheritance of the genes for minor legumin subunits is not as well established. Present evidence supports the production of the minor acidic subunits from a locus other than that for the α^M-subunits (MATTA and GATEHOUSE 1982).

Analysis of the inheritance of the Lg-3β-subunits of legumin—those which are derived from the $\alpha\beta$ subunit dimer of approximately 40×10^3 molecular weight—has suggested independent segregation of the α- and β-polypeptides of this minor species (MATTA and GATEHOUSE 1982). Such genetic behaviour, as noted above, suggests that the biosynthesis of the 40×10^3 mol. wt. subunit pair may not follow the standard pattern.

Vicilin

Structure

Vicilin from pea seeds is a heterogeneous protein that contains a number of different polypeptides, varying in molecular weight from 12×10^3 to 50×10^3 (GATEHOUSE et al. 1981; THOMSON and SCHROEDER 1978). The native protein has a molecular weight of about 1.5×10^5 and probably has a trimeric structure with subunits of molecular weight approximately 50×10^3 (GATEHOUSE et al. 1981). All native vicilin preparations seem to contain polypeptides of molecular weight 33×10^3, 19×10^3 and 12.5×10^3 in addition to the major 50×10^3 molecular weight species, together with small amounts of polypeptides of molecular weight 35×10^3, 16×10^3 and 13.5×10^3. Sequencing studies of proteins and DNA (GATEHOUSE et al. 1982, 1983; LYCETT et al. 1983) have clearly shown that the smaller polypeptides (molecular weight $< 47 \times 10^3$) are derived by proteolytic processing of a precursor polypeptide of molecular weight approximately 47×10^3; the 16×10^3 molecular weight polypeptide is probably a glycosylated form of the 12.5×10^3 molecular weight polypeptide (GATEHOUSE et al. 1982). Hence, as suggested by THOMSON et al. (1980), the fundamental subunit structure of native vicilins is simpler than their apparent polypeptide composition suggests.

Biosynthesis

Immunoprecipitation with antibody to vicilin of cell-free products of poly A-containing RNA, from developing pea seeds, results in the selective precipitation of two classes of polypeptide of molecular weights approximately 47×10^3 and 50×10^3 (CROY et al. 1980b; GATEHOUSE et al. 1981; HIGGINS and SPENCER 1981), in addition to the 70×10^3 molecular weight polypeptides of convicilin, which show immunological cross-reactivity to anti-vicilin (CROY et al. 1980b; DOMONEY and CASEY 1983). Pulse-labelling experiments (CHRISPEELS et al. 1982; GATEHOUSE et al. 1981) have shown that the smaller (47×10^3 mol. wt.) polypeptide class is 'nicked' by proteolysis to give the range of smaller polypeptides observed in vicilin purified from mature seeds.

The sequences of the mRNAs for the 50×10^3 and 47×10^3 molecular weight polypeptides show good homology over the regions where cDNA sequence data are available (LYCETT et al. 1983); despite this close similarity, there is no cross-selection of RNAs by cDNA clones for the two classes of vicilin polypeptide (CROY et al. 1982; DOMONEY and CASEY 1983). 'Northern' blot analysis shows the mRNA for both types of vicilin polypeptide to have similar sizes of about 1,750 nucleotides (DOMONEY and CASEY 1983).

Variation in vicilin structure and the genetics of vicilin

Only one extensive study of variability in vicilin subunits has been described (THOMSON and SCHROEDER 1978; THOMSON et al. 1980). Variation in the molecular weights of the 50×10^3 and of the approximately 30×10^3 polypeptides, and in the charge of 12 and 14×10^3 molecular weight polypeptides, has been observed and the inheritance of variant forms studied. Such observations show all three types of polypeptide to behave as the products of single genes which exhibit simple Mendelian inheritance. It may be assumed that the $\sim 30 \times 10^3$ molecular weight, the 12 and 14×10^3 molecular weight and the precursor polypeptides from which these are derived are all the products of the same locus. The studies of GATEHOUSE et al. (1983) show the existence of a number of copies of the genes coding for the two classes of vicilin polypeptide (of molecular weight approximately 50×10^3 and 47×10^3) and at least some of the genes for the 50×10^3 molecular weight polypeptides have been tentatively assigned to linkage group 7, close to the *r* locus (and thus also close to the legumin structural loci) (D. BOULTER, quoted in CASEY 1982). Nothing is known of the genetical relationship between the 47×10^3 and the 50×10^3 molecular weight vicilin precursor polypeptides.

Convicilin

Structure

Convicilin (CASEY and SANGER 1980; CROY et al. 1980c) is normally a minor component of pea seed protein but it can become a major constituent in certain genotypes (CASEY and SANGER 1980; CASEY et al. 1982). It has a molecular weight of about $2.2–2.8 \times 10^5$ and consists of a single size class of subunit of molecular weight approximately 70×10^3; it is therefore either a trimer or a tetramer. It is serologically closely related to vicilin but differs from it in subunit size, amino acid composition and carbohydrate content (CASEY and SANGER 1980; CROY et al. 1980c). The subunits of convicilin may or may not display heterogeneity on SDS gels, depending on the genotype under investigation; isoelectric focusing studies indicate some subunit charge heterogeneity while N-terminal amino acid analysis shows the presence of more than one end group, at least in some pea genotypes (CROY et al. 1980c).

Biosynthesis

Convicilin is not processed by internal proteolytic cleavage, but probably (in common with the vicilin precursor polypeptides of about 47 and 50×10^3 molecular weight) is synthesized as a preprotein; this can be processed by membranes isolated from dog pancreas or pea cotyledons (GATEHOUSE et al. 1981; HIGGINS and SPENCER 1981).

Antibody to SDS-denatured convicilin subunits specifically immunoprecipitates polypeptides of molecular weight approximately 70, 50 and 47×10^3 (Domoney and Casey, 1983) in agreement with the serological cross-reactivity of convicilin and vicilin (Croy et al. 1980c). Hybrid-release/translation experiments using a cDNA clone have selected a mRNA for a 70×10^3 molecular weight polypeptide, that is immunoprecipitable by antibody both to convicilin subunits and to vicilin (Domoney and Casey 1983); the clone does not, however, hybrid-select mRNAs for the vicilin precursor polypeptides. 'Northern' blot analysis of size-fractionated poly A-containing RNA, using cDNA clones for convicilin and for the two classes of vicilin polypeptide, shows that mRNA for convicilin (~2,200 nucleotides) is appreciably longer than those for the vicilin polypeptides (both about 1,750 nucleotides) (Domoney and Casey 1983). There are no reports of the structural relationship between the mRNAs for convicilin and the vicilin polypeptides.

Variation in convicilin structure and the genetics of convicilin

No variation in convicilin net surface charge or subunit charge has been reported, but variation in the molecular weight of convicilin subunits has been established (Casey and Sanger 1980; Matta and Gatehouse 1982; Thomson et al. 1978). Genetical analyses of molecular weight variants of convicilin subunits show the convicilin locus to behave as a single codominant gene (Casey 1982; Matta and Gatehouse 1982) which is not linked to the locus for legumin α^M-subunits (Casey and Sanger 1980; Matta and Gatehouse 1982). The convicilin structural locus, which is probably identical to that designated *VcA* by Thomson and Schroeder (1978) has been assigned to linkage group 2 and resides between genes *s* and *k*, 3 ± 1 map units from the *k* locus (Matta and Gatehouse 1982).

Variation in storage protein proportions

Extensive genetic variability in the amounts and proportions of pea seed proteins – which could reflect variation in the regulation of protein synthesis during development – has been reported by a number of groups (Casey et al. 1982; Davies 1980; Millerd et al. 1978; Thomson et al. 1978). At present little is known of the molecular basis of such variation (which encompasses a range of legumin: vicilin ratios of about 0.2–1.5) and such analyses will be complicated by the fact that the genotypes under examination have differing genetic backgrounds. Thus the development of near-isogenic lines differing in only a few genetic traits, including storage protein proportions, is of great importance. Davies (1980) has reported that in near-isogenic lines which differ in the nature of the *r* locus, round-seeded (*RR*) types have appreciably more legumin relative to vicilin than corresponding wrinkled-seeded (*rr*) genotypes. Thus the *r* locus, which is closely linked to the major legumin structural locus (and possibly also to that for vicilin), has an effect on legumin expression. The *r* locus also, however, influences many metabolic pathways (Coxon and Davies 1982; Kooistra 1962) and the effect on legumin synthesis may be secondary to one or more of the pleiotropic effects of the *r* locus.

Acknowledgement

We are very grateful to Agrigenetics Research Corporation, USA, for their support of some of the work described in this article.

Zusammenfassung

Die biochemische Genetik der Reserveproteine von Erbsensamen

Die Arbeit beschreibt die Genetik von Legumin, Vicilin und Convicilin, den drei Hauptreserveproteinen aus Samen der Gattung *Pisum* (Erbse), unter Berücksichtigung des genetischen Verhaltens der Proteine im Hinblick auf ihre Struktur, Biochemie und Synthese.

Краткое содержание

Биохимическая генетика запасных протеинов семян гороха

В работе рассматривается генетика легумина, вицилина и конвицилина, трёх главных форм запасного протеина из семян гороха; при этом обращается внимание на генетическое поведение протеинов в связи с их структурой, биохимией и синтезом.

Literature

Barton, K. A., J. F. Thompson, J. T. Madison, R. Rosenthal, N. P. Jarvis, and R. N. Beachy, 1982: The biosynthesis and processing of high molecular weight precursors of soybean glycinin subunits. – J. biol. Chem. **257**, 6089–6095.

Blagrove, R. J., G. G. Lilley, and R. Davey, 1980: Molecular weight of legumin from *Pisum sativum*. – Aust. J. Pl. Physiol. **7**, 221–225.

Casey, R., 1979a: Immunoaffinity chromatography as a means of purifying legumin from *Pisum* (pea) seeds. – Biochem. J. **177**, 509–520.

–, 1979b: Genetic variability in the structure of the α-subunits of legumin from *Pisum*– a two-dimensional gel electrophoresis study. – Heredity **43**, 265–272.

–, 1982: The genetics of pea seed storage proteins. – Qualitas Plantarum Plant Foods for Human Nutrition **31**, 281–295.

–, and C. Domoney, 1984: The genetics of legume storage proteins. – Phil. Trans. Roy. Soc. Lond., B **304**, 349–358.

–, and E. Sanger, 1980: Purification and some properties of a 7S seed storage protein from *Pisum* (pea). – Biochem. Soc. Trans. **8**, 658.

–, G. J. Hills, and P. Shaw, 1980: The 3-dimensional structure of legumin from *Pisum sativum*. – John Innes Ann. Rep. 20–21.

–, J. F. March, and E. Sanger, 1981a: N-terminal amino acid sequence of β-subunits of legumin from *Pisum sativum*. – Phytochemistry **20**, 161–163.

–, –, J. E. Sharman, and M. N. Short, 1981b: The purification, N-terminal amino acid sequence and some other properties of an α^M-subunit of legumin from the pea (*Pisum sativum* L.). – Biochim. Biophys. Acta **670**, 428–432.

–, J. E. Sharman, D. J. Wright, J. R. Bacon, and P. Guldager, 1982: Quantitative variability in *Pisum* seed globulins: its assessment and significance. – Qualitas Plantarum Plant Foods for Human Nutrition **31**, 333–346.

CHANDLER, P. M., T. J. V. HIGGINS, P. J. RANDALL, and D. SPENCER, 1983: Regulation of legumin levels in developing pea seeds under conditions of sulfur deficiency. – Pl. Physiol. **71**, 47–54.

CHRISPEELS, M. J., T. J. V. HIGGINS, and D. SPENCER, 1982: Assembly of storage protein oligomers in the endoplasmic reticulum and processing of the polypeptides in the protein bodies of developing pea cotyledons. – J. Cell Biol. **93**, 306–313.

COXON, D. T., and D. R. DAVIES, 1982: The effect of the r_a and r_b loci on the lipid content of the seed of *Pisum sativum*. – Theor. appl. Genet. **64**, 47–50.

CROY, R. R. D., E. DERBYSHIRE, T. G. KRISHNA, and D. BOULTER, 1979: Legumin of *Pisum sativum* and *Vicia faba*. – New Phytol. **83**, 29–35.

–, J. A. GATEHOUSE, I. M. EVANS, and D. BOULTER, 1980a: Characterisation of the storage protein subunits synthesised in vitro by polyribosomes and RNA from developing pea (*Pisum sativum* L.) I. Legumin. – Planta **148**, 49–56.

–, –, –, and –, 1980b: Characterisation of the storage protein subunits synthesised in vitro by polyribosomes and RNA from developing pea (*Pisum sativum* L.) II. Vicilin. – Planta **148**, 57–63.

–, –, M. TYLER, and D. BOULTER, 1980c: The purification and characterization of a third storage protein (convicilin) from the seeds of pea (*Pisum sativum* L.). – Biochem. J. **191**, 509–516.

–, G. W. LYCETT, J. A. GATEHOUSE, J. N. YARWOOD, and D. BOULTER, 1982: Cloning and analysis of cDNAs encoding plant storage protein precursors. – Nature, Lond. **295**, 76–79.

DAVIES, D. R., 1980: The r_a locus and legumin synthesis in *Pisum sativum*. – Biochemical Genetics **18**, 1207–1219.

DERBYSHIRE, E., D. J. WRIGHT, and D. BOULTER, 1976: Legumin and vicilin storage proteins of legume seeds. – Phytochemistry **15**, 3–24.

DOMONEY, C. A. M., 1981: Legumin synthesis and accumulation in *Pisum sativum* L. – Ph. D. Thesis, University of East Anglia.

–, and R. CASEY, 1983: Cloning and characterization of complementary DNA for convicilin, a major seed storage protein in *Pisum sativum* L. – Planta **159**, 446–453.

–,and –, 1984: Storage protein precursor polypeptides in cotyledons of *Pisum sativum* L.: identification of, and isolation of a cDNA clone for, an 80,000-M_r legumin-related polypeptide. – Eur. J. Biochem. **139**, 321–327.

GATEHOUSE, J. A., R. R. D. CROY, and D. BOULTER, 1980: Isoelectric-focusing properties and carbohydrate content of pea *(Pisum sativum)* legumin. – Biochem. J. **185**, 497–503.

–, –, H. MORTON, M. TYLER, and D. BOULTER, 1981: Characterization and subunit structures of the vicilin storage proteins of pea (*Pisum sativum* L.). – Eur. J. Biochem. **118**, 627–633.

–, G. W. LYCETT, R. R. D. CROY, and D. BOULTER, 1982: The posttranslational proteolysis of the subunits of vicilin from pea (*Pisum sativum* L.). – Biochem. J. **207**, 629–632.

–, –, A. J. DELAUNEY, R. R. D. CROY, and D. BOULTER, 1983: Sequence specificity of the post-translational proteolytic cleavage of vicilin, a seed storage protein of pea (*Pisum sativum* L.). – Biochem. J. **212**, 427–432.

GILROY, J., D. J. WRIGHT, and D. BOULTER, 1979: Homology of basic subunits of legumin from *Glycine max* and *Vicia faba*. – Phytochemistry **18**, 315–316.

GULDAGER, P., 1978: Immunoelectrophoretic analysis of seed proteins from *Pisum sativum* L. – Theor. Appl. Genet. **53**, 241–250.

HIGGINS, T. J. V., and D. SPENCER, 1981: Precursor forms of pea vicilin subunits. Modification by microsomal membranes during cell-free translation. – Plant Physiol. **67**, 205–211.

KOOISTRA, E., 1962: On the differences between smooth and three types of wrinkled peas. – Euphytica **11**, 357–373.

KRISHNA, T. G., R. R. D. CROY, and D. BOULTER, 1979: Heterogeneity in subunit composition of the legumin of *Pisum sativum*. – Phytochemistry **18**, 1879–1880.

LYCETT, G. W., A. J. DELAUNEY, J. A. GATEHOUSE, J. GILROY, R. R. D. CROY, and D. BOULTER, 1983: The vicilin gene family of pea (*Pisum sativum* L.): a complete cDNA coding sequence for preprovicilin. – Nucleic Acids Research **11**, 2367–2380.

MATTA, N. K., and J. A. GATEHOUSE, 1981: Modification of the basic subunits of pea legumin on storage. – Phytochemistry **22**, 2621–2623.
–, and –, 1982: Inheritance and mapping of storage protein genes in *Pisum sativum* L. – Heredity **48**, 383–392.
–, –, and D. BOULTER, 1981: Molecular and subunit heterogeneity of legumin of *Pisum sativum* L. (Garden Pea) – A multi-dimensional gel electrophoretic study. – J. exp. Bot. **32**, 1295–1307.
MILLERD, A., J. A. THOMSON, and H. E. SCHROEDER, 1978: Cotyledonary storage proteins in *Pisum sativum*. III. Patterns of accumulation during development. – Aust. J. Plant Physiol. **5**, 519–534.
MOREIRA, M. A., M. A. HERMODSON, B. A. LARKINS, and N. C. NIELSEN, 1979: Partial characterisation of the acidic and basic polypeptides of glycinin. – J. biol. Chem. **254**, 9921–9926.
PRZYBYLSKA, J., J. HURICH, and Z. ZIMNIAK-PRZYBYLSKA, 1979: Comparative study of seed proteins in the genus *Pisum*. IV. Electrophoretic patterns of legumin and vicilin components. – Genet. pol. **20**, 517–528.
–, –, and S. BLIXT, 1981: Variation of basic legumin components in the genus *Pisum* revealed by isoelectrofocusing. – Pisum Newsletter **13**, 44–45.
SPENCER, D., and T. J. V. HIGGINS, 1980: The biosynthesis of legumin in maturing pea seeds. – Biochemistry International **1**, 502–509.
–, –, S. C. BUTTON and R. A. DAVEY, 1980: Pulse-labelling studies on protein synthesis in developing pea seeds and evidence of a precursor form of legumin small subunit. – Plant Physiol. **66**, 510–515.
STASWICK, P. E., M. A. HERMODSON and N. C. NIELSEN, 1981: Identification of the acidic and basic subunit complexes of glycinin. – J. biol. Chem. **256**, 8752–8755.
THOMSON, J. A., and H. E. SCHROEDER, 1978: Cotyledonary storage proteins in *Pisum sativum*. II. Hereditary variation in components of the legumin and vicilin fractions. – Aust. J. Pl. Physiol. **5**, 281–294.
–, –, and W. F. DUDMAN, 1978: Cotyledonary storage proteins in *Pisum sativum* I. Molecular heterogeneity. – Aust. J. Pl. Physiol. **5**, 263–279.
–, –, and A. M. TASSIE, 1980: Cotyledonary storage proteins in *Pisum sativum*. V. Further studies on molecular heterogeneity in the vicilin series of holoproteins. – Aust. J. Pl. Physiol. **7**, 271–282.
TUMER, N. E., V. H. THANH, and N. C. NIELSEN, 1981: Purification and characterisation of mRNA from soybean seeds. – J. biol. Chem. **256**, 8756–8760.

R. CASEY, C. DOMONEY
John Innes Institute
Colney Lane
Norwich
Norfolk NR4 7UH
England

Kulturpflanze 32 · 1984 · S 109–S 116

New aspects of the subunit structure of *Vicia faba* legumin

C. HORSTMANN
Gatersleben, GDR

Summary

The 11S globulin legumin from seeds of *Vicia faba* L. var. *minor* cv. Fribo consists of acidic α- and basic β-polypeptides, one α-polypeptide is linked to one β-polypeptide via disulphide bonds to a pair. Four pairs (1–4) have been isolated as main components, and two further pairs (5, 6) as minor components of legumin. A specific α-β-combination exists in every pair. Pairs 1 and 2 (type A) are similar to each other, their α- as well as their β-polypeptides contain methionine; on the contrary pairs 3 and 4 (type B) lack methionine but are also mutually related. Further results, i.e. amino acid analyses, sequence determinations, peptide mapping, support the classification in types A and B. Deviating from the nomenclature commonly used it is proposed, to call the α-β-pairs of legumin and related 11S globulins "subunits".

The 11S storage protein of the seeds of *Vicia faba*, legumin, consists of acidic α- (MW ca. 36,000) and basic β- (MW ca. 20,000) polypeptides. In the holoprotein, one α- and one β-polypeptide are linked by S–S-bridges to an "intermediary subunit" (WRIGHT and BOULTER 1974) or a "subunit pair" (HORSTMANN 1983), and six of these pairs form the legumin molecule. Similar structures have been reported for comparable 11S proteins from other legume seeds (for a review see DERBYSHIRE et al. 1976), from seeds of other dicotyledonous plants (i.e. rapeseed, SCHWENKE et al. 1983), and even for an 11S globulin from oat seeds (BRINEGAR and PETERSON 1982). The spatial arrangement of the "intermediary subunits" in some holoproteins has been investigated by small angle *X*-ray scattering. All the 11S proteins investigated so far possess a similar spatial structure (PLIETZ et al. 1984). The α- and β-polypeptide chains of legumin are synthesized as parts of one precursor (CROY et al. 1980, BASSÜNER et al. 1983) containing one α- and one β-chain in a single polypeptide with α in its N-terminal and β in its C-terminal part (CROY et al. 1982). Mature α- and β-chains are formed from the precursor by post-translational proteolytic cleavage. α- as well as β-polypeptides exhibit heterogeneity (DERBYSHIRE et al. 1976), suggesting the occurrence of different precursors encoded by different genes, and, indeed, the existence of different precursors or their genes has been demonstrated for legumin from *Pisum* (CROY et al. 1982) and from *Vicia faba* (BASSÜNER et al. 1983, WOBUS et al. 1984). If the disulphide links between the α- and β-polypeptides are formed in the precursor before cleavage to α and β takes place, specific α-β-combinations should occur in native legumin.

Separation of the α-β-pairs of *Vicia faba* legumin has been achieved by ion exchange chromatography of legumin on DEAE-Sepharose CL-6 B in presence of 6 M urea. Four main pairs (numbered 1–4) and two minor pairs (5 and 6) could be isolated (HORSTMANN 1983). Sodium dodecylsulphate (SDS) polyacrylamide gel electrophoresis reveals for each of the pairs 1–4, without reduction of the S—S-bonds, a single band with an apparent molecular mass near 60,000; after reduction of the S—S-bonds these pairs are cleaved to their constituent α- (molecular masses near 36,000) and β-polypeptides (molecular masses near 20,000) (fig. 1). The bands

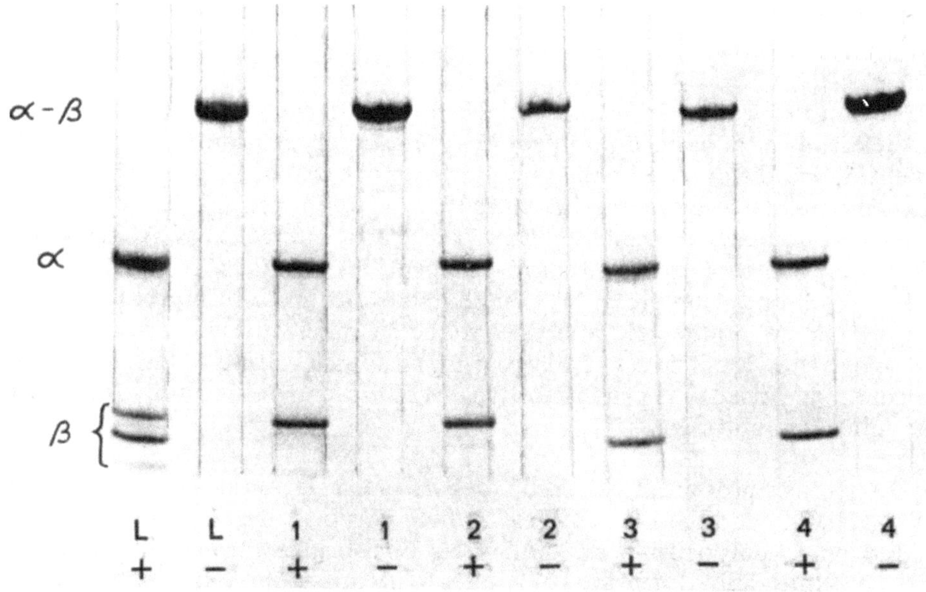

Fig. 1 SDS polyacrylamide gel electrophoresis of legumin (L) and α-β pairs 1, 2, 3, 4 with (+) and without (−) reduction with 2-mercaptoethanol

of β1 and β2 run more slowly in this gel system than those of β3 and β4. It is not known whether these differences reflect different molecular masses for β1/β2 and β3/β4, since it has been demonstrated that even a single amino acid substitution can alter the electrophoretic mobilities of polypeptides in a similar discontinous SDS polyacrylamide gel system (DE JONG et al. 1978); in a continous SDS phosphate buffer system all the β-polypeptides possess the same mobility (WRIGHT and BOULTER 1974), suggesting similar molecular masses for the β's. All the α-polypeptides have similar mobilites in the gel shown in fig. 1, but they differ in their N-terminal amino acids, which are Leu for α1 and α2, and Thr for α3 and α4. Further studies on the isolated α- and β-polypeptides (amino acid analyses, peptide maps) revealed strong similarities within each polypeptide class of pairs 1 and 2, and similarities of pairs 3 and 4, respectively, but large differences between pairs 1 and 2 on the one hand, and pairs 3 and 4 on the other hand

(HORSTMANN 1983). Differences between the pairs are also obvious after reaction with cyanogen bromide, which cleaves peptide bonds on the carboxyl side of Met residues (GROSS 1967): only α- and β-polypeptides of pairs 1 and 2 are cleaved, whereas α- and β-polypeptides of pairs 3 and 4 remain unchanged after treatment with cyanogen bromide, indicating that only pairs 1 and 2 contain Met. This result is also supported by amino acid analysis; although Met has not been quantified, only the polypeptides of pairs 1 and 2 show a Met peak. In SDS polyacrylamide gels both α1 and α2, exhibit the same pattern of three cleavage products after reaction with cyanogen bromide. A difference exists between β1 and β2, since β1 yields two and β2 three CNBr cleavage products (fig. 2), indicating

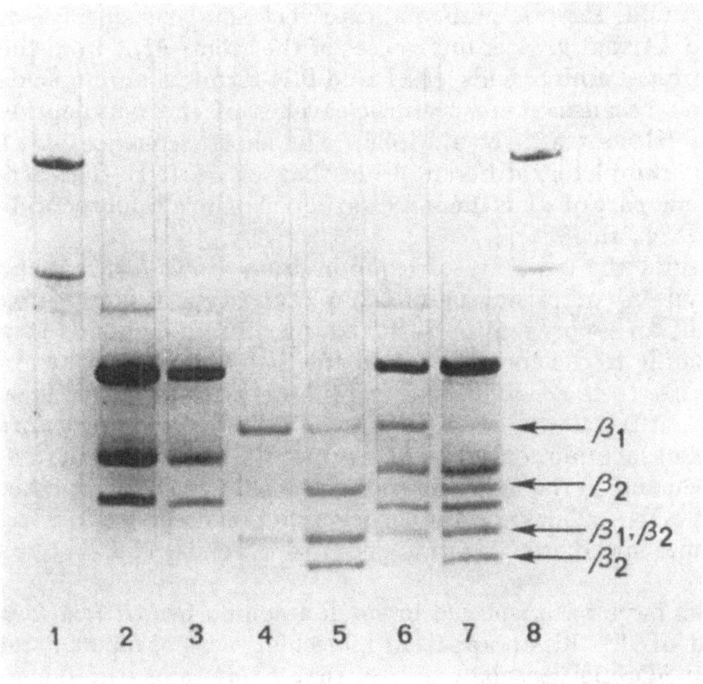

Fig. 2 SDS polyacrylamide gel electrophoretic analysis of CNBr cleavage products. Lane 1: α-β pair 1, not cleaved; lane 8: α-β pair 2, not cleaved. Lane 2–7: CNBr cleavage products of α1 (lane 2), α2 (lane 3), β1 (lane 4), β2 (lane 5), α-β pair 1 (lane 6), α-β pair 2 (lane 7). Cleavage products arising from β1 and β2, respectively, are indicated. A small cross-contamination of β1 : β2 products is due to the difficult separation of the polypeptides

that β2 contains an additional Met compared to β1. The two cleavage products of β1 have been isolated by gel filtration on Sephadex G–50 in 10 % formic acid, and OTTO et al. (1984) determined the N-terminal amino acid sequences. The product with the lower molecular mass of about 6,000 corresponds to the N-terminal part of β1, and its sequence of the first 43 amino acids differs in only two positions from the N-terminal sequence of a β-polypeptide from *Pisum*, whose sequence was deduced from the nucleotide sequence of the corresponding cDNA (CROY et al. 1982). The first 9 amino acids of the second cleavage product (molecular mass ca. 14,000) have been sequenced, they are also in full accordance with the sequence following Met at position 59 of the *Pisum* polypeptide. These results support earlier observations on the close relationship of the legumins of the two species, i.e. serological similarity (GULDAGER 1978, CROY et al. 1979), similar peptide maps of their constituent polypeptides (CASEY and HORSTMANN 1981). Therefore it is very surprising that the sequenced 43 N-terminal amino acids of β1 differ in 19 positions from those of β3/β4, the homologous polypeptides from the same protein. The 53 N-terminal amino acids of β3 sequenced so far are the same as those of β4 (61 amino acids sequenced, OTTO et al. 1984). β2 has not been sequenced yet, but results of peptide mapping suggest a primary structure very similar to β1 (HORSTMANN 1983).

Similar relationships and differences seem to exist in the group of the α-polypeptides of *Vicia faba* legumin. Peptide maps indicate very similar sequences of α1 and α2, and of α3 and α4, but greater differences of the group α1/2 from the group α3/4. Only 4 N-terminal amino acids of α1 and 9 N-terminal amino acids of α4 could be sequenced because of unspecific cleavages of the polypeptide chains during sequencing (HORSTMANN et al. 1982). The short sequence of α1 (NH_2-Leu-Arg-Glu-Gln) is completely different from that of α4 (NH_2-Thr-Ser-Ser-Glu), but the N-terminal part of α1 is the same as the one of an α^M-polypeptide of a *Pisum* legumin (CASEY et al. 1981).

According to these results the subunits of legumin from *Vicia faba* can be classified as two types: type A comprising pairs 1 and 2, and type B comprising pairs 3 and 4. In each pair an α-polypeptide is linked via disulphide bonds to a specific β-polypeptide, leading to the conclusion that the S–S-links are formed in the precursor before cleavage to the α- and β-polypeptides takes place. This view is supported by the results of BASSÜNER et al. (1983) who detected two precursors after *in vitro* biosynthesis of legumin, only one of them could be labelled by S^{35}-Met and hence must correspond to the Met-containing pairs of type A. A further support is the isolation of cDNA clones specific for legumin precursors containing Met, and other cDNA clones specific for legumin precursors lacking Met (WOBUS et al. 1984).

Two additional α-β-pairs have been isolated in small amounts from *Vicia faba* legumin: pair 5 composed of α5 with an apparent molecular mass of 55,000, and of β5 which runs faster in SDS polyacrylamide gels than β3/β4, and pair 6 consisting of α6 with an apparent molecular mass of 60,000 and of β 6 at the same position in SDS gels as β5 (fig. 3). The relationships between these pairs and between them and the type A/type B pairs have still to be investigated. A correspon ding precursor has not yet been detected by *in vitro* biosynthesis experiments, but it is interesting that a cDNA specific for a pea legumin precursor with a molecular mass of about 80,000 was isolated by CASEY and DOMONEY (1984). A precursor with a similar molecular mass could be expected for pair 6 of *Vicia* legumin.

The existence of specific α-β-pairs in legumin (and presumably in other 11 S storage proteins, for glycinin cf. STASWICK et al. 1981), which are synthesized as common precursors, poses the question of the correct nomenclature for the legumin "subunits". The term "subunit" is commonly used for the α- and β-polypeptides of legumin and corresponding polypeptides of comparable 11 S globulins from plant seeds (WRIGHT and BOULTER 1974, DERBYSHIRE et al. 1976). This apparently is a result of the application of SDS polyacrylamide gel electro-

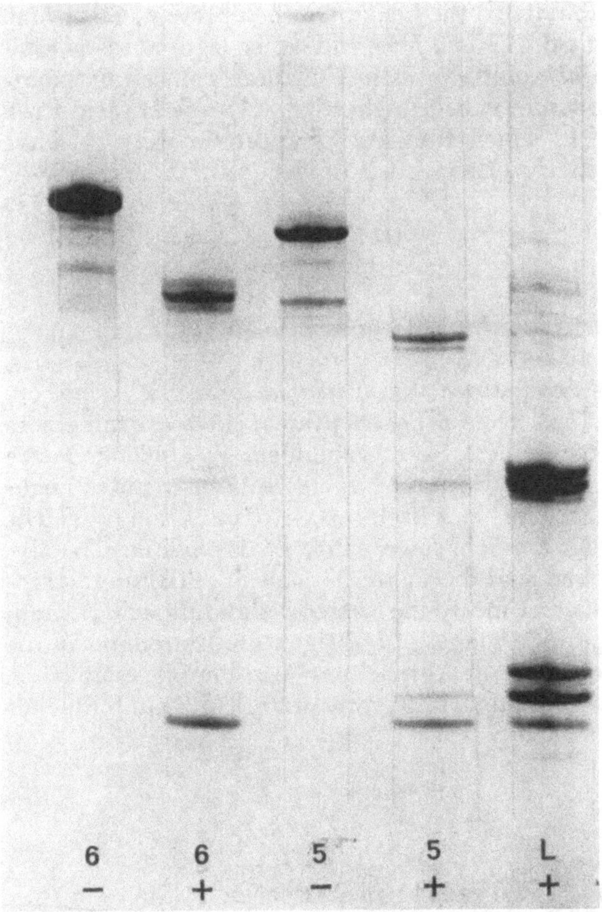

Fig. 3 SDS polyacrylamide gel electrophoresis of α-β pairs 5 and 6 with (+) and without (−) reduction with 2-mercapto-ethanol. L: legumin

phoresis to structural investigations of proteins, where every band on the gel arising from proteins after reduction of disulphide bonds is termed "subunit". According to a definition of KLOTZ et al. (1975) only "subordinate entities that interact *noncovalently* with each other to form a quaternary structure should be called subunits", and "polypeptide chains that are covalently linked by S–S bonds . . . might be called *constituent* units of a tertiary structure". Applying this definition to 11S seed proteins the term "subunit" should be used for the α-β-pairs, calling their α- and β-polypeptides "constituent polypeptides".

It is still unknown whether all α-β-pairs randomly participate in the assembly of the hexameric legumin or whether certain constraints favour the formation of distinct combinations, i.e. a fixed ratio of type A and type B subunits. Experiments to reconstitute legumin from subunits of only one type or from mixtures of different subunits resulted in every case in the formation of products with the same electrophoretic mobility as legumin in polyacrylamide gels under non-denaturing conditions, and electronmicroscopic pictures showed molecules of similar size and shape for the reconstitution products and native legumin (ADLER and HORSTMANN 1984), indicating that any subunit has the correct contact sites to associate with itself or other subunits to the hexameric holoprotein. Since Met is a nutritionally limiting amino acid of *Vicia faba* and other legume seeds, one can speculate on finding a *Vicia faba* cultivar with a legumin consisting exclusively of type A subunits or of subunits with a higher type A/type B ratio than that of the cultivar investigated (in which the ratio is approximately 1). First screening results suggest that such varieties seem to exist (MÜNTZ et al. 1984).

Zusammenfassung

Neue Aspekte der Untereinheitenstruktur von *Vicia faba*-Legumin

Das 11 S-Globulin, Legumin, aus den Samen von *Vicia faba* L. var. *minor* cv. „Fribo" besteht aus sauren α- und basischen β-Polypeptiden, ein α- ist mit einem β-Polypeptid über Disulfidbrücken zu einem Paar verbunden. Vier solcher Paare (1 bis 4) wurden als Haupt-, zwei weitere (5, 6) als Nebenkomponenten des Legumins isoliert. In jedem Paar liegt eine spezifische α-β-Kombination vor. Die Paare 1 und 2 (Typ A) sind sich sehr ähnlich, sowohl ihre α- als auch ihre β-Polypeptide enthalten Methionin, dagegen sind die Paare 3 und 4 (Typ B) methioninfrei, aber auch untereinander ähnlich. Eine Reihe weiterer Befunde, wie Aminosäureanalyse, Sequenzanalyse, "peptide maps", bestätigen die Einteilung in die Typen A und B. Es wird vorgeschlagen, in Abweichung von der bisher üblichen Nomenklatur, die α-β-Paare des Legumins und verwandter 11 S-Globuline als „Untereinheiten" zu bezeichnen.

Краткое содержание

Новые аспекты структуры субъединиц легумина конских бобов

Глобулин 11 S (легумин) из семян *Vicia faba* L. var. *minor* сорта «Фрибо» состоит из кислых α- и щелочных β-полипептидов; один полипептид α связан с полипептидом β в пару дисульфидным мостом. Четыре таких пар (от первой до четвёртой) изолировались как главные компоненты легумина, а две другие (пятая и шестая) — как побочные. В каждой паре находится специфичная α-β-комбинация. Первая и вторая пары (тип A) имеют большое сходство; их α- и их β-полипептиды содержат метионин, в то время как третья и четвёртая пара (тип B) не содержат метионина но также схожи между собой. Результаты прочих исследований, как анализ аминокислот, анализ последовательностей, „peptide maps", подтверждают правильность подразделения на типы A и B. В отличие от общепринятой номенклатуры, предлагается α-β-пары легумина и родственных 11 S глобулинов обозначать как «субъединицы».

Literature

ADLER, K., and C. HORSTMANN, 1984: Electronmicroscopic investigation of reconstitution products of isolated subunit pairs of the plant storage protein legumin from *Vicia faba*. – Proc. 3rd Seed Protein Symp. 'Genetics of Seed Proteins' (MÜNTZ, K., and C. HORSTMANN, eds.). Kulturpflanze **32**, ...–....

BASSÜNER, R., R. MANTEUFFEL, K. MÜNTZ, M. PÜCHEL, P. SCHMIDT, and E. WEBER, 1983: Analysis of *in vivo* and *in vitro* globulin formation during cotyledon development of field beans (*Vicia faba* L. var. *minor*). – Biochem. Physiol. Pflanzen **178**, 665–684.

BRINEGAR, A. C., and D. M. PETERSON, 1982: Separation and characterization of oat globulin polypeptides. – Arch. Biochem. Biophys. **219**, 71–79.

CASEY, R., and C. DOMONEY, 1984: The biochemical genetics of pea seed storage proteins. – Proc. 3rd Seed Protein Symp. 'Genetics of Seed Proteins' (MÜNTZ, K., and C. HORSTMANN, eds.). Kulturpflanze **32**, ...–....

–, J. F. MARCH, J. E. SHARMAN, and M. N. SHORT, 1981: The purification, N-terminal amino acid sequence and some other properties of an α^M-subunit of legumin from the pea (*Pisum sativum* L.). – Biochim. Biophys. Acta **670**, 428–432.

–, and C. HORSTMANN, 1981: Comparison of legumin subunits from *Pisum sativum* and *Vicia faba* by means of peptide mapping. – Abhdlg. Akad. Wiss. DDR, Abt. Math., Naturwiss., Techn. No. 5 N, 199–200.

CROY, R. R. D., E. DERBYSHIRE, T. G. KRISHNA, and D. BOULTER, 1979: Legumin of *Pisum sativum* and *Vicia faba*. – New Phytol. **83**, 29–35.

–, J. A. GATEHOUSE, I. M. EVANS, and D. BOULTER, 1980: Characterisation of the storage protein subunits synthesised *in vitro* by polyribosomes and RNA from developing pea (*Pisum sativum* L.). I. Legumin. – Planta **148**, 49–56.

–, G. W. LYCETT, J. A. GATEHOUSE, J. N. YARWOOD, and D. BOULTER, 1982: Cloning and analysis of cDNAs encoding plant storage protein precursors. – Nature **295**, 76–79

DE JONG, W. W., A. ZWEERS, and L. H. COHEN, 1978: Influence of single amino acid substitutions on electrophoretic mobility of sodium dodecyl sulfate-protein complexes. – Biochem. Biophys. Res. Commun. **82**, 532–539.

DERBYSHIRE, E., D. J. WRIGHT, and D. BOULTER, 1976: Legumin and vicilin, storage proteins of legume seeds. – Phytochemistry **15**, 3–24.

GROSS, E., 1967: The Cyanogen Bromide Reaction. – Methods Enzymol. **11**, 238–255.

GULDAGER, P., 1978: Immunoelectrophoretic analysis of seed proteins from *Pisum sativum* L. – Theor. Appl. Genet. **53**, 241–250.

HORSTMANN, C., 1983: Specific subunit pairs of legumin from *Vicia faba*. – Phytochemistry **22**, 1861–1866.

–, K. MÜNTZ, A. OTTO, R. KRAFT, und G. ETZOLD, 1982: N-terminale Sequenzen von Untereinheiten des Legumins aus *Vicia faba*. – 14. Jahrestagung der Biochemischen Gesellschaft der DDR. Neubrandenburg, 17.–19. November 1982. Abstracts, p. 179.

KLOTZ, I. M., D. W. DARNALL, and N. R. LANGERMAN, 1975: Quaternary structure of proteins. – In: The Proteins. Third Edition. Vol. I (H. NEURATH, and R. L. HILL, eds.) pp. 293–411. Academic Press, New York – San Francisco – London.

MÜNTZ, K., R. BASSÜNER, and C. HORSTMANN, 1984: Variability of Met^+ and Met^- components of the 11 S globulins from 3 different grain legumes and within the species *Vicia faba* L. – Proc. 3rd Seed Protein Symp. 'Genetics of Seed Proteins' (MÜNTZ, K., and C. HORSTMANN, eds.). Kulturpflanze **32**, ...–....

OTTO, A., R. KRAFT, and G. ETZOLD, 1984: N-terminal sequences of basic subunits of legumin from *Vicia faba* L. by solid phase sequence analysis. – Proc. 3rd Seed Protein Symp. 'Genetics of Seed Proteins' (MÜNTZ, K., and C. HORSTMANN, eds.). Kulturpflanze **32**, ...–....

PLIETZ, P., G. DAMASCHUN, D. ZIRWER, K. GAST, B. SCHLESIER, and K. D. SCHWENKE, 1984: Comparison of the structures of different 11 S and 7 S globulins by small-angle X-ray scattering, quasi-elastic light scattering and circular dichroism spectroscopy. – Proc. 3rd Seed Protein Symp. 'Genetics of Seed Proteins' (MÜNTZ, K., and C. HORSTMANN, eds.). Kulturpflanze **32**. ...–....

SCHWENKE, K. D., B. RAAB, P. PLIETZ, and G. DAMASCHUN, 1983: The structure of the 12 S globulin from rapeseed (*Brassica napus* L.). – Nahrung **27**, 165–175.

STASWICK, P. E., M. A. HERMODSON, and N. C. NIELSEN, 1981: Identification of the acidic and basic subunit complexes of glycinin. – J. Biol. Chem. **256**, 8752–8755.

WOBUS, U., H. BÄUMLEIN, R. BASSÜNER, U. HEIM, R. JUNG, K. MÜNTZ, R. PANITZ, G. SAALBACH, and W. WESCHKE, 1984: Molecular characterization of *Vicia faba* storage protein specific DNA. – Proc. 3rd Seed Protein Symp. 'Genetics of Seed Proteins' (MÜNTZ, K., and C. HORSTMANN, eds.). Kulturpflanze **32**, ...–....

WRIGHT, D. J., and D. BOULTER, 1974: Purification and subunit structure of legumin of *Vicia faba* L. (broad bean). – Biochem. J. **141**, 413–418.

C. HORSTMANN
Zentralinstitut für Genetik und Kulturpflanzenforschung
der Akademie der Wissenschaften der DDR
DDR - 4325 Gatersleben
Correnssstraße 3

Kulturpflanze 32 · 1984 · S 117–S 126

Molecular characterization of *Vicia faba* storage protein specific DNA

U. Wobus, H. Bäumlein, R. Bassüner, U. Heim, R. Jung,
K. Müntz, R. Panitz, G. Saalbach and W. Weschke
(Gatersleben, GDR)

Summary

Poly(A)$^+$RNA from membrane-bound polysomes of *Vicia faba* midmaturation stage cotyledons was used to construct a cDNA library. By means of hybrid-released translation five classes of clones were selected specifying methionine-containing preprolegumin A (6 clones), methionine-free preprolegumin B (13 clones), provicilin (17 clones), an unidentified polypeptide of 30kD (27 clones) and an also unidentified polypeptide of 27kD (1 clone). The prominent 30kD polypeptide is only detected after *in vitro* translation of poly(A)$^+$RNA but not found in the globulin fraction of cotyledons. It is not lectin.

Dot hybridization experiments demonstrated a high degree of sequence divergence between legumin A and legumin B coding genes.

Using cDNA clones four recombinant phages (3 containing legumin B gene sequences and 1 with a 30kD polypeptide gene sequence) were selected from a *V. faba* partial genomic library suggesting a family of legumin B genes with at least 10 to 20 copies and less copies for the other gene types synthesizing abundant mRNAs in mid-maturation stage cotyledons.

The interesting discovery of the existence of two clearly distinguished legumin gene types containing (type A) and lacking (type B) methionine codons not only provides a possibility to look for genotypes with an increased number of type A genes or increased synthesis of legumin A but should also be favourable in an approach to manipulate genes by reversed genetics.

Introduction

Vicia faba seed proteins have been studied for many years partly because of their relevance for animal feeding and human nutrition and partly as a model plant system since these proteins are massively synthesized during a restricted developmental period in a defined tissue (Derbyshire et al. 1976, Boulter 1981, Müntz et al. 1981). In spite of having only two major storage globulins in *Vicia* seeds detailed studies with modern analytical methods have revealed considerable polypeptide heterogeneity (Matta et al. 1981). An especially interesting example of this heterogeneity is the existence of two structural types of legumin subunit pairs (Horstmann 1983, 1984) containing (type A) and lacking (type B) methionine.

Storage protein heterogeneity may be of biological as well as of economical importance and we therefore have to ask for its underlying reasons. One possibility to do this is to analyse seed protein genes by means of recombinant DNA technology.

Results

Since storage proteins of legumes are synthesized on membrane-bound (mb) polysomes (PÜCHEL et al. 1979, BOLLINI and CHRISPEELS 1979) poly(A)$^+$RNA isolated from such polysomes is a good starting material for cDNA synthesis. If this RNA is loaded onto an urea polyacrylamide gradient gel most of it migrates in the 2kb region as one main band which may be a doublet. In addition, some minor bands can be detected (cf. Fig. 3 C in MÜNTZ et al. 1981). Translation of the same RNA in an *in vitro* wheat germ system results in a few predominant polypeptides (Fig. 1, lanes 1 and 8) the pattern of which changes gradually during development of the cotyledon (BASSÜNER et al. 1983). The two legumin precursor species are clearly distinguished by the presence or absence of methionine as indicated by *in vitro* translation experiments using either ^{35}S-methionine or ^{3}H-labelled amino acids (BASSÜNER et al. 1983, 1984a and b): preprolegumin A (ppLeA) which contains methionine and preprolegumin B (ppLeB) which is free of methionine.

The often dominating product in the polypeptide pattern synthesized *in vitro* from mb-poly(A)$^+$RNA is an up to now unidentified methionine-containing polypeptide of about 30kD (Fig. 1, lanes 1 and 8) which will be refered to as the 30kD polypeptide. It is not detectable among cotyledon extracted globulins and it remains unclear whether a minor fraction among the albumins with the same electrophoretic mobility has anything to do with it. The developmental profile of the *in vitro* synthesized 30kD polypeptide resembles that of vicilin with a clear peak in the first half of cotyledon development (BASSÜNER et al. 1983).

To construct cDNA clones we isolated total poly (A)$^+$ RNA from mb-polysomes of mid-maturation stage cotyledons and synthesized single and double stranded cDNA by the usual protocol. Insertion of ds-cDNA into the PstI-site of pBR 322 by the GC-tailing method and transformation into *E. coli* C 600 cells resulted in about 1,500 ampicillin sensitive colonies from which 82 hybridized with the original RNA under conditions allowing only abundant RNA's to react (PÜCHEL et al. 1981).

As principal method for clone identification we used hybrid-selected translation (MANIATIS et al. 1982). By checking 72 clones we identified 64 representing five different types of mRNA coding for ppLeA(met$^+$), ppLeB(met$^-$), the main vicilin precursor pVi, the 30kD polypeptide discussed above and, in addition, a minor polypeptide of about 27kD. Fig. 1 shows representative results of the *in vitro* translation experiments. Included is an experiment (lane 11) in which the *Pisum sativum* legumin cDNA clone pPs 15–75 (CHANDLER et al. 1983) was reacted with *V. faba* poly(A)$^+$RNA. It predominantly selects ppLeA(met$^+$) mRNA but also a small amount of ppLeB(met$^-$) mRNA. The degree of crosshybridization found, albeit low, was not observed with the *V. faba* ppLeA specifying cDNA clones.

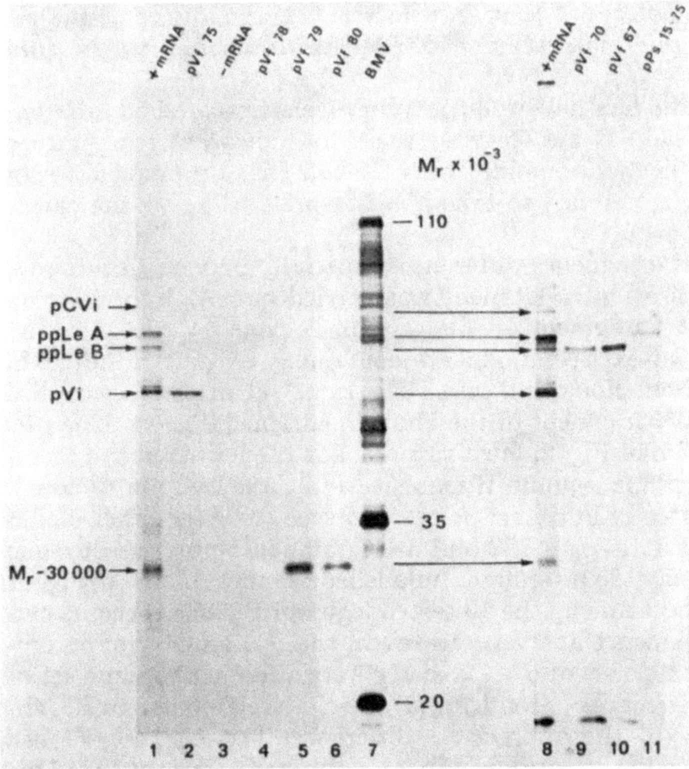

Fig. 1 Identification of cDNA clones by cell-free translation of hybrid-selected mRNA species from developing field bean (*Vicia faba* L.) cotyledons
mRNA species were isolated from cotyledonary poly(A)$^+$RNA using filter-bound plasmids (cf. MANIATIS et al. 1982) of field bean cDNA clones (designated pVf and a number) followed by translation in the wheat germ cell-free system (lanes 2, 4–6, 9, 10). For comparison, the pea cDNA clone pPs 15–75 (CHANDLER et al. 1983) bearing a legumin-specific insert, was included (lane 11)
Shown is a fluorograph obtained after SDS-gel electrophoresis (10.6 % acrylamide: lanes 1–7; 9.5 % acrylamide: lanes 8–11) of the [^{3}H]leucine labeled translation products. Lanes 1 and 8 ("+mRNA") contain translation products of poly(A)$^+$RNA whereas lane 4 ("–mRNA") shows endogenous translation products of the wheat germ system
The storage globulin-related translation products are designated as pCVi, preconvicilin; ppLe A, preprolegumin A (Met-containing); ppLe B preprolegumin B (Met-free) and pVi, previcilin (Met-free), whereas "M_r-30,000" refers to a non-identified translation product. The M_r-scale is given by BMV RNA translation products (lane).

It may be added that (a) in the case of legumin and vicilin precursor polypeptides the identity suggested by electrophoretic mobility was confirmed in immunoprecipitation experiments (results not shown) and (b) that the 30kD polypeptide is clearly not lectin since a *Pisum* lectin-specifying cDNA clone kindly provided by T. J. V. HIGGINS selects a *V. faba* lectin mRNA which directs the synthesis of a polypeptide with lower electrophoretic mobility (M_r 32,000) and different processing behavior (results not shown).

The numbers of identified clones in the different classes are as follows: 27 specify the 30kD polypeptide, 17 pVi, 13 ppLeB, 6 ppLeA and 1 the 27kD polypeptide. This distribution reflects at least in principle the abundance of the respective mRNAs in the mid-maturation cotyledon as measured by *in vitro* translation.

We were astonished by the fact how well the hybrid-selection method differentiates between legumin A and B messengers. At an hybridization temperature of 50 °C in 0.4 M salt and 65 % formamide and a GC-content of legumin DNA of about 47 % hybridization is assumed to take place about 8 °C below the calculated T_m-value.

To get a better idea of the sequence differences especially between legumin A and B clones a dot hybridization experiment was carried out. As hybridization probe an internal Pst-Hinf I fragment of the legumin B clone pVfc70 was used since hybridization with Pst-excised fragments containing GC tails resulted in unspecific cross-hybridization. For comparison, the same set of clones was also hybridized with a Pst-Hinf I fragment of the above mentioned *Pisum* clone pPs 15–75. The results are shown in Fig. 2. Most surprising is the low degree of cross-hybridization between the probe legumin B sequence and three legumin A clones. Roughly the same low degree of hybridization is also seen with the other clones tested whereas the control dots–pBR 322 and a GC-tail containing insect clone pCthBRc 19 (Wobus et al. 1980)–remain unlabelled. Under more stringent washing conditions differences among the 10 tested legumin B clones become evident, but only extensive sequence analysis can reveal the real amount of heterogeneity among the different clones of one class. Hybridization of the same set of clones with a *Pisum* legumin sequence confirms (a) the close relatedness to *V. faba* legumin A sequences and (b) the appreciable differences between the *V. faba* Met^+- and Met^--legumin-coding sequences. These results are confirmed by the partial amino acid sequences of Otto et al. (1984) since comparison of 61 amino acids from the N-terminal parts of a legumin A and a legumin B β-subunit, respectively, demonstrate only 50 % homology. Since the divergent amino acids are scattered over the sequence, complete matches of 15 and more nucleotides, necessary for cross-hybridization under stringent conditions, should be absent.

In a further attempt to characterize the cDNA clones and the respective genes the two types of legumin clones were separately hybridized in preliminary experiments to Northern blots of poly(A)$^+$RNA from different stages of cotyledon development. As expected, both clones label RNA of about 2kb and the developmental profile is principally the same as the one deduced from *in vitro* translation experiments (results not shown). This observation adds further weight to the hypothesis of a mainly transcriptionally regulated pattern of globulin biosynthesis in developing field bean cotyledons (Müntz et al. 1981, Bassüner et al. 1983).

With clearly defined cDNA clones available we started construction of genomic clones. 10 µg *V. faba* DNA fragments 15 to 20kb in length from an EcoRI partial digest were ligated to 25 µg of Charon 4A armes and packaged *in vitro*. About one million phages were obtained, amplified and stored as a partial library. On X-gal plates more than 97 % of the plaques turned out to be white and of 14 randomly choosen plaques all phages contained inserted EcoRI fragments with a mean total insert length of about 14kb. Using this value and a genome size of 13×10^6 kb (cf. Bennett et al. 1981) one million recombinant phages should represent about 30 % of the genome with 95 % probability (Clarke and Carbon 1976).

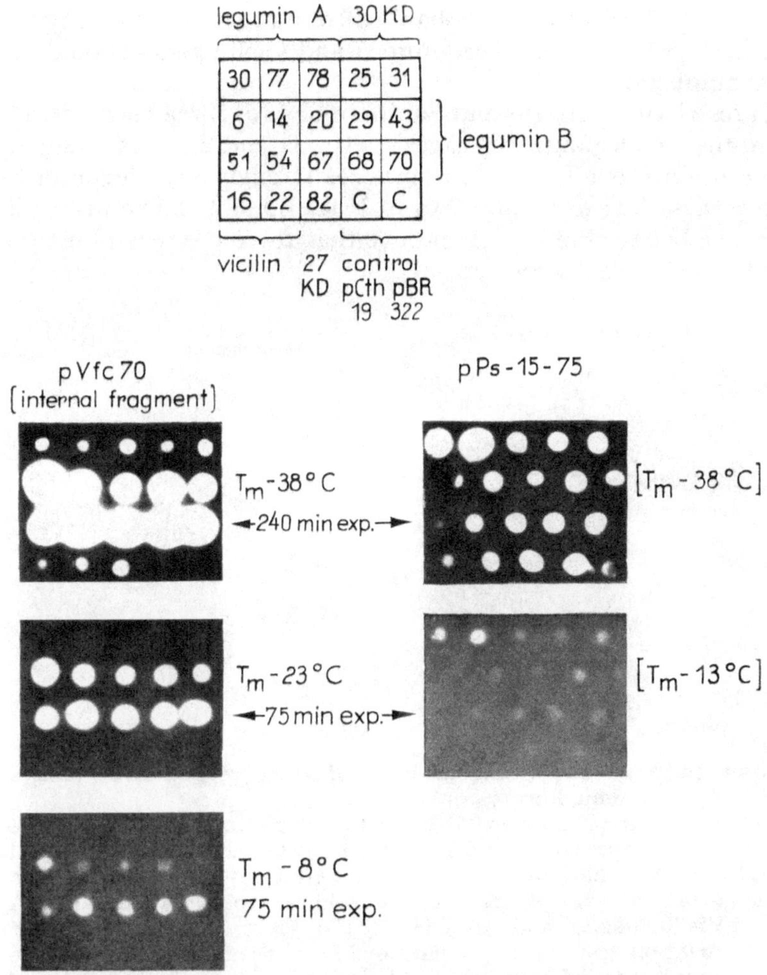

Fig. 2 Dot blots of *V. faba* cDNA clones
0.5 µg of each of the indicated clones was denatured in 0.3 N NaOH, neutralized and spotted in 12 × SSC in duplicate onto two nitrocellulose filters. After baking filters were hybridized at 0.39 M Na$^+$ and 65 °C to either an internal ^{32}P-PstI-HinfI fragment of pVfc70 or a ^{32}P-PstI-HinfI fragment of pPs 15-75 (see text for specifications) and washed stepwise under increasingly stringent conditions as indicated

The partial library was screened at a density of 10^5 plaque forming units per 14 cm plate with ^{32}P-labelled cDNA plasmids of each main class, i.e. specifying ppLeA, ppLeB, pVi and the 30kD polypeptide. Four phages were selected containing coding sequences for legumin B (λVfLeB1, 2 and 3) and the 30kD polypeptide (λVf 30.1).

Clearly, these phages represent only a certain percentage of the relevant phages present in the library since even repeated screening at low phage density does not detect all clones as demonstrated in the case of globin recombinants in the human genomic library of Maniatis et al. (Seed 1983). Therefore, the number of three seemingly different (see below) legumin B clones among the 10^6 recombinants suggests a minimal number of 10 but more probably 20 or more genes of this type. On the other hand, 30kD genes as well as legumin A and vicilin genes should be present in lower copy numbers.

Fig. 3 shows EcoRI restriction patterns and Southern blots of three selected and purified phages. From the two legumin B phages λVfLeB1 and 2 (lanes 1 and 2, respectively) in each case only one fragment hybridizes strongly with legumin B cDNA sequences. Since these fragments and that of λVfLeB3 not shown in Fig. 3 are all different in size we believe but cannot proove that the three recombinants contain sequences of three different gene copies.

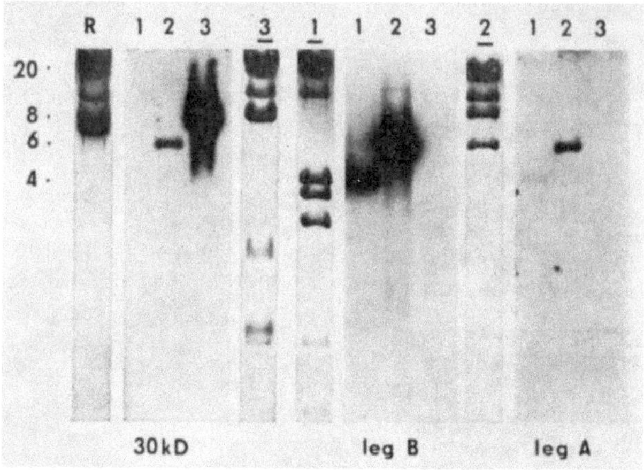

Fig. 3 EcoRI restriction patterns and Southern blots of three recombinant phages selected from a partial *V. faba* genomic library
6 μg of DNA of each recombinant phage were digested to completion with EcoRI, the samples divided into 3 parts and separated in a 0.7 % agarose gel arranged in three equal groups. After photography of the ethidium bromide stained gel DNA was transferred to nitrocellulose, the filter cutted into three pieces and hybridized to ^{32}P-labelled cDNA clones pVfc77 (legA), pVfc70 (legB) and pVfc25 (30kD) (spec. activities between $3-5 \times 10^7$ cpm/μg). Hybridization and washing conditions have been described (WOBUS et al. 1980). Numbers on the left side indicate sizes in kb derived from an EcoRI digest of Charon 4A DNA. Lane numbers indicate hybridization patterns or (if underlined) stained patterns of phages λVfLeB1 (1), λVfLeB2 (2) and λVf30.1 (3). R = staining pattern of a phage lambda EcoRI digest

Cross-hybridization of a legumin A probe and a 30kD probe to the same fragments is at least 20 times lower and that of a vicilin probe (not shown) is only detectable after long over exposure (3 days versus 4 hrs). Since this observation corresponds to principally similar results from dot hybridization and hybrid-released translation experiments we tend to believe in a low degree of real homology between legumin B and 30kD sequences.

Clone λVf 30.1 (Fig. 3, lanes 3) contains only one 8kb fragment hybridizing with cloned cDNA. All fragments of the three clones reacting with cDNA probes hybridize also with poly(A)$^+$RNA (results not shown).

For further characterization DNA of phage λVfLeB2 was hybridized to poly (A)$^+$ RNA under R-loop conditions and analyzed by electrone microscopy (Fig.4). The analysis clearly revealed a single uninterrupted R-loop of only about

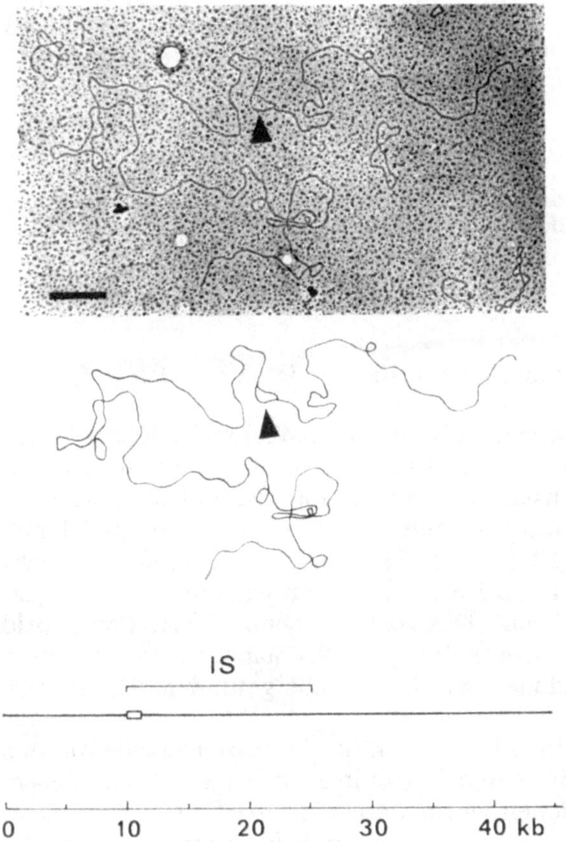

Fig. 4 Electron micrograph and tracing of a representative R loop molecule formed between clone λVfLeB2 and *V. faba* poly(A)$^+$RNA from cotyledons
The scale for both micrograph and tracing represents 2kb. The map of the recombinant molecule placed under the tracing shows the boundaries of the inserted DNA segment (IS) and the position of the R-loop on a different scale

800 bp showing that not more than one third of the expected minimal length of the gene are included in this phage. Since we used a cDNA clone for phage selection representing the 3′-half or β-subunit of the legumin B gene λVfLeB1 should contain the 3′-end of this gene together with several kb of flanking sequence. If members of the legumin B gene family are arranged in one or a few clusters as, for instance, members of a zein gene subfamily (Spena et al. 1983), a hypothetical neighbour of the λVfLeB1 gene should be at least 14 to 15kb apart since the total length of the insert is about 16kb (see Fig. 4).

To test such predictions we are presently constructing new recombinants with the main aim of selecting other legumin B clones but especially also representatives of the legumin A and vicilin gene families.

The reported experiments mainly provide the necessary groundwork for a detailed analysis of *Vicia faba* seed protein genes. Especially the existence of a dominating number of legumin genes lacking methionine provides not only a good chance to genetically improve the nutritional value of *Vicia* seed proteins by altering legumin subunit composition but this fact should also be favourable in an approach to manipulate genes directly by reversed genetics.

Acknowledgement

We wish to thank T. J. V. HIGGINS for kindly providing us with two *Pisum sativum* cDNA clones bearing legumin and lectin specific inserts, respectively.

Zusammenfassung

Molekulare Charakterisierung reserveproteinspezifischer DNA von *Vicia faba*

Poly(A)$^+$-RNA von membrangebundenen Polysomen aus *Vicia faba*-Kotyledonen mittleren Reifestadiums wurde für die Konstruktion einer cDNA-Bibliothek eingesetzt. Durch „hybrid-released translation" wurden fünf Klonklassen selektioniert, die spezifisch sind für methioninhaltiges Präprolegumin A (6 Klone), methioninfreies Präprolegumin B (13 Klone), Prävicilin (17 Klone), ein nicht charakterisiertes Polypeptid von 30kD (27 Klone) und ein weiteres nicht charakterisiertes Polypeptid von 27kD (1 Klon). Das vorherrschende 30kD-Polypeptid kann nur nach *in vitro*-Translation von Poly(A)$^+$-RNA nachgewiesen werden; in der Globulinfraktion der Kotyledonen wurde es nicht gefunden. Es ist kein Lektin.

Durch Dot-Hybridisierungsexperimente wurden große Unterschiede in den Sequenzen der Gene, die für Legumin A und Legumin B kodieren, nachgewiesen.

Durch den Einsatz von cDNA-Klonen wurden aus einer partiellen *Vicia faba*-Genbibliothek vier rekombinante Phagen (3 mit Legumin-B-Gensequenzen und 1 mit der Gensequenz für ein 30kD-Polypeptid) selektioniert, was auf eine Familie von Legumin-B-Genen mit mindestens 10 bis 20 Kopien und weniger Kopien für die anderen Gentypen hinweist, die die in Kotyledonen des mittleren Reifestadiums vorherrschenden mRNA's synthetisieren.

Die interessante Entdeckung von zwei unterschiedlichen Legumintypen, die Methionincodons besitzen (Typ A) oder nicht besitzen (Typ B), eröffnet nicht nur die Möglichkeit, nach Genotypen mit einer größeren Zahl von Typ-A-Genen oder einer erhöhten Synthese von Legumin A zu suchen, sondern stellt auch einen günstigen Ausgangspunkt für die genetische Manipulation auf molekularer Ebene dar.

Краткое содержание

Молекулярная характеристика ДНК, специфичной для запасных белков конских бобов

Поли(А)$^+$-РНК мембранно связанных полисом из котиледонов *Vicia faba* средней зрелости была взята для заложения молекулярного банка к-ДНК. Путём трансляции РНК, полученной техникой ДНК–РНК гибридизации идентифициро-

вались пять классов клонов к-ДНК, специфичных для содержащего метионин препролегумина А (6 клонов), не содержащего метионин препролегумина В (13 клонов), превицилина (17 клонов)., для одного не идентифицированного 30-кД-полипептида и для ещё одного не идентифицированного 27-кД-полипептида. Преобладающий 30-кД-полипептид выявляется только после поли(А)⁺-РНК-трансляции in vitro; во фракции глобулинов из котиледонов он не был обнаружен; это не лектин.

При помощи техники точковой гибридизации были обнаружены значительные различия в нуклеотидной последовательности генов, кодирующих легумин А и легумин В.

При применении клонов к-ДНК из частичного генного банка конских бобов были селекционированы четыре рекомбинантных фага (три со специфичными последовательностями для легумина В и один – для 30-кД-полипептида, что указывает на группу генов легумина В с 10 до 20 копиями (по меньшей мере) и с меньшим числом копий для других типов генов, на которых в котиледонах средней зрелости синтетизируется преобладающая и-РНК.

Интересное выявление двух различных типов легумина, которые имеют (тип А) или не имеют (тип В) метиониновые кодоны, открывает не только возможность поисков генотипов с большим числом генов типа А, или усиленного синтеза легумина А, но является также подходящим исходным положением для генетических манипуляций на молекулярном уровне.

Literature

Bassüner, R., R. Manteuffel, K. Müntz, M. Püchel, P. Schmidt and E. Weber, 1983: Analysis of *in vivo* and *in vitro* globulin formation during cotyledon development of field beans (*Vicia faba* L. var. *minor*). – Biochem. Physiol. Pflanzen **178**, 665–684.

–, A. Huth, R. Manteuffel and T. Rapoport, 1984a: Secretion of plant storage globulin polypeptides by *Xenopus* oocytes. Proc. 3rd Seed Protein Symp. 'Genetics of Seed Proteins' (Müntz, K., and C. Horstmann, eds.). Kulturpflanze **32**, S 215–S 218.

–, U. Wobus and T. Rapoport, 1984b: Cotranslational processing of storage globulin polypeptides of *Vicia faba* L. by signal recognition particle and microsomal membranes in a cell-free translation system. – Proc. 3rd Seed Protein Symp. 'Genetics of Seed Proteins' (Müntz, K., and C. Horstmann, eds.). Kulturpflanze **32**, S 71–S 74.

Bennett, M. D., J. B. Smith, J. Ward and G. Jenkins, 1981: The relationship between nuclear DNA content and centromere volume in higher plants. – J. Cell Sci. **47**, 91–115.

Bollini, R. and M. J. Chrispeels, 1979: The rough endoplasmic reticulum is the site of reserve-protein synthesis in developing *Phaseolus vulgaris* cotyledons. – Planta **146**, 487–501.

Boulter, D., 1981: Biochemistry of storage protein synthesis and deposition in the developing legume seed. – Adv. Botan. Res. **9**, 1–31.

Chandler, P. M., T. J. V. Higgins, P. J. Randall and D. Spencer, 1983: Regulation of legumin levels in developing pea seeds under conditions of sulphur deficiency. – Plant Physiol. **71**, 47–54.

Clarke, L. and J. Carbon, 1976: A colony bank containing synthetic ColE1 hybrid plasmids representative of the entire *E. coli* genome. – Cell **9**, 91–99.

Derbyshire, E., D. J. Wright and D. Boulter, 1976: Legumin and vicilin, storage proteins of legume seeds. – Phytochem. **15**, 3–24.

Horstmann, C., 1983: Specific subunit pairs of legumin from *Vicia faba*. Phytochem. **22**, 1861–1866.

–, 1984: New aspects of the subunit structure of *Vicia faba* legumin. – Proc. 3rd Seed Protein Symp. 'Genetics of Seed Proteins' (MÜNTZ, K., and C. HORSTMANN, eds.). Kulturpflanze **32**, S 109–S 116.

MANIATIS, T., E. F. FRITSCH and J. SAMBROOK, 1982: Molecular cloning. A laboratory manual. Cold Spring Harbor Laboratory, New York.

MATTA, N., J. A. GATEHOUSE and D. BOULTER, 1981: The structure of legumin of *Vicia faba* L. – a reappraisal. – J. exp. Bot. **32**, 183–197.

MÜNTZ, K., R. BASSÜNER, H. BÄUMLEIN, R. MANTEUFFEL, M. PÜCHEL, P. SCHMIDT and U. WOBUS, 1981: The biosynthesis of storage proteins in developing field bean seeds. In: K. MÜNTZ, R. MANTEUFFEL and G. SCHOLZ (eds.) Proc. 2nd Seed Protein Symp.: Regulation of protein biosynthesis and degradation during embryogenesis and germination of plant seeds. Akademie-Verlag, Berlin, pp. 57–72.

OTTO, A., R. KRAFT and G. ETZOLD, 1984: N-terminal sequences of basic subunits of legumin from *Vicia faba* L. by solid phase sequence analysis. – Proc. 3rd Seed Protein Symp. 'Genetics of Seed Proteins' (MÜNTZ, K., and C. HORSTMANN, eds.). Kulturpflanze **32**, S 219–S 221.

PÜCHEL, M., K. MÜNTZ, B. PARTHIER, O. AURICH, R. BASSÜNER, R. MANTEUFFEL and P. SCHMIDT, 1979: RNA metabolism and membrane bound polysomes in relation to globulin biosynthesis in cotyledons of developing field beans. – Eur. J. Biochem. **96**, 321–329.

–, H. BÄUMLEIN, P. SILHENGST and U. WOBUS, 1981: Cloning of double stranded cDNA derived from polysomal poly(A)-containing RNA from developing field bean seeds. – In: K. MÜNTZ, R. MANTEUFFEL and G. SCHOLZ (Eds.) Proc. 2nd Seed Protein Symp.: Regulation of protein biosynthesis and degradation during embryogenesis and germination of plant seeds. Akademie-Verlag, Berlin, p. 227.

SEED, B., 1983: Purification of genomic sequences from bacteriophage libraries by recombination and selection *in vivo*. – Nucleic Acids Res. **11**, 2427–2445.

SPENA, A., A. VIOTTI and V. PIRROTTA, 1983: Two adjacent genomic zein sequences: structure, organization and specific restriction pattern. – J. Mol. Biol. **169**, 799–811.

WOBUS, U., H. BÄUMLEIN, R. PANITZ, E. SERFLING and F. C. KAFATOS 1980: Periodicities and tandem repeats in a Balbiani ring gene. – Cell **22**, 127–135.

U. WOBUS, H. BÄUMLEIN, R. BASSÜNER, U. HEIM, R. JUNG, K. MÜNTZ
R. PANITZ, G. SAALBACH, W. WESCHKE
Zentralinstitut für Genetik und Kulturpflanzenforschung
der Akademie der Wissenschaften der DDR
DDR - 4325 Gatersleben
Corrensstr. 3

Kulturpflanze 32 · 1984 · S 127–S 158

Structure of 11S globulins of soybean and broad bean and its significance in the food system

T. Mori, T. Nakamura and S. Utsumi
(Kyoto, Japan)

Summary

Soybean 11S globulins (glycinins) from the seeds of 18 cultivars (soybean from Japan, U.S., China and Korea) were analyzed for their subunit compositions by electrophoreses on polyacrylamide gels under various conditions. They could be classified into five groups according to the differing molecular charges of the subunits: group I, contained 7 acidics and 8 basics; group II, 7 acidics and 7 basics; group III, 6 acidics and 7 basics; group IV, 6 acidics and 5 basics; group V, 6 acidics and 3 basics. Glycinin from soybean var. Tsuru-no-ko of the group I was fractionated on a DEAE-Sephadex column and the molecular species were investigated by electrophoretic methods. The results indicated the heterogeneity of glycinin molecular species; four molecular species with molecular weights of 375,000, 360,000, 345,000 and 340,000 were detected by PGGE. Glycinins from other cultivars also exhibited a similar heterogeneity and the extent of this heterogeneity seems to correlate with the diversity in the subunit composition.

Broad bean 11S globulins (legumins) from the seeds of six typical cultivars planted in Japan were also analyzed for their subunit compositions and examined for the molecular species. They could be classified into three groups according to the size of the seeds from the standpoint of the molecular charges, i.e., small-, medium- and large-size groups. Legumin exhibited the heterogeneity of molecular species. The heterogeneity was common to the cultivars from the standpoint of the molecular sizes of the subunits; it was deduced that five groups with nine subspecies were present in each cultivar.

The artificial 11S globulins, pseudo-glycinins and pseudo-legumins with different subunit compositions from the native 11S globulins, were formed in a reconstitution reaction which combined one each of acidics and basics of glycinin for pseudo-glycinin and one each of acidics and basics of legumin for pseudo-legumin. They were similar to the native 11S globulins; they all consisted of reconstituted intermediary subunits which were composed of acidic and basic subunits linked by disulfide bridges in a ratio of 1 : 1 and had the 6 (AB) structure. These results suggest a similarity in the structure of the acidic subunits and that of the basic subunits, and that the specificity of the interactions among subunits and among intermediary subunits in the formation of 11S-structure is not precise. This must be one of the reasons for the occurrence of the heterogeneity of glycinin and legumin molecular species. Further, hybrid-11S globulins were formed from various combinations of acidics of legumin and basics of glycinin or vice versa. The hybrid-11S globulins formed were similar to the native 11S globulins; they all consisted of six reconstituted hybrid-intermediary subunits which were composed of acidic and basic subunits linked by disulfide bridges in a ratio of 1 : 1. These observations confirm that the specificity of the interactions among subunits and among intermediary subunits in the formation of 11S-structure is not precise, and suggest that the structure of glycinin and legumin are similar to each other and that they share a common assembly mechanism.

In order to elucidate how the subunits of 11S globulins contribute to the physical properties of the gel, the textural properties of the gel made from the pseudo- and hybrid-11S globulins were investigated. The hardness of the heat-induced gels from pseudo-glycinins was different from those derived from native glycinin, depending on the acidic subunit composition. AS_5, having a larger molecular weight than the other acidic subunits, appeared to cause a significant increase in the hardness of the gel. In the case of pseudo-legumin, it was demonstrated that ASII plays an important role in increasing the hardness of the gel. The results obtained from the experiments on hybrid-11S globulins indicated that the superior gelling ability of glycinin as compared to that of legumin depends on the properties of both the acidic and basic subunits of glycinin. Further, the gel properties of glycinins were examined for each soybean cultivars from five groups having different subunit compositions each other. Hardness of the gels was different among cultivars, depending on the amount ratio of AS_5.

From the investigation on the structure and textural properties of 11S globulins described above, it was elucidated that the properties of 11S globulin at subunit level closely relate to the characteristics of gel made from the 11S globulin. Since gelation phenomena is one of the most significant functional properties with respect to usage in food systems, it may be worthwhile to take into account a cultivar of the seeds on usage of their proteins for foods. In addition to such consideration for providing desirable cultivars, breeding from the standpoint of protein composition at the level of subunit protein molecules may be a possible way to improve the quality of seed proteins.

Introduction

Soybean has a long history as a protein foodstuff in the Orient including our country. A number of soybean protein products have become established as useful ingredients in the manufacture of processed foods during the past fifteen years. The growing acceptance of these products for use in food manufacture has been prompted by their varied functional properties and good nutritional qualities. The functional properties that make soybean proteins useful in food systems are their abilities of gelation, emulsification, fiber and shred formation and so on. It is an important problem to elucidate what kind of chemical and structural properties of soybean proteins are responsible for their excellent functional properties. An understanding of these properties would further the adoption in foods of soybean proteins and other grain legume proteins similar in structure to soybean proteins. However, at present sufficient information is lacking.

The seed storage proteins of legumes including soybean contain two major components, 7S and 11S globulins (DERBYSHIRE et al. 1976), and both globulins are composed of plural subunits. Therefore, research on the physical and chemical properties of soybean proteins must be focused on the subunits in order to shed light on the problem described above. Recent years have seen a dramatic development in the elucidation of the structural aspects of 11S globulins from various kinds of seeds. Especially with regard to the 11S globulin from soybean (KOSHIYAMA 1983), several structural models of the molecule have been proposed on the basis of observations through electron microscope (SAIO et al. 1970; BADLEY et al. 1975). The model of the latter shown in Fig. 1 is generally accepted. According to

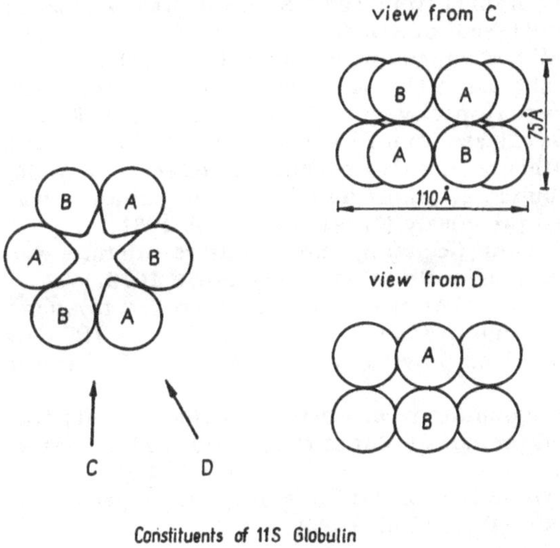

Fig. 1A Model of subunit arrangement in soybean 11S globulin. Diagram at left is top view of hexagonal rings. Side views from position C and D are shown at right

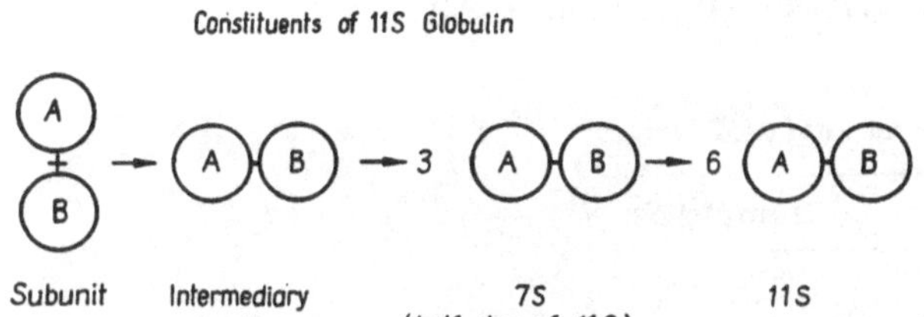

Fig. 1B Schematic representation of constituents of soybean glycinin

this model and other protein chemical analyses of the subunit structure of the soybean 11S globulin, the constituent components of the 11S globulin are regarded as shown in Fig. 1–B. In this figure, A and B refer to acidic subunits and basic subunits, respectively. The intermediary subunit (AB) is composed of disulfided acidic and basic subunits, which are linked together in specific combinations in 1 : 1 ratio (KITAMURA et al. 1976). The half-molecule (3(AB)) consists of three intermediary subunits and has the size of 7S. The 11S component consists of six intermediary subunits; that is, it has the 6 (AB) structure. Such a structure of 11S globulin shown in Fig. 1–B seems to be common to 11S-type globulins from other seeds.

Of various legumes, soybean has the highest protein content, followed by broad bean which is superior to pea, kidney bean and peanut. Broad bean has the advantage that its cultivation time is opposite to that of soybean, which makes it possible to cultivate both soybean and broad bean in the same field in a year.

This paper describes the chemical and structural properties of 11S globulins from soybean and broad bean and their relations to the functional properties of 11S globulins.

Materials and Methods

The samples of soybean seeds *(Glycine max)* examined and their country of origin are listed in Table I. The cultivars were grown at some places in Japan in 1977 where the seeds of each cultivar normally developed and matured. Six cultivars of broad bean (*Vicia faba* L.) seeds planted in Japan were obtained from Takii Seed Co., Ltd. The seeds of these cultivars differ in size, that is, the small seeds of Sanuki-Nagasaya (S.N.), Sanuki-Nagasaya-Wase (S.W.) and Gifu-Wase (G.W.), the medium seeds of Kumamoto-Churyu (K.C.) and the large seeds of Otafuku (O.) and Issun (I.). Average weights per 100 seeds of the three groups were 90, 115 and 230 g, respectively. DEAE-Sephadex A-50 and polyacrylamide gradient gel (4–30 %) were purchased from Pharmacia Co., Ltd. Ampholine (pH 3.5–10) was purchased from LKB Co. Other chemicals were guaranteed reagent grade.

Preparation of purified soybean 11S globulin (glycinin) and its native subunits was performed according to the method described previously (Mori et al. 1979, 1981).

Preparation of purified broad bean 11S globulin (legumin) and its native subunits was performed according to the method described previously (Mori and Utsumi 1979).

Polyacrylamide gel electrophoresis (PAGE) under various conditions and polyacrylamide gel electrofocusing (PGEF), and polyacrylamide gradient gel electrophoresis (PGGE) were performed according to the methods described previously (Mori et al. 1981; Utsumi and Mori 1980).

Reconstitution of pseudo- and hybrid-11S globulins from isolated native subunits was performed according to the method described previously (Mori et al. 1979, 1982; Utsumi et al. 1980, 1983).

Preparation of heat-induced gel and determination of hardness of gel were performed according to the methods described previously (Mori et al. 1982).

Table 1
List of Seed Samples Examined with Country of Origin

country	cultivar	sample No.
Japan	Tokachi Nagaha	1
	Shiro Tsuru-no-ko	2
	Rikuu No. 20	3
	Raiden	4
	Goyo Daizu	5
	Sakagami No. 2	6
	Iyo Daizu	7
	Matsuura	8
U.S.A.	Hill	9
	Hark	10
	Corsoy	11
	York	12
	Dare	13
	Ford	14
China	Bai-Hua-Zuo-Zi	15
	Tianjin-Dachingdou	16
Korea	Kinzu	17
	Huk-Tae	18

Results and Discussion

Difference in Subunit Compositions of Glycinins among Soybean Cultivars

The 11S-type globulin of soybean is called glycinin. The structural features of glycinin have been fairly elucidated up to now as described in Introduction. However, the subunit composition of glycinin has not yet elucidated sufficiently. This was investigated by several people (see references in MORI et al. 1981). Their results were different among each other, which indicates the differences in subunit compositions of glycinins among soybean cultivars.

The purified glycinins from various cultivars were analyzed by PAGE under various conditions (Figs. 2, 3, 4 and 5). All glycinins were composed of four major subunit proteins with molecular weights of 38,000, 34,800, 19,000 and 18,300 (Fig. 2). The larger two subunits and the smaller two were acidic and basic subunits, respectively. The results shown in Figs. 3 to 5 indicate that glycinins of various cultivars could be classified into two groups, one which contained an extra subunit protein (the most acidic subunit, Fig. 4). This acidic subunit appears to be linked to the basic subunit to form an intermediary subunit, not through

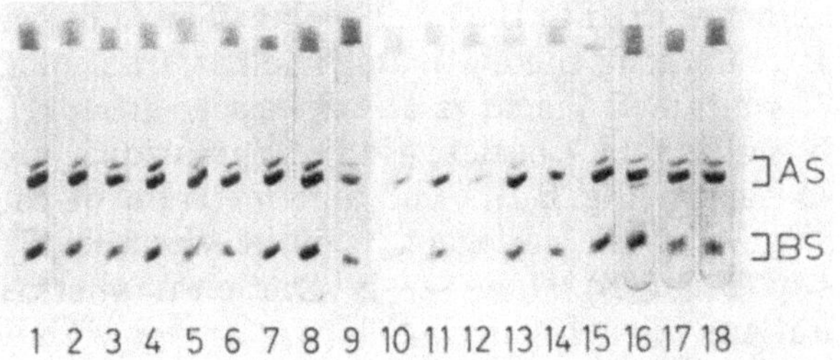

Fig. 2 SDS-PAGE in the presence of 2-ME of glycinins from samples, No. 1–18, listed in Table 1. Fifty µg of each glycinin was subjected to the electrophoresis. Migration is from top to bottom

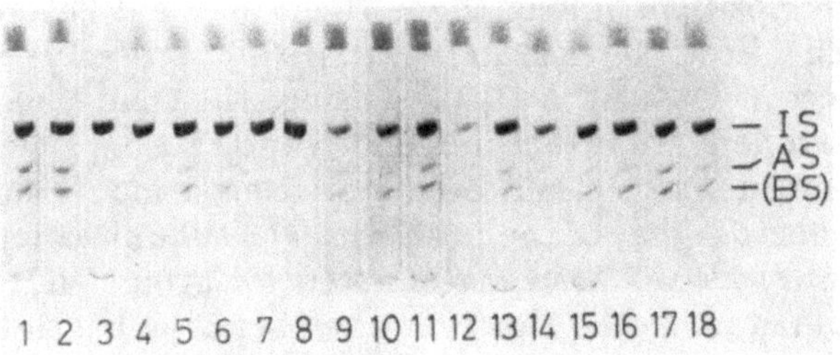

Fig. 3 SDS-PAGE in the absence of 2-ME of glycinins from the same samples described in Fig. 2. Fifty µg of each glycinin was used. Migration is from top to bottom

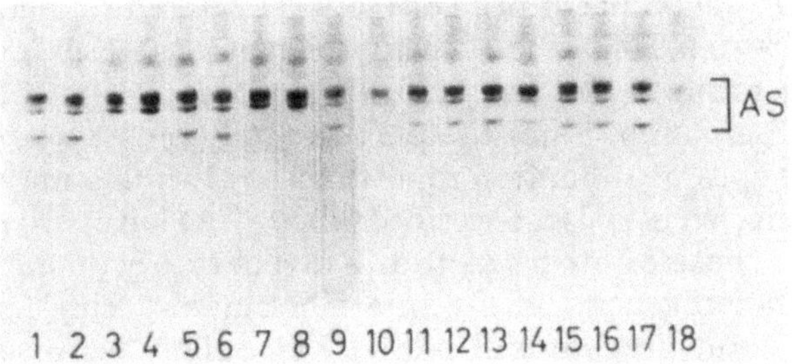

Fig. 4 Alkaline urea-PAGE in the presence of 2-ME of glycinins from the same samples described in Fig. 2. Fifty µg of each glycinin was used. Migration is from top to bottom

a disulfide bridge but through some other interactions which can be disrupted by such denaturants as sodium dodecyl sulfate (SDS) and urea (Figs. 3 and 5). Although the existence of such an intermediary subunit is rather unexpected, it may be a constituent of glycinin, since the total amount of the most acidic subunit and the basic subunit, both of which constitute the intermediary subunit,

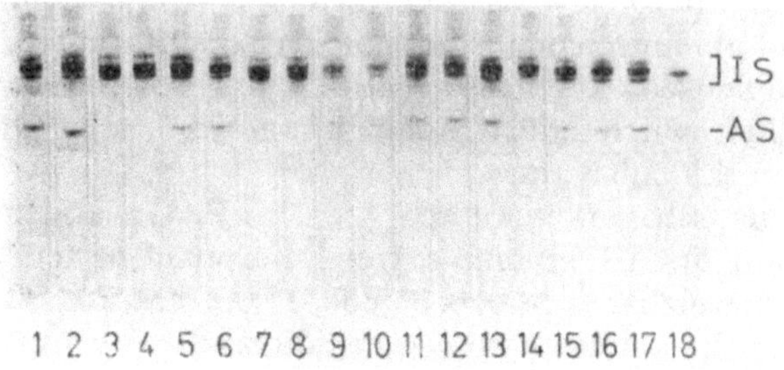

Fig. 5 Alkaline urea-PAGE in the absence of 2-ME of glycinins from the same samples described in Fig. 2. Fifty μg of each glycinin was used. Migration is from top to bottom

is comparable to the other subunits (Figs. 3 to 5). In order to obtain further details about the subunit composition of glycinins from various cultivars, glycinins were analyzed by PGEF in the presence of urea and 2-mercaptoethanol (2-ME). As shown in Fig. 6-A, the subunits of glycinins separated into two zones. One shifted to the acidic region and the other shifted to the basic region. The numbers of the acidic and basic subunits differed among the cultivars. The patterns could be sorted into five groups: group I contained 7 acidics and 8 basics; group II, 7 acidics and 7 basics; group III, 6 acidics and 7 basics; group IV, 6 acidics and 5 basics; group V, 6 acidics and 3 basics (Fig. 6-B). The amount ratio of the subunit proteins was not unity in most of the cultivars, while it was roughly similar for cultivars in the same group (Fig. 6-B). However, it is not clear whether the amount ratio of the subunit proteins in each cultivar is invariant or if they differ due to variation in physical and environmental conditions.

The variations of the subunit compositions among the cultivars may not be resulted from artifacts such as urea effect (KITAMURA et al. 1973), proteolytic action or deamidation (see references in UTSUMI et al. 1980), since the conditions for the isolation of the subunits used here are rather unfavorable to causing such artifacts, e.g., crude glycinin preparation which had been stored for a month at 5 °C did not produce results different from that of the fresh one. The subunit composition of 11S globulin among cultivars has been investigated for ground nut, lupin and pea (see references in MORI et al. 1981). However, their research has not indicated the number of the constituent subunits. Recently, it has been reported that the subunit compositions of 11S globulins of pea cultivars are related to seed phenotype, round and wrinkle (DAVIES 1980). However, the band patterns of glycinin could not be grouped according to the size and other characteristics of the seeds unlike the case of pea 11S globulin. Since the present study on soybean cultivars is not exhaustive, other types of subunit composition of glycinins may exist in soybean cultivars different from those described above. At any rate, these observations (Fig. 6) support those of others as to the diversity of subunit composition and also suggest the heterogeneity of glycinin molecular species, as has already been postulated for peanut arachin and pea 11S globulins, because

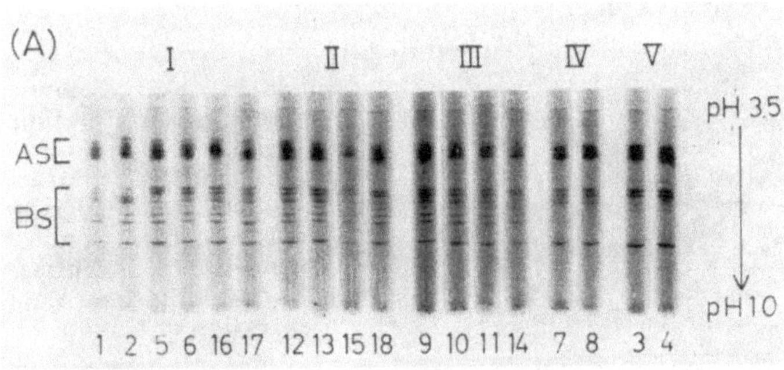

Fig. 6A PGEF of glycinins from the same samples described in Fig. 2. A hundred μg of each glycinin was used.

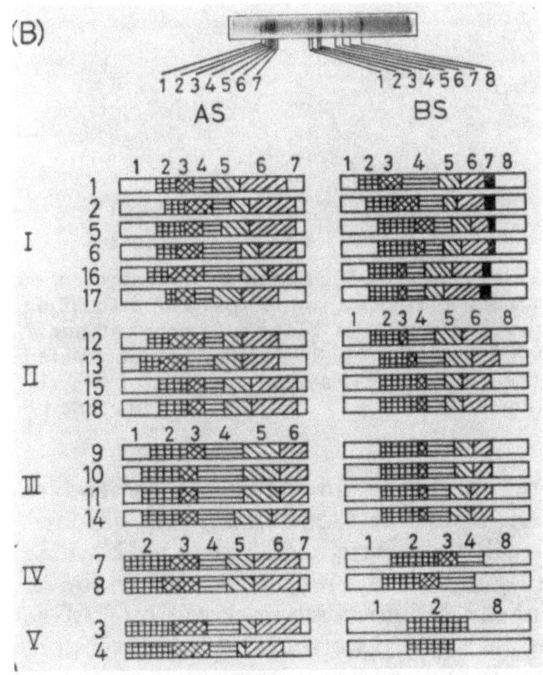

Fig. 6B The relative proportion of the acidic and the basic subunits from each cultivar was calculated from densitometric scanning of the gels and is represented schematically

more than six subunits were observed as the acidic subunits or the basic subunits in groups I, II and III, and the amount ratio of subunit proteins was not unity among the acidic subunits and among the basic subunits.

Heterogeneity of Molecular Species of Soybean Glycinin

The structural properties of glycinin have been well investigated, while glycinin was considered as a homogeneous protein with respect to molecular species. Thus, glycinin has been regarded as a single molecular species with an inherent subunit composition. However, as discussed in the above section, the results relating to the subunit compositions of glycinins from various cultivars suggest the heterogeneity of glycinin molecular species.

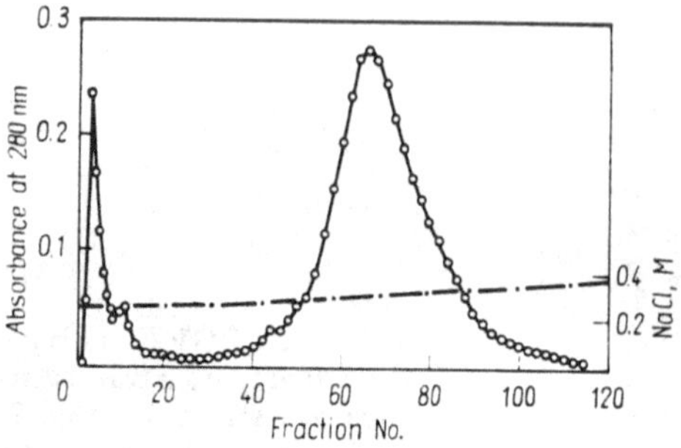

›hadex A-50 column
glycinin-rich fraction
Glycinin was equilibra-
ssium phosphate buffer
0.25 M NaCl, 10 mM
NaN$_3$ and then applied
elopment: 60 ml buffer,
Cl.—○—, absorbance at
centration of NaCl

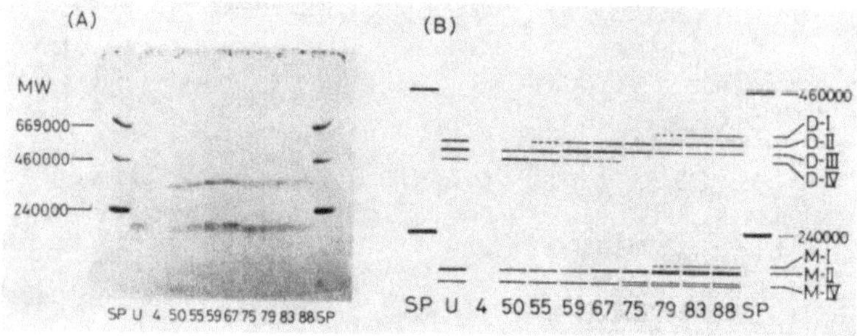

Fig. 8 PGGE of glycinin fraction from DEAE-Sephadex A-50 column. A, 20 μg protein of each fraction equilibrated with 35 mM potassium phosphate buffer (pH 7.6) containing 0.2 M NaCl, 10 mM 2-ME and 0.02 NaN$_3$ was applied on the gel. SP and U refer to standard proteins (thyroglobulin, 669,000; ferritin, 460,000; catalase, 240,000) and the unfractionated sample, respectively. The numbers under the gel are the fraction numbers of DEAE-Sephadex column chromatography. Migration is from top to bottom. B, Diagramatic representation of A

Glycinin-rich fractions obtained from var. Tsuru-no-ko by the method of cryoprecipitation were applied to a DEAE-Sephadex A-50 column. The chromatogram of glycinin was shown in Fig. 7. Glycinin eluted as a single peak which was comparatively sharper than that of legumin from broad bean (will be shown in following section). The fractions from the column were analyzed by PGGE. The electrophoretic patterns were shown in Fig. 8. Two zones were separated in all samples, the faster and slower groups of bands being monomers (half-molecules of glycinin) and dimers (glycinin), respectively. The monomer bands may be derived from the dissociation of the glycinin molecule during the electrophoresis. The unfractionated glycinin gave three bands with apparent molecular weights of 360,000 (D–II), 345,000 (D–III) and 340,000 (D–IV), of which the proportion of the middle band was higher than that of others (D–III > D–II, D–IV). Fractions 50, 59–67, 75 and 79–88 gave tow (D–IV > D–III), three (D–III > D–II, D–IV), two (D–II = D–III) and three bands (D–II > D–I (molecular weight of 375,000), D–III), respectively. The band of the largest molecular species (D–I) was not observed in the unfractionated glycinin. This may be due to the low content of this molecular species in glycinin. The results thus obtained indicate the presence of four molecular species of glycinin from the standpoint of molecular size. However, when the fractions were subjected to PAGE a similar broad band was observed in every fraction. This indicates that each glycinin molecular species has similar molecular charge, and that many more molecular species may exist from the standpoint of molecular charge than those observed from the standpoint of molecular size. The unfractionated glycinin gave two monomer bands with apparent molecular weights of 211,000 (M–II) and 199,000 (M–IV), of which the proportion of M–II was higher than M–IV. Considering the gel electrophoretic patterns of the monomers and dimers and their apparent molecular weights in fractions 50–67, the dimers D–II and D–IV may be assumed to be homodimers composed of the monomers M–II and M–IV, respectively. Similarly, the dimer D–III may be assumed to be a heterodimer made up of monomers M–II and M–IV. Since fractions 79–88 gave another monomer band with a molecular weight of 222,000 (M–I) accompanying the appearance of D–I, it may be assumed that M–I is one of the monomer components of D–I. M–IV, observed in fractions 50–67, was split into two bands in the fractions eluted after fraction 75. Similarly, M–II was split into two bands in the fractions after fraction 79. This microheterogeneity of the monomer in the fractions after fraction 75 or 79 may be partly responsible for the heterogeneity of the dimer. On the other hand, the fractions were analyzed by SDS–PAGE and by PGEF in order to elucidate their subunit compositions (data not shown). The ratio of basic subunits was different among the fractions: the more slowly glycinin eluted, the higher the content of the subunit with a molecular weight of 19,000 and the lower the content of the subunit with a molecular weight of 18,300. Differences were also observed in the subunit compositions from the standpoint of their charges among the fractions from the result of PGEF.

To elucidate whether each glycinin from a wide range of soybean cultivars exhibits the heterogeneity of molecular species, glycinins from various cultivars were analyzed by PGGE: one cultivar was selected from each of the five groups, i.e. Shiro-tsuru-no-ko, Tianjin-dachingdou, York, Ford, Iyo-daizu and Raiden from the groups I, II, III, IV and V, respectively. As shown in Fig. 9, glycinin from Shiro-tsuru-no-ko gave three dimer bands, D–II, D–III and D–IV, and two monomer bands, M–II and M–IV; Tianjin-dachingdou, three dimer bands, D–II, D–III and D–IV, and three monomer bands, M–II, M–III (molecular weight of 205,000) and M–IV; York, two dimer bands, D–II and D–III, and three monomer bands, M–II, M–III and M–IV; Iyo-daizu, two dimer bands, D–II and D–IV, and two monomer bands, M–II and M–IV; Raiden, two dimer bands, D–II and D–IV, and one monomer band, M–IV. The results show that the numbers of glycinin molecular species are different among cultivars. The extent of the heterogeneity of glycinins seems to correlate with the diversity of the subunit compositions of glycinins.

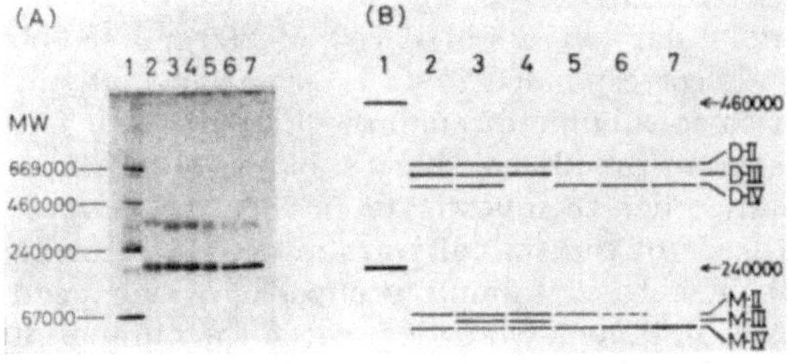

Fig. 9 PGGE of glycinins from various soybean cultivars. A, 20 μg protein of each cultivar was used. Migration is from top to bottom. Lane 1, standard proteins described in Fig. 8; Lane 2, Shiro-tsuru-no-ko; Lane 3, Tianjin-dachingdou; Lane 4, York; Lane 5, Ford; Lane 6, Iyo-daizu; Lane 7, Raiden. B, Diagramatic representation of A

Differences in Subunit Compositions of Legumins among Broad Bean Cultivars

The 11S-type globulin of broad bean is called legumin. The subunit structure of legumin has been investigated (see references in UTSUMI ET AL. 1980), however, the subunit composition has not yet been elucidated sufficiently. On the other hand, differences of subunit compositions of 11S-type globulins among cultivars have been found in several kinds of seeds as well as soybean as described in the preceding section. In order to elucidate the diversity of subunit composition among different seeds, the subunit compositions of legumins from various cultivars of broad bean planted in Japan were investigated. The legumins were analyzed by SDS-PAGE (Fig. 10). All legumins were composed of six subunit proteins,

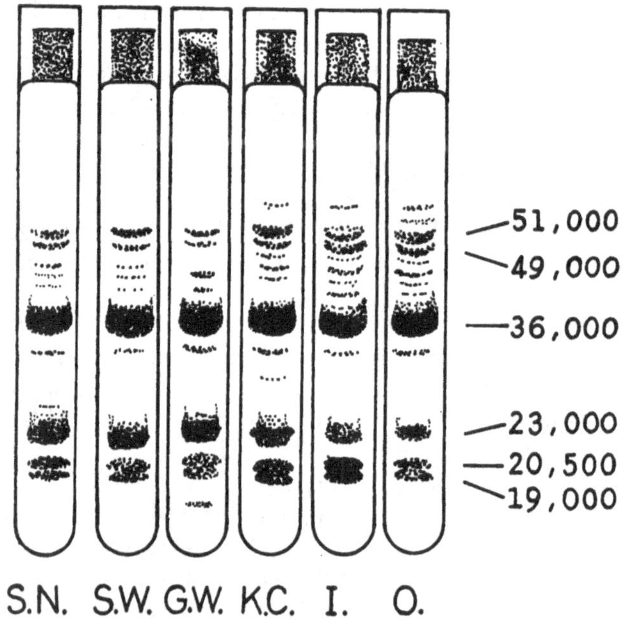

Fig. 10 SDS-PAGE of legumins from six cultivars. A hundred μg of each legumin from six cultivars was used. Gel symbols are described in the text. Migration is from top to bottom

of which three were acidic subunits with molecular weights of 51,000, 49,000 and 36,000, and three basic subunits with molecular weights of 23,000, 20,500 and 19,000. The densitometric patterns of the gels showed that the amount ratios of the acidic subunits and the basic subunits were similar among the cultivars. When the legumins were analyzed by SDS–PAGE in the absence of 2–ME (Fig. 11) three major bands with molecular weights of 61,700, 59,800 and 48,000 appeared in all legumins. These bands corresponded to an intermediary subunit in which acidic subunit linked with basic subunit through disulfide bridges. The number, proportion and molecular sizes of the intermediary subunits of legumins were similar among the cultivars. In order to investigate further the subunit compositions of legumins from the seeds of the six cultivars in detail, legumins were analyzed by PGEF (Fig. 12). The basic subunit groups, in which good separation of each band in the gels was obtained, were identical in the number of

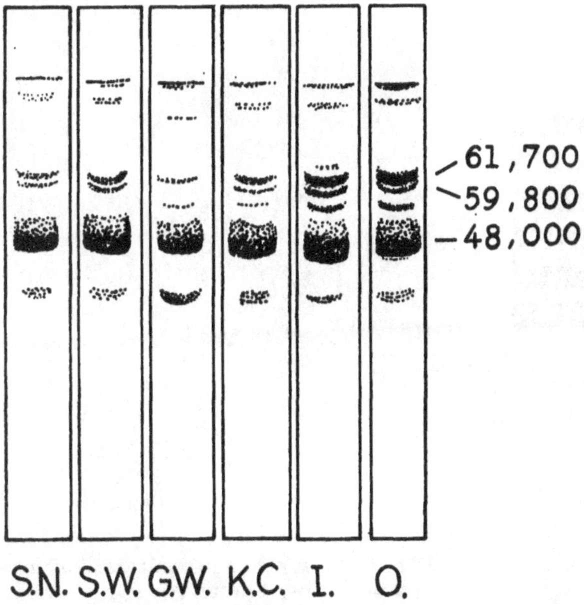

Fig. 11 SDS-PAGE in the absence of 2-ME of legumins from six cultivars. Conditions are similar to those in Fig. 10

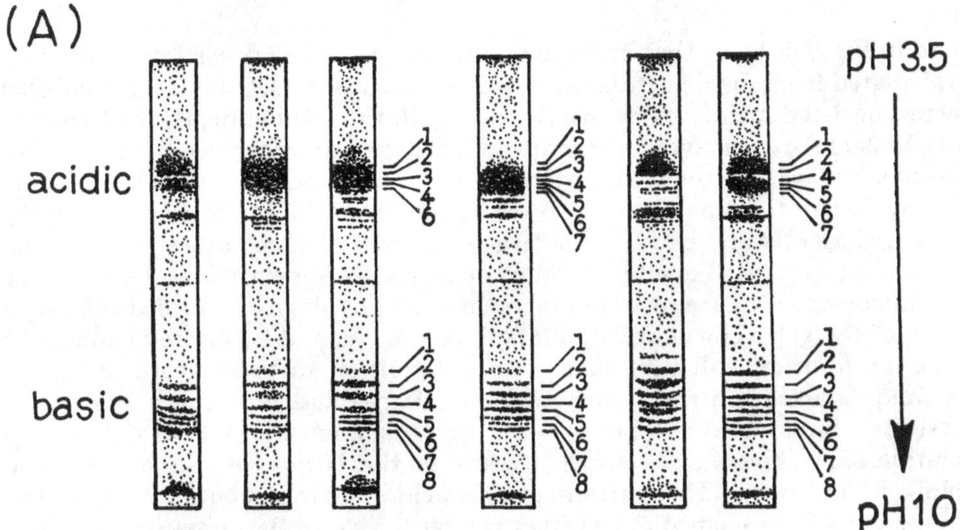

Fig. 12A PGEF of legumins from six cultivars. A hundred μg of each legumin was used.

subunits among the cultivars. However, the proportions of the subunits were different among them. Their patterns could be classified into three groups, i.e., small-, medium- and large-size groups. On the other hand, the acidic subunit groups were different both in the number and the proportion of the subunits among the cultivars. These gel electrophoretic patterns are schematically sum-

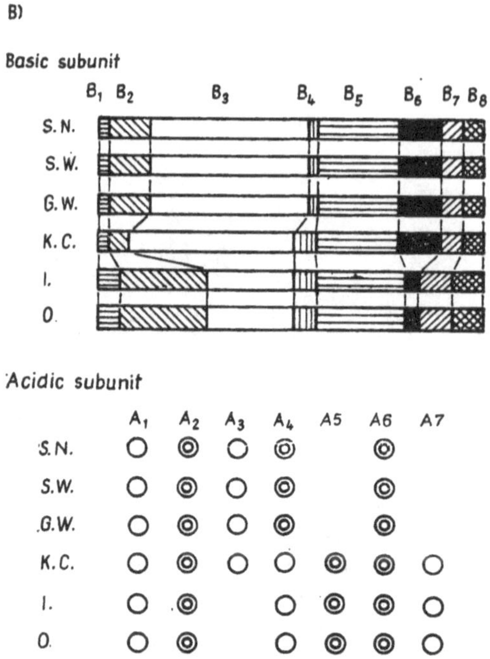

Fig. 12B Schematic representation of subunit compositions. The relative proportions of the basic subunits were calculated from densitometric scanning of the gels

marized in Fig. 12–B. In the basic subunit groups, relative proportion of subunits was calculated from the densitometric scanning of the gels. Legumins of small-size cultivars consisted mainly of B_2, B_3, B_5 and B_6. Of these subunits, B_3 predominated. In the large-size cultivars, the major subunit components were B_2, B_3 and B_5. In the medium-size cultivar, the pattern of the basic subunit composition was similar to that of the small-size cultivars except that it contains much more B_4, which is a characteristic band of the large-size group. On the other hand, in the acidic subunit groups, the subunit compositions were shown qualitatively, since it was difficult to calculate relative proportions of subunits from the densitometric scanning of the gels. The acidic subunits of A_1, A_2, A_4 and A_6 existed commonly in legumins from all cultivars examined. In addition to these acidic subunits, A_3, A_5 and A_7 existed in legumins of small-, large- and mediumsize cultivars, respectively. Thus, the number of the acidic subunits was different among the cultivars, i.e., five in the small-size, six in the large-size and seven in the medium-size cultivars. The pattern of the acidic subunit composition of the medium-size cultivar seemed to be rather similar to that of the large-size cultivars. As a whole subunit composition, the medium-size cultivar had both characteristics of the small- and large-size cultivars. The variations of the subunit compositions among the cultivars described above may not be resulted from artifacts such as urea effect, deamidation or proteolytic modification, on the basis of the following observations; (1) the isolation of subunits of legumin was performed at acidic pH (pH 6.3) and at low temperature (4 °C), (2) there was no change in the electrophoretic patterns of the subunits of legumins even if the extracts prepared using 50 mM potassium phosphate buffer (pH 7.0) or 50 mM Tris-HCl buffer (pH 7.6) each contained 1 M NaCl were incubated for 5 days at 20 °C, (3) the patterns of the subunit compositions were constant and reproducible for each cultivars. Therefore, the variations among the cultivars seem to be substantial characteristics for them. The medium-size cultivar should be regarded as a situation intermediate between the small- and the large-size cultivars, with partial representation of gene products, that is partial additive inheritance. This regard may support the conjecture that a small-size cultivar might be introduced from China into Japan for the first time and then a large-size cultivar, Otafuku, might be done.

Heterogeneity of Molecular Species of Broad Bean Legumin

The subunit structure of legumin from broad bean has been investigated by WRIGHT and BOULTER (1974) and they proposed that legumin of Triple White was composed of A_6B_6. However, they did not consider the heterogeneity of legumin molecular species. On the other hand, as described in the preceding section, the ratio of the amounts of three acidic subunits with molecular weights of 51,000, 49,000 and 36,000 was approx. 1 : 1 and legumins from various cultivars contained eight kinds of basic subunits from the standpoint of their molecular charges. These may indicate the heterogeneity of legumin molecular species. In view of these evidences, legumins from various cultivars were fractionated and their subunit compositions were investigated in order to elucidate the heterogeneity of legumin from broad bean.

Legumin from three cultivars of broad bean were prepared by sucrose density gradient centrifugation and then were applied on a DEAE-Sephadex A–50 column. The chromatograms were shown in Fig. 13. Legumin of each cultivar was eluted as a single peak. However, it was a broad peak in the range of fractions 40 to 100 (approx.). The fractions from the columns of three cultivars were analyzed

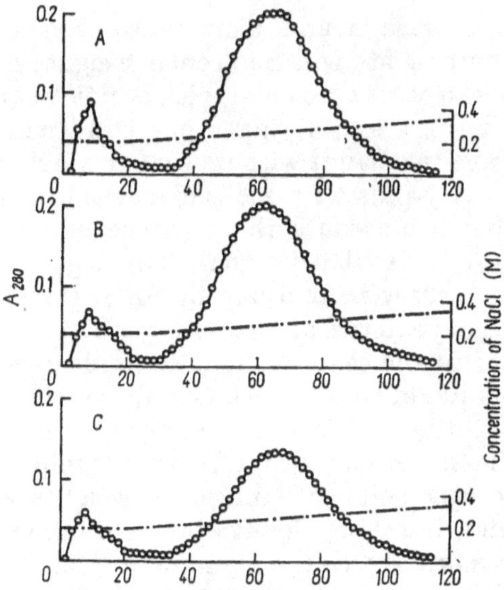

Fig. 13 DEAE-Sephadex A–50 column chromatography of legumins from three cultivars. The 11S fractions prepared by sucrose density gradient centrifugation were applied on the column for Sanuki-Nagasaya (A), Kumamoto-Churyu (B) and Otafuku (C), and then eluted. —○—: absorbance at 280 nm, —·—·—: concentration of NaCl

by various gel electrophoreses. The legumin fractions from the column of three cultivars were analyzed by PGGE in order to elucidate whether molecular species of legumin which were different in molecular weights were present in broad bean. The electrophoretic patterns of Otafuku were shown in Fig. 14. Similar patterns

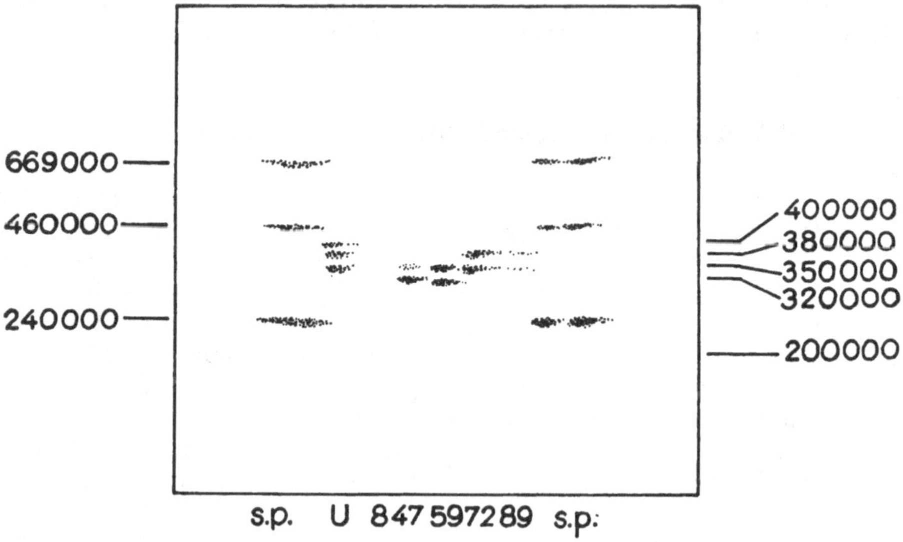

Fig. 14 PGGE of legumin fractions of Otafuku. Ten μg protein of each fraction equilibrated with 50 mM potassium phosphate buffer (pH 7.6) containing 0.2 M NaCl was electrophoresed. s.p. and U refer to the standard proteins and unfractionated sample, respectively. The numbers under the gels are the fraction numbers of the column chromatography. Migration is from top to bottom

were also obtained in the cases of Sanuki-Nagasaya and Kumamoto-Churyu. Unfractionated sample consisted of three components with molecular weights of 380,000, 350,000 and 320,000. Fraction 8 gave a very pale band with a molecular weight of around 200,000. The faster eluting fraction, 47, gave one band with a molecular weight of 320,000. Fraction 72 gave two bands with molecular weights of 380,000 and 350,000. Fraction 89 gave three bands with molecular weights of 400,000, 380,000 and 350,000. These results demonstrate the presence of four molecular species of legumin with different molecular weights. The legumin fractions eluted from the column of three cultivars were analyzed by SDS–PAGE in the presence of 2–ME. The electrophoretic patterns of Otafuku were shown in Fig. 15. Similar results were also obtained in the cases of the other cultivars. Unfractionated sample consisted of six subunits, of which the larger three bands and the smaller three bands were assigned to the acidics and the basics, respectively, as shown in the preceding section. The faster eluting fractions of legumin, fractions 44 and 47, consisted of only three components with molecular weights of 36,000, 23,000 and 20,500. On the other hand, the more slowly eluting fractions contained all of the six subunits. Thus the more slowly the legumin eluted, the more plentiful were the subunits with molecular weights of 51,000, 49,000 and

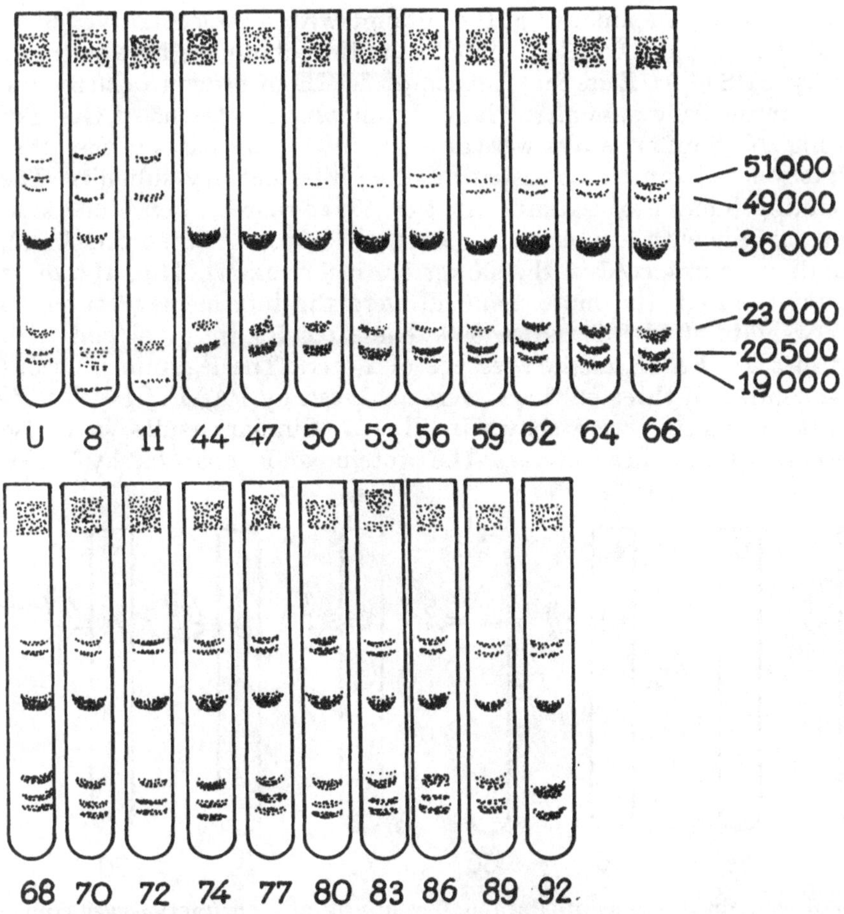

Fig. 15 SDS-PAGE of legumin fractions of Otafuku. Thirty μg protein of each fraction was used. U refers to the unfractionated sample. The numbers under the gels are the fraction numbers of the column chromatography. Migration is from top to bottom

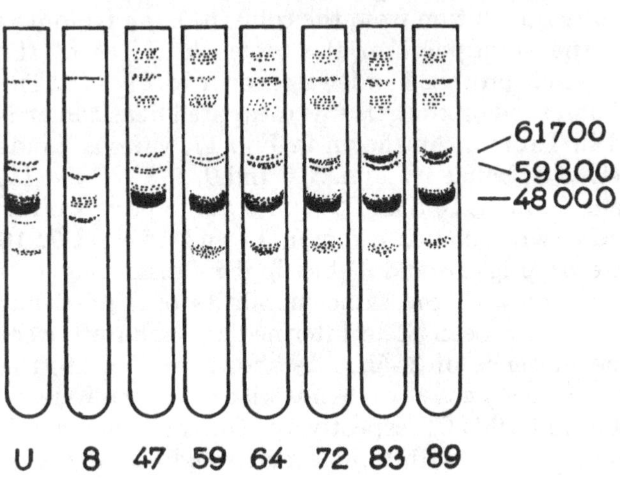

Fig. 16 SDS-PAGE of legumin fractions of Otafuku in the absence of 2-ME. The conditions are similar to those in Fig. 15

19,000, while the ratio of the amount of the subunit with a molecular weight of 20,500 decreased. The legumin fractions eluted from the column of three cultivars were analyzed by SDS–PAGE in the absence of 2–ME in order to clarify the composition of intermediary subunits. The electrophoretic patterns of Otafuku were shown in Fig. 16. Similar results were also obtained in the cases of the other cultivars. Unfractionated sample consisted of three intermediary subunits. The more quickly eluting fraction of legumin, fraction 47, consisted of only one kind of intermediary subunit with a molecular weight of 48,000. On the other hand, all of three bands were observed in the slower eluting fractions. Thus the more slowly the legumin eluted, the more plentiful were the intermediary subunits with molecular weights of 61,700 and 59,800. These results are consistent with those obtained in SDS–PAGE in the presence of 2–ME. The legumin fractions eluted from the column of three cultivars were analyzed by PAGE. The electrophoretic patterns of Otafuku were shown in Fig. 17. Similar results were also obtained in the cases of the other cultivars. The protein bands separated by PAGE

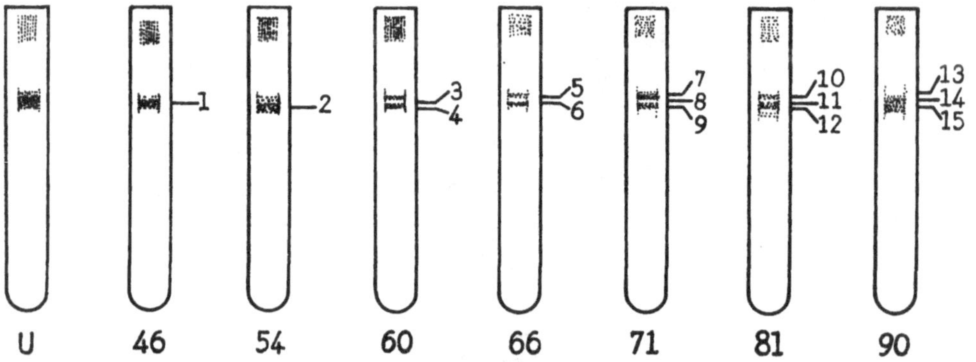

Fig. 17 PAGE of legumin fractions of Otafuku. Twenty μg of each fraction was equilibrated with 50 mM potassium phosphate buffer (pH 7.6) containing 0.2 M NaCl and then electrophoresed. The protein bands separated were numbered as described in the text

were numbered according to the order of elution from the column. Each band was analyzed by SDS–PAGE from the standpoint of the molecular sizes of the subunits. SDS–PAGE pattern of each protein band was shown in Fig. 18. The ratio of the amount of each band observed on the gels was estimated densitometrically and the results were listed in Table 2. As shown in Fig. 17, various bands had the same mobilities on PAGE, e.g., bands 2 and 4, 3, 6 and 9, 5, 8, 11 and 15, and 7, 10 and 14. However, bands 2 and 4, 3, 6 and 9, and 5 and 8 gave similar SDS–PAGE patterns, respectively, while bands 8 (or 5), 11 and 15, and 7, 10 and 14 gave fairly different patterns (Fig. 18 and Table 2). This result suggests that there are some molecular species with the same mobilities on PAGE but different subunit compositions in broad bean. The intermediary subunits were examined by SDS–PAGE in the presence of 2–ME. As shown in Fig. 19, the larger intermediary subunits, I and II, each gave two bands with molecular weights of 51,000 and 19,000, and 49,000 and 19,000, respectively. On the other hand, since the band of the intermediary subunit with a molecular weight of 48,000 is

Table 2
Relative Amounts of Subunits in Each Protein Band Separated by PAGE
The amounts were estimated densitometrically, relative to the basic subunit with molecular weight of 23,000 which was taken as unity.

Gel No.	Acidic subunits with molecular weights of			Basic subunits with molecular weights of		
	51,000	49,000	36,000	23,000	20,500	19,000
1	–	–	2.94	1	1.21	–
2	–	–	3.24	1	1.59	–
3	0.29	0.18	3.57	1	1.21	0.11
4	–	–	3.15	1	1.89	–
5	0.47	0.35	2.94	1	0.84	0.78
6	0.32	0.17	2.32	1	1.44	0.65
7	0.77	0.62	2.56	1	0.26	1.05
8	0.68	0.32	2.50	1	0.63	0.83
9	0.71	0.14	3.57	1	1.32	0.36
10	1.51	0.52	2.05	1	0.21	1.18
11	1.15	0.47	1.80	1	0.42	1.52
12	0.72	0.55	2.35	1	0.78	0.96
13	1.66	0.38	2.13	1	0.18	1.34
14	1.05	0.28	1.80	1	0.33	1.48
15	0.82	0.25	2.17	1	0.98	0.85

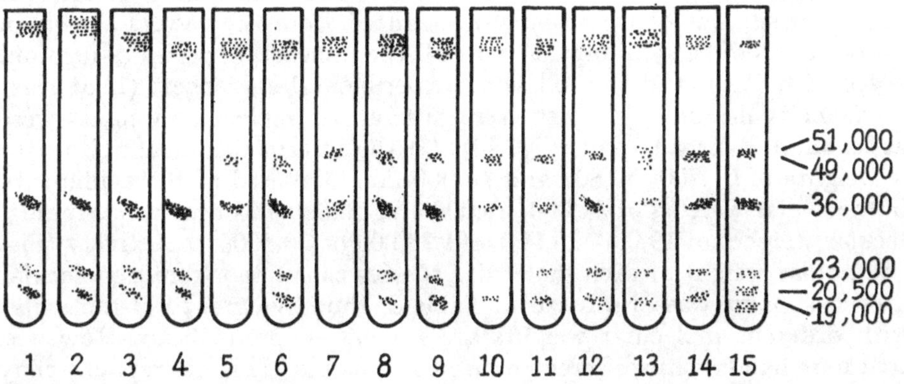

Fig. 18 SDS-PAGE of the protein bands separated by PAGE. Each stained protein band separated by PAGE (Fig. 17) was cut off and electrophoresed. Migration is from top to bottom

broad, the band was cut into two pieces, the upper (III) and the lower (IV) portions. They gave two bands with molecular weights of 36,000 and 20,500, and 36,000 and 23,000, respectively. It is of interest that the intermediary subunit composed of 36,000 and 20,500 migrated slower than that composed of 36,000 and 23,000. This may be due to the differences in compactness of the structure between these intermediary subunits.

The subunit patterns of legumins from three cultivars were similar among each other from the standpoint of the molecular sizes of the subunits (Figs. 15 and 16). Therefore, it was attempted to list the possible molecular species of

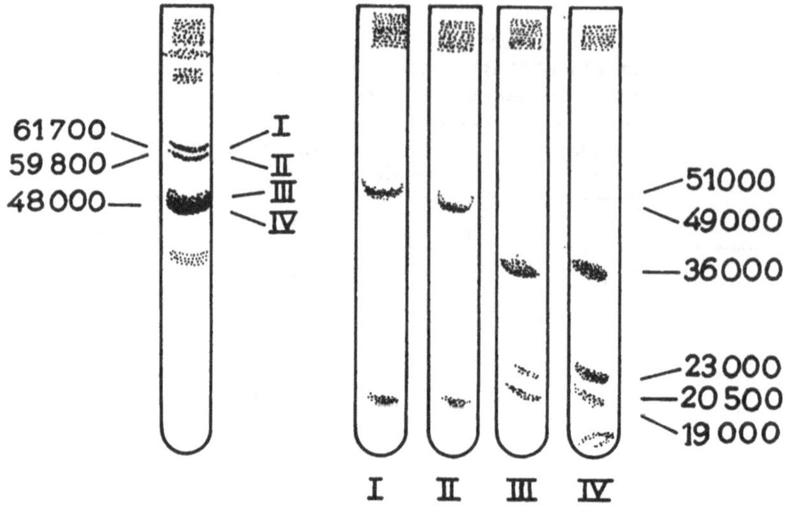

Fig. 19 Subunit compositions of intermediary subunits. Each protein band (I, II, III and IV) separated by SDS-PAGE in the absence of 2-ME was cut off and electrophoresed in the presence of 2-ME. Migration is from top to bottom

legumin which are common to the three cultivars, constructed from the subunit groups classified according to their sizes. The possible molecular species were shown in Table 3. Each subunit composition was deduced from the results shown in Table 2 on the basis of the assumption that all the molecular species of legumin are composed of 6 (AB), and the following experimental evidences: (1) at least four kinds of molecular species which were different in molecular weights were present as legumin in broad bean (Fig. 14), (2) the intermediary subunits with molecular weights of 61,700, 59,800 and 48,000 are composed of the acidic subunits with molecular weights of 51,000, 49,000 and 36,000 and the basic subunits with molecular weights of 19,000, 19,000 and 23,000 or 20,500, respectively (Fig. 19). The molecular weight of each molecular species calculated from a molecular weight of each subunit was also listed in Table 3, thus five groups of molecular species with different molecular weights are present in broad bean. However, PGGE gave four bands with different molecular weights. This discrepancy may be derived from the closeness of the molecular weights of group III (370,000) and group IV (380,000), or that they have similar apparent molecular sizes.

Legumins of Sanuki-Nagasaya, Kumamoto-Churyu and Otafuku gave eight bands as basic subunits on PGEF (Fig. 12–A), which were more than those observed on SDS–PAGE (Figs. 10 and 15). Moreover, the kinds of main bands of basic subunits were different among the three cultivars. Therefore, when the subunit compositions of legumin molecular species are represented by the subunits classified according to their charges, the number of possible molecular species of legumin must be much more than that presented in Table 3. This suggests that each of nine molecular species presented in Table 3 may have further submolecular species. However, the subunit compositions of them cannot be deduced at present. As described in the preceding section, soybean glycinin also exhibits the heterogeneity of molecular species, as well as other 11S-type globulins from

Table 3
The Possible Molecular Species of Legumin in the Broad Bean

Molecular species	Subunit composition*	Molecular weights Calculated**	Observed***
I–1	$3(A_3B_1)$ $3(A_3B_2)$	345,000	320,000
I–2	$2(A_3B_1)$ $4(A_3B_2)$		
II–1	$2(A_3B_1)$ $3(A_3B_2)$ $1(A_2B_3)$	357,000	350,000
II–2	$2(A_3B_1)$ $3(A_3B_2)$ $1(A_1B_3)$		
III–1	$2(A_3B_1)$ $2(A_3B_2)$ $1(A_2B_3)$ $1(A_1B_3)$	370,000	
III–2	$2(A_3B_1)$ $2(A_3B_2)$ $2(A_1B_3)$		
IV–1	$2(A_3B_1)$ $1(A_3B_2)$ $2(A_2B_3)$ $1(A_1B_3)$	380,000	380,000
IV–2	$2(A_3B_1)$ $1(A_3B_2)$ $1(A_2B_3)$ $2(A_1B_3)$		
V–1	$2(A_3B_1)$ $1(A_2B_3)$ $3(A_1B_3)$	396,000	400,000

* The subunits with molecular weights of 51,000, 49,000, 36,000, 23,000, 20,500 and 19,000 shown in Fig. 18 were termed A_1, A_2, A_3, B_1, B_2 and B_3, respectively.
** Calculated from molecular weight of each subunit.
*** Estimated from the results of the gradient gel electrophoresis.

ground nut and pea and 7S type-globulins from soybean, pea and lupin (see references in UTSUMI and MORI 1980; UTSUMI et al. 1981). Therefore, heterogeneity of molecular species may be an inherent property of the major storage proteins of legume seeds in general.

Formation of Pseudo-11S Globulins from Subunits of Soybean Glycinin and Broad Bean Legumin

Recent developments in investigations on the structure of the main components of seed storage protein suggest that 11S globulins are composed of six acidic (A) and six basic (B) subunits. Of various 11S globulins, those from soybean, broad bean, pea, pumpkin and sesame have been shown to have common intermediary subunits, in which the acidic and basic subunits are linked by disulfide bridges in 1 : 1 ratio and to be composed of 6 (AB) as described also in the preceding sections. In such subunit structure of 11S globulin molecule subunit composition of the globulin molecule was not unity but diverse as shown in the preceding sections. These may indicate that specificity of subunit interaction in the construction of 11S-structure is not precise. The similarity of the structure and the amino acid sequences among glycinin acidic subunits, among glycinin basic subunits, among legumin basic subunits and among basic subunits of legumins from broad bean and pea and glycinin were indicated (GILROY et al. 1979; MOREIRA et al. 1979; CASEY et al. 1981). These may permit the interchange of subunits among acidic subunits and among basic subunits in the formation of intermediary subunits and 11S-structure.

Native subunit proteins of glycinin from var. Tsuru-no-ko were fractionated by chromatography on a DEAE-Sephadex column in the presence of urea and 2–ME.

As shown in Fig. 20, glycinin was separated into six fractions. The first peak flowed through the column was composed of basic subunits with molecular weights of 19,000 and 18,300. The acidic subunits were eluted with a gradient of NaCl

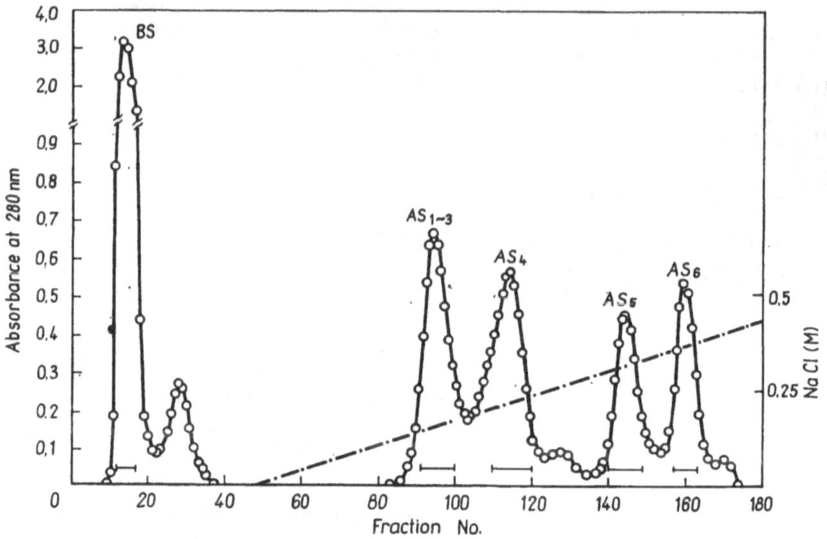

Fig. 20 Fractionation of native subunit proteins of glycinin by DEAE-Sephadex A-50 column chromatography. The purified glycinin was fractionated by the column chromatography. —○—, absorbance at 280 nm; —·—·—, concentration of NaCl. The underlinedparts were collected in each peak

and separated into four peaks, i.e., AS_{1-3}, AS_4, AS_5 and AS_6. AS_1 to AS_6 were separated by alkaline urea-PAGE and termed in the order of migration. The molecular weights of AS_{1-3}, AS_4 and AS_6 and AS_5 were 34,800 and 38,000, respectively. In order to investigate the extent of reconstitution of 11S-size complexes from the isolated native subunits, the reconstituted products from combinations of one each of glycinin acidic subunits and glycinin basic subunits, i.e., AS_{1-3} and BS, AS_4 and BS, and AS_5 and BS (the molar ratio was approx. 1 : 4 in each combination), were fractionated by sucrose density gradient centrifugation (Fig. 21). In the absorbance patterns, peaks I and II correspond to 7S (half-molecule of the 11S component) and 11S components, respectively. As shown in Fig. 21–A, the formation of 11S component from the combination of AS_{1-3} and BS was about 90 %. In the cases of the combinations of AS_4 and BS, and AS_5 and BS, it was lower (40 to 30 %) and the areas of the 7S components were larger than that of the other combination (Fig. 21–B and C). In any of the cases both free acidic subunits and free reconstituted intermediary subunits were scarcely observed. These results suggest that pseudo-glycinins which are different from the native glycinin with respect to the subunit composition can be formed from the isolated constituent subunits of glycinin except in the case of the combination of AS_6 and BS.

Native subunit proteins of legumin from var. Sanuki-Nagasaya were fractionated by chromatography on a DEAE-Sephadex column in the presence of urea and 2–ME. As shown in Fig. 22, legumin was separated into four fractions. The first peak flowed through the column and the following three peaks were eluted with a gradient of NaCl. The flowed through fraction was composed of basic subunits with molecular weights of 23,000, 20,500 and 19,000. The eluted fractions were the acidic subunits, i.e., ASI, ASII and ASIII with molecular weights of 36,000, 36,000 and 49,000 and 51,000, respectively. In order to investigate the extent of reconstitution of 11S-size complexes from the isolated native subunits of legumin, the reconstituted products from the combinations of one each of legumin acidic subunits and legumin basic subunits, i.e., ASI and BS, ASII and BS, and ASIII and BS (the molar ratio was approx. 1 : 4 in each combination), were fractionated by the sucrose density gradient centrifugation. As shown in Fig. 23–A, the extent of the 11S-size component (peak II) formed from the combination of ASI and BS was about 60 % and the complexes of the size between 7S and 11S (peak I) comprised the residual 40 %. In the case of the combinations of ASII and BS, and ASIII and BS, more than 90 % of the acidic subunits used for the reconstitution underwent reconstitution (Fig. 23–B and C). In any case both free acidic subunits and free reconstituted intermediary subunits were scarcely observed. These results suggest that pseudo-legumins which are different from the native legumin with respect to the subunit composition can be formed from the isolated constituent subunits of legumin.

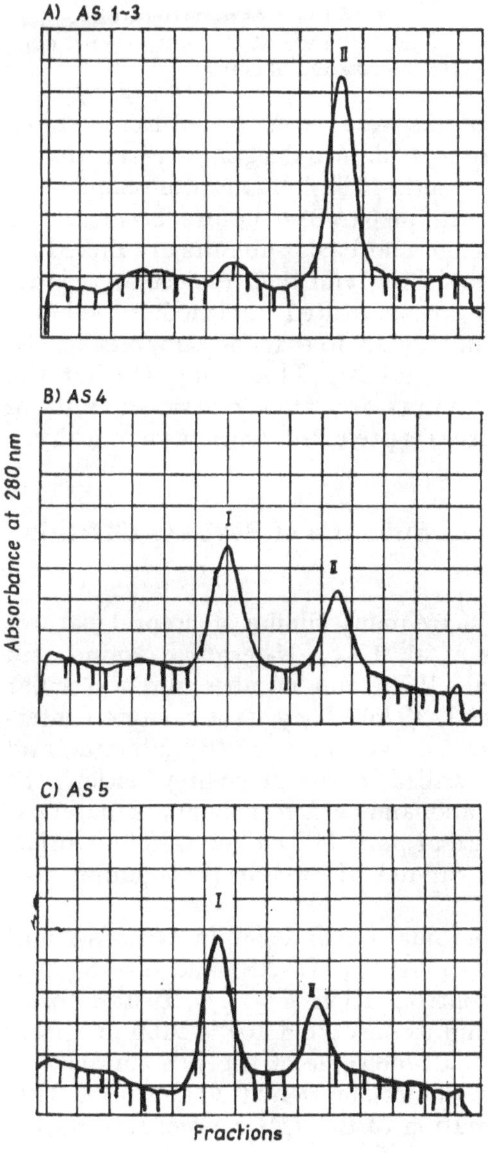

Fig. 21 Sucrose density gradient centrifugation of reconstituted products from glycinin subunits. The isolated acidic and basic subunits were combined and reconstituted in combinations of BS with AS_{1-3} (A), AS_4 (B) and AS_5 (C), and then the reconstituted products were fractionated by the sucrose gradient centrifugation. Sedimentation is from left to right

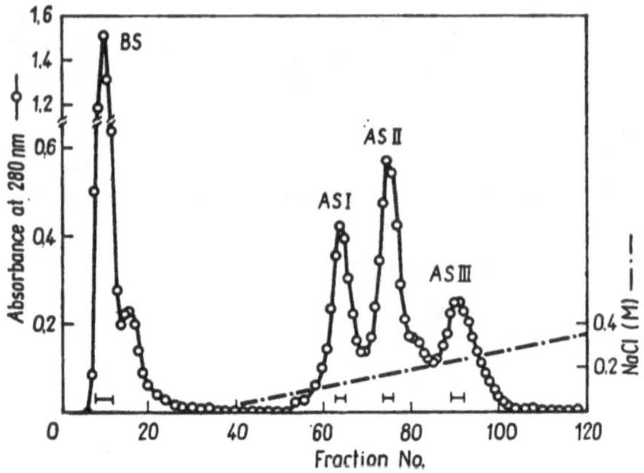

Fig. 22 Fractionation of native subunit proteins of legumin by DEAE-Sephadex A-50 column chromatography. 11S component obtained from sucrose density gradient centrifugation was fractionated by the column chromatography.—○—, absorbance at 280 nm; —·—·—, concentration of NaCl

The pseudo-glycinins and the pseudo-legumins were reconstituted from various combinations of acidic and basic subunits from glycinin and legumin, respectively, as described above. All the reconstituted pseudo-11S globulins were similar to the native glycinin and legumin with respect to the 6 (AB) structure, because they have intermediary subunits in which the acidic and basic subunits are linked by disulfide bridges in a ratio of 1 : 1 (data not shown) and their molecular weights were estimated to be six times that of the reconstituted intermediary subunit (data not shown). Moreover, hardly any unreacted free acidic subunits in the reconstituted products were found (Figs. 22 and 23). These suggest that the similarity of the structure among acidic subunits and among basic subunits of each 11S globulin; and that the 6 (AB) structure is preferred thermodynamically.

Formation of Hybrid-11S Globulins from Subunits of Soybean Glycinin and Broad Bean Legumin

Soybean glycinin and broad bean legumin share many similar structural features including molecular weights (Derbyshire et al. 1976), N-terminal amino acid sequences of basic subunits (Gilroy et al. 1979) and number and molecular weights of their subunits (Derbyshire et al. 1976). They have common intermediary subunits (AB), disulfide-bonded acidic (A) and basic (B) subunits, and are composed of 6 (AB). Moreover, as described in the preceding section, the specificity of interaction among subunits and among intermediary subunits is not precise in the formation of 11S-structure of both 11S globulins. These informations suggest that some acidic (or basic) subunit of glycinin (or legumin) may be interchanged with that of the other.

The reconstituted product from the combination of legumin basic subunit (L–BS) and each glycinin acidic subunit (G–AS_{1-3}, G–AS_4 and G–AS_5) was fractionated by sucrose density gradient centrifugation (Fig. 24). In the absorbance patterns peaks I, II and III correspond to unreacted acidic subunit and/or free intermediary subunit, 7S component (half-molecule of the 11S component, i.e., 3 (AB) structure), and 11S component (molecular weight 300,000–350,000), respectively. As shown in Fig 24, the formation of the 11S components by the

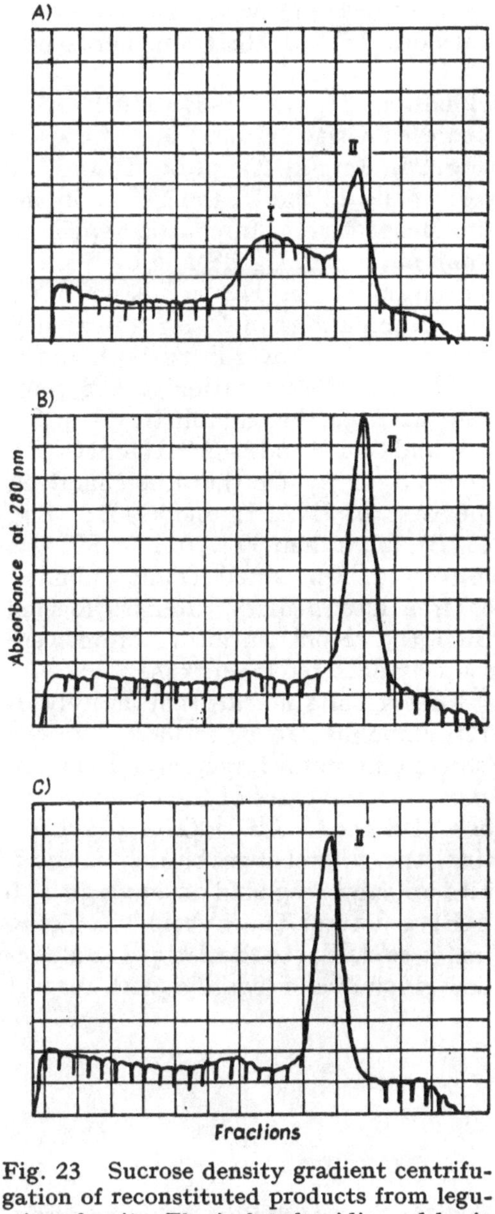

Fig. 23 Sucrose density gradient centrifugation of reconstituted products from legumin subunits. The isolated acidic and basic subunits were combined and reconstituted in combinations of BS with ASI (A), ASII (B) and ASIII (C), and then the reconstituted products were fractionated by the sucrose gradient centrifugation. Sedimentation is from left to right. Absorbance range is 0 to 0.2

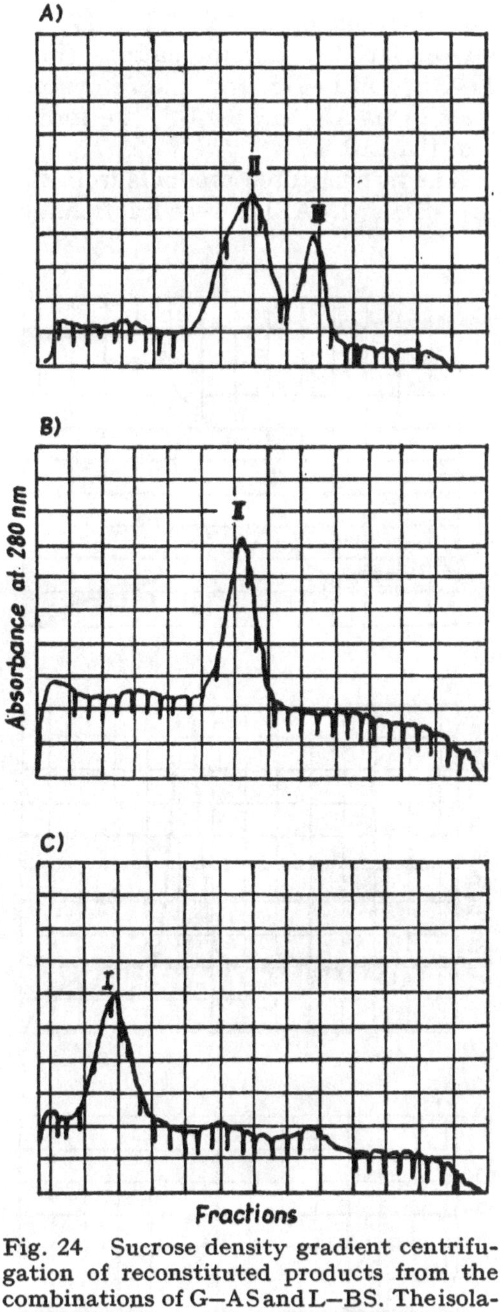

Fig. 24 Sucrose density gradient centrifugation of reconstituted products from the combinations of G–AS and L–BS. The isolated acidic and basic subunits were combined and reconstituted in combinations of L–BS and G–AS_{1-3} (A), G–AS_4 (B) and G–AS_5 (C). Sedimentation is from left to right. Absorbance range is 0 to 0.2

combinations of G–AS and L–BS was low. This was especially so in the cases of the latter two combinations (Fig. 24–B and C). From the combination of G–AS_4 and L–BS, only the 7S component was formed. On the other hand, in the case of G–AS_5 and L–BS, the formation of both 7S and 11S components was less than 10 %.

The reconstituted products from the combinations between G–BS and L–ASI, L–ASII or L–ASIII were fractionated by sucrose density gradient centrifugation (Fig. 25). In Fig. 25, peaks II and III correspond to the 7S and 11S components, respectively. In contrast with the combinations of G–AS and L–BS, the formation of the 11S components by the combinations of L–AS and G–BS was high. As shown in parts A and C of Fig. 25, the formation of 11S components from the combinations of L–ASI and G–BS and L–ASIII and G–BS were 60–70 %. On the other hand, in the case of L–ASII and G–BS (Fig. 25–B), more than 90 % of L–ASII was converted. Unreacted acidic subunits or free intermediary subunits formed from the reconstitution reaction were scarcely detected in any case.

These results indicate that the hybrid-11S globulins can be formed from the isolated subunits of glycinin and legumin except in the cases of the combination of L–BS and G–AS_4 or G–AS_5, suggesting the interchangeabilities of subunits among 11S globulins having the 6 (AB) structure. The hybrid-11S globulins were similar to the native 11S globulins, glycinin and legumin, with respect

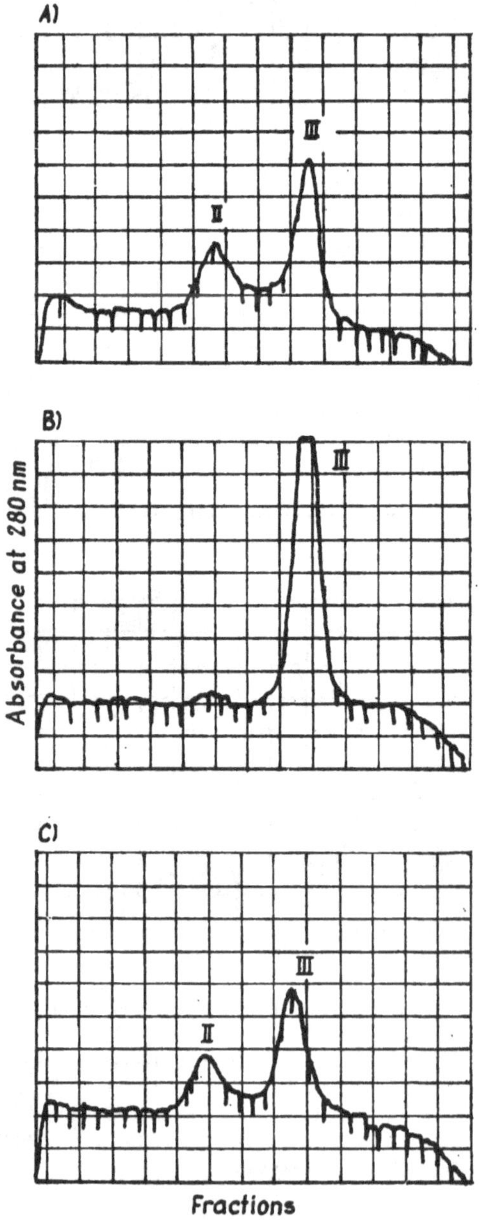

Fig. 25 Sucrose density gradient centrifugation of reconstituted products from the combinations of L–AS and G–BS. The isolated acidic and basic subunits were combined and reconstituted in combinations of G–BS and L–ASI (A), L–ASII (B) and L–ASIII (C). Other conditions are the same as in Fig. 24

to the 6 (AB) structure. There was difference in the extent of the formation of the hybrid-11S globulins from isolated subunits among the acidic subunits used, but no difference in the total yield of hybrid-7S and 11S globulins except when G–AS_5 was used (Figs. 24 and 25). These results suggest the occurrence of two types of hybrid-intermediary subunits, i.e., one can assemble into the 11S-size and the other only up to the 7S-size. In the case of the combination of G–AS_4 and L–BS, the reconstituted hybrid-intermediary subunits could assemble only up to the 7S-size. However, at present there is no evidence about the nature of these intermediary subunits. On the other hand, a comparison of the SDS–PAGE patterns of the hybrid-globulins with those of the native glycinin and legumin in the presence of 2–ME indicates that all the glycinin acidic subunits except G–AS_5 (which showed no reaction with the L–BS, Fig. 24) preferentially selected the legumin basic subunit with a molecular weight of 23,000 in the formation of the hybrid-intermediary subunits. L–ASI and L–ASII do not exhibit such specificity for one of the two glycinin basic subunits with molecular weights of 19,000 and 18,300. However, L–ASIII seems to exhibit the affinity for the glycinin basic subunit with a molecular weight of 19,000 to some extent, as is evident from the basic subunits higher presence in the hybrid-11S globulin formed from the combination of L–ASIII and G–BS, in comparison with the situation that exists in the native glycinin. These results may indicate that L–BS with a molecular weight of 23,000 is especially similar, with respect to structure, to the inherent G–BS with which G–AS_{1-3} form intermediary subunits in the native glycinin, and that G–BS with a molecular weight of 19,000 is similar to the inherent L–BS with which L–ASIII forms an intermediary subunit in the native legumin. Further studies on these points may shed light on evolutionary relationships between soybean and broad bean with respect to construction of the 6 (AB) structure and genetic relationships among the subunits of each.

Role of Constituent Subunits in Properties of Heat-Induced Gels of 11S Globulins from Soybean and Broad Bean

Gelling ability of soybean proteins is one of the most significant functional properties with respect to their usage in food systems. It is an important problem to elucidate what kind of chemical and structural properties of legume 11S globulins as well as soybean glycinin are responsible for their gelling ability. This understanding would further the adoption in foods logically and intentionally of soybean proteins and other grain legume proteins which are similar in structure to soybean proteins. However, at present sufficient information is lacking. Therefore, the physical properties of gels of glycinin, legumin, pseudo- and hybrid-11S globulins were investigated in order to elucidate how the subunits of 11S globulin contribute to the physical properties of the gel.

The pseudo-glycinins obtained from the sucrose density gradient centrifugation as described in the preceding section and native glycinin were heated to make a gel, and the hardness of the gels was determined (Fig. 26). The lowest gelation concentration for native glycinin was 2.5 %, and the gel hardness increased with increasing concentrations of protein and time of heating. The gels formed cannot

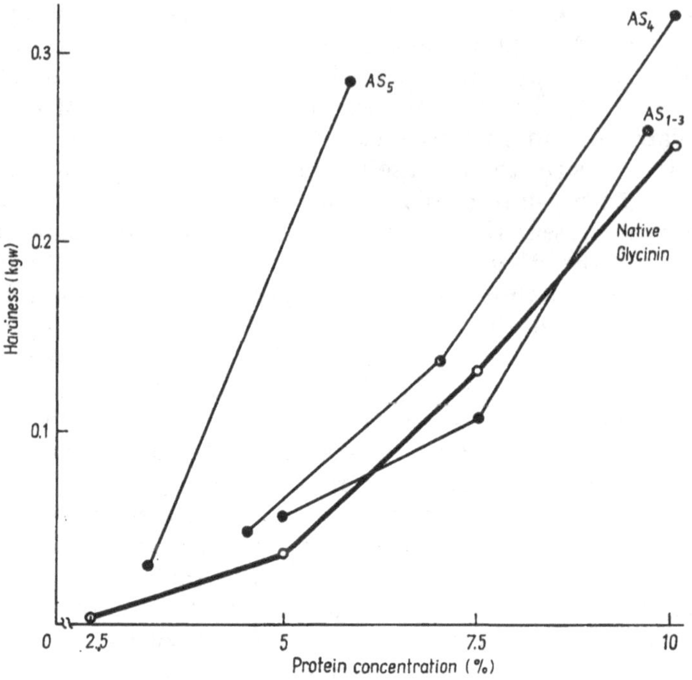

Fig. 26 Hardness of heat-induced gels of pseudo-glycinins. Native glycinin and the pseudo-glycinins were heated to make gel. The protein concentrations of the pseudo-glycinins used were as follows: combination of AS_{1-3} and BS, 5.0, 7.5 and 9.6 %; AS_4 and BS, 4.6, 7.0 and 10 %; AS_5 and BS, 3.0 and 6.0 % (w/v)

be reversed to the sol state by reheating, indicating its heat irreversibility. The gel from the pseudo-glycinin from the combination of AS_{1-3} and BS exhibited similar hardness to that from native glycinin, while that from the combination of AS_4 and BS was slightly harder. On the other hand, the gel from the pseudo-glycinin from the combination of AS_5 and BS was significantly harder than that from native glycinin. It is apparent that each gel from the pseudo-glycinins exhibits different hardness, depending on the acidic subunit composition. The result suggests that the acidic subunits contribute differently to the hardness of gel and that AS_5, having a larger molecular weight than the other acidic subunits, plays an important role for increasing the hardness of the gel. However, since some kinds of the constituent basic subunits were lacking in the pseudo-glycinin from the combination of AS_5 and BS, the possibility still remains that such basic subunits have a different contribution to the hardness of gel than the other basic subunits.

The pseudo-legumins obtained from the sucrose density gradient centrifugation as described in the preceding section and native legumin were heated to make a gel, and the hardness of the gels was determined (Fig. 27). The lowest gelation concentration for native legumin within 30 min heating was 10 %. The gel hardness increased with an increase in protein concentration and/or heating time. The higher the concentration of protein, the shorter the time required for gel formation. These tendencies coincide with those of glycinin although at the same protein concentration it tends to gel and produce a harder gel than legumin (Utsumi et al. 1982, 1983). It is noteworthy that legumin forms a transparent gel at high protein concentrations and does not form precipitates even at the low

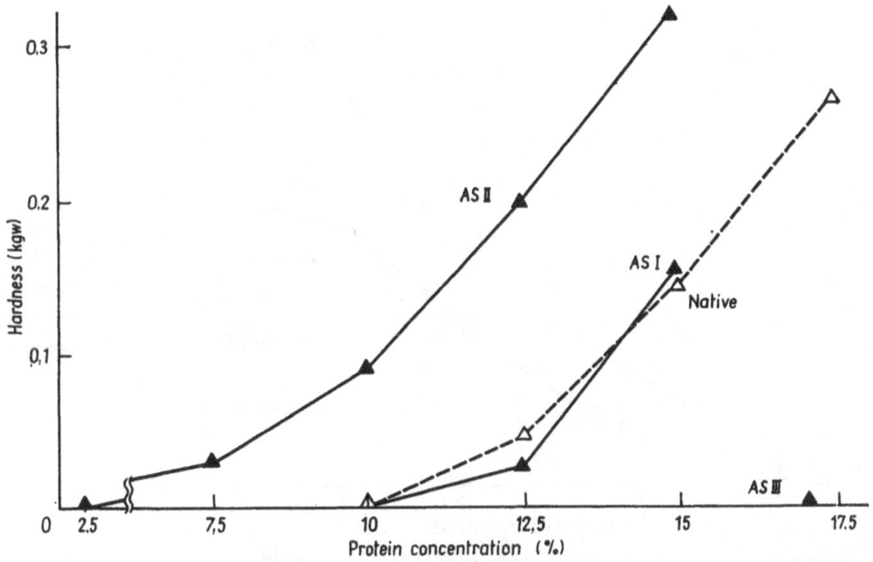

Fig. 27 Hardness of heat-induced gels of pseudo-legumins. Native legumin and the pseudo-legumins were heated to make gel

protein concentration where gel formation does not occur, although glycinin forms a turbid gel or precipitates at the low protein concentration. This difference between glycinin and legumin may be due to the difference in ease of disaggregation of their basic subunits, since free basic subunits of both glycinin and legumin are insoluble in the heating buffer. The pseudo-legumins prepared by recombining acidic and basic subunits of legumin were heated for 20 min at various protein concentrations, and the hardness of their gels formed was compared with that of the native legumin as shown in Fig. 27. The pseudo-legumin from the combination of ASIII and BS did not form a self-supporting gel at a 17.0 % concentration but formed a sol with viscosity and formed a gel when heated for 60 min. The pseudo-legumin from the combination of ASI and BS exhibited gelling ability and hardness similar to those of the native legumin. The pseudo-legumin from the combination of ASII and BS tended to form a gel at lower protein concentration and formed a significantly harder gel than did the native legumin. These results suggest that ASII of the native legumin plays an important role for determining the hardness of legumin gels. With respect to glycinin heat-induced gel, AS_5 with the highest molecular weight mainly contributes to hardness of the gel (Fig. 26). Thus, there is no such commonness between glycinin and legumin that the largest one among the acidic subunits contributes to hardness of the gel. Therefore, in order to elucidate structural factors contributing to the hardness of gel, it may be important to investigate the primary and higher structures of ASII and AS_5.

The hybrid-11S globulins obtained from the sucurose density gradient centrifugation were heated for 20 min at various protein concentrations, and the hardness of the gels obtained was compared with that of the native glycinin and legumin as shown in Fig. 28. The hybrid-globulins formed from all combinations, except that from L–ASIII and G–BS, exhibited similar gelling ability and gel hardness,

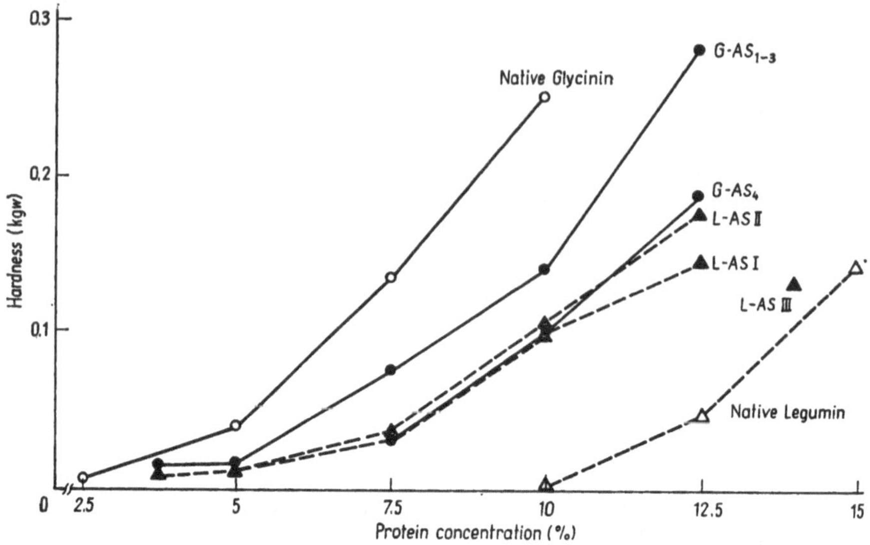

Fig. 28 Hardness of heat-induced gels of hybrid-11S globulins

the values of which were located between glycinin and legumin. However, the hardness of the gel formed from the combination of G–AS$_{1-3}$ and L–BS was slightly harder than the others. The hybrid-11S globulin from the combination of L–ASIII and G–BS did not form either a self-supporting gel or a sol below 9.6 % concentration. These results suggest that the superior gelling ability of glycinin as compared to legumin depends on the properties of both the acidic and basic subunits of glycinin.

The differences in the subunit compositions of glutenins among wheat cultivars and their relation to the bread-making quality of the flours have been studied by several people (see references in Mori et al. 1981, 1982). It has been demonstrated that the presence of a subunit of glutenin, whose molecular weight is about 145,000, correlated with bread-making quality. Concerning the 11S globulins of legume seeds, the relationship between chemical and structural properties at a subunit level and functional properties of 11S globulins is becoming to be elucidated, as described above with respect to soybean and broad bean. On the other hand, it has been demonstrated that the sulfur amino acids in glycinin from soybean cultivar CX635-1-1-1 are not distributed evenly among the various subunits, of which six acidic and four basic subunits have been distinguished (Moreira et al. 1979). Considering from these facts, it may be worthwhile to pursue the improvement of seed proteins with respect to functionality and nutritive value by breeding of conventional way and/or introducing the techniques of gene manipulation from the standpoint of subunit composition.

Zusammenfassung

Die Struktur der 11S-Globuline von Sojabohne und Ackerbohne und ihre Bedeutung im Nahrungsmittelsystem

Die Untereinheitenzusammensetzung der 11S-Globuline (Glycinine) aus den Samen von 18 Sojabohnensorten (aus Japan, USA, China und Korea) wurde in verschiedenen Polyacrylamidgelelektrophoresesystemen analysiert. Die Glycinine konnten hiernach in 5 Gruppen eingeteilt werden: Gruppe I enthielt 7 saure und 8 basische, Gruppe II 7 saure und 7 basische, Gruppe III 6 saure und 7 basische, Gruppe IV 6 saure und 5 basische, Gruppe V 6 saure und 3 basische Untereinheiten. Glycinin aus der Sojabohne var. Tsuru-no-ko (Gruppe I) wurde an einer DEAE-Sephadex-Säule getrennt und die molekularen Formen wurden gelelektrophoretisch untersucht, es wurden vier molekulare Formen mit den Molekularmassen 375,000, 360,000, 345,000 und 340,000 nachgewiesen. Glycinin aus anderen Sorten wies eine ähnliche Heterogenität auf und das Ausmaß dieser Heterogenität scheint mit der unterschiedlichen Untereinheitenzusammensetzung korreliert zu sein.

11S-Globuline (Legumine) aus Ackerbohnensamen von sechs in Japan angebauten typischen Sorten wurden ebenfalls auf ihre Untereinheitenzusammensetzung und molekulare Formen analysiert. Es konnte eine Einteilung in 3 Gruppen nach molekularer Ladung und Samengröße getroffen werden – klein-, mittel- und großsamige Gruppen.

Die Legumine verschiedener Sorten waren heterogen im Hinblick auf die molekulare Größe der Untereinheiten; fünf Gruppen mit neun Unterformen wurden in jeder Sorte gefunden.

Künstliche 11S-Globuline, Pseudoglycinine und Pseudolegumine, die sich in der Untereinheitenzusammensetzung von den nativen 11S-Globulinen unterschieden, wurden in Rekonstitutionsreaktionen durch die Kombination je einer sauren und einer basischen Untereinheit des Glycinins für Pseudoglycinin bzw. der entsprechenden Untereinheiten des Legumins für Pseudolegumin gebildet. Sie ähnelten den nativen 11S-Globulinen, alle bestanden aus über Disulfidbrücken im Verhältnis 1 : 1 zu intermediären Untereinheiten verbundenen sauren und basischen Untereinheiten und besaßen die Struktur 6 (AB). Diese Ergebnisse deuten auf ähnliche Strukturen innerhalb der sauren bzw. basischen Untereinheiten hin und zeigen, daß die Wechselwirkungen zwischen den Untereinheiten und zwischen den intermediären Untereinheiten, die zur 11S-Struktur führen, nicht spezifisch sind. Das muß einer der Gründe für die Heterogenität der molekularen Formen von Glycinin und Legumin sein. Weiterhin wurden Hybrid-11S-Globuline durch verschiedene Kombinationen saurer Untereinheiten des Legumins und basischer des Glycinins oder umgekehrt dargestellt. Die Hybridglobuline waren den nativen 11S-Globulinen ähnlich, alle bestanden aus sechs Hybridintermediäruntereinheiten, die aus über Disulfidbrücken verbundenen sauren und basischen Untereinheiten (1 : 1) zusammengesetzt waren. Diese Beobachtungen bestätigen die geringe Spezifität der Wechselwirkungen zwischen den Untereinheiten bzw. zwischen den intermediären Untereinheiten und weisen auf die strukturelle Ähnlichkeit von Glycinin und Legumin hin.

Die Textureigenschaften von aus Pseudo- und Hybridglobulinen hergestellten Gelen wurden untersucht, um den Einfluß der 11S-Globulinuntereinheiten auf die physikalischen Eigenschaften der Gele aufzuklären. Die Härten der hitzeinduzierten Gele aus Pseudoglycininen und nativem Glycinin waren unterschiedlich und abhängig von den sauren Untereinheiten. AS 5 mit einer höheren Molekularmasse als die anderen sauren Untereinheiten schien eine signifikante Erhöhung der Gelhärte zu bewirken. Für Pseudolegumin konnte gezeigt werden, daß ASII einen wichtigen Einfluß auf die Erhöhung der Gelhärte hat. Die dem Legumin überlegenen Geliereigenschaften des Glycinins werden sowohl von den sauren als auch den basischen Glycininuntereinheiten bestimmt, wie in Versuchen mit den Hybridglobulinen gezeigt wurde. Die Härte der Gele aus Glycininen von verschiedenen Sojasorten war abhängig von der Menge an AS5.

Da die Gelierfähigkeit eine der wichtigsten funktionellen Eigenschaften für den Einsatz der Proteine in Nahrungsmittelsystemen ist und von der Untereinheitenzusammensetzung der Globuline bestimmt wird, ist es wichtig, die Globulinstruktur der verschiedenen Sorten zu berücksichtigen. Züchterische Maßnahmen könnten zur Schaffung von Sorten mit der erwünschten Untereinheitenzusammensetzung der Globuline führen.

Краткое содержание

Структура 11S-глобулинов соевых и конских бобов и их значение в системах пищевых продуктов

Состав субъединиц 11S-глобулинов (глицининов) из семян 18 сортов соевых бобов (из Японии США, Китая и Кореи) анализировался в различных системах электрофореза в полиакриламидовом геле. Этим путём глицинины были подразделены на пять групп. Группа I содержит 7 кислых и 8 щелочных, группа II – 7 кислых и 7 щелочных, группа III – 6 кислых и 7 щелочных, группа IV – 6 кислых и 5 щелочных, а группа V – 6 кислых и 3 щелочных субъединиц. Глицинин из соевых бобов сорта 'Тсуру-Но-Ко' (группа I) разделялся на колонке ДЭАЭ-Сефадекс, а молекулярные формы исследовались с помощью электрофореза в геле; были обнаружены четыре молекулярных формы с молекулярной массой в 375, 360, 345, и 340 тысяч. Глицинин других сортов проявил сходную гетерогенность, причём степень этой гетерогенности, повидимому, коррелирует с различными составами субъединиц.

11S-глобулины из семян конских бобов шести типичных сортов, возделываемых в Японии, также анализировались на состав их субъединиц и их молекулярных форм. Они были подразделены на три группы по их молекулярному заряду и величине семян: мелкосемянная, среднесемянная и крупносемянная. Легумины различных сортов были гетерогенны по молекулярной величине субъединиц; внутри каждого сорта были найдены пять групп с девятью подгруппами.

Искусственные 11S-глобулины, псевдоглицинины и псевдолегумины, отличающиеся от природных 11S-глобулинов по составу их субъединиц, получались путём реконституционных реакций, комбинацией одной кислой и одной щелочной субъединицы глицинина (для псевдоглицинина) и соответствующих субъединиц легумина (для псевдолегумина). Они сходны с природными 11S-глобулинами, состояли из субъединиц, связанных дисульфидными мостами в отношении 1 : 1 в промежуточные кислые и щелочные субъединицы и обладали структурой 6 (AB). Эти результаты указывают на сходство структур внутри кислых и внутри щелочных субъединиц и говорят в пользу того, что взаимодействия между субъединицами и между промежуточными субъединицами, ведущими к структуре 11S не специфичны. Этот факт является одной из причин гетерогенности молекулярных форм глицинина и легумина. Потом были выявлены гибридные 11S-глобулины при помощи различных комбинаций кислых субъединиц легумина и щелочных глицинина (или наоборот). Гибридные глобулины были сходны с природными 11S-глобулинами, все они имели шесть гибридных промежуточных субъединиц, кислых и щелочных (1 : 1), связанных дисульфидными мостами. Эти наблюдения подтверждают низкую специфичность взаимодействий между субъединицами и между промежуточными субъединицами и указывает на структурное сходство глицинина и легумина.

Исследовались также свойства текстуры гелей, полученных из псевдоглобулинов и гибридных глобулинов чтобы выяснить влияние субъединиц 11S-глобулинов на физические свойства гелей. Твёрдости термоиндуцированных гелей из псевдоглицининов и природных глицининов были различны в зависимости от кислых субъединиц. Так субъединицы 5, с более высокой молекулярной массой, чем прочие кислые субъединицы, повидимому, были причиной достоверного повышения твёрдости геля. Для псевдолегумина было показано, что кислые субъединицы II оказывают существенное влияние на повышение твёрдости геля. Глицинин обладает лучшей способностью желатинироваться, чем легумин, что было замечено как на кислых, так и на щелочных субъединицах глицинина в опытах с гибридными глобулинами. Твёрдость гелей из глицинина различных сортов сои зависела от количества кислых субъединиц 5.

Так ка способность желатинироваться является одним из важнейших функциональных свойств для применения протеинов в системах пищевых продуктов и обусловливается составом субъединиц глобулинов, то необходимо учитывать структуру глобулинов различных сортов бобовых. Селекционные мероприятия могут привести к созданию сортов с желательным составом субъединиц глобулина.

Literature

Badley, R. A., D. Atkinson, H. Hauser, D. Oldani, J. P. Green and J. M. Stubbs, 1975: The structure, physical and chemical properties of the soy bean protein glycinin. – Biochim. Biophys. Acta **412**, 214–228.

Casey, R., J. F. March and E. Sanger, 1981: N-Terminal amino acid sequence of β-subunits of legumin from *Pisum sativum*. – Phytochemistry **20**, 161–163.

Davies, D. R., 1980: The r_a locus and legumin synthesis in *Pisum sativum*. – Biochem. Genet. **18**, 1207–1219.

Derbyshire, E., D. J. Wright, and D. Boulter, 1976: Legumin and vicilin, storage proteins of legume seeds. – Phytochemistry **15**, 3–24.

Gilroy, J., D. J. Wright, and D. Boulter, 1979: Homology of basic subunits of legumin from *Glycine max* and *Vicia faba*. – Phytochemistry **18**, 315–316.

Kitamura, K., K. Okubo and K. Shibasaki, 1973: The effects of cyanate in urea solutions on the gel electrophoresis of the subunits of soybean 11S globulin. – Agr. Biol. Chem. **37**, 1983–1984.

–, T. Takagi, and K. Shibasaki, 1976: Subunit structure of soybean 11S globulin. – Agr. Biol. Chem. **40**, 1837–1844.
Koshiyama, I., 1983: Storage proteins of soybean. – In: Seed proteins (Eds. W. Gottschalk, and H. P. Müller), pp. 427–450. Martinus Nijhoff/Dr. W. Junk Publishers, Hague.
Moreira, M. A., M. A. Hermodson, B. A. Larkins, and N. C. Nielsen, 1979: Partial characterization of the acidic and basic polypeptides of glycinin. – J. Biol. Chem. **254**, 9921–9926.
Mori, T., T. Nakamura, and S. Utsumi, 1982: Formation of pseudoglycinins and their gel hardness. – J. Agric. Food Chem. **30**, 828–831.
–, and S. Utsumi, 1979: Purification and properties of storage proteins of broad bean. – Agric. Biol. Chem. **43**, 577–583.
–, –, and H. Inaba, 1979: Interaction involving disulfide bridges between subunits of soybean seed globulin and between subunits of soybean and sesame seed globulins. – Agric. Biol. Chem. **43**, 2317–2322.
–, –, –, K. Kitamura and K. Harada, 1981: Differences in subunit composition of glycinin among soybean cultivars. – J. Agric. Food Chem. **29**, 20–23.
Saio, K., T. Matsuo, and T. Watanabe, 1970: Preliminary electron microscopic investigation on soybean 11S protein. – Agr. Biol. Chem. **34**, 1851–1854.
Utsumi, S., H. Inaba, and T. Mori, 1980: Formation of pseudo- and hybrid-11S globulins from subunits of soybean and broad bean 11S globulins. – Agric. Biol. Chem. **44**, 1891–1896.
–, –, and –, 1981: Heterogeneity of soybean glycinin. – Phytochemistry **20**, 585–589.
–, and T. Mori, 1980: Heterogeneity of broad bean legumin. – Biochim. Biophys. Acta **621**, 179–189.
–, T. Nakamura, and T. Mori, 1982: A micro-method for the measurement of gel properties of soybean 11S globulin. – Agric. Biol. Chem. **46**, 1923–1924.
–, –, and –, 1983: Role of constituent subunits in the formation and properties of heat-induced gels of 11S globulins from legume seeds. – J. Agric. Food Chem. **31**, 503–506.
–, Z. Yokoyama, and T. Mori, 1980: Comparative studies of subunit compositions of legumins from various cultivars of *Vicia faba* L. seeds. – Agric. Biol. Chem. **44**, 595–601.
Wright, D. J., and D. Boulter, 1974: Purification and subunit structure of legumin of *Vicia faba* L. (Broad bean). – Biochem. J. **141**, 413–418.

T. Mori
Research Institute for Food Science
Kyoto University
Gokasho
Uji City, Kyoto 611
Japan

Kulturpflanze 32 · 1984 · S 159—S 163

Comparison of the structures of different 11S and 7S globulins by small-angle X-ray scattering, quasi-elastic light scattering and circular dichroism spectroscopy

P. PLIETZ, G. DAMASCHUN, D. ZIRWER, K. GAST, B. SCHLESIER, and K. D. SCHWENKE
(Berlin-Buch, GDR)

Summary

The 11S globulins from sunflower seed, rape seed, and broad bean as well as the 7S globulin from french bean were investigated by small-angle X-ray scattering, quasi-elastic light scattering and circular dichroism spectroscopy. Secondary structure, molecular parameters, shape, hydration, and the structure of the proteins were determined. The structure of the 7S globulin from french bean is compared with the crystal structure of the 7S globulin from jack bean.

The 11S globulins with molar masses of (3.0–4.0)$\times 10^5$ g/Mol and the 7S globulins with molar masses of (1.3–2.0)$\times 10^5$ g/Mol are main protein fractions of different plant seeds (DERBYSHIRE et al. 1976). The 11S globulins from *Helianthus annuus* L., *Brassica napus* L. and *Vicia faba* L. as well as the 7S globulin from *Phaseolus vulgaris* were investigated by small-angle X-ray scattering (SAXS), quasi-elastic light scattering (QELS) and circular dichroism spectroscopy. The following results have been obtained for the 11S globulins (SCHWENKE et al. 1975, 1980, PLIETZ et al. 1978; 1983b, 1984):

1) The experimentally determined physical parameters of the molecules are summarized in table 1. The data show that the physical parameters of the 11S globulins from *H. annuus* and *B. napus* are identical within experimental error, however, different from those of the 11S globulin from *V. faba*. This molecule has a higher molar mass as well as larger molecular dimensions.
2) The shape of the investigated 11S globulins is nearly spherical and can be approximated by oblate ellipsiods of revolution with dimensions (11$\times$11$\times$8·8) nm, (11.2$\times$11.2$\times$8.8) nm and (12.6$\times$12.6$\times$8.82) nm, respectively.
3) A solvent shell of about 0.6 nm thickness (time-average) is fixed to the surface of the three 11S globulins in solution. This is tantamount to a solvent shell with a time-average thickness between one and two water molecules, which has to be considered as fixed to the protein surface and, therefore, influences hydrodynamic data.
4) The investigated 11S globulins consist of six structurally nearly identical subunits, which are arranged like a trigonal antiprism. The molecules have the point group symmetry 32. Therefore, the molecules have as symmetry elements a threefold rotation axis and three two-fold rotation axes oriented perpendicularly to the three-fold rotation axis (fig. 1).
5) A comparison of our experimental data with the scattering data of the 11S

globulins from *Cannabis sativa* (CLEEMANN and KRATKY 1960) and from *Glycine max.* (BADLEY et al. 1975) shows, that these globulins have the symmetry 32 too. This symmetry was also found for the 11S globulin from *Bertholletia excelsa* (SCHEPMANN et al. 1972).

Table 1

Experimentally determined molecular parameters of the 11S seed globulins from *Helianthus annuus*, *Brassica napus* and from *Vicia faba* derived from SAXS and QELS

Parameter	*Helianthus annuus*	*Brassica napus*	*Vicia faba*
Radius of gyration R_G [nm]	3.96 ±0.8	4.08 ± 0.8	4.45 ± 0.07
Maximum dimension, L [nm]	11.0 ±0.5	11.2 ± 0.5	13.0 ± 0.5
Correlation volume, V [nm^3]	410 ±40	450 ±40	683 ± 40
Surface, S [nm^2]	356 ±60	410 ±60	426 ±65
Molar mass, M [g/Mol]	(3.0 ±0.1) 10^5	(3.0 ± 0.1) 10^5	(3.5 ± 0.2) 10^5
Stokes radius, R_s [nm]	5.65 ±0.12	5.65 ± 0.15	6.3 ± 0.15
Translational diffusion coefficient, $D^0_{20,w}$[cm^2s^{-1}]	(3.78 ±0.08) 10^{-7}	(3.78 ± 0.08) 10^{-7}	(3.38 ± 0.05) 10^{-7}
Ratio of frictional coefficients f/f_0	1.28	1.28	1.36

Fig. 1 Quaternary structure of the 11S globulins

6) The amino acid residues of the investigated 11S globulins are arranged in about 14 % α-helical and in about 20 % β-sheet structure, respectively.
7) The 11S globulins from plant seeds of six different orders of dicots have a very similar secondary structure, similar shapes and the same symmetry of molecules. These results lead to the experimentally supported hypothesis, that the 11S globulins of all dicots have very similar structures.

The experimentally determined physical parameters of the 7S globulin from *Ph. vulgaris* are summarized in table 2.

The molecule has a disk-like shape and can be approximated by an ellipsoid of revolution with the axial ratio 0.3 and the dimensions (12.5×12.5×3.75) nm (PLIETZ et al. 1983c). Combining data of SAXS and QELS it can be shown that

Table 2
Experimentally determined molecular parameters of the 7S globulin from *Phaseolus vulgaris*

Parameter	*Phaseolus vulgaris*
Radius of gyration, R_G [nm]	4.05 ± 0.5
Maximum dimension, L [nm]	13.0 ± 0.5
Molar mass, M [g/Mol]	(1.45 ± 0.05) 10^5
Correlation volume, V [nm^3]	300 ±40
Surface, S [nm^2]	365 ±60
Stokes radius, R_6 [nm]	4.71 ± 0.1
Translational diffusion coefficient, $D^0_{20,w}$[cm^2s^{-1}]	(4.53 ± 0.05) 10^{-7}
Ratio of frictional coefficients f/f_0	1.34

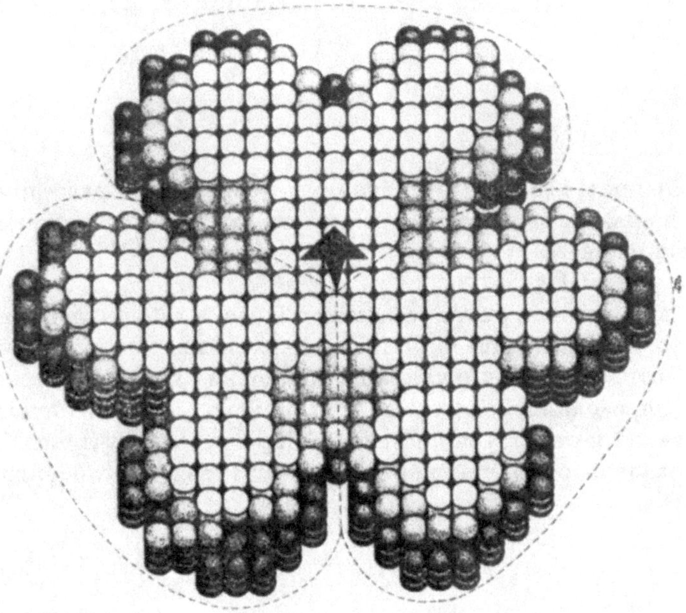

Fig. 2 Quaternary structure of the 7S globulin from *Phaseolus vulgaris*

this protein does not have a pronounced solvent shell, which is fixed to and moved with the molecule. The 7S globulin molecule has as symmetry element a three-fold rotation axis and consists of three y-shaped subunits separated by deep clefts filled with solvent (fig. 2).

The comparison of this structure with the crystal structure of the 7S globulin from *Canavalia ensiformis* (McPherson 1980) shows significant differences between the structure of the 7S globulin from *C. ensiformis* in crystal and the structure of the 7S globulin from *Ph. vulgaris* in solution. The differences are probably caused by the tryptic modification of the 7S globulin from *C. ensiformis* prior to crystallization (Plietz et al. 1983a). The secondary structure of the 7S globulin from *Ph. vulgaris* is characterized by a small portion of α-helical structure (14 %) and a marked content of β-structure (18 %).

Zusammenfassung

Vergleich der Strukturen verschiedener 11S- und 7S-Globuline durch Röntgen-Kleinwinkelstreuung, quasielastische Lichtstreuung und Zirkulardichroismus-Spektroskopie

Die 11S-Globuline von Sonnenblumen-, Raps- und Ackerbohnensamen sowie das 7S-Globulin von Buschbohnen wurden mit Hilfe der Röntgen-Kleinwinkelstreuung, der quasielastischen Lichtstreuung und der Zirkulardichroismus-Spektroskopie untersucht. Sekundärstruktur, molekulare Parameter, Gestalt, Hydratation und Struktur der Proteine wurden bestimmt. Die Struktur des 7S-Globulins der Buschbohne wird mit der Kristallstruktur des 7S-Globulins von *Canavalia ensiformis* verglichen.

Краткое содержание

Сравнение структур различных 11S- и 7S-глобулинов при помощи рассеяния рентгеновских лучей на малые углы, спектроскопии циркулярного дихроизма и квазиэластичного рассеяния света.

Глобулины 11S семян подсолнечника, рапса и конских бобов, а также глобулин 7S семян кустовой фасоли исследовались с помощью рассеяния рентгеновских лучей на малые углы, спектроскопии циркулярного дихроизма и квазиэластичного рассеяния света. Определялись: вторичная структура, молекулярные параметры, форма, гидратация и структура протеинов. Структура глобулинов 7S кустовой фасоли сравнивалась с кристаллической структурой 7S-глобулина семян *Canavalia ensiformis*.

Literature

BADLEY, R. A., D. A. ATKINSON, H. HAUSER, D. OLDANI, J. P. GREEN, and J. M. STUBBS, 1975: The structure, physical and chemical properties of the soy bean protein glycinin. – Biochim. Biophys. Acta **412**, 214–228.

CLEEMANN, J. C., and O. KRATKY, 1960: Größe, Gestalt und Solvation des Edestinmoleküls aus dem Studium der Röntgen-Kleinwinkelstreuung. – Z. für Naturforschung **15b**, 525–535.

DERBYSHIRE, E., D. J. WRIGHT, and D. BOULTER, 1976: Legumin and vicilin. Storage proteins of legume seeds. – Phytochemistry **19**, 3–24.

MCPHERSON, A., 1980: The three-dimensional structure of canavalin at 3.0 Å resolution by X-ray diffraction analysis. – J. Biol. Chem. **255**, 10472–10480.

PLIETZ, P., G. DAMASCHUN, J. J. MÜLLER, and B. SCHLESIER, 1983a: Comparison of the structure of the 7S globulin from *Phaseolus vulgaris* in solution with the crystal structure of the 7S globulin from *Canavalia ensiformis* by small-angle X-ray scattering. – FEBS Letters **162**, 43–46.

–, –, –, and K. D. SCHWENKE, 1983b: The structure of 11S globulins from sunflower and rape seed. – Eur. J. Biochem. **130**, 315–320.

–, –, D. ZIRWER, K. GAST, and B. SCHLESIER, 1983c: The structure of 7S globulin from *Phaseolus vulgaris* in solution. – Int. J. Biol. Macromol. **5**, 356–360.

–, H. DAMASCHUN, D. ZIRWER, K. GAST, K. D. SCHWENKE, W. PAEHTZ, and G. DAMASCHUN, 1978: Small-angle X-ray and quasi-elastic light scattering studies on 11S globulin from sunflower seed. – FEBS Letters **91**, 227–229.

–, D. ZIRWER, B. SCHLESIER, K. GAST, and G. DAMASCHUN, 1984: Shape, symmetry, hydration and secondary structure of the legumin from *Vicia faba* in solution. – Biochim. Biophys. Acta, **784**, 140–146.

SCHEPMAN, A. M. H., T. WICHERTJES, E. F. J. VAN BRUGGEN, 1972: Visibility of subunits in crystal of oligomeric proteins. – Biochim. Biophys. Acta **271**, 279–285.

SCHWENKE, K. D., B. RAAB, and M. SCHULTZ, 1975: Über Samenproteine. 6. Mitt. Konfor. mationsanalyse von Globulinen und Albuminen aus den Samen von Sonnenblumen (*Helianthus annuus* L.) und Raps (*Brassica napus* L.) durch Bestimmung des Circulardichroismus im fernen UV. – studia biophysica **51**, 161–172.

–, M. SCHULTZ, K.-J. LINOW, K. GAST, and D. ZIRWER, 1980: Hydrodynamic and quasi. elastic light scattering studies on the 12S globulin from rapeseed. – Int. J. Peptide Protein Res. **16**, 12–18.

P. PLIETZ, G. DAMASCHUN, D. ZIRWER, K. GAST
Zentralinstitut für Molekularbiologie der AdW der DDR
DDR - 1115 Berlin, Robert-Rössle-Straße 10

B. SCHLESIER
Zentralinstitut für Genetik und Kulturpflanzenforschung der AdW der DDR
DDR - 4325 Gatersleben, Corrensstraße 3

K. D. SCHWENKE
Zentralinstitut für Ernährung der AdW der DDR
DDR - 1505 Bergholz-Rehbrücke, Arthur-Scheunert-Allee 114–116
Akademie der Wissenschaften

Kulturpflanze 32 · 1984 · S 165–S 170

Blocks of gliadin components in wheat genetics, breeding and seed-growing

A. A. SOZINOV and F. A. POPERELYA
(Moscow and Odessa, USSR)

Scientific wheat breeding has gained high results. Further progress in creating better varieties of this culture is closely associated with the elaboration of new methods of evaluation of the genotype properties in the choice of pairs for hybridization and in the selection of valuable forms at early stages of the breeding process. The use of polymorphism of protein systems refers to long-term directions of investigations carried out in this field.

Significant polymorphism of storage proteins (MECHAM et al. 1978; PAYNE et al. 1980), amylases (SOZINOV and RYBALKA 1980; GALE et al. 1983), esterases and other enzymes (JAASKA 1978) has been found in wheat. The study of polymorphism of storage proteins – gliadins and glutenins is of the greatest interest. These proteins are coded for by polycistronic loci of chromosomes of homoeological groups 1 and 6 (SHEPHERD 1968; SOZINOV et al. 1978). They are very rich in glutamine and proline and, probably, consist of repeating sequences of amino acid residues in all cereals. Their main role is nitrogen and energy depot for the germinating embryo, though it is not ruled out that these proteins perform other functions at certain stages of metabolism.

Gliadin components (MECHAM et al. 1978; SOZINOV et al. 1975; SOZINOV and POPERELYA, 1979) have been found to be linkedly inherited as blocks. Blocks remain unchanged throughout generations. Recombinations within a block occur very rarely.

Hybridological analysis of progenies of more than 100 hybrid combinations and the use of aneuploid lines, Chinese Spring, obtained from Dr. SEARS, permitted the preparation of a catalogue of variants of blocks of gliadin components in bread winter wheats (Fig. 1).

At present there are two catalogues–for starch and polyacrylamide gels. Methods of starch gel electrophoresis worked out by our scientists in 1970 and ZILLMAN'S and BUSHUK'S modified methods of polyacrylamide gel are used (BUSHUK and ZILLMAN 1978).

It should be noted that for chromosomes of homoeological group I allelic variants of blocks practically coincide in starch and polyacrylamide gel electrophoresis; for homoeological group 6 more allelic variants have been identified using polyacrylamide gel as compared to starch gel.

The prepared catalogue of blocks serves as the basis for a genetic gliadin classification. The unit of this classification is a block of linkedly inherited electrophoretic components. The name of the block consists of gliadin symbol GLD,

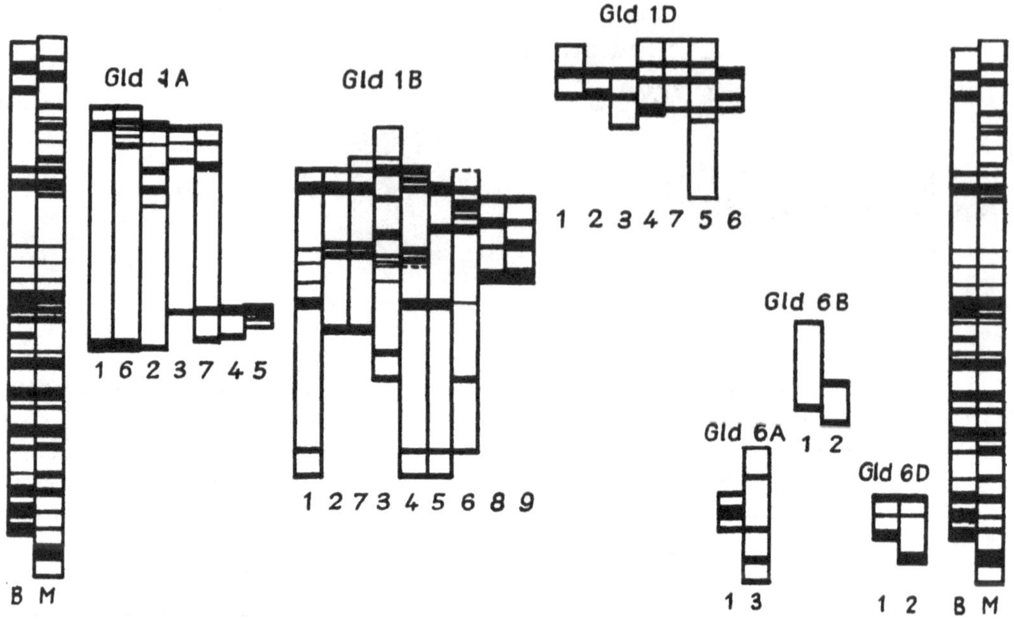

Fig. 1 Schemes of electrophoretic patterns of gliadin of Bezostaya I (B), Mironovskaya 808 (M) and identified heretofore blocks of gliadin components

number of the corresponding chromosome and index number in the catalogue. For example, GLD 1B3 is a gliadin block, controlled by the chromosome 1B and having the number 3 in the catalogue of allelic variants. The corresponding polycistronic locus is recorded as follows: GLD 1B3. Such nomenclature is suitable for breeding purposes and genetic studies since the use of letter designations restricts the number of alleles to be recorded and impedes remembering of blocks. A simple system to record gliadin genotypic formulae has been proposed (Table 1).

Blocks are characterized by stability. In modern wheat varieties blocks of aboriginal cultivars involved in hybridization 50–60 years ago are identified. For example, the GLD IBI block has been inherited by modern varieties from the aboriginal variety Krymka (Turkey).

Table 1

Genotypic formulae of gliadin of several wheat varieties

Variety	Types of blocks of components controlled by the chromosomes					
	1A	1B	1D	6A	6B	6D
Bezostaya-1	4	1	1	1	1	1
Mironovskaya-808	3	1	5	3	1	2
Rostovchanka	4	1	5	3	1	2
Odesskaya-51	2+4[a]	1	1+5[a]	3	2	1
Chaika	2	1	1	3	1	1
Kavkaz	4	3	2	1	1	1

[a] Two figures indicate varieties heterogeneous for a given locus.

Mapping of the gliadin-coding locus in the chromosome 1B has been carried out (RYBALKA and SOZINOV, 1979). Recently, PAYNE (1983) has mapped all gliadin-coding loci and loci of HMW and LMW glutenins.

The study of inheritance of gliadin components shows that designations α, β, γ and ω, while convenient, don't reflect genetics of gliadins. For example, in β-zone of different varieties we found components controlled by chromosomes 1A, 1B, 6A, 6B and 6D. On the other hand, there are blocks that consist of components localized in different gel zones. So, gliadin-coding locus may have the set of several types of gliadin genes encoding for different types of gliadin sequences.

Identification of blocks of components has substantially facilitated the determination of a relationship between protein polymorphism and variability of quantitative characters.

To this end, populations F_3, F_4 or of later generations are divided into two parts differing in the allelic state of one gliadin-coding locus and the samples are compared for a character under study. If one of allelic variants of blocks significantly prevails over another in a number of generations or in two or several combinations, it can be asserted that the variability of the quantitative character is related to the allelic state of the gliadin-coding locus.

Thus, it has been shown on many combinations that genotypes with the block GLD 1A1 are characterized by better tillering and are more productive as compared to genotypes with the GLD 1A4 block. A significant relationship of some blocks with the plant height is also observed (Table 2).

The relation of allelic variants of blocks to grain quality is most pronounced. For instance, by their influence on the sedimentation number blocks are ranged as it is shown in Table 3.

Table 2
Allelic variants of blocks of gliadin components and height of plants

Variants of blocks controlled by chromosomes 1A	1B	Height of plants (cm)	Differences
4.	1.	110.1	8.8***
4.	3.	101.5	
3.	1.	110.4	6.9**
3.	3.	103.5	

, * Significant

Table 3
Relative value of allelic blocks of gliadin components for evaluation of flour quality[a]

Gld 1A7 > Gld 1A4 > Gld 1A2 > Gld 1A5 > Gld 1A3 > Gld 1A1 ≧ Gld 1A6
Gld 1B1 > Gld 1B2 ≧ Gld 1B7 > Gld 1B5 > Gld 1B4 > Gld 1B3 > Gld 1B6
Gld 1D4 > Gld 1D5 > Gld 1D1 > Gld 1D2 > Gld 1D3
Gld 6A3 > Gld 6A1
Gld 6B2 > Gld 6B1
Gld 6D2 > Gld 6D1 > Gld 6D3

[a] Blocks associated with poorer grain quality appear on the right.

The influence of blocks on bread volume and dough rheological properties is no less essential. Thus, analysis of lines F_5 derived from hybridization of Bezostaya-I variety with the breeding line Trebishov differing in the allelic state of GLD 1A and GLD 1B loci has shown that significant differences between genotypes with GLD 1B1 and GLD 1B3 blocks made up 191.2 cm^3 by bread volume and 97.1 units by the alveograph index W; genotypes with GLD 1A4 and GLD 1A3 blocks differed by 57.2 cm^3 and 30.0 units, respectively.

Allelic variants of blocks have been found to be related to spike colour (Table 4), mass of 1,000 grains (Table 5), frost resistance, drought resistance and resistance to some diseases.

Table 4
Allelic blocks of components and spike color (Bezostaya IxChrvena Zvezda, F_4)

Blocks of components controlled by chromosomes		Number of plants		
1A	1B	Total	White	Red
4.4.	1.1.	49	49	
4.4.	1.8.	67	–	67
4.4.	8.8.	19	–	19
6.	1.1.	100	98	2
6.	1.8.	204	4	200
6.	8.8.	81	–	81

Table 5
Allelic variants of blocks of gliadin components and mass of 1,000 grains (Bezostaya I×Concho, F_6)

Compared genotypes (formulae)	Mass of 1,000 grains (g)
4. 1.1.1.1.1.–3.1.1.1.1.1.	36.9–41.1
4. 1.4.1.1.1.–3.1.4.1.1.1.	37.4–40.1
4.10.1.1.1.1.–3.10.1.1.1.1.	36.7–40.9
4.10.4.1.1.1.–3.10.1.1.1.1.	37.1–40.0
×	37.0–40.0
d	3.0**

** Significant

It should be noted that genotypic formulae of industrial varieties include a relatively small number of variants of blocks although varieties with numerous different blocks were involved in hybridization. Apparently in course of breeding genotypes with unsuitable blocks are eliminated. This is proved, in particular, by investigation of frequencies of allelic variants of blocks in populations cultivated without artificial selection (Table 6).

The findings have permitted the use of storage protein polymorphism in wheat breeding for choosing parental forms for hybridization, selecting the most valuable genotypes and eliminating unfavourable ones. Identification of blocks makes it possible to trace chromosome translocations or substitutions in interspecific hybridization. Thus, it has been shown that the genome of a number of winter durum wheat varieties includes chromosomes 1A, 1B, 6A of parental varieties.

Table 6
Frequency of blocks Gld1B1 and Gld1B3 in populations F_4 and F_5 of the hybrid Bezostaya I × Trebishov

Allele	Number of plants or progenies			Frequency		
	F_4	F_5	F_5 (p)	F_4	F_5	F_5 (p)
Gld1B1	623	474	261	0.61	0.67	0.73
Gld1B3	251	235	101	0.39	0.33	0.27

Gliadin blocks can also be used in seed-growing to prevent biological pollution, to preserve the biotypic composition and improve varieties by eliminating genotypes with undesirable properties. A new highly productive winter wheat variety Odesskaya semidwarf has been found to include two biotypes 4.1.5.1.2.1. and 4.1.1.1.2.1. Plants of the first biotype are distinguished by a high grain quality and frost resistance due to the presence of the Gld 1D5 block. Therefore, lines with the genotype 4.1.1.1.2.1. are eliminated in seed-plots.

The mechanism underlying the relationship of allelic variants of blocks with variability of many characters is unknown. It is most probable that blocks serve as effective genetic markers of individual genes or stable associations of genes determining the expression of characters. For instance, the Gld 1B3 block is a reliable genetic marker of 1B/1R substitution and the variability of some characters related with it.

Identification of blocks of components is actively used in breeding and seed-growing and gives positive results in solving numerous applied tasks.

Zusammenfassung

Die elektrophoretischen Komponenten des Gliadins werden als Blöcke vererbt. Für alle bekannten Loci, die für Gliadin codieren, wurde vielfacher Allelelismus bestätigt. Es wurden für die Chromosomen 1A, 1W, 1D, 6A, 6W, 6D entsprechend 18, 18, 8, 13, 12, 12 Blöcke von Allelvarianten identifiziert. Diese Blöcke dienen als effektive genetische Marker für Produktivitätsänderung, Frostresistenz, Trockenresistenz, Kornqualität und andere Merkmale. Das vorgeschlagene System, die Gliadin-Blöcke in der Genetik, in der Selektion und Produktion von Weizen zu nutzen, hat sich in der Praxis bewährt und wird in Selektionsprogrammen verwendet.

Краткое содержание

Электрофоретические компоненты глиадина наследуются группами в виде блоков. Множественный аллелизм подтвердился для всех известных локусов, кодирующих глиадин. В полиакриламидном гене идентифицировали 18, 18, 8, 13, 12, 12, блоков аллельных вариантов блоков для хромосом 1A, 1B, 1Д, 6A, 6B, 6Д соответственно. Эти блоки служат эффективными генетическими маркерами изменчивости продуктивности, морозоустойчивости, за сухоустойчивости, качества зерна и других признаков. Предложенная система использования глиадиновых блоков в генетике, селекции и производстве семян пшеницы успешно испытывалась в практике и применяется в селекционных программах.

Literature

Bushuk, W., and R. R. Zillman, 1978: Wheat cultivar identification by gliadin electrophoregrams. I. Apparatus, method and nomenclature. – Canad. J. Plant Sci. **58**, 505–515.

Gale, M. D., C. N. Law, A. J. Chojecki, and R. A. Kempton, 1983: Genetic control of α-amylase production in wheat. – Theor. Appl. Genet. **64**, 309–316.

Jaaska, V., 1978: NADP-dependent aromatic alcohol dehydrogenase in polyploid wheats and their relatives. On the origin and phylogeny of polyploid wheats. – Theor. Appl. Genet. **53**, 209–217.

Mecham, D. K., D. D. Kasarda, and C. O. Qualset, 1978: Genetic aspects of wheat gliadin proteins. – Biochem. Genet. **16**, 831–853.

Payne, P. I., L. M. Holt, and C. N. Law, 1981: Structural and genetical studies on the high-molecular-weight subunits of wheat glutenin. P. I. Allelic variation in subunits amongst varieties of wheat *(Triticum aestivum)*. – Theor. Appl. Genet. **60**, 229–236.

–, C. N. Law, and E. E. Mudd, 1980: Control by homoeologous group 1 chromosomes of the high-molecular-weight subunits of glutenin, a major protein of wheat endosperm. – Theor. Appl. Genet. **58**, 113–120.

–, and G. J. Lawrence, 1983: Catalogue of alleles for the complex gene loci, Glu-AI, Glu-BI, and Glu-DI which code for high-molecular-weight subunits of glutenin in hexaploid wheat. – Cereal Res. Communs. **11**, 29–35.

Rybalka, A. I., and A. A. Sozinov, 1979: Mapping of Gld 1B locus, controlling biosynthesis of reserve proteins in bread wheat. – Tsitol. Genet. **13**, 276–282 (in Russian).

Shepherd, K. W., 1968: Chromosomal control of endosperm proteins in wheat and rye. – Proc. 3rd Int. Wheat Genet. Symp. (Aust. Acad. Sci., Canberra), 86–96.

Sozinov, A. A., F. A. Poperelya, and A. I. Stakanova, 1975: Hybridological analysis as a method of study of genetic regularities of gliadin biosynthesis. – Nauchno-Tekhn. Bull. Vsesoyuznogo Selektsionno-Geneticheskogo Instituta **24**, 10–15 (in Russian).

–, A. F. Stelmakh, and A. I. Rybalka, 1978: Hybridological and monosomic analyses of gliadins in soft wheat varieties. – Genetika **24**, 1955–1967 (in Russian).

–, and F. A. Poperelya, 1979: Polymorphism of prolamines in breeding. – Vestn. Skh. Nauki **10**, 21–34 (in Russian).

–, and A. I. Rybalka, 1980: Genetic study of grain β-amylases in common wheat. – Genetika **16**, 1059–1067 (in Russian).

A. A. Sozinov
Institute of General Genetics, USSR Academy of Sciences,
Gubkin Street 3,
GSP-I Moscow B-333, USSR

F. A. Poperelya
All-Union Institute for Plant Breeding and Genetics,
Ovidiopolskaya Doroga 3
Odessa, USSR

Kulturpflanze 32 · 1984 · S 171–S 185

Combined use of two genetic systems in the development and improvement of quality protein maize

S. K. VASAL, E. VILLEGAS, C. Y. TANG, J. WERDER and M. READ
(Mexico, Mexico)

Summary

Several complex and interrelated problems are encountered in original soft endosperm *opaque-2* materials. The most important being reduced yield, chalky lustreless appearance, greater vulnerability to stored grain pests and ear rots, and slower drying. CIMMYT's strategy in developing quality protein maize (QPM) has been described. It involves combined use of the *opaque-2* gene and the genetic modifiers of the *opaque-2* locus. The genetic modifiers improve kernel phenotype and improve agronomic performance of QPM materials. The genetic modifiers are complex in inheritance and sometimes affect protein quality adversely. To facilitate breeding and attain faster progress, the need to work in homozygous *opaque-2* backgrounds has been discussed. Methodology in the development of QPM donor stocks and progress in overcoming problems associated with the *opaque-2* maize have been described. A wide array of QPM germplasm has been developed through the conversion process and the development of QPM gene pools. Breeding strategy and methodology used in the development of QPM have been briefly mentioned. Through merging and reorganisation of available QPM germplasm several new QPM pools and populations have been developed. The handling of QPM pools and populations has been described. Various uses of QPM germplasm currently available at CIMMYT have been pointed out. Renewed interest in QPM research is showing up and the QPM materials stand a good chance of commercial exploitation in some countries.

Introduction

In maize several high-lysine mutants are known which can bring about two-fold increase in the levels of lysine and tryptophan in the endosperm protein (MERTZ et al. 1964, NELSON et al. 1965, MCWHIRTER 1971). These genes exhibit several common features including increased lysine content, possessing soft chalky endosperm and deficient in total dry matter production in the grain. Though search for new and better mutants has continued over the years, the maize breeders have continued to make an extensive use of the *opaque-2* gene in developing quality protein maize materials. The use of other genes such as *floury-2* and *opaque-7* in practical maize breeding programs around the world has been negligible as these genes do not offer any advantage over the *opaque-2* gene.

As is well known, the original soft *opaque-2* materials suffer from several serious problems. The most important being reduced yield, chalky lustreless appearance, greater vulnerability to ear rot causing pathogens, increased storage losses due to stored grain pests and in losing moisture slowly following physiological maturity of the grain. The problems are complex and interrelated. Inability to solve these problems has brought frustration and decline in research interest in the breeding of quality protein maize (QPM). The benefits of *opaque-2* maize have thus been difficult to harness because of aforementioned problems that have acted as the major stumbling block in the commercial acceptance and exploitation of such materials. CIMMYT has maintained continued interest over the years. Several new research ideas and approaches have been explored to find a viable alternative that will facilitate development of QPM germplasm with inherent superior biochemical characteristics and with agronomic performance competitive to normal maize genotypes. Exploratory research work done by CIMMYT scientists in the early 70's provided convincing proof that genetic modifiers of the *opaque-2* locus could aid in remedying problems and in developing materials that will find ready acceptance with the farmers. The progress has been dramatic. Currently a wide array of QPM germplasm is available and this has revived interest in the breeding of QPM both in the national programs and in the seed industry.

Double mutant high-lysine combinations

Several high-lysine mutants reported in maize such as *opaque-2*, *floury-2* and *opaque-7* have been tried alone. Invariably they encounter problems described earlier. Double mutant combinations involving high-lysine mutants and some endosperm mutants have also been tried and produced to modify kernel properties and overcome kernel defects associated with original soft endosperm *opaque-2* materials. Some important double mutant combinations which improve the kernel modification are discussed below.

The double mutant combination involving *opaque-2* and *floury-2* had translucent appearance (NELSON 1966). At CIMMYT this double mutant combination was produced in several genetic backgrounds to improve protein content and the kernel texture. The double mutant combination failed to produce translucent kernel phenotype. Because of limited success this work was consequently stopped. In other institutions too, the success with this combination has been poor.

The other double mutant combination involving *opaque-2* and *sugary-2* also aroused considerable interest because of the translucent kernel phenotype (GARWOOD and CREECH 1972; PAEZ 1973, GLOVER and TOSELLO 1973, BAUMAN 1973, 1974). The lysine content of su_2o_2 segregates is either equal or slightly better than the soft *opaque-2* kernels.

The Purdue scientists have done a good deal of work on the breeding of su_2o_2 maize. At CIMMYT a modest effort has gone into the breeding of this type of maize. During mid 70's, *sugary-2* gene was introduced into several hard endosperm *opaque-2* maize gene pools and populations to develop *sugary-2 opaque-2* versions of these materials. On the basis of data and observations made in the field, the following general conclusions can be drawn from this work.

1. The su_2o_2 segregates vary considerably in their kernel phenotypic expression in different genetic backgrounds. Vitreousness in the kernel ranged from partial modification to complete modification.
2. In some genetic backgrounds the inner soft endosperm core resulting from the *opaque-2* gene was noticeable even in vitreous su_2o_2 segregates.
3. The su_2o_2 segregates, in general, showed intensification of endosperm colour.
4. In dent backgrounds, the su_2o_2 segregates failed to show soft cap in the dented portion of the kernel.
5. The size of the su_2o_2 segregates was generally smaller than the counterpart *opaque-2* segregates. However, considerable variation was observed between and within materials.
6. Homozygous su_2o_2 ears tended to shell more easily.
7. The su_2o_2 ears exhibited noticeable open spaces between and within the rows.
8. The protein content and quality of su_2o_2 segregates was not very different from the soft opaques (Table 1).
9. The su_2o_2 ears generally had more number of kernel rows.
10. The su_2o_2 segregates were generally lighter in kernel weight compared to counterpart soft opaques (Table 2). The kernel density of su_2o_2 segregates was, however, better.
11. In the field the su_2o_2 ears showed reduced incidence of ear rots.

Table 1
Percent protein and lysine in the whole kernel of *opaque-2* and *sugary-2*; *opaque-2* segregates recovered from different materials

Material	Protein (%)		Lysine in protein (%)	
	o_2	su_2o_2	o_2	su_2o_2
Mezcla Amarilla H.E.o_2	11.5	11.1	3.8	3.7
La Posta H.E.o_2	11.2	10.9	4.2	4.4
Pool 22 H.E.o_2	11.0	10.9	4.1	4.3
Pool 25 H.E.o_2	11.0	11.3	4.0	4.0
Pool 26 H.E.o_2	11.1	11.2	3.7	4.0

Table 2
100 Kernel weight and density comparisons of opaques and *sugary*-2 *opaque*-2 kernels in different backgrounds

Material	100 kernel weight (g)		Difference (%)	Density		Difference (%)
	o_2	su_2o_2		o_2	su_2o_2	
Mezcla Amarilla H.E.o_2	27.8	25.9	6.8	1.168	1.267	−8.48
La Posta H.E.o_2	27.3	25.7	5.9	1.214	1.262	−3.95
Pool 22 H.E.o_2	29.5	27.5	6.8	1.163	1.263	−8.60
Pool 25 H.E.o_2	28.4	25.7	9.5	1.155	1.251	−8.31
Pool 26 H.E.o_2	28.0	26.1	6.8	1.148	1.256	−9.41

Though this combination appears promising in improving kernel phenotype and its acceptability, its greatest drawback is reduced yield. At CIMMYT, a su_2o_2 composite was formed by pooling together good families exhibiting good phenotype and less difference in kernel weight from soft opaques. The recombination and selection process in this composite was continued for four cycles following formation of the composite. The data on grain yield and other agronomic characters are presented in Table 3.

Table 3
Performance of original and different cycles of selection in *sugary-2 opaque-2* composite

Cycle	Yield ton/ha	50 (%) silking	Ear height (cm)	Moisture (%)
C_0	4.59	57.4	133	37.5
C_2	5.05	56.2	124	36.0
C_4	5.13	55.7	122	35.4
L.S.D (.05)	0.55	–	3.2	–
C.V. (%)	10.9	–	7.2	–

The latest cycle of this material showed improvement over the original cycle. The latest cycle yielded 5,133 kg/ha compared to 4,590 kg/ha of the original cycle. The latest cycles were also shorter in plant and ear height, earlier in flowering and reduced moisture content. In addition to the above mentioned traits, there was also improvement in kernel phenotype, seed size and the open spaces between kernel rows became less apparent.

Even though improvements in yield have been made, the general consensus of CIMMYT scientists was that this combination will still have a yield disadvantage of 12–15 % over the normal maize genotypes. Because of this reason, the work on su_2o_2 combination has been discontinued.

Strategy involving combined use of the *opaque-2* gene and the genetic modifiers

In developing QPM materials, the CIMMYT scientists have felt strongly that the breeding approach should be such which will enable to retain the nutritional advantages of the *opaque-2* gene but which will also help circumvent problems of complex nature associated with the original soft *opaque-2* maize. Observation on partially modified kernels in *opaque-2* materials was the beginning of such a hope in the early 70's. Exploratory research work was soon initiated to gain more information on both general and basic aspects to study the feasibility of using such an approach on a large scale for the development of QPM materials.

CIMMYT Scientists were greatly stimulated by the modified *opaque-2* phenotype for more than one reason. One essential element and advantage of this approach was its ability to alter the softer kernel texture of the opaques to a more translucent phenotype with a harder texture. This will at least eliminate one major hurdle of phenotypic non-acceptability of *opaque-2* maize. Also, this will indirectly reduce vulnerability to ear rot causing pathogens and reduce losses due to insects-pests in the storage. Also during the selection for kernel phenotype possibility exists to improve kernel weight, density and yield of QPM materials.

The alteration or modification of *opaque-2* kernel phenotype results from the action of the modifying genes. Presumably there are many genes which influence kernel vitreousity in the *opaque-2* maize. Depending upon the frequency of favourable modifying genes, one would expect different degrees of kernel modification. In essence this approach then relies heavily on two genetic systems; the simple genetic system of the *opaque-2* gene to improve the levels of limiting amino acids, lysine and tryptophan and the multigene complex system of genetic modifiers superimposed on the *opaque-2* system to improve kernel characteristics and other agronomic traits. Using this strategy CIMMYT has obtained encouraging results in developing QPM materials having kernel phenotype similar to normal maize.

Modifying genes–their influence on agronomic and biochemical characteristics

As pointed out earlier one of the significant change brought about modifiers is in kernel phenotype. This phenotype is, however, altered in both regular and irregular manner giving modified kernels an appearance of banding, bridge like, scattered etc. depending upon the distribution of the vitreous endosperm. From the practical standpoint the regular type of kernel modification is of great advantage in breeding. It can be easily seen on the ear and can be gradually increased through selection until the kernel approaches normal phenotype.

The modified kernels show variation in kernel weight and density. The combined effect(s) of the *opaque-2* gene and the genetic modifiers is, however, not the same in different backgrounds. A wide range of variation in kernel weight compared to normal counterparts has been observed among and within the same material in segregating generations (Vasal et al. 1980). The variation in kernel weight of modified kernels does suggest the possibility of capitalizing this variation through recurrent selection for improving the yield.

Genetic modifiers may also influence protein quality. A decline in protein quality sometimes accompanies accumulation of modifiers in certain genotypes. Such a decrease, however, does not occur in all the families. Also the extent of decline varies considerably. The protein content, in general, is increased in translucent kernels.

Inheritance studies have shown that modifiers are controlled by a complex genetic system of polygenes. Additive gene effects seem to be more important in the expression of kernel vitreousity (Srivatnapongse et al. 1974). General combining ability effects of different modified phenotype *opaque-2* materials differ. Some materials transmit this character to other genetic backgrounds more readily than others. Such materials should serve as excellent donors of incorporating the *opaque-2* gene along with vitreous endosperm character to other agronomically promising germplasm.

The modifiers show dosage effects in triploid endosperm (Vasal 1975). In crosses involving soft opaques and modified opaques, large reciprocal differences have been observed. Most of the F_1 ears are generally soft when soft *opaque-2* parent is used as female. In contrast majority of the F_1 ears are modified or partially modified when modified opaques are pollinated by pollen of soft *opaque-2* materials.

Monitoring protein quality

Continuous monitoring of protein quality during the selection process is extremely important to accumulate favourable modifiers of kernel modification without adversely affecting the contents of lysine and tryptophan in the protein. Also, since the action of modifiers is on the endosperm phenotype, it is very essential that materials should be subjected to endosperm analyses to reduce bias from the germ and to reduce frequency of unfavourable modifiers as rapidly as possible. It may also be pointed out that in earlier stages of selection when the modification is extremely poor, one can do without analyses. However, as modification improves, the routine protein quality analyses should be initiated as rapidly as possible.

Switch-over from endosperm analyses to whole kernel analyses can be made when the kernel phenotype has become completely vitreous and that the variation within the ear has ceased to exist.

Need to work in homozygous *opaque-2* backgrounds

Genetic modifiers of the *opaque-2* locus can play a key role in overcoming problems encountered in original soft endosperm *opaque-2* materials. Since the number of modifying genes are numerous and as all of them are not necessarily favourable for improving the translucent expression without sacrificing protein quality, it is highly desirable to accumulate and increase the frequency of favourable modifiers through appropriate recurrent selection programs. To achieve this goal the best results can be obtained by working in homozygous *opaque-2* backgrounds. One can continuously select for more and more favourable modifiers for kernel phenotype and kernel weight while monitoring protein quality in each and every selection step. Incorporation of two genetic systems into any normal maize population/pool through conventional backcrossing program is somewhat laborious. Many modifications will be required for successful incorporation of the *opaque-2* gene and the modifying gene complex that goes favourably well with the *opaque-2* system. It may also be pointed out that every time one makes the ensuing backcross one starts all over again. Short cuts often necessary cannot be used without adversely affecting the accumulation of genetic modifiers. Also, before making the next backcross one is forced to advance generations at least a few times after homozygous o_2 segregates are obtained to increase the frequency of modifiers. This increases considerably the time required to obtain a true QPM version with modified endosperm of any normal material. However, through additional modifications and compromises one can make this system work. Also one can devise ways through which a fraction of the material can be handled in homozygous *opaque-2* condition for continuously increasing the frequency of modifiers and prevent dilution which normally occurs if materials with poor kernel modification are used in the backcrossing program. At least one will have to go through 3 stages; making the cross, advancing the cross to F_2 to sort out opaques and then follow some sort of recurrent selection for increasing the frequency of modifiers in homozygous *opaque-2* segregates before using in the backcrossing program.

Also, handling in homozygous *opaque-2* backgrounds has the additional advantage over the backcrossing program where one has to sort out modified *opaque-2* segregates from each segregating generation. If one is not careful or if one picks up more normal looking segregates, in segregating generations, it is quite likely that one may select so called modified opaques which are not homozygous for *opaque-2* gene. This misclassification can add to contamination and confusion. In contrast when materials are continuously handled in homozygous o_2 background, one can safely keep selecting more normal looking kernels with certainty. Also, if some contamination occurs, it can be easily rogued out through laboratory analyses.

Development of QPM donor stocks with harder texture

Before extensive efforts in the development of QPM germplasm could be undertaken using the two genetic systems, it was necessary to develop QPM donor stocks that had *opaque-2* gene and had modified phenotype kernels. Development of such donor stocks was not only slow but was also very difficult and frustrating. In addition to selection for better modifiers, the protein quality had to be maintained simultaneously. Some of the approaches that were used initially are briefly described below.

1. Accumulation of modifiers through intrapopulation schemes in some *opaque-2* materials exhibiting partially modified kernels with a reasonable high frequency. At least it took six or more generations/cycles before such materials looked fairly good with respect to kernel modification.
2. In second approach modified ears were selected and improved for kernel modification independently for a few generations. The selected ears originated from several different yellow and white *opaque-2* maize populations and were handled through selfing or sibbing. A large number of yellow and white QPM families resulted from this effort. The white and yellow QPM families were merged separately to give rise to white H.E.o_2 and yellow H.E.o_2 respectively. Following compositing, each material was further improved for modifiers for a few generations.
3. Using materials developed through the first two approaches, several promising materials from national programs and from CIMMYT's resident program were converted to QPM involving either none or a few backcrosses. Later these materials were merged in forming broadbased QPM donor stocks.

Currently, however, a wide range of QPM materials are available that meet requirements for different adaptations, maturity and kernel characteristics. The development of suitable QPM donor stocks set the stage for the development of QPM germplasm in a wide array of genetic backgrounds.

Progress in overcoming problems associated with *opaque-2* maize

It has long been realized that usefulness of QPM materials and their subsequent exploitation at the commercial production level will depend greatly upon resolving problems and making this type of maize more nearly competitive to the normal maize genotypes currently under cultivation in different agroclimatic zones. Thus in breeding of QPM germplasm at CIMMYT, major emphasis has been placed upon remedying problems through various direct and indirect criteria. The progress to date has been dramatic in agronomic performance of QPM materials without sacrificing protein quality.

1. Improving kernel phenotype

Substantial progress has occured in improving kernel phenotype of *opaque-2* maize. Many QPM materials have attained phenotypic acceptability at the ear level. Within ear variability for kernel modification has been going down consistently; there, however, still remains scope for further improvement. Data in tables 4 and 5 show improvement in kernel modification score over the cycles concomitantly with the reduction in within-ear variability for kernel modification as indicated by standard deviation and Weinberg constant values. The different

Table 4
Percentage of soft ears and rating for ear modification in different cycles of selection in some QPM Pools

Material	Cycle	Soft ears (%)	Ear modification*
White Flint QPM Pool	C_0	79.9	5.0
	C_4	28.8	3.0
	C_8	9.1	2.5
	C_{11}	4.4	1.8
Temperate × Tropical QPM (Flint)	C_0	65.7	4.2
	C_4	33.4	3.1
	C_8	10.3	2.3
	C_{12}	4.3	1.6

* Rating scale: 1-Excellent modification; 5-Completely soft.

Table 5
Mean score for kernel modification, range, standard deviation and Weinberg constant of different cycles of selection in two QPM materials

Material	Cycle	$\bar{x}$ score*	Range	S.D.	Weinberg constant
Temperate × Tropical QPM	C_0	3.7	4.8–2.9	1.07	1.60
	C_4	2.8	3.8–1.6	1.10	1.50
	C_8	2.2	3.5–1.2	0.91	1.23
CIMMYT H.E.o_2	C_0	3.4	4.7–2.7	0.95	1.40
	C_3	2.6	4.0–1.7	0.87	1.31
	C_6	2.2	2.9–1.4	0.74	1.25

* Rating scale 1–5: 1-Completely vitreous; 2–75 % vitreous; 3–50 % vitreous; 4–25 % vitreous; 5-Completely soft.

Table 6
Relative frequency of five modified kernel classes in different cycles of selection in two QPM materials

Material	Frequency of modified kernel classes* (%) Cycle	1	2	3	4	5	$\bar{x}$ score
Temperate × Tropical QPM	C_0	1.5	18.2	16.7	40.7	22.9	3.65
	C_4	6.5	44.6	17.9	23.5	7.5	2.81
	C_8	19.4	54.5	13.5	11.9	0.7	2.20
CIMMYT H.E.o_2	C_0	0.3	23.3	26.6	40.2	9.6	3.36
	C_3	4.6	53.9	21.7	19.0	0.8	2.58
	C_6	10.6	66.5	15.4	6.8	0.7	2.21

* Rating scale 1–5: 1-Completely vitreous, 2–75 % vitreous, 3–50 % vitreous, 4–25 % vitreous, 5-completely soft.

modification classes have changed dramatically over the cycles (Table 6). Modified kernels with scores of 4 and 5 have practically disappeared while the frequency of kernels with scores 1 and 2 have been continuously increasing in frequency.

2. *Narrowing down yield differences*

Yield level of QPM materials has been upgraded considerably using several approaches singly and in combination with each other. Various possibilities have been exploited including capitalizing on variation for kernel weight in segregating generations, discarding ears showing open-spaces between kernel rows, improvement through recurrent selection program and accumulation of favourable modifiers in homozygous *opaque-2* backgrounds either directly through visual selection or with partial contamination technique using normal pollen from

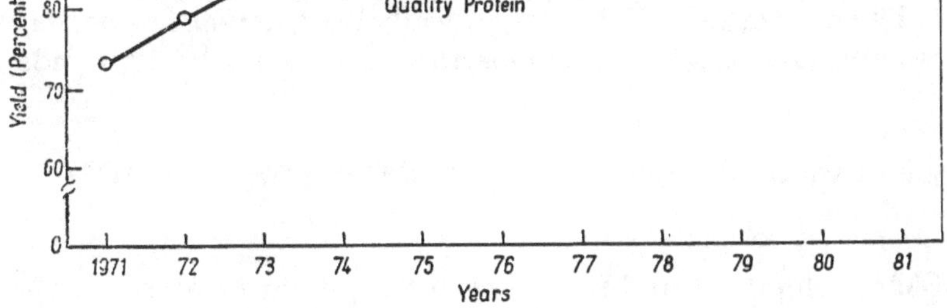

Fig. 1 Yield improvement of quality protein maize compared to the normal genotypes

materials with seed colour marker genes. Figure 1 shows continuous improvement in yield over the years compared to the normal genotypes. This yield improvement can be attributed to the improved dry matter accumulation of the grain.

3. Reduced ear rot incidence

Ear rot resistance of QPM materials has been improving gradually. This increased resistance has resulted from improvements in kernel phenotype, better drying following physiological maturity, reduced frequency of gene responsible for pericarp splitting, through population improvement involving between and within family selection both under natural and artificial conditions. Figure 2 shows ear rot incidence of QPM compared to the best check entries in different trials. It can be seen that on the average the QPM materials still show slightly higher incidence of ear rots.

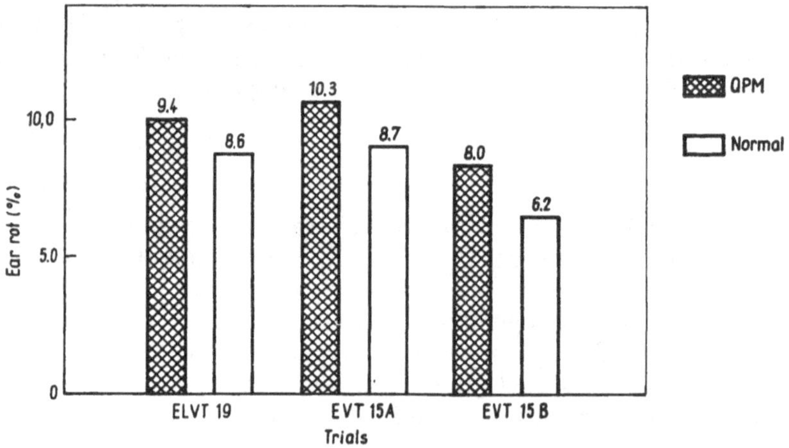

Fig. 2 Ear rot incidence of quality protein maize (QPM) compared to the best check entries (normal) in different trials

4. Losing moisture rapidly

Several known and unknown factors contribute to slower drying of *opaque-2* maize materials. Currently slower drying does not constitute any problem. Early harvesting and visual selection of faster drying ears has brought improvement in drying down ability. The data in table 7 show practically no difference in percentage moisture between normal and QPM versions in the same genetic backgrounds.

Breeding strategy and methodology used in the development of QPM germplasm

The strategy for QPM development and improvement is primarily based on the combined use of two genetic systems involving the *opaque-2* gene and the genetic modifiers. As pointed out earlier the *opaque-2* gene improves the protein quality

Table 7
Moisture percentage of normal and QPM version in the same genetic backgrounds (Poza Rica 1977 B)

Material	Moisture %	
	Normal Version	QPM Version
La Posta	29.8	28.3
Pool 25	26.0	27.0
Pool 24	28.0	26.0
Amarillo Dentado	26.0	27.0
Tuxpeño-1	27.0	29.0
Blanco Cristalino	24.0	25.0
Amarillo Cristalino	27.0	26.0
Tuxpeño Caribe	27.0	26.0
Pool 23	26.0	27.0

while the genetic modifiers help to overcome problems plaguing QPM and improve the agronomic performance of the materials.

Following development of suitable QPM donor stocks with acceptable kernel characteristics a stage was set to undertake development of QPM germplasm on a larger scale. The donor stocks were used in various ways in developing a wide array of QPM germplasm. This will be discussed in greater detail in the following paragraphs. In the breeding and improvement of QPM germplasm several criteria were used uniformly across all the materials. These are briefly mentioned below:

1. In developing QPM germplasm major thrust has been placed upon solving problems considered as major barriers in the commercial acceptance.
2. A somewhat conservative breeding strategy has been used with respect to protein quality. Emphasis has been placed on maintaining rather than on further improving the protein quality during the breeding and selection process.
3. Endosperm analyses have received preference over the whole kernel analyses in monitoring protein quality to reduce and eliminate the unfavourable modifying gene complex adversely affecting the quality protein traits.
4. Improvement in kernel modification has received high priority at all stages of the breeding effort.
5. Heavy selection pressure has been exerted against certain characters such as open spaces between kernel rows, popping tendency of the pericarp and dull modifiers during germplasm development.

In developing germplasm, the following approaches have been used:

1. Conversion of normal genotypes to QPM

A wide range of normal maize genotypes existing in CIMMYT's maize program were converted to QPM using one or more QPM donor stocks. A combination of backcross and recurrent selection was used as a breeding procedure to facilitate handling of two genetic systems (Vasal, 1979). Using improved version of recurrent parent in backcrossing, feasibility of handling materials in homozygous *opaque-2* background, flexibility in making the backcross, continuous improvement of modifiers, stability of modifiers and other agronomic characters, and finally permitting use of the products of the conversion process at various stages of the conversion program were some of the salient features of this breeding scheme. The performance of QPM versions in comparison to their normal counterparts is given in table 8. In some of the promising materials the QPM versions have registered yield ranging from 95 to 99 % of the normals. In other agronomic characters the QPM versions looked as good as the normals.

2. Development and improvement of QPM gene pools

A number of QPM gene pools with tropical and subtropical adaptation were formed through one of the following two approaches:

1. Crossing normal genotypes to several QPM donor stocks. Thereafter the F_1's were advanced to F_2 in modified ear-to-row half-sib system. Good modified QPM segregates were selected from each ear separately and recombined in half-sib crossing block. Following such recombination, the improvement of such QPM pools continued through half-sib system of breeding.
2. The second approach involved recombination of partially modified QPM families and then handling the resulting pool continuously in a half-sib breeding system.

In the formation and development of QPM gene pools diverse genetic materials and a wide array of genetic modifiers were pooled together. These QPM gene pools were handled in homozygous *opaque-2* background and received major emphasis in accumulating favourable modifiers to improve kernel phenotype and other agronomic characters. A modified ear-to-row half-sib system was followed in improving such pools. Evaluation of different cycles of selection in such pools has shown continuous improvement for several attributes (Table 9). The latest

Table 8
Comparison of normal materials and their QPM versions for yield and other agronomic traits

Material	Yield in kg/ha*		QPM as	Ear height (cm)		Days 50 % silking		Ear rots (%)	
	Normal	QPM	% of normal	Normal	QPM	Normal	QPM	Normal	QPM
Tuxpeño 1	6,187	6,152	99.4	111	110	61	60	2.2	5.4
Mix. 1 Col. Gpo. 1 ×ETO	6,117	5,681	92.9	107	105	60	58	2.9	4.5
Mezcla Amarilla	5,431	5,237	96.4	103	106	58	58	2.9	5.7
Amarillo Dentado	5,351	5,230	97.7	125	110	61	57	2.5	3.2
Tuxpeño Caribe	6,393	5,903	92.3	117	115	61	58	2.5	4.2
Ant. Rep. Dominicana	5,351	5,080	94.9	100	112	56	57	2.3	3.7
La Posta	6,470	5,903	91.2	131	122	62	59	3.1	4.0

* Pooled over years and locations.

Table 9
Comparison of original and latest cycle of selection in two QPM gene pools

Material	Cycle	Yield (kg/ha)	Days to flower	Plant height (cm)	Soft ears (%)	Ear aspect*	Endosperm** modification
Temperate × tropical	C_0	4,437	59	212	66	4.0	4.2
QPM (Flint)	C_{12}	4,920	55	192	4	2.2	1.6
Temperate white	C_0	3,445	55	204	34	3.5	3.7
QPM	C_5	3,669	55	192	10	2.3	1.9

* Rating scale 1–5 : 1-Excellent; 5-Poors.
** Rating scale 1–5 : 1-Completely modified; 5-completely soft.

cycles are shorter in plant and ear height, earlier in maturity, lesser in ear rot incidence, and better in kernel modification. Within-ear variation for kernel modification has gone down but additional cycles of selection are needed for further improvement. The yield improvement has been observed in some but not in others. The protein quality has remained practically unchanged.

Current status of QPM germplasm

For efficient handling, the merging and reorganisation of QPM germplasm has been undergoing over the past two years. From the available QPM germplasm eight tropical and six subtropical gene pools have been formed. The assigned number, name and some important phenotypic features of these pools are given in table 10. The handling of these QPM pools is done through the same procedure outlined in the previous section.

In addition to QPM pools a total of ten new advanced unit populations have also been developed. Six of the populations have tropical adaptation while the remaining four are better suited to subtropical conditions. The no. and name given

Table 10
Tropical and subtropical QPM gene pools

QPM Pool No.	Description
Pool 15 QPM	Tropical Early White Flint QPM
Pool 17 QPM	Tropical Early Yellow Flint QPM
Pool 18 QPM	Tropical Early Yellow Dent QPM
Pool 23 QPM	Tropical Late White Flint QPM
Pool 24 QPM	Tropical Late White Dent QPM
Pool 25 QPM	Tropical Late Yellow Flint QPM
Pool 26 QPM	Tropical Late Yellow Dent QPM
Pool 27 QPM	Subtropical Early White Flint QPM
Pool 29 QPM	Subtropical Early Yellow Flint QPM
Pool 31 QPM	Subtropical Intermediate White Flint QPM
Pool 32 QPM	Subtropical Intermediate White Dent QPM
Pool 33 QPM	Subtropical Intermediate Yellow Flint QPM
Pool 34 QPM	Subtropical Intermediate Yellow Dent QPM

Table 11
Formation of new QPM advanced unit populations

Population No.	Name	Adaptation	Maturity	Seed Color	Seed Texture
61	Early Yellow Flint QPM	Tropical	Early	Yellow	Flint
62	White Flint QPM	Tropical	Late	White	Flint
63	Blanco Dentado-1 QPM	Tropical	Late	White	Dent
64	Blanco Dentado-2 QPM	Tropical	Late	White	Dent
65	Yellow Flint QPM	Tropical	Late	Yellow	Flint
66	Yellow Dent QPM	Tropical	Late	Yellow	Dent
67	Templado Blanco Cristalino QPM	Subtropical	Intermediate	White	Flint
68	Templado Blanco Dentado QPM	Subtropical	Intermediate	White	Dent
69	Templado Amarillo QPM	Subtropical	Intermediate	Yellow	Flint
70	Templado Amarillo Dentado QPM	Subtropical	Intermediate	Yellow	Dent

to each population is given in table 11. These populations will be subjected to international progeny testing trials for further refinements and for improving yield dependability and adaptation. Aside from population improvement, superior fractions will be used to form site specific and across-site experimental varieties. This will aid and encourage commercial exploitation of these materials in some countries where agronomic performance and consumer acceptability of QPM as human food or as animal feed has been good. The various steps involved in the handling of QPM populations have been discussed in earlier publication (VASAL et al. 1982).

The future

Progress made at CIMMYT in solving problems encountering QPM materials and in developing a wide range of QPM germplasm suited to tropical and subtropical conditions has generated renewed enthusiasm and revived interest in breeding quality protein maize. The germplasm available at CIMMYT offers tremendous opportunities for utilization both in the national programs and in the private seed industry. The wealth of QPM germplasm currently available at CIMMYT can be used in various ways: 1) Excellent donors for the conversion program, 2) Source populations for further improvement of adaptation and other characteristics in several environments, 3) Utilization of superior fraction as experimental variety from CIMMYT's international progeny testing system, and 4) using source materials for extraction of inbred lines for developing hybrid combinations.

Some countries have already started using QPM for commercial exploitation and have reinitiated breeding efforts in this area. It is hoped that more and more countries in the near future will become interested in QPM breeding and utilization effort.

Literature

Bauman, L. F., 1973: Annual report on the inheritance and improvement of protein quality and content in maize. – Purdue University, West-Lafayette, Indiana.

–, 1974: Annual report on the inheritance and improvement of protein quality and content in maize. – Purdue University West-Lafayette, Indiana.

Garwood, D. L., and R. G. Creech, 1972: Kernel phenotypes of *Zea mays* L. genotypes possessing one to four mutated genes. – Crop Science **12**, 119–121.

Glover, D. V., and G. A. Tosello, 1973: Kernel characteristics, protein quality and biological value of the *sugary-2* mutation and its combination with *opaque-2* in *Zea mays* L. – Agronomy abstracts. p. 5.

McWhirter, K. S., 1971: A floury endosperm high-lysine locus on chromosome 10.— – Maize Genet. Coop. News Letter **45**, 184.

Mertz, E. T., L. S. Bates, and O. E. Nelson, 1964: Mutant gene that changes protein composition and increases lysine content of maize endosperm. – Science **145**, 279–280.

Nelson, O. E., 1966: *Opaque-2*, *floury-2* and high protein maize. – pp. 156–160. Proc. High Lysine Corn Conference, Corn Refiners Assn. Inc., Washington, D.C.

–, E. T. Mertz, and L. S. Bates, 1965: Second mutant gene affecting amino acid pattern of maize endosperm proteins. – Science **150**, 1469–1470.

Paez, A. V., 1973: Protein and kernel properties of modified endosperm corn involving a recessive allele at the *sugary-2* locus. – Crop Science **13**, 633–636.

Srivatnapongse, S., E. C. Johnson, S. K. Vasal, and E. Villegas, 1974: Inheritance of kernel vitreousity in *opaque*-2 maize. – SABRAO, J. **6**, 1–7.

Vasal, S. K., 1975: Use of genetic modifiers to obtain normal type kernels with the *opaque-2* gene. – In: High Quality Protein Maize, pp. 197–216, Dowden, Hutchinson and Ross, Stroudsberg, P.A.

–, E. Villegas and R. Bauer, 1979: Present status of breeding quality protein maize. – Seed Protein Improvement in Cereals and Grain Legumes, Vol. 11, pp. 127–150, (Proc. Symp. Neuherberg). IAEA, Vienna.

–, –, M. Bjarnason, M. Gelaw, and P. Goertz, 1980: Genetic modifiers and breeding strategies in developing hard endosperm *opaque-2* materials. – Improvement of quality traits of Maize for Grain and Silage Use, pp. 37–71 (Proc. Seminar, Schwäbisch Hall 1978) Martin Nijhoff, the Hague.

–, –, and C. Y. Tang, 1982: Recent advances in the development of quality protein maize germplasm at CIMMYT. Paper presented at the Research Coordination Meeting on the Use of Nuclear Techniques for Cereal Grain Protein Improvement, 6–10 December, Vienna (Unpublished).

S. K. Vasal
Centro Internacional de Mejoramiento de Maíz y Trigo (CIMMYT),
Londres 40, Apdo. Postal 6–641, Col. Juárez Deleg. Cuauhtémoc,
06600 México, D.F.

Kulturpflanze 32 · 1984 · S 187–S 188

Characterisation of winter wheat genotypes with increased crude protein content

K. Richter
(Bernburg-Hadmersleben, GDR)

As is the case with almost all species of cereals, numerous countries are making a real effort to raise the grain protein content and the protein quality in wheat without a reduction of grain yields. It is important to know for the grain breeder, which special traits are essential in efficient genotypes with increased protein content.

Already in 1969, Johnson et al. suggested trying to achieve in wheat breeding a combination of higher nitrogen uptake with greater N-translocation during grain filling. According to Sozinow et al. (1977), in the protein breeding of cereals the limitation of the grain-straw ratio should be avoided, the biomass yield increased and considered a long, active N-uptake at the roots.

In the present study, results are given from tests over three years in Hadmersleben with 14 genotypes including new wheat varieties and strains with increased grain protein content, taken from the winter wheat breeding material of the GDR. Some of the genotypes with higher protein content were obtained from combi-

Table 1
Selected traits of the tested genotypes – average values of the trials in Hadmersleben 1979–1981, extract

Varieties/ strains	Yield dt/ha	Crude protein content % (N×6.25)	Crude protein yield dt/ha	N-uptake between anthesis and maturity g/m²	N-harvest index	Biological yield at maturity g/m²	Grain-straw ratio 1 :	Harvest-index
'Alcedo' (comparison)	73.3	16.4	10.4	4.7	74.2	1,688	1.27	44.4
'Fakon' (comparison)	77.2	16.6	11.0	6.8	76.0	1,518	0.93	52.0
'Hadm. 386'	63.8	19.3	10.6	4.5	77.9	1,610	1.41	41.6
'Hadm. 374'	65.5	18.7	10.6	1.2	81.1	1,204	1.28	44.0
'Hadm. 123'	63.8	18.8	10.3	3.1	75.0	1,466	1.51	40.7
'Compal'	73.0	17.1	10.7	5.6	76.5	1,648	1.27	44.4
'Fakta'	75.3	17.5	11.3	5.3	74.0	2,057	1.42	41.4
$LSD_{5\%}$ Dunett	5.7	0.9	0.8	2.7	Diff. not signif.	227	0.14	2.8

nations with the American variety 'Atlas 66'. According to the method of Austin et al. (1977) yield traits and features of the N-metabolism were determined and conclusions discussed for the wheat breeding. In the yield tests the genotypes with the highest protein content exceeded the standard 'Alcedo' by almost 3 % protein and reached the standard in the protein yield. The varieties 'Compal' and 'Fakta' displayed 0,7–1,1 % more protein and 4–9 % higher protein yield than 'Alcedo'. According to numerous publications the protein content was correlated negatively with the grain yield, with the number of grains/ear and with the harvest index. Between the N-content of several plant fractions and the protein content no or only weak correlations existed. The grain protein yield was correlated very closely, positively with the N-uptake between anthesis and maturity, with the number of ears/m^2, grain yield and total biological yield, but not with the N-harvest index. The very efficient variety 'Fakta' differed from the most of the other tested genotypes through higher total nitrogen yield and total dry matter yield during anthesis and maturity and higher number of ears/m^2. As compared with the standard, 'Fakta' reached a further grain-straw ratio and a smaller harvest index (see table).

Literature

Austin, R. B., M. A. Ford, J. A. Edrich, R. D. Blackwell, 1977: The nitrogen economy of winter wheat. – J. Agricult. Sci. **88**, 159–167.

Johnson, V. A., P. J. Mattern, D. A. Whited, J. W. Schmidt, 1969: Breeding for high protein content and quality in wheat. – Proc. of a Panel Meeting in Röstanga, Sweden 1968, FAO, IAEA, 29–40.

Sozinow, A. A., F. A. Poperelja, A. N. Chochlov, 1977: Problemy uveličenija belkovosti zerna psenicy. – In: Problema povysenija kačestva zerna (Eds. V. N. Remeslo, A. A. Sozinow) pp. 18–30, Moscow.

K. Richter
Institut für Getreideforschung
Bernburg-Hadmersleben
der Akademie der Landwirtschaftswissenschaften der DDR
DDR - 3234 Hadmersleben

Kulturpflanze 32 · 1984 · S 189–S 191

Genotype-specific protein and isozyme patterns after direct isoelectric focusing of cotyledonous tissue of legumes

H. P. MÜLLER
(Bonn, FRG)

The analysis of the genetic structure of populations requires the discrimination between individual genotypes as near as possible to the DNA-level. Since most plant species possess large stores of variation in genes specifying isozymes, the isozyme technique has proven to be suitable for measuring the genetic variability. It satisfies the criteria for genetic markers better than any other method, because it allows the study of discrete Mendelian genes. Thus, a wider application of isozyme techniques in analysing natural and agricultural plants growing under diverse environmental conditions will provide informations on novel genetic material as well as it will be useful for basic genetic studies. For studying larger samples of seed material, a method should be available which allows rapid collection of data, and which needs only a short time and small amounts of material. In the following a method for identification and characterization of even minor differences in isoelectric points of soluble proteins of seeds, fitting with the criteria mentioned is described. It is based on the physiologically conditioned leakage of proteins within the first minutes after seed imbibition.
For analysing these soluble proteins according to their isoelectric points, small tissue sections of dry cotyledons are directly extracted on the gel during isoelectric focusing. The insertion of the tissue sections into the electric field is accomplished by placing a quarter of one cotyledon on the agarose gel (0.8 % agarose; 10 % sorbite; ampholyte mixture 2 %, pH 2–11;), mostly at a distance of 1 cm from the cathode. Before that, excess water was removed with filter paper. The gel is then placed on the precooled plate (10 °C) of a flat bed apparatus. Isoelectric focusing is performed for 20 min. with 5 mA. At the end of this time the tissue sections are removed. The electrophoretic separation runs with 5 W for 70–90 min., followed by fixation (TCA, sulfosalicylic acid, methanol, water) for 15 min. The gel is then placed for 10 min. in ethanol, afterwards dried with filter paper. The staining with Coomassie Blue takes 15 min. All these procedures are carried out at 40 °C. Destaining: 1.5 hrs. In the same manner isozymes can be detected using the usual reaction mixtures. The usefulness of the method described as a screening method for evaluating the genetic diversity within a widely distributed plant genus is shown in figure 1.

Investigations were performed on accessions of *Phaseolus vulgaris* deriving from South America, Egypt, and Germany. They included selected lines adapted to marginal environments, selected lines for disease resistence, wild types and primitive forms, mutants and high yielding varieties. Schematically drawn are the

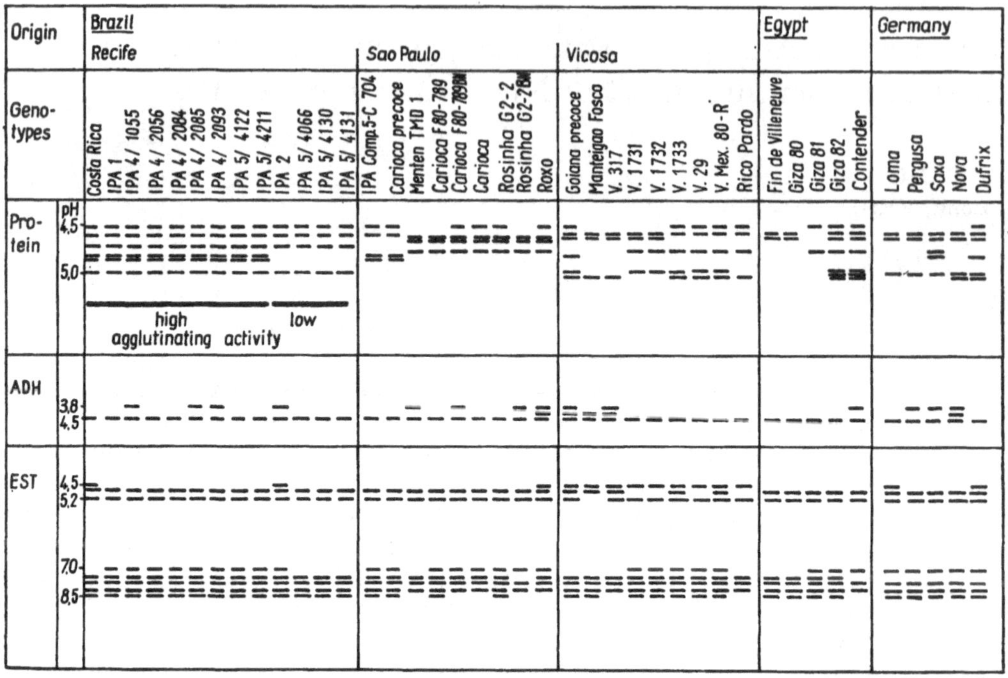

Fig. 1 *Phaseolus vulgaris*–genetic diversity

protein patterns and the isozyme patterns for ADH, the polymorphism of which sometimes can show population differentiation and adaptive differences in tolerance to waterlogging, and the non-specific esterase which was chosen because it is known that differences in the patterns can be caused by post-translational modifications either of genetical or environmental factors. In this drawing the

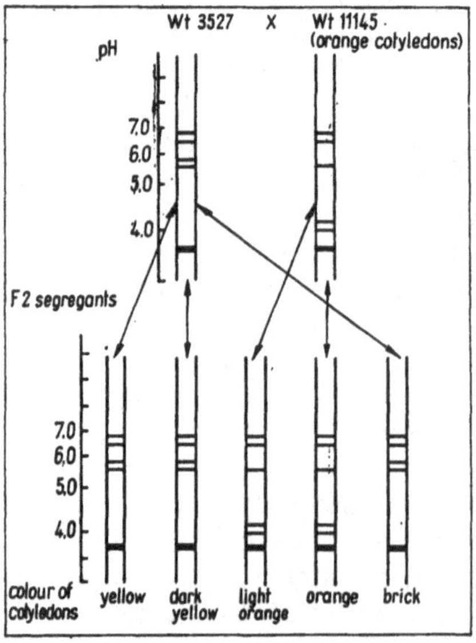

Fig. 2 Direct isoelectric focusing of tissue sections of pea cotyledons *(Pisum sativum)*; Esterase patterns (pH gradient 2–11)

different staining intensities are not considered in terms of quantitative aspects of seed protein genetics. It is only referred to numerical bandings. In combination with other physiological and biochemical traits, as indicated by the agglutination activities against human blood, the banding patterns proved to be suitable for screening.

Another example is the evaluation of genetic relationships, as shown in the inheritance of non-specific esterases in seeds of a F2-generation of *Pisum sativum* from a cross between parents having normally yellow and orange cotyledons, showing five differently coloured segregants (The seed material was kindly provided by Dr. SWIECICKI, Wiatrowo, Poland).

As indicated by the arrows, it was possible to follow the inheritance of the isozymes. Clearly, these are the first results which need more detailed confirmation.

The existing microheterogeneity between the different proteins may be helpful in selecting genotypes suitable for further investigations concerning the genetic basis of the regulation of the seed protein systems.

H. P. MÜLLER
Institute of Genetics
University of Bonn
Kirschallee 1
D - 5300 Bonn 1
FRG

Kulturpflanze 32 · 1984 · S 193–S 203

Some problems and implications in improving cereal grain protein by plant breeding

F. SCHOLZ
(Gatersleben, GDR)

Summary

Long and world-wide endeavours have proved much more difficulties than expected in breeding for higher protein yields in cereals, i.e. in converting increased protein contents of the grain into increased protein yields per unit area. The probable reasons are discussed. The real possibilities are fundamentally limited by the fact that more grain protein per unit area can only be produced if the adequate amount of additional nitrogen is available. Rough numerical calculations, based on N requirements for the protein yields, may illustrate this.

In breeding for improved protein quality, there are essential difficulties as to the impaired carbohydrate synthesis leading to reduced grain yields of the high-lysine mutants. Eventual changes in N requirement caused by changed amino acid composition of the protein are here, obviously, of at least minor importance.

Introduction

The last decades have seen most considerable yield increases in cereals, not only as to grain yields but also protein yields. This became possible by means of increasing fertilization levels, especially nitrogen, as well as by other agronomic factors. Also plant breeding, of course, has substantial share in this development, by providing varieties with higher yield potential and harvest index, improved lodging and disease resistance, etc.

As regards protein, the cereal plant is expected to produce more, in place of the same amount of carbohydrates, and that, if ever possible, without additional nitrogen. This did not look to be too difficult, as high protein contents of the grain could comparatively easily be found in collections and later on also by mutation induction. Since we know high-lysine mutants in maize, barley, and sorghum we in addition wish to improve protein quality, i.e. amino acid composition.

Apart from sporadic attempts as to protein content in former decades, intense efforts have been made for the last about fifteen years in many countries and also in international programmes to improve both protein quantity and protein quality. Decisive success in terms of commercial varieties has not yet been achieved, however. Especially as regards protein yield, genetic improvements appear to be possible on limited scale only. Looking for the reasons and in estimating the possibilities and limitations, nitrogen balance and requirement of the crop should be included into consideration.

Improving protein yields – the situation

General experience has shown that notable increases in protein content are compensated, as evidenced in reliable field trials, by at least nearly corresponding grain yield reduction. Protein yields, therefore, remain more or less unchanged, in some cases gains in protein yield in the order of about 5 per cent or so are possible. Such gains, however, are also possible and can often be observed in new varieties with increased grain yields and unchanged or even lowered protein contents. In fact, the improvements of grain yield by cereal breeding in the past decades were indeed connected with gradual improvements of protein yield (e.g. Austin et al. 1980, Focke et al. 1980). However, by means of using forms with high protein contents, i.e. by planned breeding for higher protein yield, it was not possible so far to break this trend of only gradual improvements of protein yield.

The general experience that increased protein content cannot simply be combined with high grain yield per unit area has proved true with a plenty of experiments in all cereal crops, which can not be reviewed here. We also have made this experience in early (Scholz 1960) long-term experiments with a large number of induced high-protein barley mutants which are highly isogenic to their original varieties and therefore very suitable for drawing critical conclusions.

In the course of time we were able to isolate and to test in field trials 111 different mutant lines of spring and winter barley, containing up to one-fourth more protein in the grain than the original varieties, e.g. 15 % as compared to 12 %. In no case a high-protein mutant line reached the grain yields of its original variety; there were deficits up to 10 and 20 per cent. The protein yields per unit area of some of the best mutant lines were in the order of about 5 per cent above those of their original varieties, but at the cost of reduced grain yields. This tendency proved unchanged in field trials with different, moderate or high, nitrogen levels. Later on we attempted to "break" the negative correlation between protein content and grain yield by crossing the mutants with several high-yielding varieties. After re-selection for high protein content, the yields were again inadequate, just as with the primary mutants: grain yields of 85–90 per cent and maximum protein yields of about 105 per cent as compared to the high-yielding crossing parents (Scholz 1976a, 1976b, 1984a).

The long-known negative correlation between protein content and grain yield can apparently not be due to linkage of certain genes or other genetic causes which can be "removed" by breeding. If so, i.e., if gains in protein content cannot be converted into corresponding gains in protein yield, one should look for other eventual reasons.

Implications and possibilities in improving protein yields

About 10 to 15 years ago the idea was encouraged that nitrate reductase activity (NRA) could be a limiting factor, since correlations were found between NRA and grain yield as well as plant nitrogen and grain protein in maize, wheat, and barley. Such correlations are sometimes indeed obvious, but often poor or even non-existent, so that selection for enhanced NRA soon did not appear very promising (e.g. HAGEMAN et al. 1976, HEWITT 1979). Since it is known that top yielding varieties may have low NRA, and nitrate reductase deficient mutants may have normal grain protein contents (e.g. GALLAGHER et al. 1980, OH et al. 1980), this conception had more or less to be abandoned.

One could suggest that bioenergetic implications would cause an unfavourable balance if protein content is increased. According to PENNING DE VRIES et al. (1974), 1 g of glucose can be used to produce 0.83 g carbohydrate or 0.40 g protein. On this basis, BHATIA and RABSON (1976) gave clear bioenergetic considerations for quantitative protein improvement in cereal breeding (see also RABSON et al. 1978). They have verified that each 1 per cent increase in protein content, in place of the same amount of carbohydrates, must cause an energetic loss of about 1 per cent and therefore a corresponding loss in grain yield of likewise nearly 1 per cent, because energy requirements for protein biosynthesis is about twice as high as for carbohydrates.

This inevitable cost for protein increase, however, is of an order of magnitude which is not too high and appears acceptable. In any case, this additional energy requirement cannot be the decisive reason for the situation: Namely, if there is a gain in protein content, e.g., of 3 per cent points, i.e. an increase from 12 to 15 %, one does not find a decrease of grain yield by about 3 per cent but by about 15 or 20 per cent.

In my opinion, we come closer to realistic implications if we consider the nitrogen balance and requirements of the crop. More grain protein per hectare can only be produced if more nitrogen for grain protein is available. Providing unchanged manuring conditions, the protein yield can be increased on two ways: (1) by uptake of more nitrogen from one hectare of soil and/or (2) by improved nitrogen translocation from the vegetative organs to the grain.

If nitrogen uptake from the soil is to be improved, this must more or less be compensated by higher N-fertilizing for the subsequent crop, and this would properly not meet our object. Furthermore, genetic variation of N uptake appears to be very small, at least in adapted varieties of barley and wheat (JOHNSON et al. 1968, SANDFAER and HAAHR 1975, DUBOIS and FOSSATI 1981). The reason may be that genotypes with high capacity of N uptake had always a major advantage in natural evolution as well as in selection by man (DOLL 1977). Thus, improving nitrogen uptake appears to offer extremely limited, hardly real reserves.

Better nitrogen mobilization in the vegetative organs and translocation to the grain may offer more possibilities. The high grain yield potential of the current semidwarf varieties is due to their high harvest index (grain to straw ratio), and such varieties already have a high nitrogen harvest index, i.e. a high proportion of plant protein in the grain at harvest (cf. BHATIA and RABSON 1976): About 70 % or even more of the nitrogen in the plant (above ground) is used for grain protein formation. Further improvements may be difficult. Nevertheless, there seems to exist considerable genetic variation of this trait particularly in wheat (JOHNSON et al. 1968, 1969, AUSTIN et al. 1977, DESAI and BHATIA 1978, HALLORAN 1981). Furthermore, the nitrogen content in ripe straw is still rather high, about 0.5 %, and raises much with increasing levels of N fertilizing (ANDERSEN and KØIE 1975). In other words, about 30 % or even more of the nitrogen is "wasted" in the straw, so that may still be scope for further improvements (AUSTIN 1975). More emphasis in future breeding work should be given to genetic variation of the nitrogen content of the straw.

In a broader sense, also improved harvest index itself produces an effect comparable to improved N translocation. A survey of this problem in wheat was given by KRAMER (1979). The considerable grain yield increases in the past seem primarily to be due to improved harvest indices. As the biomass production appears to be rather constant there will be, as a logical consequence, a strong positive correlation between harvest index and grain yield. This leads to a likewise strong negative correlation between grain yield and protein content, since the available nitrogen has to be redistributed to a larger biomass of grain. On the other hand, by means of a higher harvest index the protein yield will at least a little increase though protein content is decreasing, since the smaller straw proportion, in general, implies that more nitrogen is available for the grain. In agreement with this is the fact that high protein yields per unit area are usually obtained with varieties having high grain yield potentials (and high harvest indices) but sometimes rather low grain protein contents. Thus, further improvement of harvest index, as far as still possible (AUSTIN et al. 1980), may also offer certain reserves for improving protein yield.

Most of the discussed implications suggest that the genetic variation for grain protein content is indeed, according to KRAMER (1979), largely an "improper" one, because the observed genetic variation primarily is not the result of "protein genes" but is merely a derived effect of all the genes involved in the determination of plant type and in the distribution of dry matter. Also studies with ditelosomics and monosomics in wheat by METTIN et al. (1980) and BLÜTHNER and METTIN (1982) gave strong indication that the variation of grain protein content is rather a secondary effect of genes for yield and yield components as well as for several other characters.

Generally, if comparing different varieties or strains, one should be conscious that remarkable differences in both protein content and protein yield may be secondary effects of quite other characters, as maturing time, lodging resistance, disease resistance. Recently, e.g., surprisingly large effects of energy-requiring defense reactions against mildew on both yield and protein content were demonstrated in barley by SMEDEGAARD-PETERSEN and STØLEN (1981). Consequently, higher protein yields of new varieties may merely be due to improvements in disease resistance.

Nitrogen requirement for protein yield – a calculation

Estimating the possibilities for improving protein yields, one should be aware of the limitation of these potentials. If a substantial increase of protein yield is to be achieved, then one should not expect the plant to be able to do so without additional nitrogen. One cannot demand a much higher output without increasing input. If, for instance, the protein content is to be increased from 12 to 15 % and, at a grain yield level of 5,000 kg/ha, correspondingly protein yield from 600 to 750 kg/ha, then one has to pay regard to the fact that the nitrogen requirement for grain protein raises from 96 to 120 kg/ha (protein contains about 16 % N). This additional requirement is certainly not simple to be met, without additional amounts of N fertilizer.

The implications may be demonstrated in a calculation example on the nitrogen balance in cereals if breeding for higher protein yield (Table 1). This is, naturally, a simplifying calculation which, however, characterizes well the real situation (Scholz 1982, 1984b).

Table 1
Calculation on nitrogen balance and yields in cereals

Initial situation:	5,000 kg/ha grain yield and 12 % protein (for further explanation see text)					
	(1)	(2)	(3)	(4)	(5)	(6)
Protein content	12 %	15 %	15 %	15 %	15 %	12 %
Grain yield	5,000	4,000	4,567	4,375	5,000	6,250
Protein yield	600	600	685	656	750	750
N requirement for grain protein	96	96	110	105	120	120
Necessary N uptake	137	137	137	150	150	150
(if 70 % or 80 % of N into grain)	(70)	(70)	(80)	(70)	(80)	(80)

All figures in kg/ha (except last line in per cent)

The calculation proceeds from an initial, "normal" situation (first column): 5,000 kg/ha grain yield and 12 % protein in the grain, under given manuring and other environmental conditions. It is also assumed that 70 % of the uptaken N (in plant parts above ground) can be utilized for grain protein. Then, grain yield must necessarily decrease from 5,000 to 4,000 kg/ha if protein content increases from 12 to 15 % (second column). This situation can frequently be observed in performance trials.

Both higher protein and grain yields are only possible if a higher percentage of N uptake, e.g. 80 % instead of 70 %, can be utilized for grain protein, i.e., if translocation is improved (third column), or on the other hand, if N uptake itself is improved (fourth column). If both sources for additional nitrogen are effective at the same time in appropriate amount, then even a grain yield of 5,000 kg/ha with 15 % protein may be achieved, at least theoretically (fifth column).

The last column of the table shows the very high grain yield of 6,250 kg/ha. This would theoretically be possible if the higher N uptake as well as the improved

N translocation, as in the fifth column, could be utilized by a variety with 12 % instead of 15 % protein content. Then we would have adequate gains in both protein yield and total grain yield. Hence, several specialists suggest that it would be more useful to breed primarily for grain yield, trying to keep the protein content unchanged, so that the protein yield would be improved together with grain yield.

Some aspects as to improving protein quality

Only a brief outline of the current situation will be attempted, aiming also here at the more general aspect whether nitrogen balance and requirement as well as bioenergetic implications eventually may cause limitations.

About 20 years ago the first high-lysine maize mutants, *opaque* and *floury*, had been found by Mertz et al. (1964) and Nelson et al. (1965), some more mutants were added afterwards. Nearly 15 years ago the well-known 'Hiproly' barley was detected (Munck et al. 1970). A few years later followed the induced mutants from Risø (Doll et al. 1974) and several others, so that we at present know 21 different high-lysine mutants only in barley (Doll 1983), with at least 6 different loci so far identified. Also in sorghum such a mutant has been found (Singh and Axtell 1973).

All these mutants in maize, barley, and also sorghum show, more or less, the symptoms of the so-called "high-lysine syndrome" (Mertz 1976, Munck 1980): reduced synthesis of prolamins, increased synthesis of albumins and globulins with higher lysine concentration, more soluble sugars and free amino acids, etc. The most serious symptoms are impaired carbohydrate synthesis and starch accumulation, leading to smaller and shrunken grains in barley or to opaque endosperm in maize, and resulting in yield decreases in the order of 5, 10, or 20 per cent. The latter variation in yield impairment depends mainly on the extent of lysine increase.

This syndrome is always caused by one single gene. Since so many mutants with very similar effects have been identified, the probability is low to find new high-lysine mutants without deleterious side-effects and several attempts were indeed unsuccessful. New possibilities may some day be opened by means of genetic manipulation (Miflin et al. 1981). However, it is too early to comment on this as there are several constraints yet to be solved before such changes will be available for practical breeding.

Some of the existing high-lysine mutants, especially *opaque-2* in maize and 'Hiproly' in barley, have intensely been used in breeding at many places. Progress was much slower than expected initially, because of the mentioned side-effects which only became evident in the course of breeding work. In spite of these serious problems considerable progress has been made in both maize and barley, as many reports have shown. After many crosses, the yield difference between the high-lysine and normal breeding lines could little by little be reduced, by means of compensating part of the yield-decreasing side-effects by other yield components or, generally speaking, by recombination with genetic modifiers. The ultimate aim in terms of really normal-yielding high-lysine varieties has not yet been reached, however. The prospects of further advance will depend on the solution of various problems.

Calculations by BHATIA and RABSON (1976) show that the energy and substrate requirements for the synthesis of high-lysine proteins are slightly higher than those for normal proteins in both maize and barley. In three comparisons these increases are 0.1, 0.5, and 2.5 per cent. Hence, there are real increases of energy requirement but they tend to be very low and are thus of an acceptable order of magnitude.

Changed nitrogen requirement for improved protein?

The average nitrogen content of cereal grain protein varies in general around 16 %. The exact value depends on the specific amino acid composition. The proteins of high-lysine maize or barley mutants show major quantitative changes not only as to lysine, most other amino acids are also affected, some of them rather drastically. Since the N contents of the amino acids differ from about 10 % (e.g. methionine) to about 32 % (arginine), one could imagine eventual changes in overall nitrogen content. This would mean an increased or decreased nitrogen requirement for the so-called high-lysine as compared to normal protein.

Table 2

Comparable nitrogen contents of the proteins of normal and high-lysine maize and barley[1] (see also Table 3)

	Nitrogen content Per cent of protein		Relatively	
(1) Maize W64A and W64A-o_2	15.08	16.08	100	106.6
(2) Maize normal and o_2	15.57	16.39	100	105.3
(3) Barley 'Eva' and Risø 1508	15.95	16.33	100	102.4
(4) Barley C.I. 4362 (normal) and Hiproly	15.20	15.13	100	99.5

1 Calculated after data on amino acid composition from:
(1) BHATIA and RABSON (1976) after O. E. NELSON
(2) NELSON (1969)
(3) BHATIA and RABSON (1976), compiled data
(4) MUNCK (1972)

Calculations were made in two comparisons for normal and *opaque-2* maize as well as in two comparisons for normal and high-lysine barley (Table 2). The way of calculation is demonstrated in Table 3, for case No. 1 of Table 2. The N contents of the 18 amino acids, according to the molecular and atomic weights, are simply summed up, without considering peptide bonds etc. Not the exact N content of a protein can be ascertained in this way, of course, but suitable comparisons appear possible.

The results for maize, on the basis of two sources of amino acid composition data, show that about 5 or 6 per cent more nitrogen is required for *opaque-2*

Table 3
Calculation example on nitrogen contents of the proteins of normal and *opaque-2* maize (cf. case No. 1 in Table 2)

Amino acid	N content of the amino acid	Grams amino acid per 100 g protein	Grams N per 100 g protein	Grams amino acid per 100 g protein	Grams N per 100 g protein
		W64A		W64A-o_2	
Lys	0.1916	1.38	0.2644	3.53	0.6763
Try	.1372	0.26	.0357	0.67	.0919
His	.2708	2.49	.6743	3.06	.8286
Arg	.3216	2.92	.9391	4.97	1.5984
Asp	.2120	6.02	1.2762	10.31	2.1857
Glu	.1917	22.36	4.2864	18.91	3.6250
Thr	.1176	3.01	.3540	3.53	.4151
Ser	.1333	4.82	.6425	4.58	.6105
Pro	.1217	7.39	.8994	8.21	.9992
Gly	.1866	2.58	.4814	4.49	.8378
Ala	.1572	8.68	1.3645	6.88	1.0815
Val	.1196	4.64	.5549	5.06	.6052
Cys	.1166	1.55	.1807	0.86	.1003
Met	.0939	1.72	.1615	1.72	.1615
Ile	.1068	3.87	.4133	3.72	.3973
Leu	.1068	16.17	1.7270	11.08	1.1833
Tyr	.0773	4.56	.3525	3.72	.2876
Phe	.0848	5.59	.4740	4.68	.3969
	Total:		15.0818[1]		16.0821[1]
	In relative values:		**100**		**106.6**

1 100 g protein contain 15.08 and 16.08 g N, resp.; see also text

protein as compared to normal protein. This may affect a little the protein yield, of course, but in principle not the grain yield potential. The corresponding calculations for barley, for the proteins of Risø 1508 or Hiproly, resp., give even better values which are very close to those of normal protein.

Conclusions

Additional nitrogen requirement causes more or less fundamental limitation in breeding for higher protein yield. Nevertheless, there may still be some real possibilities, but one should be aware that the internal reserves of the plant are limited and utilization of these reserves is difficult because of various experimental constraints. Substantial increases in protein yield will require the use of additional amounts of nitrogen fertilizer.

On the other hand, in breeding for improved protein quality, the existing difficulties originate in the impaired carbohydrate synthesis leading to reduced grain yields of the high-lysine mutants. However, no major problems should arise here from bioenergetic reasons or from eventual changes of nitrogen requirement caused by changed amino acid composition of the storage proteins.

Zusammenfassung

Einige Probleme und Folgerungen bei der züchterischen Verbesserung des Kornproteins beim Getreide

Bei der Züchtung auf höhere Proteinerträge beim Getreide, d. h. bei der Umsetzung erhöhter Proteingehalte des Korns in erhöhte Proteinerträge je Flächeneinheit, haben sich in langen und weltweiten Bestrebungen wesentlich mehr Schwierigkeiten ergeben als erwartet. Die vermutlichen Gründe dafür werden diskutiert. Die Tatsache, daß mehr Kornprotein je Flächeneinheit nur produziert werden kann, wenn entsprechend zusätzlicher Stickstoff verfügbar ist, schränkt die realen Möglichkeiten grundsätzlich ein. Rechnerische Grobkalkulationen, auf der Grundlage des N-Bedarfs für die Proteinerträge, unterstreichen dies.

Bei der Züchtung auf verbesserte Proteinqualität bestehen wesentliche Schwierigkeiten wegen der gestörten Kohlenhydratsynthese und der dadurch verminderten Kornerträge der lysinreichen Mutanten. Mögliche Änderungen im N-Bedarf auf Grund der geänderten Aminosäurezusammensetzung des Proteins sind hier jedoch offensichtlich von zumindest untergeordneter Bedeutung.

Краткое содержание

Некоторые проблемы и следствия селекционного улучшения протеина у зерновых злаков

В селекции зерновых злаков на более высокое содержание протеина, т. е. при обращении содержания протеина зерна в повышенные урожаи белка с единицы поверхности, проводившейся в течение ряда лет во многих странах, было обнаружено больше затруднений, чем сначала предполагалось. Обсуждаются возможные причины этих затруднений. Тот факт, что большее количество зернового протеина с единицы поверхности может быть получено только тогда, когда существует соответственное количество добавочного азота — принципиально ограничивает реальные возможности. Приблизительные подсчёты на основе потребности в азоте для урожаев протеина подтверждают вышесказанное.

В селекции на улучшение качества протеина обнаруживаются существенные препятствия из за нарушенного синтеза углеводов высоколизиновых мутантов и, тем самым, заниженные урожаи зерна. Возможные изменения в потребности азота из за аминокислотного состава протеина, по всей видимости, имеют здесь только второстепенное значение.

Literature

Andersen, A. J., and B. Køie, 1975: N fertilization and yield response of high lysine and normal barley. – Agron. J. **67**, 695–698.

Austin, R. B., 1975: Economy in the use of manufactured fertilisers. – Agric. Engineer **30**, 62–65.

–, J. Bingham, R. D. Blackwell, L. T. Evans, M. A. Ford, C. L. Morgan, and M. Taylor, 1980: Genetic improvements in winter wheat yields since 1900 and associated physiological changes. – J. Agric. Sci. **94**, 675–689.

–, M. A. Ford, J. A. Edrich, and R. D. Blackwell, 1977: The nitrogen economy of winter wheat. – J. Agric. Sci. **88**, 159–167.

Bhatia, C. R., and A. Rabson, 1976: Bioenergetic considerations in cereal breeding for protein improvement. – Science **194**, 1418–1421.

Blüthner, W. D., and D. Mettin, 1982: Monosomic analysis of grain protein content and grain yield of the wheat cultivars 'Flandres Desprez' and 'Klein Aniversario'. – Biol. Zbl. **101**, 633–640.

Desai, R. M., and C. R. Bhatia, 1978: Nitrogen uptake and nitrogen harvest index in durum wheat cultivars varying in their grain protein concentration. – Euphytica **27**, 561–566.

Doll, H., 1977: Storage proteins in cereals. – In: Genetic Diversity in Plants (Eds. A. Muhammed, R. Aksel, and R. C. von Borstel), Plenum Press, New York, 337–347.

–, 1983: Barley seed proteins and possibilities for their improvement. – In: Seed Proteins: Biochemistry, Genetics, Nutritive Value (Eds. W. Gottschalk, and H. P. Müller), Martinus Nijhoff, The Hague, 207–223.

–, B. Køie, and B. O. Eggum, 1974: Induced high lysine mutants in barley. – Radiat. Bot. **14**, 73–80.

Dubois, J.-B., and A. Fossati, 1981: Influence of nitrogen uptake and nitrogen partitioning efficiency on grain yield and grain protein concentration of twelve winter wheat genotypes (*Triticum aestivum* L.). – Z. Pflanzenzüchtg. **86**, 41–49.

Focke, R., W. Porsche und K. Richter, 1980: Stand und Möglichkeiten der züchterischen Erhöhung des Rohproteingehaltes bei Weizen. – Tag.-Ber. Akad. Landwirtsch.-Wiss. DDR, Nr. 190, 117–126.

Gallagher, L. W., K. M. Soliman, C. O. Qualset, R. C. Huffaker, and D. W. Rains, 1980: Major gene control of nitrate reductase activity in common wheat. – Crop Sci. **20**, 717–721.

Hageman, R. H., R. J. Lambert, D. Loussaert, M. Dalling, and L. A. Klepper, 1976: Nitrate and nitrate reductase as factors limiting protein synthesis. – In: Genetic Improvement of Seed Proteins, Proc. Workshop Washington 1974, National Academy of Sciences, Washington, D.C., 103–131.

Halloran, G. M., 1981: Cultivar differences in nitrogen translocation in wheat. – Austral. J. Agric. Res. **32**, 535–544.

Hewitt, E. J., 1979: Primary nitrogen assimilation from nitrate with special reference to cereals. – In: Crop Physiology and Cereal Breeding, Proc. Eucarpia Workshop Wageningen 1978 (Eds. J. H. J. Spiertz, and T. Kramer), Pudoc, Wageningen, 139–155.

Johnson, V. A., P. J. Mattern, D. A. Whited, and J. W. Schmidt, 1969: Breeding for high protein content and quality in wheat. – In: New Approaches to Breeding for Improved Plant Protein, Proc. Panel Meeting Röstånga 1968, IAEA, Vienna, 29–40.

–, J. W. Schmidt, and P. J. Mattern, 1968: Cereal breeding for better protein impact. – Econ. Bot. **22**, 16–25.

Kramer, T., 1979: Environmental and genetic variation for protein content in winter wheat (*Triticum aestivum* L.). – Euphytica **28**, 209–218.

Mertz, E. T., 1976: Case histories of existing models. – In: Genetic Improvement of Seed Proteins, Proc. Workshop Washington 1974, National Academy of Sciences, Washington, D.C., 57–70.

–, L. S. Bates and O. E. Nelson, 1964: Mutant gene that changes protein composition and increases lysine content of maize endosperm. – Science **145**, 279–280.

Mettin, D., W.-D. Blüthner, G. Sternkopf, U. Buchholz und W. Drauschke, 1980:

Untersuchungen zur Lokalisierung von genetischen Faktoren für N-Aufnahme und N-Umlagerung beim Saatweizen unter Verwendung ditelosomer Linien. – Arch. Züchtungsforschg. **10**, 271–277.
MIFLIN, B. J., S. W. J. BRIGHT, and E. THOMAS, 1981: Towards the genetic manipulation of barley. – In: Barley Genetics IV, Proc. Fourth Intern. Barley Genet. Symp. Edinburgh 1981, Edinburgh University Press, Edinburgh, 919–926.
MUNCK, L., 1972: Improvement of nutritional value in cereals. – Hereditas **72**, 1–128.
–, 1980: The genetical basis for protein improvement in cereals. – In: Well-Being of Mankind and Genetics, Proc. XIV Intern. Congr. Genet. Moscow 1978, Mir Publishers, Moscow, Vol. I, Book Two, 249–260.
–, K. E. KARLSSON, A. HAGBERG, and B. O. EGGUM, 1970: Gene for improved nutritional value in barley seed protein. – Science **168**, 985–987.
NELSON, O. E., 1969: The modification by mutation of protein quality in maize. – In: New Approaches to Breeding for Improved Plant Protein, Proc. Panel Meeting Röstånga 1968, IAEA, Vienna, 41–54.
–, E. T. MERTZ, and L. S. BATES, 1965: Second mutant gene affecting the amino acid pattern of maize endosperm proteins. – Science **150**, 1469–1470.
OH, J. Y., R. L. WARNER, and A. KLEINHOFS, 1980: Effect of nitrate reductase deficiency upon growth, yield, and protein in barley. – Crop Sci. **20**, 487–490.
PENNING DE VRIES, F. W. T., A. H. M. BRUNSTING, and H. H. VAN LAAR, 1974: Products, requirements and efficiency of biosynthesis: A quantitative approach. – J. Theor. Biol. **45**, 339–377.
RABSON, R., C. R. BHATIA, and R. K. MITRA, 1978: Crop productivity, grain protein and energy: Inputs, subsidies and limitations. – In: Seed Protein Improvement by Nuclear Techniques, Proc. Meetings Baden and Vienna 1977, IAEA, Vienna, 3–20.
SANDFAER, J., and V. HAAHR, 1975: Barley stripe mosaic virus and the yield of old and new barley varieties. – Z. Pflanzenzüchtg. **74**, 211–222.
SCHOLZ, F., 1960: Versuche zur züchterischen Steigerung des Eiweißgehalts der Gerste mit Hilfe der experimentellen Mutationsauslösung. – Qual. Plant. Mater. Veget. **6**, 276–292.
–, 1976a: Zur Frage der Kombination von hohem Eiweißgehalt mit hohem Kornertrag bei Gerste. – Tag.-Ber. Akad. Landwirtsch.-Wiss. DDR, Nr. 143, 173–185.
–, 1976b: Problems of breeding for high protein yield in barley. – In: Barley Genetics III, Proc. Third Intern. Barley Genet. Symp. Garching 1975 (Ed. H. GAUL), Karl Thiemig, München, 548–556.
–, 1982: Problems and possibilities in genetic improvement of protein yields in cereals. – In: Induced Mutants for Cereal Grain Protein Improvement, Proc. Res. Coordination Meeting Nicosia 1980, IAEA-Tec. Doc. 259, IAEA, Vienna, 71–79.
–, 1984a: Some results with cross-bred descendants of induced high-protein mutants in barley. – In: Cereal Grain Protein Improvement, Proc. Res. Coordination Meeting Vienna 1982, IAEA, Vienna, 111–116.
–, 1984b: Possibilities and limiting conditions for genetic improvement of protein yield in cereals, with particular reference to nitrogen balance and requirements. – In: Cereal Grain Protein Improvement, Proc. Res. Coordination Meeting Vienna 1982, IAEA, Vienna, 269–277.
SINGH, R., and J. D. AXTELL, 1973: High lysine mutant gene *(hl)* that improves protein quality and biological value of grain sorghum. – Crop Sci. **13**, 535–539.
SMEDEGAARD-PETERSEN, V., and O. STØLEN, 1981: Effect of energy-requiring defense reactions on yield and grain quality in a powdery mildew-resistant barley cultivar. – Phytopathology **71**, 396–399.

F. SCHOLZ
Zentralinstitut für Genetik und Kulturpflanzenforschung
der Akademie der Wissenschaften der DDR
DDR - 4325 Gatersleben
Corrensstraße 3

Poster Abstracts

Kulturpflanze 32 · 1984 · S 207

The secalins of rye; chemistry, genetics, synthesis, deposition and molecular cloning

P. R. SHEWRY, M. KREIS, S. RAHMAN, S. R. BURGESS, L. TARDANI, D. BRADBERRY, J. FRANKLIN, B. G. FORDE, R. FRY and B. J. MIFLIN
(Harpenden, U.K.)

Rye seeds contain four groups of prolamin storage proteins called high molecular weight (HMW) secalins (M_r>100,000), 75K γ-secalins (M_r 75,000), ω-secalins (M_r 52,000) and 40K γ-secalins (M_r 40,000). Each group is a highly polymorphic mixture of polypeptides, the electrophoretic properties of which differ between genotypes. The four groups have been purified and characterized. The HMW secalins and ω-secalins are distinguished by a high glycine content and a low content of S-amino acids respectively. The two groups of γ-secalins have related N-terminal amino acid sequences. The chromosomal locations of the structural genes for the four groups of secalins are being studied by analysis of F_2 grain from crosses between genotypes with different secalin patterns.

Secalins are first detected in developing endosperms at about 3 weeks after anthesis, and accumulation continues almost until maturity. The 75K γ-secalins are the quantitatively predominant components, and the proportion of these increases from 45 % at 4 weeks to 56 % at maturity. Protein bodies prepared from developing endosperms contain all 4 groups of secalins.

Membrane-bound polysomes prepared from developing endosperms can be translated *in vitro* to give products corresponding in mobility on SDS-PAGE to secalins. Differential labelling experiments also indicate that the products are secalins.

Poly A^+ RNA from these polysomes was used as a template for synthesis of cDNA. Double stranded cDNA was inserted into pUC8 using Hind III linkers and cloned in *E. coli*. 932 amp^r resistant transformants with a lac z^- phenotype were identified as related to endosperm mRNAs by colony hybridization using ^{32}P-labelled cDNA as a probe. Hybrid–selected–translation was used to identify cDNA clones related to 40K and 75K γ-secalins. Other cDNA clones related to HMW secalins and ω-secalins were identified by hybridization to previously characterized cDNA clones for the homologous groups of prolamins of barley (D and C hordeins respectively). The sizes of the cDNA inserts were estimated by polyacrylamide gel electrophoresis, and restriction maps of some of the larger inserts were determined. One clone related to the 40K γ-secalins (pSc 503) has an insert of about 1390 base pairs and may represent a full-length copy of the mRNA. The nucleotide sequence of this is being determined.

P. R. SHEWRY
Rothamsted Experimental Station
Biochemistry Department
Harpenden, Herts. AL 5 2JQ
U. K.

Kulturpflanze 32 · 1984 · S 209

Genetical studies on the low-molecular-weight subunits of glutenin from wheat endosperm.

E. A. Jackson, L. M. Holt and P. I. Payne
(Cambridge, U.K.)

The low-molecular-weight (LMW) subunits of glutenin were resolved, both from each other, and from other endosperm protein groups using the two different 2-dimensional electrophoresis procedures of O'Farrell (1975; 1977). Similar analyses of several genetical lines of Chinese Spring revealed that the genes coding for LMW subunits are located on the short arms of chromosomes 1A, 1B and 1D. The segregation of LMW subunits, ω-gliadins, and γ-gliadins was studied in 800 progeny from three crosses using SDS-PAGE. No recombinations were observed, showing that the genes coding for these three protein groups are tightly linked at one locus called *Gli-1*, on each group one chromosome. Allelic variation in the major LMW glutenin subunits was studied by analysing inter-varietal chromosome substitution lines, and about 30 varieties of bread wheat. At least 6 1A, 6 1B and 4 1D variants were found. The *Gli-1* locus was mapped to the centromere (34 % R) and to the *Glu-1* locus (controlling HMW glutenin subunits) on the long arm (43 % R) showing that it is located near the end of the short arms of chromosome 1A, 1B and 1D. A line which lacks the short arm satellite of chromosome 1B does not produce those storage proteins whose genes are located at *Gli-B1*. This locus therefore occurs on the satellite, which accounts for the terminal one third of the chromosome arm.

E. A. Jackson
Plant Breeding Institute
Maris Lane, Trumpington
Cambridge CB2 2LQ
U.K.

Kulturpflanze 32 · 1984 · S 211–S 212

Chromosomal localization of *Triticum aestivum* and *Aegilops umbellulata* lectin genes

H. M. STINISSEN*, W. J. PEUMANS*, C. N. LAW** and P. I. PAYNE**
(*Leuven, Belgium; **Cambridge, UK)

Plant lectins (phytohemagglutinins) are a heterogeneous class of proteins (or glycoproteins) sharing in common their ability to recognize and bind specific sugars or sugar-containing molecules. A classical example of a seed lectin is wheat germ agglutinin (WGA). It is a dimeric protein consisting of two identical subunits of molecular weight 18,000. It has a high isoelectric point and is characterized by unusually high contents of glycine and cysteine, each exceeding 20 % on a molar basis. WGA binds specifically to N-acetylglucosamine or oligomers of this sugar, a property which is exploited in its purification by affinity chromatography.

From a genetic point of view, cereal lectins, and especially those in diploid and polyploid wheat species, represent a favourable system for a study of the expression of individual genes in complex genomes. Both in single and complex genomes, each individual genome directs the synthesis of its own lectin. In hexaploid wheat, for instance, the genomes A, B and D control the synthesis of isolectins CL_A, CL_B and CL_D respectively which can be separated by ion-exchange chromatography (PEUMANS et al. 1982).

Since it looked worthwhile to use the lectin genes as markers in plant breeding and cytogenetics, we determined their chromosomal localization. To determine the chromosomes which code for the lectin products we used the compensated nullisomic-tetrasomic collection of the wheat variety 'Chinese Spring' (CS). An analysis of these stocks revealed that the CL_A and CL_D genes are localized on chromosome 1A and 1D respectively. Furthermore, the chromosome arms bearing the lectin loci were determined using ditelocentric lines of CS. The CL_A and CL_D lectins are controlled by the long arms of chromosomes 1A and 1D. Since CS is defective for the CL_B isolectin, we used a substitution line of CS containing the 1B chromosome of 'Hope', in which CL_B is present. Here also, the lectin gene is localized on chromosome 1.

To investigate whether in other members of the Triticeae subfamily also, the lectin polypeptides are controlled by the homoeologous group 1 chromosomes, we analysed substitution and addition lines of *Aegilops umbellulata* in CS. In this species too, lectin genes are localized on chromosome 1.

It appears, therefore, that the lectin genes occur on the same chromosomes as the genes coding for ω- and γ-gliadins and on the same chromosome arms as the genes controlling the synthesis of the high-molecular-weight subunits of glutenin (PAYNE et al. 1982) and the lipopurothionins (FERNANDEZ DE CALAYA et al. 1976). So it can be concluded that group 1 chromosomes of wheat controls several groups of seed-specific proteins. Some of these proteins are exclusively expressed in the endosperm (glutenin, gliadins and lipopurothionins), whereas others (lectins) are exclusively synthesized in the primary axes. Since typical storage proteins such as gliadins and glutenins represent a very heterogeneous group of proteins, whereas lectins and purothionins have been highly conserved during the evolution and divergency of the different cereal genomes, the group 1 chromosomes apparently carries evolutionary very stable genes together with families of highly variable genes.

Literature

FERNANDEZ DE CALAYA, R., C. HERNANDEZ-LUCAS, PILAR CARBONERO and F. GARCIA OLMEDO, 1976: Gene expression in alloploids: genetic control of lipopurothionins in wheat. – Genetics **83**, 687–699.

PAYNE, P. I., L. M. HOLT, G. J. LAWRENCE, and C. N. LAW, 1982: The genetics of gliadin and glutenin, the major storage proteins of the wheat endosperm. – Qual. Plant Foods Hum. Nutr. **31**, 229–241.

PEUMANS, W. J., H. M. STINISSEN, and A. R. CARLIER, 1982: A genetic basis for the origin of six different isolectins in hexaploid wheat. – Planta **154**, 562–567.

H. M. STINISSEN,
Laboratorium voor Plantenbiochemie
KU Leuven
Kardinaal Mercierlaan, 92
3030 Leuven, Belgium

Kulturpflanze 32 · 1984 · S 213

N-terminal amino acid sequences of wheat albumins

D. LAFIANDRA (Bari, Italy and E. J.-L. LEW*, D. D. KASARDA* and N. K. FULRATH*
*Berkeley, USA)

N-terminal amino acid sequences have been determined on seed albumins extracted from three different wheat diploid species: *Aegilops squarrosa* (Syn. of *Triticum tauschii*) ssp. *eusquarrosa, Aegilops squarrosa* ssp. *strangulata* and *Triticum monococcum.*

Results showed that sequences from both subspecies of *Ae. squarrosa* were homologous with protein inhibitors of α-amylase present in hexaploid wheats.

The N-terminal amino acid sequence from *T. monococcum* showed strong homology with the recently reported sequence of a *Hordeum vulgare* trypsin inhibitor and to a lesser extent with that of *Ae. squarrosa* and α-amylase inhibitors from hexaploid wheats.

Similarity among N-terminal amino acid sequences of seed albumins from different subspecies (*eusquarrosa* and *strangulata*), species (*tauschii* and *aestivum*) and genera (*Triticum* and *Hordeum*) supports the hypothesis that divergence from a common ancestor could have originated proteins with different functions.

D. LAFIANDRA
Istituto del Germoplasma, Via Amendola 165/A
70 100 Bari
Italy

D. D. KASARDA
Western Regional Research Center, U.S.D.A.
Berkeley CA 94 710
USA

Kulturpflanze 32 · 1984 · S 215–S 218

Secretion of plant storage globulin polypeptides by *Xenopus* oocytes and legumin processing steps

R. Bassüner, A. Huth*, R. Manteuffel, and T. A. Rapoport*
(Gatersleben and Berlin-Buch, GDR)

Despite their intracellular deposition plant storage globulin polypeptides share common features with animal secretory proteins: (i) synthesis on membrane-bound polysomes (Püchel et al. 1979); (ii) occurrence of transient NH_2-terminal signal sequences (Ereken-Tumer et al. 1982); (iii) sequestration by the rough endoplasmic reticulum (Chrispeels et al. 1982a).

Totipotent oocytes from *Xenopus leavis* are regarded as a model system for studying synthesis, processing and secretion of animal secretory proteins (Lane 1981). Injecting plant mRNA coding for storage globulins we investigated the degree of homology between intracellular deposition of storage globulin polypeptides and secretion of animal proteins (Bassüner et al. 1983a). Cotyledonary poly(A)-containing RNA from field beans or french beans were coinjected with globin mRNA into oocytes, where they directed the synthesis of foreign polypeptides.

Whereas the nonsecretory globin remains inside the oocytes, the processed legumin- and vicilin-like storage globulin polypeptides are time-dependently secreted into the surrounding medium. After 48 h incubation up to $^1/_3$ of the pulse-labeled storage proteins can be recovered outside the oocytes. After fractionation of the reduced translation products on SDS-gels no acidic (α) or basic (β) chains of legumin subunits can be detected, indicating that prolegumin cleavage does not occur. Tunicamycin-mediated inhibition of the glycosylation of phaseolin polypeptides does not affect its secretion from oocytes.

Two predominating legumin subunit types of mature seeds, the methionine-containing type A (legumin A) and the methionine-free type B (legumin B) (Horstmann 1983) as well as methionine-containing (M_r 67,000) and methionine-free (M_r 64,000) primary legumin translation products (Bassüner et al. 1983b and 1984) have been described. Fig. 1 demonstrates the M_r of the mature non-reduced legumin A (M_r 58 ... 59,000) and legumin B (M_r 61 ... 62,000) in comparison to reduced polysomal legumin products obtained in a read-out cell-free translation system (prolegumin A: 66,000, prolegumin B: 62,000) and to reduced primary translation products. Prolegumins from oocytes exhibit on SDS-gels the same mobility as those from cell-free translation of polysomes (Bassüner et al. 1983a), furthermore, only the M_r-66,000 polypeptide can be labeled with [^{35}S]methionine. Therefore, the following scheme of legumin subunit processing steps can be derived:

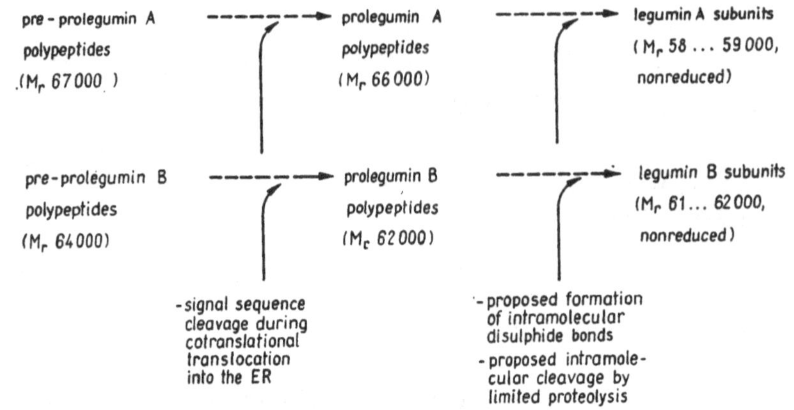

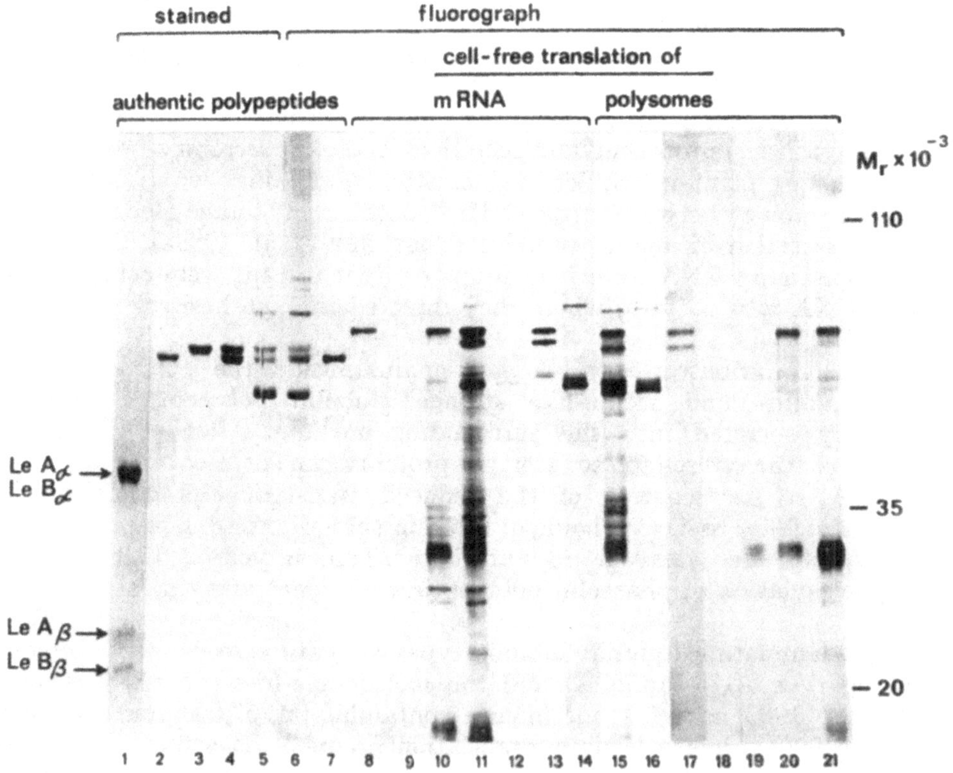

Fig. 1 Authentic storage globulin subunits from field bean cotyledons (*Vicia faba* L.) and their precursor molecules.
Total poly(A)-containing RNA and membrane-bound polysomes were isolated from immature field bean cotyledons and translated in the cell-free wheat germ system using ^{14}C-labeled amino acids. Radioactive authentic globulins were obtained by incubation of aseptic prepared cotyledons either in a liquid nutrient media supplemented with 5 ^{14}C-labeled amino acids (Asp, Glu, Gly, Phe, Ser) for 3 d or on a drop of [^{35}S]methionine for 20 h and subsequent extraction. For comparison, the samples were separated on SDS-gels (10,6 % acrylamide) together with authentic legumin A and B subunits (kindly provided by C. Horstmann)

Secretion of plant storage proteins from oocytes supports the hypothesis that the deposition of storage globulins in membrane-bounded organelles (protein bodies) of plant mesophyll cells represents an intracellular secretion process having steps in common with extracellular secretion of proteins from animal cells. Neither inhibition of glycosylation of plant storage glycoproteins nor lack of prolegumin cleavage does affect the transport of the polypeptides: i.e. secretion in oocytes (BASSÜNER et al. 1983a) or deposition in protein bodies in the plant cell (CHRISPEELS et al. 1982a and b). In accordance with these findings glycosylation and propolypeptide cleavage have been shown to be not a prerequisite for the secretion of animal secretory proteins from oocytes, too (LANE 1981, COLMAN 1982).

A definite answer concerning the signals causing storage rather than secretion in plant cells can not be given at this point.

Literature

BASSÜNER, R., A. HUTH, R. MANTEUFFEL, and T. A. RAPOPORT, 1983a: Secretion of plant storage globulin polypeptides by *Xenopus leavis* oocytes. – Eur. J. Biochem. **133**, 321–326.

–, R. MANTEUFFEL, K. MÜNTZ, M. PÜCHEL, P. SCHMIDT, and E. WEBER, 1983b: Analysis of *in vivo* and *in vitro* globulin formation during cotyledon development of field beans (*Vicia faba* L. var. *minor*). – Biochem. Physiol. Pflanzen **178**, 665–684.

–, U. WOBUS, and T. A. RAPOPORT, 1984: Cotranslational processing of storage globulin polypeptides of *Vicia faba* L. by signal recognition particle and microsomal membranes in a cell-free translation system. – Proc. 3rd Seed Protein Symp. 'Genetics of Seed Proteins' (MÜNTZ, K., and C. HORSTMANN, eds.). Kulturpflanze **32**, S 71–S 74.

Tracks represent

1,4 – authentic legumin subunits of type A (Met-containing) and type B (Met-free) (cf. HORSTMANN 1983) mixed together and reduced (track 1) or not reduced (track 4) before electrophoresis;

2,3 – authentic legumin subunits, not reduced before electrophoresis, type A (track 2) and type B (track 3);

5– 7 – globulin extracts from developing field bean cotyledons, labeled with 5 ^{14}C-amino acids (tracks 5 and 6) or with [^{35}S]methionine (track 7);

8–10 – *in vitro* translation products of poly(A)-containing RNA from developing field bean cotyledons, labeled with [^{35}S]methionine (track 10) and successively treated first with preimmune serum (track 9), then with legumin antiserum (track 8);

11–14 – *in vitro* translation products of poly(A)-containing RNA from developing field bean cotyledons, labeled with 5 ^{14}C-amino acids (track 11) and successively treated first with preimmune serum (track 12), then with legumin antiserum (track 13) and finally with vicilin antiserum (track 14);

15–18 – *in vitro* translation products of membrane-bound polysomes (read-out system containing 0.2 mM aurintricarboxylic acid) from developing field bean cotyledons labeled with 5 ^{14}C-amino acids (track 15) and successively treated first with preimmune serum (track 18), then with legumin antiserum (track 17) and finally with vicilin antiserum (track 16);

19–21 – *in vitro* read-out translation products of membrane-bound polysomes from developing field bean cotyledons labeled with [^{35}S]methionine (track 21) and successively treated first with preimmune serum (track 19) and then with legumin antiserum (track 20)

CHRISPEELS, M. J., T. J. V. HIGGINS, S. CRAIG, and D. SPENCER, 1982a: Role of the endoplasmic reticulum in the synthesis of reserve proteins and the kinetics of their transport to protein bodies in developing pea cotyledons. – J. Cell Biol. **93**, 5–14.

–, –, and D. SPENCER, 1982b: Assembly of storage protein oligomers in the endoplasmic reticulum and processing of the polypeptides in the protein bodies of developing pea cotyledons. – J. Cell Biol. **93**, 306–313.

COLMAN, A., 1982: Cells that secrete foreign proteins. – Trends Biochem. Sci. **7**, 435–437.

EREKEN-TUMER, N., J. D. RICHTER, and N. C. NIELSEN, 1982: Structural characterization of the glycinin precursors. – J. Biol. Chem. **257**, 4016–4018.

HORSTMANN, C., 1983: Specific subunit pairs of legumin from *Vicia faba*. – Phytochemistry **22**, 1861–1866.

LANE, C. D., 1981: The fate of foreign proteins introduced into *Xenopus* oocytes. – Cell **24**, 281–282.

PÜCHEL, M., K. MÜNTZ, B. PARTHIER, O. AURICH, R. BASSÜNER, R. MANTEUFFEL, and P. SCHMIDT, 1979: RNA metabolism and membrane-bound polysomes in relation to globulin biosynthesis in cotyledons of developing field beans (*Vicia faba* L.). – Eur. J. Biochem. **96**, 321–329.

R. BASSÜNER
Zentralinstitut für Genetik und Kulturpflanzenforschung,
der Akademie der Wissenschaften der DDR,
DDR - 4325 Gatersleben
Corrensstr. 3

Kulturpflanze 32 · 1984 · S 219–S 221

N-terminal sequence analysis of basic subunits of legumin from *Vicia faba* by solid-phase sequencing

A. OTTO, R. KRAFT and G. ETZOLD
(Berlin-Buch, GDR)

Introduction

Legumin, one of the major storage proteins of *Vicia faba* seeds (m.w. ca. 328,000), consists of six pairs of subunits. Each pair is composed by one (acidic) α-subunit (m.w. ca. 36,000) and one (basic) β-subunit (m.w. ca. 20,000) linked by disulphide bonds. Until now four different pairs could be isolated by ion exchange chromatography at DEAE-Sepharose and cleaved in their α- and β-components by reduction of the disulphide links and S-carboxymethylation. The N-terminal amino acid of each of the four β-subunits is Gly, but according to the sodium dodecylsulfate-polyacrylamide gel electrophoresis the four β-subunits can be subdivided into two types with different molecular weight (type A, subunit 1 and 2: higher apparent m.w.; type B, subunit 3 and 4: lower apparent m.w.). Likewise the amino acid composition of the proteins of type A (β1 and β2) and type B (β3 and β4), respectively, are very similar. Differences between the two types were also revealed by peptide mapping after proteolysis and after reaction with CNBr. Only β-subunits of type A were cleaved by CNBr, indicating that the β-subunits of type B do not contain Met (HORSTMANN 1983). The occurrence of specific pairs of basic subunits is supported by the results of the extended N-terminal amino acid sequence analysis of subunits β1, β3 and β4 (β2 in preparation) described in this contribution.

Experimental

The purified β-subunits and fragments of CNBr cleavage were obtained from Mr. C. HORSTMANN, Central Institute of Genetics and Crop Plant Research, Gatersleben, GDR. The amino acid sequences were mainly determined with a solid-phase sequencer (Rank-Hilger, England) by EDMAN degradation with phenylisothiocyanate. In all cases the proteins were coupled to 3-aminopropyl glass. β1, β3 and β4 were attached to p-phenylenediisothiocyanate activated glass under different conditions depending on the solubility of the subunits. The coupling yields were 14 % for β1, 82 % for β3 and 94 % for β4. The CNBr fragment CB β1 was immobilized by the homoserine lactone procedure (yield: 40 %).

High performance liquid chromatography of the phenylthiohydantoin derivatives of the amino acids was performed isocratically or with a gradient system on

reversed-phase columns (Spherisorb C 18, 5 μm, or LiChrosorb C 18, 10 μm). The solvent system used was acetonitrile/0.065 M ammonium acetate. The yield of the phenylthiohydantoins was calculated by a HPLC dedicated computing integrator.

Manual microsequencing was performed with 4-NN-dimethylaminoazobenzene-4'-isothiocyanate (DABITC) according to CHANG et al. 1978. In this case the amino acids were identified by thin-layer chromatography of the corresponding DAB-thiohydantoins on polyamide.

Results

The sequence of the amino acid residues of the subunit β1 and of the corresponding N-terminal CNBr fragment of β1 as well as the sequences of the subunits β3 and β4 of *Vicia faba* are shown in Fig. 1.

The comparison of the partial amino acid sequences of the β-subunits demonstrates sequence homogeneity between the first 53 positions of β3 and β4 (structural belonging to type B, lacking Met). However, the β1-subunit (type A, containing Met) shows a lot of differences indicating heterogeneity.

On the other hand the amino acid sequence of the β1-subunit of *Vicia* agrees excellently with a corresponding nucleotide sequence coding for a legumin basic subunit of *Pisum* (CROY et al. 1982). Only the amino acids at position 17 (Ser instead of Pro) and at position 27 (Gln instead of Glu) are not in accordance.

The occurrence of specific β-subunit pairs with a different amino acid sequence between each other may be explained as a result of different genes coding for legumin precursors.

Fig. 1 N-terminal amino acid sequences of β-subunits of *Vicia faba* determined mainly by solid-phase sequencing. The underlined amino acids were elucidated by the manual DABITC method. Amino acids at positions 60–68 of CB β1 were estimated from two additional CNBr cleavage products of β1, whose attachment by the homoserine lactone method was not possible (probably C-terminal fragments of β1). Hatched boxed sequences are derived from the cDNA sequence coding for a legumin basic subunit of *Pisum*. The drawn boxes indicate the position at which the sequences of the β-subunit pairs differ. Amino acid residues in parentheses are less certain

Literature

HORSTMANN, C., 1983: Specific subunit pairs of legumin from *Vicia faba*. – Phytochemistry **22**, 1861–1866.
CHANG, J. Y., D. BRAUER, and B. WITTMANN-LIEBOLD, 1978: Micro-sequence analysis of peptides and proteins using 4-NN-dimethylaminoazobenzene-4'-isothiocyanate/phenylisothiocyanate double coupling method. – FEBS Lett. **93**, 205–214.
CROY, R. R. D., G. W. LYCETT, J. A. GATEHOUSE, J. N. YARWOOD, and D. BOULTER, 1982: Cloning and analysis of cDNAs encoding plant storage protein precursors. – Nature **295**, 76–79.

A. OTTO, R. KRAFT, and G. ETZOLD,
Central Institute of Molecular Biology,
Academy of Sciences of the GDR, DDR - 1115 Berlin-Buch,
Robert-Rössle-Str. 10,
German Democratic Republic

Kulturpflanze 32 · 1984 · S 223–S 225

A gel electrophoretic study on the polypeptide structure of vicilin

B. SCHLESIER
(Gatersleben, GDR)

Introduction

The vicilin fraction of *Vicia faba* L. isolated by zonal isoelectric precipitation contains a mixture of different proteins. So far, only the main band of the acid gel system, identically with the main precipitation arc by crossed immunoelectrophoresis, is referred to vicilin (MANTEUFFEL and SCHOLZ 1975). On the basis of detailed electrophoretic investigation a general scheme for the subunit structure of vicilin-like 7S globulins of legumes is proposed in this poster.

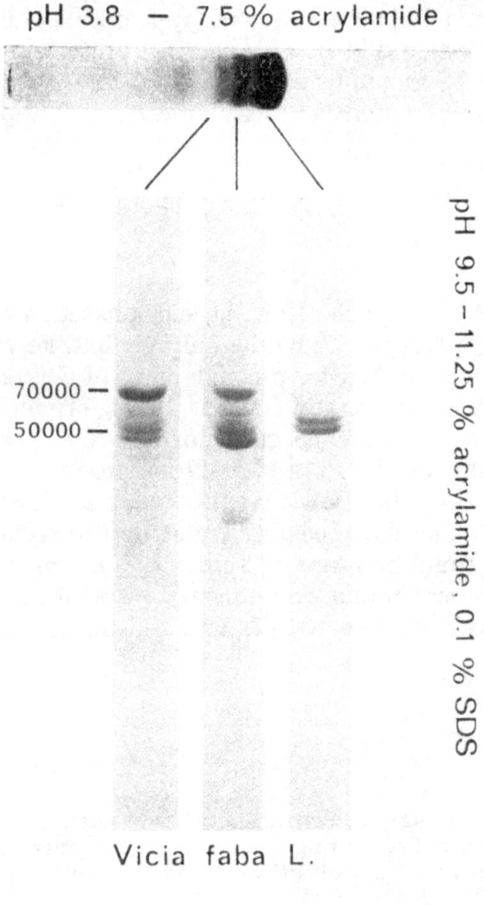

Fig. 1 Gel electrophoresis of vicilin from *Vicia faba*

Materials and Methods

Crude vicilin and legumin preparations from *Vicia faba* L., *Vicia narbonensis* L., *Pisum sativum* L. and *Phaseolus vulgaris* L. were fractionated on acid polyacrylamide gels at pH 3.8 (REISFELD et al. 1962). Individual bands or slices of these zones were cut from these gels, washed with distilled water, and reelectrophoresed on discontinuous SDS-gels (LAEMMLI 1970).

Results and Discussion

The protein pattern at pH 3.8 represents the non-dissociated globulins (SCHLESIER and SCHOLZ 1975). In addition some bands are generated by globulin association (SCHLESIER et al. 1978). In the vicilin fraction, only the polypeptide chains (ppc) 70,000 and 50,000 seem to result from translation of corresponding mRNA (GATEHOUSE et al. 1981, BASSÜNER et al. 1983) whereas smaller ppc are considered to be products of proteolysis (GATEHOUSE et al. 1981, 1982; SCHOLZ et al. 1983).

All four investigated species contain a vicilin-like 7S globulin composed of ppc of about 50,000. In *Vicia faba* L., *Vicia narbonensis* L., and *Pisum sativum* L. additional proteins appear on acid polyacrylamide gels composed of ppc 70,000 and 50,000 in varying ratios. In *Pisum sativum* L. we found a protein which consists nearly exclusively of ppc 70,000. The ppc 50,000 contamination is to be explained by vicilin association which in the non-dissociated state exhibit similar mobility as the ppc 70,000 homomer. The latter should correspond to the convicilin of *Pisum sativum* L. described by CROY et al. (1980) in accordance with the results from *Pisum fulvum* obtained by CASEY and SANGER (1980). In contrast to CROY at al. (1980) no convicilin was found in the cultivar of *Vicia faba* L. that we investigated. Since in all our investigated *Vicia* and *Pisum* species hetero-oligomeric vicilin proteins composed of ppc 70,000 and 50,000 appear concomitantly with either only 50,000 or both 50,000 and 70,000 homomeric proteins we suggest the following model of vicilin-like 7S globulins in legumes:

Vicilins represent a protein family of trimeric isomers composed of different ratios of ppc 50,000 and 70,000:

Type A	3×50,000
Type B	2×50,000+1×70,000
Type C	1×50,000+2×70,000
Type D	3×70,000

Different species do not necessarily possess all types of vicilin, e.g. phaseolin of *Phaseolus vulgaris* L. corresponding to type A seems to be the only vicilin-like 7S globulin of this species. All four types are described for β-conglycinin of *Glycine max* (L.) MERR. (THANH and SHIBASAKI 1978, SYKES and GAYLER 1981). Trimeric subunit structures have been recently proposed for 7S globulins also by other authors (THANH and SHIBASAKI 1978, CASEY and SANGER 1980, MCPHERSON 1980, PUSZTAI and STEWART 1980, SYKES and GAYLER 1981, GATEHOUSE et al. 1981, JOHNSON et al. 1982, PLIETZ et al. 1983). The real quarternary structure seems to be more complex. The number of ppc could be larger in some cases or micro-heterogeneity occurs, e.g. β-conglycinin where minor components γ and δ have been described and microheterogeneity has been detected (THANH and SHIBASAKI 1977).

Literature

BASSÜNER, R., R. MANTEUFFEL, K. MÜNTZ, M. PÜCHEL, P. SCHMIDT, and E. WEBER, 1983: Analysis of *in vivo* and *in vitro* globulin formation during cotyledon development of field beans (*Vicia faba* L. var. *minor*). – Biochem. Physiol. Pflanzen **178**, 665–684.

CASEY, R., and E. SANGER, 1980: Purification and some properties of a 7S seed storage protein from *Pisum* (pea). – Biochem. Soc. Transactions **8**, 656–657.
CROY, R. R. D., J. A. GATEHOUSE, M. TYLER, and D. BOULTER, 1980: The purification and characterization of a third storage protein (convicilin) from the seeds of pea (*Pisum sativum* L.). – Biochem. J. **191**, 509–516.
GATEHOUSE, J. A., R. R. D. CROY, H. MORTON, M. TYLER, and D. BOULTER, 1981: Characterisation and subunit structure of the vicilin storage protein of pea (*Pisum sativum* L.). – Eur. J. Biochem. **118**, 627–633.
GATEHOUSE, J. A., G. W. LYCETT, R. R. D. CROY, and D. BOULTER, 1982: The post-translational proteolysis of the subunits of vicilin from pea (*Pisum sativum* L.). – Biochem. J. **207**, 629–632.
JOHNSON, S., G. GRAYSON, L. ROBINSON, R. CHAHADE, and A. MCPHERSON, 1982: Biochemical and crystallographic data for phaseolin, the storage protein from *Phaseolus vulgaris*. – Biochemistry **21**, 4839–4834.
LAEMMLI, U. K., 1970: Cleavage of structural proteins during the assembly of the head of bacteriophage T4. – Nature **227**, 680–685.
MANTEUFFEL, R., and G. SCHOLZ, 1975: Studies on seed globulins from legumes. V. Immunoelectrophoretic control of vicilin purification by gel filtration. – Biochem. Physiol. Pflanzen **168**, 277–285.
MCPHERSON, A., 1980: The three-dimensional structure of canavalin at 3.0 Å resolution by X-ray diffraction analysis. – J. Biol. Chem. **255**, 10472–10480.
PLIETZ, P., G. DAMASCHUN, J. J. MÜLLER, and B. SCHLESIER, 1983: Comparison of the structure of the 7S globulin from *Phaseolus vulgaris* in solution with the crystal structure of 7S globulin from *Canavalia ensiformis* by small angle X-ray scattering. – FEBS Letters **162**, 43–46.
PUSZTAI, A., and J. C. STEWART, 1980: Molecular size, subunit structure and microheterogeneity of glycoprotein II from the seeds of kidney bean (*Phaseolus vulgaris* L.). – Biochem. Biophys. Acta **623**, 418–428.
REISFELD, R. A., U. J. LEWIS, and D. E. WILLIAMS, 1962: Disc electrophoresis of basic proteins and peptides on polyacrylamide gels. – Nature **195**, 281–283.
SCHLESIER, B., and G. SCHOLZ, 1975: Untersuchungen an Samenglobulinen von Leguminosen. IV. Disk-elektrophoretischer Vergleich der Globulinfraktionen einiger *Vicia*-Arten. – Kulturpflanze **23**, 157–166.
SCHLESIER, B., R. MANTEUFFEL, and G. SCHOLZ, 1978: Studies on seed globulins from legumes. VI. Association of vicilin from *Vicia faba* L. – Biochem. Physiol. Pflanzen **172**, 285–290.
SCHOLZ, G., R. MANTEUFFEL, K. MÜNTZ, and A. RUDOLPH, 1983: Low-molecular-weight polypeptides of vicilin from *Vicia faba* L. are products of proteolytic breakdown. Eur. J. Biochem. **132**, 103–107.
SYKES, G. E., and K. R. GAYLER, 1981: Detection and characterization of a new β-conglycinin from soybean seeds. – Arch. Biochem. Biophys. **210**, 525–530.
THANH, V. H., and K. SHIBASAKI, 1977: Beta-conglycinin from soybean proteins. Isolation and immunological and physicochemical properties of the monomeric forms. – Biochim. Biophys. Acta **490**, 370–384.
THANH, V. H., and K. SHIBASAKI, 1978: Major proteins of soybean seeds. Reconstitution of β-conglycinin from its subunits. – J. Agric. Food Chem. **26**, 692–695.

B. SCHLESIER
Zentralinstitut für Genetik und Kulturpflanzenforschung
der Akademie der Wissenschaften der DDR,
DDR - 4325 Gatersleben
Corrensstr. 3

Kulturpflanze 32 · 1984 · S 227–S 228

Electron-microscopic investigations of reconstitution products of isolated subunit pairs of the plant storage protein legumin from *Vicia faba*

K. Adler and C. Horstmann
(Gatersleben, GDR)

Legumin, the 11S storage protein of *Vicia faba* seeds, consists of 6 pairs of α/β-subunits. Four different pairs of disulphide linked acidic (α) and basic (β) subunits were isolated from the holoprotein (Horstmann 1983). We investigated the reconstitution of legumin molecules from subunit pairs by means of polyacrylamide gel electrophoresis and electron microscopy.

For reconstitution, subunit pairs were dissolved in 0.05 M Tris/HCl/6 M urea pH 8.0 and exhaustively dialyzed against 0.2 M Na-phosphate buffer pH 8.0, thereafter against water and freeze-dried.

The electron micrographs were taken from negatively stained specimens with a Tesla BS 500 electron microscope. The protein concentration was 0.5 mg/ml and the probes were prepared by the drop method (Haschemeyer and Myers 1973) on carbon coated grids (carbon was evaporated on mica-sheets) with 1 % uranyl acetate (UA) or 2 % phospho-tungstic acid (PTA) as contrast media.

PTA gave better contrast than UA, because the protein molecules aggregated in presence of UA (presumably because of the low pH of the UA solution).

Polyacrylamide gels of the reconstituted legumins show in all samples a band in the same position as authentic legumin and diffuse bands of non-reconstituted subunit pairs. The electron micrographs of the reconstitution products reveal molecules which do not exhibit apparent differences from the legumin molecules with regard to size and shape.

The results demonstrate that the individual subunit pairs are capable of reassociation to legumin-like molecules. Though the primary structures of the different subunit pairs exhibit major differences (Horstmann 1983, Otto et al. 1984) the pairs obviously possess the correct contact sites for the formation of a hexameric legumin.

To obtain informations on the geometrical arrangement of the subunit pairs in the legumin molecule we used the Markham technique (Markham et al. 1963) as applied by Reichelt et al. (1980) to helianthinin from sunflower seeds. The same legumin molecule shows a "subunit" structure in the copies obtained by 4-, 5-, and 6-fold rotation. By rotating different legumin molecules we could produce fine pictures of every desired subunit arrangement. We suppose that in such a high magnification range this technique does not lead to unequivocal results on the subunit arrangement.

Literature

HASCHEMEYER, R. H., and R. J. MYERS, 1973: Negative staining. – In: Principles and techniques of electron microscopy, vol. 2 (Ed. M. A. HAYAT), pp. 101–147. Van Nostrand Reinhold Company, New York, Cincinnati, Toronto, London, Melbourne.

HORSTMANN, C., 1983: Specific subunit pairs of legumin from *Vicia faba*. Phytochemistry **22**, 1861–1866.

MARKHAM, R., S. FREY, and G. J. HILLS, 1963: Methods for the enhancement of image detail and accentuation of structure in electron microscopy. – Virology **20**, 88–102.

OTTO, A., R. KRAFT, and G. ETZOLD, 1984: N-terminal sequences of basic subunits of legumin from *Vicia faba* L. by solid phase sequence analysis. – Proc. 3rd Seed Protein Symp. 'Genetics of Seed Proteins' (MÜNTZ, K., and C. HORSTMANN, eds.). Kulturpflanze **32**, S 219–S 221.

REICHELT, R., K. D. SCHWENKE, T. KÖNIG, W. PÄHTZ and G. WANGERMANN, 1980: Electron microscopic studies for estimation of the quarternary structure of the 11S globulin (Helianthinin) from sunflower seed (*Helianthus annuus* L.). – Biochem. Physiol. Pflanzen **175**, 653–663.

K. ADLER
C. HORSTMANN
Zentralinstitut für Genetik und Kulturpflanzenforschung
der Akademie der Wissenschaften der DDR,
DDR - 4325 Gatersleben,
Corrensstraße 3

Kulturpflanze 32 · 1984 · S 229

Factors affecting the reduction of protein synthesis in imbibing aged *Agrostemma githago* seeds

D. Bernhardt, R. Kosanke, K.-H. Köhler, and R. Bernhardt
(Greifswald, GDR)

In embryos of imbibing aged *Agrostemma githago* seeds the level of ^{14}C-leucine incorporation is strongly reduced corresponding to the decrease of viability of the seeds during the storage, whereas the total RNA synthesis (incorporation of ^{3}H-uridine) is affected in a lower degree. The reduction of ^{14}C-leucine incorporation is due to a diminished protein synthesis and not caused by increase of the endogenous leucine pool, which remains really constant during the long-time storage of the seeds, as was shown by application of the dansyl chloride technique.

In embryos of imbibing aged seeds the level of *de novo* synthesis of poly(A^+)RNA is remarkably high with respect to total RNA synthesis. There are, however, significant differences in the degree of polyadenylation of the poly(A^+)RNA between afterripened young and aged seed embryos, synthesized during imbibition. The amount of poly(A^+)RNA with long-chained poly(A)-sequences, eluting from the poly(U)-sepharose columns only at 50 °C or by formamide, is in embryos of aged seeds remarkably higher than in afterripened ones. When the embryos are fractionated into cell nuclei and cytoplasmic fractions, the increase of the proportion of poly(A^+)RNA with longer poly(A)-sequences is evident only in the cytoplasmic fraction, whereas in the cell nuclei the relative content of polyadenylated RNA, eluting at 20 °C, 50 °C and by formamide, changes only slightly. The amount of poly(A^+)RNA with long-chained poly(A)-sequences is high in the cell nuclei of young embryos, too. It is known, that during the ripening of mRNA for translation the length of the poly(A)-sequences is decreased to about 50 A. Evidently, this process is disturbed in the embryos of aged, long-time-stored seeds, so that the newly synthesized mRNA may be translated only by retardation or not at all.

In comparision to embryos of young afterripened seeds, aged embryos show a significant retardation of RNA transport from cell nuclei to cytoplasm. This retardation concerns the rRNA transport as well as the transport of poly(A^+)RNA.

Finally it could be shown, that during the storage of the seeds the ATP content in the embryos and their ATP formation ability using PEP and ADP are reduced corresponding to the decrease of germination rate.

It is concluded, that the decrease of protein synthesis in imbibing aged seeds is caused, among other reasons, by reduction of RNA transport from cell nuclei to cytoplasm, by disturbance of mRNA processing, and by reduction of ATP synthesis.

D. Bernhardt
Dept. Biology, Section Botany, Ernst-Moritz-Arndt University
DDR - 2200 Greifswald Grimmer Str. 80

Kulturpflanze 32 · 1984 · S 231–S 233

Seed proteins of chromosome mutants in *Pisum*

R. RAO, S. GRILLO, and L. M. MONTI
(Portici, Italy)

Introduction

Pea is one of the genetically best known species and it has been widely used in experimental mutagenesis by which many different mutants were obtained, presently included in germplasm collections. In this species, the use of the seed protein profiles obtained by electrophoresis has proved greatly useful to investigate genetic variations in induced mutants (GOTTSCHALK and MÜLLER 1974).

In *Pisum* globulins represent about 70–80 % of total seed proteins and are formed by different rates of legumin, convicilin and vicilin (BOULTER 1983) which differ in sulphur content, while albumins represent about 15–30 % of total seed proteins and contain most of the seed enzymes.

In this investigation we studied the protein fractions (both albumins and globulins) of 8 chromosome mutants of *Pisum* characterized by reciprocal translocation (obtained by LAMM, Sweden) in comparison with two genotypes having normal chromosome structure.

Crosses were performed among the two test lines and the mutant lines and some F_1 generations were analyzed.

Table 1
List of genotypes and chromosomes involved in the translocations. Protein content is also reported

Code	Line	Translocation between chromosomes	Protein (% d. m.)
1	L-III	3–7	25.0
2	L-88	4–7	18.7
3	L-112	5–7	17.6
4	L-21	5–7	23.4
5	L-SH	1–5	20.7
6	L-108	2–6	25.8
7	L-83	3–5	21.8
8	L-114	4–6	25.9
9	L-110	–	24.0
P	Proteo	–	23.1

Experimental

Genotypes used in this study are listed in table 1. The two cotyledons of each seed were separated: the one with embryo was stored for the following generation, while the other was dehulled and used for protein extraction. Globulins and albumins were separated as previously described (RAO and PERNOLLET 1981). Legumin and Vicilin were obtained through zone precipitation on a column Pharmacia 16/20, flow rate 7 ml/h. Sodium dodecyl sulphate electrophoresis (SDS-gels) were performed according to LAEMMLI (1970). Protein fractions were isofocused on 5 % polyacrylamide containing 8 M urea in a pH range 3.5–10.

Results

SDS-gels of total storage proteins and of total albumins show in the 10 genotypes both qualitative and quantitative differences among the mutants and between the mutants and the test lines. Quantitative variation is particularly evident in the convicilin region (71–73 kDalton zone) while differences in the number of bands are evident in the zone ranging about 40kD, 30–34kD and 20kD. Most of the mutant lines show a doublet in the 14kD area which is absent in both test lines.

Fig. 1 shows the result of the cross between Proteo and mutant n. 8: the F_1 cotyledon has an electrophoretic pattern which is intermediate between the two parents both for relative intensity and molecular weight of subunits.

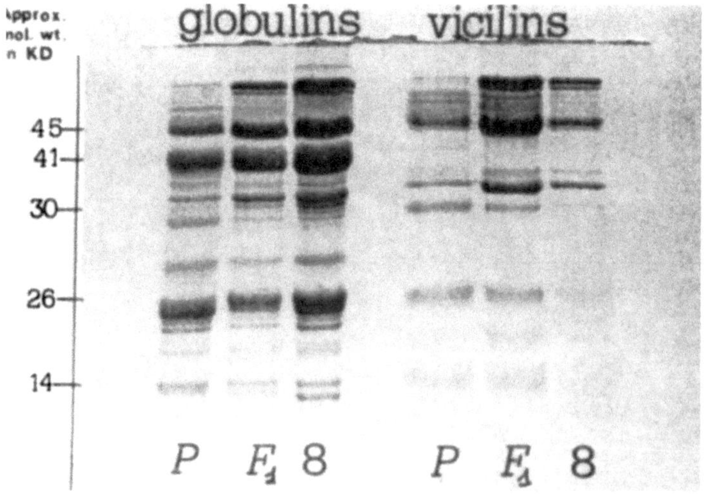

Fig. 1 SDS electrophoresis of reduced globulins and vicilins in the cotyledons of parents and F_1 (see Table 1)

Conclusions

The electrophoretic patterns here reported are similar to those described by other authors (CASEY 1979, PRZYBYLSKA et al. 1977). All the obtained profiles were highly reproducible from different protein extractions and electrophoretic analysis:

both quantitative and qualitative variations observed are genotype specific. This is also confirmed by the behaviour of the F_1 cotyledon from cross between Proteo and mutant n.8 which has an intermediate pattern. For Vicilin, focusing proved to be very reliable to evidence differences among the analyzed genotypes.

The observed variations in protein profiles could be related either to the different genetic background existing in the samples or to the presence of chromosome aberration (GOTTSCHALK and MÜLLER 1982, GRILLO and RAO 1982). The F_1 analysis gave, as expected, intermediate patterns (THOMSON and SCHROEDER 1978) and from the analysis of the F_2 generation we can identify the genetic base of the found variability.

Literature

BOULTER, D., 1981: Regulation of storage protein synthesis and deposition in developing pea seeds. – EC seminar on perspectives for peas and lupins as protein crops, (Eds. R. THOMPSON and R. CASEY). Sorrento, Italy, pp. 221–224.

CASEY, R., 1979: Genetic variability in the structure of the α-subunits of legumin from *Pisum*. A two dimensional gel electrophoresis study. – Heredity **43**, 265–272.

GOTTSCHALK, W., and H. P. MÜLLER, 1974: Quantitative and qualitative investigations on the seed proteins of mutants and recombinants of ***Pisum sativum***. – Theor. Appl. Genet. **45**, 7–20.

–, and –, 1982: Seed protein of ***Pisum*** mutants and recombinants. – Qual. Plant. Plant Foods Hum. Nutr. **31**, 297–306.

GRILLO, S., and R. RAO, 1982: Quantitative variation in protein fractions of pea mutants. – Ann. Fac. Sci. Agrar. Univ. Napoli Portici, Serie IV **16**, 1–5.

LAEMMLI, U. K., 1970: Cleavage of structural proteins during the assemblyof the head of the bacteriophage T4. – Nature **227**, 680–685.

PRZYBYLSKA, J., S. BLIXT, J. HURICH, and Z. ZIMNIAK-PRZYBYLASKA, 1977: Comparative study of seed proteins in the genus ***Pisum***. I. Electrophoretic patterns of different protein fractions. – Genetica Polonica **18**, 27–38.

RAO, R., and J. C. PERNOLLET, 1981: ***Pisum sativum*** seed globulins as biochemical markers for "wrinkled-smooth" seed character: comparison of genotypes through different protein extractions and electrophoretic procedures. Agronomie **1**, 906–916.

THOMSON, J. A., and H. E. SCHROEDER, 1978: Cotyledonary storage proteins in ***Pisum sativum***. II. Hereditary variation in components of the Legumin and Vicilin fractions. Aust. J. Plant Physiol. **5**, 281–294.

R. RAO
Università degli Studi di Napoli
Istituto di Agronomia Generale e Coltivazioni Erbacee
Cattedra di Miglioramento Genetico
Via Università, 100
80055 - Portici (NA)
Italia

Kulturpflanze 32 · 1984 · S 235

Gliadin genetics of winter wheat *Triticum aestivum*

E. V. Metakovsky, A. Yu. Novoselskaya, and A. A. Sozinov
(Moscow, USSR)

Bread wheat cultivars differ in the component composition of a polymorphic storage protein of grain–gliadin. Gliadin component composition of single F_2 grains in 36 intervarietal hybrid combinations of wheat has been studied by means of electrophoresis in polyacrylamide gel (aluminium-lactate, pH 3.1). Genetical analysis permitted the classification of more than 80 jointly inherited groups of gliadin components (blocks). Multiple allelism for all known gliadin-coding loci located on chromosomes 1A, 1B, 1D, 6A, 6B, 6D has been demonstrated: all blocks have been divided into 6 independent groups or series. Blocks of one series are controlled by alleles of one of the gliadin-coding locus and are distinguished by number, mobility and intensity of their components. Blocks consist of 1 to 8 components, blocks of one series often have not a single component with the same electrophoretic mobility. Differences observed between blocks could occur as a result of numerous changes in the gliadin-coding locus, that were not eliminated by natural selection and/or breeding. The least differences between blocks have been observed in series controlled by chromosomes 1D and 6D. The corresponding gliadin-coding loci are suggested to have had less continued period of evolution or to have greater breeding value and hence to be more conservative. Blocks of gliadin components can be used as suitable genetic markers of chromosomes of homoeological groups 1 and 6 in wheat.

E. V. Metakovsky
Institute of General Genetics,
Academy of Sciences of the USSR
Gubkin Str. 3
117 809 Moscow W-333
USSR

Kulturpflanze 32 · 1984 · S 237–S 239

Variation and genetics of electrophoretic seed albumin patterns in *Pisum**

J. Przybylska, Z. Zimniak-Przybylska, and E. Kozubek
(Poznań, Poland)

This investigation is a continuation of the earlier studies (Przybylska et al. 1977, Jakubek and Przybylska 1979, Blixt et al. 1980) on variation and genetics of electrophoretic seed albumin patterns [EP patterns] in *Pisum*.

Screening analysis of an extended *Pisum* material led to distinction of four "new" EP patterns, VI–IX; the so-far distinguished patterns are presented in Fig. 1a. Regarding distribution of EP I–IX, data obtained in this work and in the earlier investigations indicate distinctness of *P. abyssinicum* and *P. fulvum*.

The F_2 segregation data obtained for the crosses EP I×II and EP II×III provided further evidence that EP patterns are simply inherited (Tab. 1).

Table 1

F_2 segregation of EP patterns in the crosses EP I×II [A] and EP II×III [B]

A

Segregation I	I/II	II	χ^2 1 : 2 : 1	P value
179	375	170	1.1574	0.50 < P < 0.70

B

Segregation III + II/III	II	χ^2 3 : 1	P value
622	233	2.3115	0.10 < P < 0.20

Seed albumins of the accessions representing EP I–IX were fractionated on Sephadex G 100 column and the obtained peaks were analysed electrophoretically. Like in the case of the earlier investigated EP I–V, the fractions with MW ca 40,000 (fractions S2) proved to be responsible for distinction of EP VII–IX. The elution profile of the *Pisum* accession representing EP VI, lacking the major characteristic bands, showed no peak corresponding to MW ca 40,000. Electro-

* Cooperative investigations with Dr. S. Blixt from the Weibullsholm Plant Breeding Institute, Landskrona, Sweden, Prof. Dr. R. Kh. Makasheva from N. I. Vavilov All-Union Research Institute of Plant Industry, VIR, Leningrad, USSR, and Dr. Ch. Lehmann from "Zentralinstitut für Genetik und Kulturpflanzenforschung" of German Academy of Sciences, Gatersleben, GDR.

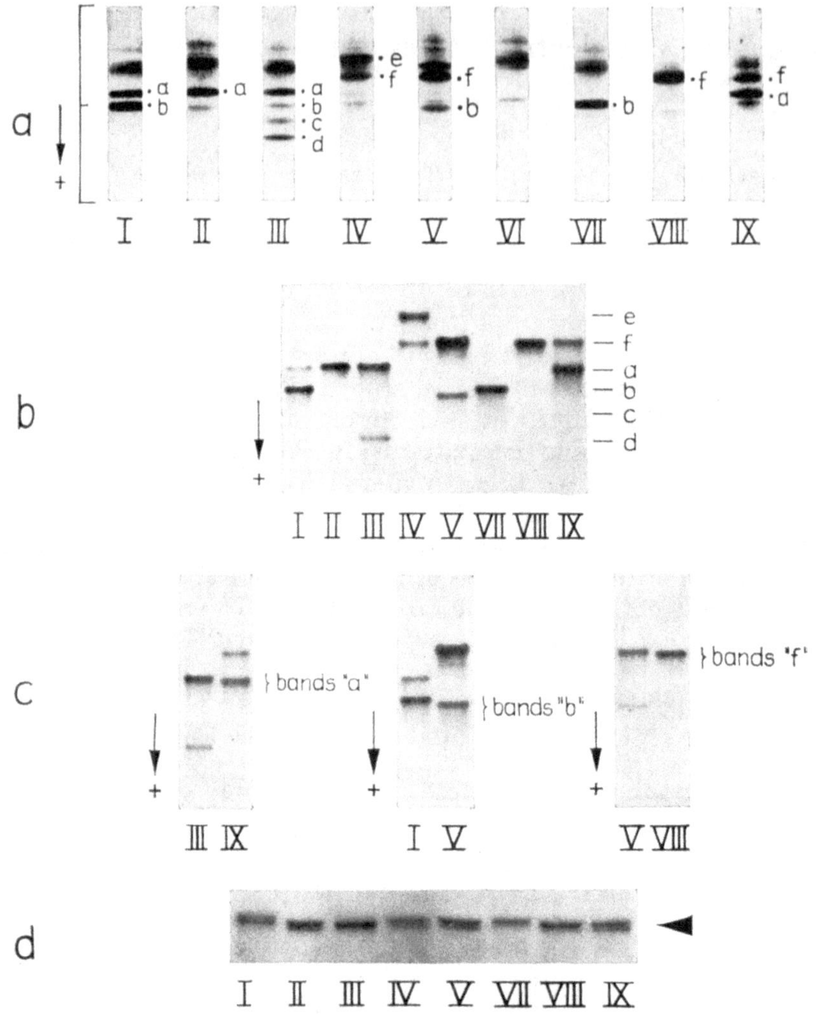

Fig. 1 Vasiation in electrophoretic seed albumin patterns in *Pisum*.
a Disc gels showing electrophoretic patterns EP I–EP IX
The patterns are produced by crude protein extracts of the following *Pisum* accessions: *P. sativum* W 110 [EP I], *P. humile* W 936 [EP II] *P. sativum* W 1490 [EP III], *P. abyssinicum* W 808 [EP IV], *P. fulvum* W 1256 [EP V], *P. sativum* W 1973 [EP VI], *P. sativum* VIR 1987 [EP VII], *P. humile* VIR 2521 [EP VIII], *P. elatius* Gat. 255 [EP IX]. The characteristic bands are indicated with letters a–f
b Slab gel showing electrophoretic patterns of fractions S2 from the *Pisum* accessions representing patterns: EP I–EP V and EP VII–EP IX. Designations as in Fig. 1 a
c Gels showing the apparently homologous bands to have slightly different electrophoretic mobilities.
Designations as in Fig. 1 a
d SDS gel [reducing conditions] of fractions S2 from different *Pisum* accessions showing subunits with MW ca 23, 000.
Designations as in Fig. 1 a

phoretic patterns of fractions S2 from the accessions representing EP I–V and VII–IX are presented in Fig. 1b.

Side-by-side electrophoresis of fractions S2 revealed variation in electrophoretic mobility of the characteristic bands which were considered homologous in the previous investigations (Fig. 1c). It should be checked whether this kind of variation is an additional element differentiating the distinguished EP patterns or whether this variation is independent.

SDS-gel electrophoresis (reducing conditions) of the specific albumins, corresponding to the characteristic bands, showed one or two polypeptide components with MW ca 23,000 (Fig. 1d).

Further investigations are necessary to understand biochemical and genetic basis of the observed variation in EP patterns.

Literature

Blixt, S., J. Przybylska, and Z. Zimniak-Przybylska, 1980: Comparative study of seed proteins in the genus *Pisum*. V. Genetics of the electrophoretic patterns I and III. – Genetica Polonica, **21**, 153–161.

Jakubek, M., and J. Przybylska, 1979: Comparative study of seed proteins in the genus *Pisum*. III. Electrophoretic patterns and amino acid composition of albumin fractions separated by gel filtration. – Genetica Polonica, **20**, 369–380.

Przybylska, J., S. Blixt, J. Hurich, and Z. Zimniak-Przybylska, 1977: Comparative study of seed proteins in the genus *Pisum*. I. Electrophoretic patterns of different protein fractions. – Genetica Polonica, **18**, 27–38.

J. Przybylska
Institute of Genetics,
Polish Academy of Sciences,
Strzeszynska 30–36
Poznań
Poland

Kulturpflanze 32 · 1984 · S 241–S 244

Isozyme patterns in cereal grain extracts in relation to the heritability of specific properties

R. Buschbeck, G. Gustavus, and R. Wilp
(Güstrow, GDR)

Determination of isozyme patterns and multiple forms of proteins obtain an increasing importance for identification of crop species and cultivars (Endo 1973; Scandalios 1977). However in practical plant breeding empirical working with morphological and cytological characterization of the breeding products is still predominant. The additional use of biochemical markers in form of isozyme patterns in combination with desirable properties can contribute to the effectiveness of the breeding process.

Using the polyacrylamide gel electrophoresis in some selected cultivars of winter rye, *Secale cereale*, and their inbred lines, the sprouting sensibility was estimated by the alpha-amylase isozyme patterns in dormant seeds. In addition the patterns of esterase and peroxidase isozymes were controlled.

In caryopsis of barley hybrids from the wild type *Hordeum bulbosum* (frost resistant, *Erysiphe* resistant) crossed with *Hordeum vulgare*, and their breeding issues of some generations the patterns of the esterase isozymes and the buffer soluble proteins were regarded.

1. Breeding of sprouting resistant rye forms

The sprouting sensibility of cereals is genetically fixed and can be influenced by the breeding process (Flintham and Gale 1980). It is possible to estimate the degree of sprouting resistance by the alpha-amylase isozyme patterns in both ripening and mature seeds of rye cultivars (Wilp 1979). In caryopsis of inbred lines of several generations (I_{12}–I_{15}) the patterns of the alpha-amylases remain constant in forms with normal cytoplasm (non restorer lines), but show reduced banding patterns in forms with male sterile cytoplasm (figure 1). The isozymes of the esterases and the peroxidases partially reflected this situation in a similar manner.

2. Species crossing with wild type barley

Crossing of *Hordeum vulgare* with the wild type *Hordeum bulbosum* is directed on the aim, to improve the resistance against frost and *Erysiphe* infection (Szigat und Wustrack 1976). Pohler und Szigat (1982) pointed out, that the

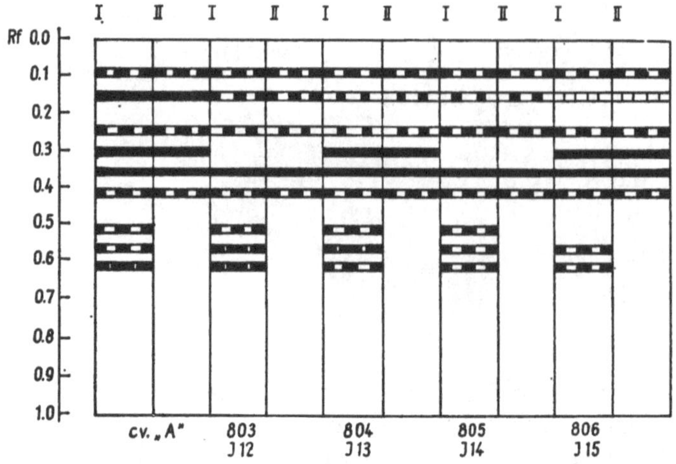

Fig. 1 A Isozyme patterns of alpha-amylases in dormant caryopsis of the cultivar "A" and the inbred lines I_{12}–I_{15}

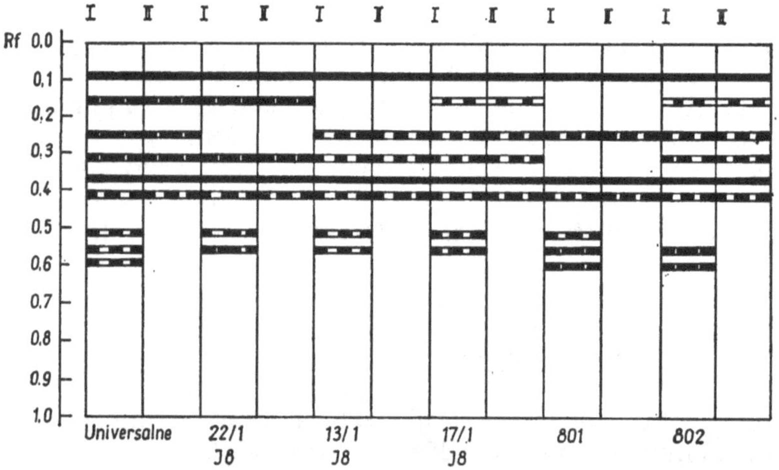

1 B Isozyme patterns of alpha-amylases in dormant caryopsis of the cultivar Universalne and the inbred lines
I = alpha- and beta amylases
II = alpha amylases

bulbosum chromosomes in the hybrid genom are eliminated, but some *bulbosum* genes obviously remain integrated. Using the isozyme technique it is possible to prove this fact (Gustavus 1982).

Figure 2 demonstrates the esterase banding patterns of *H. bulbosum, H. vulgare* and their hybrids of the F_1–F_4 generations. Only the hybrid 20001 showes the typical three bands of *H. bulbosum* about Rf 0.70, but all the hybrids developed the new "bastard bands" in Rf 0.09; 0.15 and 0.45. Crossings of *H. vulgare* with the hybrids resulted in some forms with a typical *bulbosum* morphology, the "modified barleys". They also possess the bastard bands. The physiological

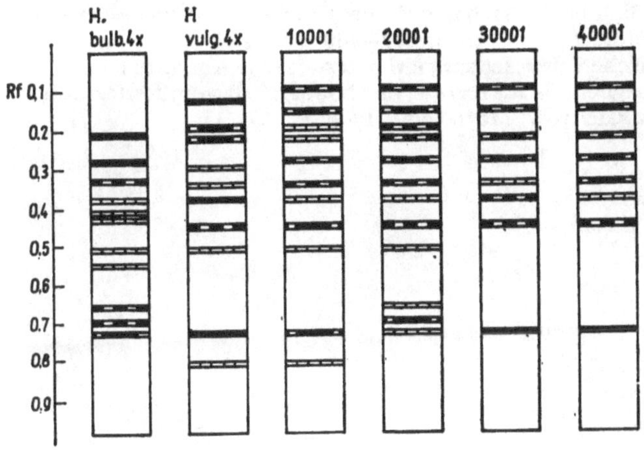

Fig. 2 Isozyme patterns of esterases from extracts of dormant caryopsis of *Hordeum bulbosum*, *H. vulgare* and the hybrids of F_1–F_4 generation (10001–40001)

properties of these plants are tending in the direction of the breeding aims, frost and *Erysiphe* resistance (SZIGAT und WUSTRACK 1976). The patterns of the buffer soluble proteins both in *H. bulbosum* and in the modified barleys developed a strong band in Rf 0.50. All these forms were not infected by *Erysiphe graminis*.

Conclusions

Isozyme patterns are reflecting the genetical situation in the crossing issue, because they originate by gene expression. Isozymes can be used as biochemical markers in the breeding process.

Coupling of distinct banding patterns with specific properties of the plants allow a direct control during the breeding process.

The sprouting sensibility of rye seems to be coupled with characteristical alpha-amylase isozymes.

Wild type properties in barley species hybrids partially can be proved by specific esterase bands in the electropherograms.

Literature

ENDO, T., 1973: Isozyme loci and a strategy of differentiation in plants. A review. – Seiken Ziho **24**, 89–104.

FLINTHAM, J. E., and M. D. GALE, 1980: The use of Gai/Rht 3 as a genetic base for low α-amylase wheats. Sprouting Symposium, England, 1979. – Cereal Research Communications **8**, 283–290.

GUSTAVUS, G., 1982: Charakterisierung von Gerstenartbastarden mit Hilfe von Isoenzymmustern. – Diss. A, Math.-Nat. Fak., Päd. Hochsch. Güstrow.

POHLER, W., and G. SZIGAT, 1982: Prüfung der Möglichkeit einer rekombinativen Genübertragung von der Wildgerste *Hordeum bulbosum* auf die Kulturgerste *H. vulgare*. I. Untersuchung der Rückkreuzung VV×BBVV. – Arch. Züchtungsforsch. (Berlin) **12**, 87–100.

SCANDALIOS, J. G., 1977: Genetic and developmental regulation of enzyme differentiation. – Proc. Confer. July 4th –9th 1977 at Halle "Regulation of developmental processes in plants"; p. 36–46.

Szigat, G., and C. Wustrack, 1976: Fertile Artbastarde bei Gerste (*Hordeum bulbosum* × ×*H. vulgare*). – Arch. Züchtungsforsch. (Berlin) **6**, 287–300.
Wilp, R., 1979: Beziehungen zwischen den Isoenzymmustern der α-Amylasen und dem Auswuchsverhalten von Winterroggen, *Secale cereale*, in Abhängigkeit vom Reifestadium. – Wiss. Zeitschr. Päd. Hochsch. Güstrow, Math.-Nat. Fak. **17**, 27–35.

R. Buschbeck, G. Gustavus, R. Wilp
Pädagogische Hochschule Güstrow
Sektion Biologie-Chemie, Bereich Botanik
DDR - 2600 Güstrow
Goldberger Straße 12

Kulturpflanze 32 · 1984 · S 245–S 248

Effect of nutrient concentration on the protein yield and amino acid composition of the protein complex in grain of maize hybrids

P. ANDONOVA, L. TYANKOVA, and T. KUDREV
(Sofia, Bulgaria)

The maize grain proteins are characterized by relatively low content of albumins and globulins, while prolamins and glutelins are presented in considerable amounts (IVANKO 1978, STEFANOV 1983). The fractional composition of grain proteins can undergo significant changes under the influence of ecological factors or genetic differences (IVANKO 1978, ANDONOVA et al. 1979, MÜNTZ et al. 1979, WOLTSON and SHAERER 1981). In our previous investigation some changes in the protein concentration, protein fractions and the ratio between them have been shown for maize hybrids grown on different nutrient media (ANDONOVA and DENCHEVA 1978, ANDONOVA and GEORGIEVA-BALABANOVA 1984).

The aim of the present investigation was to study some of the effects of different concentration of the main mineral elements on protein content, protein fractions and amino acid composition in two maize hybrids.

Material and Methods

Maize hybrids with different vegetative period, average-early hybrid Knezha 550 and late hybrid Knezha 2L 602, were grown under field conditions. The plants were irrigated, maintaining the soil moisture over 80 % of the field water holding capacity. The fertilizer doses during the different years were not the same and were calculated on the basis of the soil content of nutrient substances and the requirements of plants of basic elements for the formation of a definite yield (KUDREV and ANDONOVA 1979). Two levels of nutrient element concentration, providing for economically profitable and maximal grain yield of maize, were investigated. For three years, on the average, in maize growing for grain, the following fertilizer doses were applied: $T_1 - N_{14}P_6K_5$ and $T_2 - N_{24}P_{14}K_{14}$ kg/da, respectively (da–1,000 m^2). At the end of the vegetative period samples were taken from the grain of maize plants. Nitrogen content was determined by the method of Kjeldahl in the air-dried material. The fractional composition of protein was analyzed by consecutive extraction with distilled water, 0.5 % K_2SO_4, 80 % C_2H_5OH and 0.2 % NaOH. Each protein fraction was additionally purified by repeated precipitation and solubilization according to PLESHKOV (1968). The amino acid composition was determined on a gas-chromatograph, Pye Unicam–GCV, the amino acids were analysed as iso-butyl-hydroxy-carbonyl-methyl esters (MAKITA et al. 1976) after 72 h hydrolysis in 6 N HCL. All analyses were performed in three repetitions. The average data are presented in the tables.

Table 1
Yield, total nitrogen and protein fractions in maize grain under the influence of different concentrations of nutrients

Variants Hybrids	Yield of grain kg/da				Total nitrogen	Protein yield	Protein fractions/% N according to Kjeldahl/				
	1974	1975	1977	Average	% in grain	kg/da	Albumins	Globulins	Prolamins	Glutelins	Nonsoluble residue
Knezha 550											
T_1–$N_{14}P_6$ K_5	1,142	1,081	1,128	1,117	1.26	87.96	0.217	0.107	0.326	0.314	0.240
T_2–$N_{24}P_{14}K_{14}$	1,167	1,153	1,222	1,181	1.50	110.72	0.219	0.117	0.382	0.413	0.274
Knezha 2L 602											
T_1–$N_{14}P_6$ K_5	1,150	1,129	1,154	1,144	1.46	104.39	0.200	0.147	0.462	0.220	0.363
T_2–$N_{24}P_{14}K_{14}$	1,199	1,199	1,216	1,205	1.51	113.72	0.210	0.150	0.517	0.210	0.325
GD 5 %	45.70	38.19	18.85								
1 %	62.39	52.46	39.19								
0,1 %	84.81	72.11	52.79								

Table 2
Amino acid composition in protein fractions from maize at different concentrations of nutrients (% of the total sum)

Protein fractions	Variants	Alanine	Valine	Leucine	Isoleucine	Threonine	Serine	Proline	Aspartic acid	Hydroxyproline	Methionine	Glutamic acid	Phenylalanine	Tyrosine	Lysine
Albumins	Knezha T_1	2.12	9.21	9.12	9.86	1.24	1.90	1.30	1.42	7.28	9.82	10.77	29.00	4.42	2.54
	550 T_2	7.64	10.25	8.66	6.29	1.79	4.13	10.87	17.52	10.15	4.23	3.07	11.18	4.23	–
	Knezha T_1	2.01	8.27	8.01	5.17	1.95	3.81	8.43	19.51	19.16	14.59	1.33	6.53	1.21	–
	2L602 T_2	9.79	9.32	8.37	6.00	1.34	3.81	12.71	15.69	14.13	9.99	0.98	6.09	1.80	–
Globulins	Knezha T_1	2.02	3.96	5.41	2.75	1.92	0.98	6.98	3.30	4.93	8.83	10.32	24.42	8.94	15.25
	550 T_2	5.43	4.85	5.78	3.27	3.83	2.50	15.81	11.22	7.84	3.27	1.44	32.15	2.62	–
	Knezha T_1	4.48	7.37	10.16	6.01	4.95	0.39	9.90	13.51	19.69	9.60	1.47	10.47	2.00	–
	2L602 T_2	1.52	4.20	15.68	8.92	0.29	0.60	42.70	2.67	4.31	4.47	3.02	8.83	1.09	1.71
Prolamins	Knezha T_1	5.64	9.26	38.52	8.96	5.87	0.85	17.30	0.63	7.95	0.73	1.49	2.80	–	–
	550 T_2	1.08	1.22	2.46	1.37	1.54	0.77	3.95	4.29	9.68	10.89	5.03	29.19	18.97	10.1
	Knezha T_1	0.81	6.46	9.31	6.31	2.24	1.15	15.33	2.50	13.38	2.99	6.06	5.10	12.61	15.67
	2L602 T_2	0.94	2.28	2.69	7.57	1.47	0.74	5.39	1.68	6.46	6.51	3.73	18.43	14.79	27.30
Glutelins	Knezha T_1	1.25	8.86	12.11	5.10	1.61	1.31	34.19	1.50	9.90	2.84	2.90	15.54	2.28	0.61
	550 T_2	1.63	9.38	6.48	12.54	0.40	0.82	25.68	1.79	5.64	5.23	7.89	12.45	2.95	7.10
	Knezha T_1	8.28	6.34	8.78	3.05	2.16	1.46	28.18	6.46	16.91	10.27	1.29	5.48	1.32	–

Results and Discussion

Grain yield, total nitrogen and the protein yield increased considerably as a result of the higer fertilizing level (Table 1). Changes in protein fractions in the two hybrids were also established. In Knezha 550 the prolamins and the glutelins contents increased with the increase of total nitrogen and productivity. There were changes in prolamins of Knezha 2L 602 too, while the content of glutelins was not significantly changed at the two fertilization levels. In both the maize hybrids the changes in the content of albumins and globulins were insignificant under the same conditions.

The amino acid ratio in the protein fractions varied with the genetic specificity of both hybrids Knezha 550 and Knezha 2L 602. At the same nutrient level the variations between the hybrids in amino acid percentage of the fractions were negligible, as was the case of alanine, valine and leucine in the albumins or serine in the prolamins, but sometimes they were much pronounced (Table 2). The directions of the changes in the amino acid share were different under the influence of the mineral element level, as was the case of leucine in the glutelins which decreased to nearly 50 % in Knezha 550 and increased about two times in Knezha 2L 602. Generally, alterations in both maize hybrids from T_1 to T_2 were as follows: alanine, valine and proline increased, methionine and phenylalanine decreased in the albumins, proline increased and methionine decreased in the globulins. The share of methionine, phenylalanine, tyrosine and lysine increased, while that of valine, leucine, threonine and proline decreased in the prolamins. As regards the glutelins, the portion of valine, isoleucine, glutamic acid and lysine increased and proline and hydroxy proline decreased. The degree of the alterations was differently expressed and obviously depended on the hereditary peculiarities.

It is known that the nitrogen fertilization increases the total maize grain protein and reduces the concentrations of a number of amino acids as tryptophane, lysine, glycine, arginine and threonine (WOLTSON and SHAERER 1981). The present study has shown that when the nitrogen fertilization of maize for grain is balanced with suitable amounts of phosphorus and potassium the protein yield is enha ncedand the amino acid composition of proteins can be favourably influenced, because the essential amino acid lysine does not decrease in the basic protein fractions prolamins and glutelins. The level of the nutrients as well as the hereditary peculiarities has affected significantly the quantity of protein and amino acid composition in maize grain.

Literature

ANDONOVA, P., and S. DENCHEVA, 1978: A study on maize grain protein complex formation in the process of ripening under varying fertilizer application and irrigation treatments. – Fiziologia na rasteniata **4**, 27–35.

ANDONOVA, P., and R. GEORGIEVA-BALABANOVA, 1984: Influence of NSPKCa and Mg predomination in nutrient medium on the maize grain protein complex (investigation with ^{15}N). – Pochvosnanie i agrochimia, in press.

ANDONOVA, P., T. KUDREV, A. KATZAROV, and S. DENCHEVA, 1979: Alteration in the protein complex of grain in maize hybrids under the influence of mineral nutrition. –

Mineral Nutrition of Plants, Ed. by T. KUDREV, Publ. House of Central Cooperative Union, Sofia, 120–126.
IVANKO, ST., 1978: Formovanie proteinoveho complexu v zrne kukurice pocas dozrevania. – Biologicke prace **XXIV**, 1–178.
KUDREV, T., and P. ANDONOVA, 1979: Assessment of optimal macroelement ratios and concentrations in maize growing. – Fiziologia na rasteniata **2**, 71–80.
MAKITA, M., S. YAMAMOTO, K. SAKAI, and M. SHIRAISHI, 1976: Gas-liquid chromatography of the N-isobutyloxycarbonyl methyl esters of non-protein amino acids. – J. Chromatog. **124**, 92–96.
MÜNTZ, K., K. HAMMER, C. LEHMANN, A. MEISTER, A. RUDOLPH, and F. SCHOLZ, 1979: Variability of protein and lysine content in barley and wheat specimens from the world collection of cultivated plants at Gatersleben. Proc. IAEA/FAO Symp.: Seed protein improvement in cereals and grain legumes. **2**, 183–200. IAEA, Vienna.
PLESHKOV, B., 1968: Praktikum po biochimii rastenii. Moskva.
STEFANOV, B., 1983: Investigation of heterogeneity of glutelins in maize. – Dissertation Ph. D., Sofia.
WOLTSON, J., and G. SHAERER, 1981: Amino acid composition of grain protein of maize grown with and without pesticides and standard commercial fertilizers. – Agron. J., **73**, 611–613.

P. ANDONOVA
Institute of Plant Physiology "M. Popov"
Bulgarian Academy of Sciences
Acad. G. Bonchev str. bl. 21
Sofia 1113
Bulgaria

Kulturpflanze 32 · 1984 · S 249–S 251

Protein synthesis in *Agrostemma githago L.* seeds under stress conditions

K.-H. Köhler, U. Müller, M. Schelenz, and D. Bernhardt
Greifswald, GDR

"Priming" of seeds often improves their germination (Heydecker et al. 1973 Perl 1980). This effect is due to an imbibition of the seeds, interrupted by redrying at room temperature for one or some days before the radicle protrudes, followed by a renewed soaking or by an osmotic treatment. Very little, however, is known about the biochemical events during these stress conditions. In this paper previous investigations on RNA and DNA syntheses (Köhler et al. 1982) were completed by an analysis of protein synthesis. Moreover we compared the results of "primed" seeds to those of seeds treated osmotically with different concentrations of mannitol.

Experimental details of incorporation of radioactively labelled precursors were described by Hecker and Köhler (1979) and Köhler and Hecker (1981). Methods for germination are given by Köhler et al. (1982).

All seeds had a very high germination ability, which was not impaired by the redrying process. However, the velocity of radicle protrusion was strikingly enhanced (untreated controls started germination after 20 ... 22 h; seeds redried for 2 days had 20 % germination after 3 h, seeds redried for 7 days about 15 % after 6 h!).

The protein synthesis of such "primed" *Agrostemma* seeds, which begins at a higher level than in untreated seeds (after 2 hours only about 10 % incorporation of ^{14}C-leucine uptake) is demonstrated in fig. 1.

It should be mentioned that RNA- and DNA-syntheses in seeds imbibed for a second time are higher than after the first soaking (Köhler and Hecker 1981)!

Whereas 0.1 mol/l mannitol has nearly no influence on the germination behaviour of *Agrostemma* seeds (~96 % germination), 0.3 mol/l mannitol diminishes germination ability (only 40 % germination) and reduces germination velocity (radicle protrusion after about 60 hours). Benzylaminopurine (10^{-5} mol/l) clearly improves germination ability and velocity of osmotically stressed seeds. Due to the osmotic potential of mannitol protein synthesis is decreased and delayed compared to the untreated (H_2O) control (see fig. 2).

It can be concluded that, in contrast to vegetative parts of higher plants, in which protein synthesis is nearly totally inhibited, in embryos the protein synthesis is maintained at the same or at a somewhat decreased level even during severe stress, such as redrying or osmotic treatment. Evidently the enhanced germination velocity of "primed seeds" is correlated to a rapid increase in protein synthesis after renewed drying.

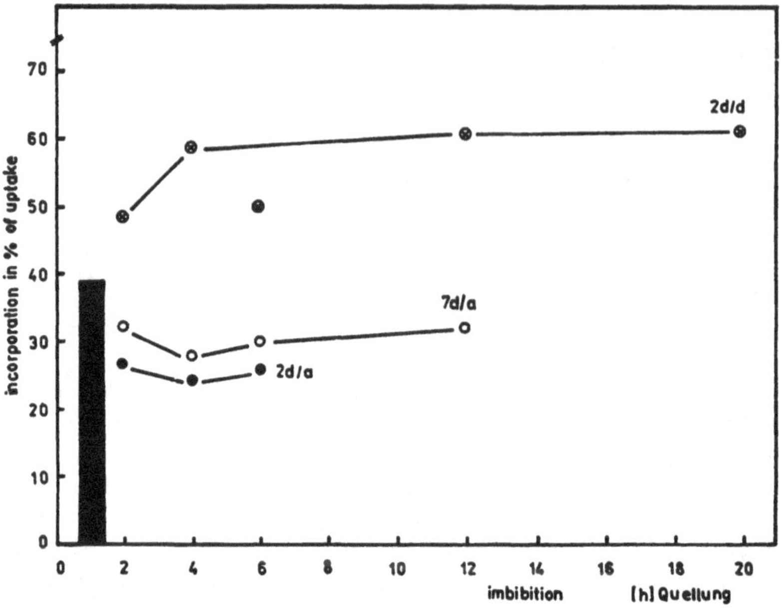

Fig. 1 Protein synthesis–measured as ^{14}C-leucine incorporation in % of uptake–in dormant (=d) and afterripened (=a) seeds after 2 and 7 days, respectively (=2d or 7d) , redried at 20 °C (the first imbibition was interrupted after 18 hours), and again soaked. The intensity of protein synthesis after 18 hours, i.e. at the interruption point, is shown by the black column

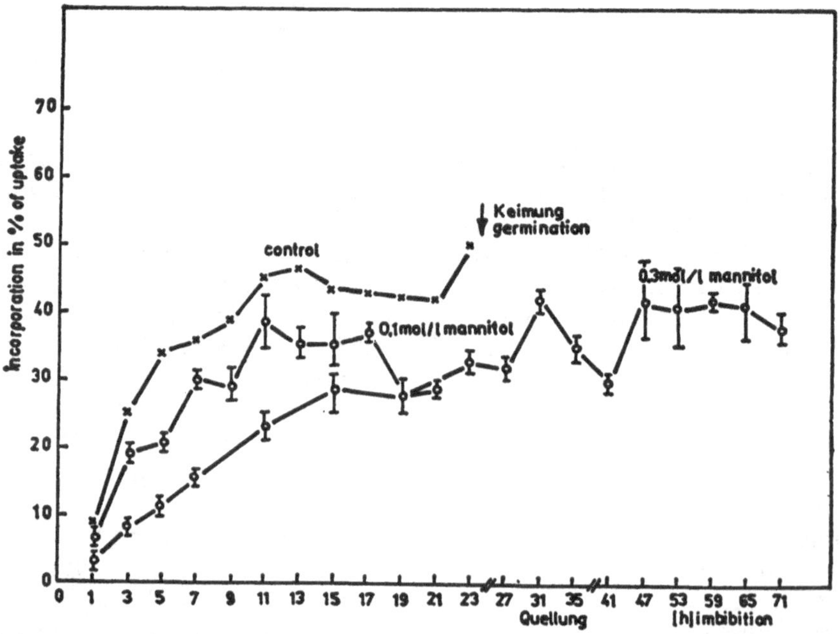

Fig. 2 Protein synthesis (see also fig. 1) in afterripened seeds, soaked with 0.1 mol/l and 0.3 mol/l mannitol, respectively. Radicle protrusion in the untreated controls (=distilled water) is indicated by the arrow

Literature

HECKER, M., and K.-H. KÖHLER, 1979: Protein and nucleic acid synthesis during imbibition of dormant and after-ripened *Vaccaria pyramidata* seeds. – Devel. Biol. **69**, 270–280.
HEYDECKER, W., J. HIGGINS, and R. L. GULLIVER, 1973: Accelerated germination by osmotic seed treatment. – Nature **246**, 42–44.
KÖHLER, K.-H., and M. HECKER, 1981: Biochemical changes in embryos of dormant and after-ripened seeds of Caryophyllaceae. – Acta Soc. Bot. Poloniae **50**, 279–288.
KÖHLER, K.-H., U. MÜLLER, M. HECKER, and D. BERNHARDT, 1982: Synthesis of proteins, RNA and DNA in dormant and afterripened Caryophyllaceae seeds. – Biol. Plant. (Praha) **24**, 354–362.
PERL, M., 1980: An ATP-synthesizing system in seeds. – Planta **149**, 1–6.

K.-H. KÖHLER
Sektion Biologie
der Ernst-Moritz-Arndt-Universität,
Wissenschaftsbereich Botanik
DDR - 2200 Greifswald
Grimmer Str. 88

Kulturpflanze 32 · 1984 · S 253–S 254

Species specific signal peptide cleavage of plant storage protein precursors in the endoplasmic reticulum

E. WEBER and A. BRANDT
Gatersleben, GDR, and Copenhagen, Denmark

Introduction

Storage protein polypeptides of plant seeds are synthesized on membrane bound polysomes of the rough endoplasmic reticulum (rER) as higher molecular weight precursors. Thus, globulins of legume cotyledons and oat caryopses as well as prolamins of corn and barley caryopses correspond to secretory proteins, e.g. a N-terminal signal peptide is cleaved off while the precursor polypeptide is vectorially discharged through the rER membrane.

We have studied the reconstitution of rough microsomes *in vitro* from barley mRNA/membrane bound polysomes (mbp) and smooth microsome membranes derived from immature barley endosperms and legume cotyledons, in order to analyse whether signal peptide cleavage takes place in hybrid microsomes reconstituted from different plant species.

Results

Initially membrane bound polysomes (mbp), mRNA, rough (RM) and smooth microsomes (SM) were isolated from cotyledons of *Vicia faba* and *Phaseolus vulgaris* as well as from endosperms of Bomi barley and mutant Risø 1508. All templates were able to programme cell-free protein synthesis in a wheat germ derived system, except the smooth microsomes. The cell-free translation of mbp was not inhibited by the cotranslational addition of SM from different plant species, whereas the mRNA translation was inhibited by 40–60 % irrespectively of the origin of SM.

Barley endosperm mbp and mRNA direct the synthesis of higher molecular weight precursor polypeptides to hordein storage proteins which are converted to the size of the native hordein polypeptides by rough microsome membranes, indicating the cleavage of a 2,000 dalton signal peptide.

Reconstitution of rough microsomes employing barley endosperm mbp or mRNA and Bomi barley smooth microsome membranes resulted in functional reconstitution as defined by the ability of the reconstituted membranes to transport and process the hordein precursor polypeptides. On the contrary, these were not transported and processed by SM from the hordein deficient barley mutant Risø 1508 as well as by SM from cotyledons of *Vicia faba* and *Phaseolus vulgaris* but remain accessible to proteolytic degradation as indicated by the almost complete digestion of the precursor polypeptides during post-translational protease treatment using sufficient enzyme concentrations.

Conclusions

Hordein precursor polypeptides are vectorially discharged into the lumen of the rough microsomes reconstituted from mRNA or mbp and smooth microsomes from Bomi barley endosperm accompanied by the signal peptide cleavage.

Transport and signal peptide cleavage of the hordein precursor polypeptides was not achieved by using SM derived from the barley mutant Risø 1508 which is deficient in hordein biosynthesis, as well as SM from cotyledons of *Vicia faba* and *Phaseolus vulgaris*.

E. Weber
Zentralinstitut für Genetik und Kulturpflanzenforschung
der Akademie der Wissenschaften der DDR
DDR - 4325 Gatersleben
Corrensstraße 3

A. Brandt
Carlsberg Laboratory
Department of Physiology
Gl. Carlsberg Vej 10
DK 2500 Copenhagen Valby

Kulturpflanze 32 · 1984 · S 255–S 257

Variability of Met⁺ and Met⁻ components of the 11S globulins from 3 different grain legumes and within the species of *Vicia faba* L.

K. Müntz, R. Bassüner, and C. Horstmann
Gatersleben, GDR

The discovery of methionine-containing (Met^+) and methionine-free (Met^-) legumin polypeptides (Horstmann 1983) as separate gene products (Bassüner et al. 1983, Wobus et al. 1984) in cotyledons of the field bean (*Vicia faba* L. var. *minor* cv. Fribo) led to the question whether an intraspecies variability of the ratio of Met^+ to Met^- legumin polypeptides exists, that eventually could be useful for plant breeders, and whether 11S globulins from other grain legumes are also composed of Met^+ and Met^- components.

The variability of the quantity of methionine in the total protein (relative Met content) of approximately 500 specimens from the Gatersleben *Vicia* collection had been screened previously (Hanelt et al. 1978). Since at present no collection of inbred *Vicia* lines was available, we started the investigation concerning a variability in the ratio of Met^+ to Met^- legumin polypeptides with 30 specimens of *Vicia faba* L. selected from the above mentioned collection, 10 of them representing the group with the highest, 10 the group with an intermediary, and the final 10 the group with the lowest relative Met content.

Defatted flour of the cotyledons from always 10 seeds per specimen was extracted according to Lichtenfeld et al. (1979). Similar protein quantities were fractionated on 12 percent SDS polyacrylamide gel slabs subsequent to globulin denaturation under reducing conditions. Stained legumine zones were densitometrically scanned under conditions that guaranteed scans to obey Lambert-Beer's law. Peak ratios of Met^+ to Met^- β-legumin polypeptides were taken to reflect the situation with total major legumin components.

The majority of genotypes from the group with high relative Met content exhibits Met^+ to Met^- β-legumin peak ratios that are statistically significant greater than one (Table 1). Contrary, no Met^+ to Met^- β-legumin peak ratio significantly smaller than one was found in the groups with middle or low relative Met content, though minor differences have been observed in this direction. In addition, variability in the electrophoretic mobility of α-legumin polypeptides was observed.

In vitro-translation of poly(A)-containing RNA from developing field bean cotyledons using [^{35}S]-methionine in the cell-free wheat germ system revealed that only preprolegumin A which represents the methionine-containing legumin polypeptide was labelled (Bassüner et al. 1983). The nonlabelling of methionine-free preprolegumin B could be explained by the lack of Met in the signal peptide as well as in the prolegumin B and by splitting off the N-terminal initiator

Table 1

Group of *Vicia* specimens with high relative Met content (HANELT et al. 1978). Statistical evaluation of the differences between peak areas (mm^2) for Met^+ and Met^- β-legumin bands from SDS-gels. a) and b) represent means from duplicates with different protein concentrations per gel. Differences were evaluated with the t-test for paired values according to Student. + – significant difference at 1 percent level, (+) – at 5 percent level; – no significant difference. % differences refer to the Met^- β-legumin peak

Specimen no.		β-legumin Met^+	Met^-	Differences %	Statistical evaluation
1	a)	26.4	15.6	69	+
	b)	29.0	19.4	49	
2	a)	20.8	13.5	54	+
	b)	25.1	18.6	35	
3	a)	24.8	19.2	29	+
	b)	26.0	22.2	17	
4	a)	28.5	20.4	40	–
	b)	not evaluable			
5	a)	27.5	15.0	83	+
	b)	25.3	17.1	48	
6	a)	25.7	14.8	71	+
	b)	33.1	18.9	75	
7	a)	11.9	14.0	–15	–
	b)	not evaluable			
8	a)	30.2	22.7	33	+
	b)	36.1	24.7	46	
9	a)	23.9	20.4	17	(+)
	b)	27.9	27.5	1.5	
10	a)	27.8	25.0	11	–
	b)	30.0	34.0	–12	

methionine in our wheat germ extract. Cell-free translation of poly(A)-containing RNA from actively globulin synthesizing developing pea (*Pisum sativum* L.) and soybean (*Glycine max* (L.) MERR.) cotyledons led to the labelling of both 11S globulin precursor polypeptides. CNBr-cleavage experiments with purified α- and β-legumin chains from peas and field beans that were performed in our laboratory revealed the total splitting of pea legumin chains whereas part of the corresponding field bean polypeptides always remained uncleaved. Since CNBr cleavage and Met were already demonstrated for all glycinin polypeptides (IYENGAR and RAVENSTEIN 1981, MOREIRA et al. 1981) the results from translation and splitting experiments agree well. In comparison of the 3 investigated grain legumes only *Vicia faba* legumin seems to be composed of Met^+ and Met^- polypeptides whereas all 11S globulin polypeptides from *Pisum sativum* and *Glycine max* contain Met residues.

Literature

BASSÜNER, R., R. MANTEUFFEL, K. MÜNTZ, M. PÜCHEL, P. SCHMIDT, and E. WEBER, 1983: Analysis of *in vivo* and *in vitro* globulin formation during cotyledon development of field beans (*Vicia faba* L. var. *minor*). – Biochem. Physiol. Pflanzen **178**, 665–684.

HANELT, P., A. RUDOLPH, K. HAMMER, H.-W. JANK, K. MÜNTZ und F. SCHOLZ, 1978: Eiweißuntersuchungen am Getreide- und Leguminosen-Sortiment Gatersleben. Teil 3: Gehalt an Rohprotein und schwefelhaltigen Aminosäuren der Ackerbohnen (*Vicia faba L.*). – Kulturpflanze **26**, 183–212.

HORSTMANN, C., 1983: Specific subunit pairs of legumin from *Vicia faba*. – Phytochemistry **22**, 1861–1866.

IYENGAR, R. B., and P. RAVENSTEIN, 1981: New aspects of subunit structure of soybean glycinin. – Cereal Chem. **58**, 325–330.

LICHTENFELD, C., R. MANTEUFFEL, K. MÜNTZ, D. NEUMANN, G. SCHOLZ, and E. WEBER, 1979: Protein degradation and proteolytic activities in germinating field beans (*Vicia faba* L. var. *minor*). – Biochem. Physiol. Pflanzen **174**, 255–274.

MOREIRA, M. A., M. A. HERMODSON, B. A. LARKINS, and N. C. NIELSEN, 1981: Comparison of the primary structure of the acidic polypeptides of glycinin. – Arch. Biochem. Biophys. **210**, 633–642.

WOBUS, U., H. BÄUMLEIN, R. BASSÜNER, U. HEIM, R. JUNG, K. MÜNTZ, R. PANITZ, G. SAALBACH, and W. WESCHKE, 1984: Molecular characterization of *Vicia faba* storage protein specific DNA. – Proc. 3rd Seed Protein Symp. 'Genetics of Seed Proteins' (MÜNTZ, K., and C. HORSTMANN, eds.). – Kulturpflanze **32**, S 117–S 128.

K. MÜNTZ, R. BASSÜNER, C. HORSTMANN
Zentralinstitut für Genetik und Kulturpflanzenforschung
der Akademie der Wissenschaften der DDR
DDR - 4325 Gatersleben
Corrensstraße 3

Kulturpflanze 32 · 1984 · S 259–S 261

Biosynthesis of Met^+ and Met^- legumin polypeptide chains during seed development of field beans (*Vicia faba* L. var. *minor* cv. Fribo)

K. MÜNTZ
Gatersleben, GDR

HORSTMANN (1983, 1984) demonstrated the occurrence of methionine-containing (Met^+) and methionine-free (Met^-) subunits of *Vicia faba* legumin, which are expressed from different legumin genes (BASSÜNER et al. 1983, WOBUS et al. 1984). Since the ratio of Met^+ to Met^- legumin polypeptides was shown to vary between different *Vicia faba* genotypes (MÜNTZ et al. 1984) the *in vivo* pattern of Met^+ and Met^- legumin subunit biosynthesis became of interest.

Cotyledon pairs were isolated from field grown *Vicia faba* seeds of different developmental stages and selected to give material of uniform cotyledon length for each harvest. Whereas one cotyledon of each pair was used for labelling with a mixture of [^{3}H]-amino acids (Amersham, 37 MBq/ml) the sister cotyledons were always incubated with [^{35}S]-methionine (Isocommerz, 45,7 GBq/mMol) applying the labelling technique of SPENCER et al. (1980). After 3 hrs incubation at room temperatur in weak illumination extraction and electrophoretic fractionation of globulins was performed as in MÜNTZ et al. (1984). Subsequent to densitometric scanning the β-legumin polypeptide region of stained gels was cut into 1 mm slices using a Mickel Automatic Gel Slicer and the radioactivity per slice was determined. The experiment was performed in duplicate; 3 identical gels were always run and used for densitometric or radioactivity evaluation.

No labelling was observed in 5 mm cotyledons (not shown here). An increasing tritium labelling took place when the cotyledon length grew to 7 and 9 mm; 12 mm cotyledons exhibited a similar percentage distribution of [^{3}H]-labelling as the 9 mm harvest. Contrary, no radioactivity peak was found in β-legumin chains of 7 mm cotyledons that were incubated with [^{35}S]-methionine. Only 9 and 12 mm cotyledons exhibited a strong rise in [^{35}S]-labelling of β-legumin (fig. 1). Since the ratio between cotyledon fresh matter and the amount of radioactivity applied per cotyledon changed in our experiments depending on the respective developmental stage, the percentage distribution as well as the ratio between the slices of corresponding gels from [^{3}H]- and [^{35}S]-labelled cotyledons were included into figure 1b. For orientation cotyledon length, fresh weight, and protein content of the different cotyledon harvests are given in fig. 1a.

Since in the native state Met^+ α-legumin is always paired with Met^+ β-legumin and Met^- α-legumin with Met^- β-legumin chains, respectively (HORSTMANN 1983), the pattern of β-legumin chain formation should reflect the situation with total major legumin components. On this basis Met^+ legumin subunit synthesis seems to be retarded in comparison to the formation of Met^- legumin chains, at least

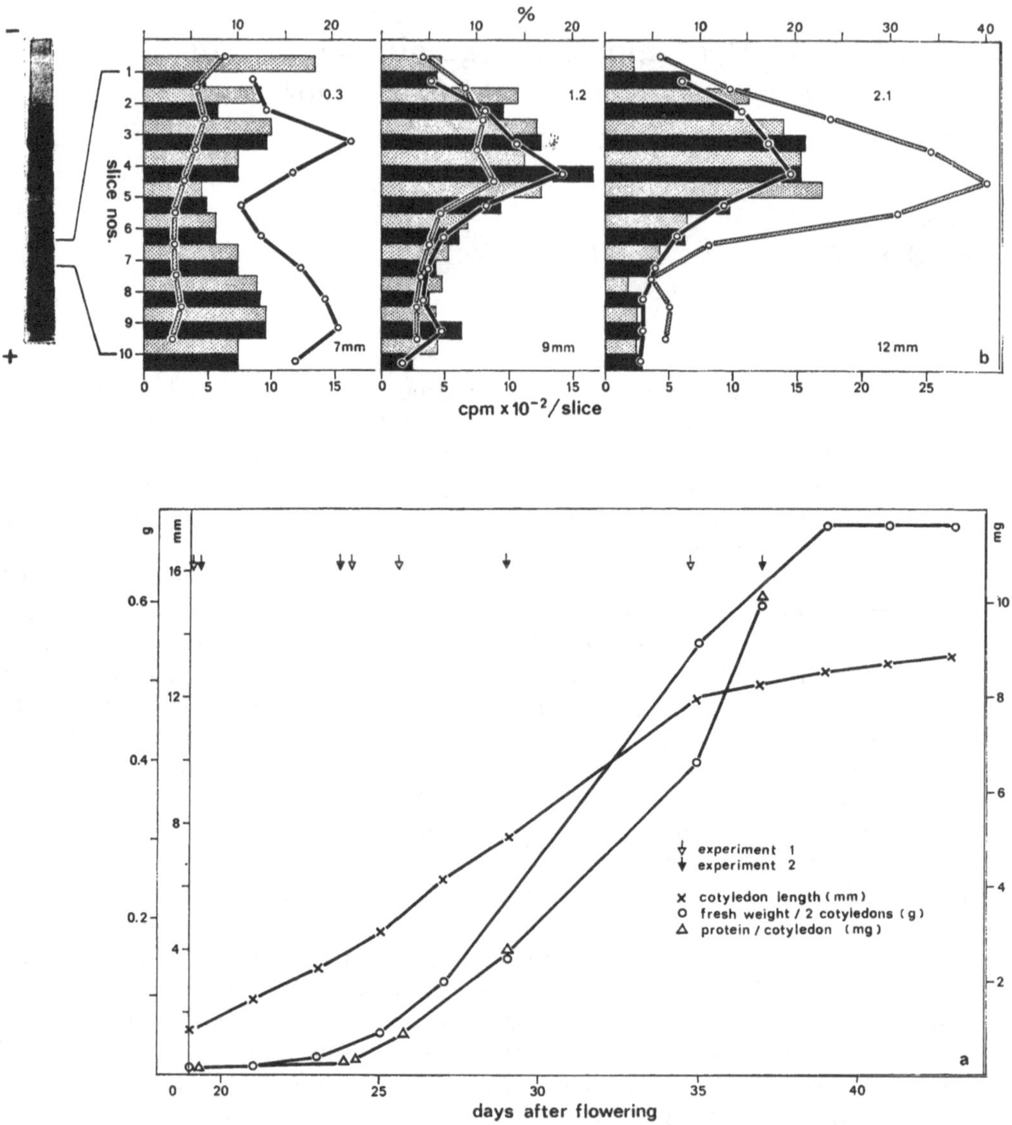

Fig. 1 Incorporation of either [^{3}H]-labelled amino acid mixture or [^{35}S]-methionine into β-legumin chains from globulins of developing field bean cotyledons. (a) Growth and protein accumulation of developing field bean cotyledons; harvests are indicated by white (1st experiment) or black arrows (2nd experiment). (b) Radioactivity and its percentage distribution found in the gel slices of the β-legumin regions; dotted lines and columns—[^{35}S]-methionine labelling, black lines and columns—[^{3}H]-amino acid labelling. The [^{35}S] to [^{3}H] radioactivity ratios are indicated for those slices exhibiting the highest radioactivity. The β-legumin bands-containing gel region that had been sliced is indicated on the gel foto on the left of the figure

in the beginning of the period of strong storage globulin formation. This retardation is reflected in the changing ratio of [^{35}S]- to [^{3}H]-incorporation into the preferentially labelled slices from the β-legumin bands of sister cotyledon pairs. Scans of the β-legumin polypeptide patterns from non-labelled cotyledons support this interpretation.

Literature

Bassüner, R., R. Manteuffel, K. Müntz, M. Püchel, P. Schmidt, and E. Weber, 1983: Analysis of *in vivo* and *in vitro* globulin formation during cotyledon development of field beans (*Vicia faba* L. var. *minor*). – Biochem. Physiol. Pflanzen **178**, 665–684.

Horstmann, C., 1983: Specific subunit pairs of legumin from *Vicia faba*. – Phytochemistry **22**, 1861–1866.

–, 1984: New aspects of the subunit structure of *Vicia faba* legumin. – Proc. 3rd Seed Protein Symp. 'Genetics of Seed Proteins' (Müntz, K., and C. Horstmann, eds.). Kulturpflanze **32**, S 109–S 116.

Müntz, K., R. Bassüner, and C. Horstmann, 1984: Variability of Met^{+} and Met^{-} components of the 11 S globulins from 3 different grain legumes and within the species *Vicia faba* L. – Proc. 3rd Seed Protein Symp. 'Genetics of Seed Proteins' (Müntz, K., and C. Horstmann, eds.). Kulturpflanze **32**, S 263–S 265.

Spencer, D., T. J. V. Higgins, S. C. Button, and R. A. Davey, 1980: Pulse-labelling studies on protein synthesis in developing pea seeds and evidence of a precursor form of legumin small subunit. – Plant Physiol. **66**, 510–515.

Wobus, U., H. Bäumlein, R. Bassüner, U. Heim, R. Jung, K. Müntz, R. Panitz, G. Saalbach, and W. Weschke, 1984: Molecular characterization of *Vicia faba* storage protein specific DNA. – Proc. 3rd Seed Protein Symp. 'Genetics of Seed Proteins' (Müntz, K., and C. Horstmann, eds.). Kulturpflanze **32**, S 117–S 128.

K. Müntz
Zentralinstitut für Genetik und Kulturpflanzenforschung
der Akademie der Wissenschaften der DDR
DDR - 4325 Gatersleben
Corrensstraße 3

List of Participants

List of Participants

BELGIUM

1 Peumans, W. — Laboratorium voor Plantenbiochemie, Katholieke Universiteit Leuven, Leuven

2 Stinissen, H. — Laboratorium voor Plantenbiochemie, Katholieke Universiteit Leuven, Leuven

BULGARIA

3 Andonova, P. — Institute of Plant Physiology „M. Popov“, Bulgarian Academy of Sciences, Sofia

CUBA

4 Travieso, A. — National Institute of Basic Research in Tropical Agronomy, Academy of Sciences of Cuba, Santiago de las Vegas

FRG

5 Feix, G. — Institut für Biologie III der Universität Freiburg, Freiburg i. Br.

6 Müller, H. P. — Abt. Biochemische Genetik der Universität Bonn, Bonn a. Rhein

GDR

7 Adler, K. — Zentralinstitut für Genetik und Kulturpflanzenforschung der AdW der DDR, Gatersleben

8 Bäumlein, H. — Zentralinstitut für Genetik und Kulturpflanzenforschung der AdW der DDR, Gatersleben

9 Bassüner, R. — Zentralinstitut für Genetik und Kulturpflanzenforschung der AdW der DDR, Gatersleben

10 Bernhardt, D. — Sektion Biologie der Ernst-Moritz-Arndt-Universität, Greifswald

11 Blüthner, W. D. — Sektion Pflanzenproduktion der Martin-Luther-Universität Halle – Wittenberg, Hohenthurm

12	Buschbeck, R.	Pädagogische Hochschule „Liselotte Herrmann", Güstrow
13	Hahne, J.	Institut für Getreideforschung der AdL der DDR, Bernburg
14	Horstmann, C.	Zentralinstitut für Genetik und Kulturpflanzenforschung der AdW der DDR, Gatersleben
15	Knoll, P.	Forschungszentrum für Molekularbiologie und Medizin der AdW der DDR, Berlin-Buch
16	Köhler, K.-H.	Sektion Biologie der Ernst-Moritz-Arndt-Universität, Greifswald
17	Kraft, R.	Zentralinstitut für Molekularbiologie der AdW der DDR, Berlin-Buch
18	Lehmann, C. O.	Zentralinstitut für Genetik und Kulturpflanzenforschung der AdW der DDR, Gatersleben
19	Lichtenfeld, C.	Zentralinstitut für Genetik und Kulturpflanzenforschung der AdW der DDR, Gatersleben
20	Jung, R.	Zentralinstitut für Genetik und Kulturpflanzenforschung der AdW der DDR, Gatersleben
21	Merbach, W.	Sektion Pflanzenproduktion der Martin-Luther-Universität Halle – Wittenberg, Halle
22	Müntz, K.	Zentralinstitut für Genetik und Kulturpflanzenforschung der AdW der DDR, Gatersleben
23	Nover, L.	Institut für Biochemie der Pflanzen der AdW der DDR, Halle
24	Ostermann, B.	Institut für Getreideforschung der AdL der DDR, Bernburg
25	Otto, A.	Zentralinstitut für Molekularbiologie der AdW der DDR, Berlin-Buch
26	Parthier, B.	Institut für Biochemie der Pflanzen der AdW der DDR, Halle
27	Plietz, P.	Zentralinstitut für Molekularbiologie der AdW der DDR, Berlin-Buch
28	Prahl, L.	Zentralinstitut für Ernährung der AdW der DDR, Bergholz-Rehbrücke
29	Richter, K.	Institut für Getreideforschung der AdL der DDR, Bernburg
30	Saalbach, G.	Zentralinstitut für Genetik und Kulturpflanzenforschung der AdW der DDR, Gatersleben
31	Scharf, K. D.	Institut für Biochemie der Pflanzen der AdW der DDR, Halle
32	Schlesier, B.	Zentralinstitut für Genetik und Kulturpflanzenforschung der AdW der DDR, Gatersleben

33 SCHOLZ, F. — Zentralinstitut für Genetik und Kulturpflanzenforschung der AdW der DDR, Gatersleben
34 TRÖBNER, W. — Sektion Biowissenschaften der Martin-Luther-Universität Halle – Wittenberg, Halle
35 WEBER, E. — Zentralinstitut für Genetik und Kulturpflanzenforschung der AdW der DDR, Gatersleben
36 WOBUS, U. — Zentralinstitut für Genetik und Kulturpflanzenforschung der AdW der DDR, Gatersleben

ITALY
37 BOLLINI, R. — Istituto Biosintesi Vegetali, C. N. R., Milano
38 GRILLO, S. — Institute of Agronomy, University of Naples, Portici
39 LAFIANDRA, D. — Germplasm Institute, Bari
40 PERRINO, P. — Germplasm Institute, Bari
41 RAO, R. — Institute of Agronomy, University of Naples, Portici

JAPAN
42 FUKAZAWA, C. — Division of Biochemistry, National Food Research Institute, Tsukuba Academic Town, Yatabe, Ibaraki
43 MORI, T. — Research Institute for Food Sciences, Kyoto University, Uji City, Kyoto

MEXICO
44 VASAL, S. K. — International Maize and Wheat Improvement Center (CIMMYT), Mexico

POLAND
45 PRZYBYLSKA, J. — Institute of Plant Genetics, Polish Academy of Sciences, Poznań
46 ZIMNIAK-PRZYBYLSKA, Z. — Institute of Plant Genetics, Polish Academy of Sciences, Poznań

SPAIN
47 GARCIA-OLMEDO, F. — Department of Biochemistry, E. T. S. Ingenieros Agronomos, Madrid

U. K.
48 BARTELS, D. — Plant Breeding Institute, Cambridge
49 CASEY, R. — John Innes Institute, Norwich
50 CROY, R. R. D. — University of Durham, Dept. of Botany, Durham
51 JACKSON, E. A. — Plant Breeding Institute, Cambridge

52	Payne, P. I.	Plant Breeding Institute, Cambridge
53	Shewry, P. R.	Rothamsted Experimental Station, Harpenden, Herts.
	USA	
54	Kasarda, D. D.	Western Regional Research Center, U. S. Dept. of Agriculture, Berkeley, California
55	Rubenstein, I.	University of Minnesota, Dept. of Genetics and Cell Biology, St. Paul, Minnesota
	USSR	
56	Egorov, Ts. A.	Institute of General Genetics, USSR Academy of Sciences, Moscow
57	Metakovsky, E. V.	Institute of General Genetics, USSR Academy of Sciences, Moscow

II. Übersichtsdarstellungen (Reviews)

Kulturpflanze 32 · 1984 · 11–34

Das Domestikationssyndrom

KARL HAMMER

(Eingegangen am 31. Januar 1984)

Zusammenfassung

Voraussetzung und Wege, die zur Domestikation führten, werden untersucht und die Ergebnisse dieses Evolutionsprozesses diskutiert. Schon von Beginn an läßt sich eine Zweiteilung in Kulturpflanzen des großflächigen Feldbaues, die über das Saatgut vermehrt werden und bei denen die Massenauslese vorherrscht, und Kulturpflanzen des kleinflächigen Gartenbaues, bei denen Formen der vegetativen Vermehrung und Individualauslese überwiegen, verfolgen. Feldbau führt zu ausgewogenen Pflanzenbeständen mit einem guten allgemeinen Ertragspotential, ohne daß sich einzelne Merkmale oder Merkmalsgruppen durch eine überdimensionale Entwicklung auszeichnen. Gartenkultur resultiert oft in beträchtlichen Merkmalsabänderungen (allometrisches Wachstum). An drei Beispielen werden Ablauf und Ergebnisse der Domestikation untersucht. *Hordeum vulgare* (primäre Kulturpflanze) und *Papaver somniferum* (sekundäre Kulturpflanze) sind Vertreter des Feldbaus. Die Bedingungen des Feldbaus bieten günstige Voraussetzungen für eine unbewußte und indirekte Domestikation von Pflanzen, die später selbst zu Kultursippen werden können (sekundäre Kulturarten). Aber auch Unkräuter, die für eine spätere Nutzung nicht in Frage kommen, werden hier indirekt domestiziert. Unter den Bedingungen der Gartenkultur dominiert die bewußte und direkte Selektion. Als Beispiel wird hier die Domestikation von *Lycopersicon esculentum* untersucht.

Obwohl der Ablauf der Domestikation in Abhängigkeit von den Arten recht kompliziert sein kann und eine Fülle von unterschiedlichen Formen hervorgebracht wird, gestattet das gesetzmäßige Auftreten von Gruppen spezifischer Kulturpflanzenmerkmale die Postulierung des Domestikationssyndroms. Eine Abgrenzung der Domestikationserscheinungen von ähnlichen Merkmalsentwicklungen bei kolonisierenden und ruderalen Arten wird vorgenommen.

Die Kulturpflanzen sind Produkt der Kulturentwicklung der Menschheit ebenso wie die Haustiere. Parallelen in der Phänomenologie bei Kulturpflanzen und Haustieren lassen auch bei diesen das Domestikationssyndrom erkennen.

Einleitung

Der Mensch hat im Laufe der Evolution in zunehmendem Maße seine Umwelt modifiziert. Besonders gravierend waren und sind seine Eingriffe in die Biosphäre. Die Reaktion der Organismen ist unterschiedlich, einige finden günstige Lebensräume in der sich verändernden Umwelt, andere, meist weniger anpassungsfähige Sippen, werden zurückgedrängt oder sterben sogar aus.

Vor etwa 10000 Jahren begann eine neue Etappe der Einflußnahme des Menschen – er begann Tiere zu zähmen und Pflanzen in Kultur zu nehmen. Damit wirkte er unmittelbar auf sie ein, indem er von Beginn an auch bewußt selektierte. Mehr oder weniger drastische Veränderungen in den Populationen waren die Folge, Kulturpflanzen und Haustiere das Ergebnis dieser Bemühungen. Der Prozeß der Überführung in den Hausstand wird als Domestikation bezeichnet. Domestizierte Pflanzen und Tiere unterscheiden sich meist in charakteristischer Weise von ihren wilden Ausgangssippen durch eine Gruppe spezifischer Merkmale. Bei den Pflanzen ist dafür der Begriff Domestikationssyndrom geprägt worden (Faegri 1981), offenbar in Anlehnung an bestimmte komplexe bestäubungsökologische Merkmalsausprägungen im Zuge der Koevolution von Bestäubern und Blumen (Faegri und van der Pijl 1979).

Das Domestikationssyndrom ist das Ergebnis eines oft Jahrhunderte oder sogar Jahrtausende währenden Evolutionsprozesses und kann im Extremfall in so ausgeprägter Form vorliegen, daß eine Kulturpflanze sich nicht ohne das Zutun des Menschen auch nur über eine Generation am Leben erhalten kann (z. B. Mais). Zur Untersuchung dieses Syndroms sind u. a. Dauer, Intensität und Qualität der Selektionsvorgänge unter dem Einfluß des Menschen zu verfolgen, außerdem Variabilität, Verbreitung und weitere botanische Merkmale der Kulturpflanzen und ihrer wildwachsenden Ausgangssippen. Diese oft nur in interdisziplinärer Zusammenarbeit möglichen Untersuchungen (vgl. Stubbe 1982, p. 11) müssen in der vorliegenden Darstellung weitestgehend auf den botanischen Aspekt reduziert bleiben. Querverbindungen zur Domestikation der Haustiere können nur gelegentlich gezogen werden.

Das Konzept von Ackerbau und Viehzucht

Archäologische Funde weisen aus, daß der Ackerbau und auch die Viehzucht an mehreren Stellen der Welt unabhängig (vgl. Merrill 1930) vor etwa 10000 Jahren entstanden sind. Dabei existiert der Mensch schon etwa 2 Millionen Jahre als Jäger und Sammler, und etwa seit 50000 Jahren gibt es den Menschen in der hochentwickelten Form des *Homo sapiens*.

Ein Charakteristikum des Übergangs zu Ackerbau und Viehzucht ist, daß es sich dabei um ein gleichsam universelles Ereignis gehandelt hat. Vor etwas mehr als 10000 Jahren lebten alle Menschen auf der Basis wildwachsender und -lebender Nahrungsquellen. Schon vor etwa 2000 Jahren ernährte sich dann ihre überwältigende Mehrheit von Ackerbau und Viehzucht. Innerhalb von nur etwa 8000 Jahren haben sich also Ackerbau und Viehzucht weltweit verbreitet. Damit stellt sich die Frage, warum der Mensch Ackerbau und Viehzucht als Strategie dem Jagen und Sammeln vorzog und warum sich diese Strategie in einer relativ kurzen Zeitspanne durchsetzte, unabhängig davon, ob das Konzept von Ackerbau und Viehzucht erfunden oder erlernt wurde. Zahlreiche Hypothesen sind zu diesem Problem aufgestellt worden, einschließlich relativ komplizierter Entwicklungsmodelle (z. B. HARRIS 1977), die nur Teilaspekte der oben erwähnten Phänomene widerspiegeln. Als Hauptursache wird von COHEN (1977) zusammenfassend das Bevölkerungswachstum und die damit einhergehende Erhöhung des Populationsdruckes angesehen, die kontinuierlich das Ökosystem beeinflussen und die Anpassung der menschlichen Kultur daran bestimmen. Das Wachstum von Jäger- und Sammler-Populationen bis zu einer gewissen Schwelle, die vielleicht als Sättigungsniveau zu bezeichnen wäre, könnte den Streß hervorgerufen haben, der Ackerbau und Viehzucht für viele menschliche Populationen notwendig machte.

Menschliche Populationen haben ein enormes Wachstumspotential. Der Hauptunterschied zwischen der biologischen Evolution und der Kulturentwicklung der Menschheit (nach VOGEL, 1983, ist der Begriff „Kulturevolution" möglichst zu vermeiden, weil anstelle der streng physisch-genetisch gebundenen Weitergabe von Informationen neue Regeln und Mechanismen getreten sind, die als Kommunikation bezeichnet werden) ist, daß Tiere und auch Pflanzen mit relativ definierten ökologischen Ansprüchen auf einen Streß durch Überbevölkerung auf lange Sicht mit einer erhöhten Sterblichkeitsrate und nur vereinzelt durch ökologische Anpassung antworten. Der Mensch mit seiner größeren Verhaltensflexibilität und ökologischen Anpassungsfähigkeit, der als einzelne Art in einem ökologischen Bereich dominiert, der in der transspezifischen Evolution der Tiere einer taxonomischen Großgruppe vom Rang eines Stammes oder einer Klasse entspricht (OSCHE 1983), kann oftmals dem Prozeß der Überbevölkerung durch entsprechende Anpassung seines Verhaltens ausweichen und trotzdem seine Population vergrößern. Hier finden sich die Wurzeln für Ackerbau und Viehzucht, und es war nicht Unwissenheit, sondern der Mangel an Notwendigkeit, der einige Völker bzw. Stämme z. T. auch heute noch davon abhält, Ackerbau und Viehzucht zu betreiben (HARLAN 1975).

Wenn auch die Anfänge von Ackerbau und Viehzucht sich nur auf einen wesentlichen Auslösefaktor, das Wachstum der menschlichen Bevölkerung, zurückführen lassen, so verkörpern sie doch kein einheitliches Konzept oder Verhalten, sondern eine Kombination von verschiedenen Verhaltensmustern. Der Vorteil des Konzepts liegt nicht primär in leichterer körperlicher Arbeit, einer höheren Lebensqualität oder einer sichereren Nahrungsgrundlage, sondern in der Erzeugung größerer Nahrungsmengen je Flächeneinheit und Zeit und damit in der Möglichkeit, dichtere Populationen zu ernähren (COHEN 1977).

Veränderung der Umwelt durch den Menschen

Der Übergang zu Ackerbau und Viehzucht vor etwa 10000 Jahren hat zu erheblichen Veränderungen der Umwelt geführt und letztendlich die Agroökosysteme hervorgebracht. Der menschliche Einfluß reicht aber naturgemäß noch weiter zurück. Die vom Menschen besiedelten Gebiete boten dann neue Räume, in denen sich Kulturfolger entfalten konnten. Während in anderen Regionen (besonders Amerika) durch die Zuwanderung von höher entwickelten Menschengruppen die Megafauna stark dezimiert wurde, erklärt beispielsweise REMMERT (1982) das Fortbestehen wesentlicher Elemente der Megafauna in Afrika aus der Möglichkeit ihrer langzeitigen Anpassung an die Existenz des Menschen.

Auch die mit menschlichen Aktivitäten einhergehenden Störungen der Vegetation, die eine Ruderalisierung zur Folge hatten, ermöglichten eine schnellere und größere Ausbreitung von sich durch kolonisierende Tendenzen auszeichnende Pflanzenarten. Später wird noch kurz auf einige Unterschiede zwischen kolonisierenden und domestizierten Pflanzen hingewiesen werden, in vielen Fällen ist es jedoch wahrscheinlich, daß erfolgreiche Kolonisatoren mit nutzbaren Eigenschaften einen Anreiz für eine Sammlung und Verwertung durch den Menschen boten (HARLAN und ZOHARY 1966). Als Beispiel kann hier die Nutzung umfangreicher, relativ dichter Bestände von Wildgetreiden in Südwestasien dienen (WRIGHT 1977). Ihre Ausbreitung vor mehr als 10000 Jahren, die sicherlich auch durch das Wirken des Menschen begünstigt wurde, läßt sich pollenanalytisch nachweisen.

Möglichkeiten der Nutzung solcher Bestände sind in einer interessanten Studie von HARLAN (1967) dargelegt worden, der eine Ernte nach alten Verfahren praktizierte und innerhalb relativ kurzer Zeit eine beträchtliche Menge Erntegut zusammenbringen konnte, das als Vorrat für nahrungsarme Zeiten gut geeignet ist.

Hier lassen sich Gedanken über Erntevölker anschließen (vgl. LIPS 1962). Ein Übergang zu einem bewußten Anbau der Pflanzen ist in den Fällen wenig wahrscheinlich, in denen die Wildpflanzenbestände in ausreichender Menge zur Verfügung stehen. Durchgreifende Veränderungen innerhalb der Populationen, im Sinne nicht mehr effektiv wirkender Verbreitungsmechanismen, wie sie für die Kulturgetreide charakteristisch sind, können aber unter diesen Bedingungen nicht erwartet werden.

In unmittelbarer Nähe der menschlichen Wohnstätten werden sich nitrophile Pflanzen angesiedelt haben, die durch die hier herrschenden Bedingungen begünstigt werden (ENGELBRECHT 1916, vgl. auch ZEVEN 1973). Von DARLINGTON (1973) werden sie als „habitation weeds" bezeichnet. Sie wurden zuweilen als Gemüsepflanzen, besonders wegen ihrer Knollen und Blätter, genutzt. Andere Arten, die zu magischen, medizinischen und technischen Zwecken Verwendung fanden (vgl. z. B. SAUER 1952) und Obstgehölze wurden bewußt dort angepflanzt oder ausgesät. Hier ist der Ursprung der gärtnerischen Kulturpflanzen zu suchen.

Es ist erstaunlich, daß schon in dieser frühen Phase die Wurzeln für Unterschiede im gärtnerischen und landwirtschaftlichen Pflanzenbau gefunden werden können, wobei sich ersterer intensiv mit relativ individuenarmen Populationen befaßt und letzterer das Management von Großpopulationen auf ausgedehnten Flächen betreibt. Besonders diese Großbestände sind zu ökologischen Räumen geworden, in denen sich für eine ganze Anzahl weiterer Arten Nischen boten, die von der Kulturpflanzenart, den Anbaumethoden (Bodenbearbeitung, Bewässerung etc.) und weiteren Selektionsdrücken dominiert werden. Sie verfügen damit über ein gewaltiges evolutionäres Potential.

Beginn der Inkulturnahme

Die bewußte Einflußnahme des Menschen begann mit der gezielten Aussaat oder dem Anpflanzen. Das wird besonders beim Verlassen traditioneller Lebensgebiete der Fall gewesen sein, wenn Material aus dem bisherigen Siedlungsgebiet mitgeführt wurde.

Hier ist es nicht notwendig, komplizierte Hypothesen ins Feld zu führen, die den Vorgang der Aussaat mythologisch zu erklären suchen (Totenkult – Begräbnis). Untersuchungen an nicht ackerbauenden Stämmen der Gegenwart zeigen, daß diese die Aussaat von Samen mit ihren Folgen durchaus kennen (Flannery 1968, Lee 1968, 1969). Mythologische Überlieferungen beinhalten der kulturellen Entwicklung angepaßte Deutungsversuche der bekannten Mechanismen und sind nicht die Auslösefaktoren. Bei manchen auf der Stufe der Jäger und Sammler lebenden Stämmen werden noch heute Samen von Wildpflanzen gesammelt und bewußt für eine spätere Nutzung ausgesät (Downs 1964). Daraus läßt sich, beim Vorhandensein entsprechender Voraussetzungen, die einen permanenten Mangel an nutzbaren Wildressourcen einschließen, ein Modell für den Beginn der Inkulturnahme ableiten. Der Zeitpunkt des Beginnes läßt sich nur ungefähr abschätzen und muß für eine Periode angenommen werden, die der von archäologischen Funden mit Pflanzenresten, die Domestikationsmerkmale zeigen (ca. 10000 Jahre alt), vorausgeht.

Die ältesten Nachweise beziehen sich auf Getreide, die in ihren Ähren gut erhaltungsfähige und erkennbare Domestikationsmerkmale besitzen. Von manchen Autoren wird den zu magischen, medizinischen und technischen Zwecken genutzten (Sauer 1952) oder den knollentragenden und Gemüsepflanzen (Harris 1969) eine frühere Inkulturnahme zugeschrieben. Die schlechte Erhaltungsfähigkeit ihrer meist fleischigen Pflanzenteile läßt aber für eine genaue Analyse nur sehr wenige brauchbare Reste erwarten.

Herausbildung der Kulturpflanzen

Die Frage nach der Entstehung der Kulturpflanzen und auch der Haustiere (vgl. Herre 1981) erfordert Überlegungen zur Populationsentwicklung. Dabei ist davon auszugehen, daß nur ein sehr geringer Teil der Ausgangspopulation in Kultur genommen wird (founder effect – Pickersgill 1981). Dieser besiedelt praktisch eine neue ökologische Nische, wobei der Mensch dafür sorgt, daß die Konkurrenz anderer Arten weitgehend ausgeschaltet wird. Für die neue Population kommt es dadurch zu einer „ökologischen Befreiung“ (ecological release – vgl. Osche 1983) und einer Erhöhung der Variabilität der Merkmale (character release – vgl. Grant 1972). Zur Illustration dieses Vorganges soll ein Beispiel dienen. *Ipomoea purpurea* wanderte von primären Standorten Zentralmexikos in die Äcker der südöstlichen USA ein und entfaltete dort eine große Variabilität der Blütenfarben (Ennos und Clegg 1983). Interessant ist, daß solche einfachen genetischen Abweichungen, wie im gegebenen Fall, beträchtliche Auswirkungen auf das Reproduktionssystem haben können.

Ein Vergleich der unter den Bedingungen der Kultur wirkenden Faktoren mit den Vorgängen der Einnischung bei Inselpopulationen bietet sich an. Hier ist eine Ursache für die große Variabilität vieler Kulturarten zu suchen.

Die Herausbildung der spezifischen Kulturpflanzenmerkmale vollzog sich schon von Anfang an unter dem Einfluß verschiedener vom Menschen bewußt oder unbewußt ausgelöster Selektionsdrücke. Besonders auffällig sind Veränderungen im morphologisch-anatomischen Bereich, die in verschiedenen Arbeiten zusammenfassend dargestellt sind (SCHWANITZ 1967, 1971, LEHMANN 1976). Dabei sind alle Organe der Kulturpflanzen, vor allem entsprechend der Nutzungsrichtung, verändert worden. Dieser Gruppe von Abänderungen stehen weniger auffällige Merkmalsabweichungen gegenüber, die physiologischer bzw. biochemischer Natur sind oder das Reproduktionssystem betreffen.

Beispielsweise unterscheidet sich nach EVANS (1980), vgl. auch SMARTT (1969), die große Formenfülle von *Phaseolus vulgaris* von der Wildsippe subsp. *aborigineus* durch folgende Merkmale oder Merkmalskomplexe. – Die Wuchsform veränderte sich drastisch. Aus einer Liane mit nicht determiniertem Wachstum entstand über verschiedene Zwischenstufen eine determiniert wachsende Buschform. Alle Sippen, von den rankenden bis zu den Buschformen, befinden sich noch in Kultur. Es erfolgte ein Übergang von ausdauernder zu einjähriger Lebensweise. Die Verringerung der Anzahl der Triebe und Blätter ging mit einer Vergrößerung der Blätter und des Stengeldurchmessers einher. Vergrößerungen gibt es auch im

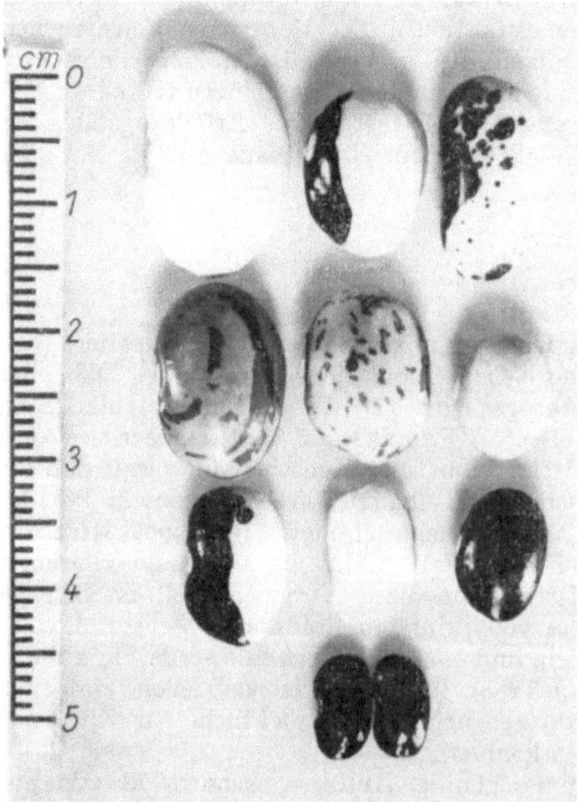

Abb. 1 Die Testafarben verschiedener Kulturformen von *Phaseolus vulgaris* sind sehr variabel. Häufig treten gefleckte Samen auf, die in der Scheckung vieler Haustierarten eine interessante Parallele haben. Die Domestikation hat zu einer Vergrößerung der Samen geführt – in der untersten Reihe zwei Samen der wildwachsenden Ausgangssippe subsp. *aborigineus*

Bereich der Hülsen, Samen und Blüten. Die Anzahl der Samen je Hülse kann erhöht sein. Als Domestikationsmerkmal bedeutsam ist die Permeabilität der Samenschale für Wasser, wodurch eine einheitliche Keimung beim Anbau garantiert ist; außerdem lassen sich solche Bohnen schneller kochen. Die Testafarben sind sehr variabel (Abb. 1). Häufig treten auch zweifarbige und gefleckte Samen auf, ein Merkmal, das Parallelen zur Scheckung vieler Haustierarten aufzeigt (vgl. HERRE 1981). Weiterhin ist eine Reduktion der Pergamentschicht der Hülse erfolgt; ihr völliges Fehlen verhindert das Platzen der Hülse bei der Reife (Abb. 2). Schließlich hat sich noch eine Änderung im photoperiodischen Verhalten vollzogen. – Auf weitere Fälle wird in den folgenden Abschnitten eingegangen.

Alle diese Merkmalsveränderungen sind Folgen der Domestikation, darüberhinaus können sie z. T. diesen Evolutionsvorgang (SCHWANITZ 1971) entscheidend verstärken oder beschleunigen. Besonders auffällig trifft das für Veränderungen im Reproduktionssystem zu. Ein Übergang von der Allo- zur Autogamie bzw. eine Verstärkung von Autogamietendenzen ist für die Domestikation vieler Arten charakteristisch (FAEGRI und VAN DER PIJL 1979: 137). Autogame Kulturpflan-

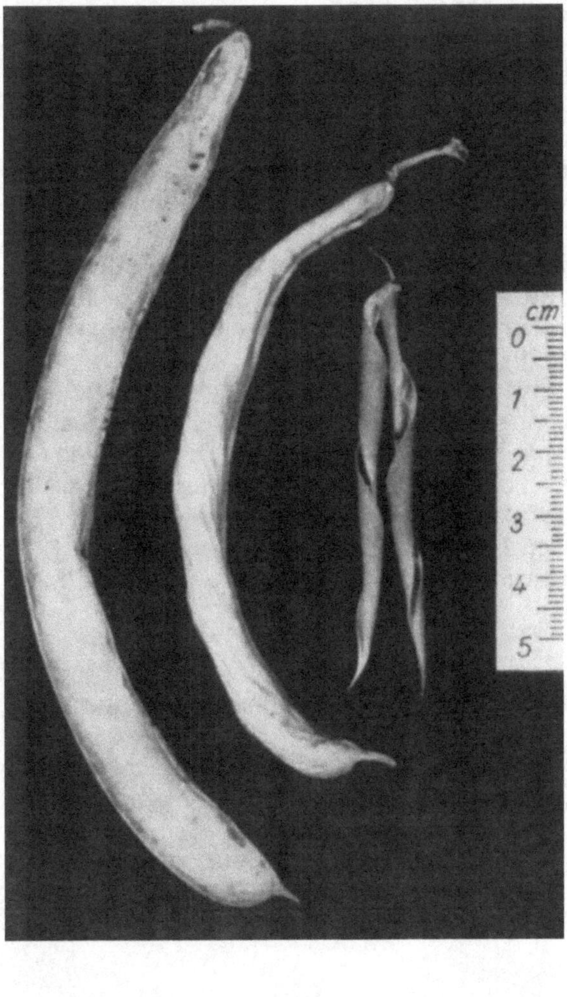

Abb. 2 Bei den Kulturformen von *Phaseolus vulgaris* ist die Pergamentschicht der Hülsen reduziert oder fehlt völlig (mittlere Hülse). Ein Platzen der Hülsen bei der Reife, wie es für die wildwachsende subsp. *aborigineus* (rechts) charakteristisch ist, wird dadurch verhindert

zen bieten dem Menschen eine bessere Selektionsgrundlage, weil gelegentlich auftretende genetisch bedingte Merkmalsabweichungen nicht nur erkannt, sondern auch relativ leicht erhalten und vermehrt werden können. Ebenso verhält es sich bei Arten, die im Laufe der Domestikation von sexueller zu vegetativer Vermehrung übergegangen sind.

Unterschiede zwischen Gartenkultur und Ackerbau

In Bemerkungen zum Reproduktionssystem klingt die schon obengenannte Zweiteilung der Kulturpflanzen in garten- (häufige vegetative Vermehrung) und ackerbaulich genutzte Arten (Vermehrung durch Samen, häufig Übergang zur Autogamie) an. Die Wurzeln dieser Zweiteilung liegen in der schon erwähnten Art der Inkulturnahme. Im Zuge der Domestikation sind so Pflanzen entstanden, die sich für einen relativ großflächigen Feldanbau eignen und in der Regel über das Saatgut reproduziert werden (seed culture – Harris 1969) und solche, die vorwiegend für eine kleinflächige Gartenkultur Verwendung finden und oft vegetativ vermehrt werden (vegeculture – Harris 1969). Diese Aufteilung trägt grundsätzlichen Charakter und hat eine globalere Bedeutung als die von Hawkes (1969) verwendete Zweiteilung in „seed agriculture“ und „root and tuber agriculture“, von denen die letztgenannte nur eine zuweilen auch großflächiger praktizierte Sonderform der Gartenkultur darstellt.

Für die großflächige Ackerkultur kam überwiegend nur eine positive **Massenselektion** in Betracht, die zu ausgewogenen Pflanzenbeständen mit einem guten durchschnittlichen Ertragspotential führte, ohne daß sich einzelne Merkmale oder Merkmalsgruppen überdimensional veränderten. Bei den Gartenkulturen überwog die **Individualselektion**, und die Folge davon waren oft beträchtliche Merkmalsabänderungen. Besonders charakteristisch für diese Gruppe ist das allometrische Wachstum (Schwanitz 1971).

Zwischen diesen beiden Systemen gibt es zahlreiche Übergänge, und auch die Kulturarten lassen sich nicht immer eindeutig zuordnen. Beispielsweise hat *Zea mays* offensichtlich ihre ersten Domestikationsschritte unter den Bedingungen der Massenauslese durchlaufen. Anschließend gaben die relativ großen Körner und ihr Zusammenstehen in Kolben, die als Selektionseinheit geeignet sind, den Anreiz für eine Individualauslese. Heute ist der Mais ganz überwiegend Feldkultur, wird aber, besonders in bestimmten Formen, auch in Gartenkultur gezogen.

Landwirtschaftliche und gärtnerische Kultur – Methoden und Ergebnisse

Nach der Inkulturnahme und der anschließenden Domestikation einer Reihe von Arten durch die ersten Pflanzenbauer wurden im Laufe der Zeit immer wieder neue Pflanzen für die Kultur erprobt, von denen sich ein Teil als Kulturarten bewährte. Dieser Prozeß dauert bis in unsere Zeit an und sein Ende ist, bedingt durch den Wechsel in den Ansprüchen des Menschen und das große Angebot an geeigneten Wildarten, nicht abzusehen.

Die unterschiedliche Dauer und Intensität der Einwirkung des Menschen auf die Kulturpflanzen, die aus den verschiedensten taxonomischen Gruppen stammen, resultieren in einer verwirrenden Vielzahl von Domestikationsvorgängen und -erscheinungen. Es soll daher versucht werden, anhand weniger ausgewählter Kulturen den Fragenkomplex zu umreißen. Die Einteilung folgt dabei dem oben vorgestellten Prinzip der Gliederung in ackerbauliche und gärtnerische Kulturmethoden.

Die Bedeutung von Kulturpflanzen-Feldbeständen mit Saatgutnutzung für die Domestikation

Feldbestände von Kulturpflanzen bilden eine ökologische Nische besonders für Arten mit einer Tendenz für ackerbauliche Standorte, und es entstanden und entstehen Unkräuter, die sich dem Selektionsdruck durch den Menschen (vor allem durch Jäten) anzupassen vermögen.

Im wesentlichen geschieht dies auf zwei Wegen (KUPCOV 1971, KUPZOW 1980): 1. Durch divergente Adaptation, indem z. B. ein Teil der Population durch hohe Vermehrungsraten den Kulturmaßnahmen ausweicht oder 2. durch konvergente Adaptation, wobei sich das Unkraut an die begleitende Kulturart anpaßt. Die sich dabei herausbildenden kulturpflanzenähnlichen Unkräuter zeigen typische Domestikationsmerkmale.

So hat die Kornrade (*Agrostemma githago*), ein Getreideunkraut, relativ ausfallfeste Kapseln, einen aufrechten, dem Getreide ähnlichen Wuchs, eine lange Bestockungsphase bei Winterformen und relativ große Samen (HAMMER et al. 1982). Diese Konvergenz (vgl. Abb. 3) sicherte *Agrostemma githago* lange Zeit den

Abb. 3 Manche Unkräuter des Ackerbaus zeigen durch konvergente Adaptation eine große Ähnlichkeit mit der Kulturpflanze, die sie begleiten. Die im Lein vorkommende Sippe der Kornrade (*Agrostemma githago* var. *linicolum*) besitzt abgerundete Samen mit einer relativ glatten Oberfläche (mittlere Reihe), die dem Lein (obere Reihe) auch in der Tausendkornmasse sehr nahe kommen (beide etwa 7 g). Die im Getreide wachsenden Sippen der Kornrade (untere Reihe) haben kantige Samen, längere Samenwarzen und eine höhere Tausendkornmasse

Fortbestand sowie meist auch einen hohen Anteil in vielen Getreidebeständen, bis es durch neuartige Saatgutreinigungsmethoden weitgehend eliminiert werden konnte und heute schon vielerorts zu den sehr seltenen Pflanzen zählt. Die Samen von *A. githago* eignen sich aufgrund ihres Gehaltes an Saponinen nicht sonderlich für den Verzehr durch den Menschen, sonst könnten die unbewußt und indirekt domestizierten Sippen, die sich auch durch ein ansprechendes Ertragsniveau auszeichnen (Hammer et al. 1982), als Kulturpflanze in größerem Umfang zum Anbau kommen. *A. githago* ist hier als Beispiel aufgeführt worden, weil sich bei ihm die Wirkung der Domestikation sozusagen in reiner Form verfolgen läßt, denn es gehört nicht in die nähere systematische Verwandtschaft der Kulturpflanzen, die es begleitet. Ein Genfluß von der Kultursippe in *A. githago* und umgekehrt ist damit nicht möglich. Dieser Effekt erschwert das Studium von Fällen nahe verwandter Unkraut- und Kulturarten.

Es gibt zahlreiche Unkräuter, die eine Kulturart imitieren. Diese Erscheinung ist unter der Bezeichnung „crop mimicry" in die Literatur eingegangen (Harlan und De Wet 1965, Wickler 1968, McNeill 1976, Barrett 1983). Wiens (1978) charakterisiert die entsprechenden Saatgutunkräuter als „seed mimics" und macht darauf aufmerksam, daß es sich hier nicht um echte Mimikry handelt, weil es keine sensorischen Wechselbeziehungen sind, die zu dieser Anpassung zwischen den beteiligten Organismen führten. Der herrschende Domestikationsdruck ist der entscheidende Wirkmechanismus.

Es ist nicht verwunderlich, daß manche dieser Unkräuter mit schon vorhandenen Domestikationsmerkmalen, bei entsprechender Eignung, selbst als Kulturpflanzen übernommen wurden. Engelbrecht (1916) wies schon zusammenfassend auf dieses Phänomen hin. Cinger (1909) lieferte überzeugende Beispiele (vgl. auch Thellung 1930), indem er die Leinunkräuter einer genaueren Untersuchung unterzog. Vavilov (vgl. die Zusammenfassung seiner Forschungen 1951) schließlich nahm die Idee auf und entwickelte die Theorie von den sekundären Kulturpflanzen.

In neueren Arbeiten gibt es zahlreiche Hinweise darauf, daß eine Reihe der Kulturpflanzen des Ackerbaus auf diese Weise entstanden ist. Die Getreide- und andere Kulturpflanzenbestände erweisen sich damit als wirksame Umwelt für die Domestikation. An zwei Beispielen sollen die Unterschiede in der Entstehung von primären (*Hordeum vulgare*) und sekundären Kulturpflanzenarten (*Papaver somniferum*) demonstriert werden. Die Herausbildung sekundärer Kulturpflanzen erfolgt überwiegend unter den Bedingungen der Massenauslese beim Ackerbau.

Hordeum vulgare

Die Gerste gehört zu den ältesten Kulturpflanzen (Zohary 1971) und ist ein typischer Vertreter der Ackerkultur. Die nächstverwandte Wildform ist *H. vulgare* subsp. *spontaneum* (Harlan 1979), deren primäre Wildstandorte im sommertrockenen Gebiet offener Eichenwälder des „Fruchtbaren Halbmondes" liegen (Zohary 1964).

Die Wildsippe zeichnet sich durch relativ große Körner aus, ein Merkmal, das ein Anreiz zum Sammeln dieses nährstoffreichen Produktes war. Pflanzen von primären Standorten haben relativ lange Antheren, die beim insgesamt autogamen

H. vulgare eine erhöhte Neigung zur Allogamie anzeigen (HAMMER 1984). Mit dem Übergang auf vom Menschen beeinflußte Standorte kam es zu graduellen Änderungen im Reproduktionssystem, wie an der Verkürzung der Antheren deutlich wird. Die Pflanzen von sekundären Standorten haben auch weniger kräftige Ähren, bei denen besonders Länge und Stärke der Grannen reduziert sind. Ihr massenhaftes Auftreten an Ruderalstandorten erhöhte den Anreiz für eine Nutzung.

Erst die bewußte Aussaat und Ernte leiteten dann den eigentlichen Domestikationsvorgang ein, denn es ist schwer vorstellbar, daß Erntemethoden allein, z. B. durch Sichel oder Sense, einen Selektionsdruck auf geringere Spindelbrüchigkeit darstellen, wie es HARLAN et al. (1973) annehmen, denn die Körner brüchiger Ähren gelangen dabei immer wieder in den Boden, während die gelegentlich auftretender Mutanten mit zäher Spindel vorwiegend geerntet und verbraucht werden. Erst mit dem Transport des Saatgutes an andere Stellen und der nachfolgenden Aussaat, vielleicht weil es hier keine Wildgerstenbestände gab, kann der oben erwähnte Populationsdruck wirksam werden, besonders natürlich bei regelmäßiger Aussaat und Ernte. Eine zähe Ährenspindel war die Folge.

Unter den neuen ökologischen Bedingungen setzte dann ein intensiver Formenbildungsprozeß ein. Er wurde gefördert durch die Ausbreitung des Gerstenanbaus in neue Regionen. Ein gutes Beispiel ist hierfür die intensive Formenentfaltung in Äthiopien, wohin der Gersten- zusammen mit dem Weizenanbau erst relativ spät gelangte.

Durch Mutation entstandene oder möglicherweise aus tibetischen Wildgersten (ÅBERG 1938, HAMMER 1984) hervorgegangene mehrzeilige Sippen erhöhten durch Bastardierungen mit zweizeiligen Gersten den Formenreichtum noch weiter.

Diese gewaltige Formenfülle kann nicht darüber hinwegtäuschen, daß durch Gründereffekte die genetische Variabilität für manche Loci beträchtlich eingeengt ist, wie NEVO et al. (1979) für verschiedene Enzyme kodierende Loci beim Vergleich mit israelischen Wildgersten nachweisen konnten. Insgesamt ist der Verlust an genetischer Variabilität nicht so beträchtlich, weil ein Genaustausch von Kultur- und Wildformen zur Evolution der Kultursippe gehörte und bis in die Gegenwart verfolgt werden kann, wie nachfolgend zu zeigen sein wird.

Mit der Kulturgerste breiteten sich allmählich auch die abgeleiteten, auf sekundären Standorten vorkommenden Sippen der subsp. *spontaneum* aus. Sie fanden besonders an Feldrändern günstige Bedingungen vor und wurden so zum Unkraut. Obwohl bei der Kultursippe die Tendenz zur Autogamie noch weiter fortgeschritten ist als bei den Unkrautgersten, gibt es doch, besonders nach witterungsbedingten Kastrationen (meist durch Trockenheit), die Möglichkeit zu spontanen Bastardierungen zwischen beiden (ZOHARY 1964, KAMM 1977, AL ALAZZEH et al. 1982), wobei häufiger ein Genfluß von der Unkraut- in die Kultursippe zu beobachten ist. Der umgekehrte Weg hat u. a. Wildsippen hervorgebracht, die als *H.* ×*lagunculiforme* bezeichnet werden (BAKHTEYEV 1964).

Es gelangen also Gene der Unkrautgersten in Kulturformen und erhöhen deren Variabilität. Unter den Selektionsbedingungen in den Kulturbeständen werden die für den Pflanzenbau ungünstigen Wildmerkmale in den Kreuzungsnachkommenschaften eliminiert. Andererseits werden Domestikationsmerkmale in die

Unkrautformen übertragen (Abb. 4). Auch sie werden schrittweise wieder zurückgedrängt, bleiben aber zu einem Teil wirksam, da sie mit einigen Selektionsvorteilen unter den vom Menschen beeinflußten ökologischen Bedingungen verknüpft sind.

Kultur- und Unkrautgersten sind gemeinsame Abkömmlinge der eigentlichen Wildsippen primärer Standorte. Unter den Selektionsbedingungen des Agroökosystems und durch Genfluß aus den Kulturformen haben sich auch bei den Unkrautgersten gewisse Domestikationsmerkmale herausgebildet (größeres Korn, kürzere und dünnere Grannen, zartere Ähren, weniger effektive Verbreitungsmechanismen für primäre Habitate). Sie sind aber nicht besonders auffällig und, im Gegensatz zur zähen Ährenspindel, oft übersehen worden.

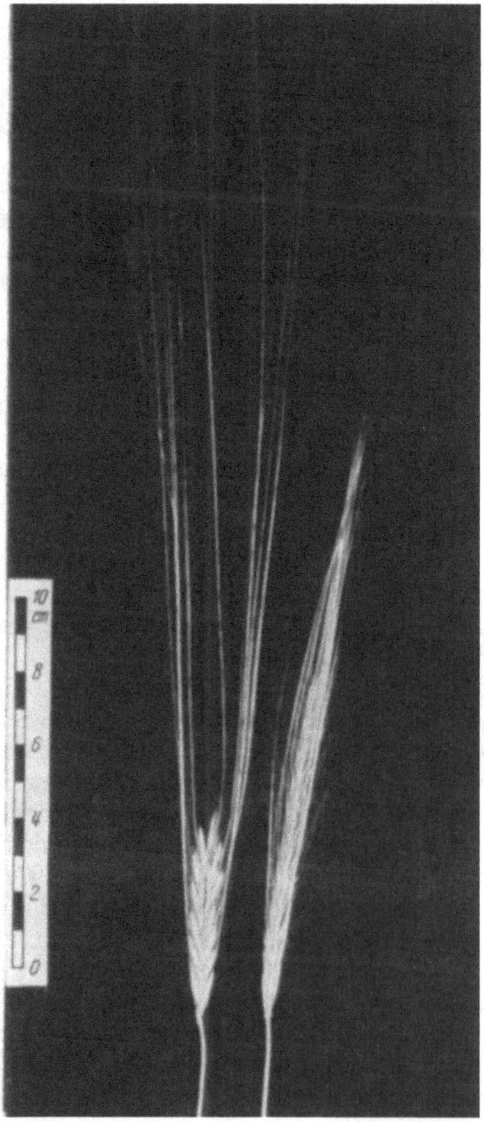

Abb. 4 *Hordeum vulgare* subsp. *spontaneum* – Form von primären Wildstandorten (links) mit langen, kräftigen Grannen und Ähren (die obere Hälfte der Ähre fehlt) und Unkrautsippe aus Libyen mit zarteren Ähren und wesentlich kürzeren Grannen. Diese Sippe zeigt Introgressionsmerkmale von Kulturgersten

Papaver somniferum

Eine gründliche Auswertung der archäologischen Befunde weist auf ein westmediterranes Entstehungsgebiet des Schlafmohns hin (Schultze-Motel 1979). Dort kommt auch heute noch die Wildsippe, *P. somniferum* subsp. *setigerum* vor. Die ältesten archäologischen Reste von Schlafmohn stammen etwa aus dem 3. Jt. v. u. Z. (vgl. Schultze-Motel 1979), einer Zeit in der der Getreidebau schon lange in diese Region vorgedrungen war. Das Agroökosystem als ökologische Großnische bot der Wildsippe offenbar günstige Entfaltungsmöglichkeiten.

Der Borstenmohn reagiert stark auf bessere Nährstoffversorgung (Abb. 5). Möglicherweise ist hier ein Parallelfall zu den sogenannten Genotrophen beim Lein (vgl. ZEVEN 1975, PATWARY 1983) gegeben.

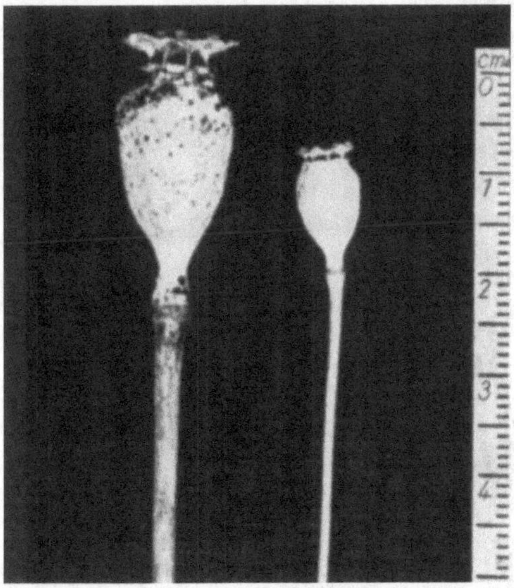

Abb. 5 Kapseln von Borstenmohn (*Papaver somniferum* subsp. *setigerum*) – Originalkapsel von einem Ruderalstandort (rechts) im Vergleich zu einer Kapsel von diesem Material, das zwei Jahre lang unter Bedingungen mit reichlicher Nährstoffversorgung angebaut wurde

Die ersten Domestikationsschritte sind damit wahrscheinlich in unmittelbarer Nähe oder innerhalb von Getreidebeständen erfolgt – sie beziehen sich auf quantitative Merkmale (Pflanzen-, Kapsel-, Samengröße) und boten den Anreiz für die Nutzung der Sippe. Der Borstenmohn ist selbstkompatibel, und es ergeben sich günstige Voraussetzungen für die weitere Selektion durch den Menschen, wobei anzunehmen ist, daß die in Größe und Färbung auffälligen Blüten eine gute Selektionsbasis boten.

Mit der Wanderung der Kultur in östliche Richtung aus dem Areal der Wildsippe heraus setzte eine intensivere Formenbildung ein, die im asiatischen Raum in der Formierung einer eigenen Subspecies gipfelte (subsp. *songaricum* – HAMMER 1981).

Der Prozeß der Domestikation wirkte sich im Entstehungsgebiet der Kulturformen auch auf die Wildsippe aus. Die ebenso wie die Kulturformen diploiden Sippen von subsp. *setigerum* (HAMMER und FRITSCH 1977) sind durch Bastardierungen mit dem Schlafmohn weitestgehend in diesen aufgegangen. Sie sind heute außerordentlich selten. Dieses Phänomen wird von HARLAN (1970) als genetische Aggression (genetic aggression) bezeichnet und läßt sich für viele andere Kulturpflanzen nachweisen. Die genetisch isolierten tetraploiden Sippen der subsp. *setigerum* konnten sich dagegen als Unkraut- und Ruderalpflanzen häufiger erhalten.

Die heutigen Formen des Schlafmohns sind in allen Organen wesentlich größer als der Borstenmohn (Abb. 6), produzieren eine größere Anzahl von Samen je Kapsel und verfügen über zahlreiche Blüten- und Samenfarbvarianten. Bei hochdomestizierten Sippen öffnen sich die Poren unter der Narbenscheibe nicht, und die Samen können nicht frei verstreut werden (Schließmohn). Diese Veränderung, die die natürliche Verbreitung der Samen unterbindet, ist in ihrer Funktion mit

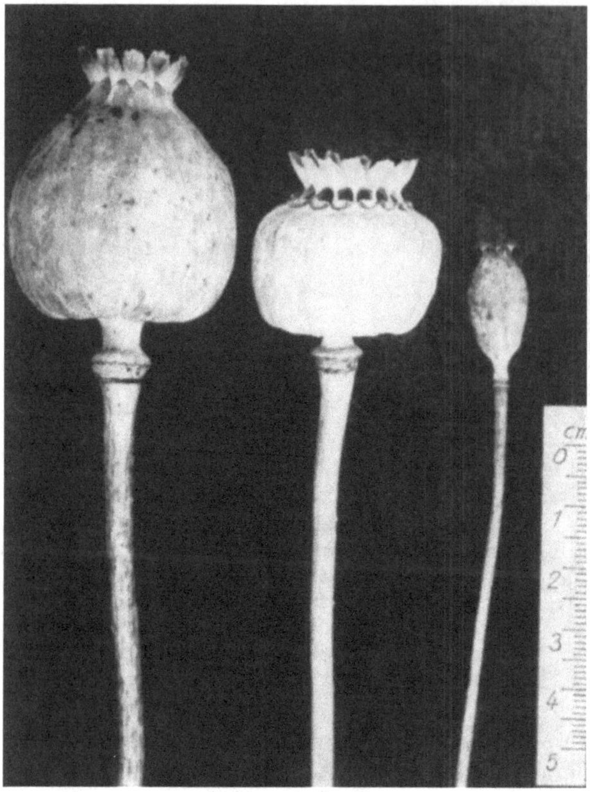

Abb. 6 Aus diploiden Sippen des Borstenmohns (rechts) sind großkapselige Schütt- (Mitte) und Schließmohne entstanden. Die Schließmohne sind hochdomestizierte Sippen, bei denen sich zur Reife die Poren unter der Narbenscheibe nicht öffnen. Eine natürliche Verbreitung ist deshalb bei ihnen sehr erschwert

der zähen Ährenspindel bei *Hordeum vulgare* vergleichbar. Im Kulturgebiet haben sich auch sekundäre Unkrautsippen, vorwiegend aus Schüttmohn-Landsorten, herausgebildet (HAMMER 1981).

Ein Großteil der gefundenen Variationen begründet sich auf Merkmale, die für eine Individualauslese leicht faßbar sind. Besonders deutlich wird das bei den Zierformen von *P. somniferum* subsp. *somniferum*, die in den Gärten weit verbreitet sind und sich durch große, farbenprächtige, oft gefüllte Blüten auszeichnen.

Beim Mohn wird deutlich, daß sich viele Kulturpflanzen nicht eindeutig dem Acker- oder dem Gartenbau zuordnen lassen, sondern daß zahlreiche Übergänge vorhanden sein können, die u. a. auch durch einen Wechsel von Massen- zu Individualselektion (Entwicklung pflanzenzüchterischer Methoden) oder umgekehrt (Übergang zum Unkrautstatus) bewirkt werden. Ein solcher Wechsel wird oft durch Veränderungen in der Nutzungsrichtung oder auch Aufgabe einer bestimmten Kultur bewirkt.

Die Umwelt gärtnerischer Kulturpflanzen als Domestikationsfaktor

Ebenso wie die ökologischen Bedingungen in den Getreidebeständen bieten auch die der Gartenkulturen Nischen für Unkräuter. Jedoch sind hier die Selektionsbedingungen sehr vielgestaltig und dadurch herrschen divergente Adaptationen vor. Eine Nutzung bestimmter, oft nitrophiler Unkrautarten und ihre anschließende gezielte Inkulturnahme ist besonders für einige Gemüsepflanzen anzunehmen. Beispielsweise wird in Mexiko und Zentralamerika das Unkraut *Jaltomata procumbens* genutzt. Damit verbunden ist eine gewisse Schonung der Art. Sie hat bereits eine Zwischenstellung zwischen Unkraut- und Kulturpflanze erreicht und zeigt demgemäß Domestikationsmerkmale (Davis und Bye 1982). Dieses Beispiel zeigt, daß unter den Bedingungen der gärtnerischen Kultur, der Domestikation oft eine Nutzung und Schonung der Unkrautarten vorhergeht, bei der Acker-

Abb. 7a Großfrüchtige Tomaten-Landsorten (Abb. 7a) im Vergleich mit *Lycopersicon pimpinellifolium* (Abb. 7b), einer Unkrautsippe, bei der die Früchte schon gewisse Nutzungsmöglichkeiten erkennen lassen

Abb. 7b

kultur diese Voraussetzungen nicht unbedingt erforderlich sind. Weitere Beispiele lassen sich in den amerikanischen Kulturpflanze-Unkraut-Komplexen (crop-weed complexes) finden (Pickersgill 1981).

An einer Kultursippe soll im folgenden der Domestikationsvorgang unter den Bedingungen der Gartenkultur (Individualauslese) untersucht werden.

Lycopersicon esculentum

Die Tomate gehört zu den alten Kulturpflanzen Amerikas. Wie bei anderen Pflanzen aus der Gruppe der Gemüse und auch der knollentragenden Pflanzen, sind die archäologischen Funde aufgrund des wenig erhaltungsfähigen Pflanzenmaterials rar. Anhand der Zentren der rezenten Formenmannigfaltigkeit lassen sich gewisse Rückschlüsse auf Kulturanfänge in Mexiko und dem westlichen Südamerika (Ekuador, Peru) ziehen.

Die Tomate ist autogam. Eine nahe verwandte Wildart, *Lycopersicon hirsutum*, ist allogam, ebenso wie *L. peruvianum*. Dieses zeigt unilaterale Kreuzungsinkompatibilität mit *L. esculentum* (Günther und Jüttersonke 1971), und es gelang, selbstkompatible Mutanten zu erzeugen (Hoffmann 1971). Beide Wildsippen kommen als Ausgangsarten für *L. esculentum* in Betracht.

Die Domestikation wird auf dem Wege der Individualauslese erfolgt sein. Sie führte zu einer allometrischen Vergrößerung vieler Pflanzenorgane, und dabei besonders der Früchte, sowie zu einer Zunahme der Selbstkompatibilität.

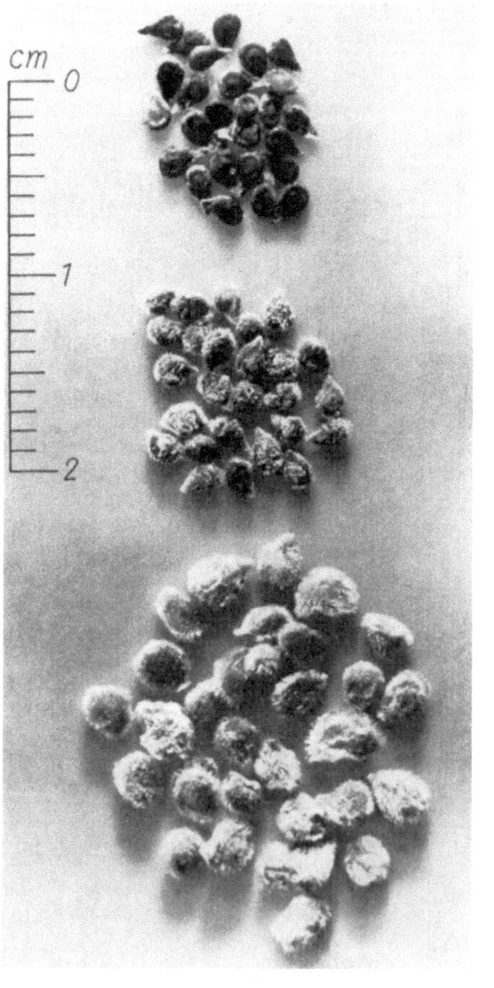

Abb. 8 Aus der Wildsippe *Lycopersicon hirsutum* (oben) sind wahrscheinlich die Unkrautsippe *L. pimpinellifolium* (Mitte) und die Kulturtomate *L. esculentum* hervorgegangen. Mit zunehmender Fruchtgröße nimmt, durch die pleiotrope Wirkung der an diesem Merkmal beteiligten Allele, auch die Größe der Samen zu

Aus den Wildsippen haben sich, über noch zur Allogamie neigende Übergangsformen, mehr oder weniger gleichzeitig die autogamen Kulturformen und die ebenso autogamen Unkrautsippen herausgebildet (Rick und Fobes 1975), von denen *L. esculentum* var. *cerasiforme* eine etwas höhere Domestikationsform repräsentiert und als Übergang zu den eigentlichen Kulturtomaten aufzufassen ist (Rick 1976). *L. pimpinellifolium* ist eine kleinfrüchtige Unkrautart (fakultativ allogame Formen dieser Art (Rick et al. 1977) stehen offenbar den zu Kultur- und Unkrautsippen überleitenden Formen recht nahe), die ebenso wie *L. esculentum* var. *cerasiforme*, durch Introgressionen aus den fortgeschrittenen Kultursippen beeinflußt wurde und wird und durch gelegentliche Einkreuzungen selbst zur Vermehrung der Formenvielfalt bei *L. esculentum* beiträgt. Wie viele Unkrautsippen zeigt sie typische Domestikationsmerkmale (Abb. 7).

Stubbe (1971) hat in Aufsehen erregenden Mutationsversuchen mit Röntgenbestrahlung weitere Domestikationsschritte nachvollziehen können und auf der Basis anschließender Individualauslese Fruchtgrößen erzielt, die manchen hochdomestizierten Sorten vergleichbar sind. Gleichzeitig wurden durch die pleiotrope Wirkung der an der Fruchtgröße und -form beteiligten Allele Samengröße und Wuchsform in Richtung der Kulturtomate verändert (Abb. 8). Es ist hier ein Modell für die Domestikation geliefert worden, das auf schon mit Domestikations-

Abb. 9 Achänen von *Helianthus annuus* – Material von einem Ruderalstandort (unten) und vom ersten bzw. zweiten Nachbaujahr unter Bedingungen mit guter Nährstoffversorgung. Die Achänengröße von Kultursorten wird erreicht. Die Verwilderung einer Kultursorte konnte damit nachgewiesen werden. Durch die Verwilderung kam es vorerst zu überwiegend phänotypischen Merkmalsabweichungen

merkmalen behaftetem Material basiert und das die Bedeutung des Unkrautstatus für die Evolution von Kulturpflanzen unterstützt.

STUBBE (1971) konnte auch die Möglichkeit der Umkehrung dieses Prozesses nachweisen und ein Modell für die Verwilderung (Abb. 9), den Prozeß des Überganges von Kulturpflanzen in den Unkrautstatus, liefern.

Im Zusammenhang mit den Ergebnissen STUBBES ist es interessant, daß LAZANYI (1979, 1983) die Ursache für Organvergrößerungen, wie sie für viele Kulturpflanzen im Zuge der Domestikation charakteristisch sind, in Tandem-Genduplikationen sieht, die im Gegensatz zur repetitiven DNA funktional sind. Aus der Duplikation von Polygenen könnten die Erscheinungen der Allometrie resultieren, die für die Domestikation typisch ist.

Schlußfolgerungen – das Domestikationssyndrom

Aus den angeführten Beispielen, für die bewußt diploide Kulturpflanzen gewählt wurden, um die Diskussion nicht durch eine Reihe weiterer mit der Polyploidie zusammenhängender Fragen (vgl. ZEVEN 1980) zu belasten, wird die Komplexität der Domestikationsvorgänge in ihrer Abhängigkeit von der Pflanzenart deutlich. Ihr Studium erfordert die Untersuchung der Kultursippen zusammen mit ihren nahe verwandten Unkraut- und Wildarten. Dabei kommt es besonders darauf an, die Abstammungsverhältnisse zu klären, denn Unkräuter können sich beispielsweise, wie oben ausgeführt wurde, von der Kultursippe ableiten, gemeinsam mit ihr aus einer Wildart hervorgegangen sein oder selbst die Kultursippe hervorgebracht haben (vgl. PICKERSGILL 1981). Zuweilen laufen alle diese Prozesse hintereinander oder sogar parallel ab.

Unkräuter müssen aber nicht in verwandtschaftlicher Beziehung zu einer Kulturart stehen, wie am Beispiel von *Agrostemma githago* aufgezeigt wurde.

Trotz der Vielgestaltigkeit der Evolutionswege, zeigen die domestizierten Pflanzen in vielen Merkmalen erstaunliche Parallelen. Das gesetzmäßige Auftreten von Gruppen spezifischer Kulturpflanzenmerkmale berechtigt die Postulierung des Domestikationssyndroms.

Wie vorangehend gezeigt werden konnte, betrifft das Domestikationssyndrom nicht nur die vom Menschen bewußt kultivierten Pflanzenarten, sondern auch weitere Pflanzen des Agroökosystems, die auf dort herrschende spezifische und allgemeine Selektionsdrücke mit konvergenter Adaptation reagieren. Weiträumige Ruderalisierungserscheinungen der menschlichen Aktionssphäre und andere oft übersehene Einflußnahmen, wie die Einführung von Honigbienen in viele Gebiete und z. T. auch ihre dortige Verwilderung, die zur Verdrängung der bodenständigen Bienenfauna führt, mit Konsequenzen für das Reproduktionssystem vieler Pflanzenarten (VOGEL 1980), bringen Veränderungen in den Wildpflanzen hervor. Hier ist eine Abgrenzung erforderlich bezüglich der Einflüsse, die den Domestikationen ähnliche Wirkungen zeigen und auch oft die Basis für sie bildeten und den eigentlichen Domestikationen.

Auf die kolonisierenden Arten in ihrer Bedeutung für die Entstehung der Kulturpflanzen wurde schon hingewiesen (vgl. auch HARLAN und ZOHARY 1966, DE WET 1981). Von BROWN und MARSHALL (1981) wird der Merkmals- und

Ursachenkomplex für diese Sippen diskutiert, der auch in wesentlichen Zügen für domestizierte Arten zutrifft, wobei graduelle Unterschiede nicht zu übersehen sind. Das Kriterium für die Abgrenzung der domestizierten Pflanzen kann also im wesentlichen nur in der nicht-konvergenten Adaptation der kolonisierenden Arten liegen. Es muß ein Respons auf die konkreten Selektionsbemühungen des Menschen unter den Bedingungen des Acker- und Gartenbaus vorliegen. Beispielsweise ist das Nichtabfallen der Früchte des in Mitteleuropa adventiven *Chenopodium pumilio* (Lhotská und Hejný 1979) kein Domestikationsmerkmal, obwohl es zweifellos durch die vom Menschen geschaffene Umwelt mit bedingt ist. Es würde aber, bei entsprechendem Bedarf, eine günstige Voraussetzung für die Inkulturnahme sein. Weitere Beispiele ließen sich anschließen (z. B. *Amaranthus bouchonii*).

Ebenso wie bei den kolonisierenden und ruderalen Sippen ist auch für manche Unkräuter das entscheidende Kriterium zur Abgrenzung gegenüber den Domestikationen ein Fehlen an konvergenten Adaptationen.

Das Domestikationssyndrom zeigen also in der Pflanzenwelt Kulturpflanzen und bestimmte Unkräuter. Der Grad seiner Ausprägung wird durch die Dauer und Intensität der vom Menschen (einschließlich der von ihm geschaffenen ökologischen Bereiche auf Feldern und in Gärten) bewirkten Selektionsdrücke bestimmt. Es kann als universelle Erscheinung eingeschätzt werden und tritt bei den verschiedensten Pflanzengruppen auf, z. B. auch bei Tangarten (van der Meer 1983) und Mikroorganismen (Stanton 1969). Es erscheint deshalb legitim, diesen Begriff auch auf die Haustiere zu übertragen, die oftmals über einen Komplex vergleichbarer Merkmalsabwandlungen verfügen (vgl. z. B. Berry 1969, Herre und Röhrs 1971, Senglaub 1976, Herre 1980, Clutton-Brock 1981). Eine Übertragung des Begriffs der Domestikation (als sog. Selbstdomestikation) auf die menschliche Population, wie das u. a. von Zeuner (1963) vorgeschlagen wird, hat schon Herre (1981) begründet zurückgewiesen.

Vor allem durch die bewußte Einflußnahme des Menschen vollzieht sich die Evolution der Kulturpflanzen und Haustiere. Es gibt zwar zahlreiche Wechselbeziehungen zwischen dem Menschen und seinen lebenden Kulturprodukten (vgl. z. B. Anderson 1967, Heiser 1981), der dominierende Faktor ist aber immer der Mensch gewesen. Haustiere und Kulturpflanzen sind Produkte der Kulturentwicklung der Menschheit (Szabó 1983), und mit der zunehmenden Beschleunigung dieser Kulturentwicklung, die sich vor allem aus der höheren Ausbreitungsgeschwindigkeit von nicht an eine sukzessive Generationsfolge (genetisches Prinzip) gebundene Informationen ergibt (Vogel 1983), z. B. durch neue Methoden in der Züchtung, werden auch ihre Veränderungen immer drastischer. Aus dem Widerspruch der schnellen Entwicklung und Ausbreitung neuer Methoden, einschließlich der dabei erzielten Ergebnisse (neue Sorten), und der an die Generationsfolge gebundenen genetischen Prinzipien erwachsen Probleme, die u. a. in einer Generosion zum Ausdruck kommen.

Die Entwicklung des Konzeptes der Domestikation basiert auf Darwins (1868) bahnbrechenden Untersuchungen, die de Candolle (1882) besonders zum Studium der wildwachsenden Ausgangssippen der Kulturpflanzen inspirierten. Erst der Vergleich zwischen den Kultur- und Wildsippen bildet die solide Basis zum Studium der Domestikationsphänomene. Nachfolgend wurden viele weitere

Informationen bekannt und besonders die Zeit nach der Wiederentdeckung der Mendelschen Vererbungsgesetze brachte neue Erkenntnisse zur Genetik und Phylogenie der Kulturpflanzen. Der Anstoß zur weiteren Durchdringung des Domestikationsphänomens wurde durch die Untersuchungen von Engelbrecht (1916) und besonders von Thellung (1925, 1930) gegeben. Thellung, als ausgezeichneter Kenner der Adventivflora, mußte genaue Kriterien zur Unterscheidung von Kultur-, Unkraut- und Wildpflanzen in ihren vielfältigen Wechselwirkungen entwickeln und lieferte ein zusammenfassendes Konzept (Thellung 1930), das die eigentliche Grundlage für eine Phänomenologie der globalen Erscheinungen der Domestikation bildet. Von Faegri (1981) wurde dafür der Begriff Domestikationssyndrom[1] eingeführt.

Summary

The domestication syndrome

Preconditions for, and manners of domestication are studied and the results of this evolutionary process discussed. Already from the beginnings of domestication a differentiation of the cultivated plants into two groups is feasible–plants from larger fields which are reproduced by seeds with predominating mass selection (seed culture) and plants from gardens often propagated vegetatively being subject mainly to individual selection (±vegeculture). Seed culture results in well balanced stands of plants producing good average yields and does not lead to excessive changes in single characters or character combinations. Garden culture often results in considerable character changes (allometric growth).

Three examples are given to demonstrate the course and results of domestication. *Hordeum vulgare* (primary crop) and *Papaver somniferum* (secondary crop) have been chosen as representatives for seed culture. The conditions effected from larger fields favour the unconscious and indirect domestication of plants which may become new crops (secondary crops) in later stages of this process. But also weeds which cannot be used by man are domesticated in this way. Conscious and direct selection is predominant under garden conditions. As an example, the domestication of *Lycopersicon esculentum* is explained.

The course of domestication may be rather complicated, depends on the plant species involved and results in a huge variability of many different forms. But the evidence of groups of specific characters in cultivated plants supports the consideration of the domestication syndrome. A delimitation from colonizing and ruderal species which may show similar characters and character combinations as domesticated plants is discussed.

Cultivated plants are a product of culture development in mankind, this is also true for domestic animals. Many domestication phenomena concerning cultivated plants parallel to those in domestic animals and, therefore, the domestication syndrome is also relevant for them.

1 frz. syndrome de la domestication, cf. J. Pernès (1983), La Recherche **14**, 910–919.

Краткое содержание

Доместикационный синдром

Рассматриваются предпосылки и пути, ведущие к доместикации и обсуждаются результаты этого эволюционного процесса. Уже с начала окультивирования можно проследить становление двух типов культурных видов: одни занимают крупные площади, размножаются посевным материалом и среди них преобладает массовый отбор; другие занимают небольшие огородные участки и среди них преобладают различные формы вегетативного размножения и индивидуальный отбор. Полеводство приводит к уравновешенным насаждениям с хорошим общим потенциалом урожайности, причём отдельные признаки или группы признаков не выделяются чрезмерным развитием. Огородные культуры часто характеризуются значительными изменениями (аллометрический рост). На примерах трёх культур исследуются процесс развития и результаты одомашнивания. *Hordeum vulgare* (первичный культурный вид) и *Papaver somniferum* (вторичный культурный вид) являются представителями полевых культур. Условия возделования на полях дают выгодные предпосылки для непреднамеренного и косвенного одомашнивания растений, которые позднее могут сами стать культурными видами (вторичные культуры). Однако, и сорные растения, о чьём использовании в дальнейшем не может быть и речи, могут быть, таким образом, одомашнены косвенным путём. В условиях огородной культуры преобладает сознательный и прямой отбор; как пример такой селекции рассматривается доместикация *Lycopersicon esculentum*.

Хотя процесс доместикации, в зависимости от характера вида, может быть очень сложным и привести к различным многочисленным формам, всё-таки закономерность появления групп специфических признаков культурных растений позволяет постулировать доместикационный синдром. Делается попытка отграничения доместикационных явлений от сходного развития признаков у колонизирующих и рудеральных видов.

Культурные растения, как и домашние животные являются продуктом развития культуры человечества. Феноменологические параллели в одомашнивании растений и животных позволяют говорить о доместикационном синдроме и у домашних животных.

Literatur

Åberg, E., 1938: *Hordeum agriocrithon* nova sp., a wild six-rowed barley. – Ann. Roy. Agr. Col. Sweden **6**, 159–216.

Al Alazzeh, A., K. Hammer, C. O. Lehmann and P. Perrino, 1982: Report on a travel to the Socialist People's Libyan Arab Jamahiriya 1981 for the collection of indigenous taxa of cultivated plants. – Kulturpflanze **30**, 191–202.

Anderson, E., 1967: Plants, Man and Life. – Univ. California Press, California.

Bakhteyev, F., 1964: Origin and phylogeny of barley. – Barley Genetics I, Proc. First Intern. Barley Genetics Symp., Wageningen 1963, 1–18.

Barrett, S. C. H., 1983: Crop mimicry in weeds. – Econ. Bot. **37**, 255–282.

Berry, R. J., 1969: The genetical implications of domestication in animals. – In: P. J. Ucko and G. W. Dimbleby (eds.), The Domestication and Exploitation of Plants and Animals, pp. 207–217. – London.

Brown, A. H. D., and D. R. Marshall, 1981: Evolutionary changes accompanying colonization in plants. – In: G. G. E. Scudder and J. L. Reveal (eds.), Evolution Today, Proc. Second Intern. Congr. Syst. Evol. Biology, pp. 351–363.
De Candolle, A., 1882: Origine des plantes cultivées. – Paris.
Cinger, N., 1909: Über die im Lein auftretenden *Camelina*- und *Spergula*-Arten und ihre Abstammung (russ.). – Trav. Mus. bot. Acad. sc. Pétersb. liv. 6, 303 pp. (zit. nach Thellung 1930).
Clutton-Brock, J., 1981: Domesticated Animals. – London.
Cohen, M. N., 1977: The Food Crisis in Prehistory. – New Haven and London.
Darlington, C. D., 1973: Chromosome Botany and the Origins of Cultivated Plants, 3rd rev. ed. – London.
Darwin, C., 1868: The Variation of Animals and Plants under Domestication, 2 vols. – London.
Davis IV, T., and R. A. Bye, 1982: Ethnobotany and progressive domestication of *Jaltomata* (Solanaceae) in Mexico and Central America. – Econ. Bot. **36**, 225–241.
Downs, J., 1964: Significance of environmental manipulation in Great Basin cultural development. – In: W. L. D'Azevedo, W. A. Davis, D. D. Fowler and W. Suttles (eds.), The Current Status of Anthropological Research in the Great Basin. – Desert Res. Inst. Tech. Rep. Ser. SH, Social Sci. and Human, Publ. No. 1.
Engelbrecht, Th. H., 1916: Über die Entstehung einiger feldmäßig angebauter Kulturpflanzen. – Geogr. Z. (Leipzig) **22**, 328–334.
Ennos, R. A., and M. T. Clegg, 1983: Flower color variation in the morning glory, *Ipomoea purpurea*. – J. Heredity **74**, 247–250.
Evans, A. M., 1980: Structure, evolution, and classification in *Phaseolus*. – In: Summerfield and Bunting (eds.), Advances in Legume Science, pp. 337–347.
Faegri, K., 1981: The social functions of botanical gardens in the society of the future. – Bot. Jahrb. Syst. **102**, 147–152.
–, and L. van der Pijl, 1979: The Principles of Pollination Ecology, 3rd rev. ed. – London.
Flannery, K. V., 1968: Archaeological systems theory and early Mesoamerica. – In: Meggers (ed.), Anthropological Archaeology in the Americas, pp. 67–86. – Washington.
Grant, P. R., 1972: Convergent and divergent character displacement. – Biol. J. Linn. Soc. **4**, 39–68.
Günther, E., und B. Jüttersonke, 1971: Untersuchungen über die Kreuzungsinkompatibilität zwischen *Lycopersicon peruvianum* (L.) Mill. und *Lycopersicon esculentum* Mill. und den reziproken Bastarden. – Biol. Zbl. **90**, 561–574.
Hammer, K., 1981: Problems of *Papaver somniferum*-classification and some remarks on recently collected European poppy land-races. – Kulturpflanze **29**, 287–296.
–, 1984: Bestäubungsökologische Merkmale und Phylogenie von *Hordeum* L. subgen. *Hordeum*. – Flora **175**, 339–344.
–, und R. Fritsch, 1977: Zur Frage nach der Ursprungsart des Kulturmohns (*Papaver somniferum* L.). – Kulturpflanze **25**, 217–227.
–, P. Hanelt und H. Knüpffer, 1982: Vorarbeiten zur monographischen Darstellung von Wildpflanzensortimenten: *Agrostemma* L. – Kulturpflanze **30**, 45–96.
Harlan, J. R., 1967: A wild wheat harvest in Turkey. – Archaeology **20**, 197–201.
–, 1970: Evolution of cultivated plants. – In: O. H. Frankel and E. Bennett (eds.), Genetic Resources in Plants – Their Exploration and Evaluation, pp. 19–32. – Oxford.
–, 1975: Crops and Man. – Madison.
–, 1979: On the origin of barley. – In: Barley: Origin, Botany, Culture, Winter Hardiness, Genetics, Utilization, Pests. – Agriculture Handbook No. 338, pp. 10–36. – Washington.
–, and J. M. J. de Wet, 1965: Some thoughts about weeds. – Econ. Bot. **19**, 16–24.
–, J. M. J. de Wet, and E. G. Price, 1973: Comparative evolution of cereals. – Evolution **27**, 311–325.
–, and D. Zohary, 1966: Distribution of wild wheats and barley. – Science **153**, 1074–1080.
Harris, D. R., 1969: Agricultural systems, ecosystems and the origins of acriculture. – In: P. J. Ucko and G. W. Dimbleby (eds.), The Domestication and Exploration of Plants and Animals, pp. 3–15. – London.

HARRIS, D. R., 1977: Alternative pathways toward agriculture. – In: C. A. REED (ed.), Origins of Agriculture, pp. 179–243. – The Hague.
HAWKES, J. G., 1969: The ecological background of plant domestication. – In: P. J. UCKO and G. W. DIMBLEBY (eds.), The Domestication and Exploitation of Plants and Animals, pp. 17–29. – London.
HEISER, C. B., 1981: Seed to Civilization. The Story of Food, 2nd ed. – San Francisco.
HERRE, W., 1980: Grundfragen zoologischer Domestikationsforschung. – Nova Acta Leopoldina n. F. **52** (241) 16 pp.
–, 1981, Domestikation. Ein experimenteller Beitrag zur Stammesgeschichte. – Naturwiss. Rundsch. **34**, 456–463.
–, und M. ROEHRS, 1971: Domestikation und Stammesgeschichte. – In: G. HEBERER (Hrsg.), Die Evolution der Organismen Bd. **2**, 2, 29–174. – Stuttgart.
HOFFMANN, W., 1971: Induktion und Analyse von selbstkompatiblen Mutanten bei *Lycopersicon peruvianum* (L.) Mill. II. Analyse der selbstkompatiblen Mutanten. – Biol. Zbl. **90**, 33–41.
KAMM, A., 1977: The range of brittle types of Cerealia barleys in Israel. – Agr. Res. Org., Volcani Center N. 165, 43 pp.
KUPCOV, A. J., 1971: Elementy obščej selekcii rastenij. – Novosibirsk.
KUPZOW, A. J., 1980: Theoretical basis of plant domestication. – Theor. Appl. Genet. **57**, 65–74.
LAZÁNYI, E., 1979: Increased plant size, heterosis and transgression as consequences of induced geneduplications in *Arabidopsis thaliana* (L.) Heynh. – Arabidopsis Inf. Serv. **16**, 107–111.
–, 1983: Átöröklés és Evolúció. – Bukarest.
LEE, R. B., 1968: What hunters do for a living or how to make out on scarce resources. – In: R. B. LEE and I. DEVORE (eds.), Man the Hunter, pp. 30–43. – Chicago.
–, 1969: !Kung Bushman subsistence: An input-output analysis. – In: VAYDA (ed.), Ecological Studies in Cultural Anthropology, pp. 47–79. – New York.
LEHMANN, CHR. O., 1976: Zu Fragen der Entstehung und Entwicklung der Kulturpflanzen. – In: H. BÖHME, R. HAGEMANN und R. LÖTHER (Hrsg.), Beiträge zur Genetik und Abstammungslehre, pp. 494–520. – Berlin.
LIPS, E., 1962: Ethnologie und Kulturpflanzenforschung. – Kulturpflanze, Beiheft 3, 191–233.
LHOTSKÁ, M., and S. HEJNÝ, 1979: *Chenopodium pumilio* in Czechoslovakia: Its strategy of dispersal and domestication. – Folia Geobot. Phytotax. **14**, 367–375.
VAN DER MEER, J. P., 1983: The domestication of seaweeds. – Bio Science **33**, 172–176.
MCNEILL, J., 1976: The taxonomy and evolution of weeds. – Weed Res. **16**, 399–413.
MERRILL, E. D., 1930: The improbability of pre-Colombian Eurasian–American contacts in the light of the origin and distribution of cultivated plants. – J. N. Y. Bot. Gard. **31**, 209–312.
NEVO, E., D. ZOHARY, A. H. D. BROWN, and M. HABER, 1979: Genetic diversity and environmental associations of wild barley, *Hordeum spontaneum*, in Israel. – Evolution **33**, 815–833.
OSCHE, G., 1983: Die Sonderstellung des Menschen in evolutionsökologischer Sicht. – Nova Acta Leopoldina n. F. **55** (253), 57–72.
PATWARY, A. K., 1983: Crosses and backcrosses with flax genotrophs. – Z. Pflanzenzüchtg. **90**, 229–235.
PICKERSGILL, B., 1981: Biosystematics of crop-weed complexes. – Kulturpflanze **29**, 377–388.
REMMERT, H., 1982: The evolution of man and the extinction of animals. – Naturwissenschaften **69**, 524–527.
RICK, CH. M., 1976: Tomato. *Lycopersicon esculentum* (Solanaceae). – In: N. W. SIMMONDS (ed.), Evolution of Crop Plants, pp. 268–273. – London and New York.
–, and FOBES, 1975: Allozyme variation in the cultivated tomato and closely related species. – Bull. Torrey Bot. Club **102**, 376–384.
–, – and M. HOLLE, 1977: Genetic variation in *Lycopersicon pimpinellifolium*: Evidence of evolutionary change in mating systems. – Plant Syst. Evol. **127**, 139–170.

Sauer, C. O., 1952: Agricultural Origins and Dispersals. – New York.
Schultze-Motel, J., 1979: Die urgeschichtlichen Reste des Schlafmohns (*Papaver somniferum* L.) und die Entstehung der Art. – Kulturpflanze **27**, 207–215.
Schwanitz, F., 1967: Die Evolution der Kulturpflanzen. – München.
–, 1971: Die Entstehung der Kulturpflanzen als Modell für die Evolution der gesamten Pflanzenwelt. – In: G. Heberer (Hrsg.), Die Evolution der Organismen **2**, 2, pp. 175–300. – Stuttgart.
Senglaub, K., 1976: Haustierkundliche Untersuchungen und Evolutionstheorie. – In: H. Böhme, R. Hagemann und R. Löther (Hrsg.), Beiträge zur Genetik und Abstammungslehre, pp. 451–470. – Berlin.
Smartt, J., 1969: Evolution of American *Phaseolus* beans under domestication. – In: P. J. Ucko and G. W. Dimbleby (eds.), The Domestication and Exploitation of Plants and Animals, pp. 451–462. – London.
Stanton, W. R., 1969: Some domesticated lower plants in Southeast Asian food technology. – In: P. J. Ucko and G. W. Dimbleby (eds.), The Domestication and Exploitation of Plants and Animals, pp. 463–469. – London.
Stubbe, H., 1971: Weitere evolutionsgenetische Untersuchungen in der Gattung *Lycopersicon*. – Biol. Zbl. **90**, 545–559.
–, 1982: Geschichte des Instituts für Kulturpflanzenforschung Gatersleben der Deutschen Akademie der Wissenschaften zu Berlin (1943–1968). – Berlin.
Szabó, A., 1983: Alkamazott Biológia a Termesztett Növények Fejlődéstörténetében. – Bukarest.
Thellung, A., 1925: Kulturpflanzeneigenschaften bei Unkräutern. – Festschrift Carl Schröter. Veröff. Geobot. Inst. Rübel, H. 3, 745–762.
–, 1930: Die Entstehung der Kulturpflanzen. – Naturwiss. u. Landwirtsch. H. 16, 91 pp.
Vavilov, N. I., 1951: Phytogeographic basis on plant breeding. The origin, variation, immunity and breeding of cultivated plants. – Chronica Botanica **13**, 1–366.
Vogel, Ch., 1983: Gibt es Vorstufen menschlicher Geschichtlichkeit bei nicht-menschlichen Primaten? – Nova Acta Leopoldina n. F. **55** (253) 79–91.
Vogel, St., 1980: Florengeschichte im Spiegel blütenbiologischer Erkenntnisse. – Rheinisch-Westfälische Akademie der Wissenschaften, Vorträge N 291, 7–40.
de Wet, J. M. J., 1981: Species concepts and systematics of domesticated cereals. – Kulturpflanze **29**, 177–198.
Wickler, W., 1968: Mimicry in Plants and Animals. – New York.
Wiens, D., 1978: Mimicry in plants. – In: M. K. Hecht, W. C. Steere and B. Wallace (eds.), Evolutionary Biology **11**. – New York and London.
Wright, H. E., 1977: Environmental change and the origin of agriculture in the Old and New worlds. – In C. A. Reed (ed.), The Origins of Agriculture, pp. 281–318. – The Hague.
Zeuner, F. E., 1963: A History of Domesticated Animals. – London.
Zeven, A. C., 1973: Dr. Th. H. Engelbrecht's views on the origin of cultivated plants. – Euphytica **22**, 279–286.
–, 1975: A possible contribution of environmentally induced changes to the domestication of plants. – Euphytica **24**, 369–370.
–, 1980: Polyploidy and domestication: The origin and survival of polyploids in cytotype mixtures. – In: W. H.Lewis (ed.), Polyploidy: Biological Relevance, pp. 385–407. – New York.
Zohary, D., 1964: Spontaneous brittle six-row barleys, their nature and origin. – Barley Genetics I, Proc. First Intern. Barley Genetics Symp., Wageningen 1963, 29–31.
–, 1971: Origin of the south west Asia cereals: Wheats, barley, oats and rye. – In: H. P. Davis, P. C. Harper and I. C. Hedge (eds.), Plant Life of South West Asia, pp. 235–263. – Edinburgh.

Dr. K. Hammer
Zentralinstitut für Genetik und Kulturpflanzenforschung
der Akademie der Wissenschaften der DDR
DDR - 4325 Gatersleben, Corrensstraße 3

Kulturpflanze 32 · 1984 · 35–65

Modellvorstellungen zur Kohlenstoff-Isotopendiskriminierung bei der Photosynthese von C_3- und C_4-Pflanzen*

MARTIN PEISKER

(Eingegangen am 6. Januar 1984)

Zusammenfassung

In der vorliegenden Arbeit werden Modelle für die Anreicherung des Kohlenstoffatoms ^{12}C gegenüber ^{13}C bei der Photosynthese von C_3- und C_4-Pflanzen beschrieben und im Zusammenhang mit Literaturdaten zum Einfluß innerer und äußerer Faktoren auf diesen Vorgang diskutiert. Bei den C_3-Pflanzen wird die Isotopendiskriminierung durch das Verhältnis der Geschwindigkeiten des Transports von CO_2 in das Blatt und der Carboxylierung des Ribulose-1,5-bisphosphats (RuBP) bestimmt. Ausdruck dieses Zusammenhangs ist eine von FARQUHAR (1980) angegebene lineare Beziehung zwischen $\delta^{13}C$-Wert und dem Verhältnis von interzellulärer und äußerer CO_2-Konzentration. Die Isotopendiskriminierung bei der C_4-Photosynthese ist ebenfalls von den Geschwindigkeiten der Diffusion von CO_2 in das Blatt und der primären Carboxylierung von Phosphoenolpyruvat (PEP) abhängig. Wegen der unvermeidlichen Verluste von CO_2 aus den Leitbündelscheidenzellen sind aber auch die sekundäre Carboxylierung von RuBP und der CO_2-Transport innerhalb des Blattes zu berücksichtigen. Ein auf dieser Grundlage entwickeltes Modell, das im Gegensatz zu einem früher beschriebenen Ansatz auch die CO_2-Abhängigkeit der Geschwindigkeitsparameter für die Carboxylierungsschritte einschließlich der Limitierung der PEP-Carboxylierung durch die PEP-Regenerationsrate berücksichtigt, wird ausführlich dargestellt. Nach diesem Modell können die wenigen bisher bekannten Befunde zum Einfluß hoher CO_2-Konzentrationen auf die Isotopendiskriminierung von C_4-Pflanzen als Hinweis darauf interpretiert werden, daß der Isotopeneffekt der RuBP-Carboxylierung bei den C_4-Pflanzen möglicherweise größer ist als bei den C_3-Pflanzen.

Einleitung

Im natürlich vorkommenden Kohlenstoff ist zu etwa 1 % das Isotop ^{13}C enthalten. Bei der photosynthetischen CO_2-Fixierung kommt es zu einer Anreicherung des leichteren Kohlenstoffatoms ^{12}C. Dieser auf geringfügigen Unterschieden in den thermodynamischen und kinetischen Eigenschaften der Moleküle mit verschiedener Isotopenzusammensetzung beruhende Vorgang wird als Isotopendiskriminierung oder Isotopenfraktionierung bezeichnet. Er führt dazu, daß die organische Substanz der Pflanzen einen geringeren Gehalt an ^{13}C besitzt als das atmosphärische CO_2, aus dem sie gebildet wird.

* Herrn Prof. Dr. K. UNGER anläßlich der 65. Wiederkehr seines Geburtstages gewidmet.

Als Maß für den ^{13}C-Gehalt einer Probe dient der $\delta^{13}C$ Wert, der durch die folgende Beziehung definiert wird:

$$\delta(‰) = \left(\frac{R_p}{R_{st}} - 1\right) \cdot 1000 \tag{1}$$

R_p und R_{st} bezeichnen die Häufigkeitsverhältnisse $^{13}C/^{12}C$ der Probe und eines Standards. Als Standard wird allgemein Pee-Dee-Belemnit (Abkürzung PDB) aus einer Kalksteinformation in South Carolina verwendet. Das Häufigkeitsverhältnis dieses Standards beträgt $R_{st} = 0.01124$ (CRAIG 1957). Bezogen darauf hatte das atmosphärische CO_2 1978 einen $\delta^{13}C$-Wert von

$$\delta_a = -7‰$$

(KEELING et al. 1979).[1] Dieser den folgenden Berechnungen zugrundeliegende Wert kann allerdings nur als Richtwert verstanden werden, der zeitlichen und örtlichen Schwankungen unterliegt. Zu erwähnen sind insbesondere ein mit dem globalen Anstieg der atmosphärischen CO_2-Konzentration verbundener zeitlicher Trend zu niedrigeren Werten (KEELING et al. 1979) und lokale Verschiebungen in der gleichen Richtung in Industriegebieten, in Gewächshäusern und innerhalb von Pflanzenbeständen (O'LEARY 1981). Aus diesem Grund können $\delta^{13}C$-Werte von Pflanzenproben, die unter verschiedenen Bedingungen erhalten worden sind, nicht ohne weiteres miteinander verglichen werden, es sei denn, die genaue Isotopenzusammensetzung der CO_2-Quelle ist bekannt. Unsicherheiten dieser Art dürften zu der beobachteten Variabilität der $\delta^{13}C$-Werte von Pflanzenmaterial beigetragen haben (LERMAN 1975).

Ein quantitatives Maß für die mit physikalischen oder biochemischen Prozessen verbundene Diskriminierung des Kohlenstoffisotops ^{13}C ist die ebenfalls in ‰ ausgedrückte Größe

$$d = \frac{\delta_q - \delta}{1 + \delta_q/1000}, \tag{2}$$

die im wesentlichen mit der Differenz der $\delta^{13}C$-Werte des Kohlenstoffs der CO_2-Quelle (δ_q) und der zu untersuchenden Probe (δ) identisch ist. Bei gegebener CO_2-Quelle ist also eine stärkere Diskriminierung von ^{13}C mit einem kleineren, d. h. mehr negativen $\delta^{13}C$-Wert gleichbedeutend.

Die Diskriminierung von ^{13}C ist bei den C_4-Pflanzen mit $\delta^{13}C$-Werten zwischen -9 und -19 ‰ geringer als bei den C_3-Pflanzen mit $\delta^{13}C$-Werten zwischen -22 und -40 ‰ (TROUGHTON 1979). Wegen dieses Unterschieds ist der $\delta^{13}C$-Wert zu einem wichtigen und häufig angewandten Differenzierungskriterium zwischen C_3- und C_4-Pflanzen geworden (TROUGHTON 1979, O'LEARY 1981). Die Differenz in den $\delta^{13}C$-Werten ist vor allem auf unterschiedliche Isotopeneffekte bei der Carboxylierung von RuBP und PEP zurückzuführen, die auch *in vitro* nachgewiesen werden konnten. Obwohl gelegentlich auch auf die Bedeutung von Diffusionsvorgängen für die Isotopendiskriminierung bei der Photosynthese hingewiesen worden ist (PARK und EPSTEIN 1960, SCHMIDT und WINKLER 1979), hat man sich bis in die jüngste Zeit mit dieser pauschalen Erklärung zufrieden gegeben (LERMAN 1975, TROUGHTON 1979). Erst in den letzten Jahren ist versucht worden,

1 Auf der südlichen Hemisphäre betrug der mittlere $\delta^{13}C$-Wert der Meeresluft 1982 nach R. J. FRANCEY -7.6‰ (s. FARQUHAR 1983).

die Isotopendiskriminierung durch geeignete Modelle quantitativ zu beschreiben und dabei sowohl den Transport von CO_2 als auch die Carboxylierungsreaktionen zu berücksichtigen (FARQUHAR 1980, O'LEARY und OSMOND 1980, VOGEL 1980, O'LEARY 1981, FARQUHAR et al. 1982b, PEISKER 1982a, FARQUHAR 1983). Im folgenden sollen die in diesen Arbeiten entwickelten Modellansätze zusammenfassend dargestellt werden. Besondere Berücksichtigung wird dabei die verbesserte Version des in einer früheren Arbeit (PEISKER 1982a) beschriebenen Modells der Isotopendiskriminierung bei C_4-Pflanzen finden. Auf die besondere Situation bei den CAM-Pflanzen, bei denen die primäre CO_2-Fixierung mit wechselnden Anteilen sowohl durch PEP- als auch durch RuBP-Carboxylierung erfolgen kann (OSMOND 1978), soll hier nicht näher eingegangen werden.

C_3 -Pflanzen

Das einfachste Modell der Isotopendiskriminierung bei der Photosynthese ist das von O'LEARY und OSMOND (1980) und von O'LEARY (1981) angegebene Zweischrittschema. Dieses Modell berücksichtigt den in der Gasphase erfolgenden CO_2-Transport zwischen der äußeren Atmosphäre und der Oberfläche der Mesophyllzellen und den Carboxylierungsvorgang, der auch die schnell verlaufenden Prozesse der Absorption des CO_2 an der Grenzschicht zwischen gasförmiger und flüssiger Phase und den Transport innerhalb der Mesophyllzellen einschließt. Der CO_2-Transport in das Blatt und in der umgekehrten Richtung wird durch die Geschwindigkeitskonstanten k_1 und k_2 beschrieben, wobei Reversibilität dieses Vorgangs angenommen werden kann: $k_1 = k_2$. Für die Carboxylierung wird näherungsweise ebenfalls eine Reaktion erster Ordnung vorausgesetzt, die durch die Geschwindigkeitskonstante k_3 charakterisiert wird. Nach diesem einfachen Modell ist die Geschwindigkeit der CO_2-Fixierung der atmosphärischen CO_2-Konzentration c_a und wegen

$$c_i = \frac{k_1}{k_2 + k_3} \cdot c_a \tag{3}$$

auch der interzellulären CO_2-Konzentration c_i proportional. Die für den Gesamtprozeß resultierende Geschwindigkeitskonstante ist durch

$$k = \frac{k_1 \cdot k_3}{k_2 + k_3} \tag{4}$$

gegeben. Als Isotopeneffekt eines bestimmten Teilprozesses definiert man das Verhältnis der Geschwindigkeitskonstanten für ^{12}C und ^{13}C:

$$E_i = \frac{k_i^{12}}{k_i^{13}} \tag{5}$$

Für den Isotopeneffekt des Gesamtprozesses erhält man

$$\frac{k^{12}}{k^{13}} = E_1 \, \frac{E_3/E_2 + k_3/k_2}{1 + k_3/k_2} \tag{6}$$

Dieses Verhältnis (genauer: sein zeitlicher Mittelwert) bestimmt die Isotopenzusammensetzung der durch die Photosynthese gebildeten organischen Substanz:

$$R = \frac{k^{13}}{k^{12}} \cdot R_a \tag{7}$$

(R_a = Häufigkeitsverhältnis $^{13}C/^{12}C$ des atmosphärischen CO_2). Die zugehörige Isotopendiskriminierung ist durch

$$d(‰) = \left(1 - \frac{k^{13}}{k^{12}}\right) \cdot 1000 \tag{8}$$

gegeben. Nach der Definition des $\delta^{13}C$-Wertes (Gleichung 1) erhält man

$$\delta(‰) = \frac{k^{13}}{k^{12}} [1000 + \delta_a(‰)] - 1000 \tag{9}$$

(PEISKER 1982a).

Numerische Werte für die Isotopeneffekte der Einzelschritte des Modells sind von O'LEARY (1981) angegeben worden. Danach gilt für die Diffusion von CO_2 in der Gasphase

$$E_1 = E_2 = 1.0044 .$$

Der Isotopeneffekt der RuBP-Carboxylierung beträgt

$$E_3 = 1.030 .$$

Nach dem Zweischritt-Modell lassen sich zwei Grenzfälle unterscheiden:

1. Der Gesamtprozeß wird allein durch den CO_2-Transport in der Gasphase begrenzt ($k_3 \gg k_2$). Man erhält nach (6) einen Gesamt-Isotopeneffekt von $k^{12}/k^{13} = E_1$ und damit eine Isotopendiskriminierung von $d = 4.4$ ‰. Der zugehörige $\delta^{13}C$-Wert beträgt $\delta = -11.4$ ‰.
2. Ausschließliche Limitierung des Gesamtprozesses durch den Carboxylierungsvorgang ($k_3 \ll k_2$) ergibt nach (6) $k^{12}/k^{13} = E_3$ und damit $d = 29.1$ ‰ sowie $\delta = -35.9$ ‰.

Nach dem Modell sind also bei den C_3-Pflanzen $\delta^{13}C$-Werte zwischen -11 und -36 ‰ zu erwarten. In guter Übereinstimmung mit dieser Voraussage liegen die experimentell bestimmten $\delta^{13}C$-Werte zwischen -22 und -40 ‰ (TROUGHTON 1979).

Mit Hilfe der Gleichungen (6) und (9) kann die Abhängigkeit des $\delta^{13}C$-Wertes vom Verhältnis der Geschwindigkeitskonstanten für die RuBP-Carboxylierung und für den CO_2-Transport berechnet werden (Abb. 1, Kurve 1). Für $k_3/k_2 = 0$ erhält man den bereits genannten Wert von -36 ‰. δ nimmt mit steigendem Verhältnis k_3/k_2 ständig zu und nähert sich dem anderen Extremwert von -11.4 ‰. Die experimentell erhaltenen $\delta^{13}C$-Werte der C_3-Pflanzen können etwa dem Bereich $k_3/k_2 < 1.5$ zugeordnet werden. Die Geschwindigkeit der Carboxylierungsreaktion ist demzufolge bei den C_3-Pflanzen immer ein wesentlicher limitierender Faktor der photosynthetischen CO_2-Fixierung.

Änderungen des Öffnungszustandes der Stomata, also Änderungen von k_2, führen bei gleichbleibender Aktivität des Photosyntheseapparates zu veränderten $\delta^{13}C$-Werten (Abb. 1, Kurve 1) und sind zugleich mit Änderungen im Verhältnis von interzellulärer CO_2-Konzentration c_i und äußerer CO_2-Konzentration c_a verbunden. Dieser Zusammenhang wird explizit durch die von FARQUHAR (1980) angegebene Beziehung

$$\delta = \delta_a - a - (b - a) \cdot \frac{c_i}{c_a} \tag{10}$$

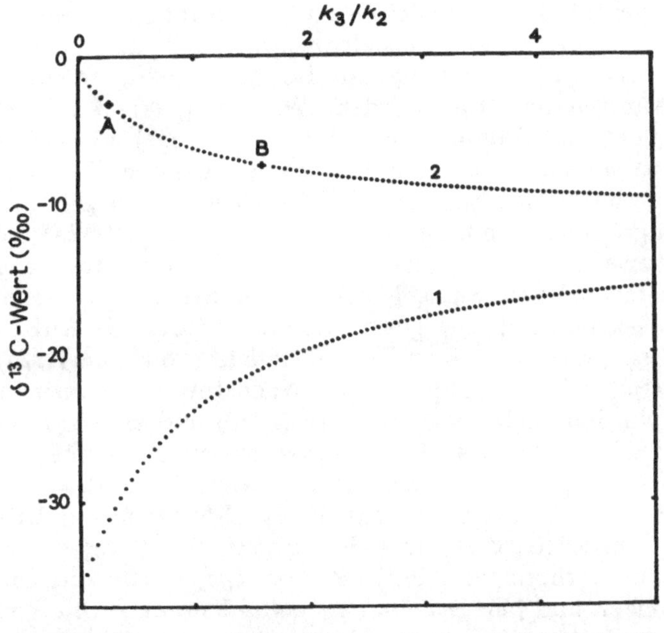

Abb. 1 Beziehung zwischen δ^{13}C-Wert und dem Verhältnis der Geschwindigkeitskonstanten für Carboxylierung und CO_2-Transport k_3/k_2 nach dem Zweischrittmodell bei RuBP-Carboxylierung (Kurve 1) und PEP-Carboxylierung (Kurve 2). Die Punkte A und B bezeichnen die bei den CAM-Pflanzen *Bryophyllum tubiflorum* und *Kalanchoë daigremontiana* gemessenen δ^{13}C-Werte des C_4-Atoms der Äpfelsäure (O'LEARY und OSMOND 1980)

ausgedrückt. a und b bedeuten die durch Diffusion in der Gasphase und durch die RuBP-Carboxylierung bedingten Diskriminierungen des Kohlenstoffatoms ^{13}C. Mit den angenommenen Werten der Isotopeneffekte E_1 und E_3 unter ist Berücksichtigung von Gleichung (8) $a = 4.4$ ‰ und $b = 29.1$ ‰ zu setzen.

In einer späteren Arbeit (FARQUHAR et al. 1982b) ist Gleichung (10) genauer begründet worden. In dieser Arbeit wurde auch gezeigt, wie mögliche Einflüsse von Licht- und Dunkelatmung auf die Isotopendiskriminierung abgeschätzt werden können. Die Autoren kommen zu dem Ergebnis, daß diese Prozesse in der Regel von untergeordneter Bedeutung sind, soweit nicht das Verhältnis c_i/c_a durch sie direkt beeinflußt wird.

Eine etwas allgemeinere Formulierung von Gleichung (10) wurde von FARQUHAR (1983) mitgeteilt. In der angegebenen Beziehung wird zwischen dem CO_2-Transport durch Diffusion im Bereich der Stomata und dem CO_2-Transport in der turbulenten Grenzschicht an der Blattoberfläche differenziert und außerdem der CO_2-Transport in der flüssigen Phase der Mesophyllzellen explizit berücksichtigt. In dieser erweiterten Form kann die Beziehung zwischen δ und c_i/c_a im Prinzip dazu dienen, die Bedeutung des Mesophyllwiderstandes als eines limitierenden Faktors des Gesamtprozesses der CO_2-Aufnahme abzuschätzen (FARQUHAR et al. 1982b). Der Einfluß der verschiedenen Teilwiderstände des CO_2-Transports auf die Isotopendiskriminierung ist auch in der Arbeit von VOGEL (1980) analysiert worden.

Gleichung (10) ist von FARQUHAR et al. (1982b) zur Herleitung einer Beziehung zwischen δ^{13}C-Wert und der Effektivität der Wassernutzung bei der Photosynthese verwendet worden. Danach ist die Isotopendiskriminierung einer C_3-Pflanze um so geringer, je größer der Wassernutzungskoeffizient der Photosynthese ist. Umgekehrt kann der δ^{13}C-Wert deshalb als Maß der Effektivität der Wassernutzung dienen (BRADFORD et al. 1983).

Die für C_3-Pflanzen vorausgesagte Zunahme des $\delta^{13}C$-Wertes mit zunehmendem Stomatawiderstand (also mit zunehmendem Verhältnis k_3/k_2; Abb. 1, Kurve 1) und abnehmendem Verhältnis c_i/c_a (Gleichung 10) ist durch experimentelle Ergebnisse verschiedener Autoren bestätigt worden. WINTER (1981) fand bei *Cicer arietinum* nach Anzucht der Pflanzen unter Wassermangel niedrigere interzelluläre CO_2-Konzentrationen und höhere $\delta^{13}C$-Werte als bei den Kontrollpflanzen. Messungen des Verhältnisses c_i/c_a und des $\delta^{13}C$-Wertes am gleichen Pflanzenmaterial wurden in größerem Umfang von FARQUHAR et al. (1982a) bei *Phaseolus vulgaris* und bei zwei Arten aus einer australischen Mangrove nach Anzucht in Lösungen mit verschiedenem Salzgehalt durchgeführt. Die erhaltene Beziehung zwischen δ und c_i/c_a läßt auf eine Diskriminierung durch die RuBP-Carboxylierung von $b = 27$ ‰ (oder auf einen Isotopeneffekt von $E_3 = 1.028$) schließen. Erhöhte $\delta^{13}C$-Werte bei C_3-Halophyten nach Wachstum oder Anzucht unter salinen Bedingungen wurden auch von GUY et al. (1980) und von ABDUL-RAHMAN und WILLIAMS (1981) beobachtet. Eine extrem starke Zunahme des $\delta^{13}C$-Wertes von -36 ‰ auf -25 ‰ ergab sich bei der Art *Puccinellia nuttalliana*, wenn das Wasserpotential der Nährlösung durch Zugabe von NaCl von -25 kPa auf -4000 kPa verringert wurde (GUY et al. 1980). Ebenfalls auf teilweisen Stomataschluß und die damit verbundene Abnahme von c_i/c_a dürfte die von WINTER et al. (1982) bei Weizen und bei *Poa pratense* gefundene Zunahme von δ nach Anzucht bei niedrigerer Luftfeuchtigkeit zurückzuführen sein. Die gleiche Ursache hatten wahrscheinlich die bei Freilandmaterial im Verlauf jahreszeitlich bedingter Trockenheitsperioden gemessenen Verschiebungen von δ zu weniger negativen Werten (WINTER und TROUGHTON 1978, WINTER et al. 1981).

Änderungen des Verhältnisses c_i/c_a können sich auch dann ergeben, wenn die biochemischen Prozesse der CO_2-Fixierung durch äußere oder innere Faktoren direkt beeinflußt werden. Eine Verringerung der CO_2-Aufnahme ist in diesem Fall durch eine Abnahme des Verhältnisses k_3/k_2 bedingt und deshalb mit einem Anstieg von c_i und mit geringeren $\delta^{13}C$-Werten verbunden (Abb. 1, Kurve 1). So lassen sich die in den innersten Jahresringen von Bäumen gemessenen relativ niedrigen $\delta^{13}C$-Werte („juvenile effect") zum Teil darauf zurückführen, daß in den frühen Entwicklungsstadien der Pflanzen die CO_2-Aufnahme durch die in den unteren Schichten des Bestandes herrschenden geringen Lichtintensitäten limitiert wird (FARQUHAR 1980, FRANCEY und FARQUHAR 1982). Experimentelle Untersuchungen zu dieser Frage führten zu widersprüchlichen Ergebnissen. Bei *Acacia farnesiana* wurden nach Anzucht bei niedrigen Lichtintensitäten tatsächlich kleinere $\delta^{13}C$-Werte erhalten als bei Lichtsättigung. Bei Freilandmaterial von *Festuca rubra* war dagegen der $\delta^{13}C$-Wert der Probe von einem dem Sonnenlicht voll exponierten Standort um 2 ‰ niedriger als der $\delta^{13}C$-Wert von im Schatten gewachsenen Pflanzen (SMITH et al. 1976). Dieses Ergebnis läßt sich auf der Grundlage der hier beschriebenen Vorstellungen zur Bedeutung von CO_2-Transport und Carboxylierung für die Isotopendiskriminierung nicht ohne weiteres erklären. Ein teilweiser Stomataschluß bei erhöhter Lichtintensität infolge von Wasserstress müßte wegen der Zunahme des Verhältnisses k_3/k_2 und der Verringerung von c_i/c_a ebenso zu einer geringeren Isotopendiskriminierung führen wie die diskutierte Beeinflussung des CO_2-Fixierungsapparates durch erhöhte Lichtintensitäten.

C_4-Pflanzen

Zweischrittmodell

Die primäre Reaktion der CO_2-Fixierung ist bei den C_4-Pflanzen die Carboxylierung von PEP. Das Substrat dieser Reaktion ist das HCO_3^--Ion. Aus diesem Grunde muß bei der Analyse der Kohlenstoffisotopendiskriminierung der C_4-Pflanzen auch der Isotopeneffekt der Hydratation des CO_2 berücksichtigt werden. Bei Vorhandensein von genügend Carbonsäure-Anhydrase stellt sich rasch ein Gleichgewicht zwischen CO_2- und HCO_3^--Konzentration ein. Bei der Modellierung der Isotopendiskriminierung kann deshalb der Isotopeneffekt dieses Vorgangs mit dem Isotopeneffekt der eigentlichen Carboxylierungsreaktion zu einem Gesamtisotopeneffekt von

$$E_3 = 0.994$$

zusammengefaßt werden (O'Leary 1981).

Die vereinfachte Beschreibung der CO_2-Aufnahme der C_4-Pflanzen durch das bereits bei den C_3-Pflanzen diskutierte Zweischrittschema ergibt die in Abb. 1, Kurve 2 dargestellte Beziehung zwischen δ^{13}C-Wert und dem Verhältnis k_3/k_2. Der Grenzfall der Limitierung des Gesamtprozesses durch die Carboxylierungsreaktion ($k_3 \ll k_2$) führt in diesem Fall zu einer Isotopendiskriminierung von $d = -6$ ‰ und mit $\delta_a = -7$ ‰ zu einem δ^{13}C-Wert von $\delta = -1$ ‰. Bei ausschließlicher Limitierung durch den CO_2-Transport ($k_3 \gg k_2$) erhält man wie bei den C_3-Pflanzen $d = 4.4$ ‰ und $\delta = -11.4$ ‰.

Das Zweischrittmodell der CO_2-Fixierung ist von O'Leary und Osmond (1980) und von Holtum et al. (1983) bei CAM-Pflanzen zur Berechnung der Isotopenzusammensetzung des C_4-Atoms der aus der PEP-Carboxylierung stammenden Äpfelsäure benutzt worden. Umgekehrt kann man aus dem δ^{13}C-Wert des C_4-Atoms der Äpfelsäure nach verschiedenen Korrekturen auf die Größe des Verhältnisses k_3/k_2 und damit nach Gleichung (3) auf das Verhältnis c_i/c_a schließen. Die in Abb. 1 auf Kurve 2 eingezeichneten Punkte geben die von O'Leary und Osmond (1980) bei zwei CAM-Pflanzen erhaltenen Resultate wieder:

A: *Bryophyllum tubiflorum* ($\delta = -3.2$ ‰; $k_3/k_2 = 0.27$; $c_i = 470$ mg m^{-3})

B: *Kalanchoë daigremontiana* ($\delta = -7.4$ ‰; $k_3/k_2 = 1.6$; $c_i = 230$ mg m^{-3})

(für $c_a = 600$ mg m^{-3}).

Nach dem Zweischrittmodell der C_4-Photosynthese hat die RuBP-Carboxylierung in den Leitbündelscheidenzellen keinerlei Einfluß auf die Isotopendiskriminierung. Es wird angenommen, daß der durch die primäre PEP-Carboxylierung einmal fixierte Kohlenstoff die folgenden Reaktionsschritte ohne eine Möglichkeit der Verzweigung durchläuft. Voraussetzung für die Gültigkeit dieser Annahme ist die vollkommene Abdichtung der Leitbündelscheidenzellen gegenüber Mesophyllzellen und äußerer Atmosphäre. In Wirklichkeit muß jedoch mit CO_2-Verlusten aus den Leitbündelscheidenzellen, also mit einem Verzweigungspunkt in der auf die PEP-Carboxylierung folgenden Reaktionssequenz, gerechnet werden (O'Leary 1981). Der Einfluß derartiger CO_2-Verluste auf die Isotopendiskriminierung bei C_4-Pflanzen ist in den Modellen von Peisker (1982a) und Farquhar (1983) untersucht worden.

Modelle mit Berücksichtigung der CO_2-Verluste aus den Leitbündelscheidenzellen

In früheren Modellen der CO_2-Kompensationskonzentration (PEISKER 1979) und der Isotopendiskriminierung (PEISKER 1982a) von C_4-Pflanzen war die Existenz von Transportwegen sowohl für einen direkten CO_2-Austausch zwischen Leitbündelscheidenzellen und äußerer Atmosphäre als auch für einen indirekten Austausch über die Mesophyllzellen angenommen worden. Dieser Ansatz soll hier durch ein Modell ersetzt werden, das den morphologischen Gegebenheiten im Blatt besser entspricht (Abb. 2). Aus der äußeren Atmosphäre (A_0) diffundiert das

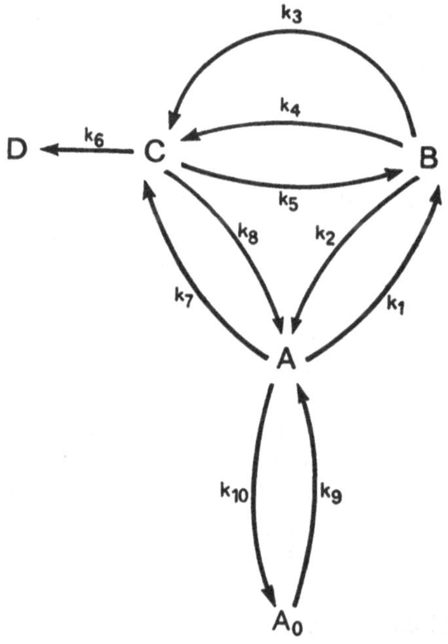

Abb. 2 Modell für die Berechnung der Isotopenzusammensetzung des fixierten Kohlenstoffs (D) bei C_4-Pflanzen. A_0, A, B, C: CO_2-Konzentrationen in der umgebenden Atmosphäre, im Interzellularraum, in den Mesophyllzellen und in den Leitbündelscheidenzellen; $k_1 = k_2$, $k_4 = k_5$, $k_7 = k_8$, $k_9 = k_{10}$: Geschwindigkeitskonstanten für den CO_2-Austausch zwischen Interzellularraum und Mesophyllzellen, Mesophyllzellen und Leitbündelscheidenzellen, Interzellularraum und Leitbündelscheidenzellen sowie äußerer Atmosphäre und Interzellularraum; k_3 und k_6: Geschwindigkeitsparameter für die PEP- und die RuBP-Carboxylierung

CO_2 zunächst in den Interzellularraum (A). Dieser CO_2-Austausch durch Grenzschicht und Stomata erfolgt in der Gasphase. Dagegen kann für den CO_2-Austausch zwischen Interzellularraum und Mesophyllzellen (B) oder Leitbündelscheidenzellen (C) sowie zwischen Mesophyll- und Leitbündelscheidenzellen ein Transport in der flüssigen Phase vorausgesetzt werden. Alle diese CO_2-Austauschvorgänge können als reversibel betrachtet werden, wenn man die Möglichkeit aktiver Transportprozesse ausschließt. Wie bei dem zuvor behandelten Zweischrittmodell werden CO_2-Flüsse und Carboxylierungsreaktionen durch Geschwindigkeitsparameter 1. Ordnung beschrieben. Die einzelnen Schritte im Modell unterliegen dabei nach O'LEARY (1981) den folgenden Isotopeneffekten:

$E_1 = E_2 = 1.000$ (CO_2-Transport in der flüssigen Phase)[2]

2 Bei diesem Wert des Isotopeneffekts für den CO_2-Transport in der flüssigen Phase ist nicht berücksichtigt worden, daß $^{13}CO_2$ um 1.1‰ weniger löslich ist als $^{12}CO_2$ (VOGEL 1980). Nach neueren Ergebnissen von M. H. O'LEARY soll außerdem der Transport von $^{13}CO_2$ in der flüssigen Phase mit einer gegenüber $^{12}CO_2$ um 0,7‰ geringeren Geschwindig-

$E_3 = 0.994$ (PEP-Carboxylierung)
$E_4 = E_5 = 1.000$
$E_6 = 1.030$ (RuBP-Carboxylierung)
$E_7 = E_8 = 1.000$
$E_9 = E_{10} = 1.0044$ (CO_2-Transport in der Gasphase)

Die Geschwindigkeitskonstante des in Abb. 2 dargestellten Gesamtprozesses ergibt sich aus den Geschwindigkeitsparametern der Einzelschritte zu

$$k = \frac{k_6 k_9 ([k_1+k_7)(k_3+k_4)+k_2 k_7]}{k_{10}[(k_2+k_3+k_4)(k_6+k_8)+k_2k_5]+k_6[(k_1+k_7)(k_3+k_4)+k_2k_7]} \quad (11)$$

Bei bekannten Werten der Parameter k_1 bis k_{10} erhält man mit dieser Gleichung zunächst k^{12}. Mit den angegebenen Isotopeneffekten E_i lassen sich nach Gleichung (5) die Geschwindigkeitsparameter k_i^{13} und daraus mit Gleichung (11) k^{13} berechnen. Der δ^{13}C-Wert ergibt sich dann nach Gleichung (9).

Im Gegensatz zu dem früher beschriebenen Modell (PEISKER 1982a) soll hier die CO_2-Abhängigkeit des die RuBP-Carboxylierung beschreibenden Parameters k_6 berücksichtigt werden:

$$k_6 = \frac{V_C}{C+K'_C} \quad (12)$$

$$K'_C = K_C\left(1+\frac{[O_2]}{K_O}\right) \quad (13)$$

(FARQUHAR und v. CAEMMERER 1982). In Gleichung (13) bedeuten K_C und K_O die Michaelis-Konstanten der RuBPCO für CO_2 und O_2 und V_C die RuBPC-Aktivität. Die CO_2-Konzentration in den Leitbündelscheidenzellen C ergibt sich nach dem Modell aus folgender Beziehung:

$$C = \frac{k_9[(k_1+k_7)(k_3+k_4)+k_2k_7]}{k_{10}[(k_2+k_3+k_4)(k_6+k_8)+k_2k_5]+k_6[(k_1+k_7)(k_3+k_4)+k_2k_7]} \cdot A_0 \quad (14)$$

(A_0 = atmosphärische CO_2-Konzentration). Man erhält eine quadratische Gleichung für k_6, deren physiologisch sinnvolle Lösung durch

$$k_6 = \frac{1}{2a_1}\left(-a_2+\sqrt{a_2^2+4a_1a_3}\right) \quad (15)$$

mit

$$a_1 = [k_{10}(k_2+k_3+k_4)+(k_1+k_7)(k_3+k_4)+k_2k_7]\,K'_C \quad (16)$$
$$a_2 = k_9[(k_1+k_7)(k_3+k_4)+k_2k_7]\,A_0+k_{10}[k_8(k_2+k_3+k_4)+k_2k_5]\,K'_C - $$
$$-[k_{10}(k_2+k_3+k_4)+(k_1+k_7)(k_3+k_4)+k_2k_7]\,V_C$$
$$a_3 = k_{10}[k_8(k_2+k_3+k_4)+k_2k_5]\,V_C$$

gegeben ist.

keit erfolgen (s. FARQUHAR 1983). Die Ergebnisse der mit dem Modell durchgeführten Rechnungen ändern sich mit dem resultierenden Isotopeneffekt von $E_1 = E_4 = E_7 = 1.0018$ nicht wesentlich.

In dem von BERRY und FARQUHAR (1978) vorgeschlagenen Modell der C_4-Photosynthese ist der direkte CO_2-Austausch zwischen Leitbündelscheidenzellen und Interzellularraum nicht explizit berücksichtigt worden ($k_7 = k_8 = 0$). Anstelle der Gleichungen (11) und (16) erhält man in diesem Fall folgende Beziehungen:

$$k = \frac{k_1 k_6 k_9 \, (k_3 + k_4)}{k_{10} \, [(k_2 + k_3 + k_4) \, k_6 + k_2 k_5] + k_6 k_1 \, (k_3 + k_4)} \tag{17}$$

$$a_1 = [k_{10} \, (k_2 + k_3 + k_4) + k_1 \, (k_3 + k_4)] \, K'_C \tag{18}$$

$$a_2 = k_1 k_9 \, (k_3 + k_4) \, A_0 + k_2 k_5 k_{10} K'_C - [k_{10} \, (k_2 + k_3 + k_4) + k_1 \, (k_3 + k_4)] \, V_C$$

$$a_3 = k_2 k_5 k_{10} V_C$$

Die CO_2-Konzentrationen im Interzellularraum, in den Mesophyllzellen und in den Leitbündelscheidenzellen können nach

$$A = \frac{k_9 [k_6 \, (k_2 + k_3 + k_4) + k_2 k_5] \, A_0}{k_{10} [k_6 \, (k_2 + k_3 + k_4) + k_2 k_5] + k_6 k_1 \, (k_3 + k_4)} \tag{19}$$

$$B = \frac{k_9 k_1 \, (k_5 + k_6) \, A_0}{k_{10} [k_6 \, (k_2 + k_3 + k_4) + k_2 k_5] + k_6 k_1 \, (k_3 + k_4)} \tag{20}$$

$$C = \frac{k_9 k_1 \, (k_3 + k_4) \, A_0}{k_{10} [k_6 \, (k_2 + k_3 + k_4) + k_2 k_5] + k_6 k_1 \, (k_3 + k_4)} \tag{21}$$

berechnet werden.

Die Notwendigkeit, in das vorliegende Modell die CO_2-Abhängigkeit von k_6 einzubeziehen, ergibt sich aus der für C_4-Pflanzen charakteristischen Anreicherung von CO_2 in den Leitbündelscheidenzellen (HATCH und OSMOND 1976). Gegenüber dem aus der Decarboxylierung der C_4-Verbindungen stammenden CO_2 ist das durch Licht- und Dunkelatmung freigesetzte CO_2 wahrscheinlich von untergeordneter Bedeutung und wird im Modell nicht berücksichtigt.

Bei den C_4-Pflanzen erfolgt der Übergang zur CO_2-Sättigung der CO_2-Aufnahmerate im allgemeinen innerhalb eines engen Bereichs der interzellulären CO_2-Konzentration. Die CO_2-Abhängigkeit der CO_2-Aufnahmerate kann deshalb genügend genau durch eine aus zwei Abschnitten bestehende Kurve approximiert werden. Bei niedrigen CO_2-Konzentrationen wird die CO_2-Aufnahme durch die PEPC-Aktivität limitiert und kann durch eine Gleichung vom Michaelis-Menten-Typ beschrieben werden. Im Bereich der CO_2-Sättigung wird der im Modell implizit enthaltene C_4-Zyklus durch die PEP-Regeneration begrenzt, für die eine konstante Maximalrate V angenommen wird. Der Geschwindigkeitsparameter k_3 nimmt also ebenso wie k_6 mit steigender CO_2-Konzentration B am Ort der Carboxylierung ab und ist durch

$$k_3 = \frac{k_3^m}{1 + \frac{B}{K_p}} \quad \text{für} \quad B < \frac{V K_p}{k_3^m K_p - V} \tag{22}$$

$$k_3 = \frac{V}{B} \qquad \text{für} \quad B \geqq \frac{V K_p}{k_3^m K_p - V}$$

gegeben (K_p = Michaelis-Konstante der PEPC für CO_2, k_3^m = Maximalwert von k_3 für $B = 0$).

In Analogie zu der die Isotopendiskriminierung der C_3-Pflanzen beschreibenden Gleichung (10) wurde von FARQUHAR (1983) für C_4-Pflanzen die folgende Beziehung abgeleitet:

$$\delta = \delta_a - a - (b_4 + b_3\Phi - a) \cdot \frac{c_i}{c_a} \qquad (23)$$

$a = 4.4$ ‰, $b_3 = 29.1$ ‰ und $b_4 = -6$ ‰ sind die für die Diffusion von CO_2 in der Gasphase, für die RuBP-Carboxylierung und für die PEP-Carboxylierung geltenden Werte der Isotopendiskriminierung. Φ ist der Teil des bei der PEP-Carboxylierung fixierten CO_2, der durch Diffusion aus den Leitbündelscheidenzellen wieder verloren geht („leakiness"). Diese Größe soll hier als Verlustkoeffizient bezeichnet werden. Konstante Werte von Φ vorausgesetzt, erhält man nach Gleichung (23) auch bei den C_4-Pflanzen eine lineare Beziehung zwischen δ^{13}C-Wert und dem Verhältnis c_i/c_a. Für $\Phi = 0.36$ ist δ von c_i/c_a unabhängig. Größere Werte von Φ führen zu einer Verringerung, kleinere Werte zu einem Anstieg von δ mit zunehmendem Verhältnis c_i/c_a (s. FARQUHAR 1983, Fig. 2).

Verglichen mit dem zuvor beschriebenen Modell erscheint Gleichung (23) weniger aussagekräftig, denn der Verlustkoeffizient Φ ist eine von zahlreichen Faktoren abhängige und damit nicht unmittelbar zugängliche Größe. Nach der Modellvariante ohne Berücksichtigung des direkten CO_2-Austauschs zwischen Leitbündelscheidenzellen und Interzellularraum (Gleichungen 17 und 18) ergibt sich Φ zu

$$\Phi = \frac{k_4\,(C - B)}{k_3 B}\,. \qquad (24)$$

Φ ist insbesondere von der Differenz der CO_2-Konzentrationen in den Leitbündelscheidenzellen und in den Mesophyllzellen abhängig, die ihrerseits durch alle Teilprozesse der C_4-Photosynthese beeinflußt werden (Gleichungen 20 und 21).

Modellparameter

1. CO_2-Transport in der Gasphase:

$k_9 = k_{10}$ ist der reziproke Wert des Gasphasenwiderstandes für die CO_2-Aufnahme. Ein für C_4-Pflanzen typischer Wert dieses Widerstandes ist 300 s m^{-1} (z. B. APEL und PEISKER 1980). Bei der Berechnung von δ^{13}C-Werten mit dem beschriebenen Modell sollen für k_9 entweder der zugehörige feste Wert von

$$k_9 = 0.0035 \text{ m s}^{-1}$$

oder veränderliche Werte zwischen 0 und 0.005 m s^{-1} angenommen werden. Auf die Unterscheidung von Grenzschichtwiderstand und Stomatawiderstand und der zugehörigen Geschwindigkeitskonstanten ist verzichtet worden, da der Grenzschichtwiderstand meist erheblich kleiner ist als der Stomatawiderstand.

2. PEP-Carboxylierung:

Der Maximalwert k_3^m des Geschwindigkeitsparameters für die PEP-Carboxylierung bei sehr niedrigen CO_2-Konzentrationen ist der reziproke Wert des Carboxylierungswiderstandes r_c. r_c ist ein Teil des intrazellulären Widerstandes für die CO_2-Aufnahme r_i, der außerdem noch den Transportwiderstand für CO_2 zwischen dem Interzellularraum und dem Ort der Carboxylierung in den Mesophyllzellen einschließt. Obwohl keine befriedigenden experimentellen Methoden zur getrennten Bestimmung der beiden Widerstandskomponenten bekannt sind, kann man wohl davon ausgehen, daß die Größe von r_i ähnlich wie bei den C_3-Pflanzen (RAVEN und GLIDEWELL 1981) vor allem durch r_c bestimmt wird. k_3^m kann somit näherungsweise als reziproker Wert von r_i erhalten werden. In 6 C_4-Arten wurden Werte von r_i zwischen 70 und 170 s m^{-1} gemessen (APEL und PEISKER 1980). Dem Mittelwert von ca. 100 s m^{-1} entspricht eine Geschwindigkeitskonstante von $k_3^m = 0.01$ m s^{-1}. In vielen Fällen sind allerdings auch niedrigere Werte von r_i erhalten worden (z. B. EDWARDS und WALKER 1983).

Im Prinzip kann k_3^m auch aus der PEPC-Aktivität V_p und der Michaelis-Konstante für CO_2 bestimmt werden:

$$k_3^m = \frac{V_p}{K_p} \tag{25}$$

Da jedoch das HCO_3^--Ion das Substrat der PEP-Carboxylierung ist, kann aus der *in vitro* erhaltenen Michaelis-Konstanten für HCO_3^- nur bei bekanntem p_H-Wert am Ort der Carboxylierung auf die Größe von K_p geschlossen werden. Unsicherheiten in der Kenntnis des intrazellulären p_H bedingen deshalb entsprechende Unsicherheiten hinsichtlich der Größe von K_p (BAUWE 1979).

In Umkehrung der Bestimmung von k_3^m aus enzymkinetischen Daten der PEPC läßt sich K_p als Produkt von V_p und r_i berechnen. Mit den von BAUWE (1979) für *Atriplex tatarica* angegebenen Werten von $r_i = 220$ s m^{-1} und $V_p = 0.66$ mg m^{-2} s^{-1} (s. Tabelle 1) erhält man

$$K_p = 150 \text{ mg m}^{-3}$$

(PEISKER und BAUWE 1984). Gesteht man diesem Wert eine gewisse Allgemeingültigkeit für C_4-Pflanzen zu, dann ergeben sich mit den bei 4 C_4-Arten der Gattung *Flaveria* gemessenen PEPC-Aktivitäten (BAUWE 1984) für k_3^m Werte zwischen 0.01 und 0.02 m s^{-1} (Tabelle 1). Als für C_4-Pflanzen repräsentative Größe soll hier

$$k_3^m = 0.015 \text{ m s}^{-1}$$

angenommen werden.

Limitierung der CO_2-Aufnahme durch die Regeneration des PEP bedeutet, daß die Maximalgeschwindigkeit V dieses Prozesses kleiner sein muß als die PEPC-Aktivität V_p. Im Mittel betrug V_p bei den in Tabelle 1 angeführten C_4-Arten 2.0 mg m^{-2} s^{-1}. Den folgenden Berechnungen liegt eine PEP-Regenerationsrate von

$$V = 1.5 \text{ mg m}^{-2} \text{ s}^{-1}$$

zugrunde. Dieser Wert ist für die von C_4-Pflanzen bekannten Sättigungsraten der CO_2-Aufnahme durchaus typisch.

Tabelle 1
Enzymkinetische Parameter von C_4-Pflanzen (*Atriplex tatarica*: BAUWE 1979, *Flaveria*-Arten: BAUWE 1984) und daraus abgeleitete Modellparameter. V_C = RuBPC-Aktivität (in mg $m^{-2} s^{-1}$); K_C = Michaelis-Konstante der RuBPCO für CO_2 (1: in der flüssigen Phase, in µM; 2: in der Gasphase, in mg m^{-3}); K_O = Michaelis-Konstante der RuBPCO für O_2 (in der flüssigen Phase, in µM); K'_C = apparente Michaelis-Konstante der RuBPCO für CO_2 bei der mit der atmosphärischen O_2-Konzentration im Gleichgewicht befindlichen Konzentration in der flüssigen Phase von 262 µM (in mg m^{-3}); k_3^m = Maximalwert des Geschwindigkeitsparameters für die PEP-Carboxylierung (in m s^{-1}); k_4^m = Maximalwert der Geschwindigkeitskonstante für den CO_2-Austausch zwischen Leitbündelscheidenzellen und Mesophyllzellen, berechnet nach Gleichung (29) mit $k_7 = 0$, $K_p = 150$ mg m^{-3}, $\gamma = 308$ mg kg^{-1} und $\Gamma_u = 2.5$ mg m^{-3} (in m s^{-1})

Art	V_C	K_C 1	K_C 2	K_O	K'_C	V_p	k_3^m	k_4^m
Atriplex tatarica	0.84	14	800	290	1500	0.66	0.0044	0.000027
Flaveria brownii	1.4	31	1800	450	2800	2.6	0.018	0.00011
Flaveria palmeri	1.1	27	1600	360	2700	2.0	0.013	0.000082
Flaveria trinervia	1.0	32	1900	340	3300	1.6	0.011	0.000067
Flaveria australasica	1.5	49	2800	430	4600	3.0	0.020	0.00012

3. *RuBP-Carboxylierung:*

Der Geschwindigkeitsparameter k_6 ist von der CO_2-Konzentration in den Leitbündelscheidenzellen C abhängig und ergibt sich aus Gleichung (15). Für die Berechnung der darin enthaltenen Funktionen a_1, a_2 und a_3 werden die RuBPC-Aktivität V_C und die apparente Michaelis-Konstante K'_C bei der atmosphärischen O_2-Konzentration von 0.275 kg m^{-3} (Gleichung 13) benötigt. In Tabelle 1 sind auch die bei *Atriplex tatarica* (BAUWE 1979) und mehreren C_4-Arten der Gattung *Flaveria* (BAUWE 1984) ermittelten Werte dieser Parameter angegeben worden. Daraus sind als für C_4-Pflanzen repräsentative Größen

$$V_C = 1.2 \text{ mg m}^{-2} \text{s}^{-1}$$

und

$$K'_C = 4000 \text{ mg m}^{-3}$$

abgeleitet worden.[3]

4. *CO_2-Transport in der flüssigen Phase:*

$k_1 = k_2$ ist der reziproke Wert des Widerstandes für den CO_2-Transport zwischen dem Interzellularraum und dem Ort der primären PEP-Carboxylierung. Dieser Widerstand ist wahrscheinlich im Vergleich mit dem Carboxylierungswiderstand von untergeordneter Bedeutung (s. o.). Für k_1 soll dementsprechend mit

$$k_1 = 0.05 \text{ m s}^{-1}$$

ein relativ großer Wert vorausgesetzt werden.

3 Bei der Wahl des angegebenen Wertes von K'_C wurde berücksichtigt, daß bei einer Reihe von C_4-Pflanzen größere Michaelis-Konstanten für CO_2 bis zu Werten von mehr als 60 µM in der flüssigen Phase gemessen worden sind (YEOH et al. 1980).

Die Parameter $k_4 = k_5$ und $k_7 = k_8$ beschreiben den CO_2-Austausch der Leitbündelscheidenzellen mit den Mesophyllzellen und dem Interzellularraum. Numerische Werte dieser Größen lassen sich mit Hilfe von Daten zur O_2-Abhängigkeit der CO_2-Kompensationskonzentration Γ abschätzen. Bei 10 C_4-Arten führte eine Erhöhung der O_2-Konzentration von 21 % auf 100 % im Mittel zu einer Zunahme von Γ um 2 mg m^{-3} (APEL und PEISKER 1978). Bei 100 % O_2 hatte also der O_2-abhängige Anteil von Γ annähernd eine Größe von $\Gamma_u = 2.5$ mg m^{-3}. Nach einem Modell für Γ bei C_4-Pflanzen (PEISKER 1979) besteht zwischen dem Widerstand r für die CO_2-Verluste aus den Leitbündelscheidenzellen und dem O_2-abhängigen Anteil Γ_u von Γ die folgende Beziehung:

$$\frac{1}{r} = \frac{V_p}{(\gamma[O_2] - \Gamma_u) \cdot \left(1 + \frac{K_p}{\Gamma_u - \lambda\,(\gamma[O_2] - \Gamma_u)}\right)} \qquad (26)$$

r und das Verhältnis λ der CO_2-Transportwiderstände zwischen Interzellularraum und Mesophyllzellen bzw. Leitbündelscheidenzellen lassen sich durch die Parameter des Modells der Isotopendiskriminierung wie folgt ausdrücken:

$$\frac{1}{r} = k_4 + k_7 + \frac{k_4 k_7}{k_1} \qquad (27)$$

$$\lambda = \frac{k_7}{k_1} \qquad (28)$$

$\gamma[O_2]$ ist der bei C_3-Pflanzen durch die Eigenschaften der RuBPCO bestimmte und von der Dunkelatmung unabhängige Teil von Γ (PEISKER 1978). Zwischen k_4, k_7 und Γ_u ergibt sich der folgende Zusammenhang:

$$k_4 = \frac{V_p}{\left(1 + \frac{k_7}{k_1}\right)(\gamma[O_2] - \Gamma_u)\left(1 + \frac{K_p}{\Gamma_u - \frac{k_7}{k_1}(\gamma[O_2] - \Gamma_u)}\right)} - \frac{k_1 k_7}{k_1 + k_7} \qquad (29)$$

Diese Beziehung (Abb. 3) drückt aus, in welchem Maße k_4 und k_7 zu der durch die CO_2-Verluste aus den Leitbündelscheidenzellen bedingten O_2-abhängigen Komponente Γ_u von Γ beitragen können. Nimmt man $k_7 = 0$ an, dann erhält man für k_4 maximale Werte. Mit $\Gamma_u = 2.5$ mg m^{-3}, $K_p = 150$ mg m^{-3}, $V_p = 2.0$ mg m^{-2} s^{-1} und dem bei einer Blattemperatur von 28 °C für die C_3-Pflanze *Phaseolus vulgaris* abgeleiteten Wert von $\gamma = 308$ mg kg^{-1} (PEISKER et al. 1981) ergibt sich bei 100 % O_2 (1.31 kg m^{-3}) $k_4^m = 0.000082$ m s^{-1}. Ist $k_7 > 0$, dann erhält man die vorgegebene Größe von Γ_u bei kleineren Werten von k_4. Der bei $k_4 = 0$ resultierende Maximalwert von k_7 ist wegen $1/k_1 > 0$ niedriger als k_4^m und nimmt mitzunehmender Größe von k_1 zu (Abb. 3). Bei den folgenden Berechnungen wird in den meisten Fällen $k_7 = 0$ und

$$k_4 = 0.00005 \text{ m s}^{-1}$$

angenommen werden.[4]

4 Bei einer Reihe von C_4-Pflanzen wurden beträchtlich höhere Michaelis-Konstanten der RuBPC für CO_2 als bei C_3-Pflanzen gefunden (YEOH et al. 1980, s. auch Tabelle 1). Deshalb muß damit gerechnet werden, daß auch der Parameter γ bei den C_4-Pflanzen größer sein kann als der hier von einer C_3-Pflanze übernommene Wert. Nach Gleichung (29) würde man dann mit dem vorgegebenen Wert von Γ_u entsprechend kleinere Werte für k_4 erhalten.

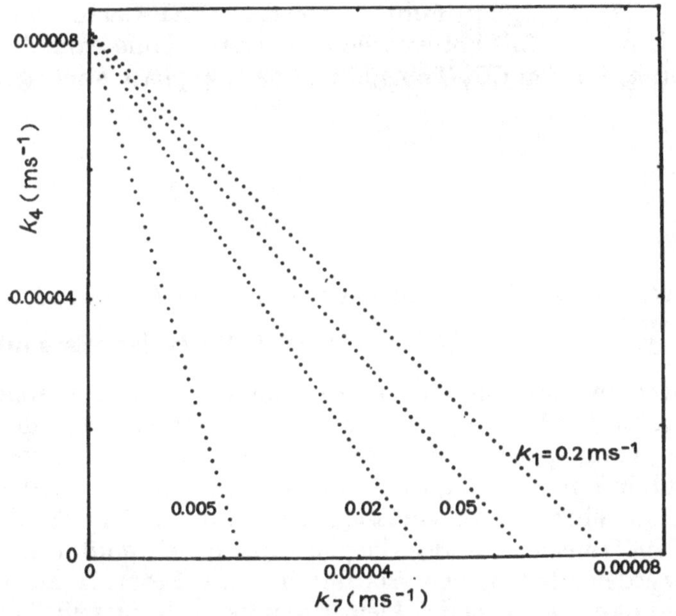

Abb. 3 Beziehung zwischen den Geschwindigkeitskonstanten für den CO_2-Austausch zwischen Leitbündelscheidenzellen und Mesophyllzellen (k_4) und zwischen Leitbündelscheidenzellen und Interzellularraum (k_7) nach Gleichung (29) bei konstanter O_2-abhängiger Komponente Γ_u der CO_2-Kompensationskonzentration und verschiedenen Werten des Parameters k_1 für den CO_2-Transport zwischen Interzellularraum und Mesophyllzellen ($\Gamma_u = 2.5\ mg\ m^{-3}$ bei $1.31\ kg\ m^{-3}$, $\gamma = 308\ mg\ kg^{-1}$, $K_p = 150\ mg\ m^{-3}$ und $V_p = 2.0\ mg\ m^{-2}\ s^{-1}$)

Quantitative Aussagen zur Bedeutung des CO_2-Transports in der flüssigen Phase lassen sich auch aus den Ergebnissen von Experimenten mit spezifischen Inhibitoren der PEPC und der decarboxylierenden Enzyme der verschiedenen Typen des C_4-Stoffwechsels ableiten. RATHNAM (1978) fand nach Applikation derartiger Inhibitoren bei Blattstreifen von 3 C_4-Arten eine beträchtliche aber nicht vollständige Hemmung der $^{14}CO_2$-Fixierung. Der nicht gehemmte Teil der $^{14}CO_2$-Fixierung erwies sich als O_2-abhängig und dürfte deshalb auf der Carboxylierung von RuBP beruhen. Diese Ergebnisse sind dahingehend interpretiert worden, daß eine direkte Fixierung von atmosphärischem CO_2 in den Leitbündelscheidenzellen signifikant zur Netto-CO_2-Aufnahme der C_4-Pflanzen beitragen soll (RATHNAM 1978). Eine solche Interpretation wird jedoch durch die in C_4-Pflanzen bei hohen O_2-Konzentrationen gefundenen niedrigen Werte von Γ ausgeschlossen (PEISKER 1978, 1979). Die als RuBP-Carboxylierung in den Leitbündelscheidenzellen zu deutende O_2-abhängige CO_2-Aufnahme war auch nur bei unterdrücktem C_4-Zyklus gemessen worden. Unter diesen Bedingungen ist offenbar ein nennenswerter CO_2-Austausch zwischen Leitbündelscheidenzellen und äußerer Atmosphäre möglich. Bei intakter C_4-Photosynthese wird dagegen der CO_2-Austausch zwischen Leitbündelscheidenzellen und der Umgebung des Blattes durch die effektive PEP-Carboxylierung in den Mesophyllzellen weitgehend eingeschränkt (APEL und PEISKER 1980). In der Terminologie des vorliegenden Modells wird der durch die Inhibitoren nicht beeinflußte Restanteil F_r der CO_2-Aufnahme vor allem durch die Parameter r (s. Gleichung 27) und k_6 bestimmt, während die PEP-Carboxylierungsrate F_p bei nicht gehemmter C_4-Photosynthese in erster Linie von der Geschwindigkeitskonstanten k_9 für den CO_2-Transport in der Gasphase abhängt. Es gilt näherungsweise

$$\frac{\frac{1}{r} \cdot k_6}{\frac{1}{r} + k_6} = \frac{F_r}{F_p} \cdot k_9$$

Mit $F_r/F_p = 0.165$ (RATHNAM 1978) und $k_9 = 0.0035 \text{ m s}^{-1}$ erhält man 0.0006 m s^{-1}. Dieser Wert stellt eine untere Grenze sowohl für $\frac{1}{r}$ als auch für k_6 dar. Es muß allerdings berücksichtigt werden, daß eine solche Abschätzung mit mehreren Unsicherheiten behaftet ist. So wird für F_r eine lineare Abhängigkeit von der äußeren CO_2-Konzentration vorausgesetzt. Weiterhin wurden mögliche unspezifische Wirkungen der Inhibitoren und mögliche Unterschiede zwischen Blattstreifen und intakten Blättern ebenso vernachlässigt wie Einflüsse der PEPC-Aktivität auf F_p und des CO_2-Transports in der Gasphase auf F_r. Es muß insbesondere damit gerechnet werden, daß in den Blattstreifen die Stomata nicht vollständig geöffnet sind und damit k_9 erheblich kleiner sein kann. In diesem Fall würde sich für die aus $\frac{1}{r}$ und k_6 resultierende Größe ein niedrigerer Wert ergeben. Veränderte Bedingungen gegenüber den intakten Blättern können sich bei den Blattstreifen auch daraus ergeben, daß CO_2 zusätzlich durch die Schnittstellen diffundieren kann. Andererseits kann nicht ausgeschlossen werden, daß der CO_2-Transport durch Grenzschicht und Stomata zum Teil in der flüssigen Phase erfolgt. Ferner gilt die für k_6 abgeleitete untere Grenze auch nur unter den Bedingungen des unterdrückten C_4-Zyklus. Bei intakter C_4-Photosynthese ist wegen der CO_2-Anreicherung in den Leitbündelscheidenzellen mit kleineren Werten zu rechnen.

Schließlich erlauben auch quantitative Angaben zur Anatomie der C_4-Pflanzen eine Abschätzung von k_4. In 6 C_4-Arten waren für den CO_2-Diffusionswiderstand zwischen Mesophyllzellen und Leitbündelscheidenzellen Werte zwischen 7000 s m^{-1} und 23000 s m^{-1} berechnet worden (APEL und PEISKER 1978). Die zugehörigen Werte von k_4 zwischen 0.00014 und 0.00004 m s^{-1} stimmen mit dem aus der O_2-Abhängigkeit von Γ abgeleiteten Schätzwert gut überein. Grundlage dieser Berechnungen ist die Annahme, daß der CO_2-Austausch zwischen Mesophyllzellen und Leitbündelscheidenzellen als Diffusion im wässrigen Medium zwischen zwei zylindrisch angeordneten Schichten beschrieben werden kann (OSMOND 1971, PEISKER 1978). Die Anwendbarkeit eines solchen Analogons auf C_4-Gräser vom NADP-ME-Typ und vom PCK-Typ ist allerdings von FARQUHAR (1983) bezweifelt worden. In diesen Gräsern enthalten die Zellwände der Leitbündelscheidenzellen eine Suberinschicht, die eine zusätzliche Diffusionsbarriere bilden könnte (HATTERSLEY und BROWNING 1981).

Zum Einfluß innerer und äußerer Faktoren auf die Isotopendiskriminierung der C_4-Pflanzen

Das Verfahren der Berechnung von $\delta^{13}C$-Werten mit vorgegebenen Zahlenwerten der Modellparameter unter Berücksichtigung der Isotopeneffekte von Einzelprozessen ist bereits in früher vorgeschlagenen Modellen der Isotopendiskriminierung von C_4-Pflanzen (PEISKER 1982a) und von C_3-C_4-Intermediärformen (PEISKER 1982b) angewandt worden. Von den früheren Ansätzen unterscheidet sich das hier beschriebene Modell für C_4-Pflanzen darin, daß die Annahme eines direkten CO_2-Austauschs zwischen Leitbündelscheidenzellen und äußerer Atmosphäre fallengelassen worden ist. Statt dessen wird nun ein CO_2-Austausch mit dem Interzellularraum angenommen, der direkt oder über die Mesophyllzellen erfolgen kann (s. Abb. 2). CO_2-Verluste aus den Leitbündelscheidenzellen führen zu einer verstärkten Isotopendiskriminierung. Dabei wirkt sich der durch die Geschwindigkeitskonstante k_7 beschriebene direkte CO_2-Austausch mit dem Interzellularraum stärker aus als der durch k_4 bestimmte indirekte Austausch über die Mesophyllzellen (Abb. 4). Der Unterschied verringert sich mit zunehmenden Werten der Geschwindigkeitskonstante k_1 für den CO_2-Transport zwischen Interzellularraum und Mesophyllzellen (Abb. 4, Kurven 2 und 3 bzw. Kurven 4 und 5). Im Gegensatz zu dem früheren Modell der Isotopendiskriminierung bei C_4-Pflanzen (PEISKER 1982a) ergibt nach dem vorliegenden Modell der Grenzfall einer ausschließlichen Limitierung des Gesamtprozesses durch den CO_2-Transport in der Gasphase ($k_9=0$) auch bei von 0 verschiedenen Werten von k_7 einen $\delta^{13}C$-Wert von -11.4 ‰ (Abb. 4, Kurven 3 und 5). Die mit $k_4=0$ und $k_7>0$ erhaltenen Kurven

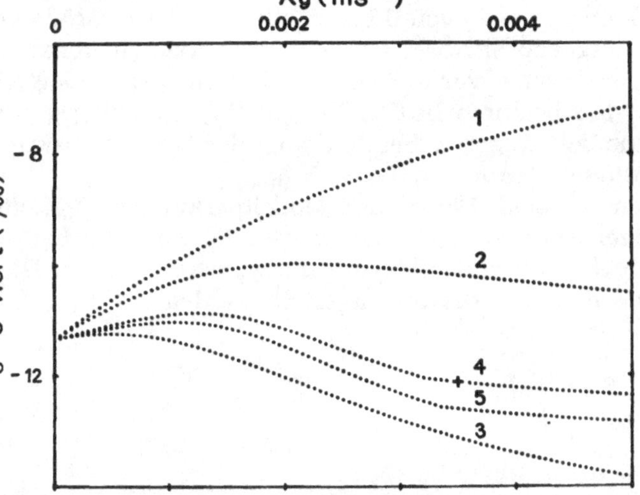

Abb. 4 Modellkurven der Abhängigkeit des $\delta^{13}C$-Wertes von der Geschwindigkeitskonstante für den CO_2-Transport zwischen äußerer Atmosphäre und Interzellularraum k_9 bei verschiedenen Werten der Parameter für den CO_2-Austausch innerhalb des Blattes. Kurve 1: $k_1=0.005$ m s^{-1}, $k_4=0$, $k_7=0$; Kurve 2: $k_1=0.005$ m s^{-1}, $k_4=0.00005$ m s^{-1}, $k_7=0$; Kurve 3: $k_1=0.005$ m s^{-1}, $k_4=0$, $k_7=0.00005$ m s^{-1}; Kurve 4: $k_1=0.05$ m s^{-1}, $k_4=0.00005$ m s^{-1}, $k_7=0$; Kurve 5: $k_1=0.05$ m s^{-1}, $k_4=0$, $k_7=0.00005$ m s^{-1}. Das + auf Kurve 4 bezeichnet den mit den Standardwerten aller Modellparameter (s. S. 52) berechneten $\delta^{13}C$-Wert

unterscheiden sich von den Kurven mit $k_4>0$ und $k_7=0$ nur quantitativ. Deshalb ist es möglich, $k_7=0$ zu setzen und direkte CO_2-Verluste in den Interzellularraum durch entsprechend größere Werte des Parameters k_4 zu berücksichtigen. In Abb. 5 ist dargestellt, wie sich in ihrer Wirkung auf den O_2-abhängigen Anteil

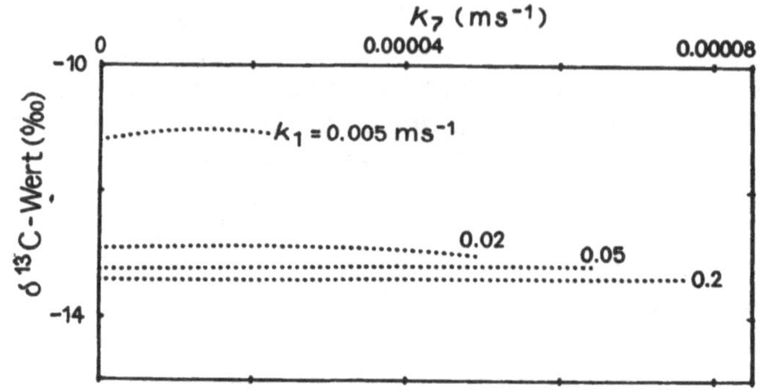

Abb. 5 Abhängigkeit des δ^{13}C-Wertes von den Parametern für den CO_2-Transport innerhalb des Blattes bei konstanter Größe der O_2-abhängigen Komponente Γ_u der CO_2-Kompensationskonzentration. Die dargestellte Zunahme der Geschwindigkeitskonstante für den CO_2-Austausch zwischen Leitbündelscheidenzellen und Interzellularraum k_7 ist nach Gleichung (29) mit einer Verringerung der Geschwindigkeitskonstante für den CO_2-Transport zwischen Leitbündelscheidenzellen und Mesophyllzellen k_4 verbunden (vgl. Abb. 3)

Γ_u der CO_2-Kompensationskonzentration einander aufhebende Änderungen der Parameter k_4 und k_7 (vgl. Abb. 3) auf die mit dem Modell berechneten δ^{13}C-Werte auswirken. Bei Abnahme des Parameters k_4 vom Maximalwert k_4^m auf 0 und gleichzeitiger Zunahme des Parameters k_7 von 0 bis zu einem von der Größe von k_1 abhängigen Maximalwert ergeben sich bei größeren k_1 praktisch konstante δ^{13}C-Werte. Auch bei kleineren Werten von k_1 ändert sich δ nur geringfügig. Es kann also festgestellt werden, daß die durch die Gleichungen (17) und (18) gegebene einfachere Modellvariante eine befriedigende Beschreibung der Isotopendiskriminierung bei der C_4-Photosynthese erlaubt.

Im vorhergehenden Abschnitt sind Werte der Modellparameter abgeleitet worden, die als für C_4-Pflanzen typisch angesehen werden können und bei den folgenden Überlegungen deshalb als Standardwerte benutzt werden sollen. Diese Parameterwerte sollen hier noch einmal zusammengestellt werden:

$k_1 = 0.05 \text{ m s}^{-1}$
$k_3^m = 0.015 \text{ m s}^{-1}$
$K_p = 150 \text{ mg m}^{-3}$
$V = 1.5 \text{ mg m}^{-2}\text{ s}^{-1}$
$k_4 = 0.00005 \text{ m s}^{-1}$
$V_C = 1.2 \text{ mg m}^{-2}\text{ s}^{-1}$
$K'_C = 4000 \text{ mg m}^{-3}$
$k_9 = 0.0035 \text{ m s}^{-1}$

Bei der atmosphärischen CO_2-Konzentration von 600 mg m^{-3} ergibt sich damit ein δ^{13}C-Wert von −12.1 ‰. Mit veränderlichen Werten des Parameters k_9 erhält man die in Abb. 4, Kurve 4 wiedergegebene Beziehung. Der Übergang von der

Limitierung des C_4-Zyklus durch die PEPC-Aktivität zur Limitierung durch die PEP-Regenerationsrate tritt hier bei $k_9 = 0.0032\ m\ s^{-1}$ auf. Der vollständige Standardsatz der Modellparameter (mit $k_9 = 0.0035\ m\ s^{-1}$) charakterisiert somit einen Zustand, bei dem die CO_2-Fixierung durch Vorgänge bei der PEP-Regeneration limitiert wird. Diese Situation ist für C_4-Pflanzen nicht ungewöhnlich, denn die CO_2-Aufnahmerate erreicht ihren Sättigungswert oft schon bei äußeren CO_2-Konzentrationen, die niedriger sind als die atmosphärische CO_2-Konzentration (z. B. CHARTIER und BETHENOT 1977).

In Übereinstimmung mit den bei $k_7 = 0$ mit dem früheren Modell erhaltenen Ergebnissen beeinflussen Änderungen von k_9, also Änderungen des Öffnungszustandes der Stomata, die Isotopendiskriminierung in Abhängigkeit von der Größe der CO_2-Verluste aus den Leitbündelscheidenzellen in unterschiedlicher Weise. Bei gegebenen Werten der die Rate der Decarboxylierung von C_4-Verbindungen in den Leitbündelscheidenzellen bestimmenden Parameter (k_3^m oder V) und der Geschwindigkeitskonstante für den CO_2-Austausch zwischen Leitbündelscheiden- und Mesophyllzellen k_4 ist der Verlauf der Beziehung zwischen $\delta^{13}C$-Wert und dem Parameter k_9 vor allem von der RuBPC-Aktivität V_C abhängig (Abb. 6).

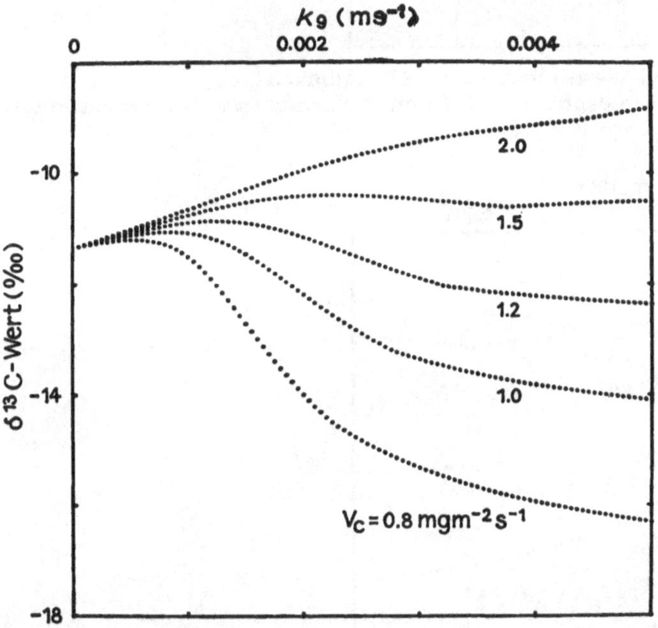

Abb. 6 Abhängigkeit des $\delta^{13}C$-Wertes von der Geschwindigkeitskonstante für den CO_2-Austausch zwischen äußerer Atmosphäre und Interzellularraum k_9 bei verschiedenen Werten der RuBPC-Aktivität V_C

Niedrige Raten der RuBP-Carboxylierung sind mit hohen CO_2-Gradienten zwischen Leitbündelscheiden- und Mesophyllzellen und entsprechend großen CO_2-Verlusten verbunden. δ nimmt in diesem Fall mit zunehmender Öffnungsweite der Stomata ab. Größere Werte von V_C führen zu kleineren CO_2-Gradienten und geringeren CO_2-Verlusten. Öffnung der Stomata bewirkt eine langsamere Abnahme des

δ^{13}C-Wertes, die bei weiterer Erhöhung von V_C in eine Zunahme übergeht. Änderungen der Decarboxylierungsrate beeinflussen δ in entgegengesetzter Richtung wie Änderungen von V_C, da die verstärkte Freisetzung von CO_2 in den Leitbündelscheidenzellen mit einem erhöhten CO_2-Gradienten und größeren CO_2-Verlusten verbunden ist (Abb. 7). Schließlich hängt der Umfang der CO_2-Verluste natürlich auch wesentlich von der Geschwindigkeitskonstante k_4 ab. Größere Werte dieses

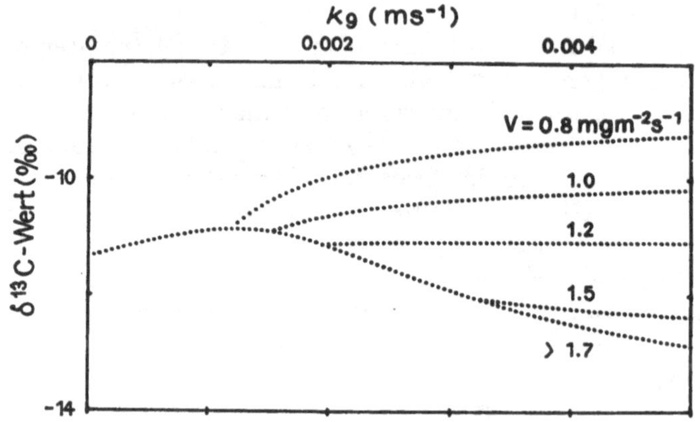

Abb. 7 Abhängigkeit des δ^{13}C-Wertes von der Geschwindigkeitskonstante für den CO_2-Austausch zwischen äußerer Atmosphäre und Interzellularraum k_9 bei verschiedenen Werten der PEP-Regenerationsrate V

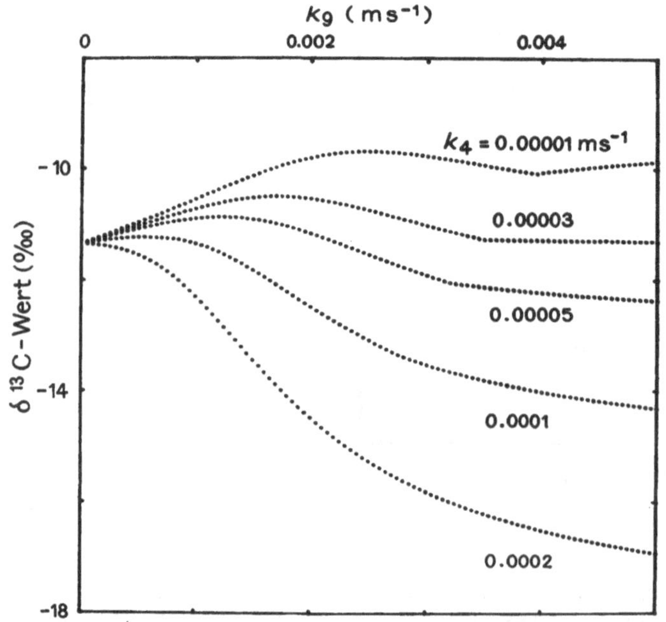

Abb. 8 Abhängigkeit des δ^{13}C-Wertes von der Geschwindigkeitskonstante für den CO_2-Austausch zwischen äußerer Atmosphäre und Interzellularraum k_9 bei verschiedenen Werten der Geschwindigkeitskonstante für den CO_2-Transport zwischen Leitbündelscheidenzellen und Mesophyllzellen k_4

Parameters führen bei zunehmender Stomataweite zu einer ähnlich starken Verringerung von δ (Abb. 8) wie niedrige Raten der RuBP-Carboxylierung.

Der Zusammenhang zwischen Isotopendiskriminierung und den CO_2-Verlusten aus den Leitbündelscheidenzellen läßt sich nach FARQUHAR (1983) auch als lineare Beziehung zwischen δ^{13}C-Wert und dem Verhältnis von äußerer und interzellulärer CO_2-Konzentration A/A_0 ausdrücken (Gleichung 23). Bei den in Abb. 8 dargestellten Kurven ergeben sich solche linearen Beziehungen erst bei größeren Öffnungsweiten der Stomata und entsprechend hohen CO_2-Konzentrationen am Ort der PEP-Carboxylierung, wenn der C_4-Zyklus durch die PEP-Regenerationsrate begrenzt wird (Abb. 9). In diesem Fall ist der Verlustkoeffizient Φ, das

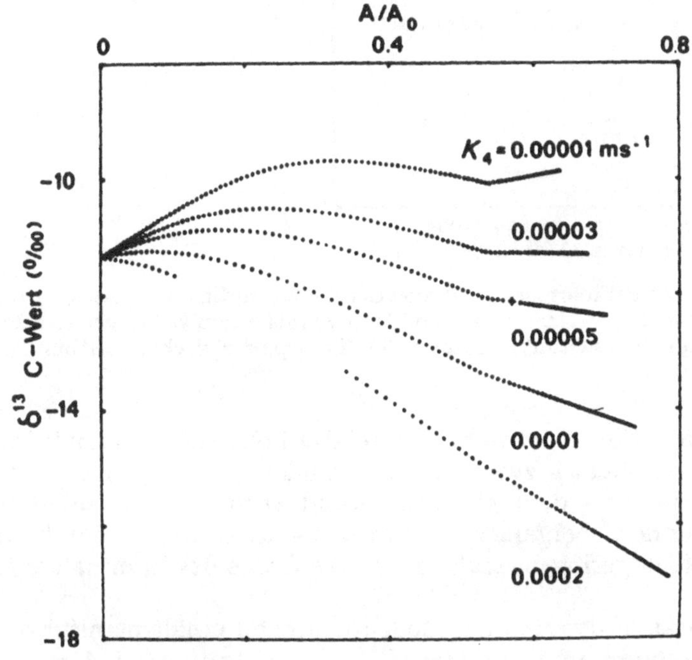

Abb. 9 Beziehung zwischen δ^{13}C-Wert und dem Verhältnis von interzellulärer und äußerer CO_2-Konzentration A/A_0 bei verschiedenen Werten der Geschwindigkeitskonstante für den CO_2-Transport zwischen Leitbündelscheidenzellen und Mesophyllzellen k_4. Die Geschwindigkeitskonstante für den CO_2-Austausch zwischen Interzellularraum und äußerer Atmosphäre wurde wie in Abb. 8 zwischen 0 und 0.005 m s^{-1} in Schritten von 0.000 05 m s^{-1} geändert

Verhältnis der CO_2-Verlustrate und der PEP-Carboxylierungsrate (Gleichung 24) nahezu konstant. Bei Limitierung des C_4-Zyklus durch die PEPC-Aktivität nimmt Φ dagegen mit steigenden Werten von k_9 ständig zu (Abb. 10). Mit $k_4 = 0.00003$ m s^{-1} erhält man bei höheren interzellulären CO_2-Konzentrationen annähernd konstante δ^{13}C-Werte von etwa -11.4 ‰ (Abb. 9) und einen Verlustkoeffizienten von $\Phi = 0.35$ (Abb. 10). Die geringfügige Abweichung von dem aus Gleichung (23) folgenden Wert $\Phi = 0.36$ (s. S. 45) erklärt sich daraus, daß die CO_2-Konzentration in den Mesophyllzellen B noch etwas niedriger ist als die

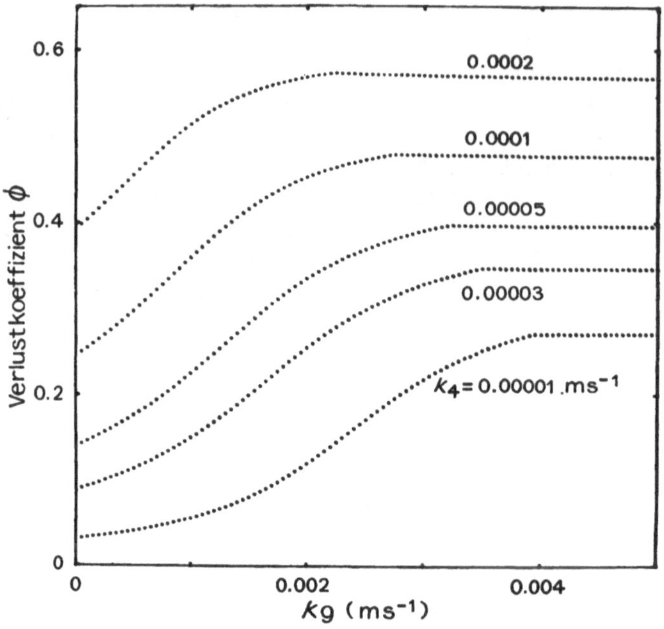

Abb. 10 Abhängigkeit des Verlustkoeffizienten Φ von der Geschwindigkeitskonstante für den CO_2-Austausch zwischen äußerer Atmosphäre und Interzellularraum k_9 bei verschiedenen Werten der Geschwindigkeitskonstante für den CO_2-Transport zwischen Leitbündelscheidenzellen und Mesophyllzellen k_4

interzelluläre CO_2-Konzentration A, obwohl eine relativ hohe Geschwindigkeitskonstante k_1 für den CO_2-Austausch zwischen Interzellularraum und Mesophyllzellen angenommen wurde. Aus dem gleichen Grund wird die interzelluläre CO_2-Konzentration, bei der der Übergang zur Limitierung des C_4-Zyklus durch die PEP-Regeneration erfolgt, in geringem Maße durch die Größe des Parameters k_4 beeinflußt (Abb. 9).

Mit den angegebenen Standardwerten der Modellparameter erhält man für das Verhältnis von interzellulärer und äußerer CO_2-Konzentration $A/A_0=0.57$ (Abb. 9). Kleinere Werte des Verhältnisses A/A_0, wie sie z. B. von WONG et al. (1979) bei Maisblättern gemessen worden sind, ergeben sich bei niedrigeren Werten von k_9. Für die Anreicherung von CO_2 in den Leitbündelscheidenzellen gegenüber der äußeren Atmosphäre sind von HATCH und OSMOND (1976) bei 2 C_4-Arten als untere Grenzen Faktoren von 2 und 5 angegeben worden. Der mit den Standardwerten der Modellparameter berechnete Anreicherungsfaktor von $C/A_0=20.4$ ist mit diesen Angaben vereinbar. Kleinere Anreicherungsfaktoren werden bei niedrigeren Werten von k_3^m oder V bzw. bei höheren Werten der Parameter k_4 und V_C erhalten. So führt eine Erhöhung von V_C auf 1.5 mg m^{-2} s^{-1} und eine gleichzeitige Verringerung von k_9 auf 0.0025 m s^{-1} zu einem δ^{13}C-Wert von -10.4 ‰ sowie zu $A/A_0=0.37$ und $C/A_0=11.6$.

In der Literatur gibt es bisher nur wenige Angaben über den Einfluß unterschiedlicher Bedingungen auf die Isotopendiskriminierung von C_4-Pflanzen. BOAG (1982) fand bei 3 C_4-Arten nach Anzucht der Versuchspflanzen unter

Na-Mangel im Vergleich zu den Kontrollpflanzen erhöhte δ^{13}C-Werte und eine Zunahme des Verhältnisses von interzellulärer und äußerer CO_2-Konzentration. Diese Wirkung des Na-Mangels ist nach FARQUHAR (1983) möglicherweise durch geringere Decarboxylierungsraten und demzufolge niedrigere CO_2-Verluste aus den Leitbündelscheidenzellen bedingt. Im Modell läßt sich eine derartige Beeinflussung des C_4-Zyklus durch geänderte PEP-Regenerationsraten oder PEPC-Aktivitäten simulieren. Eine Verringerung von V von 1.5 auf 0.8 mg m^{-2} s^{-1} führt mit den Standardparameterwerten zu einer Erhöhung des δ^{13}C-Wertes von −12.1 auf −9.4 ‰ (Abb. 7) bei einem gleichzeitigen Anstieg des Verhältnisses A/A_0 von 0.57 auf 0.71. Diese Änderungen sind von ähnlicher Größe wie die von BOAG (1982) erhaltenen Unterschiede. Eine Verringerung des Parameters k_3^m hat ebenfalls mit den Ergebnissen von BOAG vergleichbare Änderungen von δ und A/A_0 zur Folge. In diesem Fall wird mit den geänderten Parameterwerten ein Zustand erreicht, bei dem der C_4-Zyklus durch die PEPC-Aktivität und nicht mehr durch die PEP-Regenerationsrate limitiert wird.

Nach den Abb. 6 bis 8 ist zu erwarten, daß der δ^{13}C-Wert im Bereich zwischen −10 und −12 ‰ auch durch beträchtliche Änderungen der Stomataweite nur wenig beeinflußt wird. Diese Erwartung wird durch Untersuchungen von CARD et al. (1974) bestätigt. Anzucht von Pflanzenmaterial der Arten *Zea mays* und *Gomphrena globosa* in Hoagland-Lösung ergab δ^{13}C-Werte von −9.9 und −10.7 ‰. Erhöhte NaCl-Konzentrationen in der Nährlösung bewirkten demgegenüber nur geringfügige Änderungen. Die bei *Zea mays* gefundene schwache Zunahme von δ könnte durch Stomataschluß bedingt gewesen sein. Bei der Art *Spartina foliosa* wurde mit 200 mM NaCl in der Nährlösung ein δ^{13}C-Wert von −16.3 ‰ erhalten. Erhöhung der NaCl-Konzentration auf 400 und 700 mM ergab Werte von −15.7 und −16.2 ‰. Nach Abb. 6 und 8 wäre jedoch bei Stomataschluß eine eindeutige Zunahme von δ zu erwarten gewesen. Möglicherweise wurde bei dieser Salzmarschpflanze der Stomataapparat durch die erhöhten NaCl-Konzentrationen nicht beeinflußt. Es ist allerdings auch denkbar, daß ein möglicher Einfluß von Änderungen der Stomataweite auf die Isotopendiskriminierung durch gegenläufige Effekte von Änderungen im Metabolismus kompensiert worden ist. SHOMER-ILAN et al. (1975) fanden bei der Art *Suaeda monoica* nach Anzucht in 100 mM NaCl im Vergleich zu den Kontrollpflanzen eine deutliche Zunahme des δ^{13}C-Wertes von −17 auf −15.9 ‰. Bei diesem Versuch bleibt allerdings unklar, inwieweit die ebenfalls beobachtete Erhöhung der PEPC-Aktivität die Isotopendiskriminierung beeinflußt hatte.

In einer von WINTER und TROUGHTON (1978) publizierten umfangreichen Liste mit δ^{13}C-Werten von Pflanzen von Küsten- und Inlandstandorten in Israel und auf der Halbinsel Sinai sind auch Angaben über jahreszeitlich bedingte Änderungen bei mehreren C_4-Arten enthalten, die in Tabelle 2 zusammengestellt worden sind. Bei 7 dieser Arten wurde am Ende der feuchten Jahreszeit (April) ein deutlich höherer δ^{13}C-Wert gefunden als am Ende der trockenen Jahreszeit (Oktober), wobei die Differenzen 0.4 bis 2.1 ‰ betrugen. Bei 3 Arten wurden außerdem am Ende der feuchten Jahreszeit junge und alte Blätter oder Zweige miteinander verglichen. In allen Fällen ergaben sich bei den jüngeren Blättern oder Zweigen niedrigere δ^{13}C-Werte, die sich von den am Ende der Trockenperiode erhaltenen Werten relativ wenig unterschieden. Bei 3 anderen Arten änderte sich der δ^{13}C-

Tabelle 2

$\delta^{13}C$-Werte einiger C_4-Pflanzen von natürlichen Standorten des Nahen Ostens am Ende der feuchten Jahreszeit (A) und am Ende der trockenen Jahreszeit (B) nach WINTER und TROUGHTON (1978)

Art	$\delta^{13}C$-Wert (‰) A		B
Anabasis setifera Moq.	−11.8		−12.9
Atriplex halimus L.	−13.6 [1]		−14.4
	−14.8 [2]		
Calligonum comosum L'Hér.	−12.7 [1]		−14.5
	−14.2 [2]		
Hammada scoparia (Pomel) Iljin	−15.2		−15.6
Noaea mucronata (Forsk.) Aschers. et Schweinf.	−14.2		−15.3
Salsola kali L.	−12.7,	−14.5	−14.8
Suaeda fruticosa Forsk.	−12.1,	−12.2 [1]	−14.0
	−13.2,	−13.4 [2]	
Chenolea arabica Boiss.	−13.2		−13.0
Halogeton alopecuroides (Del.) Moq.	−10.8		−11.0
Seidlitzia rosmarinus Bge.	−12.5		−12.5
Suaeda monoica Forsk. ex J. F. Gmel.	−14.3		−12.8

[1] alte Blätter bzw. Zweige
[2] junge Blätter bzw. Zweige

Wert im Verlauf des Jahres kaum. Bei der Art *Suaeda monoica* schließlich wurde der höhere $\delta^{13}C$-Wert am Ende der Trockenperiode gefunden. In diesem Fall kann die beobachtete Änderung von δ durch jahreszeitlich bedingte Änderungen im Öffnungszustand der Stomata hervorgerufen worden sein (vgl. Abb. 6 und 8). Der Entwicklungsverlauf der 7 in Tabelle 2 zuerst aufgeführten Arten läßt sich demgegenüber wie folgt charakterisieren: Die in einer relativ niederschlagsreichen Periode angelegten jungen Blätter oder Zweige haben einen ziemlich niedrigen $\delta^{13}C$-Wert, der sich in der anschließenden Trockenperiode noch weiter verringern kann. Das in der darauf folgenden neuen Niederschlagsperiode einsetzende verstärkte Wachstum führt dann zu der beobachteten Zunahme des $\delta^{13}C$-Wertes. Diese läßt sich nicht auf Änderungen der Stomataweite zurückführen. Öffnung der Stomata müßte nach Abb. 6 und 8 bei Werten zwischen −16 und −12 ‰ zu einer mehr oder weniger deutlichen Verringerung von δ führen. Am wahrscheinlichsten ist wohl die Erklärung, daß mit dem Einsetzen verstärkten Wachstums mehr RuBPC synthetisiert worden ist, deren Wirkung auf den $\delta^{13}C$-Wert (Abb. 6) wesentlich stärker ausgeprägt gewesen sein müßte als eventuell gegebene Einflüsse des Stomataapparats.

Bei genügend hohen CO_2-Konzentrationen ist theoretisch zu erwarten, daß der $\delta^{13}C$-Wert der C_4-Pflanzen vor allem durch die Isotopendiskriminierung bei der RuBP-Carboxylierung bestimmt wird. Mit dem für diesen Prozeß angenommenen Isotopeneffekt von $E_6 = 1.030$ (S. 43) sollte δ sich mit steigender CO_2-Konzentration schließlich dem Grenzwert von −36 ‰ nähern (FARQUHAR 1983). Die von VOGEL (1980) bei Mais gefundene CO_2-Abhängigkeit des $\delta^{13}C$-Wertes bestätigt diese

Erwartung qualitativ. In Abb. 11 ist versucht worden, die erhaltene Beziehung mit dem hier entwickelten Modell quantitativ wiederzugeben. Dazu sind gegenüber dem Standardsatz (S. 52) zum Teil veränderte Werte der Modellparameter ange-

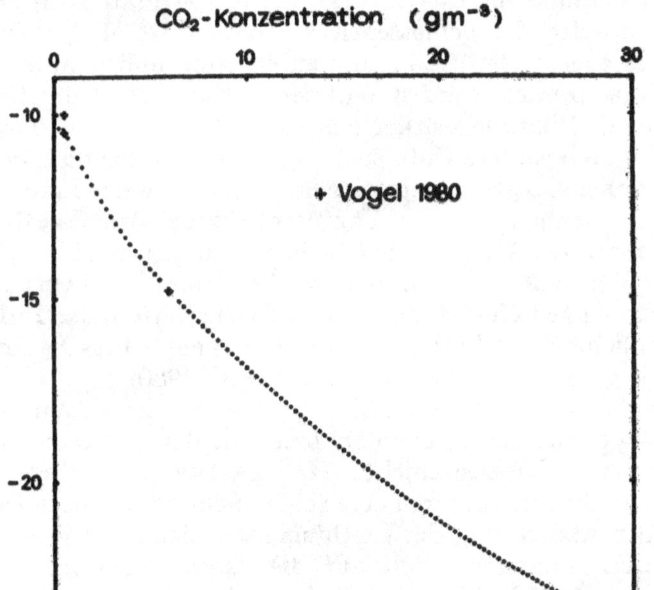

Abb. 11 **Abhängigkeit des $\delta^{13}C$-Wertes von der CO_2-Konzentration in der äußeren Atmosphäre. Für die Anpassung der Modellkurve an die bei Mais erhaltenen experimentellen Daten von Vogel (1980) wurden die folgenden von den Standardwerten (S. 52) abweichenden Werte der Modellparameter verwendet: $K'_C = 5400$ mg m^{-3}, $V_C = 2.0$ mg m^{-2} s^{-1}, $V = 0.5$mg m^{-2} s^{-1} und $k_4 = 0.000123$ m s^{-1}**

nommen worden: $K'_C = 5400$ mg m^{-3}, $V_C = 2.0$ mg m^{-2} s^{-1}, $V = 0.5$ mg m^{-2} s^{-1} und $k_4 = 0.000123$ m s^{-1}. K'_C ergab sich aus der von Yeoh et al. (1980) für Mais angegebenen Michaelis-Konstante der RuBPC für CO_2 in der flüssigen Phase von 56 μM mit $K_0 = 400$ μM (vgl. Tabelle 1). Die Anpassung an die experimentellen Daten erforderte also die Annahme einer relativ hohen RuBPC-Aktivität und einer unrealistisch niedrigen PEP-Regenerationsrate. Kleinere Werte von V_C und größere von V ergeben nach dem Modell eine geringere als die beobachtete CO_2-Abhängigkeit von δ. So erhält man mit $V_C = 1.2$ mg m^{-2} s^{-1}, $V = 1.5$ mg m^{-2} s^{-1} (s. S. 52) und $k_4 = 0.000016$ m s^{-1} bei 0.03 % CO_2 (0.54 g m^{-3}) den gleichen $\delta^{13}C$-Wert von -10.4 ‰ wie bei der in Abb. 11 wiedergegebenen Modellkurve. Bei Erhöhung der CO_2-Konzentration auf 1.5 % (27 g m^{-3}) verringert sich δ aber nur auf -15.2 ‰ gegenüber -23.3 ‰ mit den in Abb. 11 verwendeten Parameterwerten. Diese Schwierigkeit bei der Interpretation der Ergebnisse von Vogel (1980) ist möglicherweise dadurch bedingt, daß bei C_4-Pflanzen der Isotopeneffekt bei der RuBP-Carboxylierung größer sein kann als der von den C_3-Pflanzen übernommene Wert von $E_6 = 1.030$. In diesem Zusammenhang ist auch auf die ebenfalls bei Mais erhaltenen Ergebnisse von Smith und Boutton (1981) hinzuweisen. Diese Autoren fanden nach Anzucht der Pflanzen bei CO_2-Konzentrationen zwischen 0.03 und 5 % eine lineare Abnahme des $\delta^{13}C$-Wertes von -12 auf -49 ‰. Die Verwendung von technischem CO_2 mit unbekannter Isotopenzusammensetzung (s. Farquhar 1983) schränkt die Aussagekraft dieser Resultate zwar erheblich ein. Die große Spannweite der gemessenen $\delta^{13}C$-Werte ist aber auch mit niedrigen $\delta^{13}C$-Werten des zugesetzten CO_2 kaum zu erklären und könnte ebenfalls als Hinweis darauf interpretiert werden, daß der Isotopeneffekt der RuBP-Carboxylierung E_6 bei den C_4-Pflanzen stärker sein kann als bei den C_3-Pflanzen. Zur Klärung dieser Frage wären weitere Untersuchungen zur Isotopendiskriminierung der C_4-Pflanzen bei hohen CO_2-Konzentrationen wünschenswert. Eine offene Frage ist in diesem Zusammenhang, ob die Faktoren, die bei den C_4-Pflanzen erhöhte Michaelis-Konstanten der RuBPC für CO_2 bedingen (Yeoh et al. 1980), auch den Isotopeneffekt der RuBP-Carboxylierung beeinflussen (Hattersley 1982). Gäbe es eine Beziehung zwischen diesen beiden Eigenschaften der RuBPC, dann müßte mit einer ähnlich großen Variabilität des Isotopeneffektes E_6 gerechnet werden, wie sie für K_C gefunden worden ist (Yeoh et al. 1980).

C_4-Gräser vom NADP-ME-Typ und vom PCK-Typ besitzen im Gegensatz zu Gräsern vom NAD-ME-Typ und zu C_4-Dikotelydonen in den Zellwänden der Leitbündelscheidenzellen eine Suberinschicht (Hattersley und Browning 1981). Daraus ergibt sich die Frage, ob durch eine solche Schicht bedingte Unterschiede in der Qualität der Abdichtung der Leitbündelscheidenzellen gegenüber Mesophyllzellen und Interzellularraum sich auf die Isotopendiskriminierung auswirken können. Hattersley (1982) konnte sowohl in eigenen Untersuchungen als auch durch Auswertung von Angaben aus der Literatur (Brown 1977, Vogel et al. 1978, Ziegler et al. 1981) kleine aber hoch signifikante Unterschiede in den mittleren $\delta^{13}C$-Werten einiger Gräser der 3 Stoffwechseltypen nachweisen. Der höchste Mittelwert ergab sich bei den Gräsern vom NADP-ME-Typ. Es folgten die Gräser vom PCK-Typ und der niedrigste Wert wurde bei den Gräsern vom NAD-ME-Typ ohne Suberinschicht erhalten. In der gleichen Reihenfolge nahmen auch die Werte der Verlustkoeffizienten Φ zu, die von Farquhar (1983) für C_4-Monokotyledonen mit verschiedenen Decarboxylierungsmechanismen aus Unterschieden in der mittleren Quantenausbeute für die CO_2-Aufnahme (Ehleringer und Pearcy 1983) abgeleitet worden sind. Es bleibt allerdings eine offene Frage, ob diese Ergebnisse primär auf Unterschiede in der Abdichtung der Leitbündelscheidenzellen zurückzuführen sind. Diese im Modell durch den Parameter k_4 charakterisierte Eigenschaft der C_4-Pflanzen ist nur einer von mehreren Faktoren, die die CO_2-Verluste aus den Leitbündelscheidenzellen bestimmen und damit die Isotopendiskriminierung beeinflussen. Andere bedeutsame Faktoren sind vor allem die RuBP-Carboxylierungsrate (Abb. 6) und die durch PEPC-Aktivität oder PEP-Regenerationsrate (Abb. 7) limitierte Decarboxylierung von C_4-Verbindungen. Die erhaltenen Unterschiede in den $\delta^{13}C$-Werten und in den Quantenausbeuten müssen deshalb nicht notwendigerweise Unterschiede in der Güte der Abdichtung der Leitbündelscheidenzellen reflektieren, sondern können ebenso durch Besonderheiten im Metabolismus der verschiedenen Varianten des C_4-Weges der Photosynthese bedingt sein.

CO_2-Verluste aus den Leitbündelscheidenzellen erfordern einen erhöhten Auf-wand an Energie und Reduktionsäquivalenten für die Aufrechterhaltung des C_4-Zyklus und für die Synthese seiner biochemischen Komponenten. Unter diesem Aspekt sollten niedrige Verlustkoeffizienten für die Pflanze von Vorteil sein. Diesem Anspruch stehen jedoch andere für die Funktion der C_4-Photosynthese wesentliche Erfordernisse entgegen. Der Transport von Metaboliten zwischen Mesophyll- und Leitbündelscheidenzellen erfolgt ebenso wie der direkte CO_2-Austausch durch Diffusion im Symplasten (Hatch und Osmond 1976). Der diese Vorgänge limitierende Diffusionswiderstand der Plasmodesmata darf also nicht zu große Werte annehmen, damit genügend hohe Geschwindigkeiten des Metabolitentransports erreicht werden können (Hattersley und Browning 1981). Die Funktion der Suberinschicht in den Zellwänden der Leitbündelscheidenzellen bei den C_4-Pflanzen vom NADP-ME-Typ und vom PCK-Typ könnte darin bestehen, CO_2-Verluste über den Apoplasten auf ein Minimum zu reduzieren. C_4-Pflanzen ohne diese Suberinschicht haben anatomische Besonderheiten, die möglicherweise dem gleichen Ziel dienen (Hattersley und Browning 1981). Ein Mindistmaß direkten oder indirekten Gasaustausches zwischen Leitbündelscheidenzellen und Interzellularraum muß auch noch aus einem anderen Grund gewährleistet sein. Bei der C_4-Photosynthese kommt es nicht nur zu einer Anreicherung von CO_2 in den Leitbündelscheidenzellen sondern auch zu einer Erhöhung der O_2-Konzentration (Raven 1977, Berry und Farquhar 1978). Zu geringe Werte des Parameters k_4 würden die Anreicherung von O_2 begünstigen und hätten dadurch erhöhte Raten der RuBP-Oxygenierung zur Folge, die die Effektivität des Gesamtprozesses negativ beeinflussen müßten. In diesem Zusammenhang hat Farquhar (1983) auf eine mögliche Beziehung zwischen der O_2-Entwicklung in den Leitbündelscheidenzellen und dem Umfang der CO_2-Verluste hingewiesen. In den Chloroplasten der Leitbündelscheidenzellen der C_4-Pflanzen vom NADP-ME-Typ ist die Aktivität des Photosystems II und damit auch die O_2-Entwicklung mehr oder weniger stark reduziert (Edwards und Walker 1983). Damit übereinstimmend waren bei Gräsern dieses Typs sowohl aus dem mittleren $\delta^{13}C$-Wert als auch aus der mittleren Quantenausbeute die niedrigsten Werte des Verlustkoeffizienten Φ abgeleitet worden (Hattersley 1982, Farquhar 1983).

Liste der Abkürzungen

PEP : Phosphoenolpyruvat
PEPC : Phosphoenolpyruvat-Carboxylase
RuBP : Ribulose-1,5-bisphosphat
RuBPC : Ribulose-1,5-bisphosphat-Carboxylase
RuBPCO : Ribulose-1,5-bisphosphat-Carboxylase/Oxygenase
Decarboxylierende Enzyme der drei Typen von C_4-Pflanzen:
NADP-Me : NADP-abhängiges Malat-Enzym
NAD-Me : NAD-abhängiges Malat-Enzym
PCK : Phosphoenolpyruvat-Carboxykinase

Summary

Models of carbon isotope discrimination during photosynthesis of C_3 and C_4 plants

Models of fractionation of carbon atoms ^{12}C and ^{13}C during photosynthesis of C_3 and C_4 plants are described and discussed with regard to the influence of internal and external factors on this process. Isotope discrimination in C_3 plants is determined by the ratio of the rates of CO_2 transfer into the leaf and carboxylation of ribulose-1,5-bisphosphate (RuBP). This relationship can be expressed by a linear equation between $\delta^{13}C$ value and the ratio of intercellular and external CO_2 concentration as proposed by FARQUHAR (1980). Carbon isotope discrimination by C_4 photosynthesis also depends on the rates of CO_2 diffusion into the leaf and of primary carboxylation of phosphoenolpyruvate (PEP). However, because of the unavoidable losses of CO_2 from bundle sheath cells the secondary carboxylation of RuBP as well as CO_2 transfer processes within the leaf must be taken into account, too. A model based on these features is explained in some detail. In contrast to a model proposed earlier the presented model considers the dependence on CO_2 concentration of rate parameters for the carboxylation steps including limitation of PEP carboxylation by PEP regeneration rate. According to the model the spare experimental results on the influence of high concentrations of CO_2 on carbon isotope discrimination in C_4 plants may be interpreted as an indication that the isotope effect of RuBP carboxylation in C_4 plants could be higher than in C_3 plants.

Краткое содержание

Модели изотопной дискриминации углерода в фотосинтезе C_3 - и C_4 -растений

В предлагаемой работе описываются модели обогащения углеродного атома ^{12}C в сравнении с ^{13}C в фотосинтезе растений типа C_3 и C_4 и, в связи с литературным данными, обсуждается влияние внутренних и внешних факторов на этот процесс. У C_3-растений изотопная дискриминация определяется соотношением скоростей транспорта CO_2 к листьям и карбоксилирования рибулозо-1,5-дифосфата (РДФ). Выражением этой связи является, предложенная Фаркуаром (1980), линейная зависимость величины δ^{13} C от соотношения межклеточной и внешней концентрации CO_2. Изотопная дискриминация в синтезе типа C_4 зависит также от скоростей диффузии CO_2 к листу и первичного карбоксилирования фосфоэнолпирувата (ФЭП). Благодаря неизбежной потере CO_2 из клеток обкладки проводящих пучков, надо учитывать вторичное карбоксилирование РДФ и транспорт CO_2 внутри листа. Подробно описывается модель, построенная на этой основе, которая, в противоположность ранее предложенной, учитывает также параметры скорости протекания фаз карбоксилирования в зависимости от концентрации CO_2, включая лимитирование карбоксилирования ФЭП его регенеративной скоростью. Применяя описанную модель, можно немногие, до сих пор известные, данные о влиянии высоких концентраций CO_2 на изотопную дискриминацию у C_4 -растений, интерпретировать как указание на то, что изотопный эффект карбоксилирования ФЭП у C_4 -растений, возможно, оказывается более сильным, чем у C_3-растений.

Literatur

ABDULRAHMAN, F. S., and G. J. WILLIAMS, 1981: Temperature and salinity regulation of growth and gas exchange of *Salicornia fruticosa* (L.) L. – Oecologia **48**, 346–352.

APEL, P., und M. PEISKER, 1978: Einfluß hoher Sauerstoffkonzentrationen auf den CO_2-Kompensationspunkt von C_4-Pflanzen. – Kulturpflanze **26**, 99–103.

–, und –, 1980: CO_2-Versorgung der photosynthetischen Carboxylierungszentren. – In: Biophysik, Biochemie und Physiologie der Photosynthese (Hrsg. P. HOFFMANN und B. HIEKE), pp. 181–204. Colloquia Pflanzenphysiologie Nr. 3. Humboldt-Universität zu Berlin.

BAUWE, H., 1979: Biochemische und gaswechselphysiologische Untersuchungen zur Charakterisierung von Pflanzenarten mit C_3–C_4 intermediären Eigenschaften sowie zur Prüfung mathematischer Modellvorstellungen des photosynthetischen CO_2-Stoffwechsels. Dissertation A. Martin-Luther-Universität Halle – Wittenberg.

–, 1984: Photosynthetic enzyme activities and immunofluorescence studies on the localization of ribulose-1,5-bisphosphate carboxylase/oxygenase in leaves of C_3, C_4, and C_3–C_4 intermediate species of *Flaveria* (Asteraceae) – Biochem. Physiol. Pflanzen **179**, 253–268.

BERRY, J. A., and G. D. FARQUHAR, 1978: The CO_2 concentrating function of C_4 photosynthesis. A biochemical model. – In: Photosynthesis 77 (Eds. D. O. HALL, J. COOMBS, T. W. GOODWIN), pp. 119–131. The Biochemical Society, London.

BOAG, T. S., 1982: Characterization of C_4 photosynthesis in sodium-deficient plants. – Ph. D. Thesis, Australian National University, Canberra.

BRADFORD, K. J., T. D. SHARKEY, and G. D. FARQUHAR, 1983: Gas exchange, stomatal behavior, and $\delta^{13}C$ values of the *flacca* tomato mutant in relation to abscisic acid. – Plant Physiol. **72**, 245–250.

BROWN, W. V., 1977: The Kranz syndrome and its subtypes in grass systematics. – Mem. Torrey Bot. Club **23** (2), 1–97.

CARD, K. A., B. MAHALL, and J. H. TROUGHTON, 1974: Salinity and carbon isotope ratios in C_3 and C_4 plants. – Carnegie Institution Year Book **73**, 784–785.

CHARTIER, P., et O. BETHENOD, 1977: La productivité primaire à l'échelle de la feuille. – In: Les processus de la production végétale primaire (Ed. A. MOYSE), pp. 77–112. Gauthier-Villars, Paris.

CRAIG, H., 1957: Isotopic standards for carbon and oxygen and correction factors for mass spectrometric analysis of carbon dioxide. – Geochim. Cosmochim. Acta **12**, 133–149.

EDWARDS, G., and D. WALKER, 1983: C_3, C_4: mechanisms, and cellular and environmental regulation, of photosynthesis. – Blackwell Scientific Publications, Oxford – London – Edinburgh – Boston – Melbourne.

EHLERINGER, J., and R. W. PEARCY, 1983: Variation in quantum yield for CO_2 uptake among C_3 and C_4 plants. – Plant Physiol. **73**, 555–559.

FARQUHAR, G. D., 1980: Carbon isotope discrimination by plants: Effects of carbon dioxide concentration and temperature via the ratio of intercellular and atmospheric CO_2 concentrations. – In: Carbon Dioxide and Climate: Australian Research (Ed. G. I. PEARMAN), pp. 105–110. Australian Academy of Science, Canberra.

–, 1983: On the nature of carbon isotope discrimination in C_4 species. – Aust. J. Plant Physiol. **10**, 205–226.

–, M. C. BALL, S. VON CAEMMERER, and Z. ROKSANDIC, 1982a: Effect of salinity and humidity on $\delta^{13}C$ value of halophytes–evidence for diffusional isotope fractionation determined by the ratio of intercellular/atmospheric partial pressure of CO_2 under different environmental conditions. – Oecologia **52**, 121–124.

–, M. H. O'LEARY and J. A. BERRY, 1982b: On the relationship between carbon isotope discrimination and the intercellular carbon dioxide concentration in leaves. – Aust. J. Plant Physiol. **9**, 121–137.

–, and S. VON CAEMMERER, 1982: Modelling of photosynthetic response to environmental conditions. – In: Physiological plant ecology II (Eds. O. L. LANGE, P. S. NOBEL, C. B. OSMOND, and H. ZIEGLER), pp. 549–587. Encyclopedia of Plant Physiology. New Series, Vol. 12B, Springer-Verlag, Berlin – Heidelberg – New York.

FRANCEY, R. J., and G. D. FARQUHAR, 1982: An explanation of $^{13}C/^{12}C$ variations in tree rings. – Nature **197**, 28–31.
GUY, R. D., D. M. REID and H. R. KROUSE, 1980: Shifts in carbon isotope ratios of two C_3 halophytes under natural and artificial conditions. – Oecologia **44**, 241–247.
HATCH, M. D., and C. B. OSMOND, 1976: Compartmentation and transport in C_4 photosynthesis. – In: Transport in plants III (Eds. C. R. STOCKING and U. HEBER), pp. 143–184. Encyclopedia of Plant Physiology. New Series, Vol. 3. Springer-Verlag, Berlin – Heidelberg – New York.
HATTERSLEY, P. W., 1982: $\delta^{13}C$ values of C_4 types in grasses. – Austr. J. Plant Physiol. **9**, 139–154.
–, and A. J. BROWNING, 1981: Occurence of the suberized lamella in leaves of grasses of different photosynthetic types. I. In parenchymatous bundle sheaths and PCR („Kranz") sheaths. – Protoplasma **109**, 371–401.
HOLTUM, J. A. M., M. H. O'LEARY, and C. B. OSMOND, 1983: Effect of varying CO_2 partial pressure on photosynthesis and on carbon isotope composition of carbon-4 of malate from the crassulacean acid metabolism plant *Kalanchoë daigremontiana* Hamet et Perr. – Plant Physiol. **71**, 602–609.
KEELING, C. D., W. G. MOOK, and P. P. TANS, 1979: Recent trends in the $^{13}C/^{12}C$ ratio of atmospheric carbon dioxide. – Nature **277**, 121–123.
LERMAN, J. C., 1975: How to interpret variations in the carbon isotope ratio of plants: biologic and environmental effects. – In: Environmental and biological control of photosynthesis (Ed. R. MARCELLE), pp. 323–335. Dr. W. Junk b. v., Publishers, The Hague.
O'LEARY, M. H., 1981: Carbon isotope fractionation in plants. – Phytochem. **20**, 553–567.
–, and C. B. OSMOND, 1980: Diffusional contribution to carbon isotope fractionation during dark CO_2 fixation in CAM plants. – Plant Physiol. **66**, 931–934.
OSMOND, C. B., 1971: Metabolite transport in C_4 photosynthesis. – Aust. J. Biol. Sci. **24**, 159–163.
–, 1978: Crassulacean acid metabolism: a curiosity in context. – Ann. Rev. Plant Physiol. **29**, 379–414.
PARK, R., and S. EPSTEIN, 1960: Carbon isotope fractionation during photosynthesis. – Geochim. Cosmochim. Acta **21**, 110–126.
PEISKER, M., 1978: Der Einfluß von Sauerstoff auf die CO_2-Kompensationskonzentration von C_3- und C_4-Pflanzen und von Intermediärformen. – Kulturpflanze **26**, 81–98.
–, 1979: Conditions for low, and oxygen-independent, CO_2 compensation concentrations in C_4 plants as derived from a simple model. – Photosynthetica **13**, 198–207.
–, 1982a: The effect of CO_2 leakage from bundle sheath cells on carbon isotope discrimination in C_4 plants. – Photosynthetica **16**, 533–541.
–, 1982b: Theoretische Analyse der Änderungen des $\delta^{13}C$-Wertes bei der Evolution des C_4-pathway der Photosynthese. – In: Photosynthese: Regulation und Evolution (Hrsg. P. HOFFMANN und B. HIEKE), pp. 163–167. Colloquia Pflanzenphysiologie Nr. 5. Humboldt-Universität zu Berlin.
–, and H. BAUWE, 1984: Modelling carbon metabolism in C_3–C_4 intermediate species. 1. CO_2 compensation concentration and its O_2 dependence. – Photosynthetica **18**, 9–19.
–, I. TICHÁ and J. ČATSKÝ, 1981: Ontogenetic changes in the internal limitations to bean leaf photosynthesis. 7. Interpretation of the linear correlation between CO_2 compensation concentration and CO_2 evolution in darkness. – Photosynthetica **15**, 161–168.
RATHNAM, C. K. M., 1978: Metabolic regulation of carbon flux during C_4 photosynthesis. I. Evidence for parallel CO_2 fixation by mesophyll and bundle sheath cells in situ. – Z. Pflanzenphysiol. **87**, 65–84.
RAVEN, J. A., 1977: Ribulose bisphosphate carboxylase activity in terrestrial plants: significance of O_2 and CO_2 diffusion. – Curr. Adv. Plant Sci. **9**, 579–590.
–, and S. M. GLIDEWELL, 1981: Processes limiting photosynthetic conductance. – In: Physiological processes limiting plant productivity (Ed. C. B. JOHNSON), pp. 109–136. Butterworths, London – Boston – Sydney – Wellington – Durban – Toronto.
SCHMIDT, H.-L., and F. J. WINKLER, 1979: Einige Ursachen der Variationsbreite von $\delta^{13}C$-Werten bei C_3- und C_4-Pflanzen. – Ber. Deutsch. Bot. Ges. **92**, 185–191.

SHOMER-ILAN, A., S. BEER, and Y. WAISEL, 1975: *Suaeda monoica*, a C_4 plant without typical bundle sheaths. – Plant Physiol. **56**, 676–679.
SMITH, B. N., and T. W. BOUTTON, 1911: Environmental influences on $^{13}C/^{12}C$ ratios and C_4 photosynthesis. – In: Photosynthesis VI. Photosynthesis and productivity, photosynthesis and environment (Ed. G. AKOYUNOGLOU), pp. 255–262. Balaban International Science Services, Philadelphia, Pa.
–, J. OLIVER and C. MCMILLAN, 1976: Influence of carbon source, oxygen concentration, light intensity, and temperature on $^{13}C/^{12}C$ ratios in plant tissues. – Bot. Gaz. **137**, 99–104.
TROUGHTON, J. H., 1979: $\delta^{13}C$ as an indicator of carboxylation reactions. – In: Photosynthesis II (Eds. M. GIBBS and E. LATZKO), pp. 140–149. Encyclopedia of Plant Physiology. New Series, Vol. 6. Springer-Verlag, Berlin – Heidelberg – New York.
VOGEL, J. C., 1980: Fractionation of the carbon isotopes during photosynthesis. – Sitzungsberichte der Heidelberger Akad. d. Wiss., Math.-naturwiss. Klasse, Jahrgang 1980, 3. Abhdlg., 111–135.
–, A. FULS and R. P. ELLIS, 1978: The geographical distribution of Kranz grasses in South Africa. – South Afr. J. Sci. **74**, 209–215.
WINTER, K., 1981: CO_2 and water vapour exchange, malate content and $\delta^{13}C$ value in *Cicer arietinum* grown under two water regimes. – Z. Pflanzenphysiol. **101**, 421–430.
–, J. A. M. HOLTUM, G. E. EDWARDS, and M. H. O'LEARY, 1982: Effect of low relative humidity on $\delta^{13}C$ value in two C_3 grasses and in *Panicum milioides*, a C_3–C_4 intermediate species. – J. Exp. Bot. **33**, 88–91.
–, C. B. OSMOND and J. S. PATE, 1981: Coping with salinity. – In: Biology of Australian native plants (Eds. J. S. PATE and A. J. MCCOMB), pp. 88–113. University of Western Australia Press, Perth.
–, and J. H. TROUGHTON, 1978: Photosynthetic pathways in plants of coastal and inland habitats of Israel and the Sinai. – Flora **167**, 1–34.
WONG, S. C., I. R. COWAN, and G. D. FARQUHAR, 1979: Stomatal conductance correlates with photosynthetic capacity. – Nature **282**, 424–426.
YEOH, H.-H., M. R. BADGER, and L. WATSON, 1980: Variations in K_m (CO_2) of ribulose-1,5-bisphosphate carboxylase among grasses. – Plant Physiol. **66**, 1110–1112.
ZIEGLER, H., K. H. BATANOUNY, N. SANKHLA, O. P. VYAS, and W. STICHLER, 1981: The photosynthetic pathway types of some desert plants from India, Saudi Arabia, Egypt, and IRAQ. – Oecologia **48**, 93–99.

Dr. M. PEISKER
Zentralinstitut für Genetik und Kulturpflanzenforschung
der Akademie der Wissenschaften der DDR
DDR - 4325 Gatersleben
Corrensstr. 3

III. Originalarbeiten (Original Papers)

Kulturpflanze 32 · 1984 · 69–78

Rohprotein-Bestimmung durch Kjeldahl-Aufschluß und Indophenolblau-Reaktion

Bernhard Schlesier, Hans-Wolfgang Jank und Uwe Schlüter[1]

(Eingegangen am 11. Januar 1984)

Zusammenfassung

Es wird eine Methode zur Bestimmung von Stickstoff in pflanzlichem Material beschrieben. Der schnelle Aufschluß der Proben wird mit Schwefelsäure, Wasserstoffperoxid und einem Katalysator aus Kaliumsulfat und Quecksilberoxid erreicht. Ammoniak wird in aliquoten Teilen der verdünnten Aufschlüsse kolorimetrisch nach Reaktion mit Salicylat und Hypochlorit in alkalischer Lösung gemessen. Das für die Auswertung der Analysen erarbeitete Rechner-Programm wird vorgestellt.

1. Einleitung

Rohprotein-Bestimmungen zählen zu den häufigsten in der landwirtschaftlichen Forschung anfallenden Analysen (Geissler 1975). Das von Kjeldahl (1883) vorgeschlagene und seitdem vielfach verbesserte Verfahren (Bradstreet 1965, Fleck und Munro 1965, Jacobs 1965, Lange et al. 1979a) hat sich unter anderem bei der Analyse von pflanzlichem und tierischem Material bewährt. Mit der Kjeldahl-Methode wird durch den gemessenen Gesamtstickstoff ein Maß für die Proteinmenge in der Probe ermittelt. Dieser Rohprotein-Wert ist für viele Untersuchungen ausreichend. Werden Analysen des Protein-Stickstoffs benötigt, dann ist die vorherige Abtrennung des Nicht-Protein-Stickstoffs erforderlich (Gaines 1977). In der vorliegenden Arbeit wird eine Variante der Kjeldahl-Analyse beschrieben, in der die Indophenolblau-Reaktion (Berthelot 1859) zum Nachweis des Ammoniaks im Kjeldahl-Aufschluß benutzt wird. Mit dieser Methode werden seit 1978 Cerealien und Leguminosen im Rahmen eines Screening-Programms der Gaterslebener Genbank analysiert.

[1] jetzige Adresse: Dipl.-Ing. U. Schlüter, Zentralinstitut für Krebsforschung der Akademie der Wissenschaften der DDR, DDR - 1115 Berlin, Lindenberger Weg 80

2. Methoden

2.1. Gerätesysteme

Der KJELDAHL-Aufschluß wird in Reagenzgläsern (Länge 240 mm, ∅28 mm) in einem elektrisch beheizten (2,5 kW) Aluminiumblock vorgenommen. Dieser Block enthält 49 Bohrungen von 40 mm Tiefe und 30 mm Durchmesser. Zur Wärmeisolation ist der Block mit einem 15 mm dicken Luftpolster umgeben. Die Temperatur im Block wird mit einem Mantelthermoelement (2Fe-Konst, ∅2 mm)[1] gemessen und durch einen Tastbügelregler[2] bei 450 °C gehalten. Der betriebsfertige Aufschlußblock mit dem Reagenzglasständer weist die Abmessungen 500 × 320 × 200 mm³, der Regler die von 200 × 100 × 350 mm³ auf (Abb. 1).

Die Indophenolblau-Reaktion wird in einem Durchflußautomaten[3] ausgeführt. Dieses System besteht aus Probenspeicher APS 4[4], Dosierpumpe DP 2–2[4] mit PVCD-Schläuchen[3], Glaswendelsystem[3], Ultrathermostat U 10 A[3], Durchflußkolorimeter DFK 1[3] mit Wechselspannungskonstanthalter 220/0,45[5] und Kompensationsschreiber K 201[6]. Das Durchflußkolorimeter ist mit einer Halogenlampe 12 V/50 W[7], einer Fotozelle 485 GZV[8], einem Interferenzfilter FIR 656 H[6] und einer Durchflußküvette von 1 cm Schichtdicke ausgerüstet.
Die Berechnung der Analysen erfolgt auf einem Kleinrechner KRS 4201[9].

Abb. 1 Aufschlußblock

[1] VEB Mansfeldkombinat „Wilhelm Pieck", Walzwerk Hettstedt
[2] VEB Meßgerätewerk „Erich Weinert" Magdeburg
[3] VEB MLW Prüfgeräte-Werk Medingen
[4] VEB MLW Labortechnik Ilmenau [5] VEB Statron Fürstenwalde
[6] VEB C. Zeiss Jena [7] VEB NARWA Glühlampenwerk Plauen
[8] VEB Elektronische Spezialröhren Leipzig [9] VEB Kombinat Robotron Dresden

2.2. Kjeldahl-*Aufschluß*

Feste Proben (bis 200 mg) oder flüssige Proben (bis 3 ml) werden in die Reagenzgläser eingebracht und aus einem Hohlmaß mit 1,75 g ($s = \pm 0{,}084$ g, $n = 24$) Katalysatorgemisch (20 Gew.-Teile Kaliumsulfat reinst z. A. und 1 Gew.-Teil Quecksilberoxid reinst z. A.) versetzt. Nach Zugabe von 1,0 ml Wasserstoffperoxidlösung (reinst z. A., mindestens 20 % H_2O_2) und 2,00 ml ($s = \pm 0{,}004$ ml, n = 24) Schwefelsäure (reinst z. A., D = 1,84) werden die Proben mit Hilfe des Reagenzglasständers gleichzeitig in den auf 450 °C vorgeheizten Aufschlußblock gestellt. Zum Abdestillieren des Wassers wird zunächst 5 min ohne, danach 20 min mit aufgesetzten Rückfluß-Kappen erhitzt. Anschließend werden die Proben auf ca. 80 bis 100 °C abgekühlt und mit 25,0 ml 0,0005%iger Brij 35-Lösung verdünnt.

2.3. *Indophenolblau-Reaktion*

Anteile der verdünnten Proben werden in die Probenbecher des Durchflußautomaten eingefüllt. Aus jeder Probe werden 135 µl zur Analyse benötigt. Pro Stunde werden 40 Proben analysiert, wobei Probenahme- und Spülzeit gleich sind. Die Reagenzlösungen werden mit der Schlauchpumpe dosiert, Reihenfolge und Intensität des Mischens durch die Mischwendel konstant gehalten, wobei die Probe 2 min nach dem gepufferten Salicylatreagenz in das System gebracht und nach weiteren 2 min mit der Hypochloritlösung versetzt wird. Das Fließbild für Pumpe und Mischwendel ist in Abb. 2 dargestellt. Die im

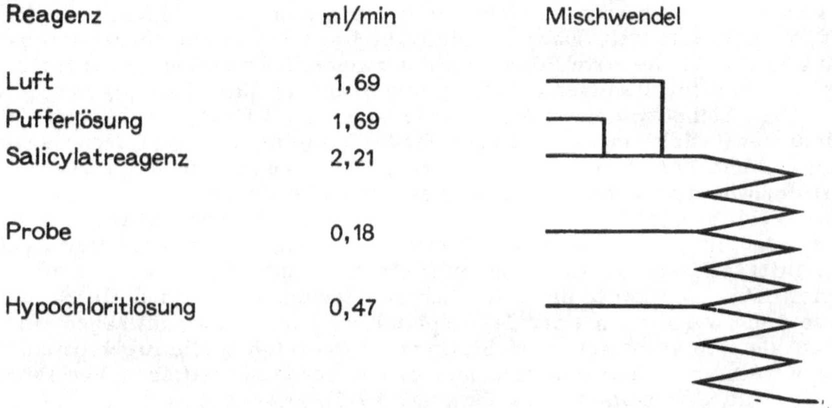

Abb. 2 Fließbild zur Indophenolblau-Reaktion

Fließsystem zur Verminderung der Probenverschleppung benötigte Luft wird zur Entfernung von Ammoniak mit 0,1 N Schwefelsäure gewaschen. Die Arbeitslösungen sind so hergestellt, daß die folgenden Endkonzentrationen im Fließsystem erreicht werden:

Pufferlösung	Kaliumchlorid (reinst z. A.)	0,133 mol/l
	K-Na-Tartrat (reinst z. A.)	0,020 mol/l
	Na_2EDTA (reinst z. A.)	0,005 mol/l
	Aceton (wasserfrei, reinst z. A.)	0,012 mol/l
Salicylatreagenz	NaOH (reinst z. A.)	0,360 mol/l
	Na-Salicylat (2. AB DDR)	0,120 mol/l
Natriumhypochloritlösung	(techn. Lösung, berechnet als aktives Chlor)	0,012 mol/l

Nach Mischen aller Reaktionspartner fließt der Ansatz bei 25 °C für 18 min durch eine Reaktionswendel. Die Farbintensität wird im Durchflußkolorimeter gemessen und mit dem Schreiber als Extinktion registriert.

2.4. *Auswertung*

Standardlösungen, die 5,00/4,00/3,00/2,00/1,00/0,50 mg Stickstoff pro 2,00 ml enthalten, werden einzeln durch Einwaage von Ammoniumsulfat reinst z. A. und Auflösen in 0,1 N Schwefelsäure, die auch als Leerwert dient, hergestellt. Von jeder Standardlösung werden parallel mindestens fünf Proben von 2,00 ml wie oben beschrieben aufgeschlossen, verdünnt und analysiert. Bei hinreichend genauer Übereinstimmung werden die Aufschlüsse jeder Konzentration vereinigt. Aliquote dieser Lösungen werden über einen Zeitraum von 14 Tagen verwendet, um die Standardreihe bei der Indophenolblau-Reaktion aufzunehmen. Die Standardreihe wird zu Beginn, nach je ca. 50 Analysen und am Ende einer Serie gemessen. Nach jeweils 20 Proben werden zwei aufeinanderfolgende Leerwerte eingesetzt und damit die Basislinie überprüft. Die Extinktionswerte werden über der Basislinie mit einem Maßstab in mm gemessen und zusammen mit Probenbezeichnung und Einwaagen in eine Tabelle eingetragen. Der Zusammenhang zwischen den Extinktionswerten der Standardreihe und den zugehörigen Gehalten an Stickstoff ist nicht exakt linear, sondern weist einen leicht konvexen Verlauf gegen die X-Achse auf. Aus diesem Grunde wird für jede Standardreihe ein Eichpolynom nach der Methode der kleinsten Quadrate berechnet (WEBER 1972). Dabei wird für alle Eichpolynome der gleiche Grad angesetzt, wobei wir mit denen zweiten Grades gute Erfahrungen gemacht haben. Die neben den Polynomkoeffizienten berechneten Standardabweichungen der Standardwerte vom Eichpolynom ermöglichen eine Aussage über die Qualität jeder einzelnen Standardkurve; über einen *F*-Test (WEBER 1972) wird angezeigt, ob aufeinanderfolgende Standardkurven in ihrer Genauigkeit hinreichend übereinstimmen. Um den Einfluß der durch das Fließsystem bedingten Drift zu reduzieren, wird für jede Probenposition durch lineare Interpolation zwischen dem führenden und dem nachfolgenden Eichpolynom ein positionsgerechtes, probenbezogenes Polynom ermittelt. Dabei ist diese Interpolation so aufgebaut, daß das erste Eichpolynom exakt für die erste Probe und das zweite Eichpolynom exakt für die letzte Probe gilt, während für die inneren Proben eine entsprechende Mischung aus dem ersten und dem zweiten Eichpolynom zur Anwendung kommt. Mit Hilfe dieser Korrekturpolynome wird dann der Extinktionswert der betreffenden Probe in den zugehörigen Gehalt an Stickstoff umgerechnet. Zusammen mit der Einwaage läßt sich hieraus der prozentuale Gehalt an Stickstoff sowie der Proteingehalt für diese Probe bestimmen.

Folgen mehrere Proben mit derselben Probennummer unmittelbar aufeinander, so werden automatisch der Mittelwert, die Standardabweichung und der Variationskoeffizient des prozentualen Stickstoffgehaltes sowie der mittlere Proteingehalt für diese Probennummer berechnet. Präzisionskontrolle und Richtigkeitskontrolle werden durch das tägliche Mitführen von zwei Proben einer Tryptophanlösung und zwei Einwaagen einer Standard-Probe aus dem zu untersuchenden Sortiment durchgeführt. Die Ergebnisse der Kontrollanalysen werden mit Kontrollkarten und den üblichen statistischen Verfahren ausgewertet (KAISER und GOTTSCHALK 1972, BARTELS 1977).

3. Ergebnisse und Diskussion

3.1. KJELDAHL-*Aufschluß*

Ein kritischer Schritt in der KJELDAHL-Methode ist die quantitative Überführung des organischen Stickstoffs in Ammoniak. Der Erfolg dieses Aufschlusses hängt wesentlich ab von Temperatur, Zeit, Katalysator und Oxidationsmitteln und es ist keine universelle Aufschlußmethode für die verschiedenartigen Substanzen zu erwarten (BRADSTREET 1965, FLECK und MUNRO 1965, LANGE et al. 1983a). Der Aufschluß in Druckgefäßen (JACOBS 1965) erschien uns für die Analyse großer Serien nicht praktikabel. Von den Varianten, die mit offenen Gefäßen arbeiten, wählten wir ein Verfahren, bei dem durch Siedepunkterhöhung der Schwefelsäure mittels Kaliumsulfat (GUNNING 1889) sowie dem Einsatz von Quecksilber (WILFARTH 1885) und Wasserstoffperoxid (KLEEMANN 1921) in kurzer Zeit auch widerstandsfähige Verbindungen mit guten Ergebnissen analysiert werden können. Quecksilber wird von mehreren Autoren (HILLER et al. 1948, WILLITS et al. 1949, KIRK 1950, MIDDLETON und STUCKEY 1951, MCKENZIE und WALLACE 1954, BAKER 1961) als der geeignetste Katalysator für den KJELDAHL-Aufschluß angesehen. Bei richtiger Handhabung sind durch den Einsatz von Wasserstoffperoxid keine Stickstoff-Verluste zu erwarten (MCKENZIE und WALLACE 1954). Unseren Aufschlußbedingungen entsprechende Verhältnisse werden auch von anderen Autoren (CONCON und SOLTESS 1973, BIETZ 1974, MONTAG 1974, KOOPS et al. 1975) sowie in der Mikro-KJELDAHL-Methode der AOAC (HORWITZ et al. 1970) verwendet. LANGE et al. 1983b arbeiten mit ähnlichen Mischungen, jedoch wesentlich längeren Aufschlußzeiten.

Wir finden unter unseren Bedingungen Tryptophan nach 20 bis 45 min Aufschlußdauer vollständig wieder. Temperaturmessungen im Aufschluß, die mit einem durch eine Glaskapillare geschützten Mantelthermoelement (2Fe-Konst, ∅ 2 mm)[1] durchgeführt wurden, zeigen, daß während der Aufschlußzeit von 25 min die kritische Temperatur von 410 °C nicht überschritten wird, wenn das Abrauchen von Schwefelsäure durch das Abdecken der Reagenzgläser mit Rückfluß-Kappen in der oben beschriebenen Weise begrenzt wird. Obwohl sich die nach dem Erhitzen der Proben in den Reagenzgläsern verbleibenden Mengen an Aufschluß von Glas zu Glas unterscheiden, haben wir auf das Auffüllen in Maßkolben verzichtet. Das Verdünnen durch Zugabe von 25,0 ml Brij 35-Lösung führt nach unseren Messungen zu maximalen Abweichungen von 1,5 %. Es ist vorteilhaft, die noch warmen Aufschlüsse zu verdünnen, weil der Aufschluß unter diesen Bedingungen bei der Zugabe der Lösung nicht kristallisiert (KOOPS et al. 1975). Die Brij 35-Lösung beeinflußt das Fließverhalten der Lösung im Durchflußsystem günstig.

3.2. Indophenolblau-Reaktion

Farbstoffe vom Typ des Indophenolblaus werden aus Ammoniak und nicht p-substituierten Phenolen in alkalischer Umgebung unter Einfluß von Oxidationsmitteln wie z. B. Hypochlorit oder Chloramin T gebildet. Zusammenfassungen der Kenntnisse über die Reaktion und neuere Untersuchungen dazu findet man bei LANGE et al. (1979a, 1980a, 1980b) und bei KROM (1980). Mit der Optimierung der Reaktionsbedingungen haben sich WEATHERBURN (1967), HARWOOD und HUYSER (1970), GEHRKE et al. (1971), GORDON et al. (1978), LANGE et al. (1980c) sowie NKONGE und BALLANCE (1982) beschäftigt. Bei geeigneter Modifikation kann die BERTHELOT-Reaktion direkt zur Ammoniak-Bestimmung im KJELDAHL-Aufschluß verwendet werden (HÄUSSLER und HAJDU 1960, FERRARI 1960, STEGEMANN und LOESCHKE 1962). Entsprechend den Variationsmöglichkeiten bei der Auswahl der Reaktionspartner und der Abhängigkeit der Farbstoffbildung von den Konzentrationen der Reaktionspartner, dem p_H-Wert, dem Temperatur-Zeit-Verhalten und dem Katalysator zur Beschleunigung der Farbstoffbildung sind eine Vielzahl von Methoden unterschiedlicher Empfindlichkeit beschrieben worden, die z. T. manuell (MANN 1963, BOHLEY 1967, KOOPS et al. 1975, FELKER

1977, NKONGE und BALLANCE 1982), z. T. in teilautomatisierten Verfahren realisiert werden (VARLEY 1966, CROOKE und SIMPSON 1971, HOLZ und KREMERS 1971, GEHRKE et al. 1973, BIETZ 1974, HAMBRAEUS et al. 1976, STEWART und RŮŽIČKA 1976, LANGE et al. 1979b).

Wir haben für die Screening-Analysen eine relativ unempfindliche Modifikation der Indophenolblau-Reaktion ausgewählt: Das früher fast ausschließlich verwendete Phenol haben wir durch das toxikologisch unbedenkliche Natriumsalicylat ersetzt (REARDON et al. 1966). Als Katalysator der Farbreaktion setzen wir Aceton ein (CROWTHER und LARGE 1956), das bei Raumtemperatur höhere Farbausbeuten liefert als Manganionen (RUSSELL 1944), die Reaktion aber nicht so empfindlich gestaltet wie Natriumnitroprussid (LUBOCHINSKY und ZALTA 1954). Wir wählten eine Reaktionstemperatur von 25 °C. Damit wird unter unseren Bedingungen eine optimale Farbreaktion ermöglicht; zugleich treten nur relativ geringe Temperaturunterschiede zwischen der Mischwendel (Verweilzeit der Lösung 6 min bei Raumtemperatur) und der Reaktionswendel auf (Verweilzeit der Lösung 18 min bei 25 °C). Widersprüchliche Angaben findet man in der Literatur zum p_H-Optimum der Indophenolblau-Bildung. Während einige Autoren (BOLLETER et al. 1961, LORENTZ 1967, WEATHERBURN 1967, ROMMERS und VISSER 1969, HARWOOD und HUYSER 1970) das Optimum mit p_H 10 bis 12 angeben, finden andere dasselbe bei p_H 13 (KOOPS et al. 1975, PYM und MILHAM 1976, KÜNSCH et al. 1977, NKONGE und BALLANCE 1982). Eigene Untersuchungen dazu (Titration, Glaselektrode GC 50[10]) bestätigen das Optimum bei p_H 13. Die zur Neutralisierung der Probe erforderliche Natronlauge ist im Reagenz enthalten. Die durch den unterschiedlichen Säureverbrauch und -verlust beim Aufschluß bedingter Differenzen im Säuregehalt der Proben werden durch eine Pufferlösung ausgeglichen. Wir verwenden dazu das Puffersystem Natronlauge/Kaliumchlorid (BATES und BOWER 1956), das im erforderlichen p_H-Bereich eine höhere Pufferkapazität aufweist als die sonst eingesetzten Phosphatpuffer (STEGEMANN und LOESCHKE 1962, MANN 1963, HARWOOD und HUYSER 1970, FELKER 1977, KÜNSCH et al. 1977). Störungen der Farbreaktion durch Quecksilber aus dem Aufschlußgemisch bzw. durch wechselnde Kationenanteile aus den Proben werden durch den Einsatz von Tartrat und Na_2EDTA unterbunden. Andere Verbindungen, die die Ammoniak-Bestimmung über Indophenolblau stören (LORENTZ 1967, PYM und MILHAM 1976, NGO et al. 1982) spielen in unserem Verfahren keine Rolle.

3.3. Auswertung

Unter den o. g. Reaktionsbedingungen werden bei der Messung der Standardreihe in einer 1 cm Küvette Extinktionen bis E = 1,3 gefunden. Die oberen Werte liegen höher, als der optimale Bereich der Informationswerte für die Spektralphotometrie in Lösungen sein sollte (GOTTSCHALK 1976). Die Leerwerte weisen eine Extinktion von E < 0,02 auf und werden über die Basislinie von den Roh-Extinktionen der Proben abgezogen. Dabei ist nach unseren Erfahrungen zu beachten, daß infolge der Probenverschleppung im Fließsystem jeweils zwei

[10] Forschungsinstitut Meinsberg

Tabelle 1
Standardkurve der Stickstoffbestimmung

Stickstoff (mg)	Peakhöhen (mm) einzelner Aufschlüsse der Standardlösungen						$\bar{x}$	s	V%
5,00	151,0	153,0	154,0	152,0	152,5	152,0	152,42	1,021	0,67
4,00	127,5	127,5	126,5	126,5	126,0	127,0	126,83	0,606	0,48
3,00	98,0	97,5	98,5	98,0	98,0	98,5	98,08	0,376	0,38
2,00	66,0	65,5	66,0	64,0	64,0	63,5	64,83	1,125	1,74
1,00	32,5	33,0	32,5	33,0	33,0	33,0	32,83	0,258	0,79
0,50	16,0	15,0	15,0	16,0	16,0	15,5	15,58	0,492	3,16
0,00	0,5	0,5	0,5	0,5	0,5	0,5			

Gleichung des Eichpolynoms: $y = b_0 + b_1x + b_2x^2$
Koeffizienten des Eichpolynoms: $b_0 = 0{,}029846$
$b_1 = 0{,}027879$
$b_2 = 0{,}000029$
Standardabweichung um das Eichpolynom: RMS = 0,04598

Leerwerte unmittelbar hintereinander eingesetzt werden müssen und dabei nur der zweite Leerwert den richtigen Meßwert liefert. Die Standardkurve verläuft nicht exakt linear (Tab. 1). Nach unseren bisherigen Untersuchungen wird dieser Mangel nicht durch die Bedingungen bei der Indophenolblau-Reaktion verursacht, da entsprechende Kurven auch auftreten, wenn stabile Farblösungen (z. B. Amidoschwarz 10B in verd. Essigsäure, Interferenzfilter SIF 625 nm) analysiert werden. Möglicherweise ist dieser Fehler durch Abweichungen im Verlaufe der Spektren von Farblösung und Filter zu erklären (BREWER 1980), denn das von uns verwendete Filter mit maximaler Durchlässigkeit bei 652 nm und einer Halbwertsbreite von 18 nm liegt nicht symmetrisch über den ungleichen Flanken des bei 654 nm gefundenen Absorptionsmaximums der Farblösung.

Das Verfahren ist für die Belange des Sreenings hinreichend zuverlässig. Bei den Analysen einer Tryptophanlösung an 22 Tagen im Verlaufe von zwei Monaten lag der Variationskoeffizient bei ±2 %. Der dabei gefundene Mittelwert $\bar{x} = 3{,}91$ mgN ($s = 0{,}08$ mgN, $n = 44$) stimmt gut mit dem Sollwert von 4,00 mgN überein, wenn man berücksichtigt, daß der Stickstoffgehalt des für die Kontroll-Lösung verwendeten Tryptophans nach Angabe des Herstellers 13,44 bis 13,92 % beträgt. Bei der Analyse von Pflanzenmaterial wird ein Variationskoeffizient von ±3 % gelegentlich überschritten. Als wesentliche Ursache dafür konnten wir die ungenügende Homogenität der zur Analyse eingesetzten Stichproben ermitteln.

Das in FORTRAN 4200 geschriebene Programm können Interessenten von H.-W. JANK erhalten.

Wir danken Frau E. SIEBERT und Frau CHR. KALLAS für die zuverlässige Mitarbeit bei der Erarbeitung der Methode und Herrn W. FANSLAU für die sorgfältige Anfertigung der Aufschlußblöcke.

Summary

Determination of crude protein by Kjeldahl digest and indophenolblue reaction

A method is described for the determination of nitrogen in plant materials. Rapid digestion of samples is obtained with sulfuric acid, hydrogen peroxide, and a mixture of potassium sulphate and mercuric oxide as a catalyst. The ammonia is determined colorimetrically in aliquots of the diluted digests after reaction with an alkaline solution of sodium salicylate and hypochlorite. The computer program is described for the evaluation of the analytical results.

Краткое содержание

Определение сырого белка методом Кельдаля и реакцией на синий индофенол

Описывается метод определения азота в растительном материале. Быстрое разложение проб достигается действием серной кислоты, перекиси водорода и катализатора, состоящего из сернокислого калия и окиси ртути. Аммиак измеряется калориметрически в аликвотах разбавленной пробы после реакции с салицилатом и гипохлоритом в щелочном растворе.

Предлагается программа обработки анализов на ЭВМ.

Literatur

Baker, P. R. W., 1961: The micro-Kjeldahl determination of nitrogen. An investigation of the effects of added salt and catalysts. – Talanta **8**, 57–71.

Bartels, H. A., 1977: Chemische Analyse. – Springer Verlag, Berlin – Heidelberg – New York.

Bates, R. G., und V. E. Bower, 1956: Alkaline solutions for p_H control. – Anal. Chem. **28**, 1322–1324.

Berthelot, M., 1859: Correspondence – 'violet d'aniline'. – Répertoire de chimie appliqué (Paris) **1**, 284.

Bietz, J. A., 1974: Micro-Kjeldahl analysis by an improved automated ammonia determination following manual digestion. – Anal. Chem. **46**, 1617–1618.

Bohley, P., 1967: Reihenbestimmungen von Stickstoff im Ultramikromaßstab. Kjeldahlveraschung und Phenol-Hypochlorit-Reaktion. – Hoppe-Seyler's Z. physiol. Chem. **348**, 100–110.

Bolleter, W. T., C. J. Bushman und P. W. Tidwell, 1961: Spectrophotometric determination of ammonia as indophenol. – Anal. Chem. **33**, 592–594.

Bradstreet, R. B., 1965: The Kjeldahl method for organic nitrogen. – Academic Press, New York – London.

Brewer, S., 1980: Solving problems in analytical chemistry. – Wiley, Chichester – New York.

Concon, J. M., und D. Soltess, 1973: Micro Kjeldahl digestion of cereal grains and other biological materials. – Anal. Biochem. **53**, 35–41.

Crooke, W. M., und W. E. Simpson, 1971: Determination of ammonium in Kjeldahl digests of crops by an automated procedure. – J. Sci. Fd. Agric. **22**, 9–10.

Crowther, A. B., und R. S. Large, 1956: Improved conditions for the sodium phenoxide-sodium hypochlorite method for the determination of ammonia. – Analyst **81**, 64–65.

Felker, P., 1977: Microdetermination of nitrogen in seed protein extracts with the salicylate-dichloroisocyanurate color reaction. – Anal. Chem. **49**, 1080.

FERRARI, A., 1960: Nitrogen determination by a continuous digestion and analysis system. – Ann. N. Y. Acad. Sci. **87**, 792–800.
FLECK, A., und H. N. MUNRO, 1965: The determination of organic nitrogen in biological materials. – Clin. Chim. Acta **11**, 2–12.
GAINES, T. P., 1977: Determination of protein nitrogen in plants. – Journal AOAC **60**, 590–593.
GEHRKE, C. W., L. L. WALL und J. S. ABSHEER, 1973: Automated nitrogen method for feeds. – Journal AOAC **56**, 1096–1105.
–, –, J. S. KILLINGLEY und K. INADA, 1971: Missouri automated method for total nitrogen in all fertilizers. – Journal AOAC **54**, 651–657.
GEISSLER, C., 1975: Die Methoden zur Rohproteinbestimmung. – Nahrung **19**, 363–377.
GORDON, S. A., A. FLECK und J. BELL, 1978: Optimal conditions for the estimation of ammonium by the BERTHELOT reaction. – Ann. Clin. Biochem. **15**, 270–275.
GOTTSCHALK, G., 1976: Standardisierung quantitativer Analysenverfahren. VI. Spektralphotometrie in Lösungen. – Z. Anal. Chem. **282**, 1–15.
GUNNING, J. W., 1889: Über eine Modification der KJELDAHL-Methode. – Z. Anal. Chem. **28**, 188–191.
HÄUSSLER, A. und P. HAJDU, 1960: Die Reststickstoffbestimmung mit der Indophenolblaureaktion. – Klin. Wschr. **38**, 874–876.
HAMBRAEUS, L., E., FORSUM, L. AMBRAHAMSSON und B. LÖNNERDAHL, 1976: Automatic total nitrogen analysis in nutritional evaluations using a block digestor. – Anal. Biochem. **72**, 78–85.
HARWOOD, J. E., und D. J. HUYSER, 1970: Some aspects of the phenol-hypochlorite reaction as applied to ammonia analysis. – Water Research **4**, 501–515.
HILLER, A., J. PLAZIN und D. D. VAN SLYKE, 1948: A study of conditions for KJELDAHL determination of nitrogen in proteins. – J. biol. Chem. **176**, 1401–1420.
HOLZ, F., und H. KREMERS, 1971: Die automatische Bestimmung des Stickstoffs als Indophenolgrün in Böden und Pflanzen. – Landwirtsch. Forsch. **24** (26/ISh), 177–191.
HORWITZ, W., P. CHICHILO und H. REYNOLDS (eds.), 1970: Official methods of analysis of the Association of Officional Analytical Chemists. 11th Ed. – AOAC, Washington.
JACOBS, S., 1965: The determination of nitrogen in biological materials. – In: Methods of biochemical analysis (ed. D. GLICK) **13**, 241–263. – J. Wiley & Sons, New York – London – Sydney.
KAISER, H., und G. GOTTSCHALK, 1972: Elementare Tests zur Beurteilung von Meßdaten. – Bibliographisches Institut AG, Mannheim.
KIRK, P. L., 1950: KJELDAHL method for total nitrogen. – Anal. Chem. **22**, 354–358.
KJELDAHL, J., 1883: Neue Methode zur Bestimmung des Stickstoffs in organischen Körpern. – Z. Anal. Chem. **22**, 366–382.
KLEEMANN, 1921: Über die Wirkungen des Wasserstoffsuperoxydes bei der Aufschließung pflanzlicher und tierischer Stoffe. – Z. f. angew. Chemie **34**, 625–627.
KOOPS, J., H. KLOMP und R. H. C. ELGERSMA, 1975: Rapid determination of nitrogen in milk and dairy products by colorimetric estimation of ammonia following an accelerated digestion procedure. – Neth. Milk Dairy J. **29**, 169–180.
KROM, M. D., 1980: Spectrophotometric determination of ammonia: A study of a modified BERTHELOT reaction using salicylate and dichloroisocyanurate. – Analyst **105**, 305–316.
KÜNSCH, U., H. SCHÄRER und A. TEMPERLI, 1977: Zur Automatisierung der Bodenanalyse. – Schweiz. landwirtsch. Forsch. **16**, 141–153.
LANGE, R., R. FRIEBE und F. LINOW, 1979a: Zur Anwendung der Methodenkombination KJELDAHL-Aufschluß/BERTHELOT-Reaktion bei der Stickstoffbestimmung in biologischen Materialien. 1. Mitt. Stand der Kenntnisse – Teile I und II. – Nahrung **23**, 255–261, 431–439.
–, –, – und B. SCHAFFNER, 1979b: 2. Mitt. Aufbau und Erprobung einer teilautomatisierten Apparatur zur routinemäßigen Bestimmung des Stickstoffs. – Nahrung **23**, 549–559.
–, J. SCHLIEMANN, – und –, 1980a: 3. Mitt. Untersuchungen über Folge- und Nebenreaktionen der Indophenolbildung. Teil I. Die Bildung von Chlorphenolen. – Nahrung **24**, 447–453.

LAMGE, R., J. SCHIEMANN und –, 1980b: 3. Mitt. Untersuchungen über Folge- und Nebenreaktionen der Indophenolbildung. Teil II. Zum Reaktionsverlauf und zur Kinetik des Indophenolabbaus. – Nahrung **24**, 907–921.
–, – und –, 1980c: 4. Mitt. Optimierung der BERTHELOT-Reaktion unter Berücksichtigung von Folge- und Nebenreaktionen der Indophenolbildung. – Nahrung **25**, 447–464.
–, – und –, 1983a: 5. Mitt. Möglichkeiten und Grenzen der simultanen Veraschung von biologischen Materialien unterschiedlicher Herkunft und differenter chemischer Zusammensetzung. – Nahrung **27**, 191–204.
–, – und –, 1983b: 6. Mitt. Die Blockaufschlußtechnik im Halbmikro- und Makromaßstab bei biologischen Feststoffen. – Nahrung **27**, 645–658.
LORENTZ, K., 1967: Mechanismus und Spezifität der Indophenolreaktion zur Ammoniakbestimmung. – Z. Klin. Chem. u. Klin. Biochem. **5**, 291–298.
LUBOCHINSKY, B., und G. P. ZALTA, 1954: Microdosage colorimetrique de l'azote ammoniacal. – Bull. Soc. Chim. biol. **36**, 1363–1366.
MANN, L. T., 1963: Spectrophotometric determination of nitrogen in total micro-KJELDAHL digests. Application of phenolhypochlorite reaction to microgram amounts of ammonia in total digest of biological material. – Anal. Chem. **35**, 2179–2182.
MCKENZIE, H. A., und H. S. WALLACE, 1954: The KJELDAHL determination of nitrogen: a critical study of digestion conditions – temperature, catalyst, and oxidizing agent. Austr. J. Chem. **7**, 55–70.
MIDDLETON, G. und R. E. STUCKEY, 1951: The standardisation of the digestion process in the KJELDAHL determination of nitrogen. – J. Pharm. Pharmacol. **3**, 829–842.
MONTAG, A., 1974: Stickstoffbestimmungen nach der KJELDAHL-Methode im Kjel-Foss-Automatic-Gerät. – Gordian **74**, 203–210.
NGO, T. T., A. P. H. PHAN, C. F. YAM und H. M. LENHOFF, 1982: Interference in determination of ammonia with the hypochlorite-alkaline phenol method of BERTHELOT. – Anal. Chem. **54**, 46–49.
NKONGE, C., und G. M. BALLANCE, 1982: A sensitive colorimetric procedure for nitrogen determination in micro-KJELDAHL digests. – J. Agric. Food Chem. **30**, 416–420.
PYM, R. V. E., und P. J. MILHAM, 1976: Selectivity of reaction among chlorine, ammonia, and salicylate for determination of ammonia. – Anal. Chem. **48**, 1413–1415.
REARDON, J., J. A. FOREMAN und R. L. SEARCY, 1966: New reactants for the colorimetric determination of ammonia. – Clin. Chim. Acta **14**, 403–405.
ROMMERS, P. J., und J. VISSER, 1969: Spectrophotometric determination of micro amounts of nitrogen as indophenol. – Analyst **94**, 653–658.
RUSSELL, J., A., 1944: The colorimetric estimation of small amounts of ammonia by the phenol-hypochlorite reaction. – J. Biol. Chem. **156**, 457–461.
STEGEMANN, H., und V. LOESCHKE, 1962: Mikrobestimmung von Stickstoff als Indophenolblau nach Chloramin-T-Oxidation. – Hoppe-Seyler's Z. physiol. Chem. **329**, 241–248.
STEWART, J. W. B., und J. RŮŽIČKA, 1976: Flow injection analysis. Part V. Simultaneous determination of nitrogen and phosphorus in acid digests of plant material with a single spectrophotometer. – Anal. Chim. Acta **82**, 137–144.
VARLEY, J. A., 1966: Automatic methods for the determination of nitrogen, phosphorus and potassium in plant material. – Analyst **91**, 119–126.
WEATHERBURN, M. W., 1967: Phenol-hypochlorite reaction for determination of ammonia. – Anal. Chem. **39**, 971–974.
WEBER, E., 1972: Grundriss der biologischen Statistik. 7. Aufl. – Fischer, Jena.
WILFARTH, H., 1885: Eine Modifikation der KJELDAHL'schen Stickstoffbestimmungs-Methode. – Chem. Central. Blatt (3. Folge) **16**, 17–19, 113–115.
WILLITS, C. O., M. R. COE und C. L. OGG, 1949: KJELDAHL determination of nitrogen in refractory materials. – Journal AOAC **32**, 118–127.

Dr. B. SCHLESIER, Dr. H.-W. JANK
Zentralinstitut für Genetik und Kulturpflanzenforschung
der Akademie der Wissenschaften der DDR
DDR - 4325 Gatersleben
Corrensstraße 3

Kulturpflanze 32 · 1984 · 79–88

Blütenbiologische Untersuchungen an der Gaterslebener Wildgersten-Kollektion (*Hordeum* L. subgenus *Hordeum*)

KARL HAMMER

(Eingegangen am 9. Januar 1984)

Zusammenfassung

In den Jahren 1980 bis 1983 wurde eine Kollektion von Wildgersten (*Hordeum* L. subgen. *Hordeum*) blütenbiologisch untersucht. Antherenlänge, -breite und Stigmatalänge wurden in die Untersuchungen einbezogen. Die Variabilität dieser Merkmale ist sehr hoch. Besonders ausgeprägt ist ihre Witterungsabhängigkeit. Im Durchschnitt der zahlreichen untersuchten Sippen über die vier Versuchsjahre lassen sich aber bestäubungsökologische Tendenzen ableiten, besonders unter Verwendung des Merkmals Antherenlänge. Die längsten Antheren hat *Hordeum spontaneum*, aber auch *H. agriocrithon* zeichnet sich durch relativ lange Antheren aus. Dagegen sind die Antheren von *H.* × *lagunculiforme* relativ kurz. Diese Sippe wird als Introgressionsprodukt von Kulturgersten in *H. spontaneum* betrachtet.

Wildgersten-Sippen mit langen Antheren besitzen eine erhöhte Neigung zur Allogamie. Sie stellen ein Problem für die sippenechte Reproduktion in der Genbank dar. Für den Vermehrungsanbau ist ihre Isolation erforderlich.

Einleitung

In den letzten Jahren wird für züchterische Zwecke (Resistenz- und Qualitätszüchtung) immer häufiger auf Wildgersten (vgl. VON BOTHMER et al. 1982) besonders aber auf Material aus dem Subgenus *Hordeum* (*Hordeum* L. sect. *Crithe* [Doell] Nevski) zurückgegriffen (AHOKAS 1979, 1981, 1982a und b, FISCHBECK 1981a, MOSEMAN et al. 1981, SEGAL et al. 1979). Mit diesen Aktivitäten ging die planmäßige Erweiterung der Wildgerstenkollektion der Genbank Gatersleben einher. Damit steht der Züchtung ein in vieler Hinsicht nutzbares Ausgangsmaterial zur Verfügung.

Für die Genbank bringt die rasche Erweiterung der Wildgersten-Kollektionen Probleme mit sich, die aus dem Wildpflanzenstatus des Materials entspringen. Bei der Reife zerfallen die Ähren; es sind daher zahlreiche Erntegänge erforderlich, um eine möglichst große Menge an Saatgut zu sichern. Erschwert wird die Ernte noch durch den unterschiedlichen Keimverzug (vgl. ÅBERG 1940) – und die damit verbundenen Reifezeitverschiebungen – selbst innerhalb einer Parzelle. Oft liegt ein gewisser Vernalisationsbedarf vor, die Pflanzen sind aber in Mitteleuropa nicht genügend winterfest. Im Zusammenhang mit der Erhaltung ergibt sich eine Reihe ganz spezifischer Fragestellungen, zu denen noch das Problem der sippenechten Reproduktion kommt, mit dem sich die vorliegende Studie befaßt.

Material und Methoden

Einbezogen in die Untersuchungen wurden die Wildgersten des Subgenus *Hordeum* – *Hordeum spontaneum* Koch, *Hordeum agriocrithon* Åberg und *Hordeum* × *lagunculiforme* (Bacht.) Bacht. ex Nikif., die alle in die engere Verwandtschaft von *Hordeum vulgare* L. s. l. gehören. Morphologische Ähnlichkeiten und Kreuzungskompatibilität sind in letzter Zeit als Argumente dafür herangezogen worden, Wild- und Kultursippen nur auf dem infraspezifischen Niveau zu trennen, wie es beispielsweise de Wet (1981) vorgeschlagen hat (*H. vulgare* L. ssp. *spontaneum* [Koch] Thell.). Dabei ergeben sich aber nomenklatorische Schwierigkeiten, so daß in der vorliegenden Bearbeitung aus praktischen Gründen die traditionellen Bezeichnungen benutzt werden.

Die blütenbiologischen Untersuchungen erfolgten in den Jahren 1980 bis 1983 an den Wildgerstenneuzugängen bzw. an den planmäßig zur Saatgutreproduktion angebauten Sippen (vgl. Tab. 1); insgesamt wurden 282 Wildgersten untersucht. Kurz vor der Anthese, wenn die Stigmata schon entfaltet waren, wurde das Gynaeceum aus je einem Mittelährchen aus dem mittleren Teil von zehn verschiedenen Ähren entnommen und auf einem Objektträger in Vaseline eingebettet. Ebenso wurde mit den Antheren verfahren. Die Vermessung von Antheren und Stigmata erfolgte nach der schon früher ausgearbeiteten Methode (vgl. Hammer 1975) bei 40facher Vergrößerung mit Hilfe eines Netzmikrometers.

Tabelle 1
Anzahl der in den einzelnen Jahren untersuchten Wildgerstensippen

Sippen	Untersuchungsjahr 1980	1981	1982	1983	insgesamt
H. spontaneum	23	21	104	60	208
H. × *lagunculiforme*	37	2	3	6	48
H. agriocrithon	9	1	13	3	26

Ergebnisse und Diskussion

In Modellversuchen mit *Hordeum vulgare* (Hammer 1975) hatte sich gezeigt, daß der Ansatz aus Fremdbefruchtung im wesentlichen durch Antheren- und Stigmatamerkmale getragen wird. Diese Merkmale lassen sich relativ leicht erfassen und sind daher zur Untersuchung eines umfangreichen Materials gut geeignet. Die nahe verwandten Wildsippen verhalten sich in ihrem Reproduktionssystem ähnlich der Kultursippe. Bei ihr liegt der Ansatz aus natürlicher Fremdbefruchtung in der Regel unter 1 % (Chaudhary et al. 1980, vgl. auch die Zusammenstellung bei Hammer 1975). Mit der Domestikation ist oft ein Trend zur Autogamie verbunden, der bei der hochdomestizierten Kulturgerste besonders ausgeprägt ist. Die verwandten Wildarten zeigen demnach eine größere Neigung zur Allogamie, die sich zwar nicht so sehr in deutlich höheren Fremdbefruchtungsraten (Fröst und Asker 1977, Brown et al. 1978a und b), sondern in einem höheren Grad an Heterozygotie in den Populationen widerspiegelt (Kahler und Allard 1981). Insgesamt können aber blütenbiologische Schlußfolgerungen für Wild- und Kulturgersten unter Benutzung der gleichen Merkmale gezogen werden.

Antherenlänge

Die Antherenlänge ist ein wichtiges Anemophiliemerkmal, weil durch sie im wesentlichen die Pollenmenge bestimmt wird. Im Mittel der Jahre liegt in diesem Merkmal *H. spontaneum* an der Spitze des Untersuchungsmaterials (Tab. 2). Damit werden die Ergebnisse früherer Untersuchungen, die an einem sehr eingeengten Material erhalten wurden, bestätigt (HAMMER 1976).

Tabelle 2

Mittlere Antherenlängen (mm)

	$\bar{x} \pm s$	Variationskoeffizient (%)	$\bar{x} + s$	Variationskoeffizient (%)
	1980		1981	
H. spontaneum	3,19 ± 0,627	19,7	3,68 ± 0,59	16,1
H. × *lagunculiforme*	2,42 ± 0,061	2,5	2,24 ± 0,24	10,7
H. agricocrithon	2,96 ± 0,462	15,6	4,05	
	1982		1983	
H. spontaneum	3,35 ± 0,67	20,1	2,90 ± 0,51	17,6
H. × *lagunculiforme*	2,08 ± 0,28	13,4	1,82 ± 0,13	7,3
H. agriocrithon	2,65 ± 0,42	16,0	2,59 ± 0,37	14,1
	insgesamt			
H. spontaneum	3,23 ± 0,67	20,8		
H. × *lagunculiforme*	2,32 ± 0,28	12,0		
H. agriocrithon	2,81 ± 0,50	18,0		

Durch die Witterungsbedingungen des Anbaujahres wird die Ausprägung der Anemophiliemerkmale stark beeinflußt, wie sich beim untersuchten Merkmal für die einzelnen Jahre verfolgen läßt. Das Jahr 1981 zeichnete sich durch günstige Blühbedingungen aus, während 1983 durch der Blühphase vorangehende Trockenheit und kühles, regnerisches Wetter während des Blühens schlechtere Voraussetzungen für die Ausbildung der Blütenorgane mit sich brachte.

H. spontaneum zeichnet sich durch eine große Variabilität der Antherenlänge aus, wie aus den Variationskoeffizienten zu ersehen ist, die für diese Sippe in jedem Jahr am höchsten sind. 1980 wurde das Maximum für die Antherenlänge bei HOR 1647 mit 4,27 mm gefunden, das Minimum bei HOR 8892 mit 1,75 mm. Die Extreme für 1981 lagen bei 5,04 mm (HOR 8523) und 2,69 mm (HOR 8528), für 1982 bei 4,98 mm (HOR 9770) und 1,97 mm (HOR 4867) und für 1983 bei 4,30 mm (HOR 8529) und 1,84 mm (HOR 9825). Diese großen Unterschiede in der Antherenlänge überdecken den vollen Bereich der für *Hordeum vulgare* gefundenen Werte (HAMMER 1975) und übertreffen ihn deutlich bei den Maximalwerten.

H. agriocrithon verfügt auch über durchschnittlich recht lange Antheren. Die relativ geringe Materialmenge aus dieser Sippe, die aus der Seltenheit ihres Vorkommens resultiert, zeigt eine beachtliche Anzahl verschiedener Typen. Die Extremwerte für 1980 lagen bei 3,13 mm (HOR 2466) und 2,63 mm (HOR 2653) – 1981 war nur eine Sippe im Anbau (HOR 8509 – 4,05 mm) – für 1982 bei 3,40 mm (HOR 3902) und 2,06 mm (HOR 1645) sowie für 1983 bei 3,01 mm (HOR 2454) und 2,02 mm (HOR 4867).

H. x *laguncuIiforme* stammt aus spontanen Kreuzungen zwischen *H. spontaneum* und mehrzeiligen Kulturgersten (Kamm 1954, 1977; Zohary 1964). In dieser Kombination erfährt die Antherenlänge etwa eine mittlere Ausprägung zwischen den Ausgangseltern. Die relativ kurzen Antheren der mehrzeiligen Kulturgersten sind für die Ausprägung dieses Merkmals bei *H.* x *laguncuIiforme* verantwortlich. Die Variationskoeffizienten (vgl. Tab. 2) zeigen die geringe Variabilität in unserem Material an, die offenbar daraus resultiert, daß bestimmte morphologische Typen von *H. vulgare* und *H. spontaneum* bevorzugt natürliche Kreuzungsprodukte hervorbringen.

Die Extremwerte von *H.* x *laguncuIiforme* lagen für 1980 bei 2,78 mm (HOR 4864) und 2,09 mm (HOR 4889), für 1981 bei 2,41 mm (HOR 3894) und 2,07 mm (HOR 4875), für 1982 bei 2,40 mm (HOR 9600) und 1,87 mm (HOR 3892) sowie für 1983 bei 2,01 mm (HOR 3886) und 1,64 mm (HOR 3897).

Antherenbreite

Die Antherenbreite ist weniger variabel als die -länge. Ein Selektionsdruck auf hohe Pollenproduktion, wie er bei Anemophilie vorhanden ist, wirkt sich vor allem auf die Antherenlänge aus; ein Ausbleiben des Selektionsdruckes beim Übergang zur Autogamie bringt im wesentlichen eine Verkürzung der Antheren mit sich und betrifft kaum deren Breite. In der Tendenz sind aber ähnliche Verhältnisse wie bei der Antherenlänge zu beobachten. *H. spontaneum* verfügt im Mittel über die breitesten Antheren und *H.* × *laguncuIiforme* zeichnet sich durch relativ schmale Staubgefäße aus (Tabelle 3).

Tabelle 3
Mittlere Antherenbreiten (mm)

	$\bar{x} \pm s$	Variationskoeffizient (%)	$\bar{x} \pm s$	Variationskoeffizient (%)
	1980		1981	
H. spontaneum	0,39 ± 0,069	17,9	0,36 ± 0,057	16,1
H. × *laguncuIiforme*	0,32 ± 0,035	10,9	0,29 ± 0,0008	9,8
H. agriocrithon	0,35 ± 0,042	12,0	0,42	
	1982		1983	
H. spontaneum	0,39 ± 0,067	17,2	0,38 ± 0,057	14,9
H. × *laguncuIiforme*	0,29 ± 0,035	12,2	0,29 ± 0,043	14,8
H. agriocrithon	0,36 ± 0,082	23,0	0,32 ± 0,083	26,3
	insgesamt 1980–1983			
H. spontaneum	0,38 ± 0,064	16,7		
H. × *laguncuIiforme*	0,31 ± 0,037	11,9		
H. agriocrithon	0,36 ± 0,069	19,4		

Stigmatalänge

Die Länge der Stigmata ist ein für die Anemophilie wichtiges Merkmal. Aber auch bei Autogamen ist eine gute Ausprägung der Stigmata nicht ohne Bedeutung, wie sich an den einfachen oder kaum gefiederten Stigmata mancher Kulturgerstensippen, die einen verminderten Ansatz zur Folge haben, zeigt (vgl. Hammer 1975). Eine Tendenz zur Autogamie wird also nur in geringem Umfang eine Reduktion der Stigmata zur Folge haben. Trotzdem gibt es auch in diesem Merkmal erhebliche Unterschiede innerhalb der Wildgerstensippen (Tabelle 4), vgl. auch die relativ hohen Variationskoeffizienten (*H. spontaneum* nimmt in der Variabilität die Spitzenstellung ein). Bei den absoluten Werten ist die Ausbildung der Narben bei *H.* x *lagunculiforme* am besten, es folgen *H. spontaneum* und *H. agriocrithon*.

Tabelle 4
Mittlere Stigmatalängen (mm)

	$\bar{x} \pm s$	Variations-koeffizient (%)	$\bar{x} \pm s$	Variations-koeffizient (%)
	1980		1981	
H. spontaneum	1,42 ±0,237	16,6	1,49 ±0,208	14,0
H. × *lagunculiforme*	1,72 ±0,232	13,5	1,92 ±0,212	11,0
H. agriocrithon	1,46 ±0,189	12,9	0,90	
	1982		1983	
H. spontaneum	1,42 ±0,307	21,6	1,69 ±0,351	20,8
H. × *lagunculiforme*	1,42 ±0,316	22,3	1,37 ±0,125	23,9
H. agriocrithon	1,44 ±0,201	14,0	1,32 ±0,328	9,5
	insgesamt 1980–1983			
H. spontaneum	1,51 ±0,325	21,6		
H. × *lagunculiforme*	1,70 ±0,28	16,8		
H. agriocrithon	1,41 ±0,212	15,0		

Korrelationen

Die Korrelationen zwischen den einzelnen blütenbiologischen Merkmalen zeigen eine ähnliche Tendenz wie die Merkmale selbst. So gibt es zwischen den Jahren erhebliche Unterschiede, die auf die unterschiedliche Witterungsabhängigkeit der blütenbiologischen Merkmale hinweisen; beispielsweise werden die Antheren in ihrer Länge besonders von den Witterungsbedingungen modifiziert, während ihre Breite relativ konstant bleibt. In der Summe der Jahre (vgl. Tabelle 5) ergibt sich aus diesem Verhalten, daß bestimmte positive Korrelationen zwar signifikant aber nur relativ schwach ausgeprägt sind (bes. Antherenlänge : Antherenbreite). Im korrelativen Verhalten ihrer Merkmale sind *H. spontaneum* und *H.* x *lagunculiforme* sehr ähnlich; hier wird die schon oben erwähnte Verwandtschaft zwischen diesen beiden Sippen sichtbar.

Tabelle 5
Korrelationsmatrix blütenbiologischer Merkmale (Summe der Jahre 1980–1983)

	Antherenlänge			Narbenlänge		
	1	2	3	1	2	3
Antherenbreite	+0,223***	+0,286*	+0,527**	+0,185**	+0,10	−0,174
Antherenlänge	+0,046	+0,373**	−0,095			

1 = *H. spontaneum*
2 = *H.* x *lagunculiforme*
3 = *H. agriocrithon*

Dagegen weisen die Merkmale von *H. agriocrithon* ein anderes korrelatives Muster auf. Damit wird im Unterschied zu anderen Befunden (ZOHARY 1960, 1971, TOVIA und ZOHARY 1962, BAKHTEYEV 1965), die Eigenständigkeit dieser Sippe betont, die im Grad in der Ausprägung ihrer blütenbiologischen Merkmale *H. spontaneum* sehr ähnlich ist.

Das Syndrom der Anemophilie (vgl. HAMMER 1976), das bei hoher Ausprägung immer deutlich positive Korrelationen zwischen den wichtigsten blütenbiologischen Merkmalen erwarten läßt, ist zwar bei den untersuchten Wildgersten noch ausgeprägter als bei der Kulturgerste (vgl. HAMMER 1975), durch die allgemeine Tendenz zur Autogamie aber ebenfalls im Rückgang begriffen. Morphologisch ist es noch klar erkennbar, in seiner Effektivität aber stark eingeschränkt.

Schlußfolgerungen

Die Untersuchung der blütenbiologischen Merkmale an einer größeren Kollektion von Wildgersten aus der Verwandtschaft des meist streng autogamen *H. vulgare*, zeigt auch bei diesen eine allgemeine Tendenz zur Autogamie und damit unter den Verhältnissen Gaterslebens eine geringe Neigung zur Fremdbefruchtung. Zur sippenechten Erhaltung beim Reproduktionsanbau machen sich daher im allgemeinen keine besonderen Isolationsmaßnahmen notwendig, und die Mehrheit der Sippen kann in den routinemäßigen Getreidereproduktionsanbau einbezogen werden. Lediglich in Jahren, deren Witterungsbedingungen die Ausbildung der blütenbiologischen Merkmale fördern – hier ergeben sich ähnliche Vorbedingungen wie bei *H. vulgare* (vgl. HAMMER 1977, GORASTEV und HAMMER 1978) – muß auf die Verwendung des Erntegutes eines Teils des Materials mit Rücksicht auf die sippenechte Erhaltung verzichtet werden. Ein Nachbau kann in diesen Fällen Aufklärung geben, weil durch den disjunktiven Anbau in Gatersleben bevorzugt zwei- und mehrzeilige Gersten benachbart stehen und die Hybriden in der F_1 erkannt werden können. Zur Einschätzung der durch die Jahreswitterung bedingten Neigung zur Allogamie ist in jedem Jahr ein ausgewähltes Sortiment von Wild- und Kulturgersten blütenbiologisch zu untersuchen.

Besondere Maßnahmen erfordern 31 Wildgersten mit einer ausgeprägten Allogamieneigung (Antherenlänge ≧ 4 mm und große Stigmata), u. a. bei *H. spontaneum* HOR 1647 und HOR 8523, bei *H. agriocrithon* HOR 2466 und HOR 8509. Während des Reproduktionsanbaus ist für sie eine Isolierung erforderlich.

Die blütenbiologischen Untersuchungen sind Bestandteil der in Gatersleben durchgeführten Reihenuntersuchungen, die Reproduktionserfordernisse erfassen und gleichzeitig eine Analyse der Variabilität des Kollektionsmaterials darstellen. Im Hinblick auf dieses Untersuchungsziel ist besonders die große Variationsbreite der geprüften Merkmale interessant. Als phylogenetische Konsequenzen ergeben sich eine Untergliederung von *H. spontaneum* in Unkraut- (mit kurzen Antheren) und Wildsippen (mit längeren Antheren), von denen sich sowohl die Unkraut- (vgl. MOLINA-CANO und CONDE 1980, MOLINA-CANO et al. 1982) als auch die Kultursippen ableiten lassen. Die Evolution von Unkrautsippen wurde durch Introgression von *H. vulgare* in die Wildpopulationen gefördert. Der umgekehrte Introgressionsweg trug in erheblichem Maße zur Formenvielfalt der Kulturgerste bei (vgl. HAMMER 1984).

H. agriocrithon verfügt über ein ähnliches Merkmalsspektrum wie *H. spontaneum*. Auch bei ihm gibt es Sippen mit ausgeprägten Anemophiliemerkmalen, die ihren Ursprung kaum in einer Kreuzung *H. spontaneum* × *H. vulgare* (mehrzeilig) genommen haben können. Es ist daher nicht auszuschließen, daß es sich bei den zur Allogamie neigenden Sippen von *H. agriocrithon* um ursprüngliche, mehrzeilige Wildgersten handelt (vgl. auch die Ergebnisse von WITCOMBE 1978 und neuere Berichte über das Vorkommen in China – KUO und ZHOU 1980, ZHOU 1981, SHAO et al. 1982, XU 1982, ZHEN 1982).

Die Ergebnisse der blütenbiologischen Untersuchungen an Wildgersten zeigen Ausgangsmaterial für die Schaffung von Hybridpartnern mit hoher Bestäubungsleistung (vgl. SCHOLZ und KÜNZEL 1981) auf, und die im Material gefundene Variabilität im Reproduktionssystem läßt Sippen mit weiteren züchterisch interessanten Eigenschaften erwarten. Ein großer Teil der in jüngster Zeit nach Gatersleben gekommenen *H. spontaneum*-Sippen verfügt beispielsweise über Mehltauresistenz in verschiedenen Ausprägungsstufen, wie sie von FISCHBECK (1981b) berichtet wurden. Das Material sollte vordringlich in phytopathologische und andere Screenings einbezogen werden.

Summary

Studies of pollination ecology in the Gatersleben collection of wild barleys (*Hordeum* L. subgenus *Hordeum*)

A collection of wild barleys (*Hordeum* L. subgen. *Hordeum*) has been studied from 1980 to 1983 in respect to pollination ecology. The characters analyzed were anther length, anther width and stigmata length. The variability of these characters is rather high. Their expression is largely dependent on weather conditions. The considerable amount of the races tested during the 4 years allowes to demonstrate some tendencies especially when using the anther length. The longest anthers can be found in *Hordeum spontaneum*, but also in some races of *H. agriocrithon*. Compared to them the anthers of *H.* × *lagunculiforme* are relatively short. This race is considered as a product of introgression from cultivated barleys into *H. spontaneum*.

Races of wild barley with long anthers show an increased tendency to allogamy. Therefore, the expected rates of outcrossing are a problem for the reproduction of this material in the gene bank. For their propagation isolation is required.

Краткое содержание

Исследования по биологии цветения в Гатерслебенской коллекции дикорастущих ячменей (*Hordeum* L. subgen. *Hordeum*)

С 1980 по 1983 гг. изучалась биология цветения дикорастущих ячменей (*Hordeum* L. subgen. *Hordeum*). Учитывались длина и ширина пыльников и длина рылец. Эти признаки показали очень высокую изменчивость. Особенно ярко выражена зависимость от погодных условий. Однако, в среднем, во время четырёхлетних наблюдений многочисленных форм, можно было заметить экологические тенденции опыления, главным образом по признаку длины пыльников. Самые длинные пыльники имеет *H. spontaneum*, но и *H. agriocrithon* выделяется относительно длинными пыльниками. Пыльники *H.* × *lagunculiforme*, напротив, оказались относительно короткими. Эта, последняя, форма рассматривается как продукт интрогрессии культурных ячменей в *H. spontaneum*.

Дикорастущие формы с длинными пыльниками обладают повышенной склонностью к аллогамии. Получение подлинных форм при репродукции в генном банке оказывается проблематичным. Размножая эти формы, их следует изолировать.

Literatur

ÅBERG, E., 1940: The Taxonomy and Phylogeny of *Hordeum* L. sect. *Cerealia* Ands. with special reference to Thibetian barleys. – Symb. bot. Upsal. VI, 2. – 156 pp.

AHOKAS, H., 1979: Cytoplasmic male sterility in barley. III. Maintenance of sterility and restoration of fertility in the ms ml cytoplasms. – Euphytica **28**, 409–419.

–, 1981: Cytoplasmic male sterility in barley. X. Distribution of MS ML – fertility restoration ability in the wild progenitor of barley in Israel. – Ann. Bot. Fenn. **18**, 313–320.

–, 1982a: Cytoplasmic male sterility in barley. XI. The ms ms 2 cytoplasm. – Genetics **102**, 285–295.

–, 1982b: Variation of kernel protein and lysine in the wild progenitor of barley. – Hereditas **96**, 29–37.

BAKHTEYEV, F. KH., 1965: Experimental data on the inheritance of some taxonomic characters in *Hordeum spontaneum*. – Züchter **35**, 365–367.

VON BOTHMER, R., N. JACOBSEN, och R. BAGGER JÖRGENSEN, 1982: Kornsläktet och användning av wilde kornarter i växtföradling. – Sveriges Utsädesförnenings Tidskrift **92**, 93–107.

BROWN, A. H. D., D. ZOHARY, and E. NEVO, 1978a: Outcrossing rates and heterozygosity on natural populations of *Hordeum spontaneum* Koch in Israel. – Heredity **41**, 49–62.

–, E. NEVO, and D. ZOHARY, 1978b: Genetic structure of *Hordeum spontaneum* in Israel. – Barley Genetics Newsletter **8**, 20–23.

CHAUDHARY, H. R., S. JANA and S. N. ADHARYA, 1980: Outcrossing rates in barley populations in the Canadian Prairies. – Can. J. Genet. Cytol. **22**, 353–360.

FISCHBECK, G., 1981a: Ergebnisse von Untersuchungen in israelischen Populationen von *Hordeum spontaneum* im Hinblick auf Gegenwartsprobleme der Resistenzzüchtung. – Ber. Arbeitst. Gumpenstein 1981, 3–12.

FISCHBECK, G., 1981b: New trends in breeding for resistance to pathogens. – Genetic Resources and Plant Breeding for Resistance to Diseases, Pests and Abiotic Environmental Conditions, Materials of IX. Congress of EUCARPIA, Leningrad, pp. 165–180.

FRÖST, S., and S. ASKER, 1977: Flavonoid patterns and polymorphisms in wild *Hordeum* species. – Hereditas **85**, 45–150.

GORASTEV, H., und K. HAMMER, 1978: Standortbedingte Unterschiede bei bestäubungsökologischen Merkmalen von Wintergersten (*Hordeum vulgare* L. s. l.). – Kulturpflanze **26**, 283–292.

HAMMER, K., 1975: Die Variabilität einiger Komponenten der Allogamieneigung bei der Kulturgerste (*Hordeum vulgare* L. s. l.). – Kulturpflanze **23**, 167–180.

–, 1976: Zur Allogamieneigung der Kulturgerste (*Hordeum vulgare* L.) – Tag.-Ber. Akad. Landwirtsch.-Wiss. DDR **143**, 221–226.

–, 1977: Fragen der Eignung des Pollens der Kulturgerste (*Hordeum vulgare* L. s. l.) für die Windbestäubung. – Kulturpflanze **25**, 13–23.

–, 1984: Bestäubungsökologische Merkmale und Phylogenie von *Hordeum* L. subgen. *Hordeum*. – Flora **175**, 339–344.

KAHLER, A. L., and R. W. ALLARD, 1981: Worldwide patterns of genetic variation among four esterase loci in barley (*Hordeum vulgare* L.). – TAG **59**, 101–111.

KAMM, A., 1954: The discovery of wild six-rowed barley and wild *Hordeum intermedium* in Israel. – Ann. Roy. Agr. Col. Sweden **21**, 287–320.

–, 1977: The range of brittle types of Cerealia barleys in Israel. – Agr. Res. Org., Volcani Center Mor. **165**. – 43 pp.

KUO PEN-CHAO, and ZHOU LI-HUA, 1980: A preliminary study on the classification and distribution of the genus *Hordeum* L. in China (Chin., Engl. summary). – Acta Phytotax. Sinica **18**, 420–426.

MOLINA-CANO, J. L., and J. CONDE, 1980: *Hordeum spontaneum* C. Koch em. Bacht. collected in Southern Morocco. – Barley Genetics Newsletter **10**, 44–47.

–, C. GOMÈZ-CAMPO and J. CONDE, 1982: *Hordeum spontaneum* C. Koch as a weed of barley fields in Morocco. – Z. Pflanzenzüchtg. **88**, 161–167.

MOSEMAN, J. G., P. S. BAENZIGER, and R. A. KILPATRICK, 1981: Genes conditioning resistance of *Hordeum spontaneum* to *Erysiphe graminis* f. sp. *hordei*. – Crop Sci. **21**, 229–232.

SCHOLZ, F., and G. KÜNZEL, 1981: Progress and problems with hybrid barley. – Barley Genetics IV, Proc. Fourth Intern. Barley Genetics Symp., Edinburgh 1981. – pp. 758–765.

SEGAL, A., G. FISCHBECK, and I. WAHL, 1979: Types of protection of wild barley, *Hordeum spontaneum* C. Koch, against *Erysiphe graminis hordei* in natural populations in Israel. – Phytoparasitica **7**, 49.

SHAO QIQUAN, LI CHANG-SEN, and SASCHAN CHIREN, 1982: Origin and evolution of cultivated barley: Wild barley from Western Szechuan and Tibet, China. – Barley Genetics Newsletter **12**, 37–42.

TOVIA, T., and D. ZOHARY, 1962: Natural hybridization between *Hordeum vulgare* and *H. spontaneum*: Corroborative evidence from progeny tests and artificial hybrids. – Bull. Res. Counc. Israel **11**, 43–45.

DE WET, J. M. J., 1981: Species concepts and systematics of domesticated cereals. – Kulturpflanze **29**, 177–198.

WITCOMBE, J. R., 1978: Two-rowed and six-rowed wild barley from the Western Himalaya. – Euphytica **27**, 601–604.

XU TINGWEN, 1982: Origin and evolution of cultivated barley in China (Chin., Engl. summary). – Acta Genetica Sinica **9**, 440–446.

ZHEN YAO, 1982: Studies on the N-banding karyotype of cultivated and wild barley (Chin., Engl. summary). – Acta Genetica Sinica **9**, 160–164.

ZHOU ZEQI, 1981: Comparison between the two-rowed wild barley of Qing-Zang Plateau and that of Israel (Chin., Engl. summary). – Acta Genetica Sinica **8**, 344–349.

ZOHARY, D., 1960: Studies on the origin of cultivated barley. – Bull. Res. Counc. Israel, Sect. D., **9**, 21–42.

ZOHARY, D., 1964: Spontaneous brittle six-row barleys, their nature and origin. – Barley Genetics I,Proc. First Intern. Barley Genetics Symp., Wageningen 1963, 29–31.
–, 1971: The fate of natural "hybrid swarms" between *Hordeum spontaneum* and *H. vulgare*. – Barley Genetics II, Proc. Second Intern. Barley Genetics Symposium, Pullman, Washington 1970, 63–64.

Dr. K. HAMMER
Zentralinstitut für Genetik und Kulturpflanzenforschung
der Akademie der Wissenschaften der DDR
DDR - 4325 Gatersleben
Corrensstr. 3

Kulturpflanze 32 · 1984 · 89–101

Rasterelektronenmikroskopische Untersuchungen an Samen der Gattung *Allium* L.

JOACHIM KRUSE

(Eingegangen am 3. Februar 1984)

Zusammenfassung

Die Skulpturmerkmale von Samenoberflächen verschiedener *Allium*-Arten wurden rasterelektronenmikroskopisch geprüft. Die Untersuchungen bestätigen die große Variabilität dieses Merkmalskomplexes. Die Ausbildung der Testaskulptur-Merkmale in verschiedenen Sektionen der Gattung sowie Probleme der infraspezifischen Variabilität werden diskutiert.

Einleitung

Der Wert von Epidermismerkmalen für die Bearbeitung systematischer Fragestellungen konnte bisher an zahlreichen Verwandtschaftskreisen nachgewiesen werden. Dieses gilt insbesondere auch für die Testamerkmale der Samen, die nicht selten wegen ihrer oft sehr spezifischen Merkmalsstruktur zur Charakterisierung von Sippen herangezogen werden.

Rasterelektronenmikroskopische Untersuchungen an Samen einiger nahe verwandter Arten der Gattung *Allium* L. wurden zunächst von BOTHMER (1974) durchgeführt. Umfassendere Samenuntersuchungen erfolgten durch PASTOR (1981), der etwa dreißig auf der Iberischen Halbinsel und den Balearen beheimatete *Allium*-Arten prüfte und dabei vielfältige Unterschiede in der Ausprägung der Testa-Oberfläche feststellen konnte. Die folgenden, an einem erweiterten Artenspektrum durchgeführten Untersuchungen dienen der weiteren Charakterisierung dieses Merkmalskomplexes. Die Fortführung dieser Arbeiten ist im Rahmen umfangreicherer taxonomischer Untersuchungen zur Verwandtschaftsstruktur der Gattung *Allium* L. vorgesehen.

Material und Methode

Das untersuchte Samenmaterial entstammt dem Kulturpflanzen-Sortiment der Genbank sowie der *Allium*-Spezialsammlung des Zentralinstituts für Genetik und Kulturpflanzenforschung Gatersleben. Die Proben der geprüften Sippen umfaßten jeweils 4–6 Samen.

Die lufttrockenen Samen wurden ohne besondere Vorbehandlung zunächst mit Kohle bedampft und anschließend im Sputtering-Verfahren mit einer Gold-Palladium-Legierung beschichtet. Die Untersuchungen erfolgten mit einem Rasterelektronenmikroskop des Typs ISI-40.

Die Zuordnung der untersuchten Arten entspricht im wesentlichen der Sektionsgliederung von Stearn (1978).

Ergebnisse und Diskussion

Die untersuchten Samenoberflächen der Gattung *Allium* L. zeigen eine ausgeprägte morphologische Mannigfaltigkeit, wie sie ähnlich auch von anderen Organkomplexen dieses Verwandtschaftskreises bekannt ist. Die Variabilität der Testamerkmale beschränkt sich dabei nicht auf Abweichungen in der Größe, Grundform und Wölbung der Zellen, sondern schließt außerdem wesentliche sippenspezifische Unterschiede in der Form und Verteilung der Oberflächenskulpturen ein.

Einem besonders einfachen Typ entsprechen die Testazellen der Samen von *Allium tricoccum* Ait. (sect. *Anguinum* G. Don ex Koch), deren obere flach gewölbte Periklinalwand vollständig glatt und ungegliedert ist (Abb. 1). Die während der Samenreife entlang der Antiklinalgrenzen auftretenden Interzellularrisse führen hier zu einer geringfügigen vertikalen Verschiebung der einzelnen Zellen gegeneinander.

Ebenso ist die Testa von *A. oreophilum* C. A. Mey., einer Art, die im System der Gattung eine relativ isolierte Stellung einnimmt, einem einfachen Skulpturtyp zuzuordnen (Abb. 2). Die stark aufgewölbten Testazellen besitzen ebenfalls eine glatte Oberfläche und einen annähernd isodiametrischen Zellumriß, der jedoch durch eine schwache S-Undulation gekennzeichnet ist. Ein spezieller Randbereich ist nur in Andeutungen erkennbar.

Die Testazellen zahlreicher *Allium*-Arten entsprechen in der Gliederung ihrer äußeren Periklinalwände dem von Barthlott und Ehler (1977) näher charakterisierten Zentralfeld-Typ, der durch die Gliederung der Epidermis-Zelloberfläche in ein mittleres Zentralfeld und ein peripheres Antiklinalfeld gekennzeichnet ist. Beide Bereiche können zusätzlich durch einen besonders hervortretenden Zentralwulst gegeneinander abgegrenzt sein. Das Antiklinalfeld zeigt dabei meist faltenartige Parallelskulpturen, die quer zur Antiklinalwand angeordnet sind, während sich das Zentralfeld durch andersartige Skulpturmuster abhebt.

Eine gewisse Gliederung der Periklinalwände konnte bei *A. nerinifolium* Bak. beobachtet werden, das zur Sektion *Calloscordon* (Herb.) Bak. gestellt wird und so ebenfalls eine gewisse Sonderstellung einnimmt (Abb. 3). Das plan bis schwach konvex entwickelte Zentralfeld besitzt eine glatte bis schwach verrucose Oberflächenskulptur. Den Rand bildet hier ein äußerst schmales Antiklinalfeld, das durch sehr kurze, quer zur Antiklinalwand orientierte Abschnitte gegliedert ist.

Eine vergleichsweise einfache Oberflächenskulptur beschreibt auch Pastor (1981) von dem zur Sektion *Anguinum* G. Don ex Koch gehörenden *A. victorialis* L. (Abb. 4). Wie aus eigenen Untersuchungen dieser Art hervorgeht, zeigt die relativ glatte Testaoberfläche hier jedoch eine deutliche Gliederung in ein relativ breites glattes Antiklinalfeld und ein in seinen Umrissen schwach gelapptes Zentralfeld mit glatter bis undeutlich granulöser Oberfläche.

Relativ glatte Skulpturen kennzeichnen auch die Samen von *A. ursinum* L.

(sect. *Ophioscorodon* [WALLR.] BUBANI), worauf auch PASTOR (1981) hinweist; jedoch ist das Zentralfeld bei dieser Art deutlich gegliedert. Dieses wird durch einen Zentralwulst begrenzt, der ein bis mehrere glatte bis faltige höckerartige Aufwölbungen umschließt. Während PASTOR (1981) nur einhöckerige Zellen erwähnt, konnten wir neben diesen auch mehr- bis zu fünfhöckerige Zellen beobachten. Selten wurde darüber hinaus auch eine höckerartige Aufgliederung des Zentralwulstes festgestellt. Im Hinblick auf die Ausbildung des Antiklinalfeldes erwies sich das untersuchte Material als heterogen. Während in einigen Fällen kein Antiklinalfeld beobachtet werden konnte, ließ sich in anderen Fällen, in Übereinstimmung mit den Befunden von PASTOR (1981), ein sehr deutlich hervortretendes Antiklinalfeld mit einer glatten Oberfläche nachweisen (Abb. 5).

Die Testazellen zahlreicher *Allium*-Arten besitzen einen durch parallele Faltenmuster auffällig gegliederten Randbereich, der von PASTOR (1981) als Interzellularregion bezeichnet wird. Gegenüber den angrenzenden Zellflächen ist dieser oft furchenartig vertieft, so daß die Testa den Eindruck einer von Interzellularrissen durchzogenen Epidermis vermittelt. Dieser Eindruck wird durch die Form der furchig umgrenzten Zellflächen, die sich lückenlos aneinanderfügen ließen, noch verstärkt. Bei vereinzelten kleineren Zellgruppen des reifen Samens können sich die Zellflächen infolge fehlender Furchen auch unmittelbar berühren. Man kann deshalb annehmen, daß diese zentralfeldartig skulpturierten Bereiche zu einem früheren Zeitpunkt der Testa-Entwicklung mit ihren Rändern aneinandergrenzen. Dennoch bleibt zunächst unklar, ob hier Trocknungsvorgänge während der Samenreife zu der Furchenbildung führen, oder ob der furchig vertiefte Randbereich durch Wachstumsprozesse entsteht und so dem Antiklinalfeld der Epidermiszelle entspricht. Für die letztere Möglichkeit spricht u. a. die Tatsache, daß neben stark vertieften auch flachrinnige oder mit der zentralen Zelloberfläche niveaugleiche Randbildungen vorkommen können. Darüber hinaus zeigen die hier auftretenden Parallelfalten die für Antiklinalfeldskulpturen charakteristische Anordnung. Da uns jedoch keine entwicklungsgeschichtlichen Untersuchungen vorliegen, die Aufschluß über den Charakter dieser vertieften Randzone geben, soll hierfür im folgenden die neutralere Bezeichnung Antiklinalbereich verwendet werden.

Ein relativ breiter und vertiefter Antiklinalbereich charakterisiert eine Gruppe von Testatypen, die sich durch eine besondere Gliederung des Zentralfeldes auszeichnen. Der Rand des Zentralfeldes besteht im einfachsten Falle aus einer breiten ringartigen Skulptur, an die sich eine ebenfalls ringartige Vertiefung anschließt. Die Mitte der Zelloberfläche ist durch die Bildung eines zentralen Höckers gekennzeichnet. Dabei kann die ringartige Vertiefung des Zentralfeldes in Abhängigkeit von der Art, aber auch innerhalb einer Testa unterschiedlich ausgeprägt sein. Oft besitzen der Zentralwulst wie auch der zentrale Höcker eine granulöse Oberfläche.

Der beschriebene Testatyp ist nach unseren Untersuchungen relativ häufig in der Sektion *Rhizirideum* G. DON ex KOCH vertreten, konnte aber auch in anderen Sektionen nachgewiesen werden. Das Vorkommen dieses Typs in verschiedenen Verwandtschaftsgruppen der Gattung *Allium* L. bestätigen auch Angaben von PASTOR (1981), der einen Typ mit zentraler Höckerbildung für eine Reihe europäisch verbreiteter Arten der Sektion *Molium* G. DON ex KOCH sowie an jeweils einer Art der Sektionen *Rhizirideum* G. DON ex KOCH und *Briseis* (SALISB.) STEARN nachwies.

Innerhalb der Sektion *Rhizirideum* G. DON ex KOCH fanden wir diesen Typ besonders eindeutig bei den Arten *A. angulosum* L. und *A. farreri* STEARN ausgeprägt (Abb. 6), während die Skulpturtypen einiger weiterer Arten Übergangs-

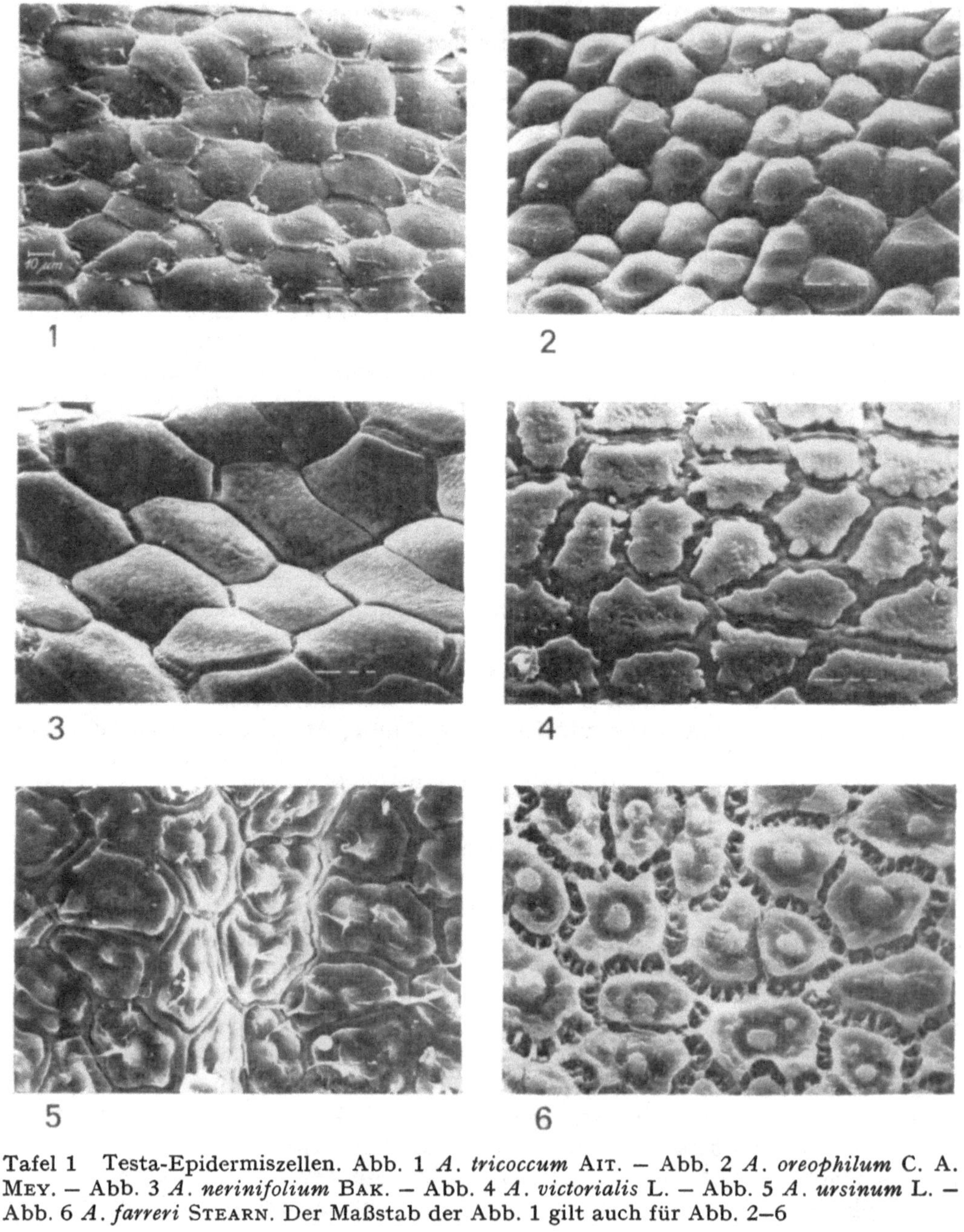

Tafel 1 Testa-Epidermiszellen. Abb. 1 *A. tricoccum* AIT. – Abb. 2 *A. oreophilum* C. A. MEY. – Abb. 3 *A. nerinifolium* BAK. – Abb. 4 *A. victorialis* L. – Abb. 5 *A. ursinum* L. – Abb. 6 *A. farreri* STEARN. Der Maßstab der Abb. 1 gilt auch für Abb. 2–6

charakter besitzen, indem der Zentralhöcker bestimmter Sippen oder auch Einzelzellen durch eine partielle oder auch vollständige Verflachung der ringartigen Vertiefung des Zentralfeldes weniger deutlich hervortritt, und sie auf diese Weise zur Bildung planer granulöser (*A. senescens* L., *A. nutans* L.) oder planer glatter Zelloberseiten (*A. albidum* Fisch.) überleiten (Abb. 7 u. 8).

Besonders deutlich ist der oben beschriebene Skulpturtyp auch bei einigen der Sektion *Molium* G. Don ex Koch nahestehenden neuweltlichen Sippen (*A. drummondii* Rgl., *A. cernuum* Roth) ausgebildet, bei denen der Zentralwulst durch eine scharf ausgeprägte Ringvertiefung vom Zentralhöcker getrennt ist (Abb. 9 u. 10).

Als morphologisch nahestehend muß eine Skulpturform angesehen werden, bei der der Zentralwulst **mehrere** (bis zu fünf) höckerartige Bildungen umschließt (*A. lineare* L., *A. senescens* L.). Die enge Formverwandtschaft beider Zelltypen findet auch in der Tatsache, daß sowohl ein- als auch mehrhöckerige Zellen in einer Testa gemeinsam vorkommen können, ihre Bestätigung (Abb. 11 u. 12). Ein entsprechender Typ von *A. senescens* L. scheint auch bei den Untersuchungen von Pastor (1981) vorgelegen zu haben, der für diese Art ebenfalls eine Variationsbreite von 1–5 Höckern je Zelle angibt.

Eine weitere Variante des oben erwähnten Grundtyps ergibt sich durch die Auflösung des Zentralwulstes in einzelne Höcker, die in ihrer Größe dem Zentralhöcker angeglichen sind (Abb. 17 u. 18). Der morphologische Zusammenhang beider Typen wird durch die Tatsache, daß beide Formen in einer speziellen Sippe einer Art gleichzeitig vertreten sein können oder daß bei einzelnen Zellen Teile des ringförmigen Zentralwulstes ungegliedert erhalten sind, bestätigt.

Wichtig erscheint die Feststellung, daß die peripheren Höcker im vorliegenden Falle entsprechend der Ringskulptur des Grundtyps unmittelbar randständig sind, während sie bei anderen mehrhöckerigen Skulpturtypen meist in deutlichem Abstand vom Rande des Zentralfeldes inseriert sind.

Pastor (1981) fand Testazellen mit einem von mehreren Höckern umgebenen Zentralhöcker bei vier Arten der Sektion *Codonoprasum* Reichenb. sowie jeweils einer Art der Sektionen *Molium* G. Don ex Koch, *Scorodon* Koch und *Rhizirideum* G. Don ex Koch. Eigene Untersuchungen konnten diesen Skulpturtyp für die Arten *A. carinatum* L., *A. pallens* L. und *A. paniculatum* L. bestätigen und darüber hinaus für *A. flavum* L. (sect. *Codonoprasum* Reichenb.) nachweisen. Dagegen kann die Auffassung des oben genannten Autors, daß die Parallelfalten der jeweils benachbarten Zellen bei den Arten der Sektion *Codonoprasum* Reichenb. sowie bei *A. senescens* L. (sect. *Rhizirideum* G. Don ex Koch) alternierend ineinandergreifen, nicht bestätigt werden. Sämtliche Proben der oben genannten Arten zeigen, daß die Falten des Antiklinalbereichs die Zentralfelder der benachbarten Zellen in gerader Linie miteinander verbinden.

Besondere Beachtung verdient die infraspezifische Variabilität der Testaskulpturen des untersuchten Verwandtschaftskreises, die hier an einer der oben erwähnten Arten näher geprüft wurde. Während die Testaskulpturen der meisten untersuchten Sippen dieser Gattung deutlich ausgeprägt sind, ergaben sich bei den Arten der Sektion *Codonoprasum* Reichenb. besondere Schwierigkeiten in der Einschätzung des Skulpturtyps, indem die für diese Gruppe charakteristischen Höckerbildungen bei einem Teil der untersuchten Herkünfte entweder nur

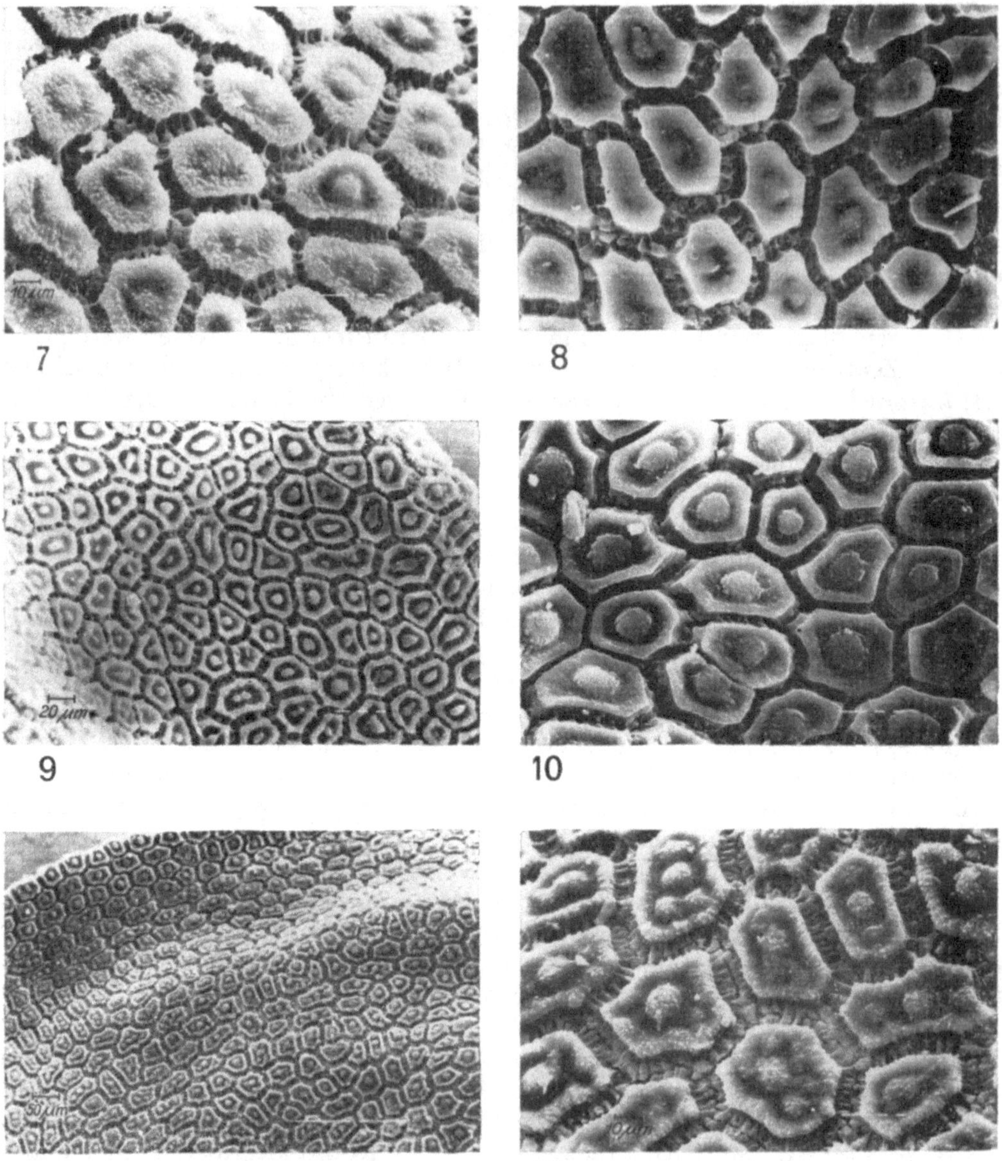

Tafel 2 Testa-Epidermiszellen. Abb. 7 *A. senescens* L. – Abb. 8 *A. albidum* Fisch. – Abb. 9 *A. cernuum* Roth – Abb. 10 *A. drummondii* Rgl. – Abb. 11 u. 12 *A. senescens* L. Der Maßstab der Abb. 7 gilt auch für Abb. 8, 10 u. 12

undeutlich oder überhaupt nicht erkennbar waren. Zum besseren Verständnis der Variabilität dieses Merkmals wurden 10 Herkünfte der Art *A. flavum* L. untersucht, wobei Typen verschiedener Wuchshöhe und unterschiedlicher Blattbereifung Berücksichtigung fanden.

Ein Teil der untersuchten Proben, bei denen das Zentralfeld der Testazellen den

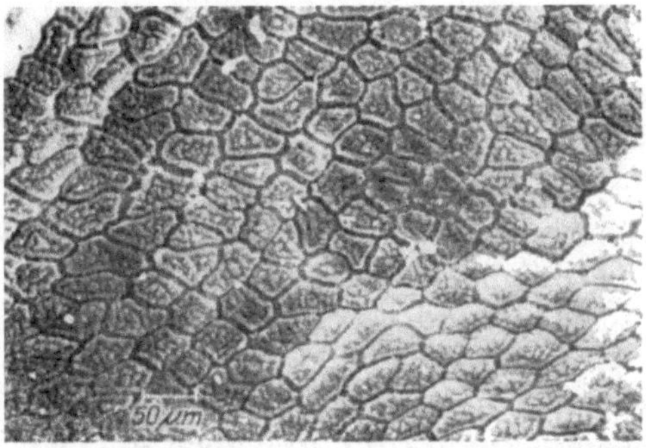

13

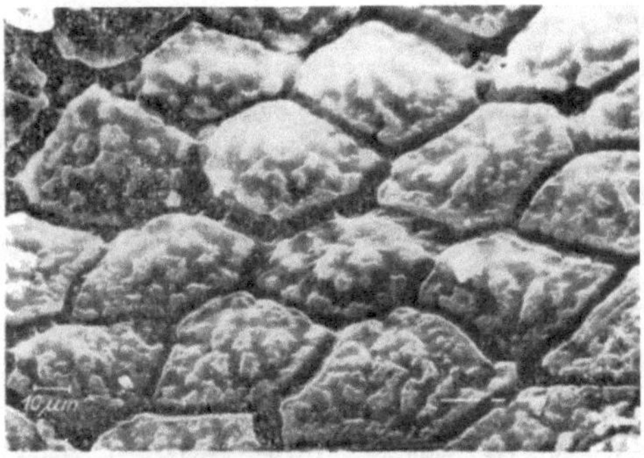

14

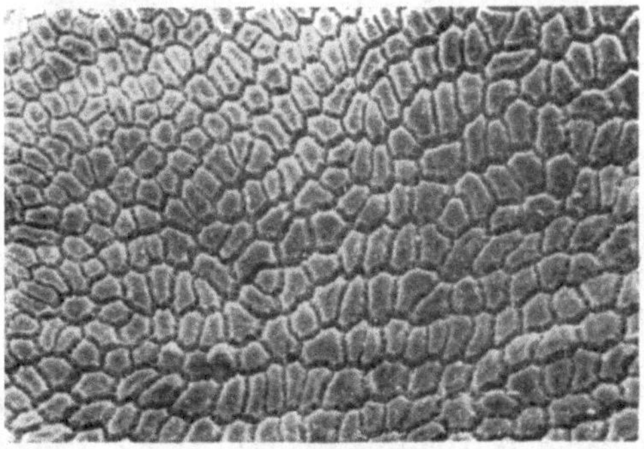

15

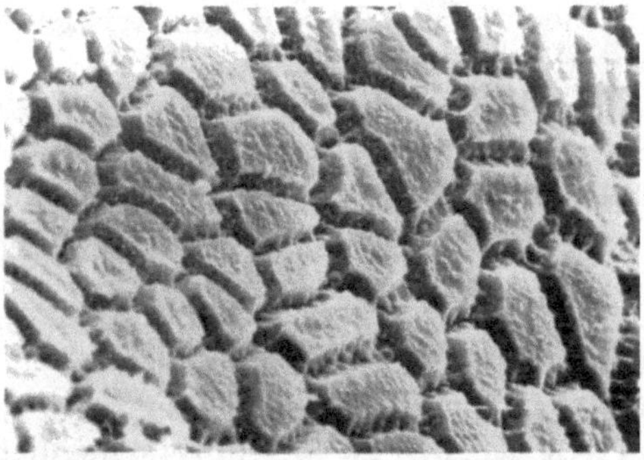

16

17

18

Tafel 3 Testa-Epidermiszellen. Abb. 13 u. 14 *A. ericetorum* THORE – Abb. 15 *A. ledebourianum* R. et S. – Abb. 16 *A. schoenoprasum* L. – Abb. 17 *A. pallens* L. – Abb. 18 *A. carinatum* L. Der Maßstab der Abb. 13 gilt auch für Abb. 15 u. 17, der auf Abb. 14 angegebene auch für Abb. 16 u. 18

furchig vertieften Antiklinalbereich wesentlich überragte, zeigte eine deutliche Höckerbildung (Abb. 19 u. 20). Daneben fanden sich Sippen, bei denen die charakteristischen Skulpturen mehr oder weniger verdeckt schienen. So konnten im Zentralbereich alle Abstufungen zu verdeckten Skulpturmustern beobachtet werden, indem die Höcker nur in undeutlichen Umrissen oder nur noch in ihrem distalen Bereich erkennbar waren (Abb. 21). Dabei scheinen die Zentralfeldskulpturen mit der relativen Erhöhung des Antiklinalbereichs undeutlicher ausgeprägt zu sein. Im Falle der vollständigen Angleichung beider Bereiche waren die spezifischen Skulpturen des Zentralfeldes weitgehend verdeckt, wobei glatte bzw. unregelmäßig gekräuselte oder grobschollige Oberflächenskulpturen auftraten (Abb. 22). Einzelne deutlicher ausgeprägte Zelloberflächen der Testa zeigten jedoch, daß auch die letzteren Formen demselben Grundtyp dieser Art zuzuordnen sind.

Ähnliche Skulpturveränderungen wurden z. T. auch im Antiklinalbereich festgestellt, so daß neben Sippen mit parallelen Faltenskulpturen auch solche mit höckerig porösen oder mehr oder weniger glatten Oberflächen beobachtet werden können.

Eine spezielle Sippe dieser Art zeichnete sich durch die extreme Förderung des Antiklinalbereichs aus (Abb. 23). Die in diesem Falle erwartungsgemäß sehr undeutlichen Zentralfeldskulpturen wurden hier vom angrenzenden Antiklinalbereich überragt. Letzterer besitzt dabei die für Antiklinalfelder typische quer zur Antiklinalwand verlaufende Faltenskulptur.

Der bei dieser Art beobachtete kontinuierliche Übergang von rinnig vertieften zu erhöhten Randbereichen mit typischen Parallelskulpturen läßt darauf schließen, daß auch die rinnigen Testavertiefungen anderer Sippen dem Antiklinalfeld entsprechen. Wie der Vergleich dieser Sippen weiterhin zeigt, ist bei einzelnen Arten mit einer gewissen innerartlichen Variabilität zu rechnen, die hier jedoch im wesentlichen auf der unterschiedlich deutlichen Ausprägung eines spezifischen Skulpturtyps beruht. Dagegen ergaben sich bisher keine Hinweise auf die Ausbildung grundlegend unterschiedlicher Skulpturmuster innerhalb einer einzelnen Art.

Ein weiterer Skulpturtyp, der hinsichtlich der Zentralfeldgliederung gewisse Anklänge an Arten der Sektion *Codonoprasum* Reichenb. zeigt, wurde bei *Allium ericetorum* Thore (sect. *Rhizirideum* G. Don ex Koch) beobachtet. Die von Pastor (1981) allgemein als granulös charakterisierte flache bis schwach gewölbte Testaoberfläche besitzt nach unseren Beobachtungen ein besonderes Verteilungsmuster (Abb. 13 u. 14). Die granulösen Skulpturen markieren dabei den nach innen unregelmäßig begrenzten, oft in kleine Abschnitte gegliederten Zentralwulst und überdecken das Zentralfeld in mehr oder weniger eckigen inselartigen Gruppen unterschiedlicher Größe und Form. Im tieferliegenden Antiklinalbereich sind z. T. parallele Falten erkennbar, die jedoch oft relativ stark verbreitert oder seitlich miteinander verschmolzen sind, so daß der Antiklinalbereich vielfach auch eine runzlige oder glatte Oberfläche aufweist.

Granulöse Skulpturen anderer Art kennzeichnen einen weiteren Typ, der besonders für eine Reihe von Arten der Sektion *Schoenoprasum* Dumort. s. l. charakteristisch ist. Die Zellen besitzen ein meist planes Zentralfeld, über das die granulösen Skulpturen relativ gleichmäßig verteilt sind (Abb. 15 u. 16). Am seitlichen Rande des erhöhten Zentralfeldes treten schwache rippenartige Bildungen auf, die in den angrenzenden Antiklinalbereich hineinragen, wobei die Rippen der jeweils benachbarten Zellen alternierend ineinandergreifen und in ihrem Umriß vermutlich einem schwachwelligen Verlauf der Antiklinalwände folgen.

Auf die alternierende Stellung dieser Randbildungen weist auch PASTOR (1981) hin, der diesen Testatyp bei zwei Arten der Sektion *Schoenoprasum* DUMORT. ermittelte. Darüber hinaus konnten wir im Antiklinalbereich einzelner Sippen mehrfach auch parallele Faltenbildungen beobachten, die dann die alternierenden Rippen mehr oder weniger verdeckten. Dieser Skulpturtyp wurde mit geringen artspezifischen Abweichungen bei folgenden Arten der Sektion *Schoenoprasum* DUMORT. beobachtet: *A. galanthum* KAR. et KIR., *A. altaicum* PALL., *A. fistulosum* L., *A. ledebourianum* R. et S. und *A. schoenoprasum* L. Ebenso kommt dieser Typ in der Sektion *Rhizirideum* G. DON ex KOCH bei *A. saxatile* BIEB. vor.

PASTOR (1981) erwähnt für *A. schoenoprasum* L. die zusätzliche Ausbildung eines kleinen zentralen Höckers, den er als Differenzierungsmerkmal zu *A. schmitzii* COUTINHO ansieht. Die von uns untersuchte Sippe von *A. schoenoprasum* L. besaß jedoch keine derartige Zusatzbildung, so daß beide Varianten möglicherweise innerhalb der Variabilitätsgrenzen der genannten Art liegen.

Ein vollständig andersartiger Typ konnte für die zur gleichen Sektion gehörende Art *A. cepa* L. nachgewiesen werden. Das Zentralfeld, an dessen Peripherie ein sehr schmaler, aber deutlich ausgeprägter Zentralwulst entwickelt ist, zeichnet sich durch dic Bildung mehrerer (etwa bis zu zehn) mittelgroßer Höcker aus (Abb. 24). Diese sind jedoch nicht, wie bei den untersuchten Vertretern der Sektion *Codonoprasum* REICHENB., randständig, sondern in deutlichem Abstand vom Rande sowie im zentralen Bereich inseriert. In dem nur flachrinnig vertieften Antiklinalbereich treten in relativ großen Abständen sehr schmale, quer zur Antiklinalwand orientierte Falten auf.

Im Gegensatz zu allen bisher hier erörterten Formen ist der folgende Testatyp durch relativ große Zellen mit einem auffällig undulierten Randbereich gekennzeichnet, wobei die randlichen Ausbuchtungen der jeweils benachbarten Zellen eng miteinander verzahnt sind (Abb. 25). Ein Antiklinalfeld ist nicht ausgebildet. Auf der Oberseite der meist mehr oder weniger aufgewölbten Zellen sind mehrere relativ große höckerartige Bildungen ausgeprägt. Dieses Skulpturmuster konnte von PASTOR (1981) an neun Arten der Sektion *Allium* L. nachgewiesen werden. Eigene Untersuchungen bestätigen die generellen Angaben über die Arten *A. ampeloprasum* L., *A. sphaerocephalon* L., *A. scorodoprasum* L. und *A. guttatum* STEV. Darüber hinaus konnten wir diesen Typ bei *A. porrum* L. feststellen.

Einzelne Artunterschiede ergeben sich z. T. durch den Undulationsgrad der Antiklinalwände, der bei vielen S-förmig ist (Abb. 26) und dann nur eine randliche Gliederung der Zellkörpers bedingt (z. B. *A. scorodoprasum* L.), während bei anderen Formen eine U-förmige Undulation sowie tiefere Einbuchtungen oder auch weiter untergliederte Ausbuchtungen der Zellen (Abb. 28) zu einer tiefgreifenden Gliederung des Zellkörpers führen können (*A. guttatum* STEV.).

Betrachtet man die relativ hohe Zahl der untersuchten Arten dieser Sektion, deren Testa ausnahmslos demselben Skulpturmuster entspricht, so läßt sich vermuten, daß hier ein Verwandtschaftskreis vorliegt, der im Gegensatz zu verschiedenen anderen Sektionen durch ein einheitliches Testamerkmal charakterisiert ist. Dieses bedeutet jedoch nicht, daß sich dieser Skulpturtyp auf die Sektion *Allium* L. beschränkt, sondern unsere Untersuchungen an Vertretern der Sektion *Megaloprason* WENDELBO (*A. giganteum* RGL., *A. stipitatum* RGL.) zeigen, daß dieser auch bei anderen Verwandtschaftskreisen vorkommen kann (Abb. 27).

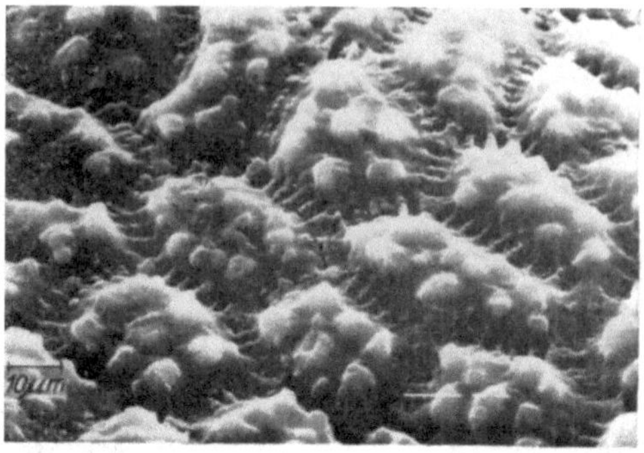

19

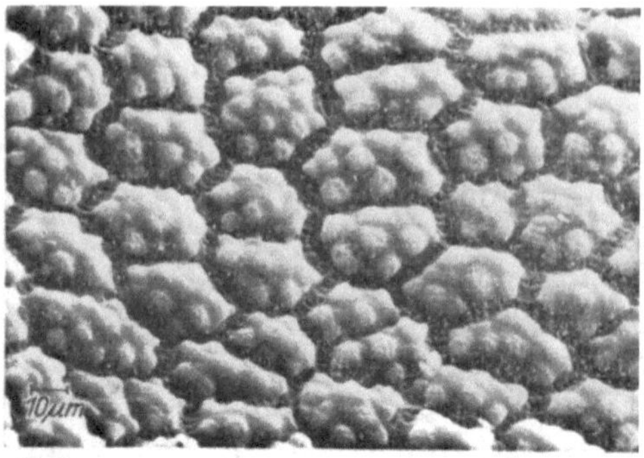

20

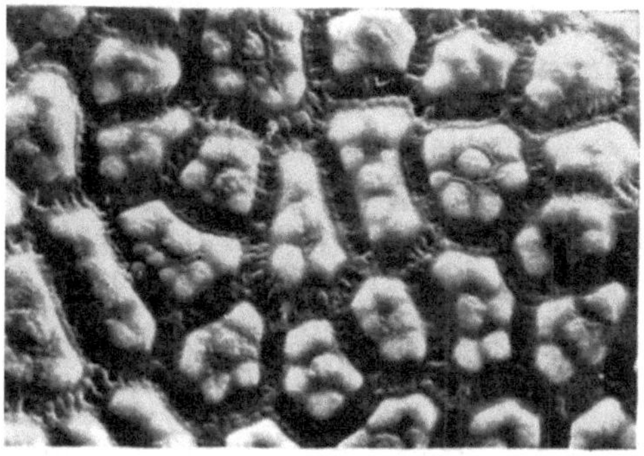

21

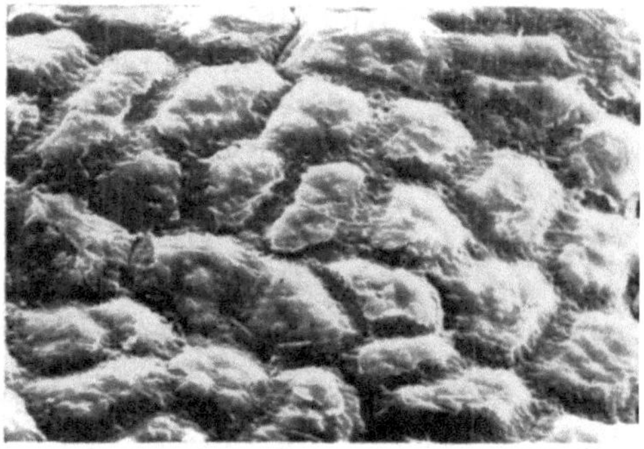

22

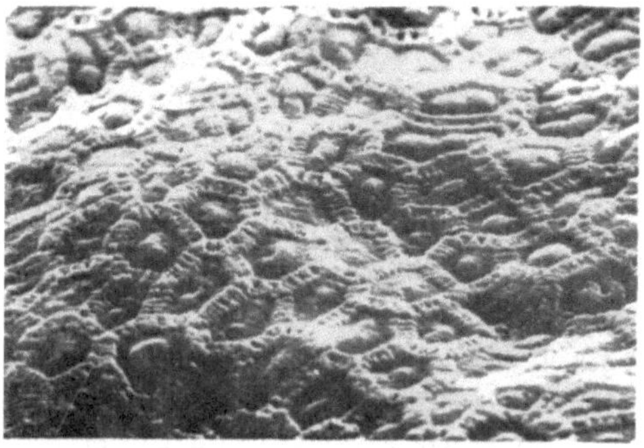

23

24

Tafel 4 Testa-Epidermiszellen. Abb. 19—23 *A. flavum* L. – Abb. 24 *A. cepa* L. Der Maßstab der Abb. 20 gilt auch für Abb. 21—24

Ein aus der Vielfalt der Gattung *Allium* L. besonders hervortretender Skulpturtyp konnte bei den auch morphologisch sehr ähnlichen Arten *A. tuberosum* ROTTL. und *A. ramosum* L. nachgewiesen werden. Die Testa dieser zur Sektion *Rhizirideum* G. DON ex KOCH gehörenden Sippen besteht aus sehr großen, langgestreckten granulösen Zellen mit einem mehr oder weniger geradlinigen bis leicht gebogenen Randverlauf (Abb. 29). Die einzelnen Grana sind in zahlenmäßig sehr

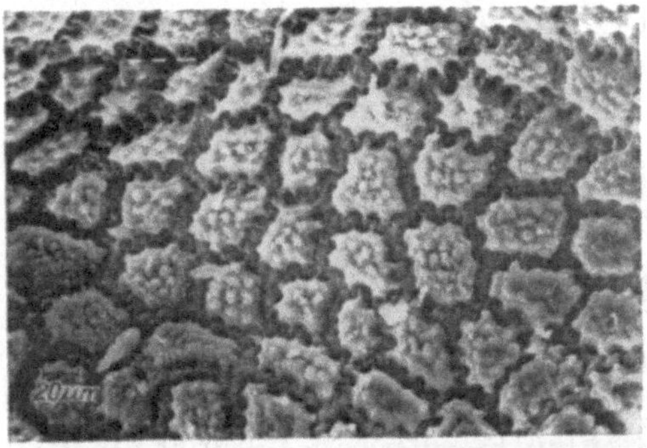

25

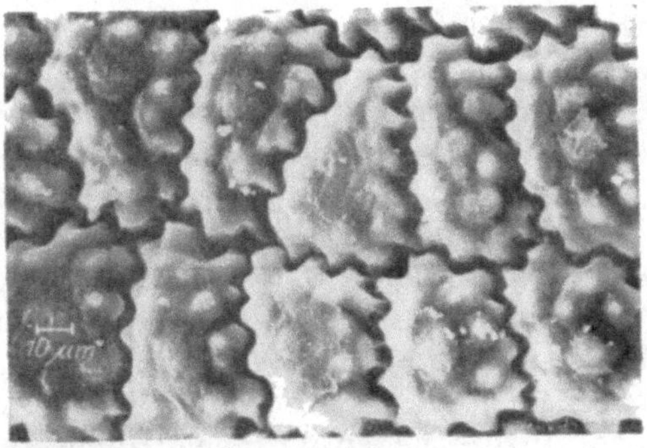

26

27

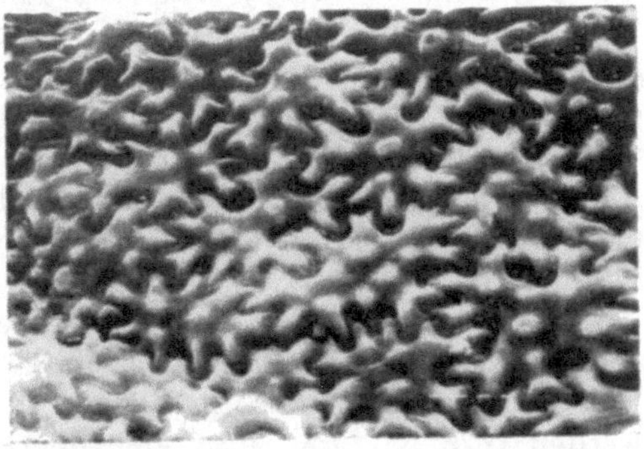

28

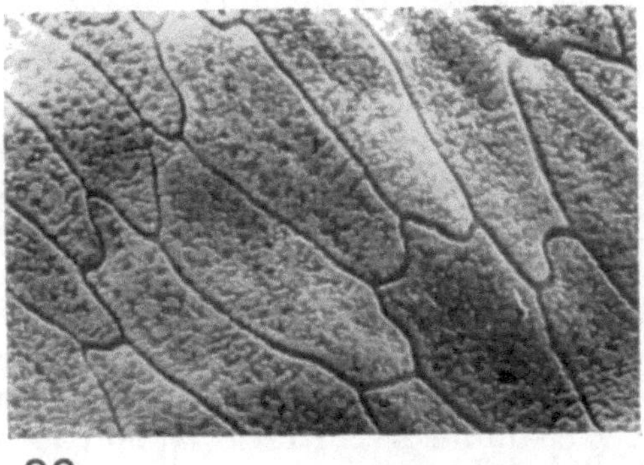

29

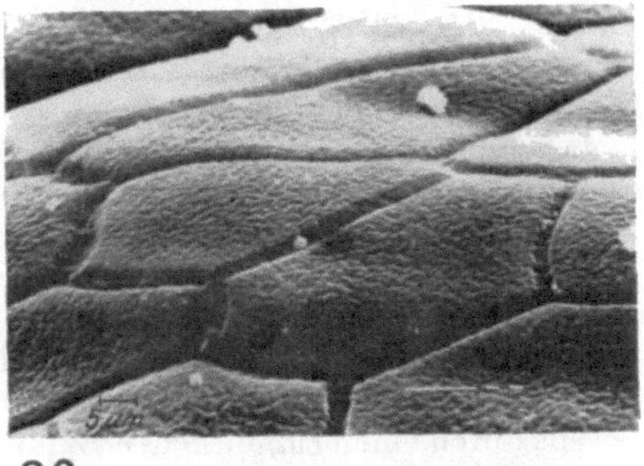

30

Tafel 5 Testa-Epidermiszellen. Abb. 25 *A. porrum* L. – Abb. 26 *A. scorodoprasum* L. – Abb. 27 *A. stipitatum* Rgl. – Abb. 28 *A. guttatum* Stev. – Abb. 29 *A. tuberosum* Rottl. – Abb. 30 *A. cyaneum* Rgl. Der Maßstab der Abb. 26 gilt auch für Abb. 27–29

unterschiedlicher Gruppierung zu Komplexen vereinigt, die, oft undeutlich gegeneinander abgegrenzt, über die gesamte Zelloberfläche verteilt sind. Das Antiklinalfeld dieses Zelltyps beschränkt sich auf einen sehr schmalen Saum.

Eine gewisse Ähnlichkeit in der Oberflächenskulptur besitzt das zur selben Sektion gehörende *A. cyaneum* Rgl. (Abb. 30). Das ebene, granulöse Zentralfeld wird hier ebenfalls von einem sehr schmalen Antiklinalfeld umgrenzt. Im Gegensatz zu den ersteren Arten sind die einzelnen Grana zu kurzen, mehr oder weniger deutlich ausgeprägten Reihen angeordnet, die durch ihren unregelmäßigen Verlauf eine schwach gekräuselte Oberflächenskulptur ergeben.

Diese Untersuchungen bestätigen die ausgeprägte Vielfalt der Testamerkmale in der Gattung *Allium* L. Neben gruppen- und artspezifischen Besonderheiten in der Wölbung, Form und Größe der Testazellen wurden Unterschiede in den Skulpturmustern sowie der relativen Größe des Zentralfeldes und des Antiklinalbereichs beobachtet. Während der Antiklinalbereich meist durch faltenartige, seltener durch glatte Skulpturen gekennzeichnet ist, treten im Bereich des Zentralfeldes neben glatten vor allem granulöse und verrucose Skulpturen auf. Weitere Besonderheiten ergeben sich durch die mögliche Bildung eines Zentralwulstes.

Bei einzelnen Arten wurde eine deutliche Variabilität der Skulpturmerkmale festgestellt, wobei verschiedene Herkünfte einer Art durch ähnliche formverwandte Skulpturtypen (*A. senescens* L.) oder durch sippenspezifische Unterschiede in der deutlichen Ausprägung eines bestimmten Skulpturmusters (*A. flavum* L.) gekennzeichnet sein können.

Mehrere der beobachteten Skulpturtypen treten bei verschiedenen Sektionen gleichzeitig auf. So ist der in der Sektion *Schoenoprasum* DUMORT. häufig vorkommende granulöse Typ auch in der Sektion *Rhizirideum* G. DON ex KOCH vertreten. Ebenso konnte der durch einen Zentralwulst und zentralen Höcker charakterisierte Testatyp sowohl in der Sektion *Rhizirideum* G. DON ex KOCH als auch an der Sektion *Molium* G. DON ex KOCH nahestehenden Arten nachgewiesen werden. Ein weiterer Typ kennzeichnet Arten der Sektionen *Allium* L. und *Megaloprason* WENDELBO. Dagegen scheinen bestimmte Muster auf einzelne Arten (z. B. *A. ursinum* L., *A. nerinifolium* BAK., *A. ericetorum* THORE, *A. oreophilum* C. A. MEY.) oder kleinere Artengruppen (z. B. *A. tuberosum* ROTTL. und *A. ramosum* L.) beschränkt zu sein, wodurch z. T. die Sonderstellung verschiedener Sippen bestätigt wird.

Andererseits läßt sich feststellen, daß mehrere Sektionen Arten mit unterschiedlichem Testatyp umfassen. Wie auch die Untersuchungen von PASTOR (1981) zeigen, trifft dieses in besonderem Maße für die sehr vielgestaltige Sektion *Rhizirideum* G. DON ex KOCH zu, in der zahlreiche Testatypen vorkommen. Eine ähnliche Heterogenität der Sektionen stellte auch FILIMONOVA (1971) hinsichtlich der Samenformen der Gattung *Allium* L. fest. Dagegen scheinen die Sektionen *Codonoprasum* REICHENB. und *Allium* L. nach den bisherigen Kenntnissen jeweils durch einen einheitlichen Skulpturtyp charakterisiert zu sein, worauf auch PASTOR (1981) hinweist. Dieses gilt auch in gewissem Umfang für die Sektion *Schoenoprasum* DUMORT., von der zahlreiche Arten eine einheitliche Testaskulptur besitzen. Die von PASTOR (1981) vermutete Einheitlichkeit dieser Sektion ist jedoch durch das andersartige Skulpturmuster von *A. cepa* L. nicht gegeben.

Die Untersuchungen machen deutlich, daß die Merkmale der Testaskulptur bei einer Reihe von Arten aussagekräftige Hinweise auf bestimmte Verwandtschaftskreise zu geben vermögen. Dabei bleibt jedoch zu klären, ob die einheitliche Skulpturform von Arten unterschiedlicher Sektionen auf Parallelentwicklungen beruht oder als Hinweis auf eine engere Verwandtschaft der entsprechenden Arten gewertet werden muß.

Summary

SEM investigations on seeds of the genus *Allium* L.

The sculpture characters of seed coats of different *Allium* species have been examined by means of SEM. The investigations confirm the great variability of this character complex. The characters of the testa sculpture in different sections of the genus as well as problems of infraspecific variability are discussed.

Краткое содержание

Исследование семян рода *Allium* L. с помощью сканирующего электронного микроскопа

Скульптурные признаки поверхности семян различных видов рода *Allium* L. изучались с помощью сканирующего электронного микроскопа. Исследования подтвердили сильную изменчивость этого комплекса признаков. Обсуждаются вопросы проявления признаков скульптуры семенной кожуры в различных секциях рода и проблемы внутривидовой изменчивости.

Die elektronenmikroskopischen Arbeiten wurden im Institut für Phytopathologie der AdL der DDR in Aschersleben durchgeführt. Der Leitung dieses Institutes danke ich für die freundliche Unterstützung. Mein besonderer Dank gilt Herrn Dr. H. B. Schmidt für die beratende Betreuung sowie Fräulein M. Müller für die technische Durchführung der elektronenmikroskopischen Arbeiten.

Literatur

Barthlott, W., und N. Ehler, 1977: Raster-Elektronenmikroskopie der Epidermis-Oberflächen von Spermatophyten. – Akad. Wiss. Lit. Mainz. Math.-Nat. Klasse. Trop. Subtrop. Pflanzenwelt **19**, 1–105.

Bothmer, R. von, 1974: Studies in the Aegean Flora. XXI. Biosystematic studies in the *Allium ampeloprasum* complex. – Opera Botanica (Lund) **34**, 1–104.

Filimonova, Z. N., 1971: Morfologija semjan sredneaziatskich vidov r. *Allium* L. – Introduksija Akklim. rast. (Taschkent, Uzbek. SSR) **8**, 111–116.

Furst, G. G., 1968: Struktura kožury semeni u raznych vidov i sortov luka. – Bjull. Glav. Bot. Sad. (Moskva) **69**, 55–60.

Pastor, J., 1981: Contribución al estudio de las semillas de las especies de *Allium* de la Península Ibérica e Islas Baleares. – Lagascalia (Sevilla) **10**, 207–216.

Stearn, W. T., 1978: European species of *Allium* and allied genera of Alliaceae: a synonymic enumeration. – Ann. Musei Goulandris **4**, 83–198.

Dr. J. Kruse
Zentralinstitut für Genetik und Kulturpflanzenforschung
der Akademie der Wissenschaften der DDR
DDR - 4325 Gatersleben
Corrensstraße 3

Kulturpflanze 32 · 1984 · 103–122

Variation of seed characters in selected *Vicia* species

PIETRO PERRINO[1], MARGARET YARWOOD[1], PETER HANELT
and G. BATTISTA POLIGNANO[1]

(Eingegangen am 10. Januar 1984)

Summary

The variability of 19 seed characters has been examined for 52 species of the genus *Vicia*. Seed circumference, relative hilum length, hilum shape and lens position could be confirmed as the most relevant characters for grouping. By cluster analyses based upon them the material has been classified into 11 groups. Discriminant analyses gave 25 % misclassifications when all 19 variables were subjected to the analysis for the determination of individual seeds. A key for determination of the species has been proposed mainly by means of quantitative seed characters; dendrograms illustrate the result of the cluster analyses and seed diagrams show schematically important seed characters.

Introduction

The genus *Vicia* with approximately 150 species (KUPICHA 1976) has several members of economic importance; the broad bean (*V. faba*) and the common vetch (*V. sativa*) are of particular interest. Other species, cultivated in certain parts of the world for human consumption, or as forage crops include *V. ervilia*, *V. benghalensis*, *V. villosa*, *V. pannonica* and *V. narbonensis*. Many others are regarded as weeds or contaminants (GUNN, 1970).

An infrageneric classification of *Vicia* has been recently proposed by KUPICHA (1976) although the rank or delimitation of many taxa is still being under question (compare also BALL, 1968). Any means of clarifying the taxonomic position would be of value to botanists and agriculturalists. In addition, a thorough evaluation of the genetic variability present within the genus is required if *Vicia* material is to be utilized efficiently in plant breeding programmes.

Vicia taxonomists have traditionally concentrated on floral and leaf morphology, paying little attention to seed characters. IANNELLI (1964) discussed the need for accurate identification of seed lots due to the increasing agricultural use of vetches and he found that species could be recognized using certain seed characters. GUNN (1970, 1971) worked more extensively on seed identification, forming seed diagrams for 100 species (1970) and a key for 36 North American

[1] Istituto del Germoplasma, Via G. Amendola 165/A, I-70 126 Bari, Italy

taxa (1971); he summarized also the pertinent literature in which seed characters had been evaluated in *Vicia* for taxonomic or identification purposes, mostly on a regional basis.

It is evident that a reliable species classification by seed characters would be a valuable time-saving technique in taxonomy, as seeds are simple to handle, durable, have a relatively long "herbarium" life, and are easy to conserve for quick cross-referencing.

It was the aim of the present work:

a) to evaluate and quantify the variation in a representative portion of the *Vicia* seed collection at Bari;
b) to determine the efficiency of GUNN's key (1971) for the identification of a larger range of species;
c) to examine the value of numerical analyses in seed identification at species level;
d) to record the seed which will subsequently be grown and investigated individually;
e) to try to present a key which could aid seed analysts in the identification of *Vicia* species.

Materials and methods

Laboratory work

The analysed material forms part of the *Vicia* collection conserved at the Germplasm Institute, Bari. 165 accessions, representing 52 species (Tab. 1) were selected. Where possible, four accessions varying in geographical origin, were chosen per species, and within each accession ten seeds were randomly selected and examined under the microscope.

Nineteen seed characters were considered (Tab. 2); these generally follow the features used by GUNN (1970, 1971). Metrical values were taken when possible; classes were used in the other cases.

Specimens authentically for each accession were checked using existing keys (BALL 1968; DAVIS and PLITMANN, 1970; SCHÄFER, 1973; IANNELLI, 1964; GUNN, 1970), on field data from seedling, adult plants, flowers, pods and also some chromatography results (unpublished data from P. PERRINO). In some cases, however, the seeds failed to germinate, and thus only seed data were available.

Table 1

List of analyzed *Vicia* species and accessions

Species	Access. No. (M. G.)	Donor
Vicia amoena FISCH. in DC.	106176	ZIGuK, Gatersleben
Vicia amurensis OETT.	106177	ZIGuK, Gatersleben
Vicia amphicarpa L.	106174	ZIGuK, Gatersleben
Vicia anatolica TURRILL	104612	BENNETT, FAO Coll. Rome
	104614	BENNETT, FAO Coll. Rome
	104618	BENNETT, FAO Coll. Rome
Vicia articulata HORNEM.	105177	Ist. Germopl., Bari
	105616	Ist. Germopl., Bari
	105776	Ist. Germopl., Bari
	105777	Ist. Germopl., Bari

Continued Table 1

Species	Access. No. (M. G.)	Donor
Vicia benghalensis L.	104746 [1]	BENNETT, FAO Coll., Rome
	104759 [1]	BENNETT, FAO Coll., Rome
	104762 [1]	BENNETT, FAO Coll., Rome
	105158 [1]	Ist. Germopl., Bari
	106140 [1]	Ist. Germopl., Bari
	104518	BENNETT, FAO Coll. Rome
	104530	BENNETT, FAO Coll. Rome
	104533	BENNETT, FAO Coll. Rome
	104535	BENNETT, FAO Coll. Rome
Vicia biennis L.	106054	Ist. Germopl., Bari
	106182	ZIGuK, Gatersleben
Vicia bithynica L.	105593	Ist. Germopl., Bari
	105648	Ist. Germopl., Bari
	105672	Ist. Germopl., Bari
	105681	Ist. Germopl., Bari
Vicia cassubica L.	106189	ZIGuK, Gatersleben
Vicia cordata WULF.	104382	BENNETT, FAO Coll. Rome
	104382	BENNETT, FAO Coll. Rome
	104383	BENNETT, FAO Coll. Rome
	104382	BENNETT, FAO Coll. Rome
	105290	Ist. Germopl., Bari
	105792	Ist. Germopl., Bari
	105970	Ist. Germopl., Bari
Vicia cracca L.	106187	ZIGuK, Gatersleben
Vicia dalmatica A. KERNER	106194	ZIGuK, Gatersleben
Vicia dasycarpa TEN.	105161	Ist. Germopl., Bari
	106267	Natl. Bur. Pl. Genet. Res., Delhi
	105601	Ist. Germopl., Bari
	105607	Ist. Germopl., Bari
Vicia disperma DC.	105621	Ist. Germopl., Bari
	106195	ZIGuK, Gatersleben
Vicia dumetorum L.	106193	ZIGuK, Gatersleben
Vicia eriocarpa (HAUSSKN.) HALACSY	106199	ZIGuK, Gatersleben
	106200	ZIGuK, Gatersleben
	106201	ZIGuK, Gatersleben
Vicia ervilia (L.) WILLD.	105205	Ist. Germopl., Bari
	106264	Natl. Bur. Pl. Genet. Res., Delhi
	105622	Ist. Germopl., Bari
	105623	Ist. Germopl., Bari
Vicia faba L.	103210	Ist. Germopl., Bari
	103242	Ist. Germopl., Bari
	103396	Svalöf AB, Sweden
	103518	Pl. Genet. Res. Centre Addis Abeba
	106262	Natl. Bur. Pl. Genet. Res., Delhi
	106263	Natl. Bur. Pl. Genet. Res., Delhi
	106152	INIA, Madrid
	106153	INIA, Madrid
	106154	INIA, Madrid
	106155	INIA, Madrid

Continued Table 1

Species	Access. No. (M. G.)	Donor
	106156	Hebr. Univ., Jerusalem
	106157	INIA, Madrid
	106158	INIA, Madrid
	106159	INIA, Madrid
	106160	INIA, Madrid
	106161	INIA, Madrid
	106162	INIA, Madrid
	106163	INIA, Madrid
	106164	INIA, Madrid
	106165	INIA, Madrid
Vicia galilaea PLITM. et ZOH.	106204	ZIGuK, Gatersleben
Vicia gigantea HOOK.	106205	ZIGuK, Gatersleben
Vicia grandiflora SCOP.	105781	Ist. Germopl., Bari
	105782	Ist. Germopl., Bari
	105783	Ist. Germopl., Bari
Vicia hirsuta S. F. GRAY	105620	Ist. Germopl., Bari
	105739	Ist. Germopl., Bari
	105771	Ist. Germopl., Bari
Vicia hybrida L.	105561	Ist. Germopl., Bari
	105562	Ist. Germopl., Bari
	105568	Ist. Germopl., Bari
	105645	Ist. Germopl., Bari
Vicia hyrcanica F. et M.	106211	ZIGuK, Gatersleben
Vicia incana GOUAN	106212	ZIGuK, Gatersleben
Vicia incisa ARCANGELI	105872	Ist. Germopl., Bari
	106028	Ist. Germopl., Bari
Vicia incisaeformis STEF.	106213	ZIGuK, Gatersleben
Vicia johannis TAMAMSCH.	106290	ZIGuK, Gatersleben
	106293	ZIGuK, Gatersleben
	106214	ZIGuK, Gatersleben
	106215	ZIGuK, Gatersleben
Vicia lathyroides L.	106218	ZIGuK, Gatersleben
Vicia lutea L.	105567	Ist. Germopl., Bari
	105565	Ist. Germopl., Bari
	105571	Ist. Germopl., Bari
	105572	Ist. Germopl., Bari
Vicia macrocarpa (MORIS) BERTOL.	106220	ZIGuK, Gatersleben
	105899	Ist. Germopl., Bari
	105969	Ist. Germopl., Bari
Vicia megalotropis LEDEB.	106222	ZIGuK, Gatersleben
Vicia melanops SIBTH. et SM.	104608	BENNETT, FAO Coll. Rome
	104736	BENNETT, FAO Coll. Rome
	105037	BENNETT, FAO Coll. Rome
	105246	Ist. Germopl., Bari
Vicia michauxii SPRENG.	106219	ZIGuK, Gatersleben
Vicia monantha RETZ.	105160	Ist. Germopl., Bari
	105160	Ist. Germopl., Bari
	105252	Ist. Germopl., Bari

Contin ued Table 1

Species	Access. No. (M. G.)	Donor
	106221	ZIGuK, Gatersleben
	105576	Ist. Germoplasm, Bari
	105651	BENNETT, FAO Coll. Rome
	105652	BENNETT, FAO Coll. Rome
Vicia narbonensis L.	105219	Ist. Germopl., Bari
	106266	Natl. Bur. Pl. Genet. Res., Delhi
	106288	ZIGuK, Gatersleben
	106289	ZIGuK, Gatersleben
	106292	ZIGuK, Gatersleben
	106294	ZIGuK, Gatersleben
	105751	Ist. Germoplasm, Bari
	105763	Ist. Germoplasm, Bari
	105773	Ist. Germoplasm, Bari
Vicia neglecta HANELT et METTIN	106231	ZIGuK, Gatersleben
Vicia orobus DC.	106188	ZIGuK, Gatersleben
Vicia pannonica CR. ssp. *striata* (BIEB.) NYM.	104732	BENNETT, FAO Coll. Rome
	104762	BENNETT, FAO Coll. Rome
	104623	BENNETT, FAO Coll. Rome
	104631	BENNETT, FAO Coll. Rome
	104633	BENNETT, FAO Coll. Rome
	104652	BENNETT, FAO Coll. Rome
	104670	BENNETT, FAO Coll. Rome
	104672	BENNETT, FAO Coll. Rome
	104742	BENNETT, FAO Coll. Rome
Vicia pannonica CR. ssp. *pannonica*	105255	Ist. Germoplasm, Bari
Vicia peregrina L.	104454	BENNETT, FAO Coll. Rome
	104696	BENNETT, FAO Coll. Rome
	104697	BENNETT, FAO Coll. Rome
	105277	Ist. Germoplasm, Bari
Vicia pisiformis L.	106241	ZIGuK, Gatersleben
Vicia pseudorobus F. & M.	106240	ZIGuK, Gatersleben
Vicia pubescens LINK	106239	ZIGuK, Gatersleben
Vicia sativa L.	104369	BENNETT, FAO Coll. Rome
	104724	BENNETT, FAO Coll. Rome
	104979	BENNETT, FAO Coll. Rome
	105159	BENNETT, FAO Coll. Rome
	105238	Ist. Germoplasm, Bari
	105697	Ist. Germoplasm, Bari
	105697	Ist. Germoplasm, Bari
	106265	Natl. Bur. Pl. Genet. Res., Delhi
	105272	Ist. Germoplasm, Bari
	105283	Ist. Germoplasm, Bari
	105337	Ist. Germoplasm, Bari
	105567	Ist. Germoplasm, Bari
Vicia sativa var. *platysperma* BARUL.	105860	Ist. Germoplasm, Bari
Vicia semiglabra RUPR. ex BOISS.	106254	ZIGuK, Gatersleben
Vicia serratifolia JACQ.	106251	ZIGuK, Gatersleben
	106252	ZIGuK, Gatersleben
	106291	Ist. Germoplasm, Bari

Continued Table 1

Species	Access. No. (M. G.)	Donor
Vicia sylvatica L.	106253	Ist. Germoplasm, Bari
Vicia tenuifolia ROTH	106255	ZIGuK, Gatersleben
	106256	ZIGuK, Gatersleben
Vicia tenuissima (BIEB.) SCHINZ et THELL.	106257	ZIGuK, Gatersleben
Vicia tetrasperma (L.) SCHREB.	106258	ZIGuK, Gatersleben
	105795	Ist. Germoplasm, Bari
	105836	ZIGuK, Gatersleben
Vicia unijuga A. BR.	106259	ZIGuK, Gatersleben
Vicia villosa ROTH	105602	Ist. Germoplasm, Bari
	105605	Ist. Germoplasm, Bari
	105706	Ist. Germoplasm, Bari

1 received as *V. atropurpurea* DESF.

Table 2

Seed characters scored for species classification

Seed

1. Shape (spherical, sub-spherical, oval, oblong, sub-lenticular, lenticular, sub-pyramidal, sub-cuboidal, cuboidal)
2. Length in mm
3. Breadth in mm
4. Thickness in mm
5. Circumference in mm
6. Background colour of seed coat
7. Presence and colour of spots on coat
8. Presence and colour of mottles on coat
9. Presence and colour of streaks on coat
10. Glabrous or velvetoid seed coat
11. Smooth or rugose seed coat

Hilum

12. Colour of center of hilum
13. Colour of border of hilum
14. Shape (oval, cuneiform, oblong, linear, circum-linear)
15. Length in mm
16. Length expressed as percentage of total seed circumference

Lens (Chalaza)

17. Distance of lens center to hilum base in mm
18. Colour
19. Degree of prominence

Statistical analysis

Computational analyses were undertaken using the Statistical Analysis System (SAS), developed by BARR et al. (1972, 1976).

The average and range of each variable for each species were calculated.

Cluster and discriminant analyses were applied to investigate species relationships. Cluster analysis, as determined by SAS, computes a distance matrix in Euclidean terms. Each observation is a cluster and, on a hierarchical basis, the two closest observations (clusters) are merged on the basis of the distances calculated from the variables under consideration.

This analysis was performed firstly by accessions and secondly, by species. Originally all nineteen seed characters were subjected to analysis. This number was successively reduced to see if there was a modification in grouping when specific seed characters were selected.

Whilst cluster analysis serves to produce groups based on a measure of similarity, subjection of data to discriminant analysis should in theory, provide a method of identification at an individual level. In practise, numerical classification using discriminant functions has not proved successful: CHISCI and MARTIN (1966) and BOGYO and PORCEDDU (1975), working on intra-specific identification reported that this technique could not discriminate well.

In the present work discriminant analysis was applied to examine if this technique could accurately identify species. To test such an approach it is a critical prerequisite to have correctly identified taxa before carrying out the analysis, so that the degree of accuracy, or efficiency, may be calculated. Again the analysis was carried out using all variables, and then a reduced number. The number of observations (individual seeds) was too great (1553) for discriminant analysis to handle in calculating the identification parameters for each species, thus the data were "calibrated": only one in four individuals were "test-data". To see if this produced a difference in classification, the analysis was repeated using the sequence of seed number 2, 6, 10 . . . instead of 1, 5, 9 . . . etc. (compare tab. 3, discrimination a resp. b). The classification criterion of SAS calculates the generalized square distance, based on individual within-group covariance matrices, and, in this case, prior probability was considered. (For further details see BARR et al., 1976; BOGYO and PORCEDDU, 1975.)

Results

Cluster analysis of accessions and that of species, using all variables, produced an overall similarity in the grouping patterns. The optimum number of groups formed was 22 with accessions, and 8 with species. However, the general picture showed that accessions of the same species preferred to cluster with themselves rather than have a wide distribution in different groups. The accessions of species traditionally considered to have a high degree of variation (BALL, 1968), for example *V. sativa,* were present in more than one cluster group.

Variation within species, based on group compilation, was low in comparison to the variation exhibited between species, and thus it was decided to concentrate on groupings, using species as the observation; amalgamation of accessions of one species gave a simplified result, but not a distortion.

It was considered relevant to determine if every variable carried equal weighting in the production of the 8 cluster groups of the 52 species. An analysis of variance was thus performed using the means of the variables for each species and the eight groups. From the mean squares, the percentage of between and within group variation for each variable was calculated. For a variable to be important in group construction, the amount of variation *between* groups must be high, and conversely, the proportion of *within* group variation must be low. Thus for group construction a high degree of constancy is necessary.

Figure 1 demonstrates that eight variables were found to be significant when an arbitrary cut-off point of 50 % within-group variation is used. The remaining eleven variables display their diversity within groups.

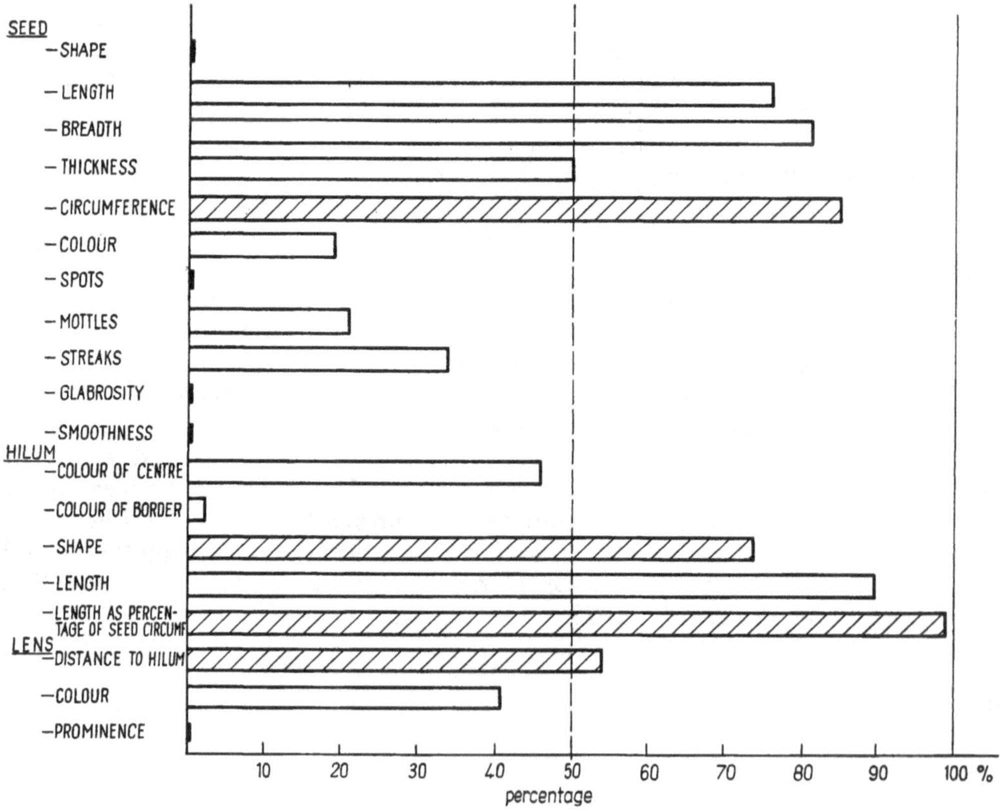

Fig. 1 Variation in percentage of 19 variables between groups

It was decided that the above eight variables, length, breadth, thickness, circumference, hilum length, likewise as a percentage of the seed circumference, hilum shape and the relative position of the lens from the hilum, could be regarded as four truly different characters, since the seed circumference summarizes seed size (length, breadth and thickness) whilst the relative hilum length as a percent-

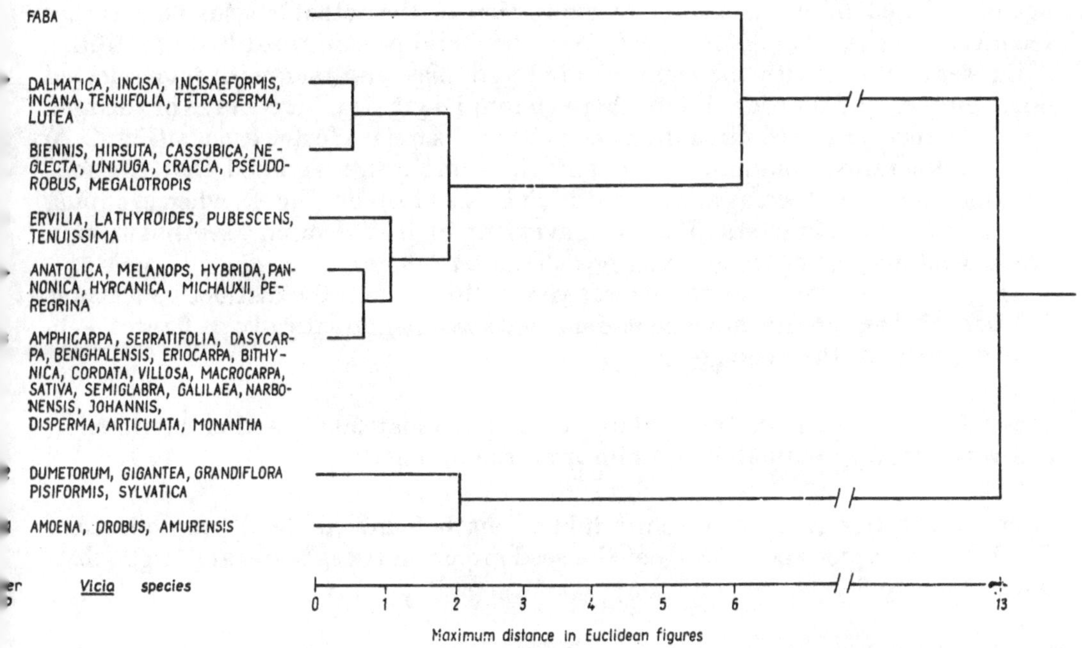

Fig. 2 Cluster groups using all 19 seed variables

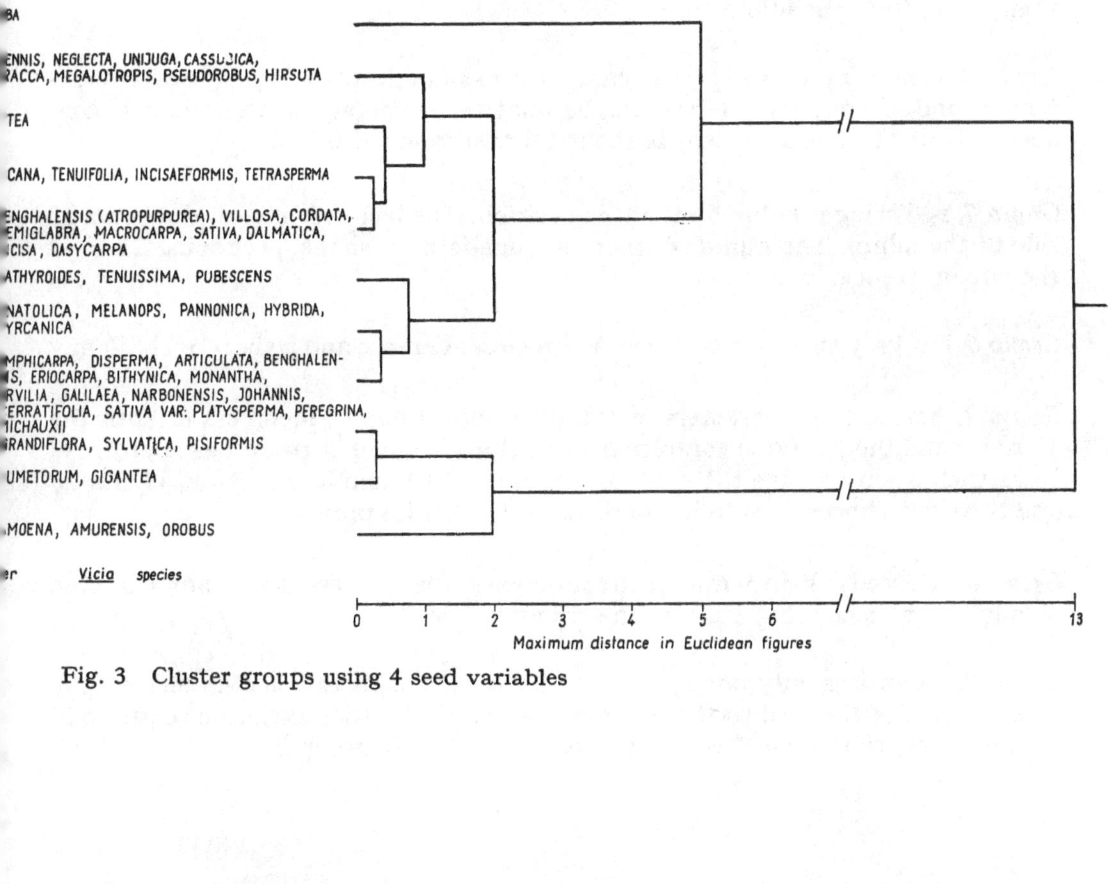

Fig. 3 Cluster groups using 4 seed variables

age of the seed circumference is an expression of the actual length. Thus earlier records on the value of distinct seed character could be confirmed (GUNN 1970).

Cluster analyses with the eight selected variables, and then with four selected ones, for i.e. lens position, hilum shape, hilum length as a percentage of the seed circumference, and seed circumference produced similar cluster groups (Fig. 2, 3) and dendrograms, obtaining practically the same result signified that the four variables could be used as accurately as eight, or nineteen (Fig. 2) when grouping species on seed characters. This is convenient in that a rapid assessment at a group level may be made utilizing few characters alone.

The optimum number of groups was eleven (Fig. 3) and the distinctive features for each of these groups have been demonstrated diagramatically in figures 4–6. Values given are the average on a group basis.

Group 1, is characterized by a hilum occupying about 50 % seed circumference, and with lens approximately 1,4 mm from the hilum.

Group 2, has the greatest absolute hilum length found in the *Vicia* collection. The hilum occupies about 75 % of the seed circumference. Seeds are larger than those of group 1.

Group 3, has a hilum occupying about 76 % of the seed circumference and a seed size similar to group 1 (10 mm circumference). In group 2 and 3 the lens is less than 1 mm from the hilum.

Group 4, consisting of 15 species shows a considerable variation in seed circumference and hilum shape, however, the relative percentage of the hilum is constant, about 11 % and the lens is about 1,4 mm from the hilum.

Group 5, is distinguishable by the lens position. The lens is found on the opposite side to the hilum. The hilum is generally cuneiform in shape, occupying 11 % of the circumference.

Group 6, has very small seeds, about 5 mm circumference and a short oval hilum.

Group 7, has certain characters in common with group 4, members of both the *V. sativa* and the *V. villosa* complex being included as well in group 7 as in group 4! Nevertheless, on average the seeds are smaller, the hilum longer (16 % cf. 11 %) and there is a shorter lens to hilum distance (1,1 cf. 1,4 mm).

Group 8, has seeds about 9 mm in circumference, the lens less than 1 mm from the oblong hilum, occupying 21 % of the circumference.

Group 9, comprises only one species, *V. lutea*. Although this has its lens on the opposite side of the seed to the hilum a character otherwise exclusive to group 5, the hilum length is about 23 % as compared to 11 % in group 5.

Group 10, is similar to group 8 (both containing besides others taxa of the *V. cracca* complex), but the hilum is longer (about 30 %) and the lens is over 1 mm from the hilum.

Group 11 is *V. faba*. This is characterized by its large dimensions; the circumference averages 45 mm thus hilum and lens values are proportionally large.

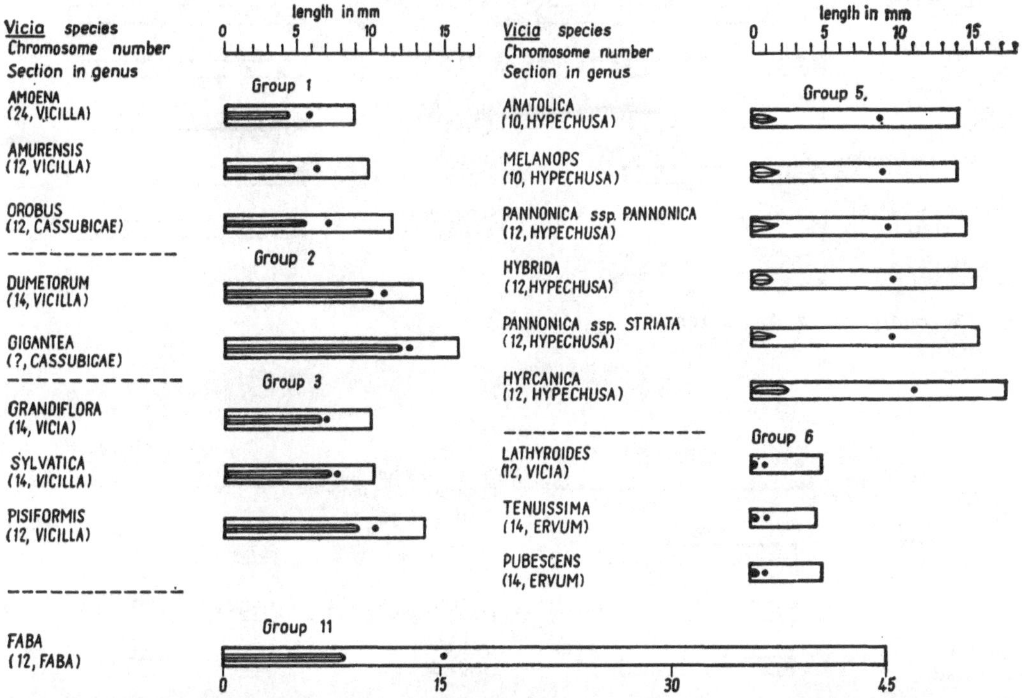

Fig. 4 Diagrams to illustrate relative seed circumference, hilum length and shape and lens position in each species by cluster groups; in brackets somatic chromosome numbers (2n) and section of the genus to which the species belongs: cluster 1–3, 5, 6, 11

Figures 4–6 also show in brackets chromosome number (2n; according to Bolchovskich et al., 1969, and earlier own studies) and section (Kupicha 1976) for each species. No clear trend is evident between groups formed by seed characters and the sections which are based mainly on floral and leaf morphology (but see e.g. group 5, made up only of taxa of sect. *Hypechusa*!). Further no distinct correlation can be recognized with respect to any seed groups and the number of chromosomes of each species within the group.

Results of discriminant analysis are summarized in Table 3. The amount of error or "misclassification" varied from 0–100 %; however, when all nineteen variables were subjected to the analysis the overall species misclassification was about 25 %. The result of the two discriminant analysis (a and b) showed little differences in total error (25 cf. 28). Over half the individuals were misclassified (51–54 %) when eight variables only were used (the selected variables were those shown to be important in cluster groupings, see figure 1).

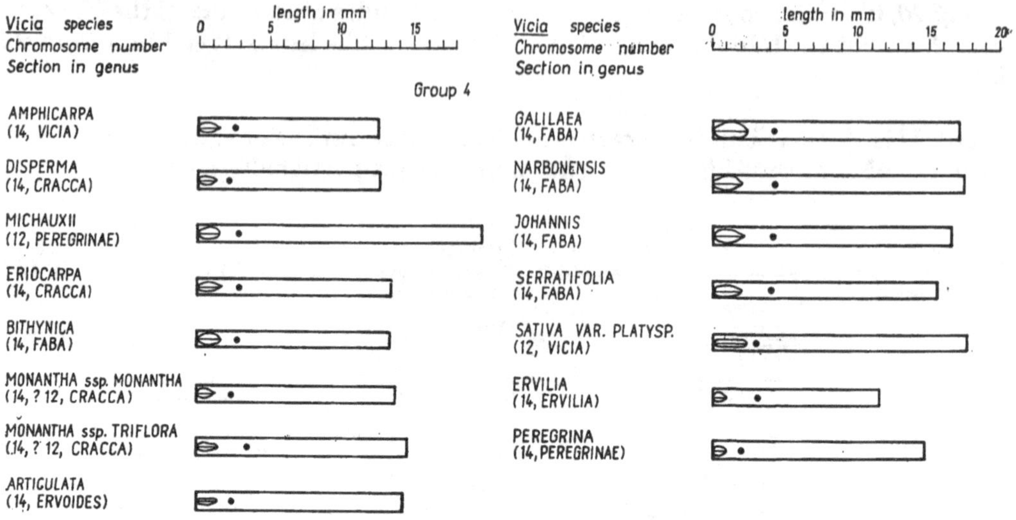

Fig. 5 compare fig. 4, cluster 4

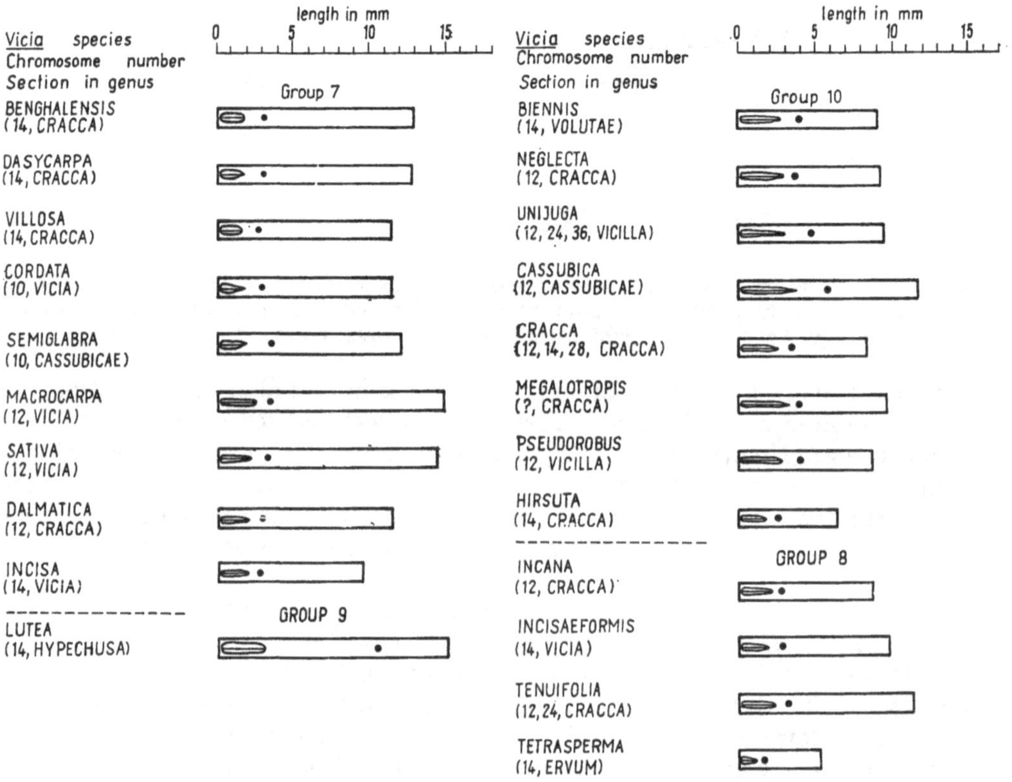

Fig. 6 compare fig. 4, clusters 7–10

The efficiency, or accuracy, of this method may thus be appreciated. It seems clear that the eight variables selected were insufficient for the analysis to correctly identify seeds (i.e. species), and the method may be regarded as a failure. With all nineteen seed characters discrimination into the correct species worked much better. Moreover, when all variables were used, the 25 % seeds misclassified were identified as species bearing certain affinities to the true one. The names given for species where misclassifications occured are shown also in Table 3. These errors may be explained mainly by the variability of species regarding seed characters partly by the taxonomic ambiguity of certain taxa. Whereever the species is clearly defined on general morphological characters, it seems to follow that also from discriminant analysis performed on seed data, that correct identification is probable. Difficulties in accurate discrimination occur in those species where discrepancies of taxonomic rank exist. This is true e.g. for the taxa of the *V. sativa*- and *V. villosa*-aggregate, here treated as species but ranked as subspecies elsewhere (e.g. by Ball 1968 in Flora Europaea).

Seed Key

Those characters more variable between groups of species were the most reliable ones for drawing a key. They were: seed circumference, relative hilum length, hilum shape and relative position of the lens to the hilum. Using relative hilum length the 52 species fell into 6 categories: very long, 75 %; long, 65 %; medium long, 50 %; medium short, 30 %; short, 20 %; and very short, 10 %. More than 50 % of the investigated species belong only to the last two categories. Particularly for those species it was hard to choose other additional characters. However, even at this level preference was given to quantitative characters (i.e. seed and hilum length, etc.). Only when species were similar on this ground qualitative characters were employed as further details for seed and species identification. Some species can be correctly identified only if the size of the sample to test is reasonable large. In dubious cases both ways of the dichotomous key should be followed. The practical value of the key is limited of course because the species considered cover only a part of the genus and represent in taxonomic and geographic respect a rather arbitrary selection. Inadequacies of the key result also from the infraspecific variability of seed characters which could not be wholly covered due to lack of material. In such cases the user of the key may be misled.

Seed key to *Vicia* species: 1 and (1), 2 and (2) etc are the 2 states of a character one of which has to be followed!

1. Hilum occupying more than 60 % of seed circumference . . . .
2. Hilum occupying 75 %, at least more than 65 % of seed circumference 3, (3)
3. Lens center about 1 mm from hilum; seed 4,5 mm in length, dark brown; hilum 10 mm in length and with a black border . . *V. dumetorum*

(3). Lens center 0,5 mm from hilum; seed 5 mm in length, dark brown and black mottled; hilum 12 mm in length and with a brown border attached at the hilum . . . . *V. gigantea*

(2). Hilum occupying about 65 % of seed circumference 4, (4)

Table 3

Discriminant Analyses – Number and Percentage of Individuals Misclassified

Species	Misclassification: Seed No.	Acc. No.	Discrimination (a) 19 Var. No.	%	8 Var. No.	%	Discrimination (b) 19 Var. No.	%	8 Var. No.	%	Classified (using misclassification for "all variables" only) as: –
Vicia amoena	9	1	2	22	3	33	1	11	2	20	*amurensis*
Vicia amurensis	10	1	2	20	2	20	1	10	4	11	*amoena, orobus*
Vicia amphicarpa	10	1	2	20	10	100	5	50	10	100	*cordata, disperma*
Vicia anatolica	30	3	6	20	21	70	9	30	19	63	*melanops, pannonica* ssp. *striata*
Vicia articulata	40	4	3	8	1	2	7	18	1	2	*serratifolia, benghalensis, lathyroides*
Vicia benghalensis (received as *atropurpurea*)	47	5	27	57	46	98	29	62	47	100	*amphicarpa, incisa, villosa, dasycarpa, cordata, bithynica, eriocarpa, sativa, disperma*
Vicia benghalensis	39	4	26	67	28	72	14	36	24	62	*villosa, dasycarpa, cordata, eriocarpa, sativa, amphicarpa*
Vicia biennis	20	2	6	30	9	45	5	25	6	30	*unijuga, cracca*
Vicia bithynica	40	4	12	30	32	80	14	35	28	70	*sativa, serratifolia, monantha, narbonensis, galilaea, ervilia, cordata*
Vicia cassubica	6	1	n.c.	n.c.	n.c.	n.c.	1	16	1	16	–
Vicia cordata	70	7	16	23	20	29	17	25	17	25	*villosa, sativa, dasycarpa, benghalensis*
Vicia cracca	10	1	3	33	3	33	6	60	5	50	*biennis, neglecta*
Vicia dalmatica	10	1	1	10	10	100	2	20	9	90	*incisaeformis*
Vicia dasycarpa	37	4	29	81	36	98	29	80	37	100	*cordata, sativa, benghalensis, eriocarpa, tetrasperma, serratifolia, villosa, dalmatica, incana*
Vicia disperma	20	2	17	85	15	75	18	90	20	100	*dasycarpa, eriocarpa, villosa, sativa, monantha*
Vici dumetorum	10	1	1	10	3	30	1	10	3	30	*pisiformis*
Vicia eriocarpa	30	3	17	57	30	100	16	53	28	93	*monantha, dasycarpa, benghalensis, villosa, sativa*
Vicia ervilia	39	4	0	0	4	10	0	0	1	3	–
Vicia faba	193	20	9	5	32	17	4	2	28	15	*villosa, narbonensis, macrocarpa*
Vicia galilaea	10	1	0	0	10	100	1	10	10	100	–
Vicia gigantea	4	1	n.c.	n.c.	n.c.	n.c.	n.c.	n.c.	n.c.	n.c.	–
Vicia grandiflora	30	3	0	0	8	27	0	0	0	0	–
Vicia hirsuta	16	3	5	31	5	31	3	19	2	20	*incana, biennis, cracca, unijuga*

Vicia incisa	20	2	7	35	16	80	7	35	16	80	*tetrasperma, incisaeformis, biennis, sativa, benghalensis*
Vicia incisaeformis	10	1	1	10	6	60	4	40	10	100	*incisa*
Vicia johannis	22	4	17	77	20	90	8	47	17	77	*bithynica, narbonensis, cordata, serratifolia*
Vicia lathyroides	10	1	0	0	4	40	0	0	7	70	–
Vicia lutea	36	4	5	14	1	3	7	19	1	3	*hyrcanica, melanops*
Vicia macrocarpa	27	3	21	78	27	100	25	92	27	100	*sativa, cordata, benghalensis, serratifolia, incisa, johannis, tetrasperma*
Vicia megalotropis	10	1	2	20	7	70	3	30	7	70	*biennis, neglecta*
Vicia melanops	39	4	16	40	37	95	22	56	38	95	*anatolica, pannonica, hybrida, cordata*
Vicia michauxii	10	1	0	0	0	0	0	0	0	0	–
Vicia monantha	31	3	22	71	31	100	21	68	31	100	*sativa, cordata, peregrina*
	30	3	1	3	16	54	4	13	22	73	*benghalensis*
Vicia narbonensis	68	8	20	29	13	19	17	25	17	19	*bithynica, serratifolia, johannis, sativa, cordata, galilaea*
Vicia neglecta	10	1	2	20	3	30	2	20	5	30	*cracca*
Vicia orobus	10	1	0	0	1	10	0	0	1	10	–
Vicia pannonica ssp. *p.*	10	1	1	10	8	80	2	20	9	80	*pannonica* ssp. *striata*
Vicia pannonica ssp. *str.*	79	8	4	5	16	20	4	5	14	20	*hyrcanica, melanops*
Vicia peregrina	38	4	1	3	5	13	6	16	5	13	*benghalensis, eriocarpa*
Vicia pisiformis	10	1	0	0	0	0	0	0	0	0	–
Vicia pseudorobus	9	1	7	78	8	88	8	88	4	88	*cracca, neglecta*
Vicia pubescens	9	1	4	44	2	22	3	33	3	22	*benghalensis, tenuissima*
Vicia sativa	112	11	26	23	21	18	25	22	19	18	*dasycarpa, monantha, articulata, incisaeformis, eriocarpa, macrocarpa, dalmatica, benghalensis*
Vicia sativa var. *platysperma*	10	1	0	0	5	50	0	0	4	40	–
Vicia semiglabra	9	1	2	22	8	88	2	22	9	88	*dasycarpa, sativa*
Vicia serratifolia	22	3	3	14	21	98	3	14	22	98	*narbonensis*
Vicia sylvatica	10	1	0	0	1	10	0	0	2	10	–
Vicia tenuifolia	19	2	1	5	11	58	3	16	11	58	*incana*
Vicia tenuissima	9	1	1	11	4	44	4	14	2	14	*dasycarpa*
Vicia tetrasperma	28	3	5	18	4	14	4	14	2	14	*incisaeformis, cordata, incana*
Vicia unijuga	9	1	1	11	3	33	2	22	8	33	*cracca*
Vicia villosa	30	3	8	36	30	100	11	37	30	100	*bithynica, macrocarpa, eriocarpa, cordata*
				$\bar{x}=25$ %		$\bar{x}=51$ %		$\bar{x}=28$ %		$\bar{x}=54$ %	

n. c. = not compared
(a) = first replicate
(b) = second replicate (see in the text for explanation)

4. Seed never more than 4 mm in length, sub-lenticular, lenticular, sub-spherical, oblong, oval; hilum never more than 8 mm in length; center of lens never more than 0,8 mm from hilum; lens raised 5, (5)
5. Seed rosy-brown, black or brown mottled, sub-lenticular or oblong; center of lens never more than 0,5 mm from hilum; hilum sometimes more than 70 % of seed circumference; sometimes remnants of funiculus along each side of hilum . . . . *V. grandiflora*
(5). Seed rosy-brown, occasionally mottled with brown paints, sub-spherical or oval; center of lens never less than 0,5 mm from hilum . *V. sylvatica*
(4). Seed never less than 4 mm in length, rosy-brown, subpyramidal, oval or sub-spherical; hilum never less than 9 mm in length; center of flat lens about 1,4 mm from hilum . . . . *V. pisiformis*
(1). Hilum occupying less than 60 % of seed circumference 6, (6)
6. Hilum occupying more than 25 % of seed circumference 7, (7)
7. Hilum occupying about 40–50 % of seed circumference 8, (8)
8. Seed never more than 4 mm in length; hilum never more than 5 mm in length 9, (9)
9. Seed rosy-brown, sub-cuboidal . . . . *V. amoena*
(9). Seed ochre or brown, sub-spherical . . . . *V. amurensis*
(8). Seed never less than 4 mm in length; hilum never less than 5 mm in length, seed rosy-brown . . . . *V. orobus*
(7). Hilum occupying about 30 % of seed circumference (in any case never less than 25 and never more than 35 %) 10, (10)
10. Seed usually more than 3 mm in length 11, (11)
11. Lens center flat 12, (12)
12. Lens center never less than 2 mm from hilum; seed mostly more than 3 mm in length, dark brown, mottled . . . . *V. cassubica*
(12). Lens center 1–2 mm from hilum; seed 3 mm or less in length, black and/or brown mottled . . . . *V. neglecta*
(11). Lens center raised 13, (13)
13. Hilum usually more than 2 mm in length, brown; seed about 3 mm long, dark brown, often brown mottled . . . . *V. megalotropis*
(13). Hilum 2 mm in length, black-brown; seed about 2 to 2,3 mm long, black-brown, dark mottled . . . . *V. unijuga*
(10). Seed usually 3 mm or less in length (but compare *V. biennis*) 14, (14)
14. Hilum usually less than 3 mm but more than 2 mm in length 15, (15)
15. Lens center never less than 1 mm from hilum and sometimes seed a little more than 3 mm in length . . . . *V. biennis*
(15). Lens center usually 1 mm or less from hilum 16, (16)
16. Seed oval, light-brown; lens raised or flat . . . . . *V. pseudorobus*
(16). Seed sub-spherical or oblong, darker (dark-brown, dark-blue or olive-brown); lens always raised . . . . *V. cracca*
(14). Hilum about 2 mm in length; seed at least 2 mm in length; funiculus usually attached at one end of hilum . . . . *V. hirsuta*
(6). Hilum occupying less than 25 % of seed circumference 17, (17)
17. Hilum occupying about 20 % of seed circumference (with few exceptions, never less than 15 % and never more than 25 %) (compare also (54): *V. melanops* and (19): *V. faba*) 18, (18)
18. Lens center on the opposite side of seed from center of hilum; seed oval, sub-spherical, sub-cuboidal, sub-linear and from 3 to 7 mm in length . . . . *V. lutea*
(18). Lens center near hilum 19, (19)
19. Seed never more than 5 mm in length 20, (20)
20. Hilum usually 2 mm and less in length 21, (21)
21. Seed coat velvet or glabrous; seed varies from 2 to 4 mm in length . . . . *V. incisa*

(21). Seed coat always glabrous; seed never more than 2 mm in length . . . . *V. tetrasperma*
(20). Hilum usually more than 2 mm in length; seed never less than 2,5 mm in length 22, (22)
22. Lens center depressed; seed about 4 mm in length; hilum sometimes more than 25 % of seed circumference . . . . *V. tenuifolia*
(22). Lens center raised; seed about 3 mm in length or somewhat more . . . . *V. incana*
(19). Seed more than 4,5 mm in length; hilum sometimes less than 15 % of seed circumference . . . . *V. faba*
(17). Hilum occupying about 10 %, more rarely up to 15 %, of seed circumference (but compare 31: *V. macrocarpa* and 43) 23, (23)
23. Lens center near hilum 24, (24)
24. Lens center 1 mm or less from hilum 25, (25)
25. Seed never more than 2–3 mm in length, lens center no more than 0,6 mm from hilum 26, (26)
26. Seed cuboidal or sub-cuboidal, creme, greyish or pink; seed coat tuberculate . . . . *V. lathyroides*
(26). Seed sub-spherical or oval; seed coat smooth 27, (27)
27. Seeds at least 2 mm in length . . . . *V. tenuissima*
(27). Seeds 2 mm or somewhat more in length . . . . . *V. pubescens*
(25). Seed usually more than 3 mm in length 28, (28)
28. Seed usually less than 5 mm in length 29, (29)
29. Hilum usually 2 mm or less in length 30, (30)
30. Hilum cuneiform or oval; seed glabrous and never less than 3 mm in length . . . . *V. disperma*
(30). Hilum oval to cuneiform, seed never less than 2,5 mm in length . . . . *V. villosa*
(29). Hilum 2 mm or more in length, hilum about 15 % of seed circumference 31, (31)
31. Seed oval, brown or mottled, never less than 3 and never more than 5 mm in length; hilum seldom less than 15 % of seed circumference . . . . *V. dalmatica*
(31). Seed sub-spherical or oblong; seed about 5 mm in length; hilum from 10 to 20 % of seed circumference, often less than 15 % . . . . *V. macrocarpa*
(28). Seed usually more than 5 mm in length 32, (32)
32. Hilum never less than 10 % of seed circumference; seed lenticular or sublenticular and never more than 7 mm in length . *V. sativa* var. *platysperma*
(32). Hilum never more than 10 % of seed circumference; seed oblong and never less than 6,4 mm in length . . . . *V. michauxii*
(24). Lens center 1 mm or more from hilum 33, (33)
33. Seed shape subpyramidal . . . . *V. ervilia*
(33). Seed shape otherwise, never subpyramidal 34, (34)
34. Hilum never more than 2 mm in length (compare also (34)) 35, (35)
35. Hilum seldom more than 10 % of seed circumference 36, (36)
36. Seed never less than 4 mm in length, sublenticular . *V. articulata*
(36). Seed sometimes less than 4 mm in length; seed glabrous velvet, subspherical . . . . *V. peregrina*
(35). Hilum seldom less than 10 % of seed circumference 37, (37)
37. Hilum oval-cuneiform and never more than 1 mm in length . . . . *V. amphicarpa*
(37). Hilum cuneiform, oval or oblong, longer 38, (38)
38. Seed never more than 5 mm in length (but compare *V. benghalensis*) . . 39, (39)
39. Lens center flat sometimes raised; lens center sometimes more than 2 mm from hilum . . . . *V. semiglabra*

(39). Lens center raised and always less than 2 mm from hilum, seeds sometimes more than 5 mm in length . . . . *V. benghalensis*
(38). Seed not seldom more than 5 mm in length 40, (40)
40. Seed with a more or less distinct black line occupying two-thirds of seed circumference; seeds seldom less than 4 mm in length . . . . *V. monantha* 41, (41)
41. Black line distinct, seeds never less than 4 mm in length, up to 6 mm . . . . *V. monantha* ssp. *triflora*
(41). Black line somewhat indistinct, seeds smaller, seldom even less than 4 mm in length; hilum shorter . . . . *V. monantha* ssp. *monantha*
(40). Seeds without this black line; sometimes less than 4 mm in length . . . 42, (42)
42. Hilum oblong; seed subspherical, brown, dark violet mottled . . . . *V. eriocarpa*
(42). Hilum oval-cuneiform; seed subspherical to sublenticular, rosy-brown, grey-black, brown mottled . . . . *V. bithynica*
(34). Hilum mostly equal or more than 2 mm in length, seldom less 43, (43)
43. Lens center usually less than 2 mm from hilum; hilum between 10 and 20 % of seed circumference 44, (44)
44. Seed often 4 and seldom more than 6 mm in length 45, (45)
45. Hilum usually less than 2,5 mm in length; seed always glabrous and smooth 46, (46)
46. Seed sub-spherical, spherical, oval, and sub-cuboidal; seed coat olivey, grey, yellow-grey, rosy brown, dark brown and occasionally with black spots . . . . *V. cordata*
(46). Seed sub-spherical, sub-lenticular, oval, spherical and sub-cuboidal; seed coat dark-brown, rosy brown, grey pink, creme, orange and mottled . . . . *V. dasycarpa*
(45). Hilum usually less but sometimes a little more than 2,5 mm in length; seed coat glabrous and/or velvet; seed sub-spherical, lenticular or sub-lenticular . . . . *V. benghalensis*
(44). Seed 5 but sometimes more than 6 mm in length (more rarely less than 5 mm); lens center sometimes more than 2 mm from hilum; seed usually rounded-rectangular, but also sub-spherical or somewhat flattened; seed coat variously coloured, monochromed to densely mottled, very variable species . . . . *V. sativa*
(43). Lens center usually 2 mm or more from hilum; hilum from 10 to 15 % of seed circumference 47, (47)
47. Lens center never more than 3 mm from hilum 48, (48)
48. Seed never more than 6 mm in length; seed coat always glabrous . . . . *V. serratifolia*
(48). Seed never less than 4,5 but often more than 6 mm in length 49, (49)
49. and (49) . . . . *V. galilaea* and *V. johannis*
(Both species are hardly to distinguish by seed characters!)
(47). Lens center from less than 2 to more than 3 mm from hilum; seed velvet or glabrous and smooth; hilum oval . . . . *V. narbonensis*
(23). Lens center on the opposite side of seed from center of hilum 50, (50)
50. Hilum oblong and never less than 2 mm in length; hilum often about 15 % of seed circumference . . . . *V. hyrcanica*
(50). Hilum seldom over 2 mm in length; hilum sometimes less than 10 % and sometimes more than 15 % of seed circumference 51, (51)
51. Hilum cuneiform and seldom oval 52, (52)
52. Lens flat; seed sub-pyramidal to subspherical . . . *V. pannonica* ssp. *pannonica*

(52). Lens somewhat raised; seed pyramidal, oblong; hilum sometimes less than 10 % of seed circumference; seed size somewhat larger . . . . *V. pannonica* ssp. *striata*

(51). Hilum cuneiform and oval 53, (53)

53. Seed spherical and seldom oval, sub-pyramidal, oblong, scarcely sub-lenticular; lens somewhat raised; hilum more oval than cuneiform, less than 2 mm, sometimes more than 15 % of seed circumference; seeds mostly 5 mm or more in length . . . . *V. hybrida*

(53). Seed sub-lenticular, sub-spherical and sub-pyramidal; seed coat greyish, pink rose, dark brown 54, (54)

54. Hilum never more than 2 mm in length; seeds mostly 4 mm, never 5 mm in length . . . . *V. anatolica*

(54). Hilum sometimes more than 2 mm in length and never less than 10 % (up to 15 %) of seed circumference; seed more than 5 mm in length (but smaller seeds may also occur) . . . . *V. melanops*

Zusammenfassung

Die Variabilität von Samenmerkmalen bei ausgewählten *Vicia*-Arten

Es wurde die Variabilität von 19 Samenmerkmalen bei 52 Arten der Gattung *Vicia* geprüft. Für eine Gruppenbildung bestätigten sich dabei der Samenumfang, die relative Nabellänge, die Nabelform und die Lage der Chalaza als besonders geeignet. Bei einer auf diesen Merkmalen aufbauenden Cluster-Analyse konnten innerhalb des Untersuchungsmaterials 11 Artengruppen ausgeschieden werden. Diskriminanz-Analysen unter Verwendung von allen 19 Variablen ergaben eine Fehlklassifikation beim Versuch der Zuordnung einzelner Samen in 25 % der Fälle. Ein Schlüssel zum Bestimmen der Arten vorzugsweise an Hand von quantitativen Samenmerkmalen wurde vorgeschlagen. Dendrogramme illustrieren die Ergebnisse der Cluster-Analysen und schematische Samendiagramme die Ausprägung der relevanten Samenmerkmale.

We thank Mrs. B. Fouquet, Gatersleben, for checking carefully the determination key.

Краткое содержание

Изменчивость признаков семян у некоторых видов *Vicia*

Исследовалась изменчивость 19 признаков семян у 52 видов рода *Vicia*. Для образования групп особенно пригодными оказались следующие признаки: величина семян, относительная длина рубчика, форма рубчика и положение халазы. Кластерный анализ, построенный на этих признаках, выявил у исследованного материала 11 групп видов. Дискриминантный анализ, при учёте всех 19 признаков дал неудачную классификацию для попытки отнесения отдельных семян к той или иной группе в 25 % случаев. Предлагается ключ для определения видов, преимущественно на основе количественных признаков семян. Дендрограммы иллюстрируют результаты кластерного анализа, а схематические диаграммы семян — выраженность их относительных признаков.

Literature

Ball, P. W., 1968: *Vicia* L. – In: Tutin, T. G. et al., Flora Europaea, vol. **2**, 129–136. – Cambridge, England.

Barr, A. J., J. H. Goodnight, J. P. Sall, and J. T. Helwig, 1972–76: SAS – A users' guide to the statistical student supply stores. – N. Carolina S. U. Raleigh. N. Carolina, 27607.

Bogyo, P. T., and E. Porceddu, 1975: The use of discriminant functions in the numerical taxonomy of *durum* wheat. – In: Scarascia-Mugnozza, G. T. (ed.), Genetics and breeding of *durum* wheat, 225–246. – Bari.

Bolchovskich, Z. V., et al., 1969: (Chromosome numbers of flowering plants). – Leningrad, Izd. Nauka, 926 S.

Chisci, G. C., and A. H. Martin, 1966: The use of discriminant functions in intraspecific classification of lucerne (*M. sativa* L.). – Genet. Agr. **20**, 37–47.

Davis, P. H., and U. Plitmann, 1970: *Vicia* L. – In: Davis, P. H. (ed.), Flora of Turkey and the East Aegean islands **3**, 274–325. – Edinburgh, Univ. Press.

Gunn, C. R., 1970: A key and diagrams for the seeds of one hundred species of *Vicia* (Leguminosae). – Proc. Internatl. Seed Test. Assoc. **35**, 773–790.

–, 1971: Seeds of native and naturalized vetches of North America. – Agr. Handb. No. **392**, Agr. Res. Serv. USDA, 42 S. – Washington.

Iannelli, P., 1964: Variety testing of vetches. – Proc. Internatl. Seed Test. Assoc. **29**, 887–907.

Kupicha, F. K., 1976: The infrageneric structure of *Vicia*. – Notes Roy. Bot. Gard. Edinb. **34**, 287–326.

Schäfer, H. I., 1973: Zur Taxonomie der *Vicia narbonensis*-Gruppe. – Kulturpflanze **21**, 211–273.

Dr. P. Hanelt
Zentralinstitut für Genetik und Kulturpflanzenforschung
der Akademie der Wissenschaften der DDR
DDR - 4325 Gatersleben
Corrensstr. 3

Kulturpflanze 32 · 1984 · 123–142

Der Genfonds an *Malus*-Arten in der DDR und seine Nutzung

Manfred Fischer[1], Gisela Mildenberger[2], Rolf Büttner[1],
Karl Hammer und Jürgen Schmidt[3]

(Eingegangen am 18. Januar 1984)

Zusammenfassung

Die ursprüngliche Naumburger Malus-Artenkollektion wurde in das Institut für Obstforschung Dresden-Pillnitz überführt. Sie umfaßt 237 Arten und Varietäten und damit die wichtigsten Vertreter der 5 Sektionen von *Malus–Malus, Sorbomalus, Chloromeles, Eriobolus, Docyniopsis* (nach Rehder 1954). Die bereits vorliegenden Untersuchungsergebnisse an dieser Kollektion über Pollenfertilität, Apomixis, Schorf-, Mehltau- und Blutlausresistenz sowie Virustoleranz werden mitgeteilt.

Auf die nationale und internationale Bedeutung dieses Genfonds, der für den Austausch zur Verfügung steht, wird hingewiesen.

Einführung

Die Erhaltung, Erweiterung und Nutzung von genetischen Ressourcen ist eine der wichtigsten Aufgaben der Pflanzenzüchtung in der Gegenwart, die sich aus der Vielfalt sowie dem Wechsel der Anforderungen an neue Pflanzensorten ergibt. Wild- und Zieräpfel als Träger entscheidender Eigenschaften fanden und finden in der Apfelzüchtung immer wieder Verwendung. Das gilt sowohl für ihre Nutzung zur Aufklärung genetischer Grundlagen als auch für ihre Einkreuzung als Resistenzdonoren.

Die Formenfülle der Arten und Varietäten innerhalb der Gattung *Malus* ist für Merkmalsstudien und genetische Untersuchungen besonders geeignet. Zahlreiche Merkmale, wie Blattform, Fruchtgröße, Kelchverhalten, Resistenzverhalten, Behaarung der Blätter, Form der Blütenblätter, sind so unterschiedlich in den einzelnen taxonomischen Gruppen der Gattung ausgebildet, daß sie deutlich bestimmt und voneinander getrennt werden können, und auch der Erbgang dieser Merkmale kann ohne Schwierigkeiten verfolgt werden. Infolge der in der Gattung stark ausgeprägten Selbststerilität sind jedoch Bastardierungen in mancherlei Weise vorgekommen, und es gibt noch immer taxonomische Probleme.

1 Institut für Obstforschung Dresden-Pillnitz der Akademie der Landwirtschaftswissenschaften der DDR, DDR-8057 Dresden, Pillnitzer Platz 2
2 Institut für Züchtungsforschung Quedlinburg der Akademie der Landwirtschaftswissenschaften der DDR, Abt. Naumburg, DDR-4800 Naumburg, Steinkreuzweg 1
3 Institut für Phytopathologie Aschersleben der Akademie der Landwirtschaftswissenschaften der DDR, DDR-4320 Aschersleben, Theodor-Roemer-Weg 4

Um solchen taxonomischen und genetischen Fragen nachgehen zu können und damit einen Beitrag zur Grundlagenforschung für die praktische Obstzüchtung zu leisten, wurde in den Jahren 1951 bis 1958 das kleine *Malus*-Wildarten-Sortiment, das in der damaligen Zweigstelle Naumburg der Biologischen Zentralanstalt bereits vorhanden war, wesentlich erweitert. Das Wildarten-Material wurde aus verschiedenen Botanischen Gärten und Arboreten Europas und Nordamerikas übernommen. Dabei war es nicht zu vermeiden, daß einige Duplikate in die Sammlung kamen. Nach sorgfältiger Prüfung wurde auf diese Duplikate verzichtet, als die Sammlung in den Jahren 1973 bis 1980 mit ihren wesentlichen Sippen in das Institut für Obstforschung Dresden-Pillnitz überführt wurde (M. FISCHER 1982). Gegenwärtig werden 237 Formen erhalten. Alle von REHDER (1954) aufgeführten Arten der fünf Sektionen sind in einer mehr oder weniger großen Anzahl Varietäten und Formen im Sortiment vertreten.

Eine Erweiterung der Kollektion geschieht im wesentlichen auf der Basis von Spontanherkünften. Auf dem Gebiet der DDR wird der Nachweis und die Sammlung von *Malus*-Wildarten in Zusammenarbeit mit den Aufgaben des Naturschutzes gelöst (vgl. SCHLOSSER 1982). Gleichzeitig ergeben sich dadurch Möglich-

Abb. 1 Neuanpflanzung von *Malus angustifolia* in Dresden-Pillnitz

keiten für eine Erhaltung in den natürlichen Habitaten. Weitere international für die Erhaltung schon erprobte Wege, wie die Anpflanzung von Obstgehölzen in Parks (MARCHENAY 1980) oder innerhalb geschützter alter Obstgärten (MARCHENAY 1981) sollen geprüft werden. Eine Zusammenarbeit zwischen Agrarhistorischen Museen und der Genbank in Gatersleben bietet sich für diese Zwecke an. Eine recht erfolgreiche Aktion zur Sammlung genetischer Ressourcen von Gemüsen und Arzneipflanzen auf dem Gebiet der DDR (HAMMER et al. 1977) könnte als Vorbild für weitere Möglichkeiten zur Erweiterung der *Malus*-Wildarten-Kollektion dienen.

Abb. 2 *Malus floribunda* verfügt über eine gute Mehltauresistenz

Die Wildartenkollektion von *Malus* (Abb. 1–4) stellt in ihrer Mannigfaltigkeit ein reichhaltiges Genreservoir für die Apfelzüchtung dar. Es wird deshalb eine Übersicht über alle *Malus*-Arten und -Varietäten gegeben, die die neu aufgebaute Sammlung im Institut für Obstforschung Dresden-Pillnitz enthält. Gleichzeitig werden alle wichtigen Ergebnisse mitgeteilt, die an diesen Sippen in den letzten Jahren bereits erarbeitet wurden.

Dieser *Malus*-Genfonds ist für den internationalen Austausch verfügbar.

Abb. 3 *Malus floribunda.* Zweig in Blüte

Abb. 4 *Malus* 'Ökonomierat Echtermeyer'

Zusammenstellung des Materials und der Untersuchungsergebnisse

Aus der Bearbeitung des Sortiments liegt bereits eine ganze Reihe Ergebnisse auf verschiedenen Forschungsgebieten vor. So gibt es Erkenntnisse im Zusammenhang mit der Taxonomie sowohl auf morphologischer als auch auf phytochemischer Grundlage, auf den Gebieten der Embryologie, Genetik, Apomixis, Pollenfertilität, der Resistenz gegen *Podosphaera leucotricha* (Ell. et Ev.) Salm, *Venturia inaequalis* (Cooke) Aderh., *Eriosoma lanigerum* Hausmann und der Frostresistenz (SEELIGER 1934, HENNING 1947, VOGLER 1957, GOLLMICK 1958a und b, WARTENBERG 1960, MILDENBERGER 1961, 1963a und b, HENKE 1963, MILDENBERGER und MIHATSCH 1968, MIHATSCH und MILDENBERGER 1981).

Das Verhalten ausgewählter *Malus*-Formen des Naumburger Sortiments hinsichtlich der Anfälligkeit gegenüber wirtschaftlich bedeutenden Virosen wurde von SCHMIDT (1979) geprüft.

Eine Zusammenstellung aller im Sortiment vorhandenen Sippen, geordnet nach dem Gliederungsvorschlag von REHDER (1954), und der Untersuchungsergebnisse an diesem Material wird im Anhang vorgelegt.

Auf methodische Einzelheiten wird an dieser Stelle verzichtet. Dazu wird auf die zitierte Literatur verwiesen.

Diskussion

Für die gezielte züchterische Verwendung von Wildarten sind nicht nur diese Arten selbst, sondern auch Tatsachen und Kenntnisse aus verschiedenen Grundlagendisziplinen, insbesondere der Pflanzengeographie, Taxonomie und Genetik, wünschenswert, die eine bessere Einschätzung der genutzten Formenkreise erlauben. Probleme der Sippenentwicklung und Formenkreisbildung und die Entstehung der Kultursorten sind z. T. noch nicht vollständig geklärt.

Von diesen Grundlagen ausgehend ergibt sich eine bestimmte Abfolge von Schritten bis zur neuen Sorte etwa in folgender Weise:
Formenkenntnis – Verwandtschaftsverhältnisse – taxonomische Gliederung der Formenkreise – stammesgeschichtliche Zusammenhänge – kritische Sichtung des Materials – Bedeutung von Merkmalen für die Gruppierung – Analyse des Materials auf verwendbare Merkmale – Entwicklung von Testmethoden für die Merkmalsprüfung – Befruchtungsbiologie – Ploidieuntersuchungen – Kreuzbarkeitstests – Einfügen von Merkmalsträgern in den Zuchtprozeß – Erbgang von Merkmalen – Entwicklung eines komplexen Zuchtprogrammes – Rückkreuzungen – Prüfung neuer Zuchtklone – neue Sorte.

Artbastardierung ist seit langem eine häufig angewandte Methode in der Obstzüchtung. In erster Linie soll damit erreicht werden, wertvolle Eigenschaften von Wildarten auf Kulturformen in Kombinationszüchtungsprogrammen zu übertragen. Vor allem handelt es sich dabei bisher um die Einkreuzung von Krankheitsresistenz, Frostwiderstandsfähigkeit oder Schwachwüchsigkeit. Der Wechsel der Anforderungen an neue Sorten bedingt die ständige Suche nach Merkmalsträgern mit neuartigen Eigenschaften. Die Kreuzungsprogramme müssen sehr langfristig angelegt werden, da Wildarten auch zahlreiche negative Eigenschaften auf ihre

Nachkommen übertragen wie Kleinfrüchtigkeit, sehr schlechte Fruchtqualität, folgernde Reife, ungünstige Wuchseigenschaften, bei Unterlagen Unverträglichkeit u. a.

Neuartige Erkenntnisse in der Befruchtungsbiologie eröffnen auch in der Obstzüchtung neue Möglichkeiten. Für den Anbau großer technologischer Einheiten sind selbstfertile Sorten von entscheidendem Vorteil. Selbstfertilität kommt innerhalb der Gattungen *Prunus*, *Fragaria* und *Cydonia* sowie bei Strauchbeerenobst vor und wird praktisch genutzt. *Malus*- und *Pyrus*-Arten sind selbststeril. Relativ häufig treten bei *Malus*-Arten aber apomiktische Vermehrungsweisen auf (insbesondere innerhalb von *M. coronaria*). Wenn es gelingt, diese Art der Vermehrung im Kreuzungsexperiment auf normalerweise sich amphimiktisch vermehrende Formen zu übertragen, sind grundsätzliche Verbesserungen in der Anbau- und Vermehrungstechnologie denkbar. Es versteht sich, daß noch langjährige Forschungsarbeiten notwendig sind, um praxisanwendbare Resultate zu erzielen.

Grundlegende blütenbiologische Untersuchungen an den *Malus*-Arten (Mihatsch und Mildenberger 1981) haben u. a. gezeigt, daß ungenügende Kreuzbarkeit teils witterungsbedingt, teils aber auch durch nicht ausreichende Pollenfertilität verursacht sein kann. Nicht an jedem Falle muß also Inkompatibilität vorliegen, wie ursprünglich vermutet. Die Ermittlung der Art und des Grades der Apomixis zahlreicher *Malus*-Arten ist die Voraussetzung für Experimente im oben genannten Sinne sowie zur Herstellung von Ausgangsmaterial für die Apfelunterlagenzüchtung.

An dem in Naumburg aufgepflanzten Material wurden zuerst von Seeliger (1934) hauptsächlich morphologische Untersuchungen vorgenommen.

Henning (1947) analysierte in seiner umfassenden Arbeit auf der Basis des Naumburger Materials eine große Anzahl von Merkmalen vor allem im Blüten- und Fruchtbereich und schätzte ihre Bedeutung für die Taxonomie ein. Sehr aufschlußreich sind die Analysen von Kreuzungsnachkommenschaften, vorwiegend zwischen Wildarten und Kultursorten, bei denen er bestimmte Dominanzverhältnisse von Merkmalen erfassen konnte. Diese Untersuchungen wurden durch ihre Vielseitigkeit grundlegend für die Arbeit am Naumburger Material. Sie wurden von Gollmick (1958a) weitergeführt, der die Art der Vernation (gerollte oder gefaltete Knospenlage der jungen Blätter), die ganzrandige oder gelappte Blattform und die Griffelzahl als Kriterium für die Nachkommenschaftsanalyse bei Kreuzungen von Kultursorte × Wildart verwandte. Später lieferte Mildenberger (1963a) am gleichen Material mit ihren morphologisch-genetischen Untersuchungen einen Beitrag zur taxonomischen Bewertung des Kelchverhaltens (abfallend/bleibend) und der Fruchtstielbehaarung auf Grund der Untersuchung von Nachkommenschaften. An dem inzwischen erweiterten und fast alle *Malus*-Formen umfassenden Sortiment wurde der spontane Befall mit den wirtschaftlich wichtigen Erregern des Apfelmehltaus und des Apfelschorfes über mehrere Jahre hinweg beobachtet (Mildenberger und Mihatsch 1968). Den mit wenigen Ausnahmen stark anfälligen Kultursorten stehen bestimmte taxonomische Gruppen von Wildarten gegenüber, die deutliche Resistenzmerkmale besitzen und sie auch vererben. Diese Untersuchungen sind von hohem züchterischem Wert, da sie grundlegend für die Resistenzzüchtung gegen wirtschaftlich bedeutsame Erreger sind. Sie ergänzen die bei Zwintzscher (1962) angeführten Befunde über die drei verschiedenen Resistenzquellen bei Schorf.

Neue Erkenntnisse aus der Virusforschung zeigen, daß Viren in unterschiedlichem Maße Ertragsdepressionen verursachen können, auch wenn die Pflanzen äußerlich symptomlos bleiben. Beim Apfel wurde eine Reihe wirtschaftlich bedeutsamer Viren nachgewiesen. In einer gründlichen Studie testete Schmidt (1979) zahlreiche Wildarten des Naumburger Sortimentes auf ihren Befall und ihre Anfälligkeit gegenüber den wichtigsten Viren. Dabei zeigte sich bei einer Reihe von *Malus*-Arten Toleranz – ein für künftige Züchtungsarbeiten beachtenswertes Ergebnis.

Zum vielfachen Einsatz von *Malus*-Arten als Resistenzdonoren kann hier nur auf die weiterführende umfangreiche Literatur verwiesen werden. Zusammenfassende Darstellungen mit Ergebnissen eigener Untersuchungen zu Schorf-, Mehltau- und Feuerbrandresistenzzüchtung finden sich bei Murawski und M. Fischer (1979), Murawski und C. Fischer (1979), C. Fischer (1979) und M. Fischer et al. (1983).

Allen diesen Arbeiten liegen letztlich Langzeitbeobachtungen am Ausgangsmaterial zugrunde, die die Notwendigkeit derartiger Untersuchungen an umfangreichen Sortimenten eindrucksvoll belegen. Sie werden in der vorliegenden Arbeit dokumentiert.

Summary

Genetic resources of *Malus* in the GDR and their utilization

The collection of *Malus* germplasm originally located in Naumburg has been transfered to the Institute of Fruit Research in Dresden-Pillnitz. It contains 237 species and forms and therewith important representatives of the 5 sections of *Malus* – *Malus, Sorbomalus, Chloromeles, Eriobolus, Docyniopsis* (according to Rehder 1954). The results already obtained on the basis of this germplasm concerning pollen fertility, apomixis, resistance to scab, mildew, woolly aphid, and virus tolerance are presented here. The national and international importance of this *Malus* collection which is available for distribution is stressed.

Краткое содержание

Генофонд видов рода *Malus* в ГДР и его использование

Коллекция видов *Malus*, которая раньше находилась в г. Наумбург, перенесена в Институт плодоводства Дрезден-Пиллниц. Коллекция состоит из 237 видов и форм и включает важнейших представителей пяти секций рода: *Malus, Sorbomalus, Chloromeles, Eriobolus, Docyniopsis* (по Редеру, 1954). Сообщаются сведения, полученные при изучении этой коллекции, о фертильности пыльцы, апомиксисе, устойчивости к парше, мучнистой росе, кровяной тле, а также о толерантности к вирусам.

Указывается на национальное и международное значение этого генофонда, образцы из которого, заинтересованные учреждения, могут получить путём обмена.

Literatur

FISCHER, C., 1979: Resistenzzüchtung gegen Schorf (*Venturia inaequalis* [Cooke] Aderh.) und Mehltau (*Podosphaera leucotricha* [Ell. et Ev.] Salm). – Tagungsber. AdL Nr. 174, 97–104.

FISCHER, M., 1982: Die Verwendung von Wildarten in der Obstzüchtung. – Naturschutzarbeit in den Bezirken Halle und Magdeburg **19**, Beiheft „Genressourcen für Forschung und Nutzung", 35–38.

–, C. FISCHER, H.-J. SCHAEFER und W. FICKE, 1983: Erste Ergebnisse der Resistenzzüchtung gegen Feuerbrand, *Erwinia amylovora* (Burril) Winslow et al. bei Kernobst. 2. Mitt. Apfel- und Birnenunterlagen. – Arch. f. Züchtungsforschung **13**, 183–191.

GOLLMICK, F., 1958a: Beobachtungen an *Malus*-Artbastarden. – Arch. f. Gartenbau **6**, 359–382.

– –, 1958b: Beobachtungen über Frostschäden an Wildäpfeln und ihren Sämlingen als Folge der strengen Frostperiode im Februar 1956. – Arch. f. Gartenbau **6**, 126–149.

HAMMER, K., P. HANELT und C. TITTEL, 1977: Sammlung autochthoner Kulturpflanzen auf dem Gebiet der DDR. – Kulturpflanze **25**, 89–99.

HENKE, O., 1963: Phytochemisch-systematische Untersuchung über Flavonoide der Gattung *Malus*. – Flora **153**, 358–372.

HENNING, W., 1947: Morphologisch-systematische und genetische Untersuchungen an Arten und Artbastarden der Gattung *Malus*. – Züchter **17/18**, 289–349.

MARCHENAY, PH., 1980: Conservation du verger cidricole. – Museum National d'Histoire Naturelle, Parc Natural Régional Normandie-Maine. – 40 pp.

–, 1981: Ethnobotanique et conservation genetique l'exemple des arbres fruitieres. – J. d'Agric. Trad. et de Bota. Appl. **28**, 85–158.

MIHATSCH, H., und G. MILDENBERGER, 1981: Beiträge zur Züchtungsforschung bei Kernobst – Untersuchungen zur Pollenfertilität, Vererbung von Fruchtmerkmalen und Mehltauresistenz. – Dissertation B, AdL der DDR.

MILDENBERGER, G., 1961: Morphologische und zytologische Studien zur Genetik und Systematik der Gattung *Malus*. – Dissertation, Friedrich-Schiller-Universität Jena.

–, 1963a: Studien zur Taxonomie der Gattung *Malus*. I. Morphologisch-genetische Untersuchungen. – Arch. f. Gartenbau **11**, 191–207.

–, 1963b: Studien zur Taxonomie der Gattung *Malus*. II. Embryologische Untersuchungen. – Arch. f. Gartenbau **11**, 493–503.

–, und H. MIHATSCH, 1968: Studien über Resistenzmerkmale bei *Malus*-Arten und -Artbastarden und ihre Beziehungen zur systematischen Ordnung. – Arch. f. Gartenbau **16**, 548–574.

MURAWSKI, H., und C. FISCHER, 1979: Schorf- und Mehltauresistenzzüchtung beim Apfel. – Arch. f. Züchtungsforschung **9**, 143–149.

–, und M. FISCHER, 1979: Beschleunigung des Zuchtfortschritts in der Obstzüchtung durch gezielte Anwendung genetischer Erkenntnisse in Verbindung mit Methoden der Kombinationszüchtung – eine internationale Aufgabe. – Tagungsber. AdL Nr. 174, 11–18.

REHDER, A., 1954: Manual of cultivated trees and shrubs. – New York.

SCHLOSSER, S., 1982: Heimische Farn- und Blütenpflanzen als Genressource für Forschung und Nutzung. – Naturschutzarbeit in den Bezirken Halle und Magdeburg **19**, Beiheft „Genressourcen für Forschung und Nutzung", 49–89.

SCHMIDT, J., 1979: Untersuchungen zum Virusbefall und zur Virusanfälligkeit von Apfelsorten und *Malus*-Formen in der Deutschen Demokratischen Republik. – Dissertation A, AdL Berlin (unveröffentlicht).

SEELIGER, R., 1934: Beobachtungen an *Malus*-Arten. I. – – Mitt. d. Dtsch. Dendrol. Ges. **46**, 1–22.

VOGLER, H., 1957: Infektionsversuche mit dem Erreger des Apfelmehltaues (*Podosphaera leucotricha* [Ell. et Ev.] Salm.) und Beobachtungen über das natürliche Auftreten der Krankheit auf *Malus*-Arten und -Sorten. Ein Beitrag zur Kenntnis des Parasit-Wirt-Verhältnisses bei den echten Mehltaupilzen *(Erysiphaceen)*. – Inauguraldissertation Friedrich-Schiller-Universität, Jena.

WARTENBERG, H., 1960: Studien am Apfelmehltau *Podosphaera leucotricha* (Ell. et Ev.) Salm. mit Berücksichtigung einer normargisch-plasmatischen Abwehrreaktion. – Phytopath. Z. **39**, 16–64.
ZWINTZSCHER, M., 1962: Die genetischen Grundlagen und die Methoden der Züchtung. – Handbuch d. Pflanzenzüchtung **6**, 651–668. – 2. Aufl.

Dr. K. HAMMER
Zentralinstitut für Genetik und Kulturpflanzenforschung
der Akademie der Wissenschaften der DDR
DDR - 4325 Gatersleben
Corrensstraße 3

Anhang

Erläuterungen zum Anhang

P	Pollenfertilität	LA	= latente Apfelviren
M	Mehltauresistenz	AM	= Apfelmosaikvirus (Spy-Verfall, chlorotische Blattfleckung, Fruchtringfleckigkeit)
S	Schorfresistenz		
B	Blutlausresistenz		
A	obligate Apomixis		
A_f	fakultative Apomixis		

Bonituren:

+ +	sehr gut	+ +	tolerant
+	gut	+	teilweise tolerant
O	mittel	–	anfällig
–	schlecht	– –	sehr anfällig
– –	sehr schlecht		nicht auswertbar
+ –	unterschiedlich in den einzelnen Jahren		
O –			

Herkunftsbezeichnungen:

A. A.	Arnold Arboretum	M. A.	Morton Arboretum
A. W.	Altenweddingen	Man.	Manchester
Be.	Berlin-Dahlem	Marb.	Marburg
Butt.	Buttstädt	Mü.	München
Darm.	Darmstadt	Müb.	Müncheberg
D. M.	Des Moines	Na.	Naumburg
Dresd.	Dresden	Roch.	Rochester
Eich.	Eichelborn	Shf.	Schraderhof
Erl.	Erlangen	Späth	Berlin/Baumsch. Späth
Geis.	Geisenheim	Ste.	Steinthaleben
Grav.	's Gravenhage	Wa.	Wageningen
Greif.	Greifswald	Ween.	Weener/Ems
Heid.	Heidelberg	Zö.	Zöschen
Kew	Kew	–	Herkunft unbekannt

Die Bezeichnung des Materials durch den Donor wurde zur Gewährleistung der Referenzmöglichkeiten in die Tabelle aufgenommen. Die Herkünfte sind unseren Beobachtungen entsprechend in das System eingeordnet.

Explanation for the appendix

P	Fertility of pollen
M	Resistance to mildew (*Podosphaera leucotricha* [Ell. et Ev.] Salm.)
S	Resistance to scab (*Venturia inaequalis* [Cooke] Aderh.)
B	Resistance to woolly aphid (*Eriosoma lanigerum* Hausm.)
A	obligate apomixis (last column)
A_f	facultative apomixis

Classification:

+ +	very good
+	good
O	middle
–	bad
– –	very bad
+ –	different in some years
O –	

LA	Latent viruses (Spy decline, chlorotic leaf spot, Apple ring spot)
AM	Apple mosaic.
+ +	tolerant
+	partially tolerant
–	susceptible
– –	highly susceptible
/	not evaluated

Обозначения

P	Фертильность пыльцы
M	Устойчивость к мучнистой росе [*Podosphaera leucotricha* (Ell. et Ev.) Salm.]
S	Устойчивость к парше [*Venturia inaequalis* (Cooke) Aderh.]
B	Устойчивость к кровяной тле [*Eriosoma lanigerum* Hausm.]
A	Апомиксис облигатный (последняя графа)
A_f	Апомиксис факультативный

Бонитировка

+ +	очень хорошо
+	хорошо
O	умеренно
–	плохо
– –	очень плохо
+ –	различно в отдельные годы
O –	

LA	латентные вирозы
AM	мозаика яблок
+ +	толерантный
+	частично толерантный
–	восприимчивый
– –	очень восприимчивый
/	нету данных

Malus-Wildsortiment

Zusammenstellung der Bewertungen verschiedener Merkmale

Sortiments-Nr.	Herkunft	Bezeichnung	P	M	S	B	LA	AM	A
Sektion I: *Malus* (*Eumalus* Zabel)									
A. Serie *Pumilae* Rehd.									
1. *M. pumila* Mill.									
194	–	*M. pumila* cv. Schwarzer Splitt	+	○	○	○			
5	Zö.	*M. pumila* var. *Sieversii*	++	–	○	–			
62	–	*M. pumila* var. *paradisiaca*	–○	○	+	–			
75	Be.	*M. pumila paradisiaca* f. *fol. aureis*		○	○	–	+	–	
451	Wa.	*M. pumila aurea*							
497	Roch.	*M. pumila*		○	+				
535	Roch.	*M. pumila*		○	○		++	–	
188	Mü.	*M. pumila*	+	++	++				
2	Zö.	*M. pumila*	+						
369	D. M.	*M. pumila ruberrima*							
80	Be.	*M. pumila plena*		○	○				
56	Be.	*M. pumila dasyphylla* f. *mitis*	++	+	+	–	+	–	
191	Be.	*M. pumila* f. *pendula*	+–	○	○	–			
31	Mü.	*M. pumila* f. *apetala*		○	○	–			
332	D. M.	*M. asiatica*		○	+				
514	Roch.	*M. adstringens*							
331	D. M.	*M. upsaliensis*	++	+	+	–	++	––	
404	Heid.	*M. orthocarpa*		–	○	–	++	+	
127	Müb.	*M.* cv. Fairy	+	+	–	–			
371	D. M.	*M. astracanica*	++	–	+	–			
124	M. A.	*M. ellerwangeriana*		○	○	–			
112	M. A.	*M. baccata* cv. Cheales Golden Gem	○	○	+	–			
113	Be.	*M. cerasifera*		–	○	○			
511	Man.	*M.* cv. Moskow Peak		○	–				
509	Man.	*M.* cv. Mercer Manitoba	+	+	○				
104	Ween.	*M. baccata flavescens*		++	++				
87	Dresd.	*M. prunifolia fructu rubro*		○	+				

26 2052/32

Sortiments-Nr.	Herkunft	Bezeichnung	P	M	S	B	LA	AM	A
105	M. A.	*M. baccata* cv. The Fairy		–	○				
70	Be.	*M. baccata* × *M. pumila*	+ +	–	–	–			
243	M. A.	*M.* cv. Wellington (bloomless)		–	○	○			
162	Müb.	*M. pumila niedzwetzkyana*	+ +	○	+	–	+	+	
196	Ween.	*M. purpurea aldenhamensis*	+ +	○	+	–			
366	D. M.	*M.* cv. Transcendent	+	○	–	–			
77	Be.	*M.* cv. Transcendent (Crab apple)	+ –	+	○	○			
118	Shf.	*M.* cv. Transcendent (Crab apple)		+ +	–	–			
61	Be.	*M.* cv. Hyslop (Crab apple)	+ +	+	–	–			
306	D. M.	*M.* cv. Hyslop		○	–	–			
313	D. M.	*M.* cv. Late Hyslop (Crab apple)	+	+	–	–			
137	Müb.	*M.* cv. Grotz' Liebling		+	○	–			
73	Be.	*M.* cv. Hohenheimer Riesling (Crab apple)		○	○	–			
78	Be.	*M.* cv. Dartt (Crab apple)	–	○	–	○			
40	Mü.	*M.* cv. Martha (Crab apple)		–	–	–			
79	Be.	*M.* cv. Löwe (Crab apple)		○	–	–			
323	D. M.	*M. prunifolia* (Crab apple)							
506	M. A.	*M.* cv. Nipissing	+ +	○	○	○			
242	M. A.	*M.* cv. Veitch's Scarlet	– ○	○	○	–			
135	–	*M.* cv. Gibbs Golden Gage (Crab apple)							
425	Grav.	*M. hybrida* cv. Scugog	+	–	+	–			
244	M. A.	*M.* cv. W. H. Anderson	○	+	○	–			
157	M. A.	*M.* cv. Marshall Oyama	+	○	+	○			
222	Ste.	*M. silvestris* cv. Steinthaleben 2	–	–	–				
321	D. M.	*M. baccata mandshurica*	– ○	+ +	–	–	+ +	–	
434	A. W.	*M. baccata pulchella*		+ +	–				
106	Ween.	*M. baccata pulchella*		+ +	–	–			
433	A. W.	*M. baccata* cv. Orange		+ +	○				
107	M. A.	*M. baccata* cv. Orange	+ –	+ +	○	–			
69	Be.	*M. denticulata*		○	+				
361	D. M.	*M. denticulata*	○ +	–	–	○			
36	Mü.	*M. ringo fastigiata bifera*		–	○				
76	Be.	*M. odorata*		–	○	–			
177	Be.	*M. prunifolia* var. *flava*		+ +	–	–	+ +	+	

Sortiments-Nr.	Herkunft	Bezeichnung	P	M	S	B	LA	AM	A
186	Wa.	*M. prunifolia* cv. Rinki	++	–	+	–	+	+	
347	D. M.	*M. armeniacaefolia*	––	++	O	–			
375	D. M.	*M. prunifolia macrocarpa*		+	–	–			
214	Be.	*M. sikkimensis*		–	O				
231	Be.	*M. spectabilis roseaplena*		O	++	O			
126	Erl.	*M. eleyi*		–	O	–	++	+	
2. *M. silvestris* Mill.									
84	Dresd.	*M. silvestris*	–+	++	++		++	+	
218	Eich.	*M. silvestris* cv. Eichelborn 3	+	++	+	–	++	+	
215	Butt.	*M. silvestris* cv. Buttstädt		++	+	–			
3. *M. prunifolia* (Willd.) Borkh.									
533	–	*M. prunifolia*					–	–	
173b	Wa.	*M. prunifolia*	+	++	–				
39	Marb.	*M. prunifolia*	O+	++	–	–			
175	Wa.	*M. prunifolia*	+	+	++	O			
370	D. M.	*M. prunifolia*	++	O	O	–			
179	M. A.	*M. prunifolia* f. *coccinea*	++	+	–	–	++	+	
406	Heid.	*M. prunifolia edulis*	++	O	–	O	+	–	
323	D. M.	*M. prunifolia lutea*		++	–	–			
173a	Wa.	*M. prunifolia* (gelb)		+	–				
174	Müb.	*M. prunifolia*							
373	D. M.	*M. prunifolia maliformis*							
374	D. M.	*M. baccata aurantiaca* (= *M. prunifolia* × *M. baccata*-Bastard)	–+	O	O	–			
358	D. M.	*M. prunifolia* (= *M. toringo* × *M. pumila*)							
314	D. M.	*M. baccata cerasiformis* (= *M. prunifolia* × *M. baccata*-Bastard)	++	O	+	O			
407	Heid.	*M. baccata* (unecht)							
113	Be.	*M. cerasifera*		+	–	–			
43	Mü.	*M. cerasifera* f. *coccinea*	++	+	–	–			
533	Roch.	*M. prunifolia*		+	O		+		
58	Be.	*M. prunifolia* var. *xanthocarpa*	+	++	–	–			

Sortiments-Nr.	Herkunft	Bezeichnung	P	M	S	B	LA	AM	A
372	D. M.	*M. maliformis*	+	○	–	–			
337	D. M.	*M. prunifolia fastigiata*	–	+	○	+	– –	/	
354	D. M.	*M. prunifolia pendula*	+ +	+ +	–	○			
183	Na.	*M. prunifolia* (freiabgeblüht)		+ +	+				
185	Na.	*M. prunifolia* (freiabgeblüht)	+	+	–				
209	Mü.	*M. prunifolia* cv. Roter Riesling		+	+ +				
264	Wa.	*M.* cv. John Dawnie		○	–	–			
601	D. M.	*M.* cv. Arrow							
547	Roch.	*M.* cv. Cashmere							
596	D. M.	*M.* cv. Chilko							
600	D. M.	*M.* cv. David							
395	D. M.	*M.* cv. Evelyn							
591	D. M.	*M.* cv. Linda							
587	D. M.	*M.* cv. Liset					– –	/	
592	D. M.	*M.* cv. Marshall Oyama							
597	D. M.	*M.* cv. Hopa							
299	Be.	*M. sitkaensis*		+	–	–			
156	M. A.	*M.* cv. Lady Northcliffe		+ +	+		–	–	
362	D. M.	*M.* cv. Dolgo		+	+				
94	Be.	*M. adstringens*		+ +	○	–			
108	M. A.	*M. baccata jackii*		+ +	+		+ +	+	
165	M. A.	*M.* cv. Ökonomierat Echtermeyer		○	–	–			
207	Be.	*M. robusta*		+ +	–	–			
205	M. A.	*M. robusta persicifolia*		+ +	+ +	+	+ +	– –	
548	Roch.	*M.* cv. Cathey		+ +	+ +				
46	Mü.	*M. orthocarpa*		+	+ +	+ +			
166	Be.	*M. orthocarpa*		+ +	○	–			
64	Be.	*M. orthocarpa*	+						
125	Geis.	*M. eleyi* hort.	○ +	+	+	+ +			
268	Wa.	*M. purpurea aldenhamensis*	○ +	+ +	○	–			
195	M. A.	*M. purpurea*	–	+	+	+ +			
315	D. M.	*M.* cv. Purple Wave	–	+	+ +	+ +			
141	Wa.	*M. hillieri*	– –	+	○				
212	Wa.	*M. sikkimensis*	– –	+ +	+	○			

Sorti-ments-Nr.	Herkunft	Bezeichnung	P	M	S	B	LA	AM	A
351	D. M.	*M. sikkimensis*	– –	+	+		–	+	A
409	A. A.	*M. sikkimensis*	– –	+	+ +				
542	Roch.	*M. sikkimensis*	– –	+	+				A ?
580	Roch.	*M. sikkimensis*							A
4. *M. spectabilis* (Ait.) Borkh.									
14	Zö.	*M. spectabilis*	○	+	–	–			
417	Grav.	*M. spectabilis albiplena*	–	○	○	○			
376	D. M.	*M. spectabilis* (single)		+ +	+	–			
130	M. A.	*M. spectabilis rosea plena* cv. Luise Dittmann	–						
33	Mü.	*M. spectabilis*							
89	Dresd.	*M. spectabilis*							
544	Roch.	*M. spectabilis*							
545	Roch.	*M. spectabilis* f. *riversii*							
402	Heid.	*M. scheideckeri*	+ +	+ +	+ +	+	+ +	+	
–	Na	*M. toringo* × *M. spectabilis* (Na 166–1)	– –	+ +	+	○	+	+	
–	Na	*M. toringo* × *M. spectabilis* (Na 166–2)	+	+ +	○				
5. *M. micromalus* Mak.									
158	Ween.	*M. micromalus*	○ +	+ +	+ +	–	–	–	
531	Roch.	*M. micromalus*							
339	–	*M. micromalus*					+ +	+	
B. Serie *Baccatae* Rehd.									
6. *M. baccata* (L.) Borkh.									
44	Darm.	*M. baccata genuina*	– +	+ +	+ +	–	+ +	+	
60	Be.	*M. microcarpa*	– +	+ +	+ +	+			
421	Grav.	*M. baccata* cv. Tanners	+ –	+ +	+ +	+			
419	Grav.	*M. baccata jackii*	+ +	+ +	+ +	+			
360	D. M.	*M. baccata columnaris*	+ +	+ +	○	+			
328	D. M.	*M. baccata edulis*	+ +	+ +	○	○			
111	M. A.	*M. baccata oblonga*	+	+ +	○				
377	D. M.	*M. baccata costata*		+ +	○	○			

Sortiments-Nr.	Herkunft	Bezeichnung	P	M	S	B	LA	AM	A
59	Be.	*M. baccata himalaica*	+ –	+ +	○	–			
396	Roch.	*M. baccata mandshurica*	– ○	+ +	–	○			
420	Grav.	*M. baccata mandshurica*	+ –	+ +	+ +	+			
521	Roch.	*M. baccata mandshurica*	–	+ +	–				
349	D. M.	*M. baccata sibirica*		+	+ +	○	+	–	
55	Be.	*M. baccata sibirica*		+ +	○	○			
467	Be.	*M. baccata sibirica*		+ +	○				
4	Zö.	*M. baccata*	–		+	○			
23	Geis.	*M. baccata*	○ +	+	○	–			
435	A. W.	*M. baccata nigra*							
322	D. M.	*M. baccata praecox*	+ –						
235	Ween.	*M. theifera*	– –	+ +	+ +	○	+ +	+	A
468	Be.	*M. theifera*	– –	+ +	+ +	+			
291	Be.	*M. theifera*							
324	D. M.	*M. flexilis*	+ +	+ +	+ +	+			
352	D. M.	*M. rockii*	+ +	○	+ +				
7. *M. hupehensis* (Pamp.) Rehd.									
341	D. M.	*M. hupehensis*	– –	+ +	+	+	– –	/	
236	Greif.	*M. hupehensis*	+ +	+ +	+	○			
8. *M. halliana* Koehne									
572	A. A.	*M. halliana spontanea*	– –	+ +	+	○	+ +	+	
9. *M. floribunda* Sieb.									
54	–	*M. floribunda*	+ –	+ +	+	+			
19	Ween.	*M. floribunda arnoldiana*	+	+	○	–	–	/	
516	Roch.	*M. arnoldiana*							
98	Be.	*M. atrosanguinea* × *M. pumila* var. *niedzwetzkyana*		+ +	+	+	+	/	

Sortiments-Nr.	Herkunft	Bezeichnung	P	M	S	B	LA	AM	A
Sektion II: *Sorbomalus* Zabel									
C. Serie *Sieboldianae* Rehd.									
10. *M. sieboldii* (Reg.) Rehd.									
27	Zö.	*M. toringo*	+ +	+ +	+ +		+	+ +	
458	Kew.	*M. sieboldii arborescens*	+ −	+ +	−	○			
353	D. M.	*M. sieboldii arborescens*	+	+	+	−			
397	Roch.	*M. sieboldii arborescens*	+	+ +	+	−	−	/	
11. *M. sargenti* Rehd.									
356	D. M.	*M. sargenti*	+	+ +	+ +	−	−	+ +	A_f
333	D. M.	*M. sargenti rosea*	− ○	○	○	+			
57	Be.	*M. sargenti*							A_f
12. *M. zumi* (Mats.) Rehd.									
248	Späth	*M. zumi*	+ −	+ +	+	○			
364	D. M.	*M. zumi*	+	+	−	−			
405	Heid.	*M. zumi*	+	+	○				
400	Roch.	*M. zumi calocarpa*	+	+	+	−			
274	M. A.	*M. zumi calocarpa*	+	+ +	+ +	−			
429	Grav.	*M. zumi calocarpa*	+	+	+ +	−			
378	D. M.	*M. zumi calocarpa*					− −	/	
21	Ween.	*M. baccata fructu flavo* (= *M. pumila* × *M. toringo*)	+ ○	○	−	−			
D. Serie *Florentinae* Rehd.									
13. *M. florentina* (Zuccagni) Schneid.									
71	Be.	*M. florentina*	+ +	+ +	+ +	−			

Sortiments-Nr.	Herkunft	Bezeichnung	P	M	S	B	LA	AM	A
E. Serie *Kansuenses* Rehd.									
14. *M. fusca* (Raf.) Schneid.									
45	Mü.	*M. fusca*	+ –	+	+	–			
289	Be.	*M. fusca*	+	+ +	+	O			
200	M. A.	*M. fusca integrifolia*	–	+ +	O	–			
121	M. A.	*M. dawsoniana*	+ O	O	+ +	–	+	– –	
15. *M. toringoides* (Rehd.) Hughes									
342	D. M.	*M. toringoides*	– –	+ +	O				A
305	D. M.	*M. toringoides macrocarpa*	– –	+	O		+	+	
239	Späth	*M. transitoria*	+ +	+ +	O				
16. *M. kansuensis* (Batal.) Schneid.									
426	–	*M. kansuensis*	+ +	+	+		+	+	
579	D. M.	*M. kansuensis* (nicht echt)	+ –	+ +	+ +	O			
F. Serie *Yunnanenses* Rehd.									
17. *M. prattii* (Hemsl.) Schneid.									
267	Wa.	*M. prattii*							
18. *M. yunnanensis* (Franch.) Schneid.									
258	Kew.	*M. yunnanensis veitchii*	+ +	+ +	+		+	+	
Sektion III: *Chloromeles* (Dcne.) Rehd.									
19. *M. platycarpa* Rehd.									
167	M. A.	*M. platycarpa*	+ O	+ +	O		–	–	A
532	Roch.	*M. platycarpa*							
293	Be.	*M. platycarpa*							A
302	D. M.	*M. platycarpa*							A

Sortiments-Nr.	Herkunft	Bezeichnung	P	M	S	B	LA	AM	A
20. *M. glaucescens* Rehd.									
317	D. M.	*M. glaucescens*	+	++	+	○	−−	/	A
21. *M. coronaria* (L.) Mill.									
522	Roch.	*M. coronaria*	+○	++	++	++			A
–	Na	*M. coronaria* × *M. coronaria* (Na 238–4/5)	−○	++	+				A
–	Na	*M. coronaria* (F_1) (Na 272–2)	+○	++	○				A
350	D. M.	*M. coronaria nieuwlandiana*	+○	++	○		−−	/	A
340	D. M.	*M. coronaria charlottae*	+○	++	○		−−	/	A
312	D. M.	*M. coronaria dasycalyx*	+−	++	+		−	/	A
348	D. M.	*M. kola*	−−	++	+				A
427	Grav.	*M. soulardii kola* (tetraploid)		++	+				A
224	Wa.	*M. soulardii*					−−	/	
593	D. M.	*M. soulardii*							A
346	D. M.	*M.* cv. Red Tip	+	++	++		−	++	A
363	D. M.	*M.* cv. Elk River		++	○		−−	/	A
330	D. M.	*M. ioensis spinosa*	−	++	○				A
557	Wa.	*M. ioensis mortonii*	○+	++	+				
325	D. M.	*M. coronaria*	−○	++	++		−	/	A
368	D. M.	*M. ioensis fimbriata*	+○	++	++		−−	/	A
298	Ec.	*M. ioensis palmeri*	−○	++	○				A
22. *M. lancifolia* Rehd.									
556	Wa.	*M. lancifolia*		++	+		−	/	A
23. *M. angustifolia* (Ait.) Michx.									
320	D. M.	*M. angustifolia*	+	++	○				A
24. *M. ioensis* (Wood) Brit.									
148	Wa.	*M. ioensis*	−○	++	+				
344	D. M.	*M. ioensis* cv. Boome	+○	++	+		−−	/	A
529	Roch.	*M. ioensis palmeri*							
225	M. A.	*M. soulardii* (= *M. ioensis* × *M. pumila*)	−−	−	○				

Sortiments-Nr.	Herkunft	Bezeichnung	P	M	S	B	LA	AM	A
343	D. M.	*M.* cv. Giant Wild	–						
134	M. A.	*M.* cv. Giant Wild Crab							
Sektion IV: *Eriolobus* (D.C.) Schneid.									
25. *M. trilobata* (Poir.) Schneid.									
428	Grav.	*M. trilobata*	–	+ +	+ +				
Sektion V: *Docyniopsis* Schneid.									
26. *M. tschonoskii* (Maxim.) Schneid.									
329	D. M.	*M. tschonoskii*		+ +	+ +		–	– –	A
240	Späth	*M. tschonoskii*	– –						A

Kulturpflanze 32 · 1984 · 143–151

Further information on farro (*Triticum monococcum* L. and *T. dicoccon* Schrank) in South Italy[1]

KARL HAMMER and PIETRO PERRINO[2]

(Eingegangen am 11. Januar 1984)

Summary

Four years of research in South Italy allow to present further information on botany, distribution, cultivation and utilization of the two farro (*T. monococcum* and *T. dicoccon*) found in the Daunia and Sannio mountains in 1980 and in the Napoletano and Lucano Appenines during missions conducted in 1981 and 1982. A detailed list of the material found and the corresponding collecting sites is given. The most frequent races are var. *vulgare* Koern. in *T. monococcum* and var. *dicoccon* and var. *rufum* Schuebl. in *T. dicoccon*.

Theoretical considerations on the genetics and evolution of the two *Triticum* species underline the importance of their "*in situ*" conservation and recall the attention of the most responsible authorities on their potentials as cultivated cereals.

Introduction

While it was thought that the two farro (*Triticum monococcum* L. and *T. dicoccon* Schrank) were not any more cultivated in Italy (PANTANELLI 1944, CIFERRI and BONVICINI 1959, ANGELINI 1965, VALLEGA 1977) some fields cultivated with these two wheats were at first found (PERRINO et al. 1981) in the territories of Castelfranco in Miscano (Benevento) and Monteleone di Puglia (Foggia). The results together with those of researches conducted in 1981 (PERRINO et al. 1982), 1982 (PERRINO and HAMMER 1983a) and 1983 (PERRINO et al. 1984) in other provinces of South Italy (see also PERRINO 1982, PERRINO and HAMMER 1982) will be mentioned and discussed here.

Methods and material

The results presented in the following paragraphs have been obtained through interviews with the farmers who provided samples of "farro medio" *(T. monococcum)* and "farro piccolo" *(T. dicoccon)*. Obviously the number of farmers interviewed and of fields and plots visited was much greater than that of the collected samples. The explored areas were

1 Preliminary studies towards a Flora plantarum agri- et horticulturae Italiae australis
2 Istituto del Germoplasma – CNR, Via G. Amendola 165/A, I–70 126 Bari, Italy

those between the Adriatic and Tirrenic sea and between 40° and 42° of North latitude. The details concerning itineraries, techniques and strategies of collecting are reported in other papers (PERRINO et al. 1981, 1982 and 1984, PERRINO and HAMMER 1983 a).

List of material and collecting sites

3. 9. 1980			
Montefalcone	No. 5802 a	*T. dicoccon*, some spikelets in *T. aestivum*	altitude: 400 m
4 km n of Castelfranco	No. 5809	*T. dicoccon*	altitude: 800 m
4. 9. 1980			
3 km e of Castelfranco	No. 5815	*T. monococcum*	altitude: 680 m
	No. 5816 b	*T. monococcum*, some spikelets in *H. vulgare*	
3 km s of Ginestra	No. 5818	*T. dicoccon*	altitude: 550 m
0,5 km s of Castelfranco	No. 5819	*T. dicoccon*	altitude: 760 m
	No. 5820	*T. dicoccon*	
4 km s–e of Castelfranco	No. 5821	*T. monococcum*	altitude: 590 m
between Monteleone di Puglia and Accadia	No. 5831 a	*T. dicoccon*	altitude: 670 m
	No. 5831 d	*T. monococcum*	
28. 8. 1981			
e of Teora near the street to S. Andrea	No. 7556 a	*T. monococcum*	altitude: 720 m
29. 8. 1981			
2 km s–w of Rapone near the road to Pescopagano	No. 7589	*T. dicoccon*	altitude: 850 m
30. 8. 1981			
Corleto Monforte	No. 7614 b	*T. dicoccon*, few spikelets in *T. durum*	altitude: 500 m
31. 8. 1981			
3 km e of Roccadàspide	No. 7619	few spikelets of *T. dicoccon* and *T. monococcum* in *Vicia faba*	altitude: 200 m
8. 9. 1982			
3 km s–e of Sicignano Scalo	No. 8618	*T. monococcum*, few spikelets in *Avena sativa*	altitude: 250 m
	No. 8619	*T. monococcum*, few spikelets in *Hordeum vulgare*	
9. 9. 1982			
3 km s–e of Brienza on the way to Marsico Nuovo	No. 8630	*T. dicoccon*	altitude: 720 m
2 km e of Brienza on the way to Marsico Nuovo	No. 8632	*T. dicoccon*	altitude: 700 m
2 km n of Satriano di Lucania	No. 8634	*T. dicoccon*	altitude: 720 m
6 km n–w of Tito	No. 8639	*T. dicoccon*	altitude: 1000 m
10. 9. 1982			
2 km before Picerno from Tito Scalo	No. 8643	*T. dicoccon*	altitude: 600 m
Montagna li Foj	No. 8649	*T. dicoccon*	altitude: 1000 m

11. 9. 1982			
Abriola	No. 8722	*T. dicoccon*	altitude: 700 m
12. 9. 1982			
Contrada Pozzi, 5 km s–e of Laurenzana	No. 8727	*T. dicoccon*	altitude: 1000 m
2 km s of Anzi, along the Basento	No. 8730	*T. dicoccon*	altitude: 550 m
13. 9. 1982			
Corleto Perticara	No. 8747	*T. dicoccon*	altitude: 800 m
14. 9. 1982			
2 km w of Viggiano			
e of Tramutola, Azienda	No. 8770	*T. dicoccon*	altitude: 800 m
sul Fondovalle	No. 8783	*T. dicoccon*	altitude: 600 m
Contrada Serra Carba	No. 8784	*T. dicoccon*	altitude: 850 m

Results

Distribution

The cultivation of the two farro in the explored area allowes the following conclusions. The two wheats are cultivated in a more or less discontinuous way along the Appeninic area (for a distribution map see PERRINO and HAMMER 1984). They have not been found in the low lands along the coast. Their altitude of cultivation is mainly comprised between 500 and 1000 m. Exceptionally, they occur as weeds in other crops in lower altitudes. As in Romania (PÉNTEK and SZABÓ 1981) *T. monococcum* has reached sometimes the status of a weed. The farro are cultivated on plateaus, basins, valleys and watersheds with more than 20 % of slope and any aspect towards rivers of primary and secondary importance.

The southern border of *T. dicoccon* cultivation seems to be the river Agri. Whereas north of the river farro could be found occasionally, in the montains south of the river there may be only very few fields. Some farmers remembered the cereal but we could not find any material. Following this direction to Calabria even the name is not known. Detailed studies in Sicily (PERRINO and HAMMER 1983b) did not give any evidence of farro cultivation, too.

An estimation concerning the amount of farmers cultivating the farro in Italy is given by PERRINO and HAMMER (1984).

Botanical classification

Triticum monococcum L.
var. *vulgare* Koern.

This is the predominant race (fig. 1). A few spikes of var. *macedonicum* Papag. could be found. Other races are very rare. DOROFEEV and KOROVINA (1979) reported var. *nigricultum* Flaksb. and var. *tauricum* Drosd. from Italy.

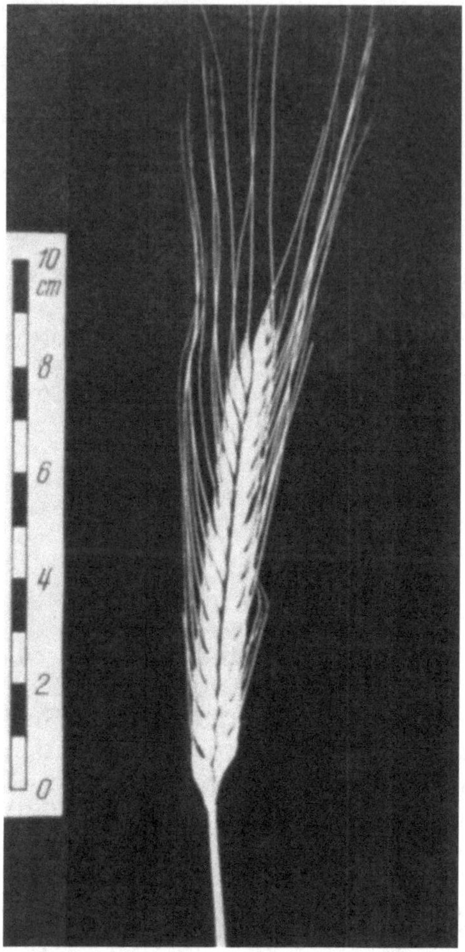

Fig. 1 Spike of *Triticum monococcum* collected near S. Andrea (Campania)

Triticum dicoccon Schrank
var. *dicoccon*

Most of the material collected belongs to this variety. There are great differences in spike size (fig. 2).

var. *rufum* Schuebl.

Occasionally among the predominantly yellow spikes. Further races may be found when studying the collected material in our reproduction fields.

Cultivation and utilization

The two wheats are sown by hand in autumn in rotation with fallow land, local maize or with other cereals like durum and bread wheats, barley and oats or with broad beans and other pulses on deep and not irrigated soils. Harvesting and threashing are fulfilled during June–July by hand, with animals or by using machines.

Fig. 2 The material of *T. dicoccon* shows great differences in spike size. Spike from a population collected near S. Andrea (Campania)–left, spike from a population collected near Castelfranco (Campania)–right. The differences between the populations exceed the infrapopulation variability by far

Cultivation techniques change according to the agricultural situations. Since farro is used exclusively for feeding animals and especially pigs, purity of seeds is often handled without care. Several farmers who do not grow farro any more have stated to have "lost" their seed because of heavy *Lolium temulentum* infestation. In our samples there was usually a great amount of other seed weeds, too, especially of *Agrostemma githago* with large seeds, which represents a progressive stage of co-domestication (Hammer et al. 1982).

Extension of cultivation is variable. In most of the cases it does not overcome half a hectare, though in some big farms it may reach one hectare (see Perrino and Hammer 1984).

Usually the production of spikelets at ripening is considered to be "good", i.e. as much as that of other cereals, and in some cases even above that of barley. One farmer told that the production was twenty times the amount that was sown.

The two farro are considered to be resistent to smut and other diseases. Only one sample of *T. dicoccon* was found to be attacked by smut (*Tilletia caries*).

According to the most old farmers, who cultivate farro since a long time, the two wheats were much more cultivated up to the second world war. The farmers who grow farro are also keeping pigs. Farro is believed to improve the fodder; the pigs would grow quicker and be more healthy. A character which would distinguish a pig fed also with farro, again according to the growers, is that of showing a more bright "hair" (coat). Another advantage would be that of being more digestible than barley.

In agricultural situations more vulnerable to the progress, probably more for human factors than for environmental ones, it seems that the cultivation of farro was suspended because of low yield.

Conservation "*in situ*"

Between the two farro still cultivated in Italy and those present in the collections of some Institutions around the world there are the following differences. The former have been conserved "*in situ*" and hence represent the result of a long and continuous selection by man but mainly by the environment (evolution under

Fig. 3 Other tetraploid wheats have contributed to the evolution of *T. dicoccon*, as can be seen from spikes collected near Montefalcone (Campania)

domestication); because these populations consist of millions of individuals one may exclude that genetic drift was as predominant as in small collecting populations. The populations contain many well balanced genotypes. Moreover, during cultivation they may have contributed to the evolution of other species of tetraploid and hexaploid cultivated wheats and vice versa (fig. 3) as well as of other wild Gramineae present in the area of cultivation. This factor alone would be a good reason to continue in cultivating the two farro or even to spread their cultivation to the most suitable environments.

Small populations of farro in collections, instead, have been preserved in the absence of natural evolution. They could have undergone a genetic drift leading to good fitness only for a limited number of environments and may be, very likely, different from the original ones.

It is easy to forsee that a study about the evolution of the two farro conserved "*in situ*" and on that of the same wheats preserved in world collections, could give results extremely interesting.

Considerations and conclusions

For more than a half century, the State, research Institutes, Seed Companies and other institutions have been competing for increasing the distribution and cultivation of varieties of wheat and other crops all over Italy but, nevertheless, the farro has not disappeared. This may be because of geographical, environmental, and social reasons and because it contains also particular features absent in other crops, i.e. competitive for the same environments or agricultural farm systems of the Appenine.

During our explorations several farms were seen, apparently with good rural buildings, but completely abondoned. In the last twenty years the reduction of the farro cultivation is more the consequence of an increase of abondoned lands mainly through emigration of people from farms to cities than to intrinsic negative features of the crop. These informations could be considered of some interest for those agronomists and other researchers responsible for the management of environments and production of the so called "marginal zones" (BONCIARELLI 1980) who precisely in the light of these discoveries could consider the possibility of redistributing the farro into rainfed zones and especially to farms with a trend toward animal husbandry.

Thanks to the work of few and unknown farmers, a prehistoric crop, cultivated on a large scale in Italy even at the beginning of our century, has continued to cope with the climate of the peninsula and exists to be included into research programmes by agronomists, plant breeders, geneticists, evolutionists and botanists. In the belief that farro had disappeared from the Italian fields they utilized for their studies material mainly from other countries.

An adequate interdisciplinary intervention at the national level could prevent a further reduction of the cultivation of the two species.

Zusammenfassung

Weitere Informationen über Farro (*Triticum monococcum* L. und *T. dicoccon* Schrank) in Süditalien

Vierjährige Studien in Süditalien erlauben es, weitere Informationen über Botanik, Verbreitung, Kultur und Nutzung der beiden Farro-Weizen (*T. monococcum* und *T. dicoccon*) zu geben, die 1980 in den Daunia und Sannio Bergen und 1981 bzw. 1982 in den Napoletano und Lucano Apenninen auf Sammelreisen gefunden wurden. Eine ausführliche Liste des Materials und der entsprechenden Sammelorte wird gebracht. Die häufigsten Sippen sind var. *vulgare* Körn. bei *T. monococcum* bzw. var. *dicoccon* und var. *rufum* Schuebl. bei *T. dicoccon*.

Theoretische Schlußfolgerungen über Genetik und Evolution der beiden Weizenarten unterstreichen die Bedeutung ihrer Erhaltung *„in situ"* und lenken die Aufmerksamkeit verantwortlicher Einrichtungen auf ihre potentiellen Möglichkeiten als Kulturgetreide.

Краткое содержание

Новые сведения о «Фарро» (*Triticum monococcum* L. и *T. dicoccon* Schrank) из Южной Италии

Четырёхлетние исследования в Южной Италии позволяют дать новые сведения о ботанике, распространении, культуре и использовании обеих пшениц «Фарро» (*T. monococcum* и *T. dicoccon*), которые были найдены во время экспедиции 1980 г. в горах Даунus и Саннио и в 1981–1982 гг. в Неаполитанских и Луканских Апеннинах. Приводится подробный список материала и местонахождений образцов. Наиболее часто встречающимися формами были: для *T. monococcum* – var. *vulgare* Körn. и для *T. dicoccon* – var. *dicoccon* и var. *rufum* Schuebl.

Теоретические выводы по генетике и эволюции обоих видов пшеницы подчёркивают значение их сохранения *in situ* и обращают внимание ответственных учреждений на их потенциальное значение как культурных хлебных злаков.

Literature

ANGELINI, F., 1965: Coltivazioni erbacee. – Facoltà di Agraria di Portici. – 829 pp.

BONCIARELLI, F., 1980: Ambiente e produzione delle zone marginali. – Rivista di Agronomia **14** (1–2), 5–16.

CIFERRI, R., e. M. BONVICINI, 1959: Revisione delle vecchie razze italiane in rapporto ai frumenti mediterranei. – Ann. Sper. Agr. **13**. – Roma.

DOROFEEV, V. F., i O. N. KOROVINA (eds.), 1979: Pšenica. – Kul't. Fl. SSSR **1**. – Leningrad, „Kolos".

HAMMER, K., P. HANELT und H. KNÜPFFER, 1982: Vorarbeiten zur monographischen Darstellung von Wildpflanzensortimenten: *Agrostemma* L. – Kulturpflanze **30**, 45–96.

PANTANELLI, E., 1944: Coltivazioni erbacee. – Facoltà di Agraria di Bari. – 325 pp.

PÉNTEK, J. és A. SZABÓ, 1981: Az alakor (*Triticum monococcum* L.) Erdélyben. – Ethnographica **92**, 259–277.

Perrino, P., 1982: Nomenclatura relativa a *Triticum monococcum* L. e *T. dicoccum* Schubler (sin. di *T. dicoccon* Schrank) ancora coltivati in Italia. – Rivista di Agronomia **16** (2), 134–137.
–, and K. Hammer, 1982: *Triticum monococcum* L. and *T. dicoccon* Schrank are still cultivated in Italy. Presence, collecting and actions. – Genetica Agraria **36**, 343–352.
–, and –, 1983a: Collection of land-races of cultivated plants in South Italy 1982. – Kulturpflanze **31**, 219–226.
–, and –, 1983b: Sicilian wheat varieties. – Kulturpflanze **31**, 227–279.
–, and –, 1984: The farro: Further information on its cultivation in Italy. utilization and conservation. – Genetica Agraria, in print.
–, – and P. Hanelt, 1981: Report of travels to South Italy 1980 for the collection of indigenous material of cultivated plants. – Kulturpflanze **29**, 433–442.
–, – and –, 1984: Collection of land-races of cultivated plants in South Italy 1983. – Kulturpflanze **32**, 207–216.
–, – and Chr. O. Lehmann, 1982: Collection of land-races of cultivated plants in South Italy 1981. – Kulturpflanze **30**, 181–190.
Vallega, V., 1977: Validità del *Triticum monococcum* nel miglioramento genetico del frumento. – Sementi Elette **23** (1), 3–8.

Dr. K. Hammer
Zentralinstitut für Genetik und Kulturpflanzenforschung
der Akademie der Wissenschaften der DDR
DDR - 4325 Gatersleben
Corrensstraße 3

Kulturpflanze 32 · 1984 · 153–196

Katalog der 1981 in der ČSSR gesammelten indigenen Kulturpflanzen-Sippen

FRANTIŠEK KÜHN[1], HARALD OHLE und KLAUS PISTRICK

(Eingegangen am 31. Januar 1984)

Zusammenfassung

Das während der Reise in ein Gebirgsgebiet der westlichen Slowakei 1981 (vgl. KÜHN et al. 1982) gesammelte Kulturpflanzenmaterial wird botanisch charakterisiert. Es handelt sich überwiegend um Landsorten, alte Sortentypen und Kultursippen von oft sehr hoher Variabilität, deren Anbau meist in raschem Rückgang begriffen ist. Charakteristische indigene Formen konnten u. a. noch bei *Triticum aestivum, Avena sativa, Papaver somniferum, Pisum sativum* und *Phaseolus vulgaris* nachgewiesen werden. Bemerkenswert ist das Vorkommen ausgesprochener Reliktkulturen (z. B. *Triticum dicoccon, Secale cereale* var. *multicaule, Panicum miliaceum, Cicer arietinum, Lathyrus sativus* und *Fagopyrum esculentum*). Die Gebirgsregion gehört zum Erhaltungsgebiet autochthoner Kulturpflanzen-Sippen in den Westkarpaten. Das rasche Fortschreiten der Generosion läßt eine baldige Fortführung der Sammeltätigkeit notwendig erscheinen.

Einleitung

Mit einer 1981 durchgeführten Sammelreise in die ČSSR wurde die Erfassung und Sammlung indigenen Kulturpflanzen-Materials in den Westkarpaten fortgesetzt (vgl. KÜHN et al. 1976, HAMMER und HANELT 1979, KÜHN et al. 1980 und HANELT et al. 1982). In den Gebirgen Biele Karpaty, Stražovské Vrchy und Javorníky (westliche Slowakei) konnten noch in beträchtlichem Umfang autochthone Sippen nachgewiesen werden (Reiseroute, chronologischen Reiseverlauf und allgemeine Ergebnisse vgl. KÜHN et al. 1982). Allerdings ist auch hier der Anbau lokaler Kulturpflanzenformen, besonders der Getreide-Landsorten, in schnellem Rückgang begriffen (Abb. 1). Im folgenden wird in katalogartiger Form eine kurze botanische Charakterisierung des Materials mit Bemerkungen zur Variabilität der Sammel-Nummern gegeben, die auf den Bonituren der Originalproben selbst und der in den vergangenen Jahren vorgenommenen Vermehrungsanbauten in Gatersleben beruht. Für jede Sippe werden Fundorte und Sammel- bzw. Herbarnummer angeführt.

Im Anhang schließt die chronologisch geordnete Originalliste der Sammlung die Zusammenstellung ab.

1 Vysoká Škola Zemědělská v Brně, Katedra botaniky a šlechtění rostlin, Brno, ČSSR

Abb. 1 Wie hier bei Čierna Lehota in den Strážovské Vrchy verschwinden in der westlichen Slowakei vielerorts die Getreide-Landsorten mit dem Auflassen der kleinen Terrassenfelder an den Gebirgshängen. Foto H. OHLE

Die Bearbeitung der Sippen verteilte sich im wesentlichen wie folgt auf die Autoren. F. KÜHN: *Commelina*; *Gramineae* p. p. (*Lolium, Bromus, Triticum, Avena, Phalaris, Echinochloa*); *Caryophyllaceae*; *Rosa*; *Pisum*; *Rhinanthus*. H. OHLE: *Allium*; *Gramineae* p. p. (*Hordeum, Secale, Panicum, Setaria, Zea*); *Fagopyrum*; *Chenopodiaceae*; *Papaveraceae*; *Umbelliferae*; *Labiatae* excl. *Elsholtzia*; *Capsicum*; *Valeriana*; *Compositae* excl. *Artemisia*. K. PISTRICK: übrige Gruppen.

Liliaceae

Allium cepa L. s. l.

Common onion-group (vgl. JONES and MANN, 1963 = var. *cepa*)

Pet'ovka (Nr. 380).

Zur eigenen Samengewinnung kultiviert, Samenträger zu mehreren an Stöcke gebunden.

Im Nachbau in Gatersleben ergaben sich braune bis bronzefarbige, runde bis leicht hochrunde Zwiebeln mit etwas ausgezogener Wurzelbasis und etwas grünlich zonierter Schnittfläche.

Aggregatum-group (= var. *aggregatum* G. DON, A. *ascalonicum* auct., non STRAND)

Vyškovec (Nr. 196), im Hausgarten S. HODULIK; Pet'ovka (Nr. 381).

aus Gartenkultur geerntetes Material starb im Nachbau jedoch ab.

Mit ihren 2–5, im Querschnitt rundlichen und von bronzefarbiger Hülle umgebenen Nebenzwiebeln ist die Kartoffelzwiebel nur einmal in Gartenkultur gefunden worden. Im Nachbauversuch in Gatersleben wurden bisher keine Blüten ausgebildet.

Allium sativum L.
var. *sativum*

Veletiny (Nr. 1), Selbstaussaat von benachbartem Knoblauchacker (Herb. 2220); Nová Bošáca (Nr. 127), Winterknoblauch; (Nr. 138), Erntegut des Bauern KRAISL, Sommerknoblauch; (Nr. 139), Winterknoblauch vom selben Bauern; Čierna Lehota (Nr. 347); Strakošovce (Nr. 594).

Die gesammelten bzw. erhaltenen Knoblauchzehen unterscheiden sich außer in der weißlich-gelblichen oder hellvioletten Farbe nicht viel. Im Nachbau in Gatersleben erbrachten die Nummern 127, 138 und 594 relativ kleine Gesamtzwiebeln (ca. 2,5 cm lang, 3,5 cm breit) sowie in der kreisartigen Anordnung und Zahl ((9)–10–(11)) der Nebenzwiebeln zeigte die Nr. 138 geringfügige Unterschiede. Brutzwiebeln wurden außer bei der Nr. 127 und 347 bei allen übrigen Knoblauchsippen beobachtet. Mit ihren sehr großen, etwa 3,5 bis 4,5 cm langen und 5 bis 6,7 cm breiten Gesamtzwiebeln mit nur 8–9 Zehen zeichneten sich die Nummern 1 und 347 als vielleicht neuere Zuchtsorten aus. Die Gesamtzwiebeln bestanden entweder aus einem Kreis median abgeflachter (Nr. 1, 127, 139, 347, 594) oder aus mehreren Kreisen transversal stark abgeflachter und gebogener Zehen (Nr. 138).

Allium porrum L.

Predpoloma (Nr. 165).

Die zur eigenen Saatguterzeugung im Hausgarten gehaltenen Pflanzen hatten nur unreife Samen und im Nachbau waren die Pflanzen sehr wenig vital.

Allium scorodoprasum L.

Nova Bošáca (Nr. 140, 208); Kroviniaci (Nr. 302); Zliechov (Nr. 463).

Fundorte des Schlangenlauchs lagen meist in der Nähe von Häusern (Hausgärten) oder er wuchs als Unkraut an Feldrändern. Über eine Nutzung der Art wurde nichts bekannt.

Commelinaceae

Commelina communis L.

Podbranč, westl. Myjava, Biele Karpaty (Nr. 88).

Einjährige Zierpflanze. Wird in den letzten Jahrzehnten in der ČSSR oft gesät. Über dem Hochblatt sind 4 Blüten. Die Blüten haben 3 weiße Kelchblätter und 3 blaue Kronblätter.

Gramineae

Lolium temulentum L.
subsp. *temulentum*

Predpoloma nördl. Nová Bošáca, Biele Karpaty (Nr. 150, 161); Bošačky östl. Vyškovec, Biele Karpaty (Nr. 243, 244d, 247g, 261); Zubák nordwestl. Púchov, Biele Karpaty (Nr. 508, 520); Hvozdnica südwestl. Bytča, Jovorníky (Nr. 577).

War in allen Fällen Saatgutunkraut in Sommergetreide (Hafer und Hafer-Gerstengemische). Die Pflanzen sind recht einheitlich, meist schwach bestockt, mit 9–11 mm breiten oberen Blättern. Die Ährchen haben bis zu 7–10 Blüten. Beim Blühen schieben sich die Staubbeutel aus den Blüten. Zur Reifezeit sind die nackten Körner zwischen Deckspelze und Vorspelze zu sehen. Die Deckspelzen sind zur Reifezeit olivenbraun, in Nr. 577 weißlich. Die Hüllspelzen sind 10–18 mm lang. Die Rhachilla ist 1,8–1,9 mm lang. Die Größe der bespelzten Körner ist ohne die Länge der Grannen 7,0–7,2×1,7–2,4×1,6–2,1 mm; die Grannen sind 10(–17) mm lang.

Bromus secalinus L.

Predpoloma nördl. Nová Bošáca, Biele Karpaty (Nr. 163, 164, Herb. 2238); Vyškovec, Biele Karpaty (Nr. 184, 205h).

Saatunkraut in Winterweizen, geht bei Aussaat im Frühjahr nicht in Blüte. Die Pflanzen sind wenig variabel. Die Blattspreiten sind 4–8 mm breit, an den Rändern bewimpert, auf der Oberseite schwach behaart oder kahl, die Knoten mit 0,1 mm langen Haaren. Die Rispenäste hatten 1(–4) Ährchen. Die Ährchen hatten meist 5–11 Blüten, waren ohne die Länge der Grannen 19–24 mm lang, die Grannen 2,5–3,5 (–6) mm lang. Die Hüllspelzen waren 5,0–6,5 und 6,0–7,0 mm, die Deckspelzen 7,3–8,0 mm lang, zum Teil unbegrannt. Die nackten Körner waren dunkelbraun.

Auf den nichtgenossenschaftlichen Äckern, die im bereisten Gebiet nur in kleinen bis mittelgroßen Feldern der meist älteren Dorfbewohner bestanden, wurden vor allem Sommergerste, Winterweizen und viel Hafer angebaut. Roggenkulturen waren recht spärlich und Mais fand sich nur sehr selten. Daß aber auch *Triticum dicoccon* SCHRANK noch örtlich im Anbau war, bewiesen Saatgutproben, die uns von einigen Dorfbewohnern zur Verfügung gestellt wurden.

Hordeum vulgare L. s. l.
Sommerformen
convar. *distichon* (L.) ALEF.
var. *vulgare* (var. *hybernum* VIB.)

Zliechov (Nr. 475).

Im besammelten Gebiet fand sich die vierzeilige Gerste nur einmal als Beimischung in Gerstengarben.

var. *nutans* (RODE) ALEF.

Veletiny (Nr. 7, 8, 10, 11, 15, 17, 23, 28) mit z. T. recht dichten (erectoiden) Ähren; Rudnik (Nr. 68d); Podzámok (Nr. 84h), als Beimengung in Saatgut von *Triticum dicoccon* SCHRANK; Nova Bošáca (Nr. 115); Predpoloma (Nr. 148, 149, 156, 157, 158); Vyškovec (Nr. 174, 175, 177); (Nr. 185), wahrscheinlich eine alte Sorte; (Nr. 188), als Beimengung in einem *Avena-sativa*-Feld; (Nr. 189h) als Beimengung in einem Weizen-Hafer-Feld; (Nr. 192a, 194), als Beimengung in einem Haferfeld; (Nr. 198), aus einer Weizen-Hafer-Saatgutprobe von S. HODULIK; Bošačky (Nr. 242, 245), aus einem Hafer-Gerstenfeld; Horná Súča (Nr. 249, 253); Kroviniaci (Nr. 285), auf einem Weizen-Hafer-Gerstenfeld; (Nr. 287, 288), mit recht dichten Ähren; Kroviniaci (Nr. 291, 301); Petrova Lehota (Nr. 306) mit ziemlich dichten Ähren; Pet'ovka (Nr. 357); Drietomská Dolina (Nr. 398, 405, 407); Brůsne bei Drietoma (Nr. 418); Zliechov (Nr. 456, 462, 469), aus eigenem Saatgutvorrat der Bäuerin VASKOVA; Zubák (Nr. 476) mit recht dichten Ähren; (Nr. 493, 494), von einem Sommergersten-Haferfeld; (Nr. 501), von abgeerntetem Sommergerste-Haferacker gesammelt; Hvozdnica (Nr. 539, 556, 567, 570, 579), aus Haferfeld; Strakošovce (Nr. 585).

Zuweilen treten neben Pflanzen mit etwas schartigen auch solche mit dichteren (erectoiden) Ähren auf, aus deren Samen sich nach Beobachtungen von HAMMER und HANELT (1979) weniger zum Lagern neigende Pflanzen als die üblichen Sommergersten ergeben können.

var. *nudum* (L.) ALEF.

Podbranč (Nr. 960).

Ein Korn wurde aus einer Probe gedroschenen Erntegutes von *Triticum dicoccon* SCHRANK ausgelesen. Als eigene Kultur wurde die Nacktgerste im Sammelgebiet nicht nachgewiesen.

Secale cereale L.
var. *cereale* (Winterform)

Podbranč (Nr. 74b, 75), Körner einer Weizensaatgutprobe von Bäuerin BATKOVA; Podbranč (Nr. 96b), aus Weizensaatgut des Bauern FILKA; Predpoloma (Nr. 141); Vyškovec (Nr. 201), Feld mit selbstvermehrter Saat, Bauer HODULIK; Bošačky (Nr. 216), im Wintergerstenfeld; Lopenik (Nr. 217, 225), im Winterweizenfeld; Kroviniaci (Nr. 283), von einem Weizen-Gerste-Haferfeld gesammelt; (Nr. 295), Winterweizenfeld; (Nr. 303), aus Getreidepuppen gesammelt; Strakošovce (Nr. 591), Winterroggenfeld.

Trotz einer gewissen Variabilität in der Ährenlänge (6,9)–10,9–(18,7) cm sind die beobachteten Merkmale des Kornes (Länge, Farbe) doch relativ einheitlich und nur selten (Nr. 591, 592, 593) treten dunkelgelbe bis rotbraun gefärbte, glasige Körner auf.

var. *multicaule* METZG.

Strakošovce (Nr. 592, 593), aus Saatgutvorräten der Ernte 1980.

Wie auch von KÜHN et al. (1979, 1980) erwähnt, wurde ein letzter Anbau von Staudenroggen in den Biele Karpaty in Talnähe oder in mittlerer Höhe an den Talhängen beobachtet. Hier lagen die aus Haselnußgebüsch hervorgegangenen Brandrodungsflächen mit Staudenroggen aber in unmittelbarer Nähe einzelner Bauerngehöfte. Der Roggen bildete etwa 170 cm hohe Pflanzen mit 15–16 cm langen Ähren aus.

Leider muß aber auch in diesem Gebirge durch Auflassen dieser alten, beschwerlichen Kultur mit einem baldigen Verschwinden des Johannisroggens gerechnet werden.

Triticum dicoccon SCHRANK
subsp. *volgense* (FLAKSB.) TZVEL.

Podbranč westl. Myjava, Biele Karpaty (Nr. 84, 96).

In beiden Fällen var. *maturatum* FLAKSB. mit Beimengung von var. *volgense* FLAKSB. Beide Varietäten sind dem Wuchs und der Morphologie nach recht einheitlich. Im oberen Teil der Ähren trägt die Ährenspindel oft unterhalb der Ährchenbasis Büschel von 3 mm langen Borsten. Emmer wird meist als Bestandteil von Blut- (und Leber-)würsten sowie zur Zubereitung von Breien und Suppen verwendet. Früher stellte man daraus auch zusammen mit zerstampften Kartoffeln eine Art Knödel („slíže“) her.

Triticum durum DESF.
subsp. *durum* var. *hordeiforme* HOST.
Zliechov, Strážovské Vrchy (Nr. 455b).

Eine Ähre in Winterweizen, in einem Gemisch von mehreren Varietäten von *Triticum aestivum*. Die Bestockung der Pflanze ist verhältnismäßig schwach, der Halm ist dünn, die Ähre relativ kurz mit langen, derben Grannen. Die Hüllspelzen sind rotbraun, sehr scharf gekielt mit recht breitem Zahn. Das Korn war im Sammelgut zum Teil ziemlich mehlig, im Nachbau 1982 glasig.

Triticum aestivum L.
subsp. *aestivum*

Meist Winterweizen; die Sommerformen sind hinter der Nummer mit + bezeichnet. Die Einreihung in Varietäten ist provisorisch; die im Sammeljahr dichten Ähren hatten in der Nachkommenschaft weniger dichte Ähren. Die gelbe Kornfarbe veränderte sich in der Nachkommenschaft oft in eine rotbraune Kornfarbe. Die Probe Nr. 587c spaltete in 2 aufeinanderfolgenden Generationen in sehr verschiedenartige Typen auf (lockerährig bis dichtährig, begrannt und unbegrannt, mit groben und mit dünnen Spelzen). An manchen Ähren war die Begrannung nur auf eine Seite der Ähre beschränkt (z. B. Nr. 204aa). Wegen der großen Anzahl der Proben führen wir hier nur die Nummern ohne Ortsnamen an (vgl. Anhang):

var. *erythrospermum* (KÖRN.) MANSF.

Nr. 25+, 84ma, 96i, 126, 183b, j, 189+ a, b, c, 195+b, 197+b, 204a, e, 209c, 224g, h, 239c–e, α, β, 263b–d, 282a, 294b, 309a, 311a, 453c, d, 454, 483c, 560eA, eB, g, k, n, 563b, 580+d–f, 581+b, 582c, 587+cB, cC, cE, g, h.

var. *ferrugineum* (ALEF.) MANSF.

69a, b, 183d, 189+d–g, 455a, aa, 560m, p, 580g, 581+c, 582d, e, 587+a.

var. *graecum* (KÖRN.) MANSF.

239f, 282b, 309b, 311c, 384b–e, 483+d, 560c, 1.

var. *milturum* (ALEF.) MANSF.

125a, b, 183e, f, 224e, 258, 455c, 560d, 580c.

var. *lutescens* (ALEF.) MANSF.

14, 24+a–c, 25+a, b, 27+a–d, 74a, aa, 80+j, 84m, 123, 142, 159a, bA, bB, 160a, b, 182, 183a, g, 195+a, 202, 203b, 204aa, b, c, ca, d, 209b, d, 215a, b, ac, 224a–d, f, 239a, b, 247c, d, 257a, b, 263a, e, 269f, 294a, d, e, 311b, 384a, 397a, 408a, 453a, b, 477+, 483+a, b, 506+, 509+a, b, 519+a, 526, 560a, b, i, α, γ, 563a, 580+a, b, 581+a, 582a, b, 587+b, cA, cD, f.

var. *alborubrum* (KÖRN.) MANSF.

125c, 560j.

var. *erythroleucon* (KÖRN.) MANSF.

382b, 383, 560qu.

var. *aureum* (LINK) MANSF.

209aA, aB, e, 215a, 247e, f, g, 294c, 382c, 384f, 560h, 563c.

var. *cinereum* (DEKAPR.) MANSF.

189g, 560o.

var. *glaucolutescens* (VATZ.) MANSF.
183h, 583a.

var. *lutescens-compactoides* (KOB.) MANSF.
247a, b.

Im Winterweizenmaterial kommt ein Typ von niedrigen Pflanzen mit breiten Blättern, dickem Halm mit verhältnismäßig kurzem oberem Internodium und dichter bis mitteldichter aufrechter Ähre mit breiter Ährenspindel vor. Die Hüll- und Deckspelzen sind meist bauchig, unbegrannt, grannenspitzig oder mit ziemlich kurzen, dicken Grannen. Die Körner sind breit, oft mehlig. Es handelt sich dabei oft um neue Hochzuchtsorten mit z.T . geringer Winterhärte. Zum Typ gehören z. B. Nr. 239b, c, 309a, 384b, c, 455a, 483c, 560i. – Außerdem treten im Material mehrere hochwüchsige Typen mit mittelstarkem Halm (bei Kultur in Gatersleben oft wenig standfest), recht schmaler Ährenspindel, lockerer Ähre, mit oft langen, biegsamen dünnen Grannen, schlanken, oft glänzenden Spelzen und mit länglichem, meist glasigem Korn auf. Bei den hochwüchsigen Weizen kann man unterscheiden: a) Banater Weizen mit gröberer, weißer, begrannter Ähre, z. T. var. *erythrospermum*, z.B. Nr. 224a, d, h, 282a. – b) ein Gebirgstyp mit langem oberem Internodium, geschlossenen Ährchen, dünneren Spelzen und dünnen, biegsamen Grannen, z. B. 560f, m, 582c, e. – c) hohe Pflanzen mit oft rotem Halm, mit langem oberem Internodium, zur Seite gebogener lockerer Ähre mit geschlossenen Ährchen, z. T. var. *miltura*, var. *lutescens*, z. B. Nr. 526, 560c, 582a, b. – d) hohe, bis fast unter die Ähre beblätterte Pflanzen mit aufrechter Ähre, z. B. Nr. 560e, 563c.

Innerhalb der Sommerweizen kann man folgende Typen unterscheiden: a) einen Gebirgstyp mit ziemlich dünnem, oft violettem Halm, schmalen Blättern, mit langen Halmknoten und über der Basis kurz behaarten Scheiden, mit lockerer, schmaler, begrannter Ähre, mit schmalen, nach oben verschmälerten, oft glänzenden Hüllspelzen, z. T. var. *ferrugineum*, z. B. Nr. 189, 578a. – b) begrannte Weizen mit aufrechter fester Ähre, die Grannen in 4+2 Längsreihen, z. B. Nr. 480c, d, 587b, e. – c) ein Typ von unbegrannten weißährigen Weizen (var. *lutescens*) mit festem Halm, stark wachsbereift, die Knoten meist kurz behaart (0,1 mm), mit mittellanger bis langer Ähre, z.B. Nr. 27a, 80, 195a, 197a, 477. – d) ein Typ mit festem Halm und dichter, am Ende keuliger Ähre mit zahlreichen Grannenspitzen am oberen Ende der Ähre, z. B. 587c.

Die Ähren hatten in manchen Proben *spelta*-ähnliche Ährenspindeln mit festen, von der Seite gesehen dreieckigen Gliedern (z. B. Nr. 587cE). In den Nr. 27b, 75aa und 509a, alle var. *lutescens*, kamen einzelne Körner ohne Embryo vor. Zur großen Variabilität der Weizen-Landsorten vgl. auch Abb. 2.

Avena fatua L.
subsp. *septentrionalis* MALZ. var. *valdepilosa* MALZ.
Hvozdnica südwestl. Bytča, Javorníky (Nr. 544).

Unkraut in einem Kartoffelacker. Ein Teil der Population hatte schwach bis mittelstark behaarte Halmknoten. Die meisten Pflanzen der Population hatten kahle Halmknoten und gehörten zu subsp. *fatua*. In der Nachkommenschaft ist die Behaarung der Knoten konstant und gut ausgeprägt. Die Pflanzen haben

Abb. 2 Ähren aus dem Nachbau einer sehr variablen Winterweizen-Landsorte, die bei Hvozdnica/Javorníky gesammelt wurde. Von links nach rechts: var. *ferrugineum*, var. *graecum*, var. *milturum*, var. *lutescens*, var. *alborubrum*, var. *erythroleucon*, var. *aureum* und var. *cinereum*. Foto G. TERPE

behaarte untere Blattscheiden. Die unteren Halmknoten sind meist, die oberen Halmknoten immer behaart. Die Deckspelzen sind weißlich, hellgrau bis braun. Die Grannen sind bis 35 mm lang. Die Körner sind ziemlich schmal. Die übrigen Merkmale entsprechen denen der subsp. *fatua*.

subsp. *cultiformis* MALZ.
var. *pseudo-culta* MALZ. subvar. *americana* MALZ.

Bošačky östl. Vyškovec, Biele Karpaty (Nr. 213).

Unkraut in einem Hafer-Gerstengemisch. Die Pflanzen sind hoch – etwa wie bei subsp. *fatua* – locker bestockt. Die Knoten sind breit und niedrig, behaart; die oberen Scheiden bleigrau; die Rispe ist groß, vom Typ lockerähriger Saathafer. Die Polster am Grunde der Rispenäste sind kurz behaart. Die Ährchen sind dick, mit breiten Hüllspelzen. Blüht chasmogam. Die Deckspelzen sind braun, am Grunde um den Kallus kurz behaart, die Rhachilla ist länger behaart. Der Rücken der Deckspelze ist kahl. Die Größe der bespelzten Frucht beträgt 16,5× ×2,8–2,9×2,6 mm. Das nackte Korn ist ockerfarbig, wenig behaart und 8,4× ×2,7×2,2 mm groß. In der Nachkommenschaft ist der Typ konstant. Von MALZEVS Beschreibung (kahle Knoten) weicht die Pflanze durch behaarte Halmknoten ab.

subsp. *fatua*
var. *fatua*

Veletiny nordwestl. Uher. Brod, Biele Karpaty (Nr. 9b, 16); Podbranč westl. Myjava, Biele Karpaty (Nr. 72, 73, 84b, d, 96e, h, j); Hrušove östl. St. Turá, Biele Karpaty (Nr. 102a); Vyškovec, Biele Karpaty (Nr. 179aa); Bošačky östl. Vyškovec, Biele Karpaty (Nr. 244k, 246); Zubák nordwestl. Púchov, Biele Karpaty (Nr. 496a); Hvozdnica südwestl. Bytča, Javorníky (Nr. 542a); Strakošovce nördl. Štiavnik, Javorníky (Nr. 586c).

var. *intermedia* (LESTIB.) LEJ. et COURT.

Nová Bošáca, Biele Karpaty (Nr. 116); Predpoloma nördl. Nová Bošáca (Nr. 144b, 147a); Vyškovec, Biele Karpaty (Nr. 179ac); Bošačky östl. Vyškovec, Biele Karpaty (Nr. 211a, 241a), Brůsne nördl. Drietoma, Biele Karpaty (Nr. 419a); Zubák nordwestl. Púchov, Biele Karpaty (Nr. 498e, 525b), Hvozdnica südwestl. Bytča, Javorníky (Nr. 564a), Strakošovce nördl. Štiavnik, Javorníky (Nr. 590b).

var. *glabrata* PETERM.

Veletiny nordwestl. Uher. Brod, Biele Karpaty (Nr. 9a, 30); Brůsne nördl. Drietoma, Biele Karpaty (Nr. 419b); Zubák nordwestl. Púchov, Biele Karpaty (Nr. 496b).

var. *vilis* (WALLR.) HAUSSKN.

Podbranč westl. Myjava, Biele Karpaty (Nr. 78f, g); Predpoloma nördl. Nová Bošáca, Biele Karpaty (Nr. 144a); Vyškovec, Biele Karpaty (Nr. 179a, b); Zubák nordwestl. Púchov, Biele Karpaty (Nr. 498e); Hvozdnica südwestl. Bytča, Javorníky (Nr. 555b).

Unkraut in verschiedenen Feldfrüchten. Die Varietät *fatua* ist meist bodenständig, die anderen Varietäten sind meist Saatgutunkräuter. Die Pflanzen der subsp. *fatua* sind hoch, meist dunkelgrün, meist unbereift, mit mittelstarken Halmen und hohen, glänzenden, unbehaarten Knoten. Die Rispen sind locker, ausgebreitet (Schlaffrispe). Die Pflanzen blühen offen. Die Hüllspelzen sind bis 29 mm lang. Die Deckspelzen sind oft klein (z. B. in Nr. 496a 13 mm lang); ihre Farbe ist braun bis grau; die grauen Typen haben manchmal helle Nerven (Nr. 78f, 144a). In den Proben von Emmer waren in einzelnen Ährchen durch hygroskopische Bewegungen der Grannen eingeschobene bespelzte Körner von *Avena fatua*.

Zwischenformen zwischen *Avena fatua* L. und *A. sativa* L.

A. fatua L. subsp. *fatua* var. *vilis* (WALLR.) HAUSSKN. subvar. *basifixa* THELL.

Veletiny nordwestl. Uher. Brod, Biele Karpaty (Nr. 12, 13, 21a); Podbranč westl. Myjava, Biele Karpaty (Nr. 70a, 71, 78d, e).

Einzelne Pflanzen als Unkraut in Getreide zusammen mit *Avena sativa*. Die Pflanzen haben einen Habitus wie *A. fatua*. Die Halmknoten sind kahl. In den Ährchen ist nur die basale Blüte begrannt und hat eine unvollständige Ablösungsstelle (Semiabszission); die zweite und die eventuell vorhandene dritte Blüte haben keine Granne und keine Ablösungsstelle. Die Deckspelzen sind meist dunkelbraun. Die Ährchen fallen weniger leicht aus als bei *A. fatua* vom normalen Typ. Die Pflanzen sind in der Nachkommenschaft konstant.

Avena fatua L. × *A. sativa* L.

Podbranč westl. Myjava, Biele Karpaty (Nr. 71); Predpoloma nördl. Nová Bošáca, Biele Karpaty (Nr. 146), Vyškovec, Biele Karpaty (Nr. 191a, 193a, b).

Pflanzen mit Merkmalen von *A. fatua* subsp. *fatua* var. *vilis* subvar. *basifixa*, deren Nachkommenschaft sich in *A. fatua*, *A. sativa* und Zwischenformen aufspaltete; als Einzelpflanzen in Getreidebeständen mit *Avena sativa* und *A. fatua* beobachtet. Zum Teil zeigten die Pflanzen einen Heterosiseffekt: in Nr. 146 waren die Pflanzen in der Nachkommenschaft bis 1,8 m hoch. In dieser Nummer waren in der Nachkommenschaft die Pflanzen vom Typ der *Avena fatua* meist bleigrau, dickkörnig.

Avena sativa L. – Fatuoid

Bošačky östl. Vyškovec, Biele Karpaty (Nr. 212).

Einzelne Pflanze in Hafer-Gerste-Gemisch. Wahrscheinlich als Mutation abgeleitet von *Avena sativa* L. subsp. *praegravis* (KRAUSE) MORDV., welche die Hauptkomponente der Population bildete. Die Pflanze bestockt sich gut, ist bleigrau, hat einen dicken Halm, niedrige, behaarte Knoten, eine mittelgroße, mitteldichte Steifrispe, mittelgroße Ährchen und breite Hüllspelzen (Ährchen 4,5 mm dick). Die Ährchen sind meist zweiblütig mit verkümmerter dritter Blüte. Alle Blüten haben eine hufeisenförmige, 1 mm breite Ablösungsstelle, welche von einem Kranz dichter, bis 1 mm langer Borsten umrandet ist. Die Rhachilla ist schwach mit kurzen Borsten besetzt. Die Deckspelze ist gelb und auf dem Rücken kahl. Jede Deckspelze hat eine hakenförmige Granne, welche im gedrehten Teil unter dem Knie schwarz ist. Die Maße des bespelzten Korns sind 16,0×3,2××2,8 mm, die des nackten Korns 9,6×2,8 mm. Die basale Blüte ist manchmal steril. Die Nachkommenschaft ist auffallend einheitlich. Die Blüten sind meist kleistogam; die 2,0–2,5 mm langen Staubbeutel sind während der Anthese meistens in den Spelzen verborgen.

Avena sativa L.

Die Proben werden hier in ein System nach den Merkmalen, die im System von MALCEV 1930 gebraucht werden, eingereiht. Die Bewertung der Merkmale ist bei KÜHN 1981 angeführt. Wegen der vielen Proben werden nur die Sammelnummern ohne Ortsnamen angeführt (vgl. Anlage).

subsp. *macrantha* (HACK.) MORDV. var. *asiatica* subvar. *iranica*

Nr. 241d, 516d, p, 517, 525a, 574j, y.

– 1. Der Typ hat behaarte Knoten, ist wenig wachsbereift, hat eine offene Rispe und schmale, unbegrannte Körner.

subsp. *praegravis* (KRAUSE) MORDV.

– 2. Typ mit behaarten Knoten und Steifrispe:

Nr. 5a–e, 6a–c, 12, 20a–d, 70a–g, 71a, 78a–c, 102b, 114a–d, 144ha, 179bb, 192b–d, 199a–d, 206h, 229a, b, 235, 241e, o, 244b, 265a–c, 269a, b, d, e, g–i, 286a–c, 489b–d, 498a, bb, c, d, f–j, 502d, e, h, 516b, 521e, f, h, j, 542c, e, f, 545a–c, 572b–g, 574l, nb, p, z, zd, 575e, 590g, h.

Im Material der häufigste Typ. Hat aufrechte, dicke, ziemlich niedrige Halme mit niedrigen, behaarten Knoten, ist standfest und hat reichlich wachsbereifte Scheiden. Die Pflanzen sind bleigrau. Die oberen Blätter haben eine lange, recht breite Scheide und eine schief angesetzte, breite Spreite. Die Rispe ist bleigrau, wachsbereift, eine Steifrispe mit recht kurzen Ästen, mitteldicht. Die Ährchen sind dick, die Hüllspelzen haben auffallende Nerven. Die Deckspelzen sind dick, grob, weiß oder gelb, die Blüten kleistogam. Oft haben die Ährchen 3 fertile Blüten, manchmal Doppelkörner. Die Proben sind oft frühreif.

– 3. Typ mit behaarten Knoten, dunkelgrün, wenig wachsbereift, mit Steifrispe:

Nr. 590ab.

– 4. Typ mit behaarten Knoten, mit Sperrispe, wachsbereift:

Nr. 71a, 176, 211c, 550a, 555a, 566a, b, 586a.

– 5. Typ mit kahlen Knoten, wachsbereift, meist mit Steifrispe:

192e, f, 269c, 502a, g, i, 512b, 516i, r, 521a, 574za, 584b, d, 590f.

– 6. Typ mit Fahnenrispe:

Nr. 241f.

Subsp. *sativa* (mit kahlen Knoten, schmalkörnig, chasmogam)

– 7. Typ mit offener, oft großer Rispe, oft hochwüchsig und wachsbereift. Deckspelzen weiß, gelb, oft dunkelbraun:

68ba, 144c–i, 154a, e, 187aa, 190f, i, l, m, n, 191c, 199e, 203a, ab, 206a–d, i, n–p, t, 241b, h–l, 244c, e, g, i, j, 516f, h, i, qu, 525aa, 574a, da, e, za, zb, 590aa.

– 8. Typ mit offener Rispe, malachitgrün, schwach wachsbereift, untere Rispenäste oberseits oft behaart, Deckspelzen am Rücken behaart, braun, unbegrannt:

Nr. 191b.

– 9. Typ mit offener Rispe, hell- bis dunkelgrün, nicht bleigrau, wenig wachsbereift, mit weißer Deckspelze:

Nr. 241n.

– 10. Typ mit offener Rispe, hell- bis dunkelgrün, nicht bleigrau, wenig wachsbereift, mit grauer Deckspelze:

190g, h, k, 191d, e, 206e, f, 241f, g, 244a, m, 513i, 514a, b, 516g, k.

– 11. Typ mit offener Rispe, hell- bis dunkelgrün, wenig wachsbereift, mit brauner Deckspelze:

512a, b, 516f.

– 12. Typ mit Fahnenrispe, mit Blatthäutchen – var. *glaberrima* (THELL.) MALZ. subvar. *contracta* (NEILR.) MALZ.:

144j, 145a, f, 155b, c, 241m, 521k, 574c.

– 13. Typ mit Fahnenrispe, ohne Blatthäutchen – var. *glaberrima* (THELL.) MALZ. subvar. *eligulata* (VAV.) MALZ.:

145fa, 190la, 145ba mit kurz behaarter Deckspelzenbasis.

subsp. *nodipilosa* (MALZ.) MORDV. (Knoten behaart, Korn schmal, chasmogam, oft spätreif).

– 14. Typ mit offener Rispe, wachsbereift:

144k–n, 154ca, cb, f, h, 178a, 199f–h, 203d, e, 206k–m, 498ba, 516b, e, g, h, na, s, t, v, w, 521l, m, 574ha, i, k, na, o, qu, r–t, v–x, zc, 590d.

– 15. Typ mit offener Rispe, hell- bis dunkelgrün, wenig wachsbereift, mit grauen Deckspelzen:

Nr. 190a–c, 513f, 514a, 516k.

– 16. Typ mit offener Rispe, hellgrün, wenig wachsbereift, Deckspelzen fast schwarz:

Nr. 21a, b.

– 17. Typ mit Fahnenrispe, mit normalen Blatthäutchen, meist malachitgrün, Deckspelzen weiß, gelb, grau – var. *glabra* MALZ. subvar. *secunda* MALZ.:

Nr. 103a, 145b, d, g, h, 155a, 187a, 190d, e, 203aa, 278c, 279, 284b, 513a–e, g, h, j, ka, kb, l, n, o, 516aa, bb, e, ga, ha, j, 521c, fa, 574g.

Zur Zeit der Expedition stand der Hafer oft noch ungemäht auf dem Felde; daher konnte eine große Anzahl von Proben gesammelt werden. Sehr oft waren die Populationen ein Gemisch von sehr verschiedenen Pflanzen; es war fast die gesamte bekannte Variabilität der Art im Sammelgut vertreten. Die Sammelnummer 574 aus einem Felde bei Hvozdnica wurde auf 30 Teilproben aufgeteilt, in einem Feld bei Zubák konnten 51 Teilproben unterschieden werden und in einem Feld bei Predpoloma 24 Teilproben. Eine ähnlich große Variabilität des Saathafers wurde auch im Jahre 1974 in der mittleren Slowakei und 1977 in der Ostslowakei festgestellt (KÜHN et al. 1976, 1980). Das Gebiet der Karpaten in der ČSSR kann daher als ein Teil des Genzentrums von *Avena sativa* angesehen werden. Aus diesem Gebiet stammt auch der älteste Getreidefund mit überwiegendem *Avena sativa*-Anteil (Červenik bei Hlohovec, ältere Eisenzeit) (TEMPIR 1966).

Unter den 1981 gesammelten Hafern traten Typen mit Semiabszission und mit Fraktur auf; meist war Semiabszission zu beobachten, Fraktur war weniger häufig (z. B. Nr. 71a, 186, 206k, 516g, 590g) und kam bei verschiedenen Unterarten des Hafers vor. Die Deckspelzen waren manchmal am Rücken lang behaart (Nr. 191b), an der Deckspelzenbasis lang (3–5 mm) behaart (Nr. 190a–c, 516f, 574zb) oder an der Deckspelzenbasis kurz (1 mm) behaart (bei der Hälfte der Proben, z. B. Nr. 241l–n, 521n) bzw. kahl (Nr. 6a–c, 20b, c, 70d, e, g, 78a, b). Die Farbe der Deckspelzen war fast schwarz (Nr. 21a, b), grau (z. B. Nr. 206e, f, 513a), braun (Nr. 146, 190f, 206a, b, d, 241b, 244c, j; gehörte manchmal zu den Hauptkomponenten der Population), aber meist gelb und weiß in verschiedenen Farbtönen. Die Begrannung war in der Nachkommenschaft oft variabel. Es kamen alle Übergänge von begrannten Typen (z. B. 192b, 206a, 241k) bis zu unbegrannten (die meisten Proben) vor. Die subsp. *praegravis* ist in niederen und offenen Lagen vorherrschend, die anderen Unterarten kommen meist höher in den Berglagen oder auf ärmeren Böden vor.

Phalaris canariensis L.

Veletiny nordwestl. Uher. Brod, Biele Karpaty (Nr. 18).

Als Vogelfutter kultiviert. Wahrscheinlich wurde Import-Saatgut verwendet. Pflanzen hoch (950–1050 mm), gut bestockt. Halme der Seitentriebe niedriger als der erste Halm.

Panicum miliaceum L.
convar. *contractum* ALEF.

Nivnice (Nr. 36, 37).

Das Saatgut der Rispenhirse stammt vermutlich aus dem staatlichen Samenhandel der Slowakei und wurde auf größerem Hangacker feldmäßig für Nahrungszwecke kultiviert.

Echinochloa crus-galli (L.) P. B.

Veletiny nordwestl. Uher. Brod, Biele Karpaty (Nr. 19).

Saatgutunkraut in *Phalaris canariensis*. Auffallend hohe Pflanzen mit aufrechten Halmen; höher als *Phalaris canariensis*. Ährchen einblütig, zum Teil zweiblütig. Die Äste der Rispe mit bis zu 4 mm langen Haaren. Hüllspelzen oft begrannt. Grannen bis 18 mm lang. Deckspelze 2,7×1,7×1,2 mm groß, graubraun, glänzend. Die Staubbeutel treten zur Blütezeit aus den Spelzen.

Setaria italica (L.) P. BEAUV.
convar. *italica*

Veletiny (Nr. 32).

Zur Vogelfuttergewinnung nur kleinflächig im Hausgarten angebaut. Das Saatgut („Senegal-Hirse") wurde im staatlichen Samenhandel gekauft.

Zea mays L.
convar. *mays*

Podbranč (Nr. 77), aus privatem Saatgutvorrat, Kolben mit Xenien, „gelber Hartmais"; Nová Bošáca (Nr. 137), aus privatem Saatgutvorrat, Haus-Nr. 390: „gelber Hartmais".

Juglandaceae

Juglans regia L.

Trenčianske Bohuslavice, Váh-Tal, Biele Karpaty (Herb. 2246).

Die Walnuß wird im Südteil des Gebiets sehr häufig kultiviert und findet in den Strážovské Vrchy noch bei über 700 m NN als Straßenbaum Verwendung. Oft mit traubenförmigen Fruchtständen (Podbranč, Trenčianské Bohúslavice).

Moraceae

Morus nigra L.

Podbranč westl. Myjava, Biele Karpaty (Herb. 2224, 2230).

Als Einzelbäume bei den Bauernhöfen am Südostrand der Biele Karpaty angepflanzt; slowakisch als ‚malina' (deutsch eigentlich = Himbeere) bezeichnet. Die reifen Scheinfrüchte werden roh als Obst verzehrt oder zur Weinbereitung verwendet. Die Schwarze Maulbeere wurde auch in den städtischen Hausgärten von Považská Bystrica beobachtet.

Polygonaceae

Fagopyrum esculentum Moench
var. *esculentum*

Nivnice (Nr. 38); Vyškovec (Nr. 180, 181), kleine Felder; Zubák (Nr. 486), niedrigwüchsige (bis 80 cm hohe) Zuchtsorte mit geringem Ertrag; (Nr. 497), wahrscheinliche Selbstaussaat aus vorjährigem Anbau in einem Kartoffelfeld; (Nr. 511), Selbstaussaat in einem Haferfeld; (Nr. 522), kleines Feld in Ortsnähe; Hvozdnica (Nr. 597), kleines Feld.

Der im Sammelgebiet beobachtete Buchweizen-Anbau wird meist auf kleinen Hangfeldern betrieben oder erscheint als Verunreinigung in selbst vermehrtem Getreidesaatgut der Bauern. Es konnte keine bedeutende Variabilität zwischen den verschiedenen Herkünften festgestellt werden und nur einmal (Nr. 597) wurden im Sammelgut geflügelte und ungeflügelte, braune oder graugestreifte Früchte gefunden. *Fagopyrum tataricum* (L.) Gaertn. wurde in den Buchweizenfeldern nicht nachgewiesen.

Caryophyllaceae

Spergula arvensis L.
var. *sativa* Boenn.

Škrátová nordöstl. Vyškovec, Biele Karpaty (Nr. 230, Unkraut in einem Haferfeld; 233, Rand eines Kartoffelfeldes).

Unkraut; wuchs in beiden Fällen zusammen mit *Spergula arvensis* var. *arvensis*. Kommt im westlichen Teil der ČSSR als Unkraut zusammen mit var. *arvensis* vor, wurde bis etwa 1975 an vielen Orten in der ČSSR als Futterpflanze, besonders als Stoppelfrucht kultiviert.

Silene gallica L.

Predpoloma nördl. Nová Bošáca, Biele Karpaty (Nr. 15, Herb. 2240); Horná Suča, Biele Karpaty (Herb. 2268).

Unkraut, welches in den höheren Lagen von Nordmähren und im östlichen Teil der ČSSR vorkommt.

Agrostemma githago L.

Winterform: Hvozdnica südwestl. Bytča, Javorníky (Nr. 561). Sommerform: Predpoloma nördl. Nová Bošáca, Biele Karpaty (Nr. 152, Herb. 2237, 2244); Zubák nordwestl. Púchov, Biele Karpaty (Nr. 503); Hvozdnica südwestl. Bytča, Javorníky (Nr. 568, 571, 576).

Saatgutunkraut, charakteristisch für Landsorten, wie auch *Bromus secalinus* und *Lolium temulentum*. Die Winterform kam in Winterweizen, die Sommerform in Hafer, Gerste und einem Hafer-Gerste-Gemisch vor. Die Pflanzen waren recht einheitlich und oft monochasial verästelt. Die Pflanzen waren oft kurz behaart mit vereinzelten langen Haaren, besonders an der Kelchbasis und unter den Kelcheinschnitten. Manchmal war der ganze Kelch reichlich behaart. Die Kelchzipfel sind kurz (25 mm) bis lang (Kelch mit Zipfeln bis 60 mm lang). Die Gabelung der Nerven unter den Kelcheinschnitten ist oft nach der 2/5 Stellung differenziert: 1. und 2. am oberen Ende gegabelt, 3. in der Mitte gegabelt, 4. an der Basis gegabelt, 5. zwei selbständige Nerven. Bei anderen Pflanzen sind die Nerven unter allen Einschnitten zu zweien verwachsen. Die Kapseln sind schmal (8,5 mm) bis breit (14 mm). Die Zähne der reifen Kapseln sind lang und schmal (Nr. 568, 571) oder kurz und breit (Nr. 561, 576). Am Grunde der Kapseln sind oft unter den Kapselzipfeln 5 braune Flecken. In Nr. 503 haben die Kapselzipfel verstärkte Mittelrippen. Die Samen sind 3,4–3,8 $\times$ 3,0–3,4 $\times$ 2,4–3,6 mm groß, schwarz, aber in Nr. 503 zum Teil tabakbraun. Nur bei einem Teil der Samen überträgt sich die braune Farbe auf die Nachkommenschaft.

Chenopodiaceae

Beta vulgaris L.
convar. *crassa* (ALEF.) HELM

Horná Suča (Nr. 262); Kroviniaci (Nr. 272); Zliechov (Nr. 436).

An den oben genannten und an weiteren Standorten wurden Samenträger an weißen Futerrüben zur Saatgutgewinnung beobachtet.

Atriplex hortensis L.
forma *rubra* ROTH.

Podbranč (Nr. 106); Vel. Košecké Podhradie (Nr. 430); Predpoloma (Herbar Nr. 2253); bei Brůsne in Bachnähe vermutlich ruderal

Die früher häufiger für ein spinatähnliches Blattgemüse verwendete Gartenmelde wurde in ihrer rotblättrigen Form jetzt als Zierpflanze in Hausgärten beobachtet oder sie wird wie auf den Feldern von Vel. Košecké Podhradie örtlich zum Vertreiben von Wühlmäusen zwischen die Gemüseparzellen gepflanzt.

forma *viridis* ALEF.

Die grünblättrige Form der Gartenmelde wurde im gleichen Ort in einem Hausgarten als Zierpflanze angetroffen.

Papaveraceae

Papaver somniferum L.
subsp. *somniferum*
convar. *somniferum*
Schüttmohn

var. *pallidum* ROTHM.

Drietomská Dolina (Nr. 402), zusammen mit den Schließmohnen var. *coerulescens* ROTHM. und var. *serenum* DANERT als Beimischung in einem kleinen Feld. Obwohl öfter Mischkulturen von Mohn angetroffen wurden, fand sich doch darin nur selten der altertümliche Schüttmohn.

Schließmohn

var. *somniferum*

Zliechov (Nr. 439).

Samenfarbe auffallend weißlich bis gelb, nur wenige blaßrosa bis violett. Die Narbenscheibe der reifen Kapsel ist flach oder trichterförmig eingesenkt.

var. *coerulescens* ROTHM.

Veletiny (Nr. 4), in einem Futterrübenfeld; Nivnice (Nr. 33), selbst vermehrtes Saatgut; Rudnik (Nr. 66), in kleiner Futterrübenparzelle, relativ kleine Kapseln; Podbranč (Nr. 93), Erntegut des Bauern FILKA; Nová Bošáca (Nr. 119), Haus Nr. 440, aus eigenem Saatgutvorrat; (Nr. 133) Haus Nr. 390, eigener Saatgutvorrat; Kapseln kugelig bis walzig; Lopenik (Nr. 227); Horná Súča (Nr. 256), Haus Nr. 96, eigenes Saatgut des Bauern KRAISL (Gemisch mit var. *serenum* DANERT); Kroviniaci (Nr. 275, 277, 290), Kapseln kugelig bis eiförmig, Narbenscheibe flach trichterförmig eingesenkt; Petrova Lehota (Nr. 308); Cierna Lehota (Nr. 337); Drietomská Dolina (Nr. 400); (Nr. 401, 402), zusammen mit var. *pallidum* ROTHM. und var. *serenum* DANERT in Hausgarten KROČILOVA; Brůsne (Nr. 417); Zliechov (Nr. 464, 468), Narbenscheibe der Kapseln am Rand meist aufwärts gebogen; Zubák (Nr. 484, 492), Kapseln walzig; Hvozdnica (Nr. 557), Kapseln mit flachen oder wenig herabgebogenen Narbenlappen.

var. *nigrum* HAYNE

Nivnice (Nr. 34, 35), im Gemisch mit var. *serenum* DANERT und var. *coerulescens* ROTHM. in kleinen Futterrübenfeldern; Kroviniaci (Nr. 276), lang ovale, große Kapseln mit Längsfurchen, Narbenlappen hoch aufgerichtet; (Nr. 290) Kapsel kugelig, Narbenscheibe flach; Petrova Lehota (Nr. 307); Pet'ovka (Nr. 371), Kapsel eiförmig, Narbenscheibe flach bis herabgebogen, Samen gelblich bis hellgrau; Zliechov (Nr. 457), Kapsel walzig, Narbenscheibe flach; (Nr. 464), Kapsel lang eiförmig; Zubák (Nr. 484), Kapseln walzig.

var. *serenum* DANERT

Nivnice (Nr. 34, 35), im Gemisch mit var. *coerulescens* ROTHM. und var. *nigrum* HAYNE, Kapseln meist spindelförmig; Horná Súča (Nr. 256) im Gemisch mit var.

coerulescens Rothm., Haus Nr. 967; Kroviniaci (Nr. 277), im Gemisch mit var. *coerulescens* Rothm.; kugelige bis birnenförmige Kapseln; Petrova Lehota (Nr. 308); Drietomská Dolina (Nr. 402), Hausgarten Krocilova, im Gemisch mit var. *coerulescens* Rothm. und var. *pallidum* Rothm. (Schüttmohn); Zliechov (Nr. 452), im Gemisch mit var. *sigillatum* Danert auf kleinen Parzellen; Zliechov (Nr. 467, 473), Kapseln klein, spindelförmig; Hvozdnica (Nr. 557).

subsp. *songaricum* Basil. (convar. *rotundilobum* Danert)
Schließmohn

var. *livens* Vess.
Veletiny (Nr. 4).

var. *rubidum* Vess.
Kroviniaci (Nr. 276), Kapseln lang spindelförmig, Narbenlappen hoch aufsteigend (Abb. 3).

var. *sigillatum* Danert
Zliechov (Nr. 452), Kapseln kugelig bis walzig.

Das hochinteressante und seltene Vorkommen dieser Mohnvarietäten in der Slowakei wurde bereits von Hammer (1981) erwähnt und ihre Verbreitung bis nach Südpolen diskutiert. Die Probe Zliechov Nr. 464 hatte dunkelviolette Kapseln.

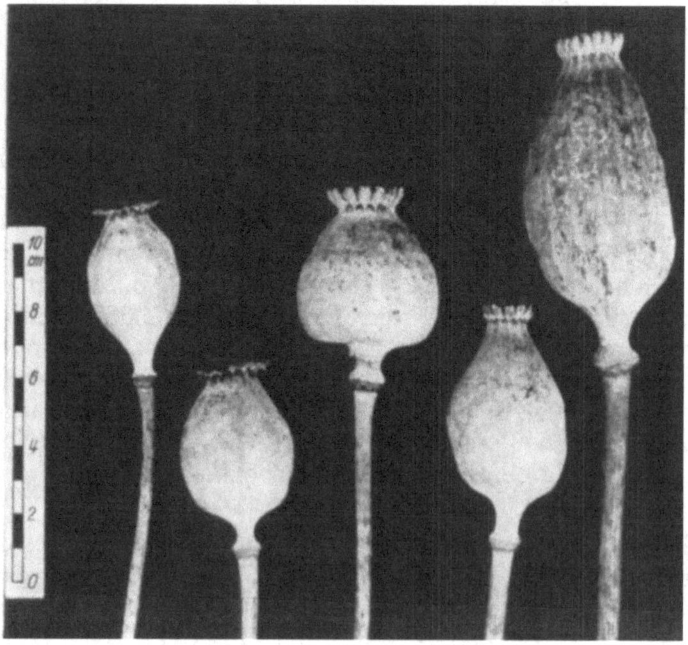

Abb. 3 Kapselform und Ausbildung der Narbenscheiben bei einigen im Gebiet häufigen Papaver somniferum-Varietäten. Von links nach rechts: var. *nigrum*, var. *somniferum*, var. *coerulescens* (2 Kapseln), var. *rubidudum*. Foto G. Terpe

Cruciferae

Brassica oleracea L.

convar. *capitata* (L.) Alef. var. *capitata* L.
Zubák nordwestl. Púchov, Biele Karpaty (Herb. 2300, 2301).

Spätreifer Strunkkohl, eine verbreitete Kulturform in den West-Karpaten (KÜHN et al. 1976, HANELT et al. 1982), wurde in einem variablen Mischbestand aus Pflanzen mit fehlender oder ± schwacher Anthocyanfärbung beobachtet. Leider konnte hier, wie von den gelegentlich angebauten Stoppelrüben (*B. rapa* L. subsp. *rapa* var. *rapa*), Kohlrüben (*B. napus* L. var. *napobrassica* (L.) RCHB.) sowie Rosenkohltypen (*B. oleracea* L. convar. *oleracea* var. *gemmifera* DC.) kein Saatgut gesammelt werden.

Brassica rapa L.
subsp. *sylvestris* (LAM.) JANCHEN

Zubák nordwestl. Púchov, Biele Karpaty (Nr. 523).

Als Unkraut in einem Buchweizenfeld. Ein weiterer Nachweis dieser seltenen Segetalform für die Slowakei (vgl. auch KÜHN et al. 1980).

Raphanus raphanistrum L.
subsp. *raphanistrum*

Vyškovec, Biele Karpaty (Nr. 166–172); Škratová nordöstl. Vyškovec, Biele Karpaty (Nr. 221–223, 231–232, Herb. 2257); Kroviniaci nördl. Horná Súča, Biele Karpaty (Nr. 292, 297–300, Herb. 2269–2270); Zubák nordwestl. Púchov, Biele Karpaty (Nr. 529–537, Herb. 2302–2303); Hvozdnica südwestl. Bytča, Javorníky (Nr. 598); Nová Bošáca, Biele Karpaty (Herb. 2242); Drietomská Dolina südl. Starý Hrozenkov, Biele Karpaty (Herb. 2284–2286).

Außer an Kalkstandorten als Unkraut häufig in Sommergetreide- und Hackfruchtkulturen. Formen mit weißen Petalen fehlten in den untersuchten Populationen, solche mit einer dunkelvioletten Aderung der blaß- bis intensiv gelben Blüten waren selten. Auch die Variabilität in den Fruchtmerkmalen war für die Unterart nicht außergewöhnlich groß (vgl. aber KÜHN et al. 1976 und HAMMER und HANELT 1979).

Raphanus sativus L.
var. *sativus*

Zliechov, Strážovské Vrchy (Nr. 470, 479).

var. *niger* KERNER

Trenčianske Bohuslavice, Váh-Tal (Nr. 108); Kroviniaci nördl. Horná Súča, Biele Karpaty (Nr. 273, Herb. 2263), Zliechov, Strážovské Vrchy (Herb. 2247).

Der Nachbau der Radieschenherkunft aus Zliechov ergab eine Population mit ± intensiv violett bzw. ± intensiv rot gefärbten runden bis plattrunden Knollen. – Runder schwarzer Rettich wird im Gebiet nicht nur Anfang August sondern auch im Frühjahr (Nr. 108, 273) ausgesät. Auch die tschechische Sorte 'Kulatá Černá' ging aus einer Landsorte hervor (BAREŠ and SEHNALOVÁ 1981).

Rosaceae

Rosa blanda AIT.

Podbranč westl. Myjava, Biele Karpaty (Herb. 2226).

Hausgarten, Zierstrauch am Zaun.

Rosa alba L.

Ilava/Váh (Herb. 2292).

Zierstrauch in einem Vorgarten, fruchtend.

Rosa turbinata AIT.

Zubák nordwestl. Púchov, Biele Karpaty (Herb. 2305).

Mehrere Ziersträucher in einem Vorgarten im mittleren Teil des Ortes, bis 5 m hoch, fruchtend, mit recht großen, gefüllten Blüten, unten spitzen, kegelförmigen Hagebutten und recht großen, kantigen Nüßchen.

Rosa centifolia L.

Podbranč westl. Myjava, Biele Karpaty (Herb. 2232); Zvratená östl. Vyškovec, Biele Karpaty (Herb. 2279); Drietomská Dolina südl. Starý Hrozenkov, Biele Karpaty (Herb. 2290, Garten bei Haus Nr. 655).

Ziersträucher in Gärten, etwa 1 m hoch, rosa- und rotblütig, gefüllt, unfruchtbar. Äste dünn, mit dünnen, meist borstenartigen Stacheln.

Sorbus domestica L.

Podbranč westl. Myjava, Biele Karpaty (Herb. 2233); Nová Bošáca, Biele Karpaty (Herb. 2252).

Einzelne Bäume in alten Obstanlagen. Die rundlichen, ca. 1,5 cm breiten Früchte (var. *pomifera* HAYNE) werden nach dem Abfallen im September gegessen und zur Bereitung eines qualitätsvollen Likörs verwendet.

Pyrus communis L. em. GAERTN.

Kroviniaci nördl. Horná Súča, Biele Karpaty (Herb. 2267).

In den Bauerngärten, an Feldrainen und in alten Obstanlagen sind überall im Gebiet noch alte Sorten von Birne, Apfel (*Malus domestica* BORKH.), Süßkirsche (*Cerasus avium* (L.) MOENCH) und Pflaume (*Prunus domestica* L.) anzutreffen (vgl. KÜHN et al. 1982). Bei Herb. 2267 handelt es sich um eine frühreife Form mit kleinen, gelben kugligen Früchten, die auch als 'Jacobs-' oder 'Stoppelbirne' bezeichnet wird (Abb. 4).

Pyrus pyraster L. BURGSD.

Predpoloma nördl. Nová Bošáca, Biele Karpaty (Herb. 2235, 2236); Kroviniaci nördl. Horná Súča, Biele Karpaty (Herb. 2265); Zliechov, Strážovské Vrchy (Herb. 2294).

Wildbirnen wurden in den Biele Karpaty mehrfach als Einzelbäume auf den Gebirgsweiden beobachtet.

Prunus cerasifera EHRH.

Nová Bošáca, Biele Karpaty (Herb. 2248).

Die Kirschpflaume wurde einmal als gelbfrüchtiger Straßenbaum festgestellt.

Prunus domestica L.
subsp. *insititia* (JUSL.) SCHNEIDER

Drietomská Dolina östl. Vyškovec, Biele Karpaty (Herb. 2282).

Niedrig geschnittene Sträucher wurden hier als Hecke am Wegrand gezogen.

Abb. 4 Alter Baumriese einer frühreifen Birnensorte bei Kroviniaci/Biele Karpaty. Foto K. PISTRICK

Leguminosae

Lotus corniculatus L.

Hvozdnica, Javorníky (Nr. 595).

Der Anbau dieser Art als Futterpflanze wurde auf einer kleinen Parzelle in Reinsaat beobachtet.

Cicer arietinum L.

Nová Bošáca, Biele Karpaty (Nr. 136).

Die Kichererbse stellt an der Nordgrenze ihres Anbauareals eine ausgesprochene Reliktkultur dar. Nur an einer Lokalität konnte noch vorjähriges Saatgut gesammelt werden. In morphologischer Hinsicht entspricht das Material Formen, die 1974 im Gebiet zwischen Krupina und Lučenec (Südslowakei) festgestellt wurden (vgl. KÜHN et al. 1976). Die hell bräunlich gelben, meist ± glatten, eulenkopfförmigen Samen zeichnen sich durch relativ geringe Größe aus (Höhe 8,0 mm, Tausendkornmasse 189,3 g).

Vicia angustifolia GRUFB.

subsp. *angustifolia*

Škratová nordöstl. Vyškovec, Biele Karpaty (Nr. 228, 234).

subsp. *segetalis* (THUILL.) GAUD.

Bošačky östl. Vyškovec, Biele Karpaty (Nr. 214); Horná Súča, Biele Karpaty (Nr. 255); Kroviniaci nördl. Horná Súča, Biele Karpaty (Nr. 264); Zubák nordwestl. Púchov, Biele Karpaty (Nr. 499).

Beide Sippen häufig als Unkraut in den Getreidefeldern des Sammelgebiets. Die Tausendkornmasse für die Herkünfte der subsp. *segetalis* liegt mit 17,3–19,7 g deutlich unter den für frühere slowakische und südpolnische Aufsammlungen angegebenen Werten.

Vicia sativa L.
convar. *sativa* var. *sativa*

Horná Súča, Biele Karpaty (Nr. 252); Kroviniaci nördl. Horná Súča, Biele Karpaty (Nr. 281); Zubák nordwestl. Púchov, Biele Karpaty (Nr. 504); Hvozdnica südwestl. Bytča, Javorníky (Nr. 540, 569); Strakošovce nördl. Štiavnik, Javorníky (Nr. 589).

Segetal in Getreidefeldern. Bei Hvozdnica (Nr. 540) zusammen mit Sommergerste kultiviert. Die Samen sind meist mittelgroß (Durchmesser 4,8–5,3 mm, Tausendkornmasse 49,0–52,5 g), nur bei Nr. 589 auffallend klein (4,2×3,8××2,7 mm, Tausendkornmasse 16 g). Bezüglich der Samenfarbe sind die Proben einheitlich hell bräunlich gelb (Nr. 589) bzw. hell graugrün, braun gesprenkelt und dunkelbraun gefleckt (Nr. 569) oder stellen Gemische von bis zu vier verschiedenen Farbvarianten dar.

Vicia faba L.
subsp. *faba* var. *equina* PERS.

Podbranč westl. Myjava, Biele Karpaty (Nr. 85, 101).

subsp. *minor* (PETERM. em. HARZ) ROTHM. var. *minor* PETERM. em. HARZ

Zliechov, Strážovské Vrchy (Nr. 459).

Die kleinkörnige Form der Ackerbohne gehört zu der in Mitteleuropa vorherrschenden, hochwüchsigen und wenig-triebigen Elatae-Gruppe der subvar. *minor*. Vereinzelt treten Samen mit hellem Nabel als Beimengung auf.

Im Material der subsp. *faba* überwiegt die subvar. *equina*. Einige Pflanzen mit leicht runzligen Hülsen müssen noch auf Merkmalskonstanz geprüft werden. Von einer Herkunft (Nr. 85) werden die Samen vor allem in grünem Zustand gekocht gegessen. Diese mittelhohe Sippe mit großer Standfestigkeit und 1–2 Hülsen je Blütenstand beansprucht züchterisches Interesse als Gemüseform für die Konservenindustrie.

Zur Anbaupraxis in den Westkarpaten vergleiche auch KÜHN et al. 1976 und HAMMER und HANELT 1979.

Lens culinaris MEDIK.

Podbranč westl. Myjava, Biele Karpaty (Nr. 86, 92); Petrova Lehota, Strážovské Vrchy (Nr. 315); Čierna Lehota, Strážovské Vrchy (Nr. 338).

Der in Zentraleuropa fast erloschene Linsenanbau konnte an einigen wenigen Lokalitäten im Sammelgebiet auf flachgründigen Kalkböden nachgewiesen werden. Die Samen sind hell graugrün bis bräunlich, vereinzelt auch ± schwach grau gefleckt und von mittlerer Größe (Durchmesser 6,0–6,3 mm).

Lathyrus sativus L.
subsp. *europaeus* ZALKIND

Moravské Lieskové nördl. Nové Mesto n. Váhom, Biele Karpaty (Nr. 45, 46, 47, Herb. 2206, 2207); Hrašné südwestl. St. Turá, Biele Karpaty (Nr. 65); Podbranč westl. Myjava, Biele Karpaty (Nr. 95); Nová Bošáca, Biele Karpaty (Nr. 118, 130, 131, Herb. 2243); Čierna Lehota, Strážovské Vrchy (Nr. 355); Zliechov, Strážovské Vrchy (Nr. 460, Herb. 2298).

Reste der früher bis nach Süddeutschland verbreiteten Kultur der Saatplatterbse wurden bereits 1974 für die südliche Slowakei angegeben (KÜHN et al. 1976). Die 1981 festgestellten Vorkommen liegen noch etwas weiter nördlich und erreichen in den Strážovské Vrchy eine Höhe von ca. 700 m NN. Der Anbau erfolgt auf kleinen Feld- oder Gartenparzellen, in Reihen am Rande der Feldstücke oder zusammen mit Kartoffeln. Mit Ausnahme einer Samenprobe (Nr. 355) stellen alle Herkünfte ein Gemisch aus Pflanzen mit blaß bläulich-violett gezeichneten und rein weißen Blüten dar. Der Zeichnung der Blüten entspricht eine braune Pigmentierung von Kiel und Nabelregion der gelblich weißen Samen. Bei einigen Nummern (118b, 131a) kommen auch auf den Seitenflächen zusätzlich leicht braun gesprenkelte Samen vor. Nach Samenmerkmalen läßt sich das Material der var. *leuco-depressus* ZALK. bzw. *marginato-depressus* ZALK. der prol. *mediterraneus* ZALK. zuordnen.

Die reifen Samen werden zur Zubereitung von Suppen genutzt.

Pisum sativum L.
subsp. *sativum*
convar. *speciosum* (DIERB.) ALEF. var. *speciosum*

Škratová nordöstl. Vyškovec, Biele Karpaty (Nr. 218a, b, Herb. 2256a, b); Horná Súča, Biele Karpaty (Nr. 251a–c); Kroviniaci nördl. Horná Súča (Nr. 280a); Zubák nordwestl. Púchov, Biele Karpaty (Nr. 505a–f, h, 510a, b, 518a, b, 524); Hvozdnica südwestl. Bytča, Javorníky (Nr. 541a–f, 554a, b, 558a, 565a–d, 575a–c, Herb. 2308).

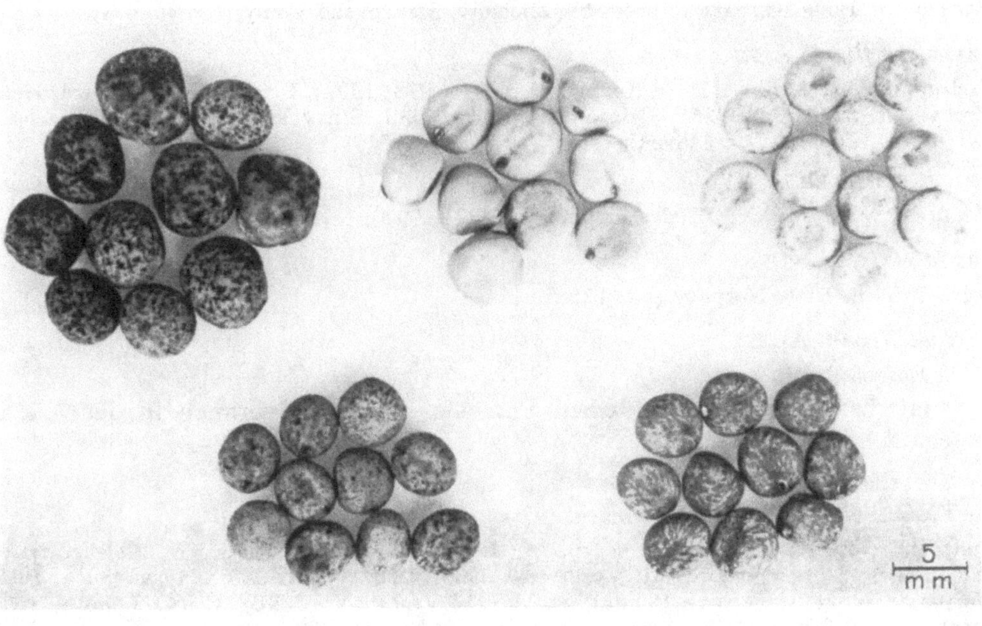

Abb. 5 Peluschken (*Pisum sativum* ssp. *sativum* convar. *speciosum*) wurden meist als Beimengung in Sommergetreide gesammelt. Außer der relativ großsamigen Form der subvar. *arvense* links oben (Nr. 218) stammen alle Samen aus dem Nachbau einer variablen Probe aus Zubák/Biele Karpaty (Nr. 505). In der Mitte oben: subvar. *caeruleum*; unten rechts: subvar. *thebaicum*. Foto G. TERPE

Die Pflanzen waren in der Samenfarbe variabel (vgl. Abb. 5). Sie können in folgende Subvarietäten eingereiht werden: Subvar. *thebaicum* (Alef.) Makash. – braun marmoriert, Nr. 251c, 281c, 505b, 510a, 558a, 565a. – subvar. *arvense* (Alef.) Makash. – violett punktiert, Nr. 218a, b, 251a, b, 505a, d, f, h, 541b–d, 565b, d, 575a, b. – Einfarbig hellgrüne bis ockerfarbige Samen (subvar. *caeruleum* (Alef.) Makash.) haben Nr. 251a, 505e, 510b, 528a, b, 524, 541a, f, 554 und 575c.

Die Peluschken waren Beimengung in Sommergetreide. Heute sind sie manchmal im Felde nicht gerne gesehen; in Hvozdnica wurden sie in zwei Feldern auf den Rain geworfen. Die Pflanzen sind auffallend hoch (0,8–1,2 m), vom Grunde an gut bestockt, oft im oberen Teil der Pflanze verästelt, mit mitteldickem Stengel, mit mittelgroßen bis großen Blüten (14–20 mm breit, kleiner in Nr. 541a) und mit recht kurzen, walzenförmigen, meist etwas von der Seite zusammengedrückten Hülsen. Die Samen sind rund bis tonnenförmig, 4–6 mm groß. Die Fahne der Blüte hat manchmal eine orangerote Basis (Nr. 505b). An den Blättchen bilden sich manchmal kurze Ranken. Die Pflanzen sind gegen Virosen wenig anfällig. Die Probe 218 weicht von den anderen Nummern durch sehr hohen Wuchs (1,5 m) ab, hat einen dicken Stengel, ist an der Basis wenig bestockt und hat walzenförmige Hülsen mit größeren violetten Samen (7 mm).

convar. *vulgare* Alef.
var. *glaucospermum* (Alef.) Gov.

Podbranč westl. Myjava, Biele Karpaty (Nr. 83b); Nová Bošáca, Biele Karpaty (Nr. 129, 132a); Petrova Lehota, Strážovské Vrchy (Nr. 325); Drietomská Dolina südl. Starý Hrozenkov, Biele Karpaty (Nr. 404b); Zliechov, Strážovské Vrchy (Nr. 46, 471).

var. *superfluens* Alef.

Podbranč westl. Myjava, Biele Karpaty (Nr. 83a, 99a, 100a), Petrova Lehota, Strážovské Vrchy (Nr. 323a, 329, 335); Drietomská Dolina südl. Starý Hrozenkov, Biele Karpaty (Nr. 404a); Zliechov, Strážovské Vrchy (Nr. 438, 451b).

var. *jesseni* Alef.

Podbranč westl. Myjava, Biele Karpaty (Nr. 89b, 99b).

var. *atrovirens* Körn.

Nová Bošáca, Biele Karpaty (Nr. 121a).

var. *vitellinum* Alef.

Nová Bošáca (Nr. 121b).

Schalerbsen wurden auf kleinen Parzellen für den Gebrauch in der Küche angebaut.

convar. *medullare* Alef.
var. *schneebergeri* Alef.

Podbranč westl. Myjava, Biele Karpaty (Nr. 83c, 99d); Nová Bošáca, Biele Karpaty (120, 121c); Škratová nordöstl. Vyškovec, Biele Karpaty (Nr. 238); Horná Súča, Biele Karpaty (Nr. 289); Petrova Lehota, Strážovské Vrchy (Nr. 323b); Čierna Lehota, Strážovské Vrchy (Nr. 341); Pet'ovka, Strážovské Vrchy (Nr. 388).

var. *waterlooense* Alef.

Podbranč westl. Myjava, Biele Karpaty (Nr. 89a, 99c), Nová Bošáca, Biele Karpaty (132b), Petrova Lehota, Strážovské Vrchy (Nr. 319), Zliechov, Strážovské Vrchy (Nr. 451a).

var. *balticum* ALEF.

Zliechov (Nr. 471b).

Markerbsen wurden auf kleinen Parzellen zur Gemüsenutzung kultiviert. Großsamig (10–11 mm) waren die Proben Nr. 89a und 120a.

Phaseolus vulgaris L.

var. *vulgaris* (Stangenbohnen)

Virides z. B. Nivnice (Nr. 44); Moravské Lieskové (Nr. 52b, c); Hrašné (Nr. 62c, 63a); Podbranč (Nr. 91a–c, 97d–j, 98b); Nová Bošáca (Nr. 122c–d, h–j, 134, 135a, b); Horná Súča (Nr. 261); Petrova Lehota (Nr. 313e, 314c, 320c, 321c, 322a); Čierna Lehota (Nr. 340, 351d, 352b); Pet'ovka (Nr. 364, 373, 391b, c); Brůsne (Nr. 415a, b, 416a, c); Vel. Košecké Podhradié (Nr. 423a, b, 428, 429, 431, 433, 434); Zliechov (Nr. 440a, 466b); Hvozdnica (Nr. 543, 546b).

Lutei z. B. Petrova Lehota (Nr. 326); Čierna Lehota (Nr. 342a, 351c).

var. *nanus* (JUSL.) ASCHERS. (Buschbohnen)

Virides z. B. Nivnice (Nr. 39, 41, 42, 43); Moravské Lieskové (Nr. 48, 49, 50a, b, 52a); Hrašné (Nr. 58b, c, 59a, b, 60); Hrušové (Nr. 104); Trenčíanske Bohuslavice (Nr. 109, 110a, b); Nová Bošáca (Nr. 122e, f, 210); Horná Súča (Nr. 259a); Kroviniaci (Nr. 274a, b); Petrova Lehota (Nr. 304, 305b, 310a, 312a, c, 313d, 317c, 320a, b, 334e); Čierna Lehota (Nr. 339a, 348); Pet'ovka (Nr. 368, 370, 376); Drietomská Dolina (Nr. 399); Brůsne

Abb. 6 Variable Bohnenprobe (*Phaseolus vulgaris* und *Ph. coccineus*) aus Podbranč/Biele Karpaty. Sie enthält u. a. im Gebiet selten gefundene zitronengelbe, schwarze und blaß grüne Samenfarbtypen der Gartenbohne. Foto G. TERPE

(Nr. 414a, c, d); Zliechov (Nr. 440b, 466c); Hor. Breznica (Nr. 482b); Zubák (Nr. 485, 488); Hvozdnica (Nr. 551, 552, 553).
Lutei z. B. Podbranč (Nr. 98a z. T.); Petrova Lehota (Nr. 327); Pet'ovka (Nr. 358, 366, 377, 387b); Drietomská Dolina (Nr. 410); Hor. Breznica (Nr. 482); Zubák (Nr. 487, 491); Hvozdnica (Nr. 546c).

Die Gartenbohne präsentiert sich als wichtigste Körnerleguminose des Gebiets in einer enormen Formenvielfalt (vgl. auch HANELT und HAMMER 1975, KÜHN et al. 1976 und 1982). Vermehrung und Bonitur des umfangreichen Materials sind noch nicht völlig abgeschlossen, so daß hier keine endgültige Einschätzung der Variabilität gegeben werden kann. Es überwiegen Trockenspeisebohnen mit Fäden.

Beim Anbau im Sammelgebiet wird oft wahrscheinlich nur nach Busch- und Stangentypen getrennt (Abb. 6). Dadurch treten schon innerhalb der Reihen und Parzellen in Hülsen- und Samenmerkmalen unterschiedliche Pflanzen auf. Nach der Auftrennung von Mischproben und der Zusammenfassung gleicher Typen vom gleichen Sammelort erhöhte sich die Zahl der Proben auf ca. 170. Allein nach der Samenfarbe lassen sich über 50 verschiedene Varianten unterscheiden (Tab. 1).

Tabelle 1
Samenfarbtypen von *Phaseolus vulgaris* (Probenzahl in Klammern hinter der Zahl der Varianten)

Grundfarbe \ Zeichnung	keine	nur 1–2 dunklere Ringe um den Nabel	flächige dunklere Zeichnung oder Fleckung
weiß	2 (32)	2 (3)	7 (12)
grün	1 (7)	–	–
gelb/ocker	1 (2)	11 (40)	3 (6)
beige/braun	1 (3)	10 (19)	12 (38)
rot/violett	–	1 (1)	2 (3)
schwarz	1 (4)	–	–
insgesamt	6 (48)	24 (63)	24 (59)

Phaseolus coccineus L.
var. *coccineus*

Podbranč westl. Myjava, Biele Karpaty (Nr. 97a, b); Petrova Lehota, Strážovské Vrchy (Nr. 328b, 333, 336b); Vel. Košecké Podhradie, Strážovské Vrchy (Nr. 424a, b, 427b).

var. *bicolor* (VELLOZO) VAN ESELT.

Petrova Lehota, Strážovské Vrchy (Nr. 314a, b, 322, 328a, 336a, Herb. 2275); Vel. Košecké Podhradie, Strážovské Vrchy (Nr. 427a, Herb. 2293).

var. *albiflorus* (DC.) BAIL.

Podbranč westl. Myjava, Biele Karpaty (Nr. 97c); Petrova Lehota, Strážovské Vrchy (Nr. 316); Čierna Lehota, Strážovské Vrchy (Nr. 350a); Vel. Košecké Podhradie, Strážovské Vrchy (Nr. 420); Zliechov, Strážovské Vrchy (Nr. 465); Hvozdnica südwestl. Bytča, Javorníky (Nr. 600).

In Umfang und Häufigkeit der Kultur wird die Feuerbohne überall im Sammelgebiet deutlich von der Gartenbohne übertroffen. Zum Teil waren Stangenformen der beiden *Phaseolus*-Arten in Mischkultur anzutreffen. Typische Buschbohnen fehlen im Feuerbohnen-Material. Allerdings wurden im Nachbau auch relativ schwach windende Formen festgestellt (Nr. 314b, 322, 465). Eine Herkunft zeichnete sich durch auffallend breite Hülsen (12,0×2,01×1,35 cm) und große Samen (23,0×14,7×10,8 cm) aus.

Umbelliferae

Petroselinum crispum (MILL.) NYM.
convar. *radicosum* (ALEF.) DANERT

var. *radicosum*

Podbranč (Nr. 82); Nová Bošáca (Nr. 128); Petrova Lehota (Nr. 330) Zliechov (Nr. 450).

Die meist in Hausgärten (z. B. Nr. 82) oder auf kleinen Gewürzpflanzenparzellen (Nr. 450) kultivierte, mehr oder weniger glattblättrige Wurzelpetersilie wird auch im Gebiet von den Bauern noch selbst vermehrt, worauf die öfter beobachteten, bis 2 m hohen Samenstände schließen ließen.

Carum carvi L.

Nová Bošáca (Nr. 124), aus dem Saatgutvorrat von Bauer OCHODNITZKY.

Anethum graveolens L.

Moravské Lieskové (Nr. 55); Petrova Lehota (Nr. 331), Pet'ovka (Nr. 379, 383); Vel. Košecké Podhradie (Nr. 426).

Der Dill wird meist auf kleinen Beeten reihenweise angebaut und wurde im östlichen Teil des Untersuchungsgebietes häufiger als im westlichen Teil (Biele Karpaty) beobachtet. Auch hier wurden die Landsorten oft noch selbst vermehrt.

Levisticum officinalis L.

Drietomská Dolina (Nr. 411), im Garten des Hauses Nr. 447.

Diese alte Gewürzpflanze wurde im Gebiet nur einmal gefunden und wirkte im Vorgarten des o. g. Hauses durch ihre Größe und glänzenden Blätter wie eine Zierstaude.

Pastinaca sativa L.
subsp. et var. *sativa*

Veletiny (Nr. 31); Drietomská Dolina (Herbar Nr. 2288), zur Saatgutgewinnung verwendet.

Labiatae

Elsholtzia ciliata (THUNB.) HYL.

Hvozdnica südwestl. Bytča, Javorníky (Herb. 2307).

In großer Menge in einem Hausgarten, vielleicht noch zu Heil- oder Gewürzzwecken kultiviert.

Salvia officinalis L.
ssp. *minor* (GMELIN) GAMS

Rudnik (Nr. 67).

Als bekannte Arznei- und Teepflanze wurde die Art nur in einem Hausgarten kultiviert gefunden („Kreuzsalbei"). Im Nachbau in Gatersleben fanden sich hellviolette (forma *rubriflora* ALEF.) und weißblütige (forma *albiflora* ALEF.) Pflanzen.

Mentha arvensis L.
ssp. *austriaca* (JACQ.) BRIQ.
Veletiny (Nr. 3), Bachufer.

Die Art wurde im Gebiet an Bachläufen häufig beobachtet.

Mentha spicata L. em. L.
Zubák (Nr. 538), im Ort am Weg (Herbar Nr. 2306).

Die Krauseminze ist hier vermutlich aus der Kultur als Arzneipflanze verwildert.

Mentha verticillata L.
Vyškovec (Herbar Nr. 2249), im Hausgarten HODULIK.

Die Art wird hier sicherlich als Teepflanze angebaut.

Rosmarinum officinalis L.

Petrova Lehota, örtlich in Hausgärten beobachtet.

Solanaceae

Solanum tuberosum L.
Hrušové östl. St. Turá, Biele Karpaty (Nr. 105, Herb. 2229); Pet'ovka, Strážovské Vrchy (Nr. 361).

Kartoffeln werden von den Bauern des Sammelgebiets wohl meist selbst vermehrt. Die Herkünfte zeigten im Nachbau am Institut für Kartoffelforschung Groß Lüsewitz ± langovale Knollen mit tiefer Augenlage (H. TIEMANN brieflich) und roter (Nr. 105) bzw. ocker Schalenfarbe. Ihre Virusresistenz war gering.

Capsicum annuum L.
Kroviniaci (Nr. 267, 268).

Von wenigen, am Feldrand verbliebenen Pflanzen waren nur unreife gelbliche bzw. kleinrundliche und dunkelviolette Früchte verfügbar. Die Samen waren unreif und keimten nicht im Nachbauversuch.

Scrophulariaceae

Rhinanthus alectorolophus (SCOP.) POLL.
var. *arvensis* (SEMLER) SCHNEIDER
Škrátová nordöstl. Vyškovec, Biele Karpaty (Nr. 226).

In Winterweizen. Saatgutunkraut auf armen Böden. Wurde auch in Wintergetreide um Vyškovec und um Podjavorník beobachtet. Die Samen haben einen schmalen, bis 0,3 mm breiten Saum.

Valerianaceae

Valeriana officinalis L.
Veletiny (Nr. 2), am Bachufer; Pet'ovka (Nr. 395) in Ortsnähe am Bach.

Diese alte Arzneipflanze fand sich überall auf den feuchten Randstreifen der Bäche, auf den Quellsümpfen und mehr oder weniger feuchten Wiesen der Talhänge der Biele Karpaty und Stražovke Vrchy.

Compositae

Inula helenium L.

Drietomská Dolina (Nr. 403), im Hausgarten der Familie KROCILOVA (Herbar Nr. 2291, 2292) und im Hausgarten HORBOVA Nr. 655 (Herbar Nr. 2289).

Vielleicht sind diese Pflanzen Relikte des früher häufigeren Anbaues als Droge für Mensch und Tier (Antiparasiticum) sowie als Küchen- und Genußmittelkraut. Ihre heutige Kultur in wenigen Exemplaren in den kleinen Wohnhaus-Vorgärten weist ihr wegen ihrer großen Blütenstände und des schönen Laubes wohl auch einen gewissen Zierwert zu.

Helianthus tuberosus L.

Podbranč (Nr. 87).

Das aus Nové Mesto n. Váhom stammende Pflanzgut von Topinambur erwies sich als wenig vital und starb in der Nachkultur ab.

Tanacetum parthenium (L.) SCHULTZ-BIP.
Drietomská Dolina, in Hausgärten zusammen mit *Ruta graveolens* L. kultiviert.

Chrysanthemum balsamita L.
Podbranč.

Diese alte südeuropäische Arznei- und Gewürzpflanze wurde in einem Hausgarten kultiviert gefunden.

Artemisia abrotanum L.

Drietomská Dolina östl. Vyškovec, Biele Karpaty (Herb. 2287).

Als Duftpflanze in einem Hausgarten kultiviert.

Lactuca sativa L.
convar. *incocta* HELM

var. *capitata* L.
Kopfsalat

Moravské Lieskove (Nr. 54); Podbranč (Nr. 81), Garten des Hauses Nr. 10; Podbranč (Nr. 90), Hausgarten Nr. 231; Petrova Lehota (Nr. 318); Pet'ovka (Nr. 394); Drietomská Dolina (Nr. 406), im Nachbau leider nicht gekeimt; Vel. Košecké Podhradie (Nr. 435); Zubák (Nr. 490); Hvozdnica (Nr. 596).

Die gesammelten Kopfsalate zeigen im Nachbau allgemein einen mittelgroßen oder manchmal auch kleinen Kopf mittlerer Festigkeit und zuweilen einen geringen Kopfschluß mit nicht immer gut deckendem Umblatt. Es zeichnet sich durch eine stark beulig-bucklige Spreite aus. Die Färbung des Kopfes ist gelblich-grün, lediglich die Nr. 54 zeigt leichte Anthocyanfärbung der Blattspreiten.

Krachsalat, Eissalat

Podbranč (Nr. 94), Garten des Bauern FILKA.

Die in Europa relativ selten kultivierten Krach- oder Eissalate fanden wir auf tiefgründigen, oft recht feuchten und doch sonnigen Hangkulturflächen der Biele

Karpaty. Im Nachbau in Gatersleben kam es zum Schossen ohne eigentliche Kopfbildung, weshalb wegen Verdachtes eines Schnittsalats eine Nachbestimmung in wiederholtem Anbau nötig ist.

var. *crispa* L.
Schnitt- oder Pflücksalat
Zliechov (Nr. 472).

Die besammelten Samenträger der in Gemüsekulturen oder zwischen Futterrüben vorhandenen Salate zeigten noch oft beachtliche basale Blattmasse, was auf gute edaphische Bedingungen im Gebiet hinweist. Die meist vorhandenen Samenträger lassen auf eine Eigenvermehrung durch die Bauern schließen.

Summary

Catalogue of indigenous taxa of cultivated plants collected 1981 in the ČSSR

The material of cultivated plants collected during a mission to western Slovakia in 1981 (cf. KÜHN et al. 1982) has been characterized botanically. It is represented mainly by land-races and old varieties which are often rather variable; their cultivation is mostly decreasing rapidly. Typical indigenous types could be discovered e.g. for *Triticum aestivum, Avena sativa, Papaver somniferum, Pisum sativum* and *Phaseolus vulgaris*. Remarkable is the occurrence if indigenous relic crops (e.g. *Triticum dicoccon, Secale cereale* var. *multicaule, Panicum miliaceum, Cicer arietinum, Lathyrus sativus* and *Fagopyrum esculentum*). The mountainous region belongs to the Western Carpathian refugial area of autochthonous taxa of cultivated plants. Because of rapidly acting genetic erosion a continuation of the collecting work seems to be necessary now.

Краткое содержание

Каталог местных форм культурных растений, собранных в 1981 г. в ЧССР

Дается ботаническая характеристика материала культурных растений, собранных во время экспедиции 1981 г. в горные области Западной Словакии (ср. Кюн с сотр., 1982). В большинстве случаев это местные сорта, старые сортотипы и культурные формы, отличающияся часто очень сильной изменчивостью; их возделование сокращается ускоренными темпами. Характерные местные формы могли быть ещё найдены, напр., для мягкой пшеницы, овса посевного, мака снотворного, гороха посевного и фасоли обыкновенной. Заслуживает внимания обнаружение явно реликтовых культур (напр.: *Triticum dicoccon, Secale cereale* var. *multicaule, Panicum miliaceum, Cicer arietinum, Lathyrus sativus, Fagopyrum esculentum*). Эта горная местность Словакии является областью сохранения автохтонных форм культурных растений в Западных Карпатах. Интенсивная генная эрозия требует скорейшего продолжения сборов исчезающих местных форм.

Den Mitarbeitern der Genbank in Gatersleben, insbesondere B. Fouquet, H. Lehmann, B. Schmidt, Chr. Schmidt, H. Selbig und K. Matthias sind wir für die bewährte Anzucht, Pflege und Bonitierung des Pflanzenmaterials zu großem Dank verpflichtet.

Literatur

Bareš, I., and J. Sehnalová, 1981: Preservation of land-races of cultivated plants in Czechoslovakia. – Kulturpflanze **29**, 67–77.

Danert, S., 1958: Zur Systematik von *Papaver somniferum* L. – Kulturpflanze **6**, 61–88.

Hammer, K., 1981: Problems of *Papaver somniferum* – classification and some remarks on recently collected European poppy land-races. – Kulturpflanze **29**, 287–296.

–, und P. Hanelt, 1979: Botanische Ergebnisse einer Reise in die VR Polen 1976 zur Sammlung autochthoner Landsorten von Kulturpflanzen. – Kulturpflanze **27**, 109–149.

Hanelt, P., und K. Hammer, 1975: Bericht über eine Reise nach Ostmähren und der Slowakei 1974 zur Sammlung autochthoner Sippen von Kulturpflanzen. – Kulturpflanze **23**, 207–215.

–, –, J. Schultze-Motel, und Władysław Kulpa, 1982: Katalog der 1978 in der VR Polen gesammelten indigenen Kulturpflanzen-Sippen. – Kulturpflanze **30**, 215–244.

Helm, J., 1956: Die zu Würz- und Speisezwecken kultivierten Arten der Gattung *Allium* L. – Kulturpflanze **4**, 130–180.

–, 1957: Versuch einer morphologisch-systematischen Gliederung der Art *Beta vulgaris* L. – Züchter **27** (5), 203–222.

Jones, H. A., and L. K. Mann, 1963: Onions and their allies. – London, New York.

Kühn, F., 1981: Problems of the infraspecific taxonomy of oat. – Kulturpflanze **29**, 267–272.

–, und K. Hammer, 1979: Das Ausklingen der Brandrodungskultur in Zentraleuropa. – Kulturpflanze **27**, 165–173.

–, –, und P. Hanelt, 1976: Botanische Ergebnisse einer Reise in die ČSSR 1974 zur Sammlung autochthoner Landsorten von Kulturpflanzen. – Kulturpflanze **24**, 283–347.

–, –, –, 1980: Botanische Ergebnisse einer Reise in die ČSSR 1977 zur Sammlung autochthoner Landsorten von Kulturpflanzen. – Kulturpflanze **28**, 183–226.

–, H. Ohle und K. Pistrick, 1982: Bericht über eine Reise in die ČSSR 1981 zur Sammlung autochthoner Sippen von Kulturpflanzen. – Kulturpflanze **30**, 245–254.

Lehmann, Chr. O., 1954: Das morphologische System der Saaterbse (*Pisum sativum* L. sens. lat. Gov. ssp. *sativum*). – Züchter **24** (11/12), 315–337.

Malcev, A., 1930: Ovsjugi i ovsy. – Bull. Appl. Bot., Leningrad, Suppl. 38, 522 pp.

Mansfeld, R., 1950: Das morphologische System der Saatgerste, *Hordeum vulgare* L. s. l. – Züchter **20**, 8–24.

–, 1951: Das morphologische System des Saatweizens, *Triticum aestivum* L. s. l. – Züchter **21**, 41–60.

Tempír, Z., 1966: Výsledky paleoetnobotanického studia pěstování zěmědelských rostlin na území ČSSR. – Vědecké Práce Československého Zemědělského Muzea, 27–144.

Dr. H. Ohle und Dipl.-Biol. K. Pistrick
Zentralinstitut für Genetik und Kulturpflanzenforschung
der Akademie der Wissenschaften der DDR
DDR - 4325 Gatersleben
Corrensstraße 3

Anlage

Originalliste des während der Reise in die westliche Slowakei 1981 gesammelten Saat- und Pflanzgut-Materials

Veletiny, nordwestlich Uher. Brod, nördliches Vorland der Biele Karpaty, 11. 8. 81

Bachufer, Grünland ca. 1 km nördlich des Ortes
1 *Allium sativum*
2 *Valeriana officinalis*
3 *Mentha arvensis* ssp. *austriaca*?
Felder ca. 1,5 km nördlich des Ortes
4 *Papaver somniferum*, Reihen in Futterrübenfeld
5 *Avena sativa*, z. T. begrannt, Haferfeld
6 *Avena sativa*, z. T. schmale Ährchen, evtl. violettspelzige Form, Haferfeld
7 *Hordeum vulgare distichon*, Beimischung in Haferfeld
8 *Hordeum vulgare distichon*, So-Gerstefeld
9 *Avena fatua*, glabrate Form aus 8
10 *Hordeum vulgare distichon*, schartige Ähren aus 8
11 *Hordeum vulgare distichon*, aufrechte Ähren aus 8
12 *Avena fatua* × *sativa*? aus 8
13 wie 12
14 *Triticum aestivum*, Wi-Form? aus 8
15 *Hordeum vulgare distichon*, So-Gerstefeld
16 *Avena fatua* mit schwarzen Deckspelzen aus 15
Felder ca. 1 km südöstlich des Ortes
17 *Hordeum vulgare distichon*, So-Gerstestoppeln
18 *Phalaris canariensis*, aus käuflichem Saatgut als Vogelfutter kultiviert
19 *Echinochloa crus-galli*, sehr hohe Pflanzen aus 18
20 *Avena sativa*, Hafer-Gerste-Weizenfeld mit Luzerne-Untersaat
21 *Avena sativa*, Schwarzhafer ?, aus 20
22 *Avena sativa*, schmalrispig, aus 20
23 *Hordeum vulgare distichon*, aus 20
24 *Triticum aestivum*, So-Form, aus 20
25 *Triticum aestivum*, begrannt, aus 20
26 *Allium vineale*, Unkraut in Hafer-Gerste-Weizenfeld
27 *Triticum aestivum*, So-Form, aus 26
28 *Hordeum vulgare distichon*, aus 26
29 *Avena sativa*, aus 26
30 *Avena fatua*, kleinspelzig mit hellen Deckspelzen, aus 26
Kleine Felder am Südostrand des Ortes
31 *Pastinaca sativa*, einzelner Samenträger, angeblich Bastard mit *Petroselinum crispum*
32 *Setaria italica*, aus käuflichem Saatgut ('Senegal-Hirse') als Vogelfutter kultiviert

Nivnice, südlich Uher. Brod, nördliches Vorland der Weißen Karpaten, 11. 8. 81
Felder ca. 1 km südöstlich des Ortes
33 *Papaver somniferum*, aus selbst vermehrtem Saatgut
34 *Papaver somniferum*, mit kleinen Kapseln zwischen Futerrüben
35 *Papaver somniferum*, Reihe zwischen Futterrüben
36 *Panicum miliaceum*, Kleine Parzelle, Saatgut aus der Slowakei
37 *Panicum miliaceum*, Stoppelacker
38 *Fagopyrum esculentum*, frisch gemähtes Feld
39 *Phaseolus vulgaris nanus*, Trockenbohne, Samen weiß mit schwarzem Nabelring
40 *Phaseolus vulgaris nanus*, Trockenbohne, Samen weiß
41 *Phaseolus vulgaris nanus*, Trockenbohne, Samen gelb mit violettem Nabelring

42 *Phaseolus vulgaris nanus*, Trockenbohne, wie 41
43 *Phaseolus vulgaris nanus*, Trockenbohne, Samen weiß mit violettem Nabelring
44 *Phaseolus vulgaris nanus*, Trockenbohne, Samen groß, weiß
57 *Hordeum vulgare distichon*, So-Gerstestoppeln

Moravské Lieskové, nördlich Nové Mesto n. Váhom, Biele Karpaty, 11. 8. 81
Felder am westlichen Ortsrand
45 *Lathyrus sativus*, Blüten weiß
46 *Lathyrus sativus*, Blüten mit bläulich-violetter Zeichnung
47 *Lathyrus sativus*, Blüten weiß
48 *Phaseolus vulgaris nanus*, Samen gelblich mit braunem und violettem Nabelring
49 *Phaseolus vulgaris nanus*, unreife Hülsen und Samen weiß, violett gefleckt
50 *Phaseolus vulgaris nanus*, Samen gelb mit hellbraunem Nabelring
51 *Phaseolus vulgaris nanus*, Gemisch verschiedener Hülsenformen und Samengrößen, Samen weiß
52 *Phaseolus vulgaris nanus*, Samen grün
53 *Allium oleraceum*? Brutzwiebeln, Wegrand am Zaun eines Obstgartens
54 *Lactuca sativa*, schmalblättriger Kopfsalat, Samenträger zwischen Futterrüben
55 *Anethum graveolens*, Samenträger zwischen Kartoffeln

Nové Mesto n. Váhom, 11. 8. 81
56 *Pisum sativum*, kleine Parzellen am östl. Stadtrand

Hrašné, südwestlich St. Turá, Biele Karpaty, 12. 8. 81
Gemüsefelder am östlichen Ortsrand
58 *Phaseolus vulgaris nanus*, Trockenbohne, Hülsen unreif violett gefleckt, Samen weiß, dunkler marmoriert
59 *Phaseolus vulgaris nanus*, Trockenbohne, Samen gelb
60 *Phaseolus vulgaris nanus*, Trockenbohne, Samen gelblich mit violettem Nabelring
61 *Phaseolus vulgaris nanus*, Trockenbohne, unreife Hülsen und Samen rot gefleckt
62 *Phaseolus vulgaris nanus*, Trockenbohne, Hülsen breit, Samen weiß und groß
63 *Phaseolus vulgaris vulgaris*, Hülsen unreif rot gefleckt, bei Reife violett gefleckt, Samen unreif rot gefleckt, bei Reife braun gefleckt
64 *Phaseolus vulgaris nanus*, Gemüsebohne, Hülsen gelb, Samen rot gefleckt
65 *Lathyrus sativus*, Blüten weiß, z. T. mit bläulich-violetter Zeichnung

Rudník, südwestlich St. Tura, Biele Karpaty, 12. 8. 81
Südwestrand des Ortes
66 *Papaver somniferum*, mit kleinen Kapseln, zwischen Futterrüben
67 *Salvia officinalis*, Hausgarten
68 *Avena sativa*, schmalkörnig

Podbranč, westlich Myjava, Biele Karpaty, 12. 8. 81
Myjava-Tal, ca. 1 km nördlich Burg Branč
69 *Triticum aestivum*, Wi-Form, begrannt, Beimischung in kleinem Wi-Weizenfeld

Ortsteil Hor. Dolina, kleine Felder im Myjava-Tal am Ostrand der Siedlung
70 *Avena sativa*, verschiedene Typen, Feldbestand
71 *Avena sativa*, braunspelzig, z. T. ca. 1,15 m hoch, aus 70
72 *Avena fatua*, hellspelzig, aus 70
73 *Avena fatua glabrata*, aus 70
74 *Triticum aestivum*, Wi-Form, Saatgutvorrat von M. Batková
75 *Triticum aestivum* und *Secale cereale*, Wi-Formen, wie 74
76 *Hordeum vulgare distichon*, mit Beimengungen, wie 74
77 *Zea mays*, Kolben mit gelben und grauen Körnern, wie 74
106 *Atriplex hortensis*, rot gefärbt. Zierpflanze am Zaun eines Hausgartens

ca. 1 km nordwestlich Burg Branč

78 *Avena sativa*, Hafer-Gerste-Weizenfeld
79 *Hordeum vulgare distichon*, aus 78
80 *Triticum aestivum*, So-Form?, aus 78

Ortsteil Podzámok

81 *Lactuca sativa*, Hausgarten der Bäuerin JANOVICHOVÁ, Podzámok Nr. 10
82 *Petroselinum crispum*, glattblättrig, wie 81
83 *Pisum sativum*, niedrig, wie 81
84 *Triticum dicoccon*, frisch ausgedroschen, K. TOMANOVÁ, Podzámok Nr. 16
85 *Vicia faba*, großsamig, wie 84

Ortsteil U Matúšu

86 *Lens culinaris*, wahrscheinlich frisches Erntegut der Bäuerin PREKOPKOVÁ, Podbranč Nr. 231
87 *Helianthus tuberosus*, wenig Pflanzenmaterial aus Nr. 231, Knollen stammen aus Nové Mesto
88 *Commelina communis*, Zierpflanze am Haus Nr. 231
89 *Pisum sativum*, hoch, Hausgarten Nr. 231
90 *Lactuca sativa*, großköpfig, Samenträger im Hausgarten Nr. 231
91 *Phaseolus vulgaris vulgaris*, Trockenbohne, Hülsen und Körner rot gefleckt, Hausgarten Nr. 231
92 *Lens culinaris*, Saatgutvorrat der Bäuerin K. PAVLÍKOVÁ
93 *Papaver somniferum*, frisches Erntegut des Bauern FÍLKA
94 *Lactuca sativa*, krausblättrig, Bauer FÍLKA
95 *Lathyrus sativus*, Saatgutvorrat des Bauern FÍLKA
96 *Triticum dicoccon*, wie 95
97 *Phaseolus vulgaris* und *coccineus*, Gemisch, wie 95
98 *Phaseolus vulgaris nanus*, Gemüsebohne, ohne Fäden, wie 95
99 *Pisum sativum*, gelb- und grünsamig, wie 95
100 *Pisum sativum*, gelbsamig, wie 95
101 *Vicia faba*, wie 95

Hrušové, östlich St. Turá, Biele Karpaty, 12. 8. 81

kleine Felder am Nordrand des Ortes

102 *Avena sativa*, kleiner Feldbestand
103 *Avena sativa*, begrannt, aus 102
104 *Phaseolus vulgaris nanus*, unreife Hülsen und Samen violett gefleckt
105 *Solanum tuberosum*?, Blätter schräg nach oben abstehend, Knolle rot

Beckov, Váh-Tal, 13. 8. 81

westexponierter Kalkfelshang südlich der Burg

107 *Allium flavum*

Trenčianske Bohuslavico, Váh-Tal, 13. 8. 81

Gemüseparzellen am südöstlichen Ortsrand

108 *Raphanus sativus niger*, Rübe schwarz, Samenträger z. T. bereits mit reifen Früchten
109 *Phaseolus vulgaris nanus*, Gemüsebohne, unreife Hülsen grün, Samen weiß
110 *Phaseolus vulgaris nanus*, Trockenbohne, frühe Sorte, junge Hülsen rot gefleckt oder grün, Samen rot gefleckt
111 *Phaseolus vulgaris nanus*, Trockenbohne, Hülsen kurz, Samen weiß

Nová Bošáca, Biele Karpaty, 13. 8. 81

Nordrand des Ortes

112 *Avena sativa*, Haferfeld mit Gerste
113 *Hordeum vulgare distichon*, aus 112
114 *Avena sativa*, verschiedene Typen aus Garben eines kleinen Haferfeldes
115 *Hordeum vulgare distichon*, Beimengung in 114

116 *Avena fatua meridionalis?*, Unkraut in Kartoffelfeld
117 *Papaver somniferum*, frisches Erntegut, A. PETRECHOVA, Nová Bošáca Nr. 441
118 *Lathyrus sativus*, Saatgutvorrat S. OCHODNITZKY, Nová Bošáca Nr. 440
119 *Papaver somniferum*, wie 118
120 *Pisum sativum*, Zuckererbse, wie 118
121 *Pisum sativum*, wie 118
122 *Phaseolus vulgaris*, verschiedenfarbige Trockenbohnen, wie 118
123 *Triticum aestivum*, Wi-Form, wie 118
124 *Carum carvi*, wie 118

Ortsteil Močariny

125 *Triticum aestivum milturum*, Feld von D. UKOVÁ östlich der Siedlung
126 *Triticum aestivum erythrospermum*, wie 125
127 *Allium sativum*, Winterknoblauch, bei der Ernte am Wegrand verlorene Zwiebel, wahrscheinlich aus dem Anbau von D. UKOVÁ, östlich der Siedlung
128 *Petroselinum crispum*, glattblättrig, Samenträger am Rande des Feldes von D. UKOVÁ, östlich der Siedlung
129 *Pisum sativum*, niedrig, grünsamig, Parzelle von L. KRAISL östlich der Siedlung
130 *Lathyrus sativus*, Blüten mit bläulich-violetter Zeichnung, wie 129
131 *Lathyrus sativus*, Saatgutvorrat von L. KRAISL, Nová Bošáca Nr. 390
132 *Pisum sativum*, niedrig, grünsamig, wie 131
133 *Papaver somniferum*, wie 131
134 *Phaseolus vulgaris nanus*, wie 131
135 *Phaseolus vulgaris vulgaris*, wie 131
136 *Cicer arietinum*, verschiedene Samentypen, 1981 nicht angebaut, wie 131
137 *Zea mays*, wie 131
138 *Allium sativum*, Sommerknoblauch, frisches Erntegut von L. KRAISL
139 *Allium sativum*, Winterknoblauch, wie 138
140 *Allium* spec., am Gartenzaun, Hausgarten L. KRAISL

Predpoloma, nördlich Nová Bošáca, Biele Karpaty, 13. 8. 81

Südhang nordöstlich des Ortes

141 *Secale cereale*, Feldbestand
142 *Triticum aestivum*, Wi-Form, einzelne Pflanzen als Beimischung in 141
143 *Spergula arvensis*, Unkraut aus 144
144 *Avena sativa*, Rispen-Hafer, Hafer-Landsorte, Feldbestand
145 *Avena sativa*, Fahnenhafer, aus 144
146 *Avena sativa*, Schwarzhafer, aus 144
147 *Avena fatua*, aus 144
148 *Hordeum vulgare distichon*, Körner gelb, aus 144
149 *Hordeum vulgare distichon*, Körner an der Basis braun, aus 144
150 *Lolium temulentum*, begrannt, Unkraut aus 144
151 *Vicia angustifolia*, Unkraut aus 144
152 *Agrostemma githago*, Unkraut aus 144
153 *Silene gallica*, Unkraut aus 144
154 *Avena sativa*, Rispenhafer, weiterer Feldbestand einer Hafer-Landsorte
155 *Avena sativa*, Fahnenhafer aus 154
156 *Hordeum vulgare distichon*, Körner an der Basis braun, aus 154
157 *Hordeum vulgare distichon*, aus 154
158 *Hordeum vulgare distichon*, aufrechte Ähren aus 154
159 *Triticum aestivum*, Wi-Form, Streifen am Rande von 154
160 *Triticum aestivum*, Wi-Form, kleiner Stoppelacker
161 *Lolium temulentum*, Unkraut aus 154
162 *Vicia sativa* ssp. *sativa*, Unkraut aus 154
163 *Bromus secalinus*, Wi-Form, am Rande von 154
164 *Bromus secalinus*, So-Form?, Unkraut aus 154
165 *Allium porrum*, zur eigenen Saatgutgewinnung, Hausgarten von A. KABELIKOVA, Predpoloma Nr. 354

Vyškovec, Biele Karpaty, ca. 500–750 m NN, 14. 8. 81

166 bis 169 *Raphanus raphanistrum*, relativ dickfrüchtige Pflanzen, Unkraut aus einem Kartoffelfeld, Hang unterhalb vom Berge Kykula

170 bis 172 *Raphanus raphanistrum*, Petalen unterschiedlich geadert, Unkraut aus 181 (Buchweizenfeld)

173 *Avena sativa*, verschiedene Formen, Haferfeld unterhalb vom Berge Kykula

174 *Hordeum vulgare distichon*, Beimengung aus 173

175 *Hordeum vulgare distichon*, Ähren z. T. schartig, So-Gerstenfeld unterhalb vom Berge Kykula

176 *Avena sativa*, Beimengung aus 175

177 *Hordeum vulgare distichon*, So-Gerstenfeld am Westhang oberhalb des Sattels

178 *Avena sativa*, Beimengung aus 177

179 *Avena fatua*, z. T. sehr dickkörnig, Unkraut aus 177

180 *Fagopyrum esculentum*, kleines Feld am Sattel, Nordosthang, 1981 letztmalig angebaut

181 *Fagopyrum esculentum*, weiteres Feld, wie 180

182 *Triticum aestivum*, So-Form?, aus 180

183 *Triticum aestivum*, Wi-Form, kleines Feld am Sattel, Nordosthang, 1981 letztmalig angebaut

184 *Bromus secalinus*, Unkraut aus 183

185 *Hordeum vulgare distichon*, Hafer-Gerstenfeld am Sattel, Südwesthang

186 *Avena sativa*, wie 185

187 *Avena sativa*, Haferfeld am Sattel, Südwesthang

188 *Hordeum vulgare distichon*, Beimengung aus 187

189 *Triticum aestivum*, z. T. var. *ferrugineum*, Weizen-Haferfeld nordöstlich vom Sattel

190 *Avena sativa*, z. T. Fahnenhafer, aus 189

191 *Avena sativa*, Schwarzhafer, aus 189

192a *Hordeum vulgare distichon*, Beimengung aus 189

192b *Avena sativa*, Haferfeld nordöstlich vom Sattel

193 *Avena sativa* × *fatua*?, aus 192b

194 *Hordeum vulgare distichon*, Beimengung in 192b

195 *Triticum aestivum*, So-Form, Feldbestand am Sattel, Nordosthang

196 *Allium cepa*, Schalotte, Hausgarten von Š. Hodulik, Vyškovec Nr. 68

197 *Triticum aestivum*, So-Weizenfeld mit Gerste und Hafer aus eigenem Saatgut von Š. Hodulik, Nordhang unterhalb vom Berge Kykula

198 *Hordeum vulgare distichon*, wie 197

199 *Avena sativa*, wie 197

200 *Vicia sativa*, Unkraut aus 197

201 *Secale cereale*, Roggenfeld, aus eigenem Saatgut von Š. Hodulik, Nordhang unterhalb vom Berge Kykula

202 *Triticum aestivum*, Wi-Form, Beimengung in 201

203 *Avena sativa*, Haferfeld, aus eigenem Saatgut von Š. Hodulik, Nordhang unterhalb vom Berge Kykula

204 *Triticum aestivum*, Wi-Form, z. T. var. *erythrospermum*, Feldbestand, aus eigenem Saatgut von Š. Hodulik, Nordhang unterhalb vom Berge Kykula

205 *Bromus secalinus*, Unkraut aus 204

206 *Avena sativa*, Gemisch, Haferfeld mit hohem Schwarzhafer-Anteil

Nová Bošáca, Biele Karpaty, 15. 8. 81

Bachgrund zwischen Ortsmitte und Ortsteil Močariny

207 *Allium oleraceum*, Straßenrand

208 *Allium* spec., Straßenrand

209 *Triticum aestivum*, Wi-Form, verschiedene Typen von einem Stoppelacker

210 *Phaseolus vulgaris*, Samen gelblich mit hellbraunem und grauem Nabelring

Bošačky, östlich von Vyškovec, Biele Karpaty, 15. 8. 81

Kleine Felder im Talgrund

211 *Avena sativa*, Gemisch, Hafer-Gerstenfeld
212 *Avena sativa*, Fatuoid, aus 211
213 *Avena sativa* ×*fatua*? aus 211
214 *Vicia angustifolia* subsp. *angustifolia*, Unkraut aus 211
215 *Triticum aestivum*, Wi-Form, Feldbestand am unteren Osthang
216 *Secale cereale*, Beimischung in 215

Lopeník, Ortsteil Škratová, nordöstlich von Vyškovec, Biele Karpaty, ca. 650–750 m NN, 15. 8. 81

217 *Secale cereale*, frisches Erntegut von O. Kročil, nordwestlich vom Lopeník-Sattel
218 *Pisum sativum speciosum*, Beimischung in einem Haferfeld aus eigenem Saatgut von O. Kročil (früher in Reinkultur angebaut), nordwestlich vom Lopeník-Sattel
219 *Vicia angustifolia*, Unkraut aus 218
220 *Avena sativa*, wie 218
221 *Raphanus raphanistrum*, Unkraut aus 218
222 *Raphanus raphanistrum*, wie 221
223 *Raphanus raphanistrum*, wie 221

Südwesthang, ca. 0,5 km nördlich vom Lopeník-Sattel

224 *Triticum aestivum*, Wi-Form, z. T. rotährig, Feldbestand
225 *Secale cereale*, Beimengung aus 224
226 *Rhinanthus alectorolophus*, Unkraut aus 224
227 *Papaver somniferum*, kleine Parzelle
228 *Vicia angustifolia*, Unkraut aus 229
229 *Avena sativa*, Feldbestand
230 *Spergula arvensis* ssp. *sativa*, Unkraut aus 229
231 *Raphanus raphanistrum*, Unkraut aus 229
232 *Raphanus raphanistrum*, wie 231
233 *Spergula arvensis*, am Rande eines Kartoffelfeldes
234 *Vicia angustifolia*, Unkraut am Rande eines Kartoffelfeldes
235 *Avena sativa*, Feldbestand
236 *Vicia angustifolia*, Unkraut aus 235
237 *Raphanus raphanistrum*, 1 Pflanze mit intensiv gelben Blüten in hellgelb blühender Population, Unkraut aus 235

Südhang, ca. 1 km nördlich vom Lopeník-Sattel

238 *Pisum sativum*, niedrig, kleine Parzelle
239 *Triticum aestivum*, Wi-Form, Garben eines Weizenfeldes
240 *Vicia angustifolia*, Unkraut aus 239

Bošačky, östlich von Vyškovec, Biele Karpaty, ca. 650 m NN, 15. 8. 81

Südhang ca. 1 km südöstlich vom Mikulčin vrch

241 *Avena sativa*, Gemisch, z. T. Schwarzhafer, Hafer-Gerstenfeld
242 *Hordeum vulgare distichon*, wie 241
243 *Lolium temulentum*, Unkraut aus 241
244 *Avena sativa*, Gemisch, z. T. Schwarzhafer, weiteres Hafer-Gerstenfeld
245 *Hordeum vulgare distichon*, wie 244
246 *Avena fatua*, grauspelzig, Unkraut aus 244
247 *Triticum aestivum*, Wi-Form, Feldbestand
248 *Allium oleraceum*, bebuschter Südhang, ca. 1,5 km südöstlich vom Mikulčin vrch

Horná Súča, Biele Karpaty, 16. 8. 81

Felder des Bauern Kramár, nordwestlich des Ortes, Horná Súča Nr. 967; nur eigenes Saatgut verwendet

249 *Hordeum vulgare distichon*, Körner z. T. an der Basis braun, Stoppelfeld

250 *Avena sativa*, verschiedene, weiß- und gelbspelzige Typen, aus Garben eines Haferfeldes
251 *Pisum sativum speciosum*, als Unkraut betrachtet, aus 250
252 *Vicia sativa* ssp. *sativa*, aus 250
253 *Hordeum vulgare distichon*, Ähren z. T. schartig oder auffallend lang, Körner z. T. an der Basis braun, aus 250
254 *Claviceps purpurea*, auf *Agropyron repens*, aus 250
255 *Vicia angustifolia*, Unkraut aus 250
256 *Papaver somniferum*, kleine Parzelle am Haus Nr. 967
257 *Triticum aestivum*, Wi-Form, Stoppeln eines Weizenfeldes
258 *Triticum aestivum milturum*, Wi-Form, in einem Strauß aus Getreideähren, Tisch in der Gastwirtschaft Horná Suča
259 *Phaseolus vulgaris nanus*, Trockenbohne, Samen gelblich mit braunem und grauem Nabelring, kleine Parzelle
260 *Phaseolus vulgaris nanus*, Samen weiß mit braunem Nabelring und braunem Längsstreifen an der Raphe, zwischen den Reihen eines Maisfeldes
261 *Phaseolus vulgaris vulgaris*, halbniedrig (Sprosse bis 2 m lang), unreife Hülsen auf grünem Grund dicht violett gefleckt, Streifen am Feldrand
262 *Beta vulgaris*, weiße Futterrübe, einzelne Samenträger

Kroviniaci, nördlich von Horná Súča, Biele Karpaty, 16. 8. 81

Westhang

263 *Triticum aestivum*, Wi-Form, z. T. begrannt, Stoppeln eines Winterweizenfeldes
264 *Vicia angustifolia segetalis*, Unkraut aus 265
265 *Avena sativa*, Gemisch, Hüllspelzen z. T. etwas violett, Haferfeld
266 *Phaseolus vulgaris nanus*, unreife Hülsen grün, später rötlich überlaufen, Samen rosa, Reihe am Feldrand
267 *Capsicum annuum*, Früchte unreif gelbgrün, länglich, wenige Pflanzen am Feldrand
268 *Capsicum annuum*, Früchte klein und rund, unreif grün, wenige Pflanzen am Feldrand
269 *Avena sativa*, zur Hälfte abgeerntetes Haferfeld
270 *Claviceps purpurea* auf *Agropyron repens*, aus 269
271 *Avena sativa* ×*fatua*?, braun begrannt, aus 269
272 *Beta vulgaris*, weiße Futterrübe, Samenträger in einem Gemüsefeld
273 *Raphanus sativus niger*, schwarzer Sommerrettich, Reihe am Feldrand
274 *Phaseolus vulgaris nanus*, Reihe am Feldrand
275 *Papaver somniferum*, kleines Feld
276 *Papaver somniferum*, Kapseln lang oval mit Längsrinnen, graublau bereift, Feldbestand zusammen mit 277
277 *Papaver somniferum*, Kapseln kugelig, blaßgrau-grün bereift, Feldbestand zusammen mit 276
278 *Avena sativa*, Gemisch, Stoppeln eines Gerste-Hafer-Weizenfeldes
279 *Avena sativa*, Fahnenhafer?, aus 278
280 *Pisum sativum speciosum*, aus 278
281 *Vicia sativa*, aus 278
282 *Triticum aestivum*, Wi-Form, aus 278
283 *Secale cereale*, aus 278
284 *Avena sativa*, aus 278
285 *Hordeum vulgare distichon*, aus 278
286 *Avena sativa*, relativ dunkelkörnig, Feldbestand
287 *Hordeum vulgare distichon*, Feldbestand
288 *Hordeum vulgare distichon*, Körner z. T. an der Basis bräunlich, Feldbestand
289 *Pisum sativum*, hochwüchsig, in einem Maisstreifen am Feldrand
290 *Papaver somniferum*, Kapseln z. T. dunkelviolett, Reihe am Feldrand
291 *Hordeum vulgare distichon*, verschiedene Typen, Feldbestand
292 *Raphanus raphanistrum*, Unkraut am Rande eines Futterrübenfeldes

293 *Phaseolus vulgaris nanus*, Trockenbohne, Samen hellbraun, Streifen am Rande eines Futterrübenfeldes
294 *Triticum aestivum*, Wi-Form, Stoppeln eines Weizenfeldes
295 *Secale cereale*, Beimengung aus 294
296 *Spergula arvensis*, Unkraut in einem Kartoffelfeld
297 bis 300 *Raphanus raphanistrum*, Potalen unterschiedlich geadert, Unkraut aus einem Kartoffelfeld
301 *Hordeum vulgare distichon*, Feldbestand
302 *Allium* spec., Erdrutsch in einer alten Obstplantage zwischen Brombeeren, Südhang
303 *Secale cereale*, Schober hinter einem Haus

Petrova Lehota, Strážovské Vrchy, 17. 8. 81

Kleine Parzellen am Osthang westlich des Ortes
304 *Phaseolus vulgaris nanus*, Blüten blaßrosa, unreife Hülsen grün, Samen graubraun
305 *Phaseolus vulgaris nanus*, Trockenbohne, Samen gelblich, um den Nabel dunkler
306 *Hordeum vulgare distichon*, Stoppeln eines So-Gerstenfeldes
307 *Papaver somniferum*, abgeerntetes Feld, nur noch wenige Kapseln vorhanden
308 *Papaver somniferum*, teilweise abgeerntetes Feld
309 *Triticum aestivum*, Wi-Form, begrannt, einzelne Ähren neben einem umgepflügten Feld
310 *Phaseolus vulgaris nanus*, unreife Hülsen gelb, Samen blaßgelb
311 *Triticum aestivum*, Wi-Form, mit dem Mähdrescher abgeerntetes Weizenfeld, nur wenige Körner
312 *Phaseolus vulgaris nanus*, Trockenbohne, Hülsen gelb, frühe Sorte
313 *Phaseolus vulgaris nanus*, Trockenbohne, Gemisch, Samen schwarz mit weißem Nabel, grau, rosa, rot gefleckt, braun mit dunkler geadertem Nabel
314 *Phaseolus vulgaris vulgaris*
315 *Lens culinaris*, erstmals angebaut, Parzelle von S. Blahó
316 *Phaseolus coccineus*, Blüten weiß, Parzelle von S. Blahó
317 *Phaseolus vulgaris nanus*, Gemisch, Samen z. T. weiß oder weiß mit roten Flecken
318 *Lactuca sativa*, Samenträger
319 *Pisum sativum*, hoch, Blüten weiß, Samen gelb
320 *Phaseolus vulgaris nanus*, Gemisch, Samen gelblich mit grauem Nabelring, gelb oder weiß mit grauer Aderung
321 *Phaseolus vulgaris nanus*, Trockenbohne, Gemisch, Hülsen z. T. violett oder dicht dunkelviolett gefleckt, dicht dunkelviolett gefleckte Hülsen gehören zu einer besonders späten Sorte, Parzelle von A. Lobnova
322 *Phaseolus coccineus*, niedrig, Hülsen rot gestreift
323 *Pisum sativum*, hoch, weißblütig, Samen gelb
324 *Phaseolus vulgaris nanus*, Samen weiß mit roter Zeichnung
325 *Pisum sativum*, niedrig, grünsamig
326 *Phaseolus vulgaris vulgaris*, Gemüsebohne
327 *Phaseolus vulgaris nanus*, Gemüsebohne, Samen weiß, rosa gefleckt
328 *Phaseolus coccineus*, Blüten rot, unreife Hülsen gelb, z. T. violett gestreift
329 *Pisum sativum*, niedrig, Blüten weiß, Samen gelb
330 *Petroselinum crispum*, glattblättrig, einzelne Samenträger auf Grünland neben einem Feld
331 *Anethum graveolens*, einzelne Samenträger auf Grünland neben einem Feld
332 *Phaseolus vulgaris vulgaris*, Gemisch, Hülsen z. T. rot gefleckt
333 *Phaseolus coccineus*, Hülsen breit
334 *Phaseolus vulgaris nanus*, Gemisch
335 *Pisum sativum*
336 *Phaseolus coccineus*, Fahne orangerot?, Flügel rosa?, Pflanzen dieser Blütenfarbe neben rot- und weißblühenden Exemplaren an einem Gartenzaun am westlichen Ortsrand

Čierna Lehota, Strážovské Vrchy, 17. 8. 81

Kleine Parzellen an den Hängen am nordwestlichen Ortsrand; Proben aus dem Ort

337 *Papaver somniferum*, kleine Parzelle

338 *Lens culinaris*, kleine Parzelle

339 *Phaseolus vulgaris nanus*, Trockenbohne, Gemisch

340 *Phaseolus vulgaris vulgaris*, Hülsen gelb mit roten Flecken, Samen weiß mit roten Flecken

341 *Pisum sativum*, Markerbse, Samen grün

342 *Phaseolus vulgaris vulgaris*, Trockenbohne, Hülsen gelb, Probe frisch geernteter Hülsen von J. MIŠAK, Čierna Lehota Nr. 109

343 *Phaseolus vulgaris nanus*, Trockenbohne, Hülsen violett gefleckt, wie 342

344 *Phaseolus vulgaris vulgaris*, Trockenbohne, Hülsen rot gefleckt, wie 342

345 *Phaseolus vulgaris*, Gemisch, rotsamig, Samenprobe von J. MIŠAK, Čierna Lehota Nr. 109

346 *Phaseolus vulgaris*, Gemisch, Samen braun oder rotviolett gefleckt, wie 345

347 *Allium sativum*, Vorrat der Bäuerin RAGANOVÁ, Čierna Lehota Nr. 112

348 *Phaseolus vulgaris nanus*, frisch ausgerissene Pflanzen von der Bäuerin RAGANOVÁ, Čierna Lehota Nr. 112

349 *Phaseolus vulgaris vulgaris*, Gemüsebohne, Samen weiß, Saatgutvorrat der Bäuerin RAGANOVÁ, Čierna Lehota Nr. 112

350 *Phaseolus vulgaris vulgaris*, Gemüsebohne, Samen weiß und groß, wie 349

351 *Phaseolus vulgaris vulgaris*, Gemüsebohne, Hülsen angeblich gelb, Samen dunkelbraun, wie 349

352 *Phaseolus vulgaris vulgaris*, Trockenbohne, Gemisch, Hülsen angeblich grün, frühe Sorte, wie 349

353 *Phaseolus vulgaris vulgaris*, Trockenbohne, Gemisch, Hülsen grün, Samen z. T. hellbraun mit braunem Nabelring, wie 349

354 *Phaseolus vulgaris vulgaris*, Trockenbohne, hoch, Hülsen angeblich gelb, Samen halb weiß/halb schwarz, wie 349

355 *Lathyrus sativus*, frisch geerntete Hülsen aus dem Garten von M. MIŠAKOVÁ, Čierna Lehota Nr. 10

Krásna Ves, Strášovské Vrchy, 17. 8. 81

356 *Linum tenuifolium*, Nordosthang am Nordwestrand von Krásna Ves, Bodenerosionsfläche auf Kalk

Pet'ovka, Strážovské Vrchy, 17. 8. 81

Kleine Felder, Nordosthang am Südende des Ortes

357 *Hordeum vulgare distichon*, schmalkörnig, Feldbestand

358 *Phaseolus vulgaris nanus*, Trockenbohne, Samen gelblich mit grauem Nabelfleck

359 *Phaseolus vulgaris nanus*, Trockenbohne, Samen weiß

360 *Phaseolus vulgaris nanus*, Trockenbohne, Samen gelb

361 *Solanum tuberosum*, selbst vermehrt

362 *Phaseolus vulgaris vulgaris*, Samen gelb

363 *Phaseolus vulgaris vulgaris*, Samen weiß

364 *Phaseolus vulgaris nanus*, Trockenbohne

365 *Phaseolus vulgaris nanus*, Trockenbohne, unreife Hülsen und Samen violett gefleckt

366 *Phaseolus vulgaris nanus*, Trockenbohne, Samen schwarz-weiß

367 *Phaseolus vulgaris nanus*, Trockenbohne, Samen weiß

368 *Phaseolus vulgaris nanus*, Trockenbohne, Samen weiß mit hellbraunen Flecken

369 *Phaseolus vulgaris nanus*, Trockenbohne, unreife Hülsen und Samen rot gefleckt

370 *Phaseolus vulgaris nanus*, Trockenbohne, Samen groß, weiß mit rotbrauner Zeichnung

371 *Papaver somniferum*

372 *Phaseolus vulgaris vulgaris*, Blüten intensiv violett, unreife Hülsen dunkelviolett, Samen weiß

373 *Phaseolus vulgaris vulgaris*, Samen ockerfarben

374 *Phaseolus vulgaris vulgaris*, Hülsen oft etwas rot marmoriert, Samen blaßbräunlich mit hellbraunem Nabelring

375 *Phaseolus vulgaris nanus*, Trockenbohne, Samen gelblich
376 *Phaseolus vulgaris nanus*, Trockenbohne, Samen hellgrün
377 *Phaseolus vulgaris nanus*, Trockenbohne, Samen weiß, dunkelviolett gefleckt
378 *Phaseolus vulgaris nanus*, Trockenbohne, Samen rosa
379 *Anethum graveolens*, Samenträger
380 *Allium cepa*, Samenträger auf einem Feld, jeweils mehrere an Stöcke angebunden
381 *Allium cepa*, Schalotte?, Zwiebeln vom Feld
382a *Triticum aestivum*, Wi-Form, meist begrannt, Körner weiß
382b *Phaseolus vulgaris nanus*, Hülsen unreif meist braun gestreift, Samen weiß
383 *Triticum aestivum ferrugineum*, Wi-Form, sehr variabel, Ähren oft kurz und dicht, Stoppeln eines Weizenfeldes
384 *Triticum aestivum*, Wi-Form, sehr variabel, Ähren oft kurz und dicht, wie 383
385 *Papaver rhoeas*, am Rande der Stoppeln eines Weizenfeldes
386 *Papaver rhoeas*, Unkraut in einer Rotkleeparzelle
387 *Phaseolus vulgaris nanus*, Trockenbohne, Gemisch, Samen z. T. weiß mit hellbrauner Zeichnung oder weiß und viereckig
388 *Pisum sativum*, mittelhoch
389 *Phaseolus vulgaris nanus*, Trockenbohne, Samen grün
390 *Phaseolus vulgaris nanus*, Gemisch, Samen z. T. rosa
391 *Phaseolus vulgaris nanus*, sehr früh, Hülsen kurz, Samen gelblich
392 *Phaseolus vulgaris nanus*, Trockenbohne, Samen braun
393 *Anethum graveolens*, Samenträger
394 *Lactuca sativa*, Pflücksalat?, Blütenstand breit, Samenträger
395 *Valeriana officinalis*, ungemähte Wiese am Bach, Südostrand des Ortes

Drietomská Dolina, südlich Starý Hrozenkov an der slowakisch-mährischen Grenze, Biele Karpaty, 18. 8. 81

Felder und Gärten am Nordosthang
396 *Claviceps purpurea* auf *Lolium multiflorum*, in 398
397 *Triticum aestivum*, Wi-Form, Streifen am Rande eines Wi-Weizen-Stoppelfeldes
398 *Hordeum vulgare distichon*, Körner z. T. an der Basis braun, Streifen am Rande des Weizenfeldes 397
399 *Phaseolus vulgaris nanus*, Samen braun
400 *Papaver somniferum*, kleine Parzelle
401 *Papaver somniferum*, Schüttmohn, mit 402 in einem kleinen Mohnfeld, Hausgarten Kročilová
402 *Papaver somniferum*, Schließmohn, mit 401 in einem kleinen Mohnfeld, Hausgarten Kročilová
403 *Inula helenium*, Höhe ca. 2 m, Hausgarten Kročilová
404 *Pisum sativum*, halbhoch, Samen gelb und grün, Hausgarten Kročilová
405 *Hordeum vulgare distichon*, kleines Stoppelfeld
406 *Lactuca sativa*, einzelne Samenträger im Hausgarten Horbova, Drietomská Dolina Nr. 655
407 *Hordeum vulgare distichon*, Stoppeln eines So-Gerstenfeldes, Hang hinter dem Haus Nr. 655
408 *Triticum aestivum*, Wi-Form, Beimengung in 407
409 *Avena sativa*, am Rande von 407
410 *Phaseolus vulgaris nanus*, Samen hellgelb mit braunem und schwärzlichem Nabelring, Rand eines Kartoffelfeldes, Hang hinter dem Haus Nr. 655
411 *Levisticum officinale*, Hausgarten Nr. 447
412 *Allium vineale*, in einem Gebüsch, Wiese an der Drietomica
413 *Papaver rhoeas*, 1 Pflanze an einem Gartenzaun im Drietomica-Tal

Brůsne, nördlich Drietoma, Biele Karpaty, 18. 8. 81

Felder im Tal nördlich der Brücke
414 *Phaseolus vulgaris nanus*, Gemisch, kleine Parzelle
415 *Phaseolus vulgaris vulgaris*, halbhoch, Hülsen unreif rot gefleckt auf gelbem Grund

416 *Phaseolus vulgaris vulgaris*, hoch und Phaseolus vulgaris nanus, Gemisch
417 *Papaver somniferum*, kleine Parzelle, größtenteils abgeerntet
418 *Hordeum vulgare distichon*, Stoppeln eines So-Gerstenfeldes aus eigenem Saatgut
419 *Avena fatua*, gelbspelzig, am Rande von 418

Vel. Košecké Podhradie, Strážovské Vrchy, 19. 8. 81
Kleine Parzellen östlich des Ortes
420 *Phaseolus coccineus*, Samen weiß
421 *Phaseolus vulgaris vulgaris*, unreife Hülsen dunkelviolett, Samen weiß mit hellbraunem Nabelring
422 *Phaseolus vulgaris vulgaris*, Samen braun gefleckt mit hellbraunem Nabelring
423 *Phaseolus vulgaris vulgaris*, Trockenbohne, unreife Hülsen gelb, Samen weiß
424 *Phaseolus coccineus*, Samen blaßviolett mit violetten Flecken und rotviolettem Nabelring
425 *Phaseolus vulgaris nanus*, Gemüsebohne, späte Sorte, unreife Hülsen gelb
426 *Anethum graveolens*, Samenträger
427 *Phaseolus coccineus*, Fahne orangerot, Flüge lrosa?
428 *Phaseolus vulgaris vulgaris*, Hülsen dicht behaart, unreif, blaß rötlich, violett gefleckt, Samen blaßviolett
429 *Phaseolus vulgaris vulgaris*, Hülsen lang und gelb, Samen weiß
430 *Atriplex hortensis*, rot gefärbt, zahlreiche Pflanzen zwischen den Parzellen, zum Abhalten von Wühlmäusen angebaut
431 *Phaseolus vulgaris vulgaris*, Hülsen meist rötlich gefleckt, Samen weiß
432 *Phaseolus vulgaris vulgaris*, Trockenbohne, Samen gelblich mit hellbraunem Nabelring
433 *Phaseolus vulgaris vulgaris*, Trockenbohne, unreife Hülsen dicht intensiv rot gefleckt, Samen violett gefleckt
434 *Phaseolus vulgaris vulgaris*, unreife Hülsen gelb, breit und mittellang
435 *Lactuca sativa*, Samenträger auf kleiner Parzelle

Zliechov, unterhalb des Berges Strážov, Strážovské Vrchy, 19. 8. 81
Kleine Felder und Gemüseparzellen am Südwesthang südöstlich des Ortes, ca. 650–750 m NN
436 *Beta vulgaris*, weiße Futterrübe, Samenträger
437 *Phaseolus vulgaris nanus*, Trockenbohne, Samen gelblich mit braunem und grauem Nabelring
438 *Pisum sativum*, halbhoch
439 *Papaver somniferum*, Samen weiß
440 *Phaseolus vulgaris nanus*, Trockenbohne, Samen weißlich mit braunem Nabelring
450 *Petroselinum crispum*, glattblättrig, großsamig, ca. 1,80 m hoher Samenträger
451 *Pisum sativum*, mittelhoch
452 *Papaver somniferum*, Kapseln kurz und breit, Samen dunkelrot-violett
453 *Triticum aestivum*, Wi-Form, Gemisch, Ähren weiß, Feldbestand
454 *Triticum aestivum*, Wi-Form, Ähren hart und weiß, aus 453
455 *Triticum aestivum ferrugineum*, Wi-Form, aus 453
456 *Hordeum vulgare distichon*, Garben eines So-Gerstenfeldes
457 *Papaver somniferum*, Gemisch, Kapseln, z. T. violett, oval oder rundlich, kleines Feld
458 *Phaseolus vulgaris vulgaris*, unreife Hülsen dunkelviolett
459 *Vicia faba*, Pflanze mit unreifen Hülsen aus einer Reihe am Feldrand
460 *Lathyrus sativus*, Blüten weiß oder mit bläulich-violetter Zeichnung, kleines Feld bei ca. 700 m NN
461 *Pisum sativum*
462 *Hordeum vulgare distichon*, Ährenfeld, Garben eines So-Gerstenfeldes
463 *Allium* spec., Brutzwiebeln, Feldrand
464 *Papaver somniferum*, Kapseln violett
465 *Phaseolus coccineus*, niedrig, Triebe meist <1 m, Blüten weiß, Reihe ohne Stützen

466 *Phaseolus vulgaris nanus*, Gemisch, unreife Hülsen z. T. rötlich
467 *Papaver somniferum*
468 *Papaver somniferum*, Kapseln z. T. violett
469 *Hordeum vulgare distichon*, Feld aus eigenem Saatgut von A. Vašková
470 *Raphanus sativus sativus*, rotes rundes Radieschen?, bei einer Pflanze zahlreiche Früchte mit stark entwickeltem Basalglied, einzelne Samenträger zwischen Gemüseparzellen, daneben 479
471 *Pisum sativum*
472 *Lactuca sativa*
473 *Papaver somniferum*
474 *Hordeum vulgare distichon*, Garben eines Sommergerstenfeldes
475 *Hordeum vulgare vulgare*, seltene Beimengung in 474
476 *Hordeum vulgare distichon*, aufrechte Ähren?, Beimengung in 474
477 *Triticum aestivum*, So-Form?, Beimengung in 474
478 *Claviceps purpurea* auf *Festuca pratensis*, aus 474
479 *Raphanus sativus niger*?, Kopf der Rübe violett, unten gelblich, einzelner Samenträger neben 470
480 *Vicia cracca*?, in Garben eines So-Gerstenfeldes
481 *Pisum sativum*

Hor. Breznica westlich Púchov, Biele Karpaty, 20. 8. 81

Westhang nördlich des Ortes
482 *Phaseolus vulgaris nanus*, Trockenbohne, unreife Hülsen gelb, Samen weiß, dunkelviolett gefleckt
483 *Triticum aestivum*, z. T. begrannt, verlorene Ähren am Wegrand

Zubák, nordwestlich Púchov, Biele Karpaty, 20. 8. 81

Felder und Gemüseparzellen nördlich des Ortes
484 *Papaver somniferum*, Kapseln violett
485 *Phaseolus vulgaris nanus*, Trockenbohne, Samen hellbraun
486 *Fagopyrum esculentum*, schlechter Ansatz, wahrscheinlich aus dem Verkehr gezogene Zuchtsorte
487 *Phaseolus vulgaris nanus*, Trockenbohne, unreife Hülsen rot gestreift, Samen rot gefleckt
488 *Phaseolus vulgaris nanus*, Trockenbohne, unreife Hülsen gelb, Samen gelbbraun mit braunem und grauem Nabelring
489 *Avena sativa*, z. T. begrannt, Stoppeln eines Haferfeldes
490 *Lactuca sativa*, einzelne Samenträger zwischen Futterrüben
491 *Phaseolus vulgaris nanus*, Gemisch, Samen z. T. hellbraun
492 *Papaver somniferum*, Samen hellbraun, einzelne Kapseln von einer abgeernteten Parzelle

Felder südwestlich des Ortes
493 *Hordeum vulgare distichon*, Körner z. T. an der Basis bräunlich, Stoppeln eines So-Gersten-Haferfeldes
494 *Hordeum vulgare distichon*, Körner z. T. an der Basis bräunlich, Stoppeln eines So-Gerstenfeldes
495 *Avena sativa*
496 *Avena fatua*, kleine Ährchen, Unkraut in einem Kartoffelfeld
497 *Fagopyrum esculentum*, vom Vorjahr?, seltene Beimengung in einem Kartoffelfeld
498 *Avena sativa*, dick- und schmalkörnige Typen, von Vögeln und Wild stark geschädigtes Feld am Waldrand, ob zur Wildfütterung angebaut?
499 *Vicia angustifolia*, Grasweg in einem Fichtenforst westlich des Ortes
500 *Papaver rhoeas*, Milchsaft an der Luft rötend, Grabenrand am Wege westlich des Ortes

Osthang in einem Seitental südlich der Ortsmitte
501 *Hordeum vulgare distichon*, Stoppeln eines So-Gersten-Haferfeldes

502 *Avena sativa*, wie 501
503 *Agrostemma githago*, aus 501
504 *Vicia sativa* subsp. *sativa*, aus 501
505 *Pisum sativum speciosum*, aus 501
506 *Triticum aestivum lutescens*, So-Form?, 1 Ähre aus 501
507 *Vicia sativa* dicksamig, aus 501

Nordosthang südlich der Ortsmitte

508 *Lolium temulentum*, aus dem Haferfeld 516
509 *Triticum aestivum*, So-Form? Beimengung in 516
510 *Pisum sativum speciosum*, aus 516
511 *Fagopyrum esculentum*, vom Vorjahr?, Beimengung in 516
512 *Avena sativa*, Schwarzhafer, aus 516
513 *Avena sativa*, Fahnenhafer, aus 516
514 *Avena sativa grisea*, aus 516
516 *Avena sativa*, verschiedene Typen, Rispe z. T. ausgebreitet, Hüllspelzen z. T. begrannt oder violett überlaufen, Haferfeld
517 *Avena sativa macrantha*, aus 516
518 *Pisum sativum speciosum*, Garben des Haferfeldes 521
519 *Triticum aestivum*, So-Form?, Beimengung in 521
520 *Lolium temulentum*, Unkraut aus 521
521 *Avena sativa*, Garben eines Haferfeldes
522 *Fagopyrum esculentum*, Buchweizenfeld oberhalb der Häuser
523 *Brassica rapa*, Unkraut aus 522
524 *Pisum sativum speciosum*, aus 522
525 *Avena fatua*, Unkraut aus 522
526 *Triticum aestivum*, Wi-Form, Stoppeln eines Wi-Weizenfeldes, 2 Pflanzen am Rande
527 *Phaseolus vulgaris nanus*, Trockenbohne, unreife Hülsen gelb, Samen blaßgelb mit dunklem Nabelring, Reihe am Feldrand
528 *Sinapis arvensis*, Schoten borstig, Unkraut aus 522
529 bis 531 *Raphanus raphanistrum*, Petalen schwarzviolett geadert, Unkraut aus einem Kartoffelfeld am Nordwesthang oberhalb des Ortes
532 bis 537 *Raphanus raphanistrum*, Petalen unterschiedlich geadert, Unkraut aus 522
538 *Mentha spicata*, Straßenrand im Dorf

Hvozdnica, südwestlich Bytča, Javorníky, 21. 8. 81

Nordosthang oberhalb des Ortes

539 *Hordeum vulgare distichon*, So-Gersten-Saatwickenfeld
540 *Vicia sativa* subsp. *sativa*, wie 539
541 *Pisum sativum speciosum*, aus 539
542 *Avena sativa*, Beimischung in 539
543 *Phaseolus vulgaris nanus*, Trockenbohne, unreife Hülsen grün mit roten Streifen, Reihe am Feldrand
544 *Avena fatua septentrionalis*, Unkraut aus einem Kartoffelfeld
545 *Avena sativa*, z. T. begrannt, Stoppeln eines Haferfeldes
546 *Phaseolus vulgaris nanus*, Trockenbohne, unreife Hülsen z. T. rötlich gefleckt, Samen weiß, z. T. grau marmoriert
547 *Phaseolus vulgaris nanus*, Trockenbohne, Samen hellbraun
548 *Phaseolus vulgaris nanus*, Trockenbohne, unreife Hülsen dunkelviolett gestreift
549 *Phaseolus vulgaris nanus*, Trockenbohne, unreife Hülsen gelb, Samen ockerfarben
550 *Avena sativa*, z. T. begrannt, Garben eines Haferfeldes
551 *Phaseolus vulgaris nanus*, Trockenbohne, Samen braun, Nabelfleck dunkelbraun
552 *Phaseolus vulgaris nanus*, Trockenbohne, Samen gelb mit braunem Nabelring
553 *Phaseolus vulgaris nanus*, Trockenbohne, Gemisch

Felder an den Hängen des Seitentales westlich des Ortes
554 *Pisum sativum speciosum*, aus 555
555 *Avena sativa*, Garben eines So-Gersten-Haferfeldes
556 *Hordeum vulgare distichon*, wie 555
557 *Papaver somniferum*, Samen hellgrau
558 *Pisum sativum speciosum*, Stoppeln eines So-Gerstenfeldes
559 *Phaseolus vulgaris nanus*, unreife Hülsen violett
560 *Triticum aestivum*, Wi-Form, sehr variabel, Ähren weiß oder braun, begrannt oder unbegrannt, z. T. schmalährig, Feldbestand
561 *Agrostemma githago*, Unkraut aus 560
562 *Vicia angustifolia segetalis*, Unkraut aus 560
563 *Triticum aestivum*, Wi-Form, Samtweizen?, aus 560
564 *Avena sativa*, Beimengung in einem So-Gerstenfeld
565 *Pisum sativum speciosum*, an den Rand geworfen, evtl. z. T. bei der Arbeit gegessen, aus 566
566 *Avena sativa*, Stoppeln eines Hafer-So-Gerstenfeldes
567 *Hordeum vulgare distichon*, wie 565
568 *Agrostemma githago*, Unkraut aus 566
569 *Vicia sativa sativa*, aus 570
570 *Hordeum vulgare distichon*, Körner z. T. mit brauner Basis, Hafer-So-Gerstenfeld
571 *Agrostemma githago*, Unkraut aus So-Gerstenfeld
572 *Avena sativa*, Beimengung in 571
573 *Vicia angustifolia*, Unkraut aus 571
574a *Avena sativa*, z. T. Fahnenhafer, aus 574b
574b *Avena sativa*, sehr variabel, z. T. in Schwaden gemähtes Haferfeld
575 *Pisum sativum speciosum*, aus 574b
576 *Agrostemma githago*, Unkraut aus 574b
577 *Lolium temulentum*, Unkraut aus 574b
578 *Avena sativa – fatua*, Zwischenform, aus 574b
579 *Hordeum vulgare distichon*, Beimengung in 574b

Štiavnik, Javorníky, 21. 8. 81

580 *Triticum aestivum*, So-Form, Feldbestand, östliches Seitental, ca. 2 km nördlich des Ortes
581 *Triticum aestivum ferrugineum, lutescens, erythrospermum*, So-Formen, Feldbestand östlich des Baches im Talgrund nördlich des Ortes

Strakošovce, nördlich Štiavnik, Javorníky, 21. 8. 81

582 *Triticum aestivum*, So-Form, Gemisch, z. T. var. milturum und ferrugineum, Feldbestand westlich des Baches im Talgrund
583 *Triticum aestivum*, So-Form, Ähren grau, wohl z. T. krank, aus 582
584 *Vicia angustifolia*, Unkraut aus 582
585 *Hordeum vulgare distichon*, Pflanzen z. T. ca. 1,50 m hoch, Beimengung in 582
586 *Avena sativa*, Beimengung in 582
587 *Triticum aestivum*, Gemisch, Ähren z. T. grau, Garben eines So-Weizenfeldes am Hang östlich des Ortes
588 *Vicia angustifolia*, Unkraut aus 587
589 *Vicia sativa sativa*, aus 587
590 *Avena sativa*, Gemisch, Feldbestand am Hang östlich des Ortes
591 *Secale cereale*, Wi-Form, Stoppeln eines Winterroggenfeldes am Hang östlich des Ortes
592 *Secale cereale* var. *multicaule*, 1980 zuletzt geerntet, Saatgutvorrat von M. Plávsky, Strakošovce Nr. 584
593 *Secale cereale* var. *multicaule*, 1980 zuletzt geerntet, Saatgutvorrat von J. Kollárik, Strakošovce Nr. 588
594 *Allium sativum*, Abfallhaufen an einer Bachböschung im Ort

Hvozdnica, südwestlich Bytča, Javorníky, 21. 8. 81
Südwesthang oberhalb des Ortes
595 *Lotus corniculatus*, kleine Parzelle
596 *Lactuca sativa*, Samenträger in einem Vorgarten am Ortsrand
597 *Fagopyrum esculentum*, Gemisch, Früchte braun und geflügelt oder grau mit dunkelbraunen Streifen und ungeflügelt, Garben eines Buchweizenfeldes
598 *Raphanus raphanistrum*, Unkraut aus 597
599 *Sinapis arvensis*, Unkraut aus 597
600 *Phaseolus coccineus*, Blüten weiß, Zaun eines Hausgartens am Ortsrand

Kulturpflanze 32 · 1984 · 197–206

Report on a travel to the Socialist People's Libyan Arab Jamahiriya 1983 for the collection of indigenous taxa of cultivated plants

PIETRO PERRINO[1], G. BATTISTA POLIGNANO[1], KARL HAMMER and CHRISTIAN O. LEHMANN

(Eingegangen am 23. Januar 1984)

Summary

The mission was carried out within the Mediterranean Programme of the IBPGR in April/May 1983 with participants from the Socialist People's Libyan Arab Jamahiriya (Agricultural Research Centre, Tarabulus), the Istituto del Germoplasma, Bari, and the Zentralinstitut für Genetik und Kulturpflanzenforschung, Gatersleben. In Tripolitania, the Fezzan- and Ghadāmis-areas 316 samples, mostly cereals and vegetables, have been collected. In Tripolitania and the Ghadāmis-area the land-races have been nearly completely replaced by new varieties. In the Fezzan genetic erosion is in progress. There, several cereal- and vegetable-land-races could still be collected. The land-races may be useful for breeding programmes.

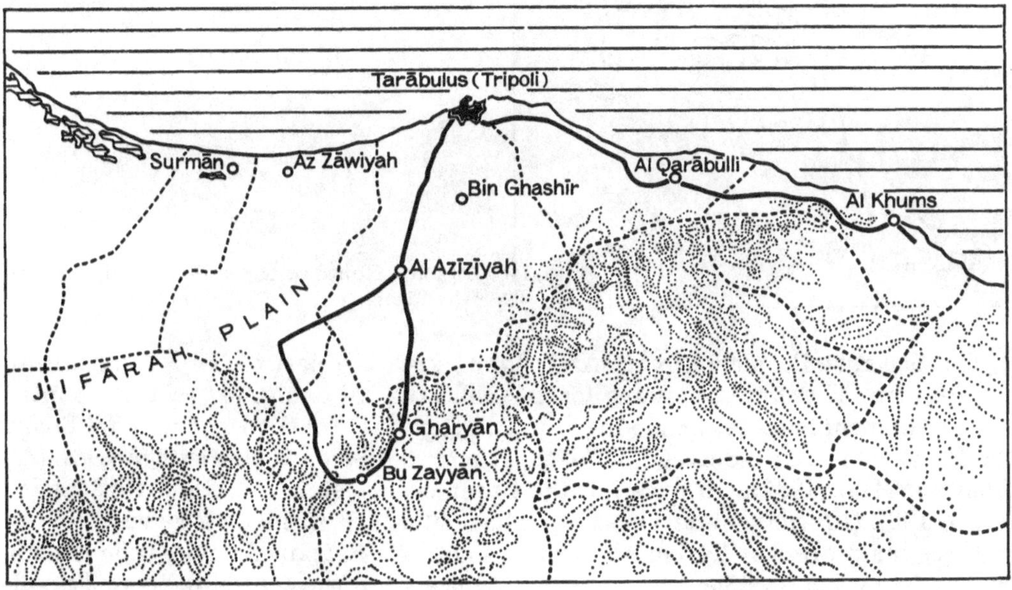

Fig. 1 Collecting route in Tripolitania

1 Istituto del Germoplasma, Via G. Amendola 165/A, I-70126 Bari, Italy.

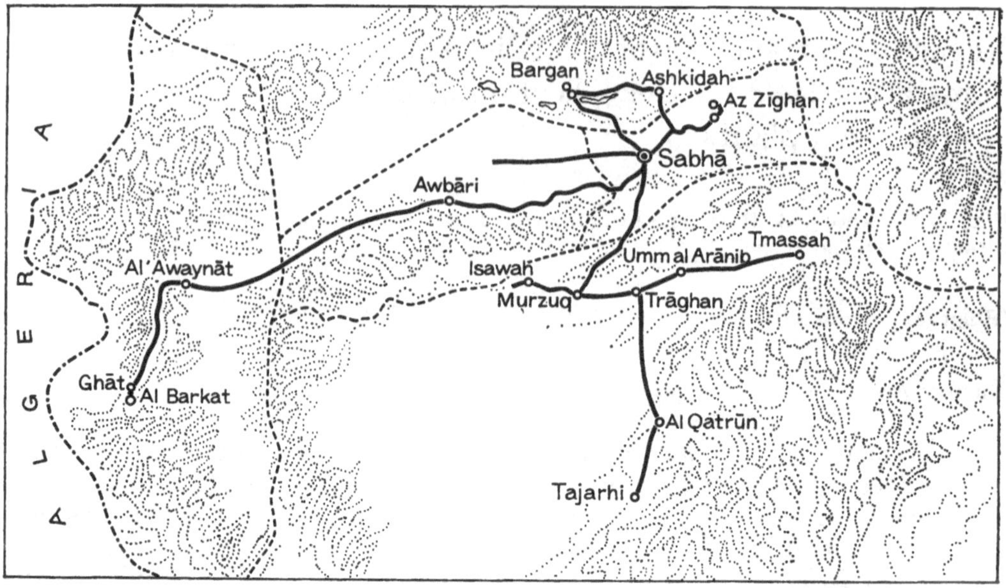

Fig. 2 Collecting route in the Fezzan, including the Ghāt oases

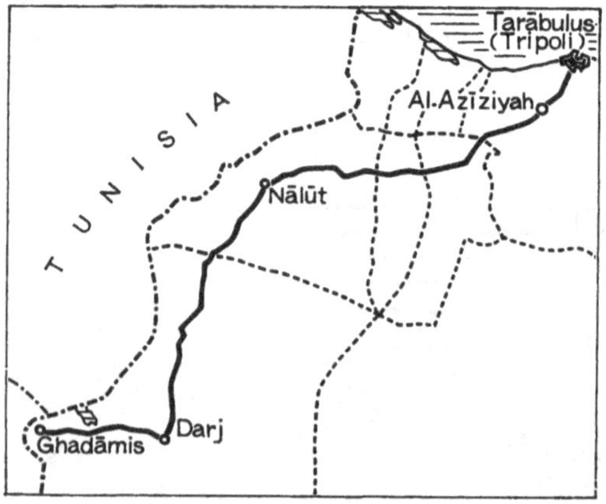

Fig. 3 Route of the collecting mission to Ghadāmis

The Socialist People's Libyan Arab Jamahiriya is a part of the Mediterranean gene centre. Therefore, missions for the collecting of landraces in this country have been included into the Mediterranean Programme of the International Board for Plant Genetic Resources (IBPGR). The first mission took place in 1981 (AL ALAZZEH et al. 1982a, b) covering the Cyrenaica. A second mission was carried out in 1982 (LEHMANN and HAMMER 1983a, b) which included a general survey of three main areas – Cyrenaica, Tripolitania and Fezzan. At that time genetic erosion had proceeded very fast in the Cyrenaica and the central parts of Tripolitania. Consequently, the third mission had to be concentrated to the marginal parts of Tripolitania and to the Fezzan area.

The travel was organized by the Agricultural Research Centre (ARC) in

Tarābulus and the research station of the ARC in Sabhā. We are thankful to Dr. A. Maddur, Director General, ARC, for his effective support of the mission.

Participants were, as consultants of the IBPGR, Dr. P. Perrino, Dr. G. B. Polignano (Bari) and Dr. Chr. O. Lehmann, Dr. K. Hammer (Gatersleben) together with Libyan colleagues who joined the mission as guides, for helping and for being trained in collecting activities.

Table 1
Material collected during the mission

Crop	Number of accessions	Sum
Triticum aestivum	56	
Triticum durum	5	
Hordeum vulgare	48	
Avena sativa	3	
Zea mays	11	
Sorghum bicolor	19	
Pennisetum americanum	10	
Wild grasses	9	
Cereals and grasses		161
Vicia faba	12	
Vicia spp.	9	
Trigonella foenum-graecum	7	
Pisum sativum	6	
Vigna sinensis	6	
Lens culinaris	3	
Lupinus sp.	1	
Legumes		44
Allium cepa	12	
Allium sativum	2	
Allium sp.	1	
Capsicum annuum	10	
Lycopersicon esculentum	7	
Solanum melongena	2	
Cucumis melo	7	
Cucurbita pepo	3	
Other Cucurbitaceae	8	
Beta vulgaris	5	
Abelmoschus esculentus	6	
Hibiscus sabdariffa	1	
Corchorus olitorius	4	
Brassica raba var. *rapa*	7	
Other vegetables	8	
Spice-, medicinal- and oil-crops	28	
Vegetables incl. spice-, medicinal- and oil-crops		111
Total sum		316

The collections were done from April 21st to April 22nd in Tripolitania (fig. 1), from April 25th to May 2nd in the Fezzan (fig. 2) including the Ghāt oases–for this part two teams were mounted–and from May 4th to May 5th in the Ghadāmis area (fig. 3). Altogether 316 accessions could be collected (table 1). Most of the material comes from the Fezzan.

Tripolitania

21. 4. Tarābulus–Al Azīziyah–Gharyān area–back to Tarābulus via Jifārah plain, collecting in the Gharyān area;

22. 4. Tarābulus–Al Khums–back to Tarābulus, collecting on the way near Al Qarābūlli and Al Khums.

The results of the rapidly acting genetic erosion could be observed in this area. Only some land-races could be collected in addition to the material of the last year's mission (LEHMANN and HAMMER 1983a). The collections include local types of barley, field beans and peas as well as wild relatives of cultivated plants like *Aegilops* spp. (fig. 4) and *Vicia* spp. New varieties are progressively spreading and have replaced the landraces even in remote areas. Since most of the wheats in the mountainous parts were still unripe, the team moved to Fezzan.

Fig. 4 Collecting *Aegilops* sp. in the Gharyān area (Tripolitania)

Fezzan area, including Ghat oases

25. 4. Sabhā–Samnū–Az Zīghan–back to Sabhā, collecting in Az Zīghan.
For the following exploration of the area two collecting teams were mounted.
First team (Dr. PERRINO, Dr. HAMMER):
26. 4. Sabhā–Awbāri–Al 'Awaynāt, collecting in the Wādī al Ḥayāh and around Al 'Awaynāt;
27. 4. Al 'Awaynāt–Ghāt–Al Barkat–back to Al 'Awaynāt, collecting in Ghāt and Al Barkat;
28. 4. Al 'Awaynāt–Awbāri–Sabhā, collecting around Awbāri and in the Wādī al Ḥayāh;
1. 5. Sabhā–Gabrahon–Sabhā, collecting in Gabrahon.
Second team (Dr. POLIGNANO, Dr. LEHMANN):
26. 4. Sabhā–Murzuq–Tsāwah–back to Murzuq, collecting around Spartat, Aqār, Umm al Ḥamām, Wādī 'Atabah;
27. 4. Murzuq–Qāwāt–Trāghan–Umm al Arānib–Misquwīn–Zuwaylah–Tmassah–back to Murzuq, collecting on the way;
28. 4. Murzuq–Trāghan–Al Qaṭrūn–Tajarhi–back to Murzuq, collecting in Al Qaṭrūn and Tajarhi;
29. 4. Murzuq–Sabhā, collecting in the Godwa area;
1. 5. Sabhā–Wādī ash Shāṭi–back to Sabhā, collecting in the Wādī ash Shāṭi
Though also in this area new varieties, especially of cereals, are spreading from large farming projects, many old land-races could be collected.

Four- and six-rowed barleys are characteristic for the oases in the Sahara (CHEVALIER 1932). ERROUX (1956) and AL-JIBOURI (1966) reported also about some two-rowed barleys in the Fezzan area. Our last year's mission had not been successful in collecting those types, but this year they could be detected in some places. They belong to *Hordeum vulgare* L. s. l. convar. *deficiens* (Steud.) Mansf., a race which is very rare in Africa, except Ethiopia, and have yellow, naked kernels (var. *nudideficiens* Körn.). As in other cereals, e.g. sorghum and maize, the *deficiens* barleys could have been introduced long ago to the Fezzan from southern or south-eastern areas. ERROUX (1956) has also described a *deficiens* barley with violett kernels from the Fezzan, which could not be found by us. But there was a four-rowed barley with naked, violett kernels (convar. *vulgare* var. *violaceum* Körn.). More races will be detected during propagation of the material in our fields because of the variability in some of the spike- and in the grainsamples.

The wheats collected mostly belong to the special oases type (cf. LEHMANN and HAMMER 1983a). The spikes are pubescent and bear inflated awns (fig. 5 and 6). These wheats are well adapted to the extreme environmental conditions in the oases and may be useful in breeding programmes concerning stress resistance (fig. 7). They are praized by the farmers because of their good flour quality. The wheats of the Fezzan have been studied by CIFERRI and GARAVINI (1941) and by ERROUX (1954, 1972). Comparing these surveys with our results and also considering the compilation made by AL-JIBOURI (1966) we can conclude that most of the wheat types known from the Fezzan are now collected and will be maintained in gene banks. But some types today occured only as rare admixtures in larger wheat fields.

Fig. 5 Typical constituents of a wheat field in Sabhā (Fezzan)

Fig. 6 Magnification from a wheat spike collected in Sabhā (Fezzan) showing the hairy, inflated glumes. This trait is characteristic for many wheat land-races from the oases (×6)

Fig. 7 The cereals from the Fezzan are expected to be resistant to environmental stress. The fields are supplied with water by irrigation tubes or canals. Partly harvested fields with wheat and naked barley in Az Zīghan (Fezzan)

In the Fezzan many vegetables and also spice-, medicinal- and oil-crops are grown (cf. table 1). Predominantly they are land-races (fig. 8). Genetic erosion seems to be progressing only slowly in these crops. But also in the far remote places new varieties are introduced. This is also true for trees (fig. 9).

Estonishing is the amount of different crops. Apart from those mentioned in table 1, *Portulaca oleracea, Coriandrum sativum, Anethum graveolens, Raphanus sativus, Eruca sativa, Petroselinum crispum, Nicotiana rustica, Lawsonia inermis* etc. could be collected.

Especially the Ghāt oases turned out to be a refuge for variable old land-races. The material collected from this area deserves a special study.

Fig. 8 Shallow beds are prepared for planting with vegetables. Some drought resistant *Cucurbitaceae* and sorghums are already growing–Az Zīghan (Fezzan)

Fig. 9 Formerly indigenous *Tamarix* spp. were planted to prevent wind erosion–Az Zīghan (Fezzan). Now they have been mostly replaced by Australien *Casuarina* spp.

Ghadāmis area

4. 4. Tarābulus–Sīnāwan–Darj–Ghadāmis, collecting 5 km west of Ghadāmis and near the Algerian border;
5. 5. Ghadāmis–Nālūt–Shakshūk–Al Azīziyah–Tarābulus, collecting along the road.

In the Ghadāmis area genetic erosion was found to be very high (fig. 10). Only some samples, including *Triticum durum*, could be collected.

Fig. 10 **Protection against wind erosion by fences constructed of leaves of *Phoenix* in the Ghadāmis oases. Mainly new varieties are cultivated**

This and the preceding missions covered the most promising areas of Libya and many old land-races could be collected. Only some marginal parts, including the Kufrah region, are still promising for indigenous material of cultivated plants. Further collections, preferably on a local basis, should be mounted to complete the land-race collections in this country. The material collected is indispensable as initial material for breeding highly adapted new varieties in Libya.

Zusammenfassung

Bericht über eine Reise in die Sozialistische Libysche Arabische Volksjamahiriya 1983 zur Sammlung autochthoner Sippen von Kulturpflanzen

Die Reise erfolgte im Rahmen des Mediterranean Programme des IBPGR im April/Mai 1983 mit Teilnehmern aus der Sozialistischen Libyschen Arabischen Volksjamahiriya (Agricultural Research Centre, Tarābulus), des Istituto del

Germoplasma, Bari, und des Zentralinstituts für Genetik und Kulturpflanzenforschung, Gatersleben. In Tripolitanien, dem Fezzan und dem Gebiet von Ghadāmis wurden 316 Proben, meist von Getreiden und Gemüsen, gesammelt. In Tripolitanien und dem Gebiet von Ghadāmis sind die Landsorten fast vollständig von neueren Sorten verdrängt worden. Im Fezzan schreitet die Generosion fort. Hier konnten noch zahlreiche Getreide- und Gemüselandsorten gesammelt werden. Die Landsorten können für Züchtungsprogramme nützlich sein.

Краткое содержание

Отчёт об экспедиции в Социалистическую Народную Ливийскую Арабскую Джамахирию 1983 г. для сборов автохтонных форм культурных растений

Экспедиция проводилась в рамках Средиземноморской программы ИБПГР в апреле–мае 1983 г. с участниками из Ливийской Джамахирии (Сельскохозяйственный н.-и. центр в Триполи), Института зародышевой плазмы в Бари и Центрального института генетики и культурных растений в Гатерслебене. В Триполитании, Феццане и в области Гадамиса были собраны 316 образцов, главным образом зерновых злаков и овощей. В Триполитании и в области Гадамиса местные сорта почти полностью вытеснены новыми сортами. В Феццане генная эрозия тоже прогрессирует, но здесь можно было собрать ещё многочисленные местные сорта зерновых злаков и овощей. Местные сорта могут быть полезными в селекционных программах.

Literature

Al Alazzeh, A., K. Hammer, Chr. O. Lehmann, and P. Perrino, 1982a: Report on a travel to the Socialist People's Libyan Arab Jamahiriya 1981 for the collection of indigenous taxa of cultivated plants. – Kulturpflanze **30**, 191–202.

–, –, –, and –, 1982b: Collecting in Libya. – Plant Genetic Resources – Newsletter **49**, 22–23.

Al-Jibouri, H. A., 1966: Description and distribution of wheat and barley varieties grown in Libya. – FAO, Rome. – 38 pp.

Chevalier, A., 1932: Les productions végétales du Sahara et de ses confins nord et sud. – Rev. Bot. Appl. Agr. Trop. **12**, 669–919.

Ciferri, R., e M. Garavini, 1941: I Cereali dell' Africa Italiana. III. I frumenti oasicoli del Sahara libico. – Bibliotheca Agraria Coloniale 19. – Firenze, Regio Istituto Agronomico per l'Africa Italiana. – 155 pp.

Erroux, J., 1954: Les blés du Fezzan. – Bull. Soc. Hist. Nat. Afr. Nord **45**, 302–317.

–, 1956: Les orges du Fezzan. – Bull. Soc. Hist. Nat. Afr. Nord **47**, 74–83.

–, 1972: Blés indigènes Nord-Africains. – Plant Genetic Resources – Newsletter **28**, 17–23.

Lehmann, Chr. O., und K. Hammer, 1983a: Bericht über eine Reise in die Sozialistische Libysche Arabische Volksjamahiriya 1982 zur Sammlung autochthoner Sippen von Kulturpflanzen. – Kulturpflanze **31**, 207–217.

–, und –, 1983b: Cereal and vegetable collection in Libya. – Plant Genetic Resources – Newsletter **55**, 26–27.

Dr. K. Hammer and Dr. Chr. O. Lehmann
Zentralinstitut für Genetik und Kulturpflanzenforschung
der Akademie der Wissenschaften der DDR
DDR - 4325 Gatersleben, Corrensstraße 3

Kulturpflanze 32 · 1984 · 207–216

Collection of land-races of cultivated plants in South Italy 1983

PIETRO PERRINO[1], KARL HAMMER and PETER HANELT

(Eingegangen am 3. Februar 1984)

Summary

In continuation of collecting missions in 1980, 1981 and 1982a further mission has been carried out in South Italy jointly by staff members of the Istituto del Germoplasma, Bari, and the Zentralinstitut für Genetik und Kulturpflanzenforschung, Gatersleben, in July 1983 for studying and collecting indigenous forms of cultivated plants and their wild relatives. 277 samples, mainly of cereals, grain legumes and vegetables have been collected. The material represents highly variable old land-races; this is especially true for wheat, maize, french beans, faba beans and *Brassica* spp. In some fields of the Sila Mts. introgressions from *Secale montanum* into cultivated rye could be observed. The results of this mission stress the necessity for continuing the exploration of plant genetic resources in South Italy.

Introduction

The joint explorations of plant genetic resources in South Italy–begun in 1980–have been continued in 1983 by the Istituto del Germoplasma in Bari and the Zentralinstitut für Genetik und Kulturpflanzenforschung in Gatersleben within the frame of the agreement between the Consiglio Nazionale delle Ricerche (C.N.R.) of Italy and the Academy of Sciences of the GDR. Regarding the importance and the general agricultural situation of this region one should compare the earlier travel reports (PERRINO et al. 1981a, 1982a, PERRINO and HAMMER 1983a).

The recent mission could confirm the observations made during the last years of the still existing small-scale cultivation of old varieties and land-races of cereals and leguminous species and especially of different garden crops but likewise also of the advanced stage of gene-erosion for the whole of the collecting region and for distinct crops. Especially in Calabria this process is enhanced by the widerspread rural emigration and the frequent abandonment of former farms.

The collecting work has been concentrated in 1983 on the southern parts of the region Basilicata and on northern and central Calabria (fig. 1) which means a

1 Istituto del Germoplasma, Via G. Amendola 165/A, I-70126 Bari, Italy

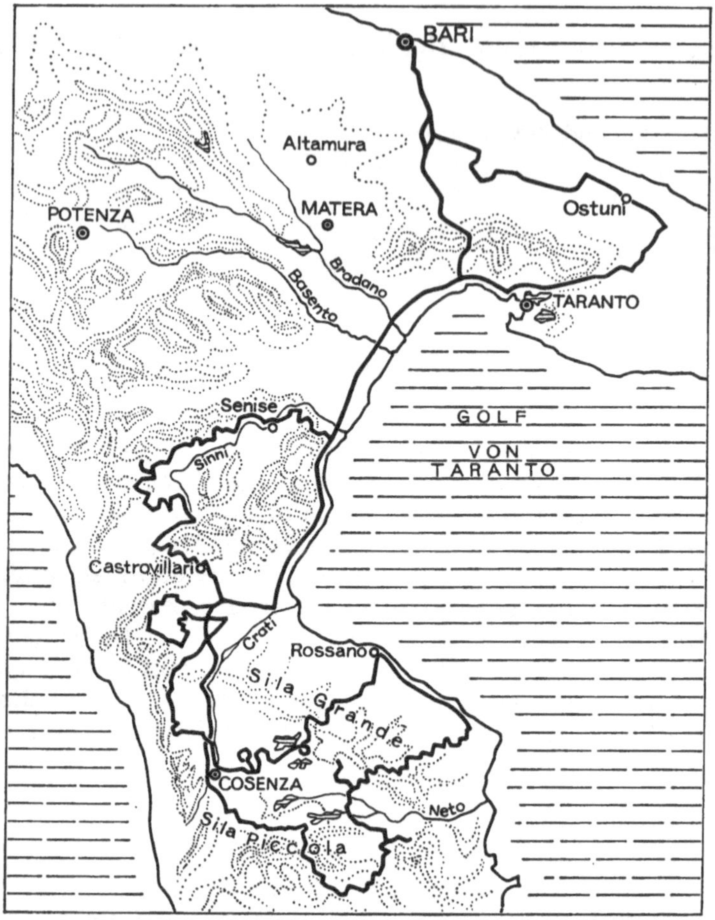

Fig. 1 Route of the collecting mission

systematic continuation of the earlier activities in southward direction. The exploration has been done mostly in altitudes between 400 and 1200 m NN within the belt, in which the original vegetation is represented by deciduous broadleaved forests, *Quercus* ssp.–*Castanea vesca* in the lower, *Fagus sylvatica* associations in the uppermost parts.

The mission has been again organized by the Bari institute, participants having been the authors of this paper. The field work took place from July 18th to July 30th, the detailled itinerary is summed up in the following (see also fig. 1):

18. 7. Bari – Mòttola – Metaponto – Policoro – Senise – Chiaramonte;
19. 7. Chiaramonte – Fardella – Episcopia – Agromonte – Castellùccio Superiore;
20. 7. Castellùccio Superiore – Castellùccio Inferiore – Rotonda – Viggianello – Contrada – Zarafa – Prastio – Tempa Pertugio – Rifugio;
21. 7. Rifugio – Morano Calabro – Castrovillari – Firmo – Acquaformosa – S. Donato – Roggiano Gravina – Altomonte;
22. 7. Altomonte – S. Marco Argentano – Mongrassano – Cavallerizzo – Cerzeto – S. Martino – S. Giacomo di Cerzeto – Rota Greca – Diretto – Latarico – Montaldo Uffugo – Cosenza;

23. 7. Cosenza – Spezzano – Celico – S. Pietro in Guarano – Camigliatello;
24. 7. Camigliatello – Cava di Melis – Longobucco – Destro – Cropalati – Mirto;
25. 7. Mirto – Cariati – Crucoli – Torretta – Cirò – Umbriatico – Perticaro – Verzino – Savelli;
26. 7. Savelli – Pino Grande – S. Giovanni in Fiore – Lago Ampollino – Germano;
27. 7. Germano – S. Giovanni i. F. – Lago Ampollino – Cotronei – Roccabernarda – Petilia Policastro – Foresta – Mesoraca – Filippa – Petronà – Filippa;
28. 7. Filippa – Petronà – Cerva – Sersale – Zagarise – Magisano – S. Pietro – S. Giovanni Albi – Albi – Villaggio Mancuso – Villaggio Mancise – Spineto;
29. 7. Spineto – Rogliano – Cosenza – Sibari – Policoro – Metaponto – Taranto – Carovigno;
30. 7. Carovigno – Ostuni – Alberobello – Putignano – Bari.

Results

Previous missions in South Italy (Perrino et al. 1981a and b, Perrino et al. 1982a and b, Perrino and Hammer 1983a and b) have shown that this area is still rich in plant genetic resources. The mountainous areas provide suitable niches for the preservation of old land-races. This impression from the former missions was suppoιted by the observations made this year in southern parts of Basilicata (fig. 2) and in some areas of Calabria, where the traditional agricultural

Fig. 2 Traditional type of agriculture guarantees the preservation of old land-races, southern Basilicata (near Fardella)

practice is still maintained. For many sites in Calabria a severe genetic erosion could be observed, mainly because a large amount of farms have been abandoned by rural emigration (fig. 3).

Fig. 3 In Calabria farms are often left and the fields are abandoned (E of Rogliano). By this process genetic erosion is speeded up

Altogether the collections comprise 277 accessions. The detailed list of the material collected is given on p. 479.

In addition some samples of further species have been collected for special purposes of the participating institutes.

The procedure of handling, reproducing and evaluating the collected material has been described elsewhere (see Perrino and Hammer 1983a). It is expected that during the reproduction and evaluation of the material there will be detected and separated some further interesting morphological variants.

Crop	Number of accessions	Sum
Wheat, *Triticum* spp.	22	
Maize, *Zea mays*	21	
Cultivated and Wild Rye, *Secale cereale* and *S. montanum*	18	
Barley, *Hordeum vulgare*	13	
Oats, *Avena* spp.	8	
Aegilops spp.	6	
Lolium spp.	2	
Cereals incl. Wild Grasses		90
French Beans, *Phaseolus vulgaris*	31	
Faba Beans, *Vicia faba*	14	
Lupins, *Lupinus* spp.	13	
Chick Pea, *Cicer arietinum*	9	
Peas, *Pisum sativum*	4	
Other Pulses	12	
Legumes		83
Brassica spp.	28	
Onions, *Allium* spp.	5	
Cucumbers, Melons, *Cucumis* spp.	7	
Lettuce, *Lactuca sativa*	5	
Tomato, *Lycopersicon esculentum*	8	
Peppers, *Capsicum annuum*	9	
Other Vegetables, Medicinal and Spice Plants	42	
Vegetables incl. some Medicinal and Spice Plants		104
Total Sum		277

Some remarkable findings are briefly summarized in the following:

- As in previous missions the variability of *Triticum* land-races was rather high. In some fields *T. durum* and *T. aestivum* were grown together, sometimes with *Secale cereale* as an admixture (fig. 4); *T. dicoccon* could not be observed during this mission. Apparently, the recent cultivation of this relic crop does not pass over the Agri river. On the southern slope of this river basin the farmers still knew emmer wheat but it has not been cultivated already for a long time. Proceeding further to the south and entering into Calabria we could not even find some knowledge concerning this old wheat species (see also HAMMER and PERRINO 1984).
- In the Sila Mts. *Secale cereale* is a widespread cereal (fig. 5). Newly introduced varieties are grown and competite with old markedly variable land-races. Their variability partly results from introgressions of the wild growing *Secale montanum* which often inhabits areas near the cereal fields or even at their borders (fig. 6). Some introgressed spikes of *S. cereale* could be found and also some spikes from *S. montanum* possessing characters of *S. cereale*. These observations can be compared with earlier ones by ZOHARY (1964) who reported introgressed populations between *S. cereale* and *S. montanum* from the Central Anatolian

Fig. 4 Variable cereal population containing several botanical varieties of *Triticum aestivum* and *T. durum* but also some spikes of *Secale cereale*

Fig. 5 In the Sila Mts. rye is a rather common cereal. Field of rye near Germano under a plantation of apple trees

Fig. 6 ***Secale montanum*** **growing at the edge of a wheat field near Camigliatello (Sila Mts.) In this area introgressions from this species into *Secale cereale* could be found**

plateau, growing there most commonly at the edges of cultivation. Possible introgressions between the two species have been discussed recently (PERRINO et al. 1982a) when studying material collected from the slopes of Mt. Etna (Sicily). The newly collected material deserves a special study.

- Some spikes of *Triticale* were collected from a wheat field. According to informations of the farmers *Triticale* has been cultivated in this area in experimental plots some twenty years ago. Today it is a rare admixture of some wheat fields in the Sila Mts.
- As in other regions of South Italy *Phaseolus vulgaris* proved to be very variable. But also *Vicia faba* in characterized by a considerable variability. Of special interest are dark seeded and small grained varieties of this crop which are cultivated as green manure.
- Weedy types of *Lupinus graecus* grew abundantly in some fields of the Sila Mts., they are characterized by comparatively large seeds (this species has not been reported from South Italy, cf. GLADSTONES 1974).
- Again variable early ripening land-races of maize have been collected. Forms of hard maize with very small cobs and grains resembling puff maize have been collected for the first time during our missions.
- Rather many old races of vegetable crops are maintained by the farmers in their gardens, in which very often a high crop diversity is to be found mainly by means of frequent intercropping (fig. 7); here e.g. different indigenous varieties of

Fig. 7 Various intercroppings, e.g. maize and beans, and selfmade seed production in a garden of the mountain valley east of Cosenza. These conditions favour the preservation of many old land-races, especially vegetables

Fig. 8 Tubiferous *Brassica rapa*, a rare root-crop in South Italy, from the garden shown in fig. 7

Brassica spp. (leafy races of *Br. rapa*, primitive cultivars of *Br. oleracea*, variable broccolis etc.) are still rather often cultivated (fig. 8).

The collections have confirmed that South Italy is an important refuge for land-races. The exploration of this area will be continued in the following years.

Zusammenfassung

Sammlung von Kulturpflanzen-Landsorten in Süditalien 1983

In Fortführung von Sammelreisen in den Jahren 1980, 1981 und 1982 wurde im Juli 1983 eine weitere Reise nach Süditalien gemeinsam von Mitarbeitern des Istituto del Germoplasma, Bari, und des Zentralinstituts für Genetik und Kulturpflanzenforschung, Gatersleben, unternommen, die dem Studium und der Sammlung indigener Formen von Kulturpflanzen und ihrer verwandten Wildsippen diente. 277 Proben, besonders von Getreiden, Körnerleguminosen und Gemüsen, wurden gesammelt. Das Material besteht aus hochvariablen alten Landsorten, besonders von Weizen, Mais, Gartenbohnen, Ackerbohnen und *Brassica*-Arten. In einigen Feldern des Sila-Gebirges konnten Introgressionen von *Secale montanum* in Kulturroggen beobachtet werden. Die Ergebnisse dieser Sammelreise unterstreichen die Notwendigkeit zur Weiterführung der Exploration pflanzengenetischer Ressourcen in Süditalien.

Краткое содержание

Сборы местных сортов культурных растений в Южной Италии в 1983 г.

Начатые в 1980 г. и продолженные в 1981 и 1982 гг. экспедиционные сборы, смогли осуществится и в 1983 г. Экспедиция в Южную Италию была предпринята для дальнейшего изучения и сборов автохтонных форм культурных растений и их сородичей совместно сотрудниками Института зародышевой плазмы в Бари и Центрального института генетики и культурных растений в Гатерслебене. Было собрано 277 образцов, главным образом, зерновых злаков, зерновых бобовых и овощей. Материал сборов представлен сильно изменчивыми старыми местными сортами, большинство которых относится к пшенице, кукурузе, огородной фасоли, конским бобам и видам *Brassica*. На некоторых полях в горах Сила наблюдалась интрогрессия между *Secale montanum* и культурной рожью. Результаты экспедиции говорят в пользу того, что необходимо и дальше продолжить изыскания генетических растительных ресурсов Южной Италии.

Literature

GLADSTONES, J. S., 1974: Lupins of the Mediterranean region and Africa. – Technical Bulletin Nr. 26, Dept. Agr. Western Australia, 48 pp.

PERRINO, P., and K. HAMMER, 1983a: Collection of land-races of cultivated plants in South Italy 1982. – Kulturpflanze **31**, 219–226.

– and –, 1983b: Collecting in southern Italy. – Plant Genetic Resources – Newsletter **54**, 26–27.

– and –, 1984: Further information on farro (*Triticum monococcum* L. and *T. dicoccon* Schrank) in South Italy. – Kulturpflanze **22**, in print.

–, – and P. HANELT, 1981a: Report of travels to South Italy 1980 for the collection of indigenous material of cultivated plants. – Kulturpflanze **29**, 433–442.

–, – and –: 1981b: Collecting in southern Italy. – Plant Genetic Resources – Newsletter **42**, 24.

–, – and CHR. O. LEHMANN, 1982a: Collection of land-races of cultivated plants in South Italy 1981. – Kulturpflanze **30**, 181–190.

–, – and –: 1982b: Collecting in southern Italy. – Plant Genetic Resources – Newsletter **49**, 32–33.

ZOHARY, D., 1964: Spontaneous brittle six-row barleys, their nature and origin. – Barley Genetics I, Proc. First Intern. Barley Genetics Symp., pp. 27–31. – Wageningen 1963.

Dr. K. HAMMER and Dr. P. HANELT
Zentralinstitut für Genetik und Kulturpflanzenforschung
der Akademie der Wissenschaften der DDR
DDR - 4325 Gatersleben
Corrensstr. 3

Kulturpflanze 32 · 1984 · 217–227

Report of a collecting mission to the Georgian SSR 1983 for the study of indigenous material of cultivated plants

RUSUDAN K. BERIDZE [1], REINHARD FRITSCH, KLAUS PISTRICK and IRINA M. SAKVARELIDZE [1]

(Eingegangen am 13. Januar 1984)

Summary

In the summer 1983 the joint activities for the collection of autochthonous material of cultivated plants have been continued in the outermost western regions of Georgia (Abkhazia, Megrelia). Altogether 434 accessions could be obtained, more than one third of them being grain legumes, and one fourth vegetables and spice plants, resp. The remaining part includes mainly maize. Gene-erosion is very advanced especially for cereals and some grain legumes besides *Phaseolus vulgaris* for which an astonishing genetic variability could be confirmed within the local material. Some details of the accessions regarding botanical interest and breeding value have been outlined. The collecting work should be continued.

Introduction

The cooperation between the Institute of Botany of the Academy of Sciences of the Georgian SSR and the Zentralinstitut für Genetik und Kulturpflanzenforschung of the Academy of Sciences of the GDR on the field of cultivated plant research was continued. A third travel had been organized into hitherto unexplored parts of western Georgia in order to study and collect indigenous material of cultivated plants and their wild relatives in a region whose population played a more active role in domestication of indigenous useful plants than other peoples did and which belongs to one of the Vavilovian gene centres (see also SAKVARELIDZE et al. 1969, BERIDZE et al. 1982, BERIDZE et al. 1983). The mission went to the Akhazian ASSR (rayons Suchumi, Gul'ripš, Očamčira, and Gudauta) and to the historical province Megrelia (rayons Zugdidi, Calendžicha, Čochorocku, Gegečkori, Chobi, Cchakaja, and Poti). The collection took place in settlements and villages from the lowlands of the Inguri, Chobi and Rioni river systems and near the border of the Black Sea up to the mountain valleys (400–650 m above sea level), for details see tab. 1 and fig. 1.

The accessions were mostly obtained from the seed reserves of the farmers or were collected in their gardens (figs. 2–6). Sometimes they were bought on local

1 Institute of Botany, Academy of Science of the Georgian SSR, Department of Cultivated Flora, 380007 Tbilisi, USSR

Table 1

Chronology of the travel to the Georgian SSR for the collection of indigenous material of cultivated plants (August 22 to September 10, 1983)

22. 8.	Departure by plane from Berlin to Tbilisi via Moscow.
23. 8.	Excursion by car in western direction to the Kura valley near Mccheta and Uplisciche.
24. 8.	Final preparations for the trip; visit of the Botanical Institute in Tbilisi.
25. 8.	Departure by plane from Tbilisi to Suchumi.
25. 8. – 31. 8.	Main basis at Suchumi; from here travels by car.
26. 8.	Visit of the Botanical Garden and the seed control laboratory at Suchumi.
27. 8.	Travels in northeastern direction to the villages Kelasuri, Besleti, Achalsopeli, Cebel'da (ca. 400–550 m above sea level) and Lata.
28. 8.	Visit of the bazaar at Suchumi for obtaining seeds of local races of vegetables and spices; travels in the surroundings of the town.
29. 8.	Travels in southeastern direction to Gul'ripš-Mestnikova, Pšapi, Atara Abchazkoe and Atara Armjanskoe.
30. 8.	Travels in northwestern direction, via Novyj Afon, to the villages Primorskoe, Verchnjaja Mcara, Srednjaja Mcara and Orlinoe gnezdo.
31. 8.	Travels in northeastern direction to the villages Kada, Azanta (ca. 600 m) and Cebel'da.
1. 9.	Departure by car from Suchumi to Zugdidi.
1. 9. –7. 9.	Main basis at Zugdidi; from here daily travels by car.
2. 9.	Travels in southwestern direction to Kvemokachati, Achalkachati, Darčeli, Anaklia and Koki.
3. 9.	Travels in southeastern direction to the kolchoz Kolchida and the villages Narazeni, Staryj Abastumani, Abastumani, Urta and Čitackari.
4. 9.	Visits of the bazaars at Zugdidi and Poti for obtaining seeds of local races of vegetables and spices; collecting in the villages Schepi and Chamiskuri in southern and southeastern direction of Zugdidi.
5. 9.	Travels in northeastern direction, via Calendžicha, to the villages Skuri, Čkvaleri, Džvari, Poccho and Saberio.
6. 9.	Travels in northeastern direction to Dzichaškari and via Čchorocku to the villages Osindale (ca. 650 m) and Muchuri.
7. 9.	Travels in eastern direction, via Gegečkori, to the villages Taleri and Doberazeni (ca. 350–400 m).
8. 9.	Return by car from Zugdidi to Tbilisi.
9. 9.	Managing the necessary permissions for the export of the material.
10. 9.	Return by plane from Tbilisi to Dresden.

markets. Some of these farmers are highly interested in growing old land races and such crop species which are commonly out of use now. However, many of the younger people do not know of the existence of those ancient cultivated plants.

The troublesome work of organization of this third exploration trip had been done again by Mrs. Dr. Beridze. Further participants have been from Tbilisi Mrs. I. Sakvarelidze, scientific coworker, and Mr. K. Chelašvili, biochemist, and from Gatersleben Dr. R. Fritsch and K. Pistrick.

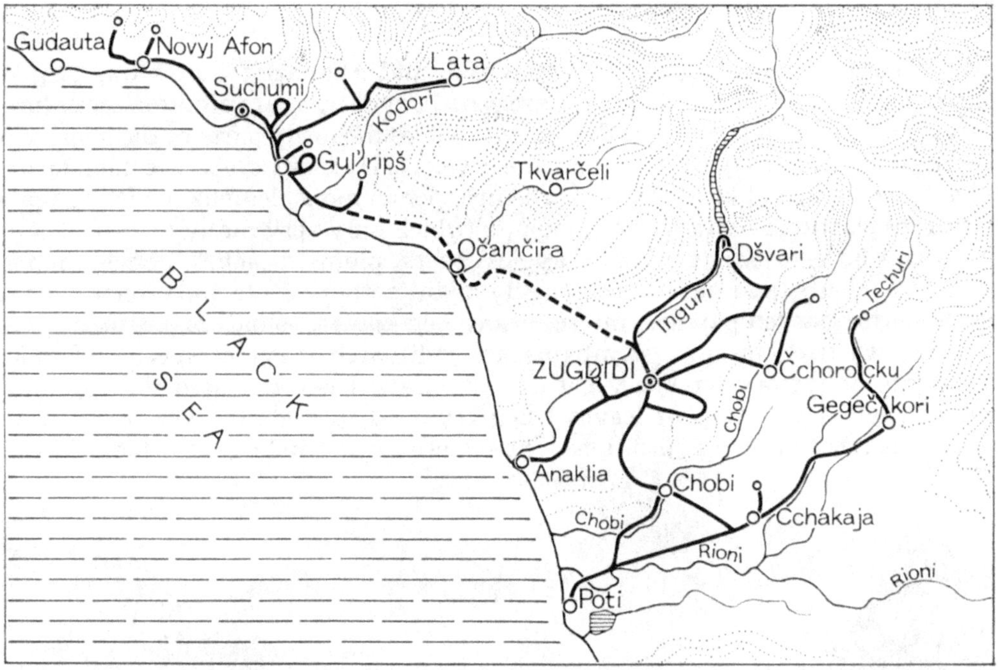

Fig. 1 Travel route of the collecting mission 1983 to Western Georgia
Drawing R. PERTACK

Fig. 2 Landscape at Gul'ripš near the Black Sea with mandarin-plantations and gardens at the slopes
Photo R. FRITSCH

Results

In Abkhazia, the area south of the Bzybskij and Abkhazkij chains was visited. There, cultivated fields and gardens are mostly small and are situated on natural terraces of the mountains or in the valleys, where the lower parts of the slopes are preferred to the stony areas along the rivers. The main crops are mandarins, tobacco and maize, but in the orchards figs, persimmons, lemons (which are protected by plastic cover in winter), loquats (which only very rarely bear fruit) and feijoa can be seen besides grapes, apples, pears, plums, peaches, pomegranates, quinces, walnuts, and hazel nuts. Mostly as intercroppings in the orchards or in neighbouring garden patches, maize, beans, cucumbers, tomatoes, eggplants, red peppers, watermelons and pumpkins are cultivated. Several species of spice plants are obligatory for every garden, too. Laurel is often used for bordering hedges and gives a crop of leaves when trimmed in autumn. Tea plants were locally seen, not only in plantations, but sometimes also in house gardens.

Fig. 3 A house garden at Kelasuri near Suchumi, mandarins, maize and *Phaseolus* beans in the foreground Photo R. FRITSCH

In Megrelia however, tea plantations are the dominant aspect of landscape (fig. 4) instead of the characteristic slopes covered with mandarins in Abkhazia (fig. 2). They are absent only in very wet areas of the lowlands and in the higher mountains above 400 m. Maize is the only other main crop but tobacco and mandarins are also seen. Tung (*Aleurites fordii*) is more or less often grown besides tea plantations or intermixed with them. The gardens are resembling in most

Fig. 4 Tea plantations protected by tree rows of *Cryptomeria japonica* at Narazeni near Zugdidi
Photo K. PISTRICK

Fig. 5 A traditional granary called 'nalia' or 'bereli' in an orchard at Skuri, northeast of Zugdidi
Photo K. PISTRICK

characters to those ones in Abkhazia. Here, maize, *Phaseolus* pole beans and squashes are much more often intercropped.

Altogether 434 accessions have been collected, several samples of *Phaseolus* beans, soya beans and partly also maize consist of different morphological strains which will be separately propagated and maintained. Nearly all accessions were divided into two parts, so that the institutes in Tbilisi and Gatersleben share an almost identical set of our collection. A detailed catalogue of the material is given in tab. 2. Additional herbarium specimens have been collected, too.

Fig. 6 Red pepper, Phaseolus-beans, pumpkins, maize and cucumbers in a house garden at Skuri
Photo R. FRITSCH

In all visited areas a high diversity of garden crops could be observed. Mostly the old land races are still in cultivation and a considerable part of the collecting work has devoted to them. As already observed during the first and second collection missions, especially the common beans are represented by an astonishing array of different variants not only in regard to morphological characters of seeds and pods, but also in regard to ripening time, yield characters and cooking

Table 2
Catalogue of the material collected in Abkhazia and Megrelia 1983

crop	number of accessions	
Zea mays, maize	33	
Setaria italica, foxtail millet	2	
Sorghum sp., sorghum millet	1	
maize and millets altogether		36 accessions
Phaseolus vulgaris, common bean	147	
Ph. coccineus, runner bean	1	
Glycine max, soya bean	11	
Pisum sativum, pea	7	
Vicia faba, broad bean	4	
grain legumes altogether		170 accessions
Cucurbita sp., squash	28	
Cucumis sp., cucumber, melon	12	
Citrullus lanatus, water melon	7	
Brassica sp., cabbages	19	
Beta vulgaris, beets	4	
Lactuca sativa, salad	3	
Spinacia oleracea, spinach	1	
Rumex rugosa, garden sorrel	2	
Raphanus sativus, radish	7	
Daucus carota, carrot	3	
Lycopersicon esculentum, tomato	10	
Solanum melongena, eggplant	1	
wild taxa of *Daucus* and *Raphanus*	8	
vegetables and cucurbits altogether		105 accessions
Allium sativum, garlic	26	
A. cf. ampeloprasum, great headed garlic	8	
A. porrum, leek	12	
A. cepa, onion	9	
wild species of *Allium*	4	
Apium graveolens, celery	8	
Trigonella caerulea, sweet trefoil	8	
Ocimum basilicum, basilic	7	
Coriandrum sativum, coriander	6	
Tagetes erecta, marigold	6	
Petroselinum crispum, parsley	5	
Foeniculum vulgare, fennel	4	
Capsicum annuum, red pepper	4	
Lepidium sativum, garden cress	3	
Satureja hortensis, savory	3	
Anethum graveolens, dill	2	
spice plants altogether		115 accessions
Nicotiana sp., tobacco	1	
further wild taxa	7	8 accessions
	sum total	434 accessions

quality. In Megrelia we found an interesting polymorphism of garlic accessions which was not seen in other parts of western Georgia (fig. 7). Besides the common or 'tasteful' garlic, which itself is very polymorphous in regard to form and

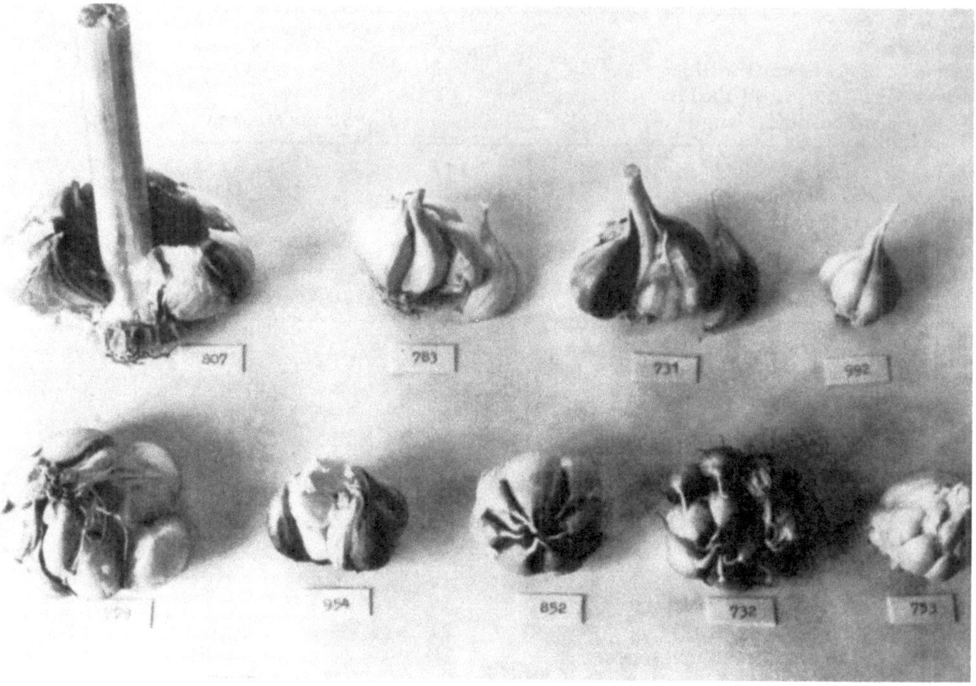

Fig. 7 Variability of Megrelian garlic accessions Photo R. Fritsch

number of cloves per bulb and to characters and colour of bulb coats, there is a second type called 'turkuli niori' (foreign garlic) characterized by greater cloves with thick, yellowish brown bulb coats and a much more penetrating, unpleasant odour and a more bitter, sharp taste. The cloves of those both types are used for seasoning. The fresh green leaves and stalks of a third type of plants, however, are used – and highly prized – for seasoning, and commonly not the cloves. It may well be that some of the third type plants do not belong to *Allium sativum* but to the great headed garlics *A. ampeloprasum* L. *s. l.*

Some other garden plants deserve also special interest. Among Armenian people in Abkhazia, a special form of a leafy cabbage called 'lachaná' is cultivated for its leaves which are mostly eaten raw or cooked. In most other parts of western Georgia and among other people in Abkhazia, it is replaced by a primitive kind of kohlrabi or turnip kale with a rather insignificantly fleshy-swollen stem the leaves of which are used in the same way and the peeled stems of which are pickled eaten, too. Another spinach-like vegetable is a broad leaved form of *Rumex rugosa* which is sometimes grown. A special dish is provided from the fresh leaves of a plant called 'cicmati' which seems to be a species of *Brassica.*

In the past, the cultivation of wheat, barley, oat and rye had never greater importance and is now fully ceased in the areas visited. Foxtail millet (*Setaria*

Fig. 8 *Setaria italica*, a main cereal of the past in Georgia, still cultivated in a small scale at Doberazeni, north of Gegečkori
Photo R. Fritsch

italica) was a main cereal in the past and is still extremely rarely cultivated (fig. 8) and used for human food. The main staple diet is maize. Most fields of it are sown by the rather modern Georgian dent corn cultivar 'Adžametis tetri' (white maize from Adžamethi) introduced there in the thirties. Sometimes local races of white flint corn are also grown, but only a few accessions of the old indigenous races with red or yellow kernels could be obtained.

Some farmers still use a mixture of maize and soya flour for preparing a special kind of noodles which are highly prized for their taste, but soya beans are not often cultivated and mostly used for feeding livestock. Broad beans and peas are rarely to be seen, and mostly only their unripe seeds are taken for soups. Their use as grain legumes have been passed over to *Phaseolus* beans.

A detailed analysis of the collected material will be given later, when the accessions will have been examined in the Gatersleben gene bank collection and after comparison with the other strains from the Georgian SSR collected during the former years.

The collecting mission has shown again, that gene erosion has reached a very critical stage in the regions visited. Field crops, especially cereals, foxtail millet, and indigenous forms of maize, are highly endangered. The same is true for several grain legumes. The great diversity among the indigenous races of garden crops makes them an interesting tool for plant breeding purposes and for basic research on cultivated plants. That's why the continuation of the collecting missions to other parts of Georgia seems to be essential.

Once again we have to stress the friendly hospitability of the Georgian farmers and of other informants and the valuable support given by the local administrations and the Academy of Sciences of the Georgian SSR.

Zusammenfassung

Bericht über eine Reise in die Georgische SSR 1983 zur Sammlung indigenen Materials von Kulturpflanzen

Im Sommer 1983 wurde eine weitere gemeinsame Reise zur Sammlung autochthonen Kulturpflanzenmaterials unternommen, die in die westlichen Teile von West-Georgien (Abchasien, Megrelien) führte. Es konnten insgesamt 434 Proben gesammelt werden, von denen mehr als ein Drittel Körnerleguminosen und je etwa ein Viertel Gemüse- und Gewürzpflanzen waren. Getreide war fast ausschließlich durch Mais vertreten. Die Gen-Erosion ist bei Getreide und einigen Körnerleguminosen bereits sehr weit fortgeschritten, während bei *Phaseolus vulgaris* wiederum eine erstaunliche Vielfalt an Formen festgestellt werden konnte. Besonders interessantes Material von Knoblauch, Blattkohlformen und einigen Spinatgemüsen wird kurz vorgestellt. Die Sammeltätigkeit sollte fortgesetzt werden.

Краткое содержание

Отчёт о поездке в Грузинскую ССР в 1983 г. для сборов местного материала культурных растений

Летом 1983 г. состоялась ещё одна совместная экскурсия для сборов автохтонного материала культурных растений, предпринятая в западные части Западной Грузии (Абхазия, Мегрелия). Всего собрано 434 образца, из которых больше трети были зерновые бобовые и, приблизительно равные части овощных и пряных растений. Зерновые злаки представлены почти только кукурузой. Генная эрозия у хлебных злаков и некоторых зерновых бобовых достигла уже высокой степени, в то время как фасоль обыкновенная опять оказалась представленной удивительным многообразием форм. Кратко описываются особенно интересные формы чеснока, листовой капусты и некоторых «шпинатных» овощей. Предполагается продолжение сборов местного материала.

Literature

BERIDZE, R. K., P. HANELT, and J. KRUSE, 1982: Report of a travel to the Georgian SSR 1981 for the collection of indigenous material of cultivated plants. – Kulturpflanze 30, 203–213.

–, –, and R. FRITSCH, 1983: Report of a collecting mission to the Georgian SSR 1982 for the study of indigenous material of cultivated plants. – Kulturpflanze 31, 173–184.

SAKVARELIDZE, I. M., V. N. KANDELAKI, and I. T. KARUMIDZE, 1969: (Flora of cultivated plants of Kartli – composition, origin and evolution. – In Georgian language. Unpublished research report.) Tbilisi.

Dr. R. FRITSCH and Dipl.-Biol. K. PISTRICK
Zentralinstitut für Genetik und Kulturpflanzenforschung
der Akademie der Wissenschaften der DDR
DDR - 4325 Gatersleben
Corrensstraße 3

Kulturpflanze 32 · 1984 · 229–243

Literatur über archäologische Kulturpflanzenreste (1982/1983)

JÜRGEN SCHULTZE-MOTEL

(Eingegangen am 9. Januar 1984)

Zusammenfassung

In Fortsetzung von Bibliographien über archäologische Kulturpflanzenreste wurde die einschlägige Literatur der Jahre 1982/1983 zusammengestellt und der Inhalt der erfaßten 78 Arbeiten kurz kommentiert.

Einleitung

Hiermit wird die Bibliographie über archäologische Kulturpflanzenreste fortgesetzt (vgl. SCHULTZE-MOTEL 1983). Es wurde versucht, die entsprechende Literatur der Jahre 1982 und 1983 zu erfassen. Bisher übersehene Titel aus früheren Jahren oder solche, die nicht eher beschafft werden konnten, sind ebenfalls aufgenommen worden.

Wieder erhielt ich durch Frau Bibliothekarin V. SCHNEIDER und den wissenschaftlichen Mitarbeiter Herrn W. MATTHIAS (beide am Landesmuseum für Vorgeschichte Halle/S.) uneigennützige Unterstützung bei der Ermittlung und Bereitstellung der Literatur. Auch Frau Dr. MARIA HOPF (Römisch-Germanisches Zentralmuseum Mainz) und Herr Dr. H.-H. MÜLLER (Zentralinstitut für Alte Geschichte und Archäologie Berlin) übermittelten zahlreiche Zitate. Allen Genannten sage ich für ihre wertvolle Unterstützung meinen herzlichen Dank.

Es folgen die Einzelangaben zu den alphabetisch angeordneten Arten. Hierbei wurden folgende Abkürzungen benutzt: N = Neolithikum, Bz = Bronzezeit, Ez = Eisenzeit, Rz = Römerzeit, Ma = Mittelalter.

Amaranthus lividus L.
Welzheim, Rz, KÖRBER-GROHNE et al. 1983; Münsterhof in Zürich, 12.–13. Jh., JACQUAT et al. 1982; Heidelberg, 15./16. Jh., MAIER 1983; Duisburg, Ma, KNÖRZER 1983a.

Amygdalus communis L., Mandelbaum
Heidelberg, 15./16. Jh., MAIER 1983.

Anethum graveolens L., Dill
Welzheim, Rz, KÖRBER-GROHNE et al. 1983; Münsterhof in Zürich, 12.–13. Jh.,

Jacquat et al. 1982; Heidelberg, 15./16. Jh., Maier 1983; Lübeck, Ma, Meyer und Neugebauer 1982.

Apium graveolens L., Sellerie
Karthago, Tunesien, Rz (1.–2. Jh. u. Z.) und byzantinisch (um 600), van Zeist und Bottema 1983; Welzheim, Rz, Körber-Grohne et al. 1983; Duisburg, Ma, Knörzer 1983a; Lübeck, 13. und 17./18. Jh., Lynch und Paap 1982.

Atriplex hortensis L., Gartenmelde
Welzheim, Rz, Körber-Grohne et al. 1983.

Avena fatua L., Flughafer, und *Avena sativa* L., Saathafer
Kastanas, Griechenland, frühe Ez, Kroll 1982; Niedergurig, Kr. Bautzen, jüngere Bz, Nötzold 1982a; Dresden-Nickern, Aunjetitzer Kultur, Nötzold 1982b; Crosskirk near Thurso in Caithness, Schottland, Ez, Dickson 1979; Pod bei Bugojno, Zentralbosnien, Ez, Kučan 1983; Răcătău, Rumänien, Latène, Cârciumaru 1983; Pizzica-Pantanello, Italien, 3. Jh. v. u. Z., Costantini 1983; Welzheim, Rz, Körber-Grohne et al. 1983; Leuben, Kr. Oschatz, Römische Kaiserzeit, Nötzold 1982a, b; Magdeborn, Lkr. Leipzig, slawisch, Nötzold 1982a; Krivina, Bulgarien, 4.–10. Jh. u. Z., Hajnalová 1982; Machrins, Island of Colonsay, Schottland, ca. 800 u. Z., Dickson 1982; Monastyrek, mittleres Dnepr-Gebiet, 8.–10. Jh., Paškevič und Petrašenko 1982; Hont, Komitat Nográd, Ungarn, 10. Jh., Hartyányi 1983; Sornzig, Kr. Oschatz, frühdeutsch, 12./13. Jh., Nötzold 1982b; Duisburg, Ma, Knörzer 1983a; Lübeck, Ma, Meyer und Neugebauer 1982; Stift Lilienfeld, Niederösterreich, Grab von 1429, Sackl 1980; Smeerenburg, Spitzbergen, 17. Jh., van Wijngaarden-Bakker und Pals 1981.

Avena sterilis L.
Pod bei Bugojno, Zentralbosnien, Ez, Kučan 1983.

Beta vulgaris L.
Duisburg, Ma, Knörzer 1983a.

Brassica napus L., Raps, oder *Brassica rapa* L., Rübsen
Lübeck, Ma, Meyer und Neugebauer 1982; Lübeck, 17./18. Jh., Lynch und Paap 1982.

Camelina sativa (L.) Crantz, Leindotter
Kastanas, Griechenland, frühe Ez, Kroll 1982; Monastyrek, mittleres Dnepr-Gebiet, 8.–10. Jh., Paškevič und Petrašenko 1982; Lübeck, 13. Jh., Lynch und Paap 1982; Duisburg, Ma, Knörzer 1983a.

Cannabis sativa L., Hanf
Lübeck, 13. Jh., Lynch und Paap 1982; Šarišské Michalany, ČSSR, Ma, Šiška und Hajnalová 1983; Inowrocław, Polen, Ma, Iwaszkiewicz et al. 1979.
Opravil (1983) berichtet über die Anbaugeschichte des Hanfes.

Capsicum annuum L., Paprika
Ladenburg, 15./16. Jh., MAIER 1983.

Carum carvi L., Kümmel
Duisburg, Ma, KNÖRZER 1983a.

Cicer arietinum L., Kichererbse
Pizzica-Pantanello, Italien, 3. Jh. v. u. Z., COSTANTINI 1983.

Citrullus lanatus (THUNB.) MATSUMURA et NAKAI
Semna, Sudan, ca. 1500–1480 v. u. Z., VAN ZEIST 1983b.

Coriandrum sativum L., Koriander
Welzheim, Rz, KÖRBER-GROHNE et al. 1983; Heidelberg und Ladenburg, 15./16. Jh., MAIER 1983.

Cucumis melo L., Melone
Nidderau-Heldenbergen, Main-Kinzig-Kreis, Rz, BAAS 1982; Karthago, Tunesien, byzantinisch, ca. 600 u. Z., VAN ZEIST und BOTTEMA 1983.

Cucumis sativus L., Gurke
Münsterhof in Zürich, 12.–13. Jh., JACQUAT et al. 1982.

Cyperus esculentus L., Erdmandel
Semna, Sudan, ca. 1500–1480 v. u. Z., VAN ZEIST 1983b.

Daucus carota L., Gartenmöhre
Heidelberg, 15./16. Jh., MAIER 1983; Duisburg, Ma, KNÖRZER 1983a.

Fagopyrum sagittatum GILIB., Buchweizen
Duisburg, Ma, KNÖRZER 1983a.

Ficus carica L., Feigenbaum
Semna, Sudan, ca. 1500–1480 v. u. Z., VAN ZEIST 1983b; Kastanas, Griechenland, Bz, frühe Ez, KROLL 1982; Pizzica-Pantanello, Italien, 4.–3. Jh. v. u. Z., COSTANTINI 1983; Karthago, Tunesien, punisch (4. Jh. v. u. Z.), Rz (1.–2. Jh. u. Z.) und byzantinisch (ca. 600), VAN ZEIST und BOTTEMA 1983; Nidderau-Heldenbergen Main-Kinzig-Kreis, Rz, BAAS 1982; Welzheim, Rz, KÖRBER-GROHNE et al. 1983; Duisburg, Ma, KNÖRZER 1983a; Lübeck, Ma, MEYER und NEUGEBAUER 1982; Lübeck, 13. und 17./18. Jh., LYNCH und PAAP 1982; Heidelberg und Ladenburg, 15./16. Jh., MAIER 1983.

Foeniculum vulgare MILL., Fenchel
Münsterhof in Zürich, 12.–13. Jh., JACQUAT et al. 1982; Heidelberg und Ladenburg, 15./16. Jh., MAIER 1983.

Fragaria vesca L., Wald-Erdbeere
Lübeck, Ma, MEYER und NEUGEBAUER 1982; Duisburg, Ma, KNÖRZER 1983a; Münsterhof in Zürich, 12.–13. Jh., JACQUAT et al. 1982; Heidelberg und Ladenburg, 15./16. Jh., MAIER 1983; Lübeck, 17./18. Jh., LYNCH und PAAP 1982.

Hordeum vulgare L., Gerste
Skoroszowice, Gem. Strzelin, Polen, N, KLICHOWSKA 1981; Aartswoud, Niederlande, N, PALS 1983; Eberdingen-Hochdorf, Kr. Ludwigsburg, N, KÜSTER 1983a, b; Sarup, Dänemark, N, JØRGENSEN 1982; Federseeried, spätes N, BLANKENHORN und HOPF 1982; Ostiano (Cremona), Italien, frühe Bz, NISBET 1982; Otomani, Oarța de Sus und Sărata Monteoru, Rumänien, Bz, CÂRCIUMARU 1983; Tiszaalpár-Várdomb, Ungarn, Bz, HARTYÁNYI 1982; Kastanas, Griechenland, Bz, frühe Ez, KROLL 1982; Niedergurig, Kr. Bautzen, jüngere Bz, NÖTZOLD 1982a; Hahnefeld, Kr. Riesa, und Bautzen (Proitzschenberg), jüngere Bz, NÖTZOLD 1982b; Wiesens, Stadt Aurich, Ldkr. Aurich, jüngere Bz, BEHRE 1983b; Harkenbleck, Gem. Hemmingen, Ldkr. Hannover, jüngere Bz, KULLIG 1983, BEHRE 1983b; Dresden-Nickern, Aunjetitzer Kultur, NÖTZOLD 1982b; Histria und Popești, Rumänien, Ez, CÂRCIUMARU 1983; Pod bei Bugojno, Zentralbosnien, Ez, KUČAN 1983; Crosskirk near Thurso in Caithness, Schottland, Ez, DICKSON 1979; Lauffen am Neckar, Kr. Heilbronn, Frühlatènezeit, ca. 400 v. u. Z., PIENING 1983; Altenritte, Kr. Kassel und Wehren/Fritzlar, Latène, HAHNER 1982; Pizzica-Pantanello, Italien, 4.–3. Jh. v. u. Z., COSTANTINI 1983; Titelberg, Luxemburg, Latène und Rz, ROWLETT et al. 1982; Welzheim, Rz, KÖRBER-GROHNE et al. 1983; Nidderau-Heldenbergen, Main-Kinzig-Kreis, Rz, BAAS 1982; Leuben, Kr. Oschatz, Römische Kaiserzeit, NÖTZOLD 1982a, b; Magdeborn, Lkr. Leipzig, slawisch, NÖTZOLD 1982a; Krivina, Bulgarien, 4.–10. Jh. u. Z., HAJNALOVÁ 1982; Monastyrek, mittleres Dnepr-Gebiet, 8.–10. Jh. u. Z., PAŠKEVIČ und PETRAŠENKO 1982; Machrins, Island of Colonsay, Schottland, ca. 800 u. Z., DICKSON 1982; Lübeck, Ma, MEYER und NEUGEBAUER 1982; Duisburg, Ma, KNÖRZER 1983a; Sornzig, Kr. Oschatz, frühdeutsch, 12./13. Jh., NÖTZOLD 1982b; Lübeck, 13. und 17./18. Jh., LYNCH und PAAP 1982; Stift Lilienfeld, Niederösterreich, Grab von 1429, SACKL 1980; Smeerenburg, Spitzbergen, 17. Jh., VAN WIJNGAARDEN-BAKKER und PALS 1981.

SHEWRY et al. (1982) untersuchten die Zusammensetzung der Proteine und Aminosäuren bei rezenten und subfossilen Gerstenkörnern. Anhaltspunkte zur Bestimmung und Datierung archäologischer Proben ließen sich aus den Ergebnissen nicht gewinnen.

Humulus lupulus L., Hopfen
Haithabu bei Schleswig, 9.–10. Jh., BEHRE 1983a; Lübeck, 12./13. und 17./18. Jh., LYNCH und PAAP 1982.

Juglans regia L., Walnußbaum
Karthago, Tunesien, Rz (1.–2. Jh. u. Z.) und byzantinisch (ca. 600), VAN ZEIST und BOTTEMA 1983; Welzheim, Rz, KÖRBER-GROHNE et al. 1983; Duisburg, Ma, KNÖRZER 1983a; Lübeck, Ma, MEYER und NEUGEBAUER 1982; Heidelberg, 15./16. Jh., MAIER 1983.

Lathyrus sativus L., Saat-Platterbse
Kastanas, Griechenland, frühe Bz, KROLL 1982; Tiszaalpár-Várdomb, Ungarn, Bz, HARTYÁNYI 1982.

Lens culinaris MEDIK., Linse
Kastanas, Griechenland, Bz, frühe Ez, KROLL 1982; Tiszaalpár-Várdomb, Ungarn, Bz, HARTYÁNYI 1982; Kalakača, Jugoslawien, Ez, MEDOVIĆ 1978; Pod bei Bugojno, Zentralbosnien, Ez, KUČAN 1983; Pizzica-Pantanello, Italien, 3. Jh. v. u. Z., COSTANTINI 1983; Nidderau-Heldenbergen, Main-Kinzig-Kreis, Rz, BAAS 1982; London, Rz, STRAKER 1983.

Linum usitatissimum L., Lein
Weyregg am Attersee, Oberösterreich, N, SCHMIDT 1983; Eberdingen-Hochdorf, Kr. Ludwigsburg, N, KÜSTER 1983a, b; Aartswoud, Niederlande, N, PALS 1983; Kastanas, Griechenland, frühe Bz, frühe Ez, KROLL 1982; Tiszaalpár-Várdomb, Ungarn, Bz, HARTYÁNYI 1982; Crosskirk near Thurso in Caithness, Schottland, Ez, DICKSON 1979; Welzheim, Rz, KÖRBER-GROHNE et al. 1983; Leuben, Kr. Oschatz, Römische Kaiserzeit, NÖTZOLD 1982a, b; Lübeck, Ma, MEYER und NEUGEBAUER 1982; Duisburg, Ma, KNÖRZER 1983a; Inowrocław, Polen, Ma, IWASZKIEWICZ et al. 1979; Monastyrek, mittleres Dnepr-Gebiet, 8.–10. Jh., PAŠKEVIČ und PETRAŠENKO 1982; Münsterhof in Zürich, 12.–13. Jh., JACQUAT et al. 1982; Lübeck, 13. und 17./18. Jh., LYNCH und PAAP 1982.

Malus sylvestris (L.) P. MILLER und *Malus domestica* BORKH., Apfelbaum
Nidderau-Heldenbergen, Main-Kinzig-Kreis, Rz, BAAS 1982; Duisburg, Ma, KNÖRZER 1983a; Münsterhof in Zürich, 12.–13. Jh., JACQUAT et al., 1982; Heidelberg und Ladenburg, 15./16. Jh., MAIER 1983.

Mespilus germanica L., Mispel
Nidderau-Heldenbergen, Main-Kinzig-Kreis, Rz, BAAS 1982.

Morus nigra L., Schwarzer Maulbeerbaum
Duisburg, Ma, KNÖRZER 1983a; Münsterhof in Zürich, 12.–13. Jh., JACQUAT et al. 1982; Heidelberg und Ladenburg, 15./16. Jh., MAIER 1983.

Olea europaea L., Ölbaum
Pizzica-Pantanello, Italien, 4.–3. Jh. v. u. Z., COSTANTINI 1983; Karthago, Tunesien, punisch (4. Jh. v. u. Z.), Rz (1.–2. Jh. u. Z.) und byzantinisch (um 600), VAN ZEIST und BOTTEMA 1983.

Panicum miliaceum L., Rispenhirse
Hain, Kr. Borna, jüngere Linienbandkeramik, NÖTZOLD 1982a; Eberdingen-Hochdorf, Kr. Ludwigsburg, N, KÜSTER 1983b; Kastanas, Griechenland, Bz, frühe Ez, KROLL 1982; Bautzen (Proitzschenberg), jüngere Bz, NÖTZOLD 1982b; Harkenbleck, Gem. Hemmingen, Ldkr. Hannover, jüngere Bz, KULLIG 1983, BEHRE 1983b; Cîndești, Rumänien, Bz (?), CÂRCIUMARU 1983; Kalakača, Jugoslawien, Ez, MEDOVIĆ 1978; Pod bei Bugojno, Zentralbosnien, Ez, KUČAN 1983;

Tăşad, Brad, Popeşti, Barboşi und Ocniţa, Rumänien, Ez, Cârciumaru 1983; Lauffen am Neckar, Kr. Heilbronn, Frühlatènezeit, ca. 400 v. u. Z., Piening 1983; Nidderau-Heldenbergen, Main-Kinzig-Kreis, Rz, Baas 1982; Welzheim, Rz, Körber-Grohne et al. 1983; Krivina, Bulgarien, 4.–10. Jh. u. Z., Hajnalová 1982; Monastyrek, mittleres Dnepr-Gebiet, 8.–10. Jh., Paškevič und Petrašenko 1982; Münsterhof in Zürich, 12.–13. Jh., Jacquat et al. 1982; Lübeck, 13. und 17./18. Jh., Lynch und Paap 1982; Heidelberg, 15./16. Jh., Maier 1983; Duisburg, Ma, Knörzer 1983a; Inowrocław, Polen, Ma, Iwaszkiewicz et al. 1979; Smeerenburg, Spitzbergen, 17. Jh., van Wijngaarden-Bakker und Pals 1981.

Papaver somniferum L., Schlafmohn
Eberdingen-Hochdorf, Kr. Ludwigsburg, N, Küster 1983a, b; Federseeried, spätes N, Blankenhorn und Hopf 1982; Kastanas, Griechenland, frühe Bz, frühe Ez, Kroll 1982; Răcătău und Popeşti, Rumänien, Latène, Cârciumaru 1983; Karthago, Tunesien, punisch (4. Jh. v. u. Z.), Rz (1.–2. Jh. u. Z.) und byzantinisch (ca. 600), van Zeist und Bottema 1983; Welzheim, Rz, Körber-Grohne et al. 1983; Duisburg, Ma, Knörzer 1983a; Münsterhof in Zürich, 12.–13. Jh., Jacquat et al. 1982; Heidelberg und Ladenburg, 15./16. Jh., Maier 1983.

Petroselinum crispum (Mill.) Nym., Petersilie
Eberdingen-Hochdorf, Kr. Ludwigsburg, N, Küster 1983a, b.

Phoenix dactylifera L., Dattelpalme
Semna, Sudan, ca. 1500–1480 v. u. Z., van Zeist 1983b.

Pistacia lentiscus L., Mastixstrauch
Karthago, Tunesien, punisch (4. Jh. v. u. Z.) und Rz (1.–2. Jh. u. Z.), van Zeist und Bottema 1983.

Pisum sativum L., Erbse
Hain, Kr. Borna, jüngere Linienbandkeramik, Nötzold 1982a; Eberdingen-Hochdorf, Kr. Ludwigsburg, N, Küster 1983a, b; Tiszaalpár-Várdomb, Ungarn, Bz, Hartyányi 1982 (auch *P.* cf. *elatius*); Großsteinberg, Kr. Grimma, mittlere Bz, Nötzold 1982b; Bautzen (Proitzschenberg), jüngere Bz, Nötzold 1982b; Cîrlomăneşti, Rumänien, Ez, Cârciumaru 1983 (*Pisum* cf. *elatius*); Pod bei Bugojno, Zentralbosnien, Ez, Kučan 1983; Pizzica-Pantanello, Italien, 4.–3. Jh. v. u. Z., Costantini 1983; Nidderau-Heldenbergen, Main-Kinzig-Kreis, Rz, Baas 1982; Krivina, Bulgarien, 4.–10. Jh. u. Z., Hajnalová 1982; Magdeborn, Lkr. Leipzig, slawisch, Nötzold 1982a; Duisburg, Ma, Knörzer 1983a; Münsterhof in Zürich, 12.–13. Jh., Jacquat et al. 1982; Smeerenburg, Spitzbergen, 17. Jh., van Wijngaarden-Bakker und Pals 1981.

Prunus avium L., Süßkirsche
Nidderau-Heldenbergen, Main-Kinzig-Kreis, Rz, Baas 1982; Lübeck, Ma, Meyer und Neugebauer 1982; Duisburg, Ma, Knörzer 1983a; Münsterhof

in Zürich, 12.–13. Jh., JACQUAT et al. 1982; Lübeck, 13. Jh., LYNCH und PAAP 1982; Heidelberg und Ladenburg, 15./16. Jh., MAIER 1983.

Prunus cerasus L., Sauerkirsche
Duisburg, Ma, KNÖRZER 1983a; Münsterhof in Zürich, 12.–13. Jh., JACQUAT et al. 1982; Lübeck, 13. Jh., LYNCH und PAAP 1982; Heidelberg, 15./16. Jh., MAIER 1983.

Prunus domestica L., Pflaume
Nidderau-Heldenbergen, Main-Kinzig-Kreis, Rz, BAAS 1982; Welzheim, Rz, KÖRBER-GROHNE et al. 1983; Duisburg, Ma, KNÖRZER 1983a; Münsterhof in Zürich, 12.–13. Jh., JACQUAT et al. 1982; Lübeck, 13. und 17./18. Jh., LYNCH und PAAP 1982; Heidelberg und Ladenburg, 15./16. Jh., MAIER 1983.

Prunus persica (L.) BATSCH, Pfirsich
Welzheim, Rz, KÖRBER-GROHNE et al. 1983; Duisburg, Ma, KNÖRZER 1983a; Inowrocław, Polen, Ma, IWASZKIEWICZ et al. 1979; Lübeck, 17.–18. Jh., LYNCH und PAAP 1982.

Punica granatum L., Granatapfelbaum
Karthago, Tunesien, punisch (4. Jh. v. u. Z.), Rz (1.–2. Jh. u. Z.) und byzantinisch (ca. 600), VAN ZEIST und BOTTEMA 1983.

Pyrus communis L., Birnbaum
Nidderau-Heldenbergen, Main-Kinzig-Kreis, Rz, BAAS 1982; Duisburg, Ma, KNÖRZER 1983a; Münsterhof in Zürich, 12.–13. Jh., JACQUAT et al. 1982; Heidelberg und Ladenburg, 15./16. Jh., MAIER 1983.

Rumex scutatus L., Schildblättriger Ampfer
Welzheim, Rz, KÖRBER-GROHNE et al. 1983.

Secale cereale L., Roggen
Cîndeşti, Rumänien, Bz (?), CÂRCIUMARU 1983; Cîrlomăneşti und Popeşti, Rumänien, Ez, CÂRCIUMARU 1983; Lauffen am Neckar, Kr. Heilbronn, Frühlatènezeit, ca. 400 v. u. Z., PIENING 1983; Welzheim, Rz, KÖRBER-GROHNE et al. 1983; Krivina, Bulgarien, 4.–10. Jh. u. Z., HAJNALOVÁ 1982; Monastyrek, mittleres Dnepr-Gebiet, 8.–10. Jh., PAŠKEVIČ und PETRAŠENKO 1982; Hont, Komitat Nográd, Ungarn, 10. Jh., HARTYÁNYI 1983; Duisburg, Ma, KNÖRZER 1983a; Lübeck, Ma, MEYER und NEUGEBAUER 1982; Lübeck, 13. und 17./18. Jh., LYNCH und PAAP 1982; Smeerenburg, Spitzbergen, 17. Jh., VAN WIJNGAARDEN-BAKKER und PALS 1981.

Setaria italica (L.) PAL. BEAUV. oder *S. viridis* (L.) PAL. BEAUV.
Lauffen am Neckar, Kr. Heilbronn, Frühlatènezeit, ca. 400 v. u. Z., PIENING 1983; Duisburg, Ma, KNÖRZER 1983a.

Triticum aestivum L., Saatweizen (incl. *Tr. compactum* HOST)
Hain, Kr. Borna, jüngere Linienbandkeramik, NÖTZOLD 1982a; Eberdingen-

Hochdorf, Kr. Ludwigsburg, N, KÜSTER 1983a, b; Sarup, Dänemark, N, JØRGENSEN 1982; Aartswoud, Niederlande, N, PALS 1983; Štúrovo, ČSSR, N, HAJNALOVÁ 1983; Federseeried, spätes N, BLANKENHORN und HOPF 1982; Ostiano (Cremona), Italien, frühe Bz, NISBET 1982; Otomani, Sărata Monteoru und Cîndeşti, Rumänien, Bz, CÂRCIUMARU 1983; Tiszaalpár-Várdomb, Ungarn, Bz, HARTYÁNYI 1982; Ussat, Ariège, Frankreich, Bz, ERROUX 1982; Lehma, Kr. Altenburg, jüngere Bz, NÖTZOLD 1982a; Niedergurig, Kr. Bautzen, jüngere Bz, NÖTZOLD 1982a; Bautzen (Proitzschenberg), und Hahnefeld, Kr. Riesa, jüngere Bz, NÖTZOLD 1982b; Kastanas, Griechenland, frühe Bz, KROLL 1982; Cîrlomăneşti, Tăşad und Popeşti, Rumänien, Ez, CÂRCIUMARU 1983; Lauffen am Neckar, Kr. Heilbronn, Frühlatènezeit, ca. 400 v. u. Z., PIENING 1983; Wehren/Fritzlar, Latène, HAHNER 1982; Pizzica-Pantanello, Italien, 4.–3. Jh. v. u. Z., COSTANTINI 1983; Titelberg, Luxemburg, Latène und Rz, ROWLETT et al. 1982; London, Rz, STRAKER 1983; Welzheim, Rz, KÖRBER-GROHNE et al. 1983; Nidderau-Heldenbergen, Main-Kinzig-Kreis, Rz, BAAS 1982; Leuben, Kr. Oschatz, Römische Kaiserzeit, NÖTZOLD 1982a, b; Krivina, Bulgarien, 4.–10. Jh. u. Z., HAJNALOVÁ 1982; Monastyrek, mittleres Dnepr-Gebiet, 8.–10. Jh., PAŠKEVIČ und PETRAŠENKO 1982; Hont, Komitat Nográd, Ungarn, 10. Jh., HARTYÁNYI 1983; Magdeborn, Lkr. Leipzig, slawisch, NÖTZOLD 1982a; Duisburg, Ma, KNÖRZER 1983a; Mutzschen, Kr. Grimma, Ma um 1430, NÖTZOLD 1982a; Smeerenburg, Spitzbergen, 17. Jh., VAN WIJNGAARDEN-BAKKER und PALS 1981.

UDAČIN (1982) berichtete über das Vorkommen eines Weizens in Tadžikistan, der große Ähnlichkeit mit *Triticum antiquorum* HEER hat. JACOMET und SCHLICHTHERLE (1983) setzten sich mit dem Problem des kleinen Pfahlbauweizens von OSWALD HEER auseinander.

Triticum dicoccon SCHRANK, Emmer

Hain, Kr. Borna, jüngere Linienbandkeramik, NÖTZOLD 1982a; Eberdingen-Hochdorf, Kr. Ludwigsburg, N, KÜSTER 1983a, b; Sarup, Dänemark, N, JØRGENSEN 1982; Aartswoud, Niederlande, N, PALS 1983; Štúrovo, ČSSR, N, HAJNALOVÁ 1983; Altenburg bei Burgerroth, Gde. Aub, Unterfranken, spätes N, HOPF 1982b; Federseeried, spätes N, BLANKENHORN und HOPF 1982; Ussat, Ariège, Frankreich, Bz, ERROUX 1982; Ostiano (Cremona), Italien, frühe Bz, NISBET 1982; Kastanas, Griechenland, Bz, frühe Ez, KROLL 1982; Tiszaalpár-Várdomb, Ungarn, Bz, HARTYÁNYI 1982; Otomani, Oarţa de Sus und Sărata Monteoru, Rumänien, Bz, CÂRCIUMARU 1983; Lehma, Kr. Altenburg, jüngere Bz, NÖTZOLD 1982a; Hahnefeld, Kr. Riesa, jüngere Bz, etwa 1200–1000 v. u. Z., NÖTZOLD 1982b; Harkenbleck, Gem. Hemmingen, Ldkr. Hannover, jüngere Bz, KULLIG 1983, BEHRE 1983b; Dresden-Nickern, Aunjetitzer Kultur, NÖTZOLD 1982b; Cîrlomăneşti, Tăşad und Grăniceşti, Rumänien, Ez, CÂRCIUMARU 1983; Pod bei Bugojno, Zentralbosnien, Ez, KUČAN 1983; Grabanlage des 'Anch-Hor, Ägypten, 26. Dynastie, LIESE-KLEIBER und SCHLICHTHERLE 1982; Lauffen am Neckar, Kr. Heilbronn, Frühlatènezeit, ca. 400 v. u. Z., PIENING 1983; Pizzica-Pantanello, Italien, 4.–3. Jh. v. u. Z., COSTANTINI 1983; Titelberg, Luxemburg, Latène und Rz, ROWLETT et al. 1982; Welzheim, Rz, KÖRBER-GROHNE et al. 1983; Leuben, Kr. Oschatz, Ältere Römische Kaiserzeit, NÖTZOLD 1982b;

Monastyrek, mittleres Dnepr-Gebiet, 8.–10. Jh., PAŠKEVIČ und PETRAŠENKO 1982; Mutzschen, Kr. Grimma, Ma um 1430, NÖTZOLD 1982a.

Triticum durum DESF., Hartweizen
Kastanas, Griechenland, frühe Ez, KROLL 1982; Cîrlomăneşti, Rumänien, Ez, CÂRCIUMARU 1983.

Triticum macha DEKAPR. et MENABDE
Georgien, UdSSR, N, GORGIDZE 1979 (vgl. auch BREGADZE 1982).

Triticum monococcum L., Einkorn
Hain, Kr. Borna, jüngere Linienbandkeramik, NÖTZOLD 1982a; Štúrovo, ČSSR, N, HAJNALOVÁ 1983; Eberdingen-Hochdorf, Kr. Ludwigsburg, N, KÜSTER 1983a, b; Sarup, Dänemark, N, JØRGENSEN 1982; Aartswoud, Niederlande, N, PALS 1983; Federseeried, spätes N, BLANKENHORN und HOPF 1982; Kastanas, Griechenland, Bz, frühe Ez, KROLL 1982; Otomani und Sărata Monteoru, Rumänien, Bz, CÂRCIUMARU 1983; Tiszaalpár-Várdomb, Ungarn, Bz, HARTYÁNYI 1982; Großsteinberg, Kr. Grimma, mittlere Bz, NÖTZOLD 1982b; Harkenbleck, Gem. Hemmingen, Ldkr. Hannover, jüngere Bz, KULLIG 1983, BEHRE 1983b; Bautzen (Proitzschenberg), und Hahnefeld, Kr. Riesa, jüngere Bz, NÖTZOLD 1982b; Lehma, Kr. Altenburg, jüngere Bz, NÖTZOLD 1982a; Dresden-Nickern, Aunjetitzer Kultur, NÖTZOLD 1982b; Kalakača, Jugoslawien, Ez, MEDOVIĆ 1978; Pod bei Bugojno, Zentralbosnien, Ez, KUČAN 1983; Cîrlomăneşti und Tăşad, Rumänien, Ez, CÂRCIUMARU 1983; Lauffen am Neckar, Kr. Heilbronn, Frühlatènezeit, ca. 400 v. u. Z., PIENING 1983; Titelberg, Luxemburg, Latène und Rz, ROWLETT et al. 1982; Welzheim, Rz, KÖRBER-GROHNE et al. 1983; London, Rz, STRAKER 1983; Leuben, Kr. Oschatz, ältere römische Kaiserzeit, NÖTZOLD 1982b.

Triticum spelta L., Spelz
Štúrovo, ČSSR, N, HAJNALOVÁ 1983; Federseeried, spätes N, BLANKENHORN und HOPF 1982; Kastanas, Griechenland, Bz, frühe Ez, KROLL 1982; Otomani und Sărata Monteoru, Rumänien, Bz, CÂRCIUMARU 1983; Cîrlomăneşti und Tăşad, Rumänien, Ez, CÂRCIUMARU 1983; Pod bei Bugojno, Zentralbosnien, Ez, KUČAN 1983; Lauffen am Neckar, Kr. Heilbronn, Frühlatènezeit, ca. 400 v. u. Z., PIENING 1983; Welzheim, Rz, KÖRBER-GROHNE et al. 1983; Titelberg, Luxemburg, Rz, ROWLETT et al. 1982; Monastyrek, mittleres Dnepr-Gebiet, 8.–10. Jh., PAŠKEVIČ und PETRAŠENKO 1982; Heidelberg, 15./16. Jh., MAIER 1983.

Vicia ervilia (L.) WILLD., Erve
Kastanas, Griechenland, Bz, frühe Ez, KROLL 1982; Pod bei Bugojno, Zentralbosnien, Ez, KUČAN 1983; Pizzica-Pantanello, Italien, 3. Jh. v. u. Z., COSTANTINI 1983; Krivina, Bulgarien, 4.–10. Jh. u. Z., HAJNALOVÁ 1982.

Vicia faba L., Ackerbohne
Kastanas, Griechenland, frühe Bz, frühe Ez, KROLL 1982; Lehma, Kr. Altenburg, jüngere Bz, NÖTZOLD 1982a; Bautzen (Proitzschenberg), jüngere Bz,

NÖTZOLD 1982b; Pod bei Bugojno, Zentralbosnien, Ez, KUČAN 1983; Pizzica-Pantanello, Italien, 3. Jh. v. u. Z., COSTANTINI 1983; Nidderau-Heldenbergen, Main-Kinzig-Kreis, Rz, BAAS 1982; Welzheim, Rz, KÖRBER-GROHNE et al. 1983; Smeerenburg, Spitzbergen, 17. Jh., VAN WIJNGAARDEN-BAKKER und PALS 1981.

Vicia sativa L., Saatwicke
Nidderau-Heldenbergen, Main-Kinzig-Kreis, Rz, BAAS 1982.

Vitis vinifera L., Weinstock
Semna, Sudan, ca. 1500–1480 v. u. Z., VAN ZEIST 1983b; Kastanas, Griechenland, Bz, frühe Ez, KROLL 1982; Pizzica-Pantanello, Italien, 4.–3. Jh. v. u. Z., COSTANTINI 1983; Karthago, Tunesien, punisch (4. Jh. v. u. Z.), Rz (1.–2. Jh. u. Z.) und byzantinisch (ca. 600), VAN ZEIST und BOTTEMA 1983; Welzheim, Rz, KÖRBER-GROHNE et al. 1983; Nidderau-Heldenbergen, Main-Kinzig-Kreis, Rz, BAAS 1982; Duisburg, Ma, KNÖRZER 1983a; Münsterhof in Zürich, 12.–13. Jh., JACQUAT et al. 1982; Lübeck, 13. und 17./18. Jh., LYNCH und PAAP 1982; Heidelberg und Ladenburg, 15./16. Jh., MAIER 1983.

Zea mays L., Mais
DE WET (1981) befaßte sich mit der Entstehung und Domestikation des Maises.

Ziziphus spina-christi (L.) WILLD.
Semna, Sudan, ca. 1500–1480 v. u. Z., VAN ZEIST 1983b.

Zum Schluß seien zusammenfassende Arbeiten bzw. solche allgemeinen und methodischen Inhaltes besprochen.

Experimente zur Wirksamkeit von „Samen-Maschinen" sind von VAN DER VEEN (1983) unternommen worden.

SCHLICHTHERLE (1983) hat mikroskopische Untersuchungen an neolithischen Gefäßinhalten aus Schweizer Pfahlbauten vorgenommen und dazu auch Verkohlungsversuche durchgeführt.

Eine kurze Übersicht älterer neolithischer Kulturpflanzenfunde zwischen Rhein und Seine wird von HEIM (1983) gegeben.

Die Problematik der Getreidegruben hat NOVÁKI (1983) behandelt.

Mit der Interpretation archäologischer Pflanzenreste aus verschiedenen Ländern setzten sich CLARKE et al. (1983), HILLMAN (1983a), JONES (1983), KNÖRZER (1983b), LUNDSTROM-BAUDAIS (1983), PAAP (1983), SHAY (1983), VAN VILSTEREN (1983) und WILLERDING (1983a) auseinander.

KOSINA (1983) untersuchte die Morphologie von Weizen-Körnern, um Anhaltspunkte für die Bestimmung einiger Arten zu gewinnen und HILLMAN (1983b) prüfte die Ährenachsen tetraploider und hexaploider Weizen auf Unterscheidungsmerkmale. Auch KISLEV (1983) beschäftigte sich mit methodischen Problemen bei archäologischen Nacktweizen aus dem Nahen Osten.

Die archäologischen und urkundlichen Nachweise von Pflanzen aus dem mittelalterlichen England hat GREEN (1983) zusammengestellt.

NEUSTUPNÝ und DVOŘÁK (1983) haben ein Modell der Ernährung urgeschichtlicher Bauern entwickelt.

Ein umfassender Katalog vor- und frühgeschichtlicher Kulturpflanzen aus dem nördlichen Deutschland ist von Hopf (1982a) veröffentlicht worden. Willerding (1983b) gibt einen Überblick über den ältesten Ackerbau in Niedersachsen.

Van Zeist (1983a) beschreibt den Gebrauch von Pflanzen in früherer Zeit. Grünert et al. (1982) und Hoffmann (1982) behandeln die Entstehung von Pflanzenanbau und Haustierhaltung.

Die Begründung und Entwicklung des frühen europäischen Ackerbaus ist von Jarman et al. (1982) ausführlich dargestellt worden.

Forni (1983) versucht eine einheitliche Theorie für die Entstehung der Domestikation zu entwickeln.

Summary

Literature on archaeological remains of cultivated plants (1982/1983)

The publications of 1982 and 1983 on archaeological remains of cultivated plants were put together and the contents of the 78 references briefly discussed.

Краткое содержание

Литература об археологических находках остатков культурных растений (1982/1983 гг.)

Продолжение библиографии работ об археологических находках остатков культурных растений. Дается список соответствующей литературы за 1982 и 1983 годы. Кратко комментируется содержание 78 работ.

Literatur

Baas, J., 1982: Kultur- und Nutzpflanzen aus römischen Siedlungsgruben in Nidderau-Heldenbergen (Main-Kinzig-Kreis). Ein Beitrag zur Geschichte unserer Kulturpflanzen. – Saalburg Jahrbuch **38**, 110–119.

Behre, K.-E., 1983a: Zur Geschichte des Bieres nach Fruchtfunden und schriftlichen Quellen. – Papers and Abstracts, VI Symposium, International Work Group for Palaeoethnobotany, Groningen, 30 May–3 June 1983, 118.

–, 1983b: Zwei jungbronzezeitliche Getreidefunde aus Niedersachsen. – Nachrichten aus Niedersachsens Urgeschichte **51**, 1982, 281–292.

Blankenhorn, B., und M. Hopf, 1982: Pflanzenreste aus spätneolithischen Moorsiedlungen des Federseerieds. – Jahrb. Röm.-German. Zentralmus. Mainz **29**, 75–99, Taf. 1–3.

Bregadze, N. A., 1982: Očerki po agroėtnografii Gruzii. – Mecniereba, Tbilisi, 264 pp.

Cârciumaru, M., 1983: Consideraţii paleoetnobotanice şi contribuţii la agricultura geto-dacilor. Considérations paléoethnobotaniques et contributions a l'agriculture chez les Géto-Daces. – Thraco-Dacica **4** (1–2), 126–134, Fig. 1–4.

Clarke, A., L. Cubis, and B. Gott, 1983: Progress and problems of Australian palaeoethnobotany. – Papers and Abstracts, VI Symposium, International Work Group for Palaeoethnobotany, Groningen, 30 May–3 June 1983, 120.

Costantini, L., 1983: Bioarchaelogical [sic] research at Pizzica-Pantanello. (5 unnumbered pages [32–36]). In: The Territory of Metaponto 1981–1982. – The University of Texas at Austin, [51 pp.].

Dickson, C. A., 1982: Cereals and charcoal from Machrins, Colonsay. (Appendix 3: pp. 275–276). In: J. N. G. Ritchie, Excavations at Machrins, Colonsay. – Proc. Soc. Antiq. Scot. **111**, 1981, 263–281.

Dickson, J. H., 1979: The botany of brochs. In: Early Man in the Scottish Landscape (Scottish Archaeological Forum **9**), 62–65.

Erroux, J., 1982: Analyse de grains de céréales provenant de la grotte des Églises Ussat (Ariège). – Bull. Soc. Préhist. Ariège **37**, 152–156.

Forni, G., 1983: From pyrophytic to domesticated plants: the palaeontological-linguistic evidences for a unitary theory on the origin of domestication. – Papers and Abstracts, VI Symposium, International Work Group for Palaeoethnobotany, Groningen, 30 May–3 June 1983, 121.

Gorgidze, A. D., 1979: [Living relic of cultivated wheat]. – Bull. Acad. Sci. Georgian SSR **96** (2), 425–428. (Ref.: Plant Breeding Abstracts **53**, 1983, no. 2872).

Green, F. J., 1983: The archaeological and documentary evidence for plants from the Medieval period in England. – Papers and Abstracts, VI Symposium, International Work Group for Palaeoethnobotany, Groningen, 30 May–3 June 1983, 1–36.

Grünert, H., et al. 1982: Geschichte der Urgesellschaft. – VEB Deutscher Verlag der Wissenschaften, Berlin, 360 pp. (Die Entstehung von Pflanzenanbau und Haustierhaltung: pp. 167–208).

Hahner, A., 1982: Die Latènezeit in Niederhessen. – Führer vor- u. frühgesch. Denkmäl. **50**, 136–158.

Hajnalová, E., 1982: Archäobotanische Funde aus Krivina (pp. 207–236, Taf. 80–85). In: Iatrus-Krivina. Spätantike Befestigung und frühmittelalterliche Siedlung an der unteren Donau. Bd. 2. Ergebnisse der Ausgrabungen 1966–1973. – Schriften zur Geschichte und Kultur der Antike **17**, Akademie-Verlag, Berlin, 242 pp., 87 Taf.

–, 1983: Paleobotanické neolitické nálezy zo Štúrova. Paläobotanische Funde aus dem Neolithikum in Štúrovo. – Slovenská Archeológia **31**–1, 199–218.

Hartyányi, B. P., 1982: A Tiszaalpár-Várdomb bronzkori lakóteleprõl származó mag- és termésleletek. Die von der bronzezeitlichen Wohnsiedlung Tiszaalpár-Várdomb stammenden Korn- und Fruchtfunde. – Cumania VII. Archaeologia, Bács-Kiskun Megyei Múzeumok Közleményei, Acta Museorum ex Comitatu Bács-Kiskun, Kecskemét, 133–164, 271–286.

–, 1983: Kora Árpád-Korból származó búza a Hont-i ispánsági várból. Aus der frühen Árpádenzeit stammender Weizen der Honter Gespannschaftsburg. – Magyar Mezögazdasági Múzeum Közleményei 1981–1983, 95–112.

Heim, J., 1983: Apports récents de la paléobotanique à la connaissance de l'importance des activités culturales (agricoles) des néolithiques anciens entre Rhin et Seine. – Dissertationes Archaeologicae Gandenses **21**, Progrès récents dans l'étude du Néolithique ancien, Brugge, 62–70.

Hillman, G. C., 1983a: Interpretation of archaeological plant remains: the application of ethnographic models from Turkey. – Papers and Abstracts, VI Symposium, International Work Group for Palaeoethnobotany, Groningen, 30 May–3 June 1983, 37–84.

–, 1983b: Criteria for distinguishing rachis remains of 4x and 6x naked wheats. – Papers and Abstracts, VI Symposium, International Work Group for Palaeoethnobotany, Groningen, 30 May–3 June 1983, 124.

Hoffmann, E., 1982: Zu den Anfängen von Pflanzen- und Tierproduktion im Vorderen Orient. – Ethnogr.-Archäol. Z. **23**, 639–650.

Hopf, M., 1982a: Vor- und frühgeschichtliche Kulturpflanzen aus dem nördlichen Deutschland. – Römisch-Germanisches Zentralmuseum Mainz, 108 pp., 94 Taf. (Kataloge vor- und frühgeschichtlicher Altertümer **22**).

–, 1982b: Botanische Untersuchungen in Getreideabdrücken im keramischen Fundmaterial der Grabungen 1919–1921 (pp. 512–513). In: D. R. Spennemann, Die spätneolithische Höhensiedlung auf dem Altenberg bei Burgerroth, Gde Aub, Unterfranken. – Frankfurt a. M.

Iwaszkiewicz, M., M. Klichowska, and M. Sobociński, 1979: Materiały przyrodnicze. [Natural materials.] In: Zaplecze gospodarcze konwentu OO. Francizkanów w Inowrocławiu. [The basis of supplies of the monastery of the Franciscan order in Inowrocław]. Poznań, 171–177.

JACOMET, S., and H. SCHLICHTHERLE, 1983: Der kleine Pfahlbauweizen Oswald Heers – Bemerkungen zur Morphologie alter und neuer Nacktweizenfunde. – Papers and Abstracts, VI Symposium, International Work Group for Palaeoethnobotany, Groningen, 30 May–3 June 1983, 125.

JACQUAT, C., B. PAWLIK, und W. SCHOCH, 1982: Die mittelalterlichen Pflanzenfunde (pp. 267–278). In: Der Münsterhof in Zürich. – Schweizer Beiträge zur Kulturgeschichte und Archäologie des Mittelalters **10**.

JARMAN, M. R., G. N. BAILEY, and H. N. JARMAN (eds.), 1982: Early European Agriculture. Its Foundations and Development. – Cambridge University Press, Cambridge etc., X + 283 pp.

JONES, G., 1983: Interpretation of archaeological plant remains: ethnographic models from Greece. – Papers and Abstracts, VI Symposium, International Work Group for Palaeoethnobotany, Groningen, 30 May–3 June 1983, 85–116.

JØRGENSEN, G., 1982: Korn fra Sarup. Med nogle bemaerkninger om agerbruget i yngre stenalder i Danmark. Cereals from Sarup. With some remarks on plant husbandry in Neolithic Denmark. – Kuml 1981, 221–231.

KISLEV, M., 1983: Botanical evidence for ancient naked wheats in the Near East. – Papers and Abstracts, VI Symposium, International Work Group for Palaeoethnobotany, Groningen, 30 May–3 June 1983, 139–160.

KLICHOWSKA, M., 1981: Odciski zbóż i chwastów zachowane na neolitycznej polepie i ceramice ze Skoroszowic, gmina Strzelin, stanowisko 1. Getreide- und Unkrautabdrücke auf der neolithischen Keramik und dem gebrannten Lehm aus Skoroszowice, gem. Strzelin, Fundst. 1 (pp. 118–125). In: W. WOJCIECHOWSKI, Wczesnoneolityczna osada w Skoroszowicach. – Acta Universitatis Wratislaviensis, Studia Archeologiczne **12**.

KNÖRZER, K.-H., 1983a: Mittelalterliche Pflanzenfunde unter dem Alten Markt in Duisburg (pp. 78–87). In: Duisburg im Mittelalter, Begleitschrift zur Ausstellung, 4. September bis 27. November 1983. – Stadtarchiv und Niederrheinisches Museum der Stadt Duisburg, 141 pp.

–, 1983b: Aussagemöglichkeiten paläoethnobotanischer Latrinenuntersuchungen. – Papers and Abstracts, VI Symposium, International Work Group for Palaeoethnobotany, Groningen, 30 May–3 June 1983, 126–127.

KÖRBER-GROHNE, U., M. KOKABI, U. PIENING, und D. PLANCK, 1983: Flora und Fauna im Ostkastell von Welzheim. – Forschungen und Berichte zur Vor- und Frühgeschichte in Baden-Württemberg **14**, Konrad Theiss Verlag, Stuttgart, 149 pp., 1 Beilage.

KOSINA, R., 1983: Morphology of the crease of wheat caryopses and its usability for an identification of some species. – Papers and Abstracts, VI Symposium, International Work Group for Palaeoethnobotany, Groningen, 30 May–3 June 1983, 128.

KROLL, H., 1982: Pflanzliche Großreste vom Siedlungshügel bei Kastanas. – Jahrb. Röm.-German. Zentralmus. Mainz **26**, 1979, 229–239.

KUČAN, D., 1983: Kulturpflanzenfunde aus Pod bei Bugojno/Zentralbosnien (Hallstatt- und La Tène-Zeit). – Papers and Abstracts, VI Symposium, International Work Group for Palaeoethnobotany, Groningen, 30 May–3 June 1983, 129.

KÜSTER, H., 1983a: Rekonstruktionsversuche zur neolithischen Landwirtschaft nach botanischen Funden aus Eberdingen-Hochdorf (Kreis Ludwigsburg). – Archäol. Korrespondenzblatt **13**, 37–39.

–, 1983b: Neolithic plant remains from Eberdingen-Hochdorf (South Germany). – Papers and Abstracts, VI Symposium, International Work Group for Palaeoethnobotany, Groningen, 30 May–3 June 1983, 130.

KULLIG, C.-G., 1983: Eine Grube der Jüngeren Bronzezeit mit Getreideresten bei Harkenbleck, Gemeinde Hemmingen, Ldkr. Hannover. – Nachrichten aus Niedersachsens Urgeschichte **51**, 1982, 277–279.

LIESE-KLEIBER, H., und H. SCHLICHTHERLE, 1982: Untersuchung von Erdproben aus den Beeten des Lichthofes (XI. Anhang = p. 291). In: M. BIETAK und E. REISER-HASLAUER, Das Grab des 'Anch-Hor, Obersthofmeister der Gottesgemahlin Nitokris. – Wien.

LUNDSTROM-BAUDAIS, K., 1983: Interpretation of macroplant remains from a Neolithic

lakeshore site in France. – Papers and Abstracts, VI Symposium, International Work Group for Palaeoethnobotany, Groningen, 30 May–3 June 1983, 131.
Lynch, A., und N. Paap, 1982: Untersuchungen an botanischen Funden aus der Lübecker Innenstadt (Ein Vorbericht). – Lübecker Schriften zur Archäologie und Kulturgeschichte **6**, 339–360, Taf. 35.
Maier, U., 1983: Nahrungspflanzen des späten Mittelalters aus Heidelberg und Ladenburg nach Bodenfunden aus einer Fäkaliengrube und einem Brunnen des 15./16. Jahrhunderts. – Forschungen und Berichte der Archäologie des Mittelalters in Baden-Württemberg **8**, 139–183.
Medović, P., 1978: Naselja gvozdenog doba u Jugoslovenskom Podunavlju. Die relative Chronologie der Südlagen der älteren Eisenzeit im jugoslawischen Donaugebiet. – Beograd, 112 pp.
Meyer, D., und M. Neugebauer, 1982: Archäologisch-baugeschichtliche Beobachtungen und Teiluntersuchungen im Haus Engelsgrube 56 und seinen Nachbarhäusern in Lübeck. Mit einem botanischen Beitrag von Norbert Paap. – Lübecker Schriften zur Archäologie und Kulturgeschichte **6**, 163–183, Abb. 1, 47–51, Taf. 17.
Neustupný, E., Z. Dvořák, 1983: Výživa pravěkých zemědělců: model. Nutrition of prehistoric farmers: a model. – Památky Archeologické **74**/1, 224–257.
Nisbet, R., 1982: I resti vegetali macroscopici di Ostiano, S. Salvatore (antica Età del Bronzo). – Preistoria Alpina **18**, 217–223.
Nötzold, T., 1982a: Vor- und frühgeschichtliche karpologische Kulturpflanzenreste aus Sachsen. – Arbeits- und Forschungsberichte zur sächsischen Bodendenkmalpflege Beiheft **17**, Beiträge zur Ur- und Frühgeschichte Teil II, 445–460.
–, 1982b: Vor- und frühgeschichtliche karpologische Kulturpflanzenreste aus Sachsen II. – Arbeits- und Forschungsberichte zur sächsischen Bodendenkmalpflege **24**/**25**, 159–172.
Nováki, G., 1983: Régészeti és paleoethnobotanikai adatok a „gabonásvermek" kérdéséhez. Archäologische und paleoethnobotanische Angaben zur Frage der „Getreidegruben". – Magyar Mezögazdasági Múzeum Közleményei 1981–1983, 57–94.
Opravil, E., 1983: Z historie šíření konopě seté (*Cannabis sativa* L.). Aus der Geschichte der Verbreitung des Anbaus von Hanf. – Archeologické rozhledy **35**, 206–213.
Paap, N., 1983: Botanical investigations in Amsterdam. – Papers and Abstracts, VI Symposium, International Work Group for Palaeoethnobotany, Groningen, 30 May–3 June, 1983, 132.
Pals, J. P., 1983: Plant remains from Aartswoud, a Neolithic settlement in the coastal area. – Papers and Abstracts, VI Symposium, International Work Group for Palaeoethnobotany, Groningen, 30 May–3 June 1983, 133.
Paškevič, G. O., V. O. Petrašenko, 1982: Zemlerobstvo i skotarstvo v Seredn'omu Podniprov'i v VIII–X st. Zemledelie i skotovodstvo v Srednem Podneprov'e v VIII–X vv. – Archeologija **41**, 46–62.
Piening, U., 1983: Verkohlte Pflanzenreste der Frühlatènezeit von Lauffen am Neckar, Kreis Heilbronn. – Fundberichte aus Baden-Württemberg **8**, 47–54.
Rowlett, R. M., A. L. Price, and M. Hopf, 1982: Differential grain use on the Titelberg, Luxembourg. – J. Ethnobiol. **2** (1), 79–88.
Sackl, H., 1980: Pharmakognostische Aspekte der Mittelalter-Archäologie. – Österr. Apothek.-Ztg. **34**, 985–987.
Schlichtherle, H., 1983: Mikroskopische Untersuchungen an neolithischen Gefäßinhalten aus Hornstaad, Yverdon und Burgäschisee-Süd. – Naturwissenschaftliche Untersuchungen zur Ermittlung Prähistorischer Nahrungsmittel. Ein Symposionsbericht, hrsg. von H. Müller-Beck und R. Rottländer, Univ. Tübingen, Archaeologica Venatoria e. V., 39–61.
Schmidt, R., 1983: Pollen und Großreste aus der neolithischen Station Weyregg I am Attersee, Oberösterreich. – Fundberichte aus Österreich **21**, 1982, 157–169, Taf. 1.
Schultze-Motel, J., 1983: Literatur über archäologische Kulturpflanzenreste (1981/1982). – Kulturpflanze **31**, 281–297.
Shay, C. T., 1983: Comparison of archaeobotanical data with literary evidence in Upper Fort Garry, Manitoba. – Papers and Abstracts, VI Symposium, International Work Group for Palaeoethnobotany, Groningen, 30 May–3 June 1983, 134.

SHEWRY, P. R., M. A. KIRKMAN, S. R. BURGESS, G. N. FESTENSTEIN, and B. J. MIFLIN, 1982: A comparison of the protein and amino acid composition of old and recent barley grain. – New Phytol. **90**, 455–466.
ŠIŠKA, S., E. HAJNALOVÁ, 1983: Stredoveké obydlie a depot semien konopy siatej zo Šarišských Michalian. Mittelalterliches Wohnobjekt und ein Hanfsamendepotfund aus Šarišské Michalany. – Študijné Zvesti Archaeologického Ústavu Slovenskej Akadémie Vied zo Nitra, 303–317.
STRAKER, V., 1983: Roman London: Carbonized grain from the Forum and the Thames waterfront. – Papers and Abstracts, VI Symposium, International Work Group for Palaeoethnobotany, Groningen, 30 May–3 June 1983, 135.
UDAČIN, R. A., 1982: O vozmožnom suščestvovanii *Triticum antiquorum* HEER v naši dni. On the possible existance [sic] of *Triticum antiquorum* HEER nowadays. – Bjull. vses. Inst. Rast. im. N. I. VAVILOVA **119**, 72–73.
VEEN, M. VAN DER, 1983: Seeds and 'seed-machines'. – Circaea **1** (2), 61–62.
VILSTEREN, V. T. VAN, 1983: The medieval village of Dommelen: a case study for the interpretation of charred seeds from postholes. – Papers and Abstracts, VI Symposium, International Work Group for Palaeoethnobotany, Groningen, 30 May–3 June 1983, 136.
WET, J. M. J. DE, 1981: Origin and domestication of maize. – Technical Communication, Department of Agriculture and Fisheries, South Africa No. **172**, 14–20 (Proc. Fourth So. Afr. Maize Breed. Symp.).
WIJNGAARDEN-BAKKER, L. H. VAN, and J. P. PALS, 1981: Life and Work in Smeerenburg. The bio-archaeological aspects. – Proceedings of the International Symposium Early European Exploitation of the Northern Atlantic 800–1700, Arctic Centre, University of Groningen, Netherlands, 133–151 (I. P. P. Publicatie 300).
WILLERDING, U., 1983a: Paläo-ethnobotanische Befunde und schriftliche sowie ikonographische Zeugnisse in Zentraleuropa. – Papers and Abstracts, VI Symposium, International Work Group for Palaeoethnobotany, Groningen, 30 May–3 June 1983, 161–166.
–, 1983b: Zum ältesten Ackerbau in Niedersachsen (pp. 179–219). In: Frühe Bauernkulturen in Niedersachsen. – Archäologische Mitteilungen aus Nordwestdeutschland, Beiheft **1**.
ZEIST, W. VAN, 1983a: Plantengebruik in vroeger tijd (pp. 10–36). In: H. F. LINSKENS et al. (eds.), Plantenveredeling en genenmanipulatie. – Pudoc Wageningen.
–, 1983b: Fruits in foundation deposits of two temples. – J. Archaeol. Sci. **10**, 351–354.
–, and S. BOTTEMA, 1983: Palaeobotanical studies of Carthage. A comparison of microscopic and macroscopic plant remains. – CEDAC Carthage Bulletin **5**, 18–22.

Dr. J. SCHULTZE-MOTEL
Zentralinstitut für Genetik und Kulturpflanzenforschung
der Akademie der Wissenschaften der DDR
DDR - 4325 Gatersleben
Corrensstraße 3

Kulturpflanze 32 · 1984 · 245–259

Taxonomy and evolution of cultivated plants: Literature review 1982/1983

JÜRGEN SCHULTZE-MOTEL, REINHARD FRITSCH, KARL HAMMER,
PETER HANELT, JOACHIM KRUSE, HELGA I. MAASS,
HARALD OHLE, and KLAUS PISTRICK

(Eingegangen am 17. Januar 1984)

Summary

Important papers on taxonomy and evolution of cultivated plants published in 1982 and 1983 were compiled and briefly discussed.

Introduction

In continuation of our literature reviews on taxonomy and evolution of cultivated plants (SCHULTZE-MOTEL et al. 1983) we present here the papers on this topic published in 1982 and 1983 as well as some publications of former years being overlooked in our earlier papers.

Amaranthaceae

Amaranthus L.
The relationship between the two basic numbers ($n=16$ and $n=17$) in the grain group of the genus was investigated by an interspecific dibasic cross of *A. hypochondriacus* L. ($n=16$) × *A. hybridus* L. ($n=17$). From meiotic analysis the origin of $n=17$ from $n=16$ by primary trisomy is suggested (PAL et al. 1982).

Chenopodiaceae

Beta vulgaris L.
BURENIN and GAVRILJUK (1982) have proposed Asia Minor, Transcaucasia, and Central Asia as regions for the primary domestication of *Beta vulgaris* L. The domesticated races reached the Mediterranean area where they intercrossed with primitive local types. Under Mediterranean conditions highly domesticated races arose. The progenitor of cultivated beets is *Beta maritima* L. (*Beta atriplicifolia* ROUY and *B. orientalis* ROTH are considered as subspecies of *B. maritima*).

Cruciferae

Brassica L.
GRAY (1982) discussed the origin and application of the name broccoli. The history and evolution of broccoli (*B. oleracea* L. var. *italica* PLENCK) is considered in relation to cauliflower (*B. oleracea* L. var. *botrytis* L.) and its diversification into annual and biennial types is discussed.

The knowledge of the evolution of the different types of cauliflower has been useful in identifying sources of reduced susceptibility to cabbage root fly (*Delia radicum* L.), and may also be useful in the search for reduced susceptibility to clubroot (*Plasmodiophora brassicae* WOR.). CRISP (1982) argued that knowledge of the phylogeny of types within each crop species is of great importance in the exploitation of genetic resources.

The origin and evolution of amphidiploid *Brassica* species has been investigated by mutational analysis of the chloroplast genome (PALMER et al. 1983). A detailed chloroplast DNA phylogeny enables identification of the maternal parent for most of the accessions examined.

From records of the classical Chinese literature and from comparisons of artificial hybrids it has been suggested that Chinese cabbage (*Br. pekinensis* [LOUR.] RUPR.) originated from hybrids between turnip (*Br. rapa* L.) and pak-choi (*Br. chinensis* L.). Chinese local varieties of Chinese cabbage have been grouped into 6 basic forms using morphological, ecolcgical and agronomic data (LI 1981).

A genealogical diagram of Korean varieties of Chinese cabbage produced between 1900 and 1970 has been presented, most of them being hybrids between Japanese and Chinese local varieties (PYO 1981).

Lepidium L.
GIRENKO (1982) dealt with productivity characters and their variability in *L. sativum* L. The accessions were classified into 5 varietal types.

Raphanus L.
Four species of the genus were examined by electrophoresis of water soluble proteins. The accession of *R. sativus* L. var. *sativus* showed the greatest genetic similarity when compared with an Italian population of *R. landra* MORETTI ex DC. (LEWIS-JONES et al. 1982).

Dioscoreaceae

Dioscorea L.
The volume "Yams. Ignames" edited by MIÈGE and LYONGA (1982) comprises mainly lectures on food-yams. These inform on plant improvement, agronomic studies and taxonomy, plant protection, postharvest biology and technology and economics of production and mechanization. In this taxonomical contributions MIÈGE treats some characters (morphology of bulbils, chromosome numbers, foliar epidermis, sections of leaf-stalk), which enable a better intraspecific discrimination of *D. bulbifera* L. including a determination key of yams of the Ivory Coast made

by BUFFARD-MOREL. Besides, he reports on chemotaxonomical investigations of the *D. cayenensis* LAM.–*D. rotundata* POIR. complex and completes his information about this group by remarks and indications in an appendix.

Gramineae

Echinochloa PAL. BEAUV.
DE WET et al. (1983b) reviewed the relatively recent domestication of Sawa millet *E. colona* (L.) LINK in India, where it remains an important cereal only across the tropical and subtropical parts. Four morphological races are recognized, although these do not have geographical, ecological or ethnological unity.

Hordeum L.
SHAO et al. (1982) concluded on the basis of morphological, cytological and genetical investigations that *H. spontaneum* K. KOCH is the eldest ancestor, *H. lagunculiforme* BACHT. represents the second stage and *H. agriocrithon* ÅBERG the third stage, while the cultivated barley denotes the final stage of evolution of *H. sativum* L. They reported natural occurrences of *H. spontaneum* over a wide area of western and southwestern China and of *H. lagunculiforme* in many places in southwestern China.

Hordeum spontaneum C. KOCH was morphometrically analyzed (BAUM and BAILEY 1983). As conclusion of these studies it was stated that *H. spontaneum* is highly variable but shows no segregates which would allow the recognition of infraspecific taxa. In another paper (BAUM 1983) the relationships within *Hordeum* L. subgen. *Hordeum* are studied. Three morphs ought to be recognized as taxonomic entities: *H. spontaneum* C. KOCH, *H. distichum* L. and *H. vulgare* L. In a hypothesis *H. lagunculiforme* BACHT. is postulated as the progenitor of this group.

Paspalum L.
DE WET et al. (1983a) found it difficult to distinguish wild, weed and cultivated strains of *P. scrobiculatum* L., because they merge in all morphological characters studied. The species was apparently domesticated across its range of present-day cultivation in India some 3000 years ago.

Setaria PAL. BEAUV.
KAWASE and SAKAMOTO (1982) examined phenol color reaction in a total of 376 strains of *S. italica* (L.) PAL. BEAUV. collected from different areas throughout Eurasia. The 50 strains that showed positive reaction occurred more frequently in the lower altitudinal regions of Asia, while the 319 strains with the negative phenotype occurred in almost all the regions. Genetic analyses showed that the phenol color reaction is controlled by a single gene, and that the positive phenotype is dominant.

MAISAYA and GORGIDZE (1980) analyzed hybrids between Georgian *S. italica* (L.) PAL. BEAUV. and the wild species *S. viridis* (L.) PAL. BEAUV. and *S. ketzchovelii* MEN. et ER. The cross compatibility data indicated that the cultivated Georgian millet is closely related both to the wild species and that the differences seen result from long evolution under different ecological and cultural conditions.

In France it is tried to reestablish the culture of *S. italica* (BRABANT et al. 1981). The authors discussed the probable domestication of that species some 6000–7000 years ago in China.

Sorghum ADANS.
IVANJUKOVIČ (1983) gives a review of the species of the series Nervosa of the genus *Sorghum.*

Triticum L. and *Aegilops* L.
A new form of *Triticum petropavlovskyi* UDACZ. et MIGUSCH. was described by BOGUSLAVSKIJ (1982) as ssp. *mexicana.*

SAITO et al. (1982) carried out experiments to estimate the relationship between eight *Triticum* species on the basis of leaf rust susceptibility.

Character variability and evolutionary aspects of land-races of *T. aestivum* L., *T. durum* DESF. and *T. turgidum* L. from Sicily are discussed by PERRINO and HAMMER (1983).

GEORGIEV et al. (1983) investigated species of different ploidy level in the genera *Triticum* and *Aegilops* by means of nuclear test and karyological analyses. The results show that *T. boeoticum* BOISS. em. SCHIEM. and *T. monococcum* L. cannot be genome donors in tetra- and hexaploid wheats.

BARISASHVILI and GORGIDZE (1979) analysed the genome composition in hybrids between *T. spelta* L., *T. macha* DEK. et MEN., and *T. karamyschevii* NEVSKI on the one side and *Ae. tauschii* NEVSKI on the other side by means of gel electrophoresis and double immunodiffusion techniques. It is supposed that *T. macha* DEK. et MEN. is a hybrid of *T. karamyschevii* NEVSKI and *Ae. tauschii* NEVSKI. Carbonized plant material and plant imprints from Georgia 5000–6000 years B. C. have been examined by GORGIDZE (1979). Besides husk types naked-grained wheats have been found indicating a wheat cultivation in Georgia at that time. The possibility that *T. spelta* L. was a component of a heterogeneous population of *T. macha* DEK. et MEN. is discussed.

The restriction fragment patterns of chloroplast DNAs of the Emmer-Dinkel and Timopheevi groups as well as the *Aegilops kotschyi–Ae. variabilis* complex have been compared with those of the putative donors of cytoplasm (*Triticum urartu* TUM. and the *Aegilops* species of the *Sitopsis* group) (TSUNEWAKI and OGIHARA 1983). The patterns of Emmer and Dinkel wheat revealed a close relationship to *Aegilops longissima* SCHWEINF. Furthermore the patterns of the Timopheevi group of wheat and *Ae. aucheri* BOISS. were identical. The donor of the *Ae. variabilis* cytoplasm seems to be *Ae. bicornis* (FORSSK.) JAUB. et SPACH. or *Ae. searsii* FELDMAN et KISLEV ex HAMMER. BOWMAN et al. (1983) examined the chloroplast DNA of seven *Triticum* species and six *Aegilops* species. The comparison of the DNA types indicated that only *T. dicoccoides* KÖRN. belongs to the same cytoplasmatic class as *T. aestivum* L. em. FIORI et PAOL., whereas the species of *Sitopsis* group and *T. urartu* TUM. are characterized by other cytoplasmatic types and therefore do not seem to be the donors of the B genome.

Cross experiments with telosomic lines of *Triticum aestivum* L. em. FIORI et PAOL. and *Aegilops speltoides* TAUSCH showed no references to the origin of the B genome of polyploid wheats (ALONSO and KIMBER 1983).

Investigations of the C-banding patterns of 11 diploid *Aegilops* species confirmed the existing principal genome groups (TEOH and HUTCHINSON 1983). It is proposed to give *Ae. uniaristata* VIS. a separate symbol and to term *Ae. umbellulata* ZHUK. and *Ae. caudata* L. with different genome symbols. *Aegilops speltoides* TAUSCH and *Ae. sharonensis* EIG are regarded as the possible donors of the B genome of polyploid wheat species.

MIGUŠOVA and CHAKIMOVA (1982) recommend the division of *Ae. crassa* BOISS. into three species (*Ae. crassa* BOISS. s. str., *Ae. vavilovii* (ZHUK.) CHENNAV. and *Ae. trivialis* (ZHUK.) MIGUSH. et CHAK.).

A guide to the species of *Aegilops* was prepared by WITCOMBE (1983). The nomenclature used is not up to date.

Zea L.

Seed protein analyses of 17 accessions of *Z. mays* L., *Z. luxurians* (DURIEU) BIRD, *Z. perennis* (HITCHC.) REEVES et MANGELSD. and *Z. diploperennis* ILTIS, DOEBLEY et GUZMÁN (MASTENBROEK et al. 1981) confirmed the earlier suggested taxonomy of sections *Zea* and *Luxuriantes* DOEBLEY et ILTIS based on morphology, but analysis of isoenzymes did not completely correspond with the morphological classification.

GALINAT and PASUPULETI (1982) included a review on several aspects of maize evolution in their discussion on the potential value of *Z. diploperennis* for maize improvement.

Juglandaceae

KOMANIČ (1980) gives some remarks to the history of walnut cultivation from ancient times, about variability, genetic resources, genecenters, interspecific and intergeneric hybridizations, karyology of distant-hybrids and breeding for yield and resistance.

Leguminosae

Serological studies of the seed proteins (KONAREV et al. 1983) of different species and genera of the tribe Vicieae have confirmed more or less the established classification of this group; it is interesting to note the intermediate position of *Vicia narbonensis* L. between *V. faba* L. (here recognized as genus *Faba* MILL.) and other species of *Vicia*. The same situation is true for *V. ervilia* (L.) WILLD. which links *Vicia* L. and *Lens* MILL.; the complex of *V. sativa* L. proved to be rather homogenous also regarding protein serology.

Arachis L.

Arachis villosa BENTH. or a similar species and *A. batizocoi* KRAP. et GREG. have been proposed as genome donors for the tetraploid *A. hypogaea* L. by means of data from chromosome morphology (KIRTI et al. 1983).

KLOZOVÁ et al. (1983) investigated the relationship of 12 species by comparison of seed protein patterns. Their results agree quite well with recent classification schemes of this genus.

Cajanus DC.
Based on morphological studies within the subtribe Cajaninae BENTH. in connexion with results of interspecific hybridizations, of cytological and chemotaxonomical examinations VAN DER MAESEN (1981) comes to the conclusions that *Atylosia* WIGHT et ARN. is congeneric with *Cajanus*. A taxonomical revision is announced.

A new study on the evolution of *C. cajan* (L.) MILLSP. has been made by KRISHNA and REDDY (1982). They suggest that this species is phylogenetically nearest to *Atylosia cajanifolia* HAINES. Thus, they confirm VAN DER MAESEN'S opinion that the latter species is the putative progenitor of pigeon pea.

Cicer L.
RAVI and RAVI (1981) analysed the isozyme patterns in six annual species with regard to the origin of *C. arietinum* L.

Glycine L.
In a genetical study KORSAKOV (1982) has analysed wild and cultivated stocks of the VIR collection. Three centres of origin of forms of the genus have been distinguished: East Africa, Australia, and South-eastern Asia, whereat the first should be the eldest centre and the last the youngest.

Medicago L.
SMALL and LEFKOVITCH (1982) examined 14 agronomically important chemical characters and 12 morphological traits of 527 stocks of *M. sativa* L. grown at the same site. Analyses of variance of chemical variables showed that only S and Ca content differed significantly among subspecies. The authors discussed some factors as possible explanation of the differences and concluded that the two chemical variables proved to have very limited value as taxonomic discriminators.

ISELY (1983) has critically reviewed recently proposed classifications of the *M. sativa* aggregate and suggested a rather simplistic own one in which *M. sativa* L. is considered in a broad sense, including diploid and tetraploid wild and semi-wild populations as well as all cultivated or escaped ones derived completely or only in part from the original *M. sativa* (that means also the hybrid derivates known often as *M.×varia* MART.).

Trigonella L.
INGHAM (1981) examines the phytoalexins of 41 species after inoculation with *Helminthosporium* and distinguishes three groups, linking the genus to *Medicago* L. and *Factorovskya* EIG, to *Melilotus* MILL., and to *Trifolium* L.

Vicia L.
Rather unconvincingly and without consideration of the data already available in the literature the traditional genus *Vicia* has been split into *Ervum* L., *Ervilia* LINK, *Vicia* L. s. str. and *Bona* MEDIK. (STANKEVIČ 1982), in consequence some well-known cultivated plants received uncommon names (*Vicia ervilia* L. = *Ervilia sativa* LK.; *Vicia narbonensis* L. = *Bona narbonensis* (L.) MEDIK. etc).

Different seed protein profiles within the *V. sativa* L. complex could not be related to any specific karyotype, chromosome number or traditional taxon in contrary to the situation in other leguminous genera. This variability of the proteins may be important for the ecological flexibility of the group and has probably originated by interkaryotypic hybridization (LADIZINSKY and WAINES 1982).

An informal classification of Western European cultivars of *V. faba* L. has been elaborated based upon 3 discontinuous characters (flower colour, testa and hilum colour) thus avoiding the difficulties of using continuous characters, e.g. seed size, which has been used hitherto as main character for the delimitation of the botanical varieties of this species (HIGGINS et al. 1981). A variable, comparatively large incompatibility between 4 so-called subspecies (*'minor'*, *'major'*, *'paucijuga'*, *'pliniana'*) of *V. faba* has been the result of infraspecific hybridization experiments (LE GUEN 1983). Different varieties of one group show sometimes different compatibility patterns in crossings with another group. Thus incompatibility barriers are not strongly correlated with infraspecific groupings.

Vigna SAVI and *Phaseolus* L.

Recent results on the delimitation of the genera *Phaseolus* and *Vigna* and the transfer of the Old World cultivated taxa known hitherto as *Phaseolus* species to the genus *Vigna* has been reviewed from the literature by BUDANOVA (1982).

The fine structure of the pollen exine of *Vigna angularis* (WILLD.) OHWI et OHASHI has been studied in detail for the cultivated and the Japanese wild taxa known as var. *nipponensis* (OHWI) OHWI et OHASHI. Both proved to be quite similar in this respect, conforming earlier speculations on their evolutionary relationship (OHASHI and TAKAHAGI 1981).

In contrary to other data SEM studies of the testa have indicated *V. umbellata* (THUNBG.) OHWI et OHASHI (incorrectly called *V. calcarata* in the paper) as the putative ancestor of *V. radiata* (L.) WILCZ. (JAIN and BABU 1982). But flavonoid analyses revealed again the great similarity between the cultivated variety of this species and its spontaneous one var. *sublobata* (ROXB.) VERDC. (being clearly distinguished from *V. mungo* (L.) HEPPER), thus further investigations seem to be necessary for the Oriental Vignas (ISHIKURA et al. 1981).

In order to get a better understanding of the domestication of the common bean (*Phaseolus vulgaris* L.) the large collections of wild strains of this species from CIAT, Colombia, have been studied morphologically. South American accessions (var. *aborigineus* (BURK.) BAUD.) stand distinctly apart from the Mexican wild types, the Mexican weedy types of the species possessing a somewhat intermediary position but in general being more similar to the Mexican wild forms. Thus the hypothesis of independent domestication of the species in South and Meso-America may be substantiated (VANDERBORGHT 1983).

The 231 cultivars of the bush bean, *Ph. vulgaris* L. var. *nanus* (JUSL.) ASCHERS., officially released in France have been described in detail and classified into many informal groups and subgroups, called groups, sections and classes (ANONYMUS 1983), diagnostic characters being colour and strings of the pod, pod shape and size, seed coat colour, foliage colour, size of the flower bracts and seed shape. The main cultivar groups are characterized by green pods with strings, green stringless pods, waxy pods with strings resp. waxy stringless pods and the further subdivision uses shape of pods and seed colour.

Malvaceae

Gossypium L.
SHADMANOV and DAUTOVA (1981) have crossed *Gossypium hirsutum* L. with the diploids *G. sturtii* F. MUELL., *G. raimondii* ULBR. and *G. trilobum* HUTCHINS. et GHOSE. The hybrids were polyploidized with colchicine. Electrophoretic patterns of *G. hirsutum*×*G. raimondii* were similar to *G. hirsutum*. This result suggests that *G. raimondii* was involved in the evolution of 4x cottons.

Martyniaceae

Proboscidea parviflora (WOOT.) WOOT. et STANDL.
A form of this species showing clearly the influence of domestication originated in southern Arizona–northern Sonora. It was described by BRETTING (1982) as var. *hohokamiana*.

Pandanaceae

Pandanus PARKINS.
STONE (1982) analyzed the taxonomy of the widespread and variable species *P. tectorius* PARKINS. in Australia, indicating that it occurs there in several varieties and forms, and that several recently proposed species should be reduced to synonymy or to a minor taxonomic rank.

Pedaliaceae

Sesamum L.
RANGASWAMY and RATHINAM (1982) found mutagen induced male sterility in the M_2 of two varieties of *S. indicum* L. and their hybrid to be monogenically recessive.

Rosaceae

Armeniaca SCOP.
A new variety, *Armeniaca vulgaris* LAM. var. *glabra* is described, the principal distinguishing character of which is its smooth fruit (SHENGXIANG and CHONGLIAN 1982).

Aronia MEDIK.
The cultivated blackfruited chokeberry (*Aronia mitschurinii* SKVORTS. et MAITULINA) arose in the nursery of I. V. MITSCHURIN and is characterized by apomixis and tetraploidy in contrast to the wild diploid and amphimict *A. melanocarpa* (MICHX.) ELLIOTT. The possible mechanism of its origin is discussed (SKVORCOV et al. 1983).

Malus MILL.
PONOMARENKO (1982) has given new aspects on the historical process of domestication of *Malus sieversii* (LEDEB.) ROEMER in the mountains of Middle Asia. He also suggests the West European and East European secondary gene-centres as localities for new complex populations of intra- and interspecific apple-hybrids.

Prunus L.
From morphological and cytogenetical studies in spontaneous hybrids of *Prunus divaricata* LEDEB. and *P. spinosa* L. BAIAŠVILI (1981) developed the theory that cultivated plum is a product of such hybridizations in combination with autopolyploidy.

Sanguisorba L.
Wild populations of *Sanguisorba officinalis* L. were subdivided by using morphological and biochemical (polyphenolics in rootstocks) characters into some groups. One group with accelerated phenological development gives rise to mountain biotypes, and another group of forms with large leaves, flowers and a high content of polyphenolics in roots is of interest in economy and plant breeding (ORIŠČENKO and TICHONOVA 1981).

Rubiaceae

Cinchona L.
One of the most important of the early botanists in southern America (Colombia), JOSÉ CELESTINO MUTIS, published in 1793 under the title "El Arcano de la Quina" his knowledge concerning the tropical genus *Cinchona*. KIRKBRIDE (1982) found 4 species validly published: *Cinchona lancifolia, C. oblongifolia, C. cordifolia* and *C. ovalifolia.*

Coffea L.
The hypothesis that *Coffea canephora* PIERRE diverged from *C. congensis* FROEHN., whereafter the latter species differentiated into *C. arabica* L. and *C. eugenioides* MOORE is supported by means of analysis and comparisons with chloroplast- and mitochrondrial DNA in 9 taxa of the genus *Coffea* (BERTHOU et al. 1983).

Sapindaceae

Litchi SONN.
LIANG (1981) traced the different stages of historical development of the cultural techniques of the lychee (*Litchi chinensis* SONN.) which originated in southern China and is known from literature for more than 2000 years.

Solanaceae

Datura L.
A revision of the sections *Dutra* BERNH., *Ceratocaulis* BERNH. and *Datura*, which contains also cultivated races, is presented on the basis of a living collection (HAMMER et al. 1983).

Lycopersicon MILL.
Electrophoretic patterns and genetical analysis of Fraction 1 protein in *Lycopersicon* species incl. *Solanum pennellii* CORRELL indicate that the evolution within the genus proceeds by gene mutation rather than by alloploidy (CAMMAERTS and JACOBS 1982).

Petunia JUSS.
Petunia×*hybrida* hort. ex VILM. (*P.*×*atkinsiana* D. DON) is a hybrid of the putative parents *P. axillaris* (LAM.) BRITTON, STERN et POGGENBURG (white-flowered) and *P. integrifolia* (HOOK.) SCHINZ et THELLUNG (purple-flowered). A revision of the wild species is given (WIJSMAN 1982).

Solanum L. (tuber-bearing species)
The hypothesis was investigated of *Solanum ajanhuiri* JUZ. et BUK. being a hybridogenic taxon of *S. stenotomum* JUZ. et BUK.×*S. megistacrolobum* BITT. (HUAMÁN et al. 1982). There is strong evidence from crossing experiments that the Yari group of *S. ajanhuiri* could be an F_1 *S. stenotomum*×*S. megistacrolobum* hybrid, the Ajawiri group seems to be a backcross of an F_1 hybrid to *S. stenotomum*. The hypothesis is supported by these results. As a consequence, the hybridogenic taxon should be designated as *Solanum*×*ajanhuiri* JUZ. et BUK. SCHMIEDICHE et al. (1982) crossed the tetraploid *Solanum acaule* BITT. with diploid species to obtain triploid *S.*×*juzepczukii* BUK. The resynthesis could be easily made. As the best diploid parent *S. goniocalyx* JUZ. et BUK. was found. Therefore, this species is proposed as genome donor of *S.*×*juzepczukii*. The pentaploid *S.*×*curtilobum* JUZ. et BUK. was not successfully resynthesized. As a result of these crosses tetraploids occurred containing at least one genome from *S. acaule*. This may be a good method for the transfer of *S. acaule* germ plasm, which is resistant to frost, into the tetraploid cultivated gene pool.

Solanum L. (fruit-bearing species)
The evolution of the *Solanum nigrum* L. complex is discussed by RAO (1982). Diploid, tetraploid and hexaploid races can be found. The polyploids are suggested to have occurred under agricultural conditions and are highly adapted to new environmental conditions. Therefore, the diploids become progressively more restricted.

Theaceae

Camellia L.
Based on cytological investigations CHERIAN and STEPHEN (1981) suggest to consider *C. sinensis* (L.) KUNTZE and *C. assamica* (J. W. MASTERS) W. WIGHT as distinct species.

TANI and NAKAGAHRA (1981) studied isozymes of 28 strains and varieties of *C. sinensis* and distinguished 17 groups.

Vitaceae

Vitis L.
PIRMAGOMEDOV (1982) reported the occurrence of *Vitis vinifera* L. ssp. *sylvestris* (GMEL.) BEGER in the Tabasaran region of Daghestan, and classified this material into 4 taxa (var. *typica* NEGR., var. *aberrans* NEGR., var. *balcanica* NEGR., and var. *tabasaranica* NEGR.). Local varieties of cultivated grape, similar in some characters, probably originated from these forms.

Zusammenfassung

Taxonomie und Evolution der Kulturpflanzen: Literaturübersicht 1982/1983

Wichtige Arbeiten über Taxonomie und Evolution der Kulturpflanzen aus den Jahren 1982 und 1983 wurden zusammengestellt und kurz kommentiert.

Краткое содержание

Таксономия и эволюция культурных растений: обзор литературы за 1982–1983 гг.

Дается список работ по таксономии и эволюции культурных растений, вышедших в 1982 и 1983 годах. Работы кратко комментируются.

Literature

ALONSO, L. C., and G. KIMBER, 1983: A study of genomic relationships in wheat based on telocentric chromosome pairing – I. – Z. Pflanzenzüchtung (J. Plant Breed.) **90**, 23–31.

ANONYMUS, 1983: Description et essai de classification des variétés de Haricot Nain. – Inscrites au catalogue officiel Français et/ou bénéficiant d'un certificat d'obtention végétale au 30. 11. 1981, 232 pp. – I. N. R. A., Versailles.

BAIAŠVILI, E. I., 1981: K izučeniju proischoždenija kul'turnoj slivy. – Soobšč. AN Gruz. SSR **101** (1), 121–124.

BARISASHVILI, M. A., and A. D. GORGIDZE, 1979: [Origin of the wheat *Triticum macha* DEK. et MEN.]. – Bull. Acad. Sci. Georgian SSR **95** (2), 409–412.

BAUM, B. R., 1983: Morphometric relationships in *Hordeum vulgare* (Triticeae, Poaceae). II. *Hordeum agriocrithon, H. distichum, H. lagunculiforme, H. spontaneum,* and *H. vulgare.* – Can. J. Bot. **61**, 2023–2031.
–, and L. G. BAILEY, 1983: Morphometric relationships in *Hordeum vulgare* (Triticeae, Poaceae). I. *Hordeum spontaneum.* – Can. J. Bot. **61**, 2015–2022.
BERTHOU, F., C. MATHIEU, and F. VEDEL, 1983: Chloroplast and mitochondrial DNA variation as indicator of phylogenetic relationships in the genus *Coffea* L. – Theor. Appl. Genet. **65**, 77–84.
BOGUSLAVSKIJ, R. L., 1982: Novaja botaničeskaja forma geksaploidnoj pšenicy. The new botanical form of hexaploid wheat. – Bjull. vses. Inst. Rast. im. N. I. VAVILOVA **119**, 73–74.
BOWMAN, C. M., G. BONNARD, and T. A. DYER, 1983: Chloroplast DNA variation between species of *Triticum* and *Aegilops.* Location of the variation on the chloroplast genome and its relevance to the inheritance and classification of the cytoplasm. – Theor. Appl. Genet. **65**, 247–262.
BRABANT, P., J. BELLIARD, G. METAILIE, E. N. VAN, S. POIRIER, B. POIRIER, et J. PERNES, 1981: Données préliminaires pour la reintroduction et la culture du millet *Setaria* en France. – J. Agric. tradit. Bot. appl. **28**, 309–328.
BRETTING, P. K., 1982: Morphological differentiation of *Proboscidea parviflora* ssp. *parviflora* (Martyniaceae) under domestication. – Amer. J. Bot. **69**, 1531–1537.
BUDANOVA, V. I., 1982: Novoe v sistematike rodov *Phaseolus* L. i *Vigna* SAVI. Some new data on the systematics of the genera *Phaseolus* L. and *Vigna* SAVI. – Trudy prikl. bot., genet. selekc. **72** (1), 16–20.
BURENIN, V. I., i I. P. GAVRILJUK, 1982: Sistematika i filogenija roda *Beta* L. – Trudy prikl. bot. genet. selekc. **72** (3), 3–12.
CAMMAERTS, D., and M. JACOBS, 1982: Support for the evolution of Fraction 1 protein in the genus *Lycopersicon* by gene mutation. – Z. Pflanzenphysiol. **106**, 251–256.
CHERIAN, T. T., and J. STEPHEN, 1981: Cytology of tea. – Cytologia, Tokyo **46**, 767–772.
CRISP, P., 1982: The use of an evolutionary scheme for cauliflowers in the screening of genetic resources. – Euphytica **31**, 725–734.
GALINAT, W. C., and C. V. PASUPULETI, 1982: *Zea diploperennis*: II. A review on its significance and potential value for maize improvement. – Maydica **27**, 213–220.
GEORGIEV, S., M. ANASTASSOVA-KRISTEVA, and H. NICOLOFF, 1983: Activity of rRNA cistrons as a tool for identification of genomes in the genus *Triticum* L. – Biol. Zbl. **102**, 41–47.
GIRENKO, M. M., 1982: Priznaki produktivnosti i ich izmenčivost u kress-salata (*Lepidium sativum* L.). – Trudy prikl. bot., genet. selekc. **72** (3), 79–84.
GORGIDZE, A. D., 1979: [Living relic of cultivated wheat]. – Bull. Acad. Sci. Georgian SSR **96** (2), 425–428.
GRAY, A. R., 1982: Taxonomy and evolution of Broccoli (*Brassica oleracea* var. *italica*). – Econ. Bot. **36**, 397–410.
HAMMER, K., A. ROMEIKE, und C. TITTEL, 1983: Vorarbeiten zur monographischen Darstellung von Wildpflanzensortimenten: *Datura* L., sectiones *Dutra* BERNH., *Ceratocaulis* BERNH., et *Datura.* – Kulturpflanze **31**, 13–75.
HIGGINS, J., J. L. EVANS, and P. J. REED, 1981: Classification of Western European cultivars of *Vicia faba* L. – J. nat. Inst. agric. bot. **15** (3), 480–487.
HUAMÁN, Z., J. G. HAWKES, and P. R. ROWE, 1982: A biosystematic study of the origin of the cultivated diploid potato, *Solanum* ×*ajanhuiri* JUZ. et BUK. – Euphytica **31**, 665–676.
INGHAM, J. L., 1981: Phytoalexin induction and its chemosystematic significance in the genus *Trigonella.* – Biochem. System. and Ecology **9**, 275–281.
ISELY, D., 1983: Classification of alfalfa (*Medicago sativa* L.) and relatives. – Iowa State J. Res. **57** (3), 207–220.
ISHIKURA, N., M. IWATA, and S. MIYAZAKI, 1981: Flavonoids of some *Vigna*-plants in Leguminosae. – Bot. Mag., Tokyo **94** (1035), 197–205.
IVANJUKOVIČ, L. K., 1983: Obzor vidov serii Nervosa roda *Sorghum* (Poaceae). A review of the species of the series Nervosa of the genus *Sorghum* (Poaceae). – Bot. Ž. **68**, 1086–1089.

Jain, N. C., and C. R. Babu, 1982: Seed coat polymorphism in *Vigna calcarata* and its evolutionary significance. – Seed Sci. Technol. **10** (3), 451–456.

Kawase, M., and S. Sakamoto, 1982: Geographical distribution and genetic analysis of phenol color reaction in Foxtail Millet, *Setaria italica* (L.) P. Beauv. – Theor. Appl. Genet. **63**, 117–119.

Kirkbride, J. H., Jr., 1982: The *Cinchona* species of José Celestino Mutis. – Taxon **31**, 693–697.

Kirti, P. B., M. Bharati, U. R. Murty, and N. G. P. Rao, 1983: Chromosome morphology in three diploid species of *Aratis* [!] and its bearing on the genomes of groundnut (*Arachis hypogaea* L.). – Cytologia **48**, 139–151.

Klozová, E., J. Švachulová, J. Smartt, E. Hadač, V. Turková, and V. Hadačová, 1983: The comparison of seed protein patterns within the genus *Arachis* by polyacrylamide gel electrophoresis. – Biologia Plantarum, Praha **25**, 266–273.

Komanič, I. G., 1980: Biologija, kul'tura, selekcija greckogo orecha. – Kišinev (Moldavskaja SSR), Štiinca, 143 pp.

Konarev, V. G., A. M. Tarlakovskaja i I. P. Gavriljuk, 1983: Immunochimičeskoe izučenie belkov semjan predstavitelej triby Vicieae v svjazi s voprosami differenciacii vidov i rodov. – Bot. Ž. **68**, 1153–1159.

Korsakov, N. I., 1982: Geografičeskie očagi formoobrazovanija i gomologičeskie rjady v nasledstvennoj izmenčivosti priznakov roda *Glycine* L. (Geographical centers of origin and homologous series in hereditary variability of characters in the genus *Glycine* L.). – Trudy prikl. bot., genet. selekc. **72** (1), 3–15.

Krishna, T. G., and L. J. Reddy, 1982: Species affinities between *Cajanus cajan* and some *Atylosia* species based on esterase isoenzymes. – Euphytica **31**, 709–713.

Ladizinsky, G., and G. Waines, 1982: Seed protein polymorphism in *Vicia sativa* agg. (Fabaceae). – Plant Syst. Evol. **141**, 1–5.

Le Guen, J., 1983: Incompatibilité unilatérale chez *Vicia faba* L. I. Analyse globale de croisements intraspécifiques entre quatre sous-espèces. – Agronomie **3** (5), 443–449.

Lewis-Jones, L. J., J. P. Thorpe, and G. P. Wallis, 1982: Genetic divergence in four species of the genus *Raphanus*: Implications for the ancestry of the domestic radish *R. sativus*. – Bot. J. Linn. Soc. **18**, 35–48.

Li, C. W., 1981: The origin, evolution, taxonomy and hybridization of Chinese cabbage (pp. 3–10). In: N. S. Talekar, and T. D. Griggs (eds.), Chinese cabbage. Proc. first internatl. symp. – Tainan, Taiwan.

Liang, J., 1981: Le litchi, origine, utilisation et développement de sa culture. – J. Agric. tradit. Bot. appl. **28**, 259–270.

Maesen, L. J. G. van der, 1981: Taxonomy of *Cajanus*. – Proc. Internatl. Workshop on Pigeonpeas, 15–19 Dec. 1980, International Crops Research Institute for the Semi-Arid Tropics, Patancheru, India, vol. **2**, 9–13.

Maisaya, I. I., and A. D. Gorgidze, 1980: Establishing the genome composition of Georgian *Setaria* millet. – Bull. Acad. Sci. Georgian SSR **99**, 169–172. (Ref.: Plant Breeding Abstr. **53**, 1983, no. 3028).

Mastenbroek, I., C. E. Cohen, and J. M. J. de Wet, 1981: Seed protein and seedling isozyme patterns of *Zea mays* and its closet relatives. – Biochem. Syst. Ecol. **9**, 179–183. (Ref.: Plant Breeding Abstr. **53**, 1983, no. 2986).

Miège, J., and S. N. Lyonga (eds.), 1982: Yams. Ignames. – Clarendon Press Oxford, XVIII + 411 pp.

Migušova, È. F., i A. G. Chakimova, 1982: K voprosu o sistematike *Aegilops crassa* Boiss. – Bjull. vses. Inst. Rast. im. N. I. Vavilova **119**, 75–76.

Ohashi, H., and H. Takahagi, 1981: Pollen morphology of *Vigna angularis* (Leguminosae). – Bot. Mag., Tokyo **94** (1034), 177–180.

Oriščenko, N. D., i V. L. Tichonova, 1981: Vnutrividovaja izmenčivost *Sanguisorba officinalis* L. – Rast. resursy **17**, 390–395.

Pal, M., R. M. Pandey, and T. N. Khoshoo, 1982: Evolution and improvement of cultivated amaranths. – J. Heredity **73**, 353–356.

Palmer, J. D., C. R. Shields, D. B. Cohen, and T. J. Orton, 1983: Chloroplast DNA evolution and the origin of amphidiploid *Brassica* species. – Theor. Appl. Genet. **65**, 181–189.

PERRINO, P., and K. HAMMER, 1983: Sicilian wheat varieties. – Kulturpflanze **31**, 227–279.
PIRMAGOMEDOV, P. M., 1982: Dikorastuščij vinograd Tabasaranskogo rajona Dagestana i ego svjaz s mestnymi sortami. – Bjull. vses. Inst. Rast. im. N. I. VAVILOVA **126**, 47–50.
PONOMARENKO, V. V., 1982: Proischoždenie i rasprostranenie kul'tury jabloni – *Malus domestica* BORKH. – Bjull. vses. Inst. Rast. im. N. I. VAVILOVA **126**, 7–12.
PYO, H. K., 1981: Historical observations on the cultivar development of Chinese cabbage in Korea (pp. 41–47). In: N. S. TALEKAR, and T. D. GRIGGS (eds.), Chinese cabbage. Proc. first internatl. symp. – Tainan, Taiwan.
RANGASWAMY, M., and M. RATHINAM, 1982: Mutagen induced male sterile lines in sesame. – Ind. J. Genet. Plant Breeding **42** (2), 142–143. (Ref.: Plant Breeding Abstr. **53**, 1983, no. 3302).
RAO, G. R., 1982: A hypothesis on polyploidy and evolution of *Solanum nigrum* L. – Science & Culture **48**, 206–207.
RAVI, S. B. (BALA RAVI, S.), and P. B. RAVI (PADMA BALA RAVI), 1981?: Phylogenetic relationships in the genus *Cicer* and the origin of chick pea (*Cicer arietinum* L.) on the basis of isozymes (Abstract). – In: Fourth Internat. SABRAO congress, 4–8 May 1981, at Univ. Kebangsaan Malaysia and Federal Hotel, Kuala Lumpur, p. 4, IARI Reg. Sta., Hyderabad, India. (Ref.: Plant Breeding Abstr. **53**, 1983, no. 3581).
SAITO, H., M. TANAKA, and N. ISHIDA, 1982: Origin and phylogenetical differentiation of wild tetraploid wheats inferred from their susceptibility to leaf rust (preliminary note). Susceptibility of the genus *Triticum* to two races of *Puccinia recondita* f. sp. *tritici*. – Japan. J. Breed. **32**, 274–281.
SCHMIEDICHE, P. E., J. G. HAWKES, and C. M. OCHOA, 1982: The breeding of the cultivated potato species *Solanum* ×*juzepczukii* and *S.* ×*curtilobum*. II. The resynthesis of *S.* ×*juzepczukii* and *S.* ×*curtilobum*. – Euphytica **31**, 695–707.
SCHULTZE-MOTEL, J., R. FRITSCH, K. HAMMER, P. HANELT, J. KRUSE, H. OHLE, and K. PISTRICK, 1982: Taxonomy and evolution of cultivated plants: Literature review 1981/1982. – Kulturpflanze **31**, 299–316.
SHADMANOV, R. K., and E. S. DAUTOVA, 1981: Electrophoretic characteristics of the seed proteins of cotton amphidiploids. – Uzbekistan Biologija Žurnali No. 1, 66–69 (Ref.: Plant Breeding Abstr. **53**, 1983, 377, no. 4081).
SHAO, QIQUAN, LI CHANG-SEN, and BASCHAN CHIREN, 1982: Origin and evolution of cultivated barley: Wild barley from Western Szechuan and Tibet, China. – Barley Genetics Newsletter **12**, 37–42.
SHENGXIANG, S., and P. CHONGLIAN, 1982: A new botanical variety of apricot. (Chin.). – Acta hort. Sinica **9** (1), 58.
SKVORCOV, A. K., JU. K. MAJTULINA, i JU. N. GORBUNOV, 1983: O meste, vremeni i vozmožnom mechanizme vozniknovenija kul'turnoj černoplodnoj Aronii. – Bjull. Mosk. obšč. ispyt. prir. biol. **88** (3), 88–96.
SMALL, E., and L. P. LEFKOVITCH, 1982: Agrochemotaxometry of alfalfa. – Can. J. Plant Sci. **62**, 919–928.
STANKEVIČ, A. K., 1982: O sistematičeskom položenii nekotorych sekcij roda *Vicia* L. – Trudy prikl. bot., genet. selekc. **72** (1), 21–27.
STONE, B. C., 1982: *Pandanus tectorius* PARKINS. in Australia: a conservative view. – Bot. J. Linn. Soc. **85**, 133–146.
TANI, H., and M. NAKAGAHRA, 1981: Application of isozyme analysis to classification of tea plants, *Camellia sinensis* (L.) O. KUNTZE. – In: Annual Report 1980, Division of Genetics, National Institute of Agricultural Sciences, Yatabe, Japan, 44–45. (Ref.: Plant Breeding Abstr. **53**, 1983, no. 3274).
TEOH, S. B., and J. HUTCHINSON, 1983: Interspecific variation in C-banded chromosomes of diploid *Aegilops* species. – Theor. Appl. Genet. **65**, 31–40.
TSUNEWAKI, K., and Y. OGIHARA, 1983: The molecular basis of genetic diversity among cytoplasms of *Triticum* and *Aegilops* species. II. On the origin of polyploid wheat cytoplasms as suggested by chloroplast DNA restriction fragment patterns. – Genetics **104**, 155–171.
VANDERBORGHT, T., 1983: Evaluation of *Phaseolus vulgaris* wild types and weedy forms. – Plant Genetics Resources Newsletter **54**, 18–25.

Wet, J. M. J. de, K. E. Prasada Rao, M. H. Mengesha, and D. E. Brink, 1983a: Diversity in Kodo Millet, *Paspalum scrobiculatum*. – Econ. Bot. **37**, 159–163.
–, –, –, –, 1983b: Domestication of Sawa Millet (*Echinochloa colona*). – Econ. Bot. **37**, 283–291.
Wijsman, H. J. W., 1982: On the interrelationships of certain species of *Petunia*. I. Taxonomic notes on the parental species of *Petunia hybrida*. – Acta bot. Neerl. **31**, 477–490.
Witcombe, J. R., 1983: A guide to the species of *Aegilops* L. Their taxonomy, morphology and distribution. – IBPGR Secretariat Rome, 74 pp.

Dr. J. Schultze-Motel, Dr. R. Fritsch, Dr. K. Hammer, Dr. P. Hanelt,
Dr. J. Kruse, Dr. H. I. Maass, Dr. H. Ohle and K. Pistrick
Zentralinstitut für Genetik und Kulturpflanzenforschung
der Akademie der Wissenschaften der DDR
DDR - 4325 Gatersleben
Corrensstraße 3

Kulturpflanze 32 · 1984 · 261–272

Beiträge zur Kenntnis der Vaucheriaceae XXIV. Über eine in die *pseudogeminata*-Gruppe gehörende Art aus dem Harz

ALFRED RIETH

(Eingegangen am 1. November 1983)

Zusammenfassung

Es wird eine *Vaucheria* aus dem Hochharz nach Freilandfunden 1981 und 1982, sowie an Hand von Kulturen beschrieben, die in ihren Merkmalen eine gewisse Mittelstellung zwischen *V. pseudogeminata* DANGEARD und *V. alaskana* BLUM einnimmt. Die zahlreich beobachteten Prolifikationen im Sexualorganbereich werden analysiert.

Einleitung

Die Pflanzen, die dieser Untersuchung zugrunde liegen, wurden im Hochharz Anfang Juli 1981 und Ende Juli 1982 am Rande des vom Forsthaus Plessenburg zu den Zeterklippen (SW Ilsenburg) führenden Weges, etwa 300 m oberhalb der als „Stern" bezeichneten Wegespinne, in ungefähr 775 m Meereshöhe, als kleine, grüne, Moosprotonema ähnliche, wenig auffallende Beläge gesammelt. Sie wuchsen auf feuchtem Granitverwitterungssand mit beigemischter organischer Substanz, meist Fichtennadeln. Die mikroskopische Untersuchung zeigte, daß es sich vorwiegend um nur vegetativ vorliegende *Vaucheria*-Thalli handelte. Als sie begannen, mit Bachwasser aus demselben Gebiet angesetzt, Mitte bis Ende August Sexualorgane zu bilden, konnte festgestellt werden, daß neben spärlich vertretener *Vaucheria prolifera* DANGEARD eine andere Spezies vorherrscht, deren Zuordnung einige Schwierigkeit bereitet, da sie Merkmale sowohl von *V. pseudogeminata*, die ich 1975 im Harz nachweisen konnte und *V. alaskana*, bisher für Mitteleuropa nicht bekannt, aufweist. Die Population soll daher kurz beschrieben und diskutiert werden.

Beobachtungen

Nach der Thallusfadendicke ((13,0–) 23,5–36,5 (–47,0) µm) (Abb. 1a) gehört die vorliegende Form zu den dünnfädigeren Arten. Zuweilen treten undulierte Thallusbezirke auf (Abb. 3). Die mehr oder minder lang ((96,0–) 130,0–309,0 (–468,0) µm) (Abb. 2) gestielten, in einen schneckenartig eingerollten, 1–1¼ Umgänge beschreibenden Antheridialast – an dem das Antheridium selbst mit

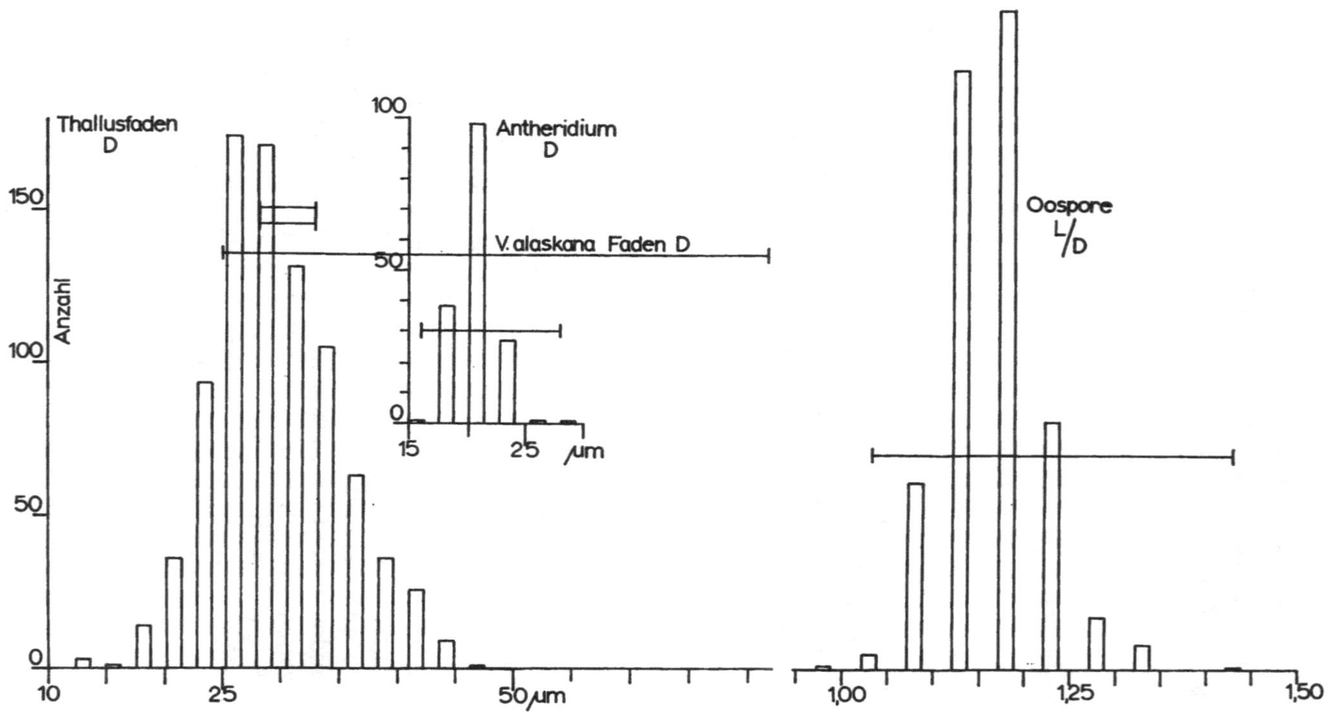

Abb. 1a Harzmaterial. Thallusfaden-Durchmesser, Antheridiendurchmesser (an der Querwand gemessen), Oosporen Verhältnis Länge zu Breite. Als Querstriche ist die Variationsspanne wie sie für *V. alaskana* in der Literatur angegeben ist, eingetragen. Außerdem ist als leerer Querbalken die von DANGEARD in der Originaldiagnose genannte Spanne des Thallusfadendurchmessers für *V. pseudogeminata* vermerkt

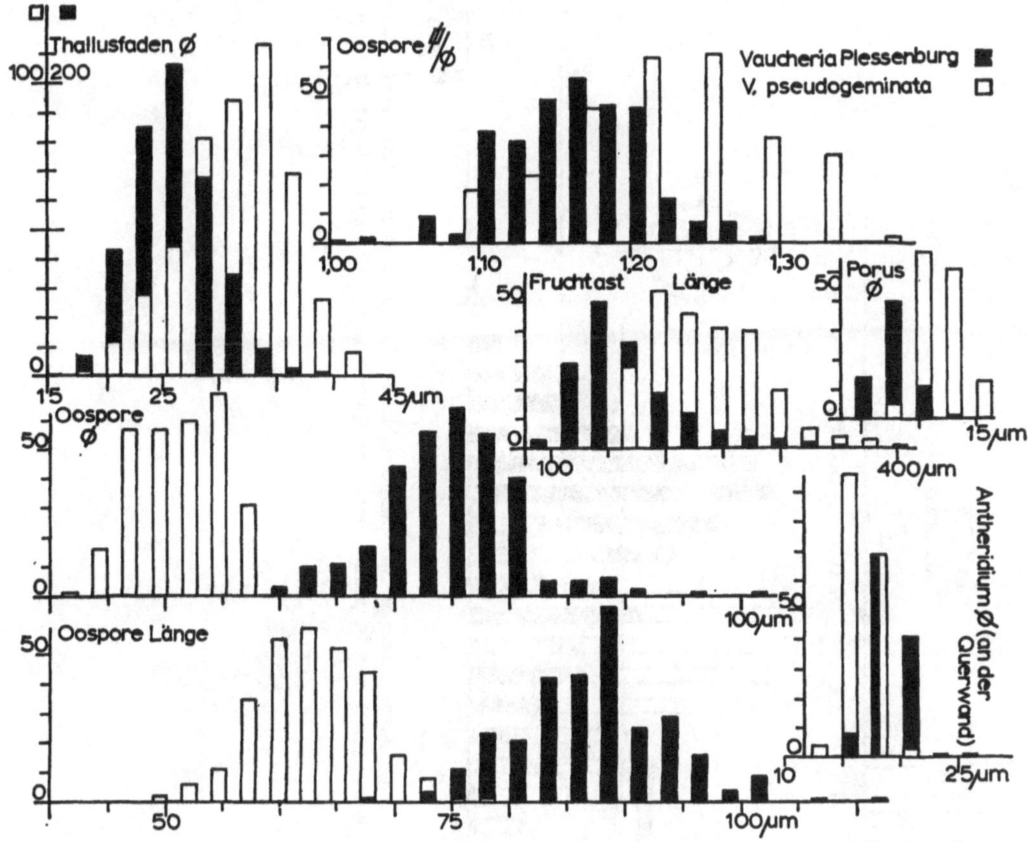

Abb. 1 b *V. pseudogeminata* und *V. spec.* aus dem Harz: Größenvergleich nach Kulturen unter gleichen Bedingungen

ungefähr 1/2 Umgang beteiligt ist – endenden, bisexuellen Fruchtäste sitzen dem Mutterfaden seitlich an. Sie tragen lateral, unterhalb des männlichen Astes kurz gestielt ein, oder seltener zwei gegenständige, aufrechte Oogonien. Die Medianebene des oder der Oogonien steht senkrecht auf der Symmetrieebene des Antheridialastes, dessen Scheitelhöhe etwa der der Oogonien entspricht (Abb. 3, a–c; Abb. 7, Fig. 19, 19a). Charakteristisch ist die Schnabelung des ♀ Hüllorgans, die der Art eindeutig eine Stellung in der Pseudogeminata Gruppe (RIETH 1980) zuweist. Seine Ausbildung ist in dem Harzmaterial etwas variabel. Wie die Figuren 5–17 der Abbildung 4 zeigen, ist der Schnabel in der Masse der Fälle jedoch kurz hakig gekrümmt, mit relativ enger, durch die Krümmung bedingt, abwärts zum Tragfaden zu gerichteter Befruchtungsöffnung. In einem größeren, über 1000 Sexualorganstände umfassenden Material treten allerdings auch Oogonien mit fast kaum gebogenen, nicht abwärts, sondern schräg aufwärts gerichteten Schnäbeln auf (Abb. 4, Fig. 1–4), die an die Verhältnisse bei *Vaucheria pseudogeminata* erinnern, aber im Unterschied zu dieser verhältnismäßig kleinporig sind (siehe zum Vergleich die Abbildungen von *V. pseudogeminata* bei RIETH 1975 (Abb. 3 u. 4) und 1980 (Fig. 25)).

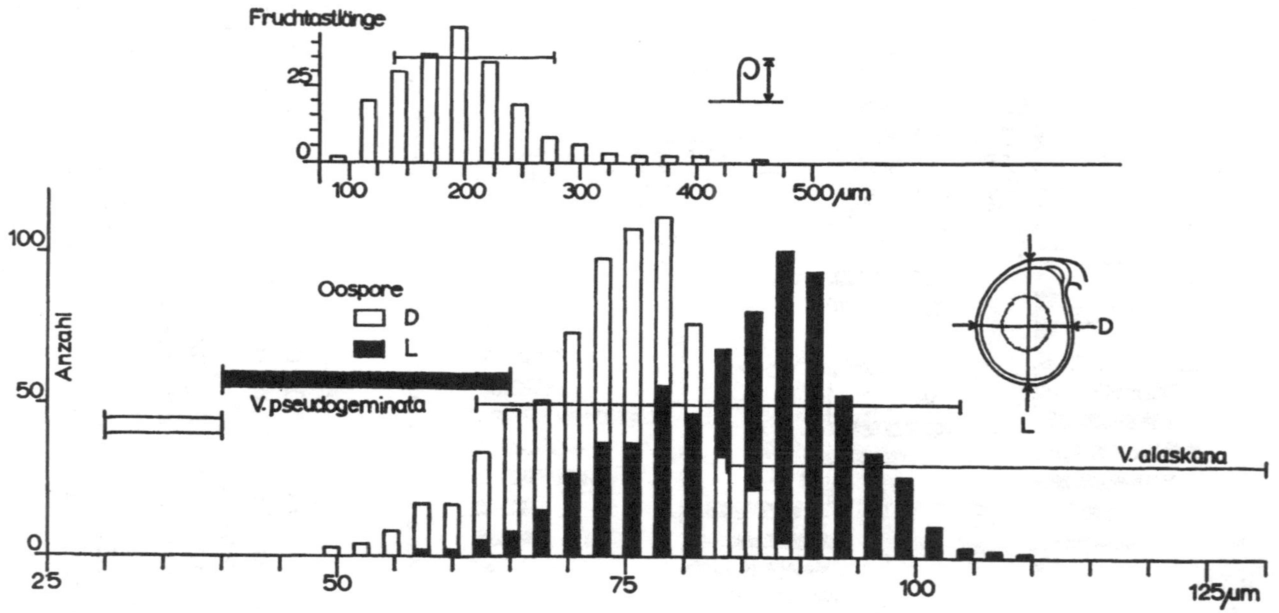

Abb. 2 Harzmaterial. Fruchtastlänge, sowie Oosporenlänge und Oosporenbreite. Die entsprechenden Werte nach der Literatur für *V. alaskana* (einfache Querstriche) und *V. pseudogeminata* (Querbalken) sind zum Vergleich eingetragen

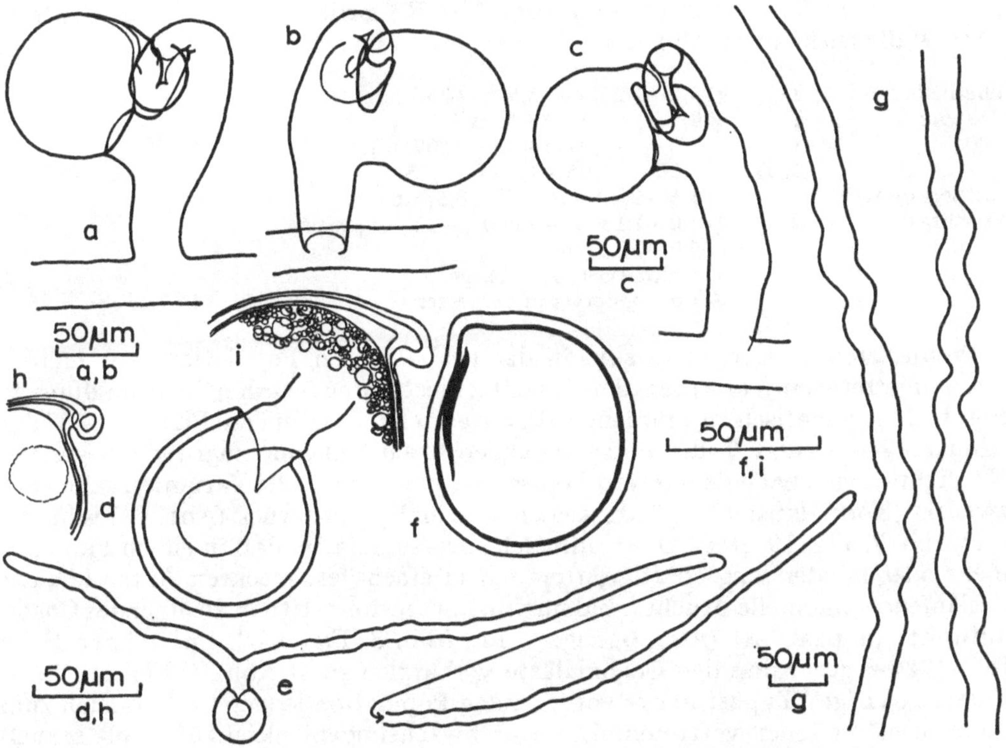

Abb. 3 Fig. a–c: Die in dem Harzmaterial häufigste Fruchtastausbildung; Fig. d, e: Keimende Oospore; Fig. f: Keimrißlage (nach Laktophenolbaumwollblaufärbung); Fig. g: Undulierte Thallusfadenbereiche

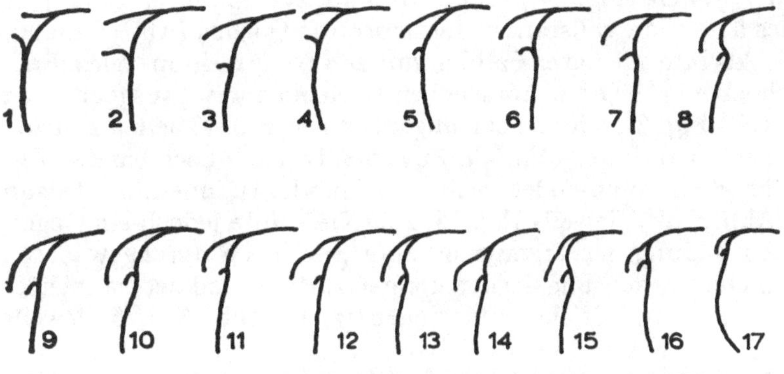

Abb. 4 Fig. 1–17: Variation der Schnabelausbildung im Harzmaterial

Die Oospore füllt das Oogon mit Ausnahme des Schnabels völlig aus. Nur zuweilen finden sich Fälle, in denen sich die Spore mehr oder minder weit auch in den Schnabel erstreckt (Abb. 3, Fig. h u. i). Sie enthält bei der Reife einen rötlich-gelb gesäumten, rötlich-braunen Zentralfleck und ist von einer mäßig dicken, nicht ornamentierten, zweischichtigen Wand, deren innere Schicht sich mit Laktophenolbaumwollblau tingiert, umgeben. Der Keimriß entsteht an der Ventralseite. Maße (siehe auch Abb. 1!):

Thallusfaden	D	(13,0 –) 23,5 – 36,5 (– 47,0) μm
Oospore	D	(49,5 –) 62,5 – 88,5 μm
	L	(57,0 –) 70,0 – 100,0 (– 109,0) μm
	L/D	(1,00 –) 1,06 – 1,28 (– 1,35)
Antheridium₊	D	15,5 – 21,0 – 26,0 (– 28,5) μm
Fruchtast	L	(96,0 –) 130,0 – 309,0 (– 468,0) μm
Schnabelporus	D	5,0 – 10,5 μm
	D =	Durchmesser; L = Länge
	+	An der Querwand gemessen

Proliferierende Fruchtäste sind in der beobachteten Population sehr häufig. Die aufgetretenen Varianten der Fruchtastgestaltung wurden in Abbildung 5, Fig. 1–22 schematisch zusammengefaßt dargestellt, während die Figuren 1–22 der Abbildungen 6 und 7 die entsprechenden, dem Schema zugrunde liegenden Konfigurationen nach dem mikroskopischen Bild zeigen. Alle Varianten gehen auf zwei als „Normalausbildung“ anzusehende Grundtypen zurück (Abb. 6, Fig. 1, 1a und Abb. 7, Fig. 19, 19a). Diese unterscheiden sich darin, daß in einem Falle der mehr oder minder langschäftige, terminal in einen geschneckten Antheridialast auslaufende bisexuelle Fruchtstand nur ein kurzgestielt seitlich ansitzendes Oogon aufweist, „unpaar“ ist (Abb. 6, Fig. 1, 1a; Abb. 3, Fig. a–c), im anderen Falle jedoch zwei gegenständige Oogonialäste vorhanden sind (Abb. 7, Fig. 19, 19a). Dieser „paarige“ Typ ist in der vorliegenden Population jedoch im Vergleich zum „unpaaren“ seltener vertreten. Die Durchwachsungen, gleichgültig ob sexuell determiniert oder vegetativ, entspringen, einer gewissen Gesetzmäßigkeit folgend, nur an bestimmten Punkten des Fruchtastes. So finden sie sich an Stelle eines mehr oder minder in der Ausbildung gehemmten ♀ Organs. Dabei wird in einem Extremfall statt des völlig unterdrückten Oogons ein neuer Sexualorganstand gebildet (Abb. 5 und 7, jeweils Fig. 14), im anderen sprießt ein neuer Fruchtast aus dem Schnabel des äußerlich vollständig ausgeformten Oogons (Abb. 5 und 6, jeweils Fig. 10 u. 11). Als intermediärer Fall kommt es vor, daß ein in seiner Entwicklung „steckengebliebenes“ ♀ Organ am Rücken zu einem normal ausgebildeten Oogon (Abb. 5, 6, jeweils Fig. 2), oder zu einem ganzen neuen bisexuellen Fruchtstand durchwächst (Abb. 5, 6, jeweils Fig. 8). Zuweilen ist auch noch ein zweites, nicht zur Funktionsfähigkeit kommendes, mehr oder minder rudimentäres Oogon zwischengeschaltet (Abb. 5 u. 7, jeweils Fig. 15, 16). Nie wurde jedoch ein Oogon gefunden, das nach Ausbildung der Trennwand zum Stiel noch durchgewachsen wäre. An den gleichen Durchwachsungsstellen können statt sexuell determinierter Proliferationen auch vegetative Triebe entspringen (z. B. Abb. 5 u. 6, jeweils Fig. 6, 7, 9, 10).

Im Bereich des Antheridialastes entsteht ein Proliferationsprozeß bevorzugt am Scheitel (Abb. 5 u. 6, jeweils Fig. 3, 11, 14), gelegentlich auch dicht an der Trennwand zum Antheridium (Abb. 5 u. 6, jeweils Fig. 3, 4, 5, 11; Abb. 5 u. 7, jeweils Fig. 14), jedoch bildet sich dabei nie direkt ein Oogon, sondern entweder ein vegetativer Ast (Abb. 5 u. 6, jeweils Fig. 3, 11) oder ein bisexueller Stand (Abb. 5 u. 6, jeweils Fig. 4; Abb. 5 u. 7, jeweils Fig. 14) bzw. ein Stand, der nur das Antheridium und an Stelle eines funktionsfähigen Oogons einen vegetativen Ast ent-

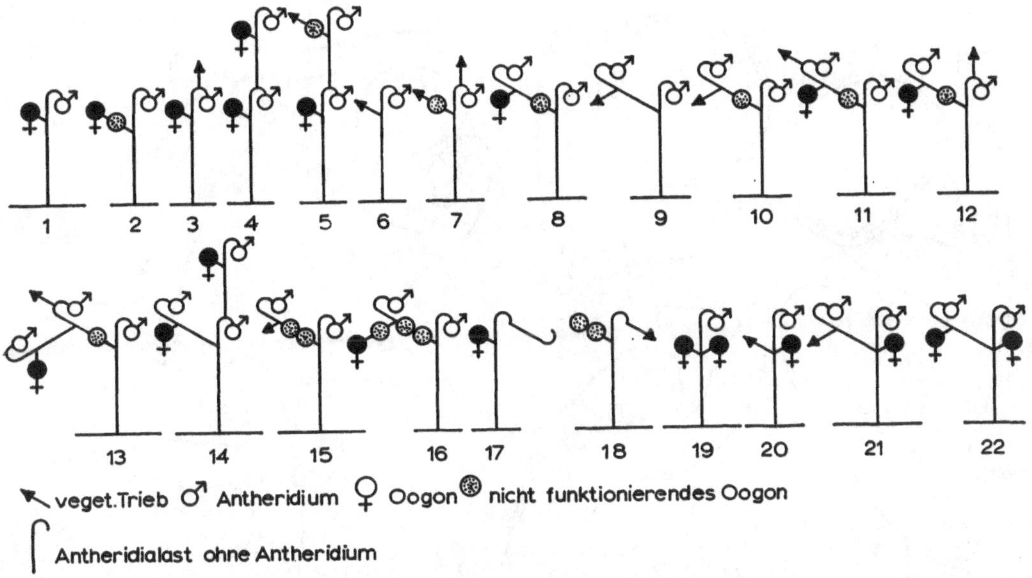

Abb. 5 Fig. 1–22: Die gefundenen Varianten der Morphologie im Sexualorganbereich, schematische Übersicht. Fig. 1 und Fig. 18 stellen die „Normalausbildung" dar

hält (Abb. 5 u. 6, jeweils Fig. 5). Sehr selten wächst der Antheridialast selbst, statt ein Antheridium zu bilden, entweder vegetativ (Abb. 5 u. 7, jeweils Fig. 18), oder zu einem – allerdings rudimentär bleibenden – Fruchtstand weiter (Abb. 5 u. 7, jeweils Fig. 17).

Die Entscheidung, ob die untersuchte Population innerhalb der Pseudogeminata-Gruppe zu *V. pseudogeminata* oder zu *V. alaskana* gestellt werden soll (*Vaucheria adunca* JAO 1939, an die zu denken wäre, hat wesentlich kleinere Oosporen und längeren Schnabel), ist bei kritischer Einschätzung der geschilderten Merkmale nicht einfach, vor allem auch deshalb, weil gute, die Variabilitätsbreite des Habitus erfassende Abbildungen nach lebenden Pflanzen für *V. alaskana* nicht vorliegen. Manche Darstellungen, wie die Mikrophotographien bei SANTRA und ADHYA (1976), sind völlig unbrauchbar. BLUM (1972) nennt folgende Unterscheidungsmerkmale:

	V. pseudogeminata	*V. alaskana*
Oogonschnabel	kegelförmig	eine kurze Röhre
Durchmesser	23 – 31 µm	8 – 14 µm
Oogon L/D	1,32 – 1,41	1,18 – 1,20

Seine Angaben über die Beschaffenheit der Oosporenwand bei *V. alaskana* sind etwas widersprüchlich. In der Originaldiagnose (BLUM 1953) steht: „. . . its wall composed of two layers, thin and colorless" was mit den Verhältnissen bei unserem Material übereinstimmt. 1972 aber findet sich: „. . . with a conspicuous wall, the outer wall sometimes thickened with a pattern of reticulate ridges". Leider findet sich keine Abbildung dieser 1972 geschilderten Oosporenornamentierung.

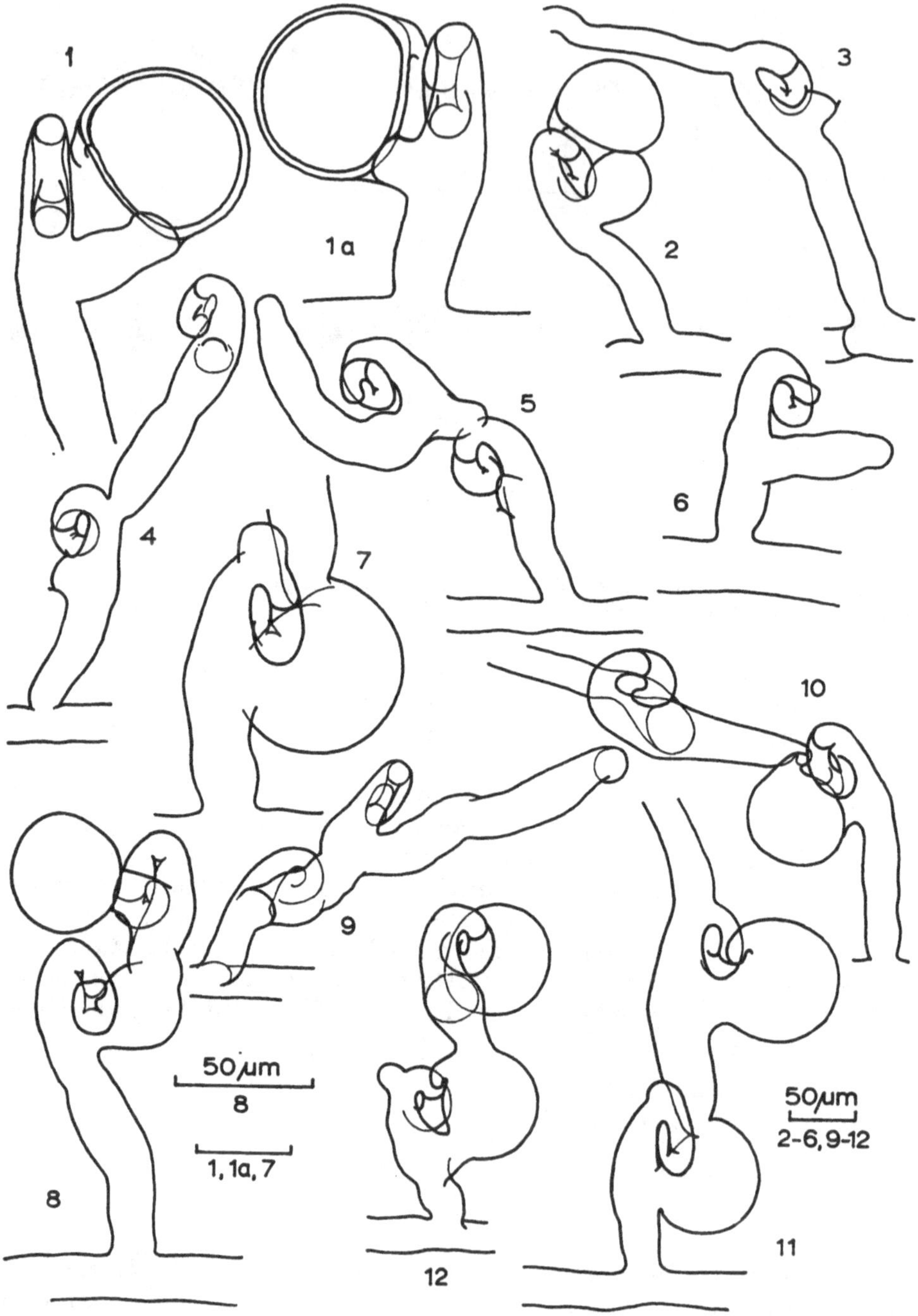

Abb. 6 und Abb. 7 Fig. 1–17: Die den entsprechend nummerierten Figuren des Schemas Abb. 5 zugrunde liegenden Konfigurationen nach dem mikroskopischen Bild. Alle Figuren nach lebenden Pflanzen gezeichnet

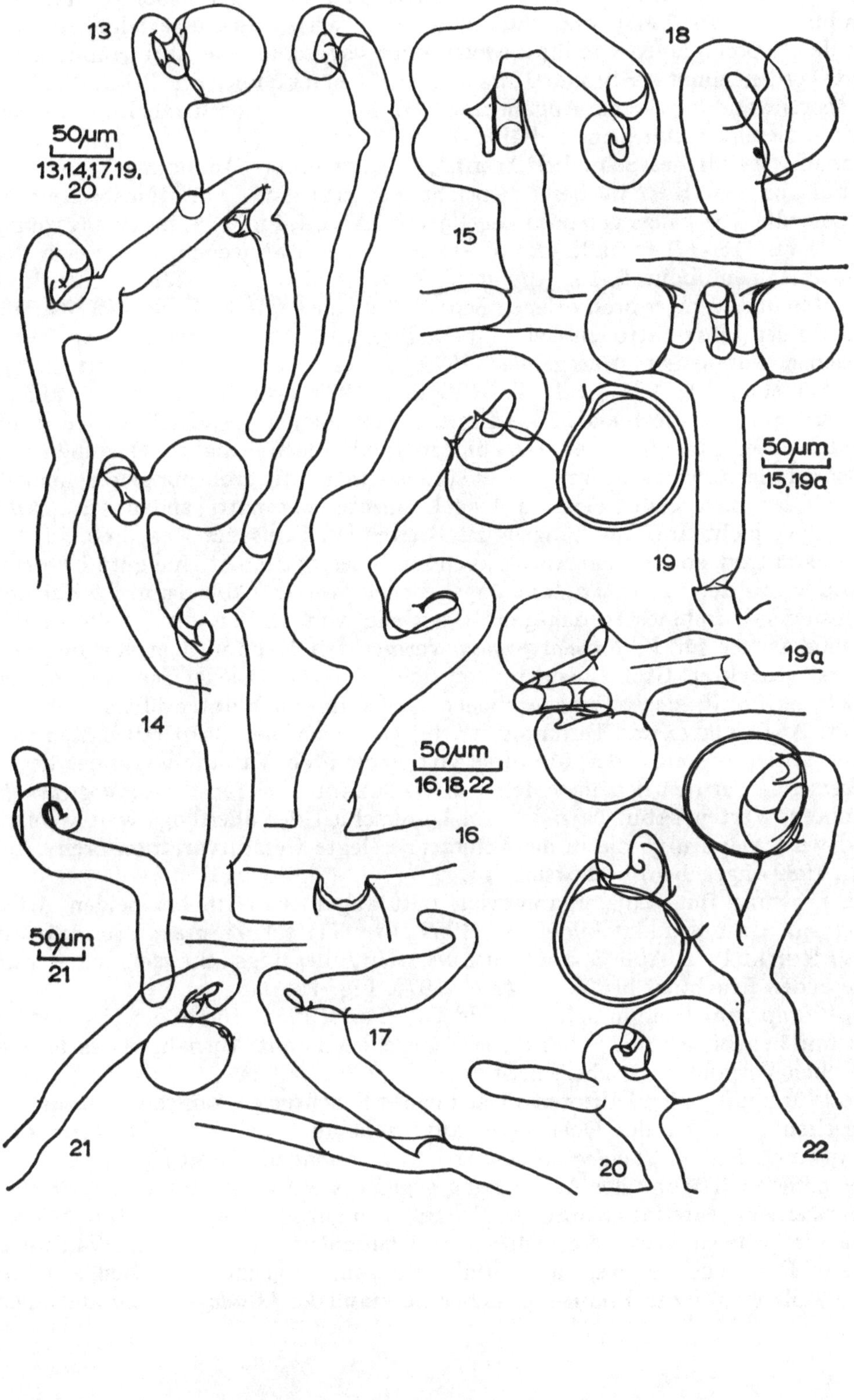
13
50µm
13,14,17,19,
20
18
15
50µm
15,19a
19
19a
14
50µm
16,18,22
16
50µm
21
17
21
20
22

Unser seinerzeit, allerdings auch nur an fixierten Proben, untersuchtes Material aus China (RIETH 1963) wies durchweg recht derbe Thalluswandungen auf. Vollreife Oosporen lagen von ihm jedoch nicht vor, sodaß über ihre Hülle nichts gesagt werden konnte. SIMONS (1974 und 1977) macht über die Beschaffenheit der Sporenwandung keine Angaben. Seine Figur, augenscheinlich in beiden Arbeiten dieselbe, zeigt keine auffällig derbe Wand.

Was die Gestalt des Schnabels betrifft, so zeigt unsere Abbildung 4, daß das Material aus dem Harz in dieser Hinsicht eine erhebliche Variationsbreite aufweist, die alle Übergänge von breit kegelförmig (Abb. 4, Fig. 1–4) bis zu eng röhrig (Abb. 4, Fig. 13–17) umfaßt, die Masse der Fälle liegt jedoch im Bereich der Konfigurationen Figur 5–13, Abbildung 4, zeigt also einen stärker zurückgekrümmten und weniger breit offenen Schnabel als die typische *V. pseudogeminata*. Im Laufe der Jahre hatte ich Gelegenheit, Populationen der letzteren von 5 verschiedenen Fundorten (Harzgebiet: 1973, Gegend Roßtrappe (RIETH 1975); 1974, Selketal; 1979, Bodetal; V. R. Polen 1975: Nordrand der Beskiden; Bundesrepublik Deutschland: 1983 Bayern, Starnberger Seegebiet) an Kulturen zu untersuchen. In all diesen Herkünften fanden sich lediglich Oogonien mit typisch schräg aufwärts bis waagerecht stehendem, relativ großsporigem Schnabel. Nach Art der vorliegenden Form zurückgekrümmte, röhrenartig endende Schnäbel traten dort nicht auf. Im Längen- zu Breitenverhältnis der Oosporen, in den Fruchtastlängen, sowie beim Antheridiendurchmesser besteht eine gute Übereinstimmung mit den für *V. alaskana* angegebenen Werten (Abb. 1a u. 2). Für den Thallusfaden-Durchmesser dagegen liegen die von DANGEARD (1939) in der Originaldiagnose für *V. pseudogeminata* vermerkten Maße in dem von unserem Material gedeckten Größenbereich, der sich seinerseits wieder nur im unteren Bereich der für *V. alaskana* angegebenen Größe, ihn nach unten überschreitend, bewegt. Auch Länge und Durchmesser der Oogonien und Oosporen halten sich an der unteren Grenze der für diese Art genannten Variationsspanne, liegen andererseits aber deutlich über den von DANGEARD l. c. für *V. pseudogeminata* genannten Werten (Abb. 1a u. 2), ein Unterschied der allerdings weit weniger scharf wird, wenn man die in der Literatur belegte Gesamtvariationsbreite (bei RIETH 1980 angegeben) berücksichtigt.

Versprossung im Sexualorganbereich tritt augenscheinlich bei beiden Arten häufig auf. CHRISTENSEN bildet sie (1969, Fig. 11) für *V. pseudogeminata* ab, ebenso RIETH (1975, Abb. 3, a–c), SIMONS stellt, allerdings sehr klein, einen proliferierenden Fruchtast bei *V. alaskana* (1974, Fig. 4b) dar.

Im Biotop „auf Erde in lichten Wäldern" (BLUM 1953) dürfte der Fundort im Harz mit den bisher für *V. alaskana* angegebenen (mit Ausnahme des niederländischen Vorkommens) übereinstimmen.

Die Würdigung aller Faktoren weist meines Erachtens darauf hin, daß der im Renneckenberggebiet des Oberharzes augenscheinlich häufigen *Vaucheria* eine Stellung zwischen *V. pseudogeminata* und *V. alaskana* zukommt.

Zur näheren Klärung der Abgrenzung gegen erstere wurde Ende August 1983 im Harz erneut Material gesammelt. Es fand sich auf sonst noch vegetationsloser, schwarzer Schwemmerde eines durch Holzeinschlag aufgewühlten Waldweges (Unterer Renneckenbergweg) als auffallende isolierte grüne Fleckchen, gebildet von jeweils wohl einer Pflanze. Gleichzeitig stand *V. pseudogeminata* aus einem

Fund im Juli 1983 im Gebiet der bewaldeten Hügel des Starnberger (Würm-)See Ostufers zur Verfügung. (Es dürfte sich um den ersten Nachweis der Art in Bayern handeln). Beide wurden in der Folge parallel nebeneinander unter gleichen Bedingungen kultiviert. Dabei traten die Unterschiede klar hervor. Die *V. pseudogeminata* wies durchweg den breit kegelförmigen, schräg aufwärts gerichteten Schnabel mit großem Porus auf, stimmte im Habitus völlig mit den von mir 1975 und 1980 gegebenen Abbildungen und auch mit den Größenverhältnissen dieser Art überein. Die fragliche Spezies aus dem Harz aber zeigte deutlich die vorstehend beschriebene abweichende Schnabelgestaltung und Unterschiede in den Meßwerten, besonders klar größere Oosporen und engeren Schnabelporus (Abb. 1b, Tab. 1). Eine Zuordnung zu *V. pseudogeminata* kann danach kaum in Frage kommen. Die Entscheidung andererseits, ob eine Einbeziehung in den Formenkreis von *V. alaskana* möglich ist, fällt aus den oben geschilderten Gründen derzeit schwer. Einstweilen sei sie – mit allem Vorbehalt – zu dieser Art gestellt, ich neige aber zur Vermutung, daß eine neue Art vorliegt. Sollte sich erweisen, daß die von Blum angeführte „verdickte äußere Sporenwand mit einem Muster vernetzter Rippen“ nicht nur „sometimes“, sondern regelmäßig bei *V. alaskana* auftritt, wäre ein klares Unterscheidungsmerkmal zu der von mir beschriebenen Pflanze gegeben.

Tabelle 1
Vaucheria pseudogeminata und *Vaucheria spec.* aus dem Hochharz. Grössen nach Messungen an Parallelkulturen unter gleichen Bedingungen

V. pseudogeminata		
Thallusfaden	D	18,0 – 33,0 – 41,5 µm
Oospore	D	41,5 – 51,0 – 60,0 µm
	L/D	1,08 – 1,23 – 1,39
	L	49,5 – 62,5 – 78,0 µm
Antheridium	D	13,0 – 17,0 – 21,0 µm
Fruchtast	L	171,0 – 411,0 – 494,0 µm
Schnabelporus	D	7,8 – 15,8 µm
V. spec.		
Thallusfaden	D	(15,5 – 21,0 – 26,0 – 34,0 (– 39,0) µm
Oospore	D	(60,0 –) 67,5 – 75,5 – 80,5 (– 101,5) µm
	L	(67,5 –) 75,5 – 87,0 – 96,0 (– 118,0) µm
	L/D	(1,00 –) 1,10 – 1,21 (– 1,29)
Antheridium	D	15,5 – 18,0 – 26,0 µm
Fruchtast	L	88,5 – 156,0 – 234,0 (– 338,0) µm
Schnabelporus	D	5,0 – 8,0 – 13,0 µm

Summary

Contributions to the knowledge of the Vaucheriaceae XXIV. A species belonging to the *pseudogeminata*-group from the Harz mountains

A *Vaucheria*-species is described from the Harz Mountains which, within the *pseudogeminata* group holds a position between *V. pseudogeminata* and *V. alaskana*. Parallel cultivation of *V. pseudogeminata* and the Harz-*Vaucheria* under identical conditions revealed, that there exist clear cut differences between these two. A discrimination to *V. alaskana*, for reasons discussed in the paper, is not so easy. So, for the time being–with all reservations–the Harz material is assigned to this species, but supposed, that it may be a different, new species.

Краткое содержание

Материалы для изучения Vaucheriaceae. XXIV. Об одном виде водоросли с Гарца, принадлежащей к группе *pseudogeminata*

Описывается вид *Vaucheria* с Высокого Гарца по находкам 1981 и 1982 гг., а также по культивированным водорослям, которые по своим признакам занимают промежуточное положение между *V. pseudogeminata* DANGEARD и *V. alaskana* BLUM Анализируются многочисленные пролиферации в области половых органов.

Literatur

BLUM, J. L., 1953: The racemose Vaucheriae with inclined or pendent oogonia. – Bull. Torr. Bot. Club **80** (6), 478–497.

–, 1972: Vaucheriaceae: in: North American Flora II, 8. – The New York Bot. Garden, 1–64.

CHRISTENSEN, T., 1969: *Vaucheria* collections from VAUCHER's region. – Kongl. Danske Vidensk. Selskab Biol. Skrifter **16** (4), 3–36.

DANGEARD, P., 1939: Le genre *Vaucheria* spécialement dans la région du sud-ouest de la France. – Le Botaniste **XXIX**, 183–265.

JAO, CHIN-CHIH, 1939: Studies on the Freshwater algae from China III. New Chlorophyceae from Hopeh, Chekiang and Szechwan. – Sinensia **X**: 147–160.

RIETH, A., 1963: Die Algen der chinesisch-deutschen biologischen Sammelreise durch Nord- und Nordostchina 1956. – Limnologica **1** (4): 287–313.

–, 1975: Beiträge zur Kenntnis der Vaucheriaceae XVIII. *Vaucheria pseudo-geminata* DANGEARD 1939 und eine bemerkenswerte Vaucheriengesellschaft aus dem Harz. – Hercynia N. F. **12**, 160–170.

–, 1980: Vaucheriales in: Süsswasserflora von Mitteleuropa **4**, Xanthophyceae 2. Jena.

SANTRA, S. C., and ADHYA, T. K., 1976: Vaucheriaceae of eastern Himalayas (India). – Nova Hedwigia **XXVII**, 655–659.

SIMONS, J., 1974: Further remarks on the genus *Vaucheria* in the Netherlands. – Acta Bot. Neerl. **23** (4), 399–413.

–, 1977: De nederlandse *Vaucheria*-soorten. – Vetensch. mededelingen K. N. N. V. **120**, 1–32.

Prof. (em.) Dr. A. RIETH
Zentralinstitut für Genetik und Kulturpflanzenforschung
der Akademie der Wissenschaften der DDR
DDR - 4325 Gatersleben
Corrensstraße 3

IV. Das Institut im Jahre 1983
(The Institute in 1983)

Kulturpflanze 32 · 1984 · 275–289

A. Jahresberichte der Bereiche

1. Molekular- und Zellgenetik

Auf dem Gebiet der **Gentechnik bei Bakterien** gelang zusätzlich zum Alpha-Amylasegen aus *Bacillus licheniformis* die Klonierung der entsprechenden, aber nicht homologen Gene aus zwei anderen *Bacillus*-Stämmen. Die auf verschiedene Plasmidvektoren übertragenen *amy*-Gene bilden den Ausgangspunkt für Untersuchungen zur Regulation der Enzymsynthese und zur Analyse von Teilsequenzen der Gene mit Kodierungsfunktion für eine hitzestabile bzw. eine hitzelabile Alpha-Amylase.

Parallel dazu konnten Versuche zur Reisolierung des Gens für das Exoenzym Beta-Glucanase von *Bacillus amyloliquefaciens* aus einer Genbank des Phagen Lambda Charon 4A erfolgreich durchgeführt werden. Das Gen *bgl* wurde sowohl in den *E. coli* Plasmidvektor pBR 322 als auch in Plasmiden aus *Streptococcus* spec. rekloniert und zur Expression gebracht. In diesem Falle ist das Beta-Glucanasegen aus einer grampositiven Bakterienart in dem gram-negativen Empfänger *E. coli* aktiv. Dieses Ergebnis entspricht bisherigen Befunden, wonach Gene grampositiver Bakterien, im Gegensatz zur Situation beim recA-Gen, auch in gram-negativen aktiv sind. Durch Transposon Tn5-Mutagenese gelang es, das Strukturgen *bgl* in einem 3.6 kb-DNA-Fragment zu lokalisieren.

Die nunmehr verfügbaren Gene amy^+ bzw. bgl^+ bilden die Grundlage für die Herstellung gentechnisch manipulierter Produktionsstämme. Die 1982 begonnenen Arbeiten zur Übertragung und Nutzung von Plasmiden aus *Streptococcus* spec. in *Bacillus subtilis* waren erfolgreich. Die Plasmide besitzen in *B. subtilis* eine ausreichend hohe Stabilität und eine genügend große Kopienzahl.

Die Arbeiten über das recA-Gen von *Proteus mirabilis* und dessen Wirkung in gram-negativen und gram-positiven Bakterien (*E. coli*, *P. mirabilis* bzw. *B. subtilis*) wurden abgeschlossen. Nach *in vitro*-Übertragung des recA-Gens auf einen Plasmidvektor mit der Fähigkeit, sich sowohl in gram-positiven, als auch in gram-negativen Bakterienarten zu vermehren, konnte in recE-Mutanten von *B. subtilis* die phänotypische Aktivität des heterologen aus *P. mirabilis* stammenden recA-Gens in *B. subtilis* nachgewiesen werden. Die genetische Aktivität eines Gens aus gram-negativen Bakterien in grampositiven *B. subtilis* steht im Widerspruch zur bisher allgemein festgestellten Inaktivität von Genen gram-negativer Bakterien in *B. subtilis*.

Im Rahmen der **Untersuchungen zur Induktion von Gen- und Chromosomenmutationen** in Zellkulturen des Chinesischen Hamsters und Wurzelspitzenmeristemen von *Vicia faba* wurden weitere Versuche zur Adaptation der Zellen an mutagene Wirkungen durch Vorbehandlung mit niedrigen Dosen mutagener Agenzien, die selbst keine Mutationen hervorrufen („adaptive response"), weitergeführt. Aus einer Vielzahl von Daten kann geschlossen werden, daß bei Hamsterzellen, analog zu *E. coli*, mutagene und clastogene Adaptation auf die Induktion separater, adaptiver Reparatursysteme zurückzuführen ist. Dadurch konnte erstmalig der Nachweis erbracht werden, daß Säugerzellen *in vitro* auf die Einwirkung sehr niedriger Mutagendosen pleiotrop reagieren können.

In Wurzelspitzenmeristemen von *Vicia faba* konnte nachgewiesen werden, daß Proteinsynthesehemmung die „adaptive response" bei der Induktion von Chromosomenaberrationen verhindert. Zum gleichen Ergebnis führten *in vitro*-Versuche mit N-methyl-N'-nitro-N-nitrosoguanidin (MNNG). Vorbehandlung von V79-Zellen mit MNNG reduzierte die Mengen der untersuchten DNA-Methylierungsprodukte, gemessen unmittelbar und mehrere Stunden nach Alkylierung mit Methylnitrosoharnstoff. Dieser Effekt trat nicht auf, wenn Cycloheximid als Proteinsynthesehemmer gegeben wurde.

Untersuchungen zur Verteilung induzierter Chromatidenaberrationen nach Einwirkung verschiedener Mutagene auf die Translokationsmutante T140 von *Vicia faba* ergaben eine außergewöhnliche Häufung von Chromatidentranslokationen in der Nukleolar-Organisatorregion. 30 % aller Aberrationen traten in der NOR auf. Die Befunde sind das bisher extremste Beispiel positionsabhängiger Häufung von Chromatidenaberrationen in einem definierten Genomteil.

Weiterhin wurde eine Methode entwickelt, mit deren Hilfe durch Kreuzung bestimmter Translokationskaryotypen gezielt trisome und andere aneuploide Pflanzen von *Vicia faba* hergestellt werden können.

Im Verlauf der Arbeiten zur **Analyse von Struktur und Expression** eukaryotischer Gene in den polytänen Riesenchromosomen der Dipteren-Art *Chironomus thummi* wurde die Analyse des rekombinanten Lambda-Phagen BRc-1 weitergeführt. Die Sequenzanalyse von fünf Repetitionseinheiten der BRc-DNA erbrachte eine hochgradige Identität mit der bereits bekannten Sequenz der 249 Basenpaare langen Repetitionseinheit. Die erste Analyse eines weiteren Phagenklons BRc-2 ergab Hinweise für die Existenz einer Variante der bisher nachgewiesenen Repetitionseinheit.

In Zusammenarbeit mit dem Institut für Zytologie und Genetik der AdW der UdSSR in Novosibirsk wurde die Klonierung der DNA des BRa, der einen bisher nicht analysierten zellspezifischen BR-Typ von *Chironomus* repräsentiert, im Mikromaßstab durchgeführt. Durch *in situ*-Hybridisierung mit polytänen Speicheldrüsenchromosomen konnten mehrere BRa-spezifische Klone nachgewiesen werden, deren Feinstrukturanalyse noch aussteht.

Zur weiteren **Analyse der Samenprotein-Gene der Ackerbohne** *Vicia faba* wurden in enger Zusammenarbeit mit der Abt. Eiweißstoffwechsel 64 der 72 verfügbaren cDNA Klone durch ein modifiziertes Hybrid-Selektionsverfahren identifiziert. Dabei ergaben sich 5 Gruppen von Plasmiden, die Gensequenzen folgender Produkte enthalten: Präprolegumin A (Methionin-haltig) – 6 Klone; Präprolegumin B (Methionin-frei) – 13 Klone; Prävicillin – 17 Klone; 30kD-Protein unbekannter Funktion – 27 Klone; 27 kD-Protein unbekannter Funktion – 1 Klon. Die identifizierten Plasmide wurden und werden durch Restriktions-

und Sequenzanalyse weiter charakterisiert sowie für analytische Arbeiten eingesetzt. Nach umfangreichen Vorarbeiten konnte im Lambdaphagen-Vektor Charon 4a eine Genbank hergestellt werden. Zwei Versuchsserien erbrachten jeweils ca. 1×10^6 rekombinante Phagen, die insgesamt zwei Drittel des *V. faba*-Genoms mit 95 % Wahrscheinlichkeit repräsentieren sollten. Screening von bislang 1×10^6 Phagen mit cDNA-Plasmiden spezifisch für Legumin A, Legumin B, Vicillin und 30kD-Protein führte zur Isolation von 3 Phagen mit DNA-Sequenzen von Legumin B-Genen und einem Phagen mit einer Gen-Sequenz des 30kD-Proteins. Durch elektronenmikroskopische Analyse von R-loops erfolgte der Nachweis, daß VfLe-1 nur etwa 800 bp des 3'-Endes eines Legumin B-Gens enthält. Die weitere Charakterisierung sowie Selektion zusätzlich rekombinanter Phagen mit Samenprotein-Gensequenzen wurde begonnen.

Die Untersuchungen zur **Genetik und Biochemie der Nitratreduktion bei höheren Pflanzen** wurden fortgesetzt. Durch somatische und sexuelle Hybridisierungen konnten die Allelieverhältnisse der bisher vorliegenden nitratreduktase (NR)-defekten Varianten von *Nicotiana tabacum* bestimmt werden. Außerdem wurden die NR-Varianten von *N. tabacum* mit den von Marton et al. (1982) isolierten NR-Varianten von *N. plumbaginifolia* somatisch hybridisiert. Diese interspezifischen Zellhybriden erlaubten, funktionelle Analogien zwischen den NR-spezifischen Genen beider Nicotiana-Arten festzustellen. Es wurde gefunden, daß nicht-analoge NR-Varianten einander auch in interspezifischen Kombinationen vollständig komplementieren. Durch diese Untersuchungen wurden bisher vier funktionell verschiedene NR-spezifische Gene identifiziert, und zwar außer dem Strukturgen des NR-Apoproteins *(nia)* noch drei Gene, die die Funktion des Molybdän-Cofaktors kontrollieren *(cnxA, cnxB, cnxC)*. Diese Gene liegen im haploiden Genom von *N. plumbaginifolia* in je einem Exemplar, bei der amphidiploiden Art *N. tabacum* jedoch (wie im Falle des *nia*-Gens gezeigt wurde) dupliziert vor.

Die drei *cnx*-Genloci konnten biochemisch genauer charakterisiert werden. Dazu mußten neue Methoden entwickelt werden, die es gestatten, zwischen Defekten im Molybdän-Cofaktor und solchen im Molybdän-Processing zu unterscheiden. Es wurde gefunden, daß bei den *cnxA*- und *cnxB*-Mutanten die Strukturkomponente des Cofaktors vorhanden ist, ihre Dimerisierungsfunktion erfüllt und *in vitro* durch Molybdat reaktiviert werden kann. Die bisherigen Ergebnisse deuten darauf hin, daß das *cnxA*-Gen für ein Molybdän-carrier-Protein und das *cnxB*-Gen für eine molybdänspezifische Chelatase codiert. Das *cnxC*-Genprodukt ist dagegen wahrscheinlich an der Synthese der Strukturkomponente des Cofaktors beteiligt. (Analoge Untersuchungen wurden auch bei *cnx*-Typ-Mutanten von *Neurospora crassa* vorgenommen.)

Verschiedene *nia*-Doppelmutanten von *N. tabacum* wurden durch Selektion von Nitratverwertern in Protoplastenkulturen hinsichtlich des Auftretens von spontanen Revertanten analysiert. Bei allen fünf bisher untersuchten Mutanten traten unter den aus Mesophyll-Protoplasten angezogenen Mikrokolonien NR^+-Typen mit einer Häufigkeit von rund 10^{-5} auf. Ein Teil dieser Revertanten-Klone wurde zu fertilen NR^+-Pflanzen regeneriert. Die Spaltungsanalyse der Samennachkommenschaft ergab, daß erwartungsgemäß jeweils nur ein Allel an einem *nia*-Locus revertiert war.

Durch Kreuzung von *nia*-Doppelmutanten mit Wildtyp-Pflanzen von *N. tabacum* konnten Genotypen gewonnen werden, die sich hinsichtlich der Anzahl der aktiven NR-Strukturgene unterscheiden. Diese wurden benutzt, um die Gendosisabhängigkeit der NR-Aktivität zu untersuchen. Die Ergebnisse zeigen, daß in den frühen Keimlingsstadien die NR-Aktivität der Gendosis proportional ist und somit keiner „feedback"-Kontrolle unterliegt. In späteren Entwicklungsstadien gewinnen dagegen Regulationsmechanismen bestimmenden Einfluß und führen zu einer vollständigen Kompensation der Gendosiseffekte. Pflanzen mit nur einem *nia*$^{+}$-Gen wiesen eine genau so hohe NR-Aktivität auf wie Wildtyp-Pflanzen, die vier *nia*$^{+}$-Gene besitzen. Daß Tabakblätter eine unerwartet hohe NR-Synthese-Kapazität besitzen, die den normalerweise zur Aufrechterhaltung des regulierten NR-Niveaus erforderlichen Wert weit überschreitet, konnte auch mit Hilfe von „leaky" *nia*-Mutanten demonstriert werden. Beispielsweise beträgt bei der Mutante *Nia46* die *in vivo*-Aktivität in jungen Keimlingen weniger als 1 %, in Blättern älterer Pflanzen aber rund 60 % des Wildtyp-Wertes.

Im Rahmen der ***in vitro*-Kultivierung pflanzlicher Zellen und Gewebe** konnten die Arbeiten zur Gewinnung eines Verfahrens zur Regeneration von mindestens 10 Pflanzen/cm^2 aus Gewebeexplantaten der Kartoffel abgeschlossen werden. Insgesamt wurden 25 dihaploide Klone, 5 Sorten und 4 diploide Wildformen geprüft. Von 8 dihaploiden Klonen konnten mehr als 10 (z. T. bis zu 30) Pflanzen/cm^2 Explantatoberfläche regeneriert werden, von 4 der 5 Sorten nur wenig mehr als 10, von einer Sorte und von allen 4 Wildformen nur erheblich weniger als 10. Internodien waren als Explantate besser geeignet als Blattstückchen. Die Regeneratpflanzen zeigten nach Topfen in Erde wenig karyotypische Veränderungen, aber es war zu erkennen, daß bei vielen Dihaploiden (erwünschte) Chromosomenverdopplungen stattgefunden haben, worauf auch vorläufige cytologische Untersuchungen hinweisen. Diese Ergebnisse und die Genotypidentität der Regeneratpflanzen werden gegenwärtig im Institut für Kartoffelforschung Groß-Lüsewitz untersucht.

Zur Induktion und Selektion gegenüber *Phytophthora*-Kulturfiltraten resistenter Zellinien und regenerierter Pflanzen aus Protoplastenkulturen der Kartoffel wurden weitere Versuche durchgeführt. Aus Mesophyllprotoplasten zweier dihaploider Genotypen konnten bei Kultivierung in bzw. auf toxinhaltigem Nährmedium bis jetzt 19 Pflanzen regeneriert werden. Ihre Resistenz gegenüber der Toxinpräparation wird gegenwärtig untersucht. Vermutlich handelt es sich nicht bei allen auf dem Toxinmedium überlebenden und regenerierenden Zellkolonien um Mutantenkolonien. Da die Toxinpräparation, wie anzunehmen ist, aus mehreren aktiven Komponenten besteht, ist die Wahrscheinlichkeit, daß die regenerierten Pflanzen ihr gegenüber völlig resistent sind, ziemlich gering.

Die Versuche zur Auffindung einer aus Zellkultur regenerationsfähigen Leguminose ergaben bei *Medicago sativa* und *Trifolium repens* positive Ergebnisse. Bei beiden Objekten gelang bei allen untersuchten Sorten, sowohl direkt aus meristematischen und nicht meristematischen Explantaten als auch aus dem Primärkallus beider Explantatgruppen, die Regeneration fertiler Pflanzen.

Die parallel dazu begonnenen Versuche an Körnerleguminosen führten zu fundierten Aussagen über deren begrenzte Regenerationsfähigkeit. Bei *Pisum sativum* konnten aus meristematischen und aus drei Blattexplantaten wurzellose oder nur spärlich bewurzelte Sprosse gewonnen werden, die zum Teil auch zur Blüte kamen, aber nicht in Erdkultur überführt werden konnten. Bei *Phaseolus vulgaris* entstand aus zwei Apikalmeristemen und aus einem Blattexplantat je eine fertile Regeneratpflanze. Alle anderen Versuchsansätze führten zur Bildung von Kallus, aus dem sich bei *Phaseolus vulgaris* gelegentlich Sproßknospen oder Wurzeln, bei *Pisum sativum* lediglich Wurzeln, differenzierten. Bei *Glycine max* und bei beiden Lupinearten kam es zur Kallusbildung ohne Regeneration.

Im **Mutagenitäts-Testlaboratorium** wurden die Arbeiten zum Nachweis induzierter Mutationen im LDH-X(C_3)Locus von Spermien der Hausmaus mit Hilfe monospezifischer Antikörper weitergeführt. Nachdem bereits im Vorjahr gezeigt werden konnte, daß Äthylnitrosoharnstoff (ENU) Punktmutationen in diesem Gen induziert, wurde im Jahre 1983 die Dosis-Wirkungs-Kurve vollständig erarbeitet und dabei eine nichtlineare Dosis-Wirkungs-Beziehung nachgewiesen.

In Dosisfraktionierungsexperimenten wurde gefunden, daß im Vergleich zu einer akuten Behandlung mit 180 mg/kg ENU eine fraktionierte Behandlung mit 9×20 mg/kg bei 24 Stunden Intervallen eine signifikant niedrigere Frequenz an LDH-X Varianten induziert, die durch eine Verlängerung des Intervalls auf 7 Tage noch einmal deutlich verringert wird.

Das Vorliegen eines experimentellen „no-effect-levels" (unter 60 mg/kg ENU), die deutliche Nichtlinearität der Dosis-Wirkungs-Kurve und die Ergebnisse der Dosisfraktionierungsexperimente sprechen eindeutig für den Ablauf von DNA-Reparaturprozessen in den Spermatogonien der Maus.

Die Induktion von LDH-X-Mutanten durch EMS (100 und 300 mg/kg), Dimethylsulfat (25 und 50 mg/kg) und Methylnitrosoharnstoff (20, 40 und 60 mg/kg) konnte nicht nachgewiesen werden, während Mitomycin C in diesem Test positiv ist.

Diese Ergebnisse stimmen mit den Daten des Spezifischen-Locus-Testes überein und zeigen, daß der LDH-X-Test sensitiv für Mutagene ist, die in den Spermatogonien Mutationen induzieren, während Verbindungen mit postspermatogonialer Spezifität nicht nachgewiesen werden können.

Die Untersuchungen zur transplazentalen Induktion von Chromosomenaberrationen, SCE und Fellfarbmutationen wurden fortgesetzt und zu einem vorläufigen Abschluß gebracht. Dabei konnte die im Vorjahr getroffene Feststellung weiter erhärtet werden, daß der Fellfleckentest (Mammalian-Spot-Test) mit hoher Spezifität induzierte Genmutationen nachweist, während Chromosomenmutationen und Non-Disjunktion als Ursache für die Induktion von Farbflecken ausgeschlossen werden können.

Neben diesen Arbeiten wurden die Routineprüfungen von Produkten der chemischen Industrie und die zytogenetischen Untersuchungen zur Analyse des genetischen Risikos der Exposition gegenüber Chemikalien am Arbeitsplatz an Aıbeitern der chemischen Industrie fortgesetzt.

Die Arbeiten auf dem Gebiet der **Entwicklungsbiologie der Säuger** konzentrierten sich auf die Schwerpunkte (1) Gentransfer bei Säugern, (2) Befruchtungsbiologie und (3) Differenzierung *in vitro* kultivierter Teratokarzinom- und embryonaler Stammzellen. Beim Gentransferprogramm bei Säugetieren konnten ganz wesentliche Fortschritte erzielt werden. Sie betreffen in erster Linie die Etablierung der Mikroinjektionstechnik von DNA in den Pronucleus von Zygoten der Hausmaus, den Transfer der so behandelten Zygoten in den Ovidukt pseudogravider Wirtsmütter und eine sich daran anschließende erfolgreiche ontogenetische Entwicklung. In Kooperation mit dem Zentralinstitut für Molekularbiologie der AdW der DDR wurde ein Plasmid, welches das HSV-TK-Gen trägt, auf diesem Wege in die Maus übertragen und seine Integration in den Zellkern sowie die Transmission in die F_1-Generation nachgewiesen.

Auch die Versuche zum Transfer neuer genetischer Eigenschaften in den Säuger mit Hilfe *in vitro* kultivierter Teratokarzinom- und embryonaler Stammzellen wurden weitergeführt. Dabei konnten aus einer embryonalen Zellinie 9 6-Thioguanin-resistente Klone isoliert werden. Dabei wurden 3 mit einer stark reduzierten HGPRT-Aktivität als Mutanten klassifiziert.

Die mit diesen Klonen durchgeführten histologischen, biochemischen und karyologischen Untersuchungen zeigten jedoch, daß während der Mutantenselektion eine Änderung des Differenzierungsstatus und damit eine Reduktion der Entwicklungspotenz eintritt.

Weitere Arbeiten mit den embryonalen Stammzellinien betreffen daher Versuche zum Gentransfer mit Hilfe der Calzium-Präzipitationstechnik. Die Charakterisierung der bisher isolierten Transformanten steht noch aus.

Obwohl die aus Blastozysten isolierten *in vitro* kultivierten Stammzellen sowohl *in vitro*, als auch nach subkutaner Injektion *in vivo* ein breites Differenzierungsspektrum zeigen, konnte eine Beteiligung dieser Zellen nach Injektion in die Blastozyste an der Embryonalentwicklung und der Bildung von Chimären bisher nicht nachgewiesen werden.

Die Eignung dieser Zellen für die Übertragung genetischer Information in den Säuger ist damit sehr fraglich geworden.

Auf dem Gebiet der Befruchtungsbiologie wurden das Hamsteroozyten-Penetrationssystem als Test zur Analyse von Teilschritten des Befruchtungsprozesses aufgebaut und mit der Isolierung und Aufarbeitung von 17000 ovulierten Oozyten Voraussetzungen für die Untersuchung der Membranreifung der Eizelle geschaffen.

2. Molekularbiologische Grundlagen der pflanzlichen Stoffproduktion

Die Arbeiten über den **Eiweißstoffwechsel in Samen** von Körnerleguminosen konzentrierten sich weiter auf die Strukturanalyse an Globulinen der Ackerbohne (*Vicia faba* L. var. *minor*), auf die innerzellulären Vorgänge von Synthese, Transport und Speicherung der Reserveproteine, sowie auf die Anwendung gentechnischer Methoden zur Aufklärung von Primärstruktur und Regulation der Synthese von Globulinen.

Nach der Entdeckung von methionin-haltigen (Met^+) und methionin-freien (Met^-) Untereinheiten des Legumins wurden in Zusammenarbeit mit dem Zentralinstitut für Molekularbiologie der AdW der DDR die N-terminalen 50 bis 60 Aminosäurereste von rein dargestellten β-Leguminpolypeptidketten sequenziert. Met^+- und Met^--β-Leguminketten unterscheiden sich in 50 Prozent der Aminosäurereste, wobei in homologer Position zum Met einer Met^+-Kette bei Met^--Ketten das Met fehlte. Met^+-β-Legumin der Ackerbohne ist dem in der Literatur beschriebenen β-Legumin der Erbse (*Pisum sativum* L.) hochgradig homolog. Die beiden polymorphen Formen der Met^+-Leguminuntereinheiten (Legumin-Typ A) unterscheiden sich im Met-Gehalt der β-Leguminpolypeptidketten. Untersuchungen über die minor-Komponenten des Legumins, die sich durch α-Leguminketten mit relativen Molekularmassen (M_r) von 55000 bis 60000 auszeichnen, über eine eventuelle züchterisch nutzbare genetische Variabilität im Verhältnis von Met^+- zu Met^--Leguminuntereinheiten sowie zum Verlauf der Synthese der beiden Legumin-Typen während der Samenentwicklung wurden begonnen. Das Vorkommen spezifischer mRNA bzw. gesonderter Gene für die polymorphen Leguminuntereinheiten ist durch die Selektion von 6 Met^+-Legumin-spezifischen und 13 Met^--Legumin-spezifischen Clonen aus einer cDNA-Bank sowie durch die erste Selektion von Met^--Legumin-spezifischen genomischen DNA-Sequenzen der Ackerbohne belegt worden (vgl. diesen Bericht S. 276). Die mit der Maxam-Gilbert-Methode vorgenommene DNA-Sequenzanalyse eines Abschnittes aus einem Met^+-cDNA-Clon betraf den C-terminalen Abschnitt der α-Legumin- und den N-terminalen Abschnitt der zugehörigen β-Leguminkette mit dem dazwischen liegenden Pentapeptid-linker des gemeinsamen precursors und ergab Übereinstimmung mit den Resultaten der Aminosäuresequenzanalyse.

Die detaillierte Untersuchung der Untereinheitenzusammensetzung gelelektrophoretisch trennbarer Vicilinbanden bei verschiedenen Körnerleguminosen führte unter Einbeziehung von Daten aus der Literatur über die Soja-Globuline zu einer einheitlichen Strukturhypothese für die vicilin-artigen 7 S-Reserveglobuline bei Leguminosen. Die Holoproteine stellen sich als Trimere dar, die in unterschiedlichen Verhältnissen aus zwei Typen von Untereinheiten aufgebaut sind. Grundsätzlich kommen folgende Kombinationen dieser beiden Untereinheitentypen (M_r 70000 und 50000) vor: A) 3×70000; B) (2×70000)+(1×50000); C) (1×70000)+(2×50000); D) 3×50000. Während man bei *Pisum sativum* L. alle vier Formen der Vicilin-Holoproteine antrifft, kommen bei *Vicia faba* L. nur die Formen B bis D in merklichen elektrophoretisch nachweisbaren Mengen vor, und bei *Phaseolus vulgaris* L. beschränkt sich das 7 S-Globulin auf die Form D. Auf der Grundlage reiner Präparate dieses Globulins (Phaseolin) konnte nun nach den Quartärstrukturmodellen für 11 S-Globuline erstmals auch ein solches Modell für ein 7 S-Globulin im Zentralinstitut für Molekularbiologie der AdW der DDR durch Röntgenkleinwinkelstreuungsanalyse der Moleküle in Lösung aufgestellt werden. Aus der erwähnten cDNA-Bank wurden 17 Clone für Vicilin 50000 mittels 'hybrid selected translation' im Weizenkeimsystem isoliert.

Die Verfügbarkeit der cDNA-Clone gestattete die spezifische Isolierung von mRNA für Legumin- und Vicilinuntereinheiten. Derartige mRNA wurde in Kombination mit Hundepankreas-SRP (signal recognition particle) bzw. mit diesen SRP und K-RM (potassium washed rough microsomes) aus Hundepankreas[1] zur *in vitro*-Translation im zellfreien Weizenkeimsystem eingesetzt. Während SRP die Translation blockierten, erfolgte in Anwesenheit von SRP und K-RM sowohl Translation als auch Abspaltung der Signalsequenz und vektorielle Sequestierung der Globulinpolypeptidketten in das Lumen der ER-Membranen.

1 Für die Bereitstellung von SRP und K-RM aus Hundepankreas danken wir Dr. T. A. Rapoport, Zentralinstitut für Molekularbiologie, Berlin-Buch, der Akademie der Wissenschaften der DDR.

Dies belegt den sekretorischen Charakter der Reserveproteinbildung und die Funktion tierischer SRP und K-RM mit pflanzlicher mRNA *in vitro*. Die Beteiligung des Golgi-Apparates am innerzellulären Transfer von Reserveproteinpolypeptidketten vom ER zu den Aleuronkörnern wurde mit immunohistochemischen Techniken sowohl für Vicilin als auch für Legumin in Zusammenarbeit mit dem Institut für Biochemie der Pflanzen der AdW der DDR bewiesen (s. auch S. 284 dieses Berichtes). Schließlich gelang es in Kooperation mit dem Carlsberg-Institut, Kopenhagen (Dänemark), durch homo- und heterologe Kombination der dissoziierten Polysomen und Membranen des RER von Gerstenendosperm und *Vicia faba*-Kotyledonen bei der *in vitro*-Translation im zellfreien Weizenkeimsystem zu zeigen, daß korrektes processing der Hordeinpolypeptidketten nur im homologen System zu erfolgen scheint.

Die Ergebnisse der Forschungsarbeiten über den Eiweißstoffwechsel bei der Reserveproteinbiosynthese in pflanzlichen Samen wurden auf dem 3. internationalen Samenproteinsymposium vorgestellt, das unter dem Thema ‚Genetik der Samenproteine' vom 31. 8. bis 2. 9. 1983 in unserem Institut abgehalten wurde. Der Fortschrittsbericht dieses Symposiums ist in diesem Band der ‚Kulturpflanze' zu finden (S. S 1–S 268.)

Im Rahmen **ertragsphysiologischer Untersuchungen** an Getreide wurde an einem Testsortiment von 10 Sommergerstensorten der Temperaturabhängigkeit der Jungpflanzenentwicklung nachgegangen. Moderne Hochzuchtsorten weisen bei niedrigen Temperaturen (4 °C) gegenüber alten Sorten eine beschleunigte Wurzelentwicklung auf. Die Bildung der Wurzelmasse ist positiv korreliert mit den Bestockungsraten dieser Sorten unter Freilandbedingungen.

Bei der Suche nach Frühselektionsverfahren für die Getreidezüchtung wurde geprüft, ob die Selektion nach Anzucht bei erhöhtem CO_2-Gehalt (1500 vpm) in Phytokammern zu positiven Ergebnissen führt. Mehrere aus einem M_2-Ramsch von Sommergerste auf diese Weise selektierte Linien brachten im Freiland einen signifikanten Mehrertrag gegenüber der Ausgangsform. Das Ergebnis bedarf noch einer mehrjährigen und mehrortigen Prüfung.

In Fortführung der Untersuchungen zur Evolution des C_4-pathways der Photosynthese wurden kinetische Messungen an der Phosphoenolpyruvat-Carboxylase durchgeführt. Es wurden art- und photosynthesetypspezifische Unterschiede in den kinetischen Eigenschaften dieses Enzyms nachgewiesen. Während das Enzym aus *Flaveria cronquistii* (C_3) und *F. pubescens* (C_3–C_4) niedrige Km-Werte für PEP (Phosphoenolpyruvat) mit Hillkoeffizienten ~ 1 aufweist, zeigt dieses Enzym aus den C_4-Arten *F. palmeri, F. trinervia* und *F. australasica* relativ geringe PEP-Affinität bei sigmoidaler v/S Charakteristik (v = Reaktionsgeschwindigkeit, S = Substratkonzentration).

Bezüglich der Ribulosebisphosphat-Carboxylase/Oxygenase konnte eine hohe evolutionäre Konstanz der O_2-Bindungseigenschaften des aktiven Zentrums belegt werden. Im Gegensatz dazu sind die CO_2-Bindungseigenschaften dieses Enzyms relativ flexibel und korrelieren mit dem Photosynthesetyp.

Es zeichnet sich heute die Vorstellung ab, daß die bisher bekannten Intermediärformen sich im Laufe der Evolution in unterschiedlichem Maße dem C_4-Typ angenähert haben. Die intermediären *Flaveria*-Arten sind in dieser Hinsicht am weitesten in der Evolution fortgeschritten, d. h., am „C_4-ähnlichsten".

Die in der Abt. Ertragsbildung entwickelten mathematischen Modellvorstellungen tragen diesem schrittweisen Übergang vom C_3- zum C_4-Typ Rechnung und gestatten eine quantitative Abschätzung der Rolle einzelner Komponenten (Enzymaktivitätsmuster, Kompartimentierung) bei der Ausprägung des C_4-Syndroms.

Auf dem Gebiet des **Mutantenstoffwechsels** bei höheren Pflanzen wurde gefunden, daß die Aminosäure Nicotianamin = (2S : 3'S : 3''S)-N-[3-(3-Amino-3-carboxypropylamino)-3-carboxypropyl]-azetidin-2-carboxylsäure, deren Synthese in der Tomatenmutante *chloronerva* blockiert ist, die anomal hohe Eisenaufnahme dieser Mutante auf den Wert der Wildform reguliert, wenn sie in einer Konzentration von 10^{-5} M in der Nährlösung enthalten ist oder auf die jüngsten Blätter aufgepinselt wird. Wahrscheinlich ist das Nicotianamin an einem System beteiligt, das die Aufnahme von Eisen und vielleicht auch von anderen Schwermetallen im Sinne einer negativen Rückkoppelung regelt.

Es besteht eine positive Korrelation zwischen der Konzentration des der Nährlösung zugegebenen bzw. auf die Blätter aufgetragenen Nicotianamins und dem Längenwachstum der Mutantenwurzel. Diese Wirkung ist unabhängig von der Mineralstoffzufuhr. Nicotianamin löst in Zellkulturen des Wildtyps innerhalb von 2 Stunden eine Senkung der Mitoserate um 17 % aus; bei der Mutante wird sie geringfügig erhöht.

In Zusammenarbeit mit dem Institut für Organische Chemie und Biochemie der Tschechoslowakischen Akademie der Wissenschaften wurden früher bereits einmal durchgeführte Untersuchungen über die Verbreitung des Nicotianamins im Pflanzenreich mit neuer Fragestellung und verbesserter Methodik wiederholt. Dabei wurde Nicotianamin erstmals in niederen Pflanzen (in Pilzen, Moosen, Lebermoosen, Flechten) gefunden, teilweise nur in Spuren *(Pellia epiphylla)*, gelegentlich aber auch in erheblichen Konzentrationen *(Polyporus squamosus)*. Negative Resultate ergaben Bakterien, einzellige Algen und Bäckerhefe. Bestätigt wurde der im Gegensatz zu den verwandten Amaryllidaceen und Iridaceen extrem niedrige Nicotianamingehalt in Liliaceen *(Allium cepa)*. Von dieser Ausnahme abgesehen war kein Zusammenhang zwischen der systematischen Stellung der Untersuchungsobjekte und der in weiten Grenzen schwankenden Nicotianaminkonzentration erkennbar, die wahrscheinlich von ökologischen und physiologischen Faktoren erheblich beeinflußt wird.

Nicotianamin-Analysen während verschiedener Entwicklungsstadien von *Lycopersicon esculentum*, *Vicia faba* und *Phaseolus vulgaris* ließen keine Bevorzugung oder Benachteiligung bestimmter Organe erkennen, wenn man vom Rückgang der Nicotianaminkonzentration mit fortschreitendem Blattalter absieht. Die Samen der Körnerleguminosen besitzen im Vergleich zu Tomatensamen einen ca. 10mal höheren Nicotianamingehalt. Ein Zusammenhang mit der Eisenkonzentration war in den untersuchten Organen nicht feststellbar.

Die Durchmusterung des Gaterslebener Kulturpflanzensortiments wurde durch Rohprotein- und Lysinanalysen an 761 Weizen-, 30 Gersten- und 68 Maisformen fortgesetzt.

Im Vordergrund der vielseitigen Anwendung **serologischer Methoden** standen Untersuchungen zur Entwicklung hochempfindlicher Nachweistechniken für Antigene bzw. Antikörper. Immuntransfertechniken zur Identifizierung und

Charakterisierung von Proteinen, von Glycoproteinen bzw. ihrer proteolytischen Spaltprodukte nach gelelektrophoretischer Trennung wurden etabliert, spot-Teste an festen Trägern in Form von RiA (Radioimmunoassay) und EiA (Enzymimmunoassay) zum schnellen Antigen- bzw. Antikörpernachweis erprobt und ihre Spezifität ermittelt.

Die Testempfindlichkeiten von ELISA-Techniken (enzyme-linked immunosorbent assay) im Mikrotiterplattenverfahren wurden unter Verwendung von Hilfsseren und Einsatz unterschiedlicher Enzym-Konjugate vergleichsweise getestet und ein competition-ELISA zur Bestimmung von serologischen Proteinverwandtschaften aufgebaut. Zur Steigerung von Testsensitivität und -spezifität wurde das Antiserensortiment komplettiert, und neben der Proteinmarkierung durch radioaktive Isotope wurden unterschiedliche Konjugationsverfahren zur Enzymkopplung sowie lösliche Immunkomplexe bzw. Lectin-Enzymkomplexe zum Antigen- bzw. Antikörpernachweis eingesetzt.

Die etablierten Techniken und die produzierten Immunreagenzien fanden bei der Bearbeitung folgender Probleme Einsatz:

- Nachweis der Beteiligung der Dictyosomen am Transport der Reserveproteine in reifenden *Vicia faba*-Kotyledonen durch serohistologische Markierung von Ultradünnschnitten und Identifizierung des Organelleninhaltes im Immuntransfer.
- Untersuchung der Heterogenität des Reserveproteins Vicilin durch Nachweis der serologischen Verwandtschaft der Untereinheiten und Bestimmung ihres Glycosilierungsgrades durch Affinitätschromatographie an Lectin-Sorbentien und Immuntransfer.
- Testung der serologischen Verwandtschaft der Viciline verschiedener Leguminosenarten zur Ermittlung eines möglichen Rezipienten für die Transformation durch Reserveproteingene aus *Vicia faba*.
- Testung alloenzymspezifischer Antikörper gegen die GPI-Isoenzyme (Glucose-phosphat-Isomerase-) der Hausmaus durch ELISA.
- Untersuchung zur Funktion und Anordnung der Untereinheiten des ATPase-Komplexes der Thylakoidmembran durch Einsatz von anti-Untereinheiten-Antikörper und Immuntransfer.
- Serologischer Nachweis der Expression des rec-A Proteins in bakteriellen Klonen.
- Untersuchungen zur serologischen Identifizierung der Nitratreduktase durch Immuntransfer und EiA.
- Serologische Identifizierung und Charakterisierung pflanzlicher Hitzeschockproteine durch Immuntransfer.

Die **Elektronenmikroskopie** wurde u. a. zur Darstellung von rekonstituierten Legumin-Molekülen aus biochemisch isolierten Untereinheiten-Paaren im Rahmen der Untersuchungen über Reserveproteine der Ackerbohne eingesetzt. Dabei konnte gezeigt werden, daß die verschiedenen Untereinheiten-Paare trotz erheblicher Unterschiede in der Primärstruktur noch über die charakteristischen Bindungsstellen verfügen müssen, da die rekonstituierten Holoproteinmoleküle in Größe und Form dem Molekül des nativen Legumins entsprechen. Im Fall des Legumins (M_r 360000) lieferte die Markham-Technik zur Darstellung von symmetrischen Untereinheiten-Strukturen keine eindeutigen Ergebnisse mehr.

Liposomen werden in der Gentechnik als Übertragungs-Vehikel von genetischem Material (DNA- oder RNA-Molekülen) verwendet. Sie werden im Labor künstlich aus Lipiden präpariert. Unter Anwendung von rasterelektronenmikroskopischen Techniken und Gefrierätzverfahren gelang die Charakterisierung dieser künstlichen Lipidpartikel, wobei die Rasterelektronenmikroskopie die Darstellung von äußerer Form, Größe und Beschaffenheit, die Gefrierätztechnik eine Untersuchung des inneren Aufbaues aus ein- oder mehrschichtigen Hüllen gestattete.

Zur Weiterentwicklung der Untersuchungstechniken konnte vor allem die Herstellung von Platin/Kohle-Abdrücken bei der Gefrierätztechnik verbessert und eine automatische Regelung des Bedampfungsvorganges mit selbsttätiger Abschaltung nach Erreichen einer vorgegebenen Schichtdicke aufgebaut werden.

Ein wesentlicher Aspekt der Untersuchungen über **Biogenese, Struktur und Funktion des Photosyntheseapparates** war die Aufklärung der Spektralformen des Chlorophylls und deren Bedeutung für die photosynthetische Energieumwandlung. Durch spektroskopische Analyse von Pigmentgemischen, isolierten Chlorophyll-Proteinen, Teilstrukturen der Photosynthesemembranen sowie ergrünenden Pflanzen konnte unter Anwendung der Hauptkomponentenanalyse in Verbindung mit mathematischen Methoden der Spektrenzerlegung nachgewiesen werden, daß sich die Spektren der Antennenchlorophylle *in vivo*, die ca. 99 % des Blattchlorophylls ausmachen, durch zwei verschiedene Pigmentkollektive im Rahmen der Meßgenauigkeit erklären lassen. Die weitergehende Strukturierung der Spektren in Form von Subbanden ist danach als Ergebnis der Pigment-Pigment-Wechselwirkung innerhalb spektral homogener Struktureinheiten zu betrachten und nicht als Ausdruck einer strukturell-funktionellen Inhomogenität.

Bei der weiterführenden Analyse der Chlorophyll a/b-Proteine mono- und dikotyledoner Pflanzen durch verbesserte gelelektrophoretische Methoden konnten 4 verschiedene Apoproteine nachgewiesen werden, die in den Granabereichen der Thylakoidmembranen angereichert und vermutlich mit Photosystem II strukturell und funktionell gekoppelt sind. Die Untersuchungen ergaben außerdem Hinweise für die Existenz von zwei verschiedenen Chlorophyll a/b-Protein Populationen, die selektiv von der Thylakoidmembran abgelöst werden können und deshalb unterschiedlich gebunden oder lokalisiert sein müssen.

Untersuchungen an der ATP Synthetase (CF_1–CF_0) der Chloroplasten mit Untereinheiten-spezifischen Antikörpern führten zu der Schlußfolgerung, daß die Komponenten I und II des H^+-transportierenden Teilkomplexes CF_0 von essentieller Bedeutung für die ATP Synthese sind. Für Untereinheit I, die wenigstens teilweise mit einem Chl a/b-Protein strukturell identisch ist, wird eine Funktion bei der Wechselwirkung mit PS II angenommen. Im Hinblick auf die Funktion des Enzymsektors CF_1 führten die Untersuchungen zu der Hypothese, nach der dieser Protein-Komplex aus zwei Teilstrukturen aufgebaut ist, die weitestgehend unabhängig voneinander für die ATP Synthese und Hydrolyse verantwortlich sind.

Im Rahmen einer erweiterten **Anwendung mathematischer und physikalischer Methoden** wurden für die spektralen Untersuchungen an den Chlorophyllen in Zusammenarbeit mit dem VEB Carl Zeiss Jena Verfahren zur Korrektur von Fluoreszenzspektren im roten und nahen infraroten Bereich unter Verwendung von Laserfarbstoffen entwickelt.

Neben der Erarbeitung verschiedener allgemein nutzbarer und statistischer Programme konzentrierten sich die Bemühungen um erweiterte Anwendung der Rechentechnik auf die Fertigstellung des Programmsystems Sequenzanalyse, das aus etwa 20 Einzelprogrammen besteht.

Die Nutzungsmöglichkeiten der vorhandenen Rechentechnik wurden durch Anschluß eines Graphikdisplays und einer Tastatur an den KRS 4201 sowie einer Lochbandperipherie an den Geräterechner SILEX verbessert. Dies war mit der Bereitstellung von Software für die Bedienung dieser peripheren Geräte verbunden.

Die Kapazität für **chemisch-präparative Arbeiten** wurde wesentlich erweitert. Neben der Reinigung von Chemikalien für spezifische Zwecke standen Naturstoffisolierungen sowie Synthesen bestimmter organischer und anorganischer Verbindungen für den Forschungsprozeß im Mittelpunkt der Arbeiten, die vor allem auf Stoffe für die Zell- und Gewebekultur bei Pflanzen, auf Hilfsstoffe für Polyacrylamidgelelektrophorese-Systeme, für die Szintillationsspektrometrie, für die Aminosäureanalyse und verschiedene spektroskopische Verfahren gerichtet waren.

3. Angewandte Genetik, Taxonomie und Genbank der Kulturpflanzen

Auf dem **Gebiet der angewandten Genetik bei Kulturpflanzen** wurden die Arbeiten zu überwiegend längerfristigen Aufgaben mit verschiedenen speziellen Untersuchungen fortgesetzt.

Im Rahmen der Apomixisforschung bei Futtergräsern wurden Nachkommenschaften aus speziellen Bestäubungsversuchen (freie oder isolierte Abblüte bzw. Kreuzung) mit Art- und Gattungsbastarden bestimmter Kombinationen zur Beurteilung des Fortpflanzungsmodus weiter geprüft. Die Ergebnisse erhärten die früheren Hinweise, daß bei mehreren 84-chromosomigen Nachkommen von drei F_1-Bastarden aus *Festuca gigantea* ($2n=42$) $\times$ *F. arundinacea* ($2n=42$) relativ häufig diploide (z. T. auch haploide) Parthenogenese vorkommt. Die Nachkommen sind allerdings nicht muttergleich und auch untereinander verschieden. In cytologischen und embryologischen Untersuchungen wurde in Übereinstimmung damit festgestellt, daß es bei den Pflanzen dieser Kombination zur Restitutionskernbildung zumindest überwiegend in Meiose II kommt, offenbar gefolgt von parthenogenetischer Entwicklung der Eizelle. Die Frage, inwieweit sich die vorhandene Tendenz zur Parthenogenese mit Restitutionskernbildung in Meiose I verbinden läßt, ist noch offen. – Ein im Zusammenhang mit der Apomixisforschung aufgebautes Sortiment von rund 1700 F_1-Art- und Gattungsbastarden zwischen verschiedenen *Lolium*- und *Festuca*-Arten wurde hinsichtlich einer Reihe von Ertrags- und Leistungsmerkmalen geprüft, ein Teil davon wird zur züchterischen Nutzung abgegeben.

In den Untersuchungen zur Entwicklung genetischer Systeme der Hybridsaatguterzeugung bei Gerste, insbesondere auf der Grundlage chromosomal-genisch bedingter männlicher Sterilität, wurden weitere Fortschritte beim Aufbau und der Prüfung von sog. BTT-Formen (balancierte tertiäre Trisomie) erzielt. Auf breiter Materialbasis, insbesondere mit Hilfe eines großen Translokationssortiments, ist es bisher gelungen BTT-Formen zu entwickeln, die auf insgesamt 20 verschiedene Kombinationen (bei 4 verschiedenen *msg*-Genen) zurückgehen. Die Vitalität der BTT-Pflanzen verschiedener Formen ist außerordentlich variabel, das Spektrum reicht von subletal bis hin zu annähernd voll vital, abhängig vom jeweils beteiligten translozierten Zusatzchromosom. Die cytogenetische Stabilität, d. h. vor allem enge Kopplung zwischen Translokation und *msg*-Gen sowie geringe Pollentransmission des Zusatzchromosoms, ist in der Mehrzahl relativ gut. – Spezielle Mutationsversuche zur Induktion geeigneter, mit *msg*-Genen eng gekoppelter Marker haben einen Stand erreicht, der das Problem der zusätzlichen genetischen Markierung von BTT-Systemen als lösbar erscheinen läßt. Darüber hinaus erfolgten Untersuchungen zu alternativen Möglichkeiten, sowohl mit Hilfe von „male gamete eliminator"-Mutationen als auch mit Hilfe cytoplasmatischer männlicher Sterilität.

In Leistungsprüfungen mit Sojabohnen, u. a. unter Beteiligung an einer internationalen Prüfung, wurden Sorten aus nördlichen Anbauländern, Zuchtstämme, Mutantenlinien und Sortimentsherkünfte hinsichtlich ihrer Anbau- und züchterischen Eignung untersucht. Weitere Mikroleistungsprüfungen bestätigten, daß proteinreiche Linien in der Mehrzahl ertragsunterlegen sind, fettreiche Linien dagegen den Standard meist übertreffen. Die hohen Kornerträge des Jahres 1983, ausgezeichnet durch außerordentlich warme und trockene Witterung während der Vegetationszeit, weisen darauf hin, daß unter hiesigen Bedingungen nicht die Wasserversorgung, sondern die Temperatursumme der begrenzende Faktor für hohe Kornerträge ist. Die eingeleiteten genetischen Untersuchungen zur Temperaturtoleranz erscheinen deshalb besonders aktuell.

Die Bearbeitung der Sortimente von Cucurbitaceen und Mais wurde fortgeführt, einschließlich von Untersuchungen zur Variabilität des Kornproteingehalts bei Mais. Weitergeführt wurden auch Untersuchungen an der Gartenerdbeere im Zusammenhang mit dem offensichtlich genetisch determinierten Schadbild der sog. Frühjahrsbuntheit.

Das im Rahmen der Forschungsvorhaben zur **Taxonomie und Evolution von Kulturpflanzen** im Vorjahr abgeschlossene Manuskript der mehrbändigen Neuauflage des Verzeichnisses landwirtschaftlicher und gärtnerischer Kulturpflanzen der Erde von R. MANSFELD wurde termingemäß dem Akademie-Verlag, Berlin, übergeben, zusätzlich wurde an der Zusammenstellung von Nachträgen gearbeitet. Für die parallel dazu laufend bearbeiteten Bibliographien zur Literatur über Taxonomie und Evolution sowie zur Archäologie von Kulturpflanzen wurden weitere Folgen zusammengestellt.

Neu aufgenommen wurde ein längerfristiges Programm zur Untersuchung taxonomischer und abstammungsgeschichtlicher Probleme in der Gattung *Allium* L., insbesondere ihrer Kulturpflanzen enthaltenden Verwandtschaftskreise. In Fortführung früherer Arbeiten zum Aufbau einer lebenden Spezialkollektion wurde ein etwa 300 Sippen umfassendes Freilandsortiment etabliert. Die Untersuchungen bezogen sich vor allem auf die Ermittlung taxonomisch relevanter Merkmalsstrukturen. Raster-elektronenmikroskopische Ermittlungen der Feinstruktur der Samentesta ergaben in dieser Hinsicht wertvolle Hinweise für Verwandtschaftsbeziehungen; ebenso erwiesen sich Untersuchungen zur Wuchsform, zur Struktur der Speicherorgane, zur Anatomie der Laubblätter (besonders der Leitbündelanordnung) sowie zur Karyologie der Sippen als aussichtsreich. Karyologische Studien wurden vor allem im Verwandtschaftskreis von *A. cepa* L. durchgeführt, wobei verschiedene Karyotyp-Gruppen abgegrenzt werden konnten. Eine chorologisch-ökologische Analyse dieser Gruppe wurde abgeschlossen, wesentlich gefördert durch einen Studienaufenthalt am Botanischen Institut der Akademie der Wissenschaften der Tadžikischen SSR in Dušanbe.

In Weiterführung der numerisch-taxonomischen Arbeiten bei Kulturpflanzen wurde die neuentwickelte Methode zur Ermittlung von Klassifikationsdendrogrammen mit Hilfe wahrscheinlichkeits-theoretischer Informationsanalysen zur Abarbeitung der Programme an die BESM6-Anlage angepaßt, die Berechnungen wurden mit unterschiedlichen Merkmalssätzen durchgeführt. Beim Vergleich der Resultate mit denen auf der Basis von Ähnlichkeitskoeffizienten und Cluster-Analysen ergab die Kombination Gower-Koeffizient/UPGMA-Clusteranalyse eine etwa den Befunden der wahrscheinlichkeits-theoretischen Informationsanalysen gleichwertige Gliederung des Untersuchungsmaterials (*Triticum* und *Aegilops* spp.).

Die taxonomische Bearbeitung der Gattung *Raphanus* L. wurde mit Wuchsform-Analysen, anatomischen Studien der Speicherorgane und Früchte, Herbarstudien und Untersuchungen von Wildformen am Standort fortgesetzt. – Die Aufbewahrung der umfangreichen Archivsammlungen konnte durch den Einbau von Hebelschubanlagen in 3 Sammlungssälen wesentlich verbessert werden. Die Archive wurden durch Material der lebenden Kollektionen und der Sammelreisen ergänzt.

Zur Erfassung pflanzengenetischer Ressourcen und entsprechend einer Kooperationsvereinbarung mit dem Botanischen Institut der Akademie der Wissenschaften der Georgischen SSR in Tbilisi konnte eine weitere Sammelreise in westliche Teile (Abchasien, Megrelien) dieser Unionsrepublik durchgeführt werden. Insgesamt 434 Saat- und Pflanzgutproben wurden dort gesammelt, darunter viele Körnerleguminosen und Gemüse-Sippen (vgl. Bericht in diesem Band). Durch Beteiligung an der diesjährigen Sammelreise nach Süd-Italien und auch von anderen Auslandsaufenthalten wurde weiteres Saatgutmaterial für die Genbank eingebracht. Die botanische Bearbeitung des Materials der Sammelreise in die Slowakische SR 1981 wurde abgeschlossen und die der früheren Reisen in die Georgische SSR fortgeführt.

Für die Sektion Geobotanik und Phytotaxonomie der Biologischen Gesellschaft der DDR wurden die Vortragstagung „Vegetationswandel in der DDR" in Leipzig, 26.–28. 8. 1983, und die Gründungsversammlung der Arbeitsgruppe Morphologie mit Vortragstagung in Halle, 17. 11. 1983, organisiert.

Der Bestand der **Genbank der Kulturpflanzen** nahm im Jahre 1983 um 1888 Muster zu und beträgt 53959 Sippen:

Getreide und Gräser	27533
Hülsenfrüchte	11805
Gemüse	7025
Heil- und Gewürzpflanzen, Industriepflanzen	4875
Mutanten (*Antirrhinum, Glycine, Lycopersicon*)	2721

Im nationalen und internationalen Samentausch wurden 12155 Proben, davon an Einrichtungen in der DDR 2419, abgegeben. Für die Genbank und Forschungsgruppen im Institut gingen 1698 Proben ein.

Die im Jahre 1981 begonnene Sammlung genetischer Ressourcen der Kulturpflanzen in der Sozialistischen Libyschen Arabischen Volksjamahiriya im Rahmen des Mediterranean Programme des International Board for Plant Genetic Resources wurde im April/Mai fortgesetzt. Zwei Mitarbeiter des Zentralinstituts waren daran als Konsultanten beteiligt. Die Sammelgebiete lagen in Tripolitanien und in der Region um Sabha. Es konnten 316 Muster von Landsortenpopulationen, in der Mehrzahl von Weizen, Gersten und Gemüsen, gesammelt werden. Sie stellen ein wertvolles Ausgangsmaterial für die Entwicklung einer künftigen eigenständigen Pflanzenzüchtung in der Volksjamahiriya dar.

Die im Jahre 1980 in Süditalien begonnene Sammlung von Material indigener Kulturpflanzen und ihrer wildwachsenden verwandten Arten wurde im Rahmen der Vereinbarung der Akademie der Wissenschaften der DDR mit dem Consiglio Nazionale delle Ricerche weitergeführt. Die Sammelgebiete der von Mitarbeitern des Istituto del Germoplasma in Bari im Juli organisierten Unternehmungen lagen im südlichen Teil der Provinz Basilicata und in der Provinz Calabria. Diese Sammlungen brachten 276 Muster von Getreide-, Hülsenfrucht- und Gemüse-Landsortenpopulationen. Sie sind eine wertvolle Bereicherung der Kollektionen in den Genbanken in Bari und Gatersleben.

In Österreich erfolgte im Juli eine Sammlung autochthoner Hülsenfrucht-Landsorten gemeinsam mit Dr. R. SCHACHL von der Landwirtschaftlich-chemischen Bundesversuchsanstalt in Linz, und zwar im Rahmen des Abkommens der Akademie der Wissenschaften der DDR und der Österreichischen Akademie der Wissenschaften. Im Burgenland, in der Steiermark und in Kärnten wurden 122 Muster gesammelt, überwiegend von *Phaseolus vulgaris* und *Phaseolus coccineus*. – Erste Berichte über die Sammelreisen werden in diesem Band publiziert.

Die Bearbeitung des Materials der Genbank richtete sich hauptsächlich auf die Vermehrung und Charakterisierung des in den vergangenen Jahren gesammelten Materials von genetischen Kulturpflanzenressourcen sowie auf die Evaluierung von Krankheitsresistenz- und Qualitätseigenschaften (Eiweiß) in den Kollektionen von Getreiden und Gemüsen. Abgeschlossen wurden genbankspezifische Untersuchungen der Kollektionen von Wildgersten (Subgen. *Hordeum*), niederländischen Bohnenlandsorten *(Phaseolus vulgaris)* und Stechapfel *(Datura)* sowie die methodischen Untersuchungen für die Anwendung chemischer Agenzien zur Verlangsamung des Alterungsprozesses und Reduzierung der Chromosomenschäden bei der Lagerung des Saatgutes. – Die Einlagerung von Samen in das Samenkühllagerhaus wurde planmäßig fortgeführt.

Im Rahmen der im Jahre 1982 begonnenen Kooperation mit den Agrarhistorischen Museen in der DDR bei der Sammlung und Erhaltung der genetischen Ressourcen von Obst fand am 30. und 31. Mai eine Arbeitsberatung statt. In ihrem Ergebnis konnte die Zusammenarbeit wesentlich aktiviert werden.

Der Genbank wurde innerhalb des FAO/UNDP/IBPGR European Co-operative Programme Genetic Resources (ECP/GR) die Funktion einer Leiteinrichtung für Gerste übertragen. Dazu fand am 18. und 19. Mai eine Beratung der „Arbeitsgruppe Gerste“ statt, bei der die Richtlinien für die Arbeit bis zum Jahre 1985 festgelegt wurden. Es steht u. a. die Aufgabe, einen Datenspeicher für die in Europa vorhandenen genetischen Gerstenressourcen aufzubauen.

B. Kolloquien; Ausstellungen; Vortragsabende

19. 1. 1983 Prof. Dr. H. SAEDLER, Max-Planck-Institut für Züchtungsforschung, Köln/BRD: Transponierbare Elemente im Genom von *Zea mays*.

25. 1. 1983 Dr. M. METZLAFF, Wissenschaftsbereich Genetik der Martin-Luther-Universität Halle – Wittenberg, Halle: Molekulare Grundlagen der Genom-Plastom-Wechselwirkung in höheren Pflanzen.

1. 3. 1983 Dr. E. SERFLING, Zentralinstitut für Genetik und Kulturpflanzenforschung der AdW der DDR, Gatersleben: Zur Expression der Gene eukaryotischer Organismen. Zugleich Vortrag anläßlich der Verteidigung der Dissertation im Rahmen des Promotionsverfahrens B.

8. 3. 1983 Dr. G. KÜNZEL, Zentralinstitut für Genetik und Kulturpflanzenforschung der AdW der DDR, Gatersleben: Zytogenetische Ergebnisse im Rahmen der Hybridforschung bei Gerste.

28. 3. 1983 Dr. J. CULLUM, Department of Biochemistry, Institute of Science and Technology, University of Manchester, Manchester/Großbritannien: Jumping genes.

29. 3. 1983 Dr. Dr. med. habil. U. EHLING, Institut für Genetik der Gesellschaft für Strahlen- und Umweltforschung, Neuherberg/BRD: Dominante Katarakte bei Mäusen – Modelluntersuchungen für den Menschen.

5. 4. 1983 Prof. Dr. R. B. GOLDBERG, Department of Biology, University of California, Los Angeles, Cal./USA: Organization and expression of developmentally regulated genes in soybean.

13. 4. 1983 P. STEPHAN, Zentralinstitut für Genetik und Kulturpflanzenforschung der AdW der DDR, Gatersleben: Bevölkerungsbiologische Untersuchungen in Ditfurt, einem Dorf im Nordharzvorland, im 17. und 18. Jahrhundert. Zugleich Vortrag anläßlich der Verteidigung der Dissertation im Rahmen des Promotionsverfahrens A.

13. 4. 1983 Dr. J. SCHÖNEICH, Zentralinstitut für Genetik und Kulturpflanzenforschung der AdW der DDR, Gatersleben: Gefährdung des menschlichen Erbgutes durch mutagene Chemikalien. Zugleich Vortrag anläßlich der Verteidigung der Dissertation im Rahmen des Promotionsverfahrens B.

19. 4. 1983 Dr. B. MICHEEL, Zentralinstitut für Krebsforschung der AdW der DDR, Berlin-Buch: Monoklonale Antikörper – Herstellung und Anwendungsmöglichkeiten.

28. 4. 1983 Dr. E. KLOZOVÁ, Institut für experimentelle Botanik der Tschechoslovakischen Akademie der Wissenschaften, Praha/ČSSR: Protein characters in plant taxonomy.

5. 5. 1983 Dr. CH. T. HARMS, Versuchsstation Eschikon des Instituts für Pflanzenbau, Eidgenössische Technische Hochschule Zürich, Eschikon/Schweiz: Komplementationsanalyse aminosäureanalogresistenter Karottenzellen durch Protoplastenfusion.

10. 5. 1983 Dipl.-Biol. J. SCHOTTE, Zentralinstitut für Genetik und Kulturpflanzenforschung der AdW der DDR, Gatersleben: Einfluß exogener Faktoren auf den photosynthetischen Elektronentransport und die Thylakoidmembranstruktur. Zugleich Vortrag anläßlich der Verteidigung der Dissertation im Rahmen des Promotionsverfahrens A.

18. 5. 1983 Dr. J. H. JØRGENSEN, Risø National Laboratory, Roskilde/Dänemark: Barley mildew resistance in Denmark: Epidemiology, resistance breeding, durability of resistance, and new strategies.

24. 5. 1983 Dipl.-Biol. E. HÜTTNER, Zentralinstitut für Genetik und Kulturpflanzenforschung der AdW der DDR, Gatersleben: Zytogenetische Untersuchungen an Lymphozyten des peripheren Blutes zur Analyse des genetischen Risikos beruflich exponierter Chemiearbeiter. Zugleich Vortrag anläßlich der Verteidigung der Dissertation im Rahmen des Promotionsverfahrens A.

26. 5. 1983 Prof. Dr. S. IGALI, National Research Institute for Radiobiology and

Radiohygiene 'Frederic Juliot-Curie', Budapest/Ungarische VR: Mutagenic interactions of ionizing radiation and chemicals.

7. 6. 1983 Dr. K. Adler, Zentralinstitut für Genetik und Kulturpflanzenforschung der AdW der DDR, Gatersleben: Die Elektronenmikroskopie in den Biowissenschaften – Entwicklung und Stand der Untersuchungstechniken.

21. 6. 1983 Doz. Dr. Th. Börner, Wissenschaftsbereich Genetik der Humboldt-Universität zu Berlin, Berlin: Wird Zytoplasmatische männliche Sterilität durch Mitochondriengene verursacht?

28. 6. 1983 Dr. J. C. Wray, Department of Biochemistry and Microbiology, University of Saint Andrews, Saint Andrews/Großbritannien: Structure of nitrate reductase in higher plants.

6. 9. 1983 Dr. V. Römheld, Institut für Pflanzenernährung der Universität Stuttgart-Hohenheim, Stuttgart-Hohenheim/BRD: Mobilisierung, Aufnahme und Verlagerung von Eisen bei unterschiedlich Fe-effizienten Genotypen.

13. 9. 1983 Dr. R. Kulessa: Möglichkeiten der Laborautomatisierung im Zentralinstitut für Genetik und Kulturpflanzenforschung Gatersleben.

20. 9. 1983 Dr. A. Rosenthal, Sektion Chemie der Humboldt-Universität zu Berlin, Berlin: Chemische Synthese von natürlichen und abgewandelten DNA-Fragmenten und ihre Anwendung in der Molekularbiologie.

23. 9. 1983 Dipl.-Math. H. Knüpffer, Zentralinstitut für Genetik und Kulturpflanzenforschung der AdW der DDR, Gatersleben: Nutzung der EDV zur Dokumentation in Genbanken – Aufbau eines Kleinrechner-Datenspeichersystems für die Gaterslebener Genbank der Kulturpflanzen. Zugleich Vortrag anläßlich der Verteidigung der Dissertation im Rahmen des Promotionsverfahrens A.

27. 9. 1983 Dr. P. Apel, Zentralinstitut für Genetik und Kulturpflanzenforschung der AdW der DDR, Gatersleben: Ertragsphysiologische Aspekte der Getreidezüchtung. Zugleich Vortrag anläßlich der Verteidigung der Dissertation im Rahmen des Promotionsverfahrens B.

28. 9. 1983 Dr. A. Czeizel, Abteilung Humangenetik des Staatlichen Instituts für Hygiene, Budapest/Ungarische VR: Self-poisoning as a model for the study on mutagenicity of chemicals in human beings.

11. 10. 1983 Dr. E. Brecht, Zentralinstitut für Genetik und Kulturpflanzenforschung der AdW der DDR, Gatersleben: Anwendung der Fluoreszenzspektroskopie in der biologischen Forschung.

18. 10. 1983 Prof. Dr. O. Stølen, Department of Crop Husbandry and Plant Breeding, Royal Veterinary and Agricultural University, Kopenhagen/Dänemark: Experiments and problems in developing multiline varieties for mildew resistance in barley.

24. 10. 1983 Prof. Dr. F. C. Kafatos, Biological Laboratories, Harvard University, Cambridge, Mass./USA: Molecular evolution of genes and gene families.

28. 10. 1983 Prof. Dr. J.-E. Edström, Europäisches Laboratorium für Molekularbiologie, Heidelberg/BRD: Ultramicrotechniques in gene analysis and gene expression studies.

22. 11. 1983 Dr. K. Becker und Dr. A.-M. Wobus, Zentralinstitut für Genetik und Kulturpflanzenforschung der AdW der DDR, Gatersleben: Genetische Manipulation am Säugerembryo – Möglichkeiten und Ergebnisse.

29. 11. 1983 Prof. Dr. K. Apel, Botanisches Institut und Botanischer Garten der Christian-Albrechts-Universität Kiel, Kiel/BRD: Die lichtabhängige Ergrünung der Gerste.

13. 12. 1983 Dr. G. Scholz, Zentralinstitut für Genetik und Kulturpflanzenforschung der AdW der DDR, Gatersleben: Siderophore bei Mikroorganismen und Pflanzen.

Bereichskolloquien

20. 1. 1983 Dr. H. SOMMER, Max-Planck-Institut für Züchtungsforschung, Köln/BRD: *Tam* 1, ein transponierbares Element am *nivea*-Locus von *Anthirrhinum majus*.

7. 2. 1983 Dr. R. FRITSCH, Dr. K. HAMMER, Dr. P. HANELT und Dr. CHR. O. LEHMANN: Berichte über die Kulturpflanzen-Sammelreisen im Jahr 1982 (Georgische SSR, Libyen, Italien).

8. 2. 1983 Dr. R. BASSÜNER: 'Signal recognition particle' – eine dritte ribosomale Untereinheit?

21. 2. 1983 Dr. G. EITNER, Dr. B. ADLER und Dr. J. HOFEMEISTER: RecA-Gen und Genprodukt von *Proteus mirabilis*.

24. 2. 1983 Dr. K. GRÖBER, Dr. F. MATZK und Dr. M. ZACHARIAS (vorgetragen von Dr. K. GRÖBER): Über die Einbeziehung cytogenetisch und blütenbiologisch geeigneter Kulturpflanzen zur Intensivierung der Apomixisforschung.

22. 3. 1983 Dr. H. BAUWE: Untersuchungen an C_3-C_4-intermediären Arten der Photosynthese.

2. 5. 1983 Prof. Dr. D. LAU, Institut für Getreideforschung der AdL der DDR, Hadmersleben: Stand und Perspektiven der Sommergerstenzüchtung in der DDR.

16. 5. 1983 Dr. N. N. KOLESNIKOV, Institute of Cytology and Genetics, Academy of Sciences of the USSR, Novosibirsk/UdSSR: Fundamental and applied research works of the Institute of Cytology and Genetics of the Academy of Sciences of the USSR in Novosibirsk.

16. 5. 1983 Prof. Dr. A. SCHINZEL, Institut für Medizinische Genetik der Universität Zürich, Zürich/Schweiz: Teratogenese beim Menschen. Klinische, epidemiologische und pharmakogenetische Aspekte.

26. 5. 1983 Dr. S. ASKER, Institut für Genetik der Universität Lund, Lund/Schweden: Untersuchungen zur Fortpflanzungsweise bei *Potentilla* und *Hordeum*.

30. 5. 1983 Dipl.-Biol. C. HÜBNER, Martin-Luther-Universität Halle – Wittenberg, Sektion Biowissenschaften, Wissenschaftsbereich Genetik, Halle: Untersuchungen zur positionsabhängigen und Aberrationstyp-spezifischen Mutagenempfindlichkeit der NOR bei *Vicia faba*.

14. 6. 1983 Dipl.-Biol. S. BRANTL, Martin-Luther-Universität Halle – Wittenberg, Sektion Biowissenschaften, Wissenschaftsbereich Genetik, Halle: Streptokokkenplasmide in *Bacillus subtilis*.

27. 6. 1983 Prof. Dr. K. MÜNTZ: Stand der Erforschung der Gene für Reserveproteine pflanzlicher Samen.

12. 7. 1983 Dr. W. MERBACH, Martin-Luther-Universität Halle – Wittenberg, Sektion Pflanzenproduktion, Wissenschaftsbereich Physiologie und Ernährung der Kulturpflanzen, Halle: Stickstoffumsatz und symbiontische N_2-Fixierung bei Körnerleguminosen.

22. 9. 1983 Dr. R. BRAUN: Vergleichende Analyse des durch alkylierende Agenzien transplazental induzierten Mutationsspektrums in Embryonalzellen der Hausmaus.

26. 9. 1983 Dr. J. B. SCHVARTZMAN, Instituto de Biologia Celular, Madrid/Spanien: Persistence of DNA-damage.

27. 9. 1983 Dr. A. MEISTER: Aufklärung der spektroskopisch unterscheidbaren Pigmentformen.

27. 9. 1983 Dr. M. PEISKER: Modellvorstellungen zur photosynthetischen CO_2-Fixierung und die Evolution des C_4-Weges der Photosynthese.

20. 10. 1983 Prof. Dr. O. STØLEN, Department of Crop Husbandry and Plant Breeding, Royal Veterinary and Agricultural University, Kopenhagen/Dänemark: Agriculture in the nordic countries with special emphasis on Denmark.

24. 10. 1983 Dr. J. PUSTELL, Biological Laboratories, Harvard University, Cambridge, Mass./USA: Analysis of DNA sequence homology by computer: Fact, fantasy, and inference.

23. 11. 1983	Prof. Dr. J. BOGDANOV, Institut für allgemeine Genetik der AdW der UdSSR, Moskva/UdSSR: Advances and prospects in elekctronmicroscopic karyotyping of meiotic cells in plants and animals.
24. 11. 1983	Dr. J. SCHÖNEICH und Dr. E. HÜTTNER (vorgetragen von Dr. J. SCHÖNEICH): Analyse des genetischen Risikos von Arbeitern der β-Naphtal- und Chromat-Produktion des VEB Chemiekombinat Bitterfeld.
24. 11. 1983	Dr. E. WEBER: Kompartimentierung von Biosynthese und Speicherung der Reserveglobuline in Ackerbohnensamen.
29. 11. 1983	Dr. K. GRÖBER: Cytologische und embryologische Untersuchungen zum Fortpflanzungsverhalten ausgewählter Bastarde von Futtergräsern.
29. 11. 1983	Dr. K. HAMMER und Dr. CHR. O. LEHMANN (vorgetragen von Dr. K. HAMMER): Die Bearbeitung ausgewählter Genbank-Kollektionen: Wildgersten, niederländische Bohnen-Landsorten, Stechapfel.
30. 11. 1983	Prof. Dr. M. SELLYEI, Cytogenetic Laboratory of the Pathology Department, Robert Károly Hospital, Budapest/Ungarische VR: Experiences with reverse sister chromatid differentiation.
14. 12. 1983	Dr. C. TITTEL: Behandlungen von Samen mit chemischen Agenzien zur Verlangsamung des Alterungsprozesses.
20. 12. 1983	Prof. Dr. K. MÜNTZ und Dr. CHR. HORSTMANN: Methioninhaltiges und methioninfreies Legumin aus Ackerbohnensamen.

Ausstellungen

15. 6. bis 1. 7. 1983	Ausstellung der Zirkel Keramik und Kindermalen des Zentralinstituts Gatersleben.
16. 9. bis 30. 9. 1983	SABINE LESSIG, Leipzig: Malerei & Grafik.

Vortragsabende

13. 1. 1983	A. HÜNEKE, Halle: Die Farbstürme Emil Noldes.
19. 1. 1983	Ragtime-Programm mit Klavierstücken und Liedern von Scott Joplin: CHRISTA MIHN, Gesang, GERHARD ERBER, Klavier.
17. 2. 1983	Dr. H. KNAPPE, Harzmuseum Wernigerode: Indien – Ein weiter Weg ins Paradies.
16. 3. 1983	„Lieder und Tänze auf die Lauten". Musik in Europa zur Zeit der Reformation und Nachreformation: MONIKA ROST, Renaissance-Laute und Gitarre, JÜRGEN ROST, Gitarre.
13. 4. 1983	Konzert mit dem Nasdala-Quartett: CAROLA NASDALA, I. Violine, ANDREAS SCHULIK, II. Violine, FRANK STRAUCH, Viola, KLEIF CARNARIUS, Violoncello.
21. 4. 1983	Dr. P. HANELT: Eindrücke aus Georgien – Landschaften, Bauten, Geschichte.
17. 5. 1983	Festliches Chorkonzert aus Anlaß des Karl-Marx-Jahres: Chor der Thomas-Müntzer-Oberschule, Aschersleben, und Solisten. Leitung: MARTIN FEHRE.
18. 5. 1983	Kabarett-Abend „Die Taktlosen".
15. 6. 1983	Konzert des Hausmusikkreises des Zentralinstituts Gatersleben.
30. 6. 1983	Dr. F. SCHOLZ und Dr. CHR. O. LEHMANN: Zwischen Donau und Alpen – Eindrücke aus Österreich.
21. 9. 1983	Klavierabend mit WINFRIED APEL.
29. 9. 1983	TH. GROSS, Radebeul: Folklore in Rumänien – Maramureş und Oaş.
13. 10. 1983	CH. WASSER, Berlin: Pablo Picasso – Das Leben eines Malers in Wort, Bild und Musik.
26. 10. 1983	Konzert mit dem Kammertrio Berlin: JOCHEN AMELN, Violoncello, AXEL WILCZOK, Violine, GUNTHER ANGER, Klavier.

17. 11. 1983 Dr. v. GYNZ-REKOWSKI, Wernigerode: Wernigerode und der Harz: Eine Wanderung durch die Stadt und über die Berge.

30. 11. 1983 Lieder, Balladen, Chansons mit und von WERNER BERNREUTHER.

8. 12. 1983 Prof. Dr. J. IRMSCHER, Zentralinstitut für Alte Geschichte und Archäologie der AdW der DDR, Berlin: Der Vesuvausbruch vom Jahre 79 und der Untergang der Städte Pompeji, Herculaneum und Stabiae.

C. Veröffentlichungen aus dem Institut 1983

ADLER, KLAUS and KLAUS MÜNTZ: Origin and development of protein bodies in cotyledons of *Vicia faba*. Proposal for an uniform mechanism. Planta 157 (1983) S. 401–410.

APEL, PETER: Ertragsphysiologische Aspekte der Getreidezüchtung. Diss. B, Akad. Wiss. DDR, Berlin (1983).

– Jungpflanzenentwicklung und Einzelpflanzenertrag bei Sommergerste. Kulturpflanze 31 (1983) S. 167–172.

BASSÜNER, RONALD, ANTJE HUTH, RENATE MANTEUFFEL and TOM A. RAPOPORT: Secretion of plant storage globulin polypeptides by *Xenopus laevis* oocytes. Eur. J. Biochem. 133 (1983) S. 321–326.

–, RENATE MANTEUFFEL, KLAUS MÜNTZ, MANFRED PÜCHEL, PETER SCHMIDT and ERNST WEBER: Analysis of *in vivo* and *in vitro* globulin formation during cotyledon development of field beans (*Vicia faba* L. var. minor). Biochemie u. Physiol. Pflanzen 178 (1983) S. 665–684.

BAUWE, HERMANN: Comparative phylogenetic age of C_3–C_4 intermediate species of *Moricandia* determined by isoelectric focusing and amino acid composition of small subunit of ribulose 1,5-bisphosphate carboxylase/oxygenase. Photosynthetica 17 (1983) S. 442–449.

BERIDZE, R. K., P. CHANEL'T [PETER HANELT] i I. KRUZE [JOACHIM KRUSE]: O naučnom sotrudničestve v izučenii kul'turnoj flory Gruzii. Izv. Akad. Nauk GSSR, Ser. Biol. 9 (1983) Nr. 5, S. 1–3.

–, PETER HANELT and REINHARD FRITSCH: Report of a collecting mission to the Georgian SSR 1982 for the study of indigenous material of cultivated plants. Kulturpflanze 31 (1983) S. 173–184.

BORRISS, RAINER: Bacillen für Medizin und Industrie. Spectrum 14 (1983) Nr. 9, S. 24–25.

BRAUN, ROLF, HANS-PETER JÄKEL and JÖRG SCHÖNEICH: *In vitro* metabolism and mutagenic activity of isoniazid-correlation with hydrazine formation. Mutat. Res. 113 (1983) S. 236.

–, JÖRG SCHÖNEICH, HEINRICH V. MALLING and AFTAB A. ANSARI: Sensitivity of somatic tissues and spermatogonia of the mouse for induction of point mutations by alkylating agents. Abstracts book 13th Annual Meeting of the European Environmental Mutagen Society, Montpellier/Frankreich (1983) S. I–4–2.

–, JÖRG SCHÖNEICH, HEINRICH V. MALLING and AFTAB A. ANSARI: Sensitivity of somatic tissues and spermatogonia of the mouse for induction of point mutations by alkylating agents. Abstracts Symposium "Current problems of environmental mutagenesis", Poznań/VR Polen (1983) S. 32.

–, JÖRG SCHÖNEICH, LUDWIG WEISSFLOG and WOLFGANG DEDEK: Activity of organophosphorus insecticides in bacterial tests for mutagenicity and DNA repair – direct alkylation versus metabolic activation and breakdown. II. 0,0-dimethyl-0-(1,2-dibromo-2,2-dichloroethyl)-phosphate and two O-ether derivatives of trichlorfon. Chem.-Biol. Interact. 43 (1983) S. 361–370.

FALUDI-DÁNIEL, ÁGNES, OSCAR SCHMIDT, ANDRZEJ SZCZEPANIAK and OTTO MACHOLD: Distribution of characteristic membrane proteins in granum and stroma thylakoids. Eur. J. Biochem. 131 (1983) S. 567–570.

GERSTER, TH., J. BANERJI, J. DE VILLIERS, C. A. GEHRING, L. OLSON, D. PICARD, F. WEBER, EDGAR SERFLING and W. SCHAFFNER: Cloned genes are preferentially activated either by transcriptional 'enhancers' or by template DNA replication. Ab-

stracts 15th Annual Meeting of the Union of the Swiss Society for Experimental Biology, Fribourg (1983) S. 38.
GRAFE, REINHARD and ANDREAS J. MÜLLER: Complementation analysis of nitrate reductase deficient mutants of *Nicotiana tabacum* by somatic hybridization. Theoret. & appl. Genetics 66 (1983) S. 127–130.
GRÖBER, KURT und MANFRED ULRICH: Chloroplastenverdünnung als Ursache der Frühjahrsbuntheit bei der Gartenerdbeere (*Fragaria ananassa* DUCH.). Arch. Zücht.-Forsch. 13 (1983) S. 205–216.
– und MARTIN ZACHARIAS: Über Nutzungsmöglichkeiten der apomiktischen Samenausbildung bei ausgewählten Kulturpflanzen. Kulturpflanze 31 (1983) S. 145–163.
HAMMER, KARL und PETER HANELT: Von Alphonse de Candolle zur Genbank. Spectrum 14 (1983) Nr. 3, S. 32.
–, ANNELIESE ROMEIKE und CLAUS TITTEL: Vorarbeiten zur monographischen Darstellung von Wildpflanzensortimenten: *Datura* L., sectiones *Dutra* BERNH., *Ceratocaulis* BERNH. et *Datura*. Kulturpflanze 31 (1983) S. 13–75.
HANELT, PETER, JÜRGEN SCHULTZE-MOTEL and CHARLES E. JARVIS: (715) Proposal to conserve *Triticum aestivum* L. (1753) against *Triticum hybernum* L. (1753) (Gramineae). Taxon 32 (1983) S. 492–498.
HEINDORFF, KONRAD, OTTO AURICH, ARND MICHAELIS and RIGOMAR RIEGER: Genetic toxicology of ethylenediaminetetraacetic acid (EDTA). Mutat. Res. 115 (1983) S. 149–173.
HOFEMEISTER, JÜRGEN, M. ISRAELI-RECHES and D. DUBNAU: Integration of plasmid pE194 at multiple sites on the *Bacillus subtilis* chromosome. Molecular & gen. Genetics 189 (1983) S. 58–68.
HORSTMANN, CHRISTIAN: Specific subunit pairs of legumin from *Vicia faba*. Phytochemistry 22 (1983) S. 1861–1866.
HÜTTNER, EDITH: Zytogenetische Untersuchungen an Lymphozyten des peripheren Blutes zur Analyse des genetischen Risikos beruflich exponierter Chemiearbeiter. Diss. A, Akad. Wiss. DDR, Berlin (1983).
JÄKEL, HANS-PETER, ANNA M. WOBUS, CHRISTINE BLOCH, NATASCHA JÄKEL and JÖRG SCHÖNEICH: A micromethod for the determination of alkaline phosphatase in mammalian cells. Biomed. Biochim. Acta, Berlin 42 (1983) S. 1123–1128.
JUNG, RUDOLF: Chemische Synthese der hypothetischen ancestralen Sequenz Ca-modulierter Eiweiße und analoger langkettiger Peptide mit potentieller Ca-Bindungsfähigkeit. Diss. A, Martin-Luther-Univ. Halle – Wittenberg, Halle (1983).
KAINA, BERND: Studies on adaption of V79 Chinese hamster cells to low doses of methylating agents. Carcinogenesis 4 (1983) S. 1437–1443.
– Cross-resistance studies with V79 Chinese hamster cells adapted to the mutagenic or clastogenic effect of N-methyl-N'-nitro-N-nitrosoguanidine. Mutat. Res. 111 (1983) S. 341–352.
–, KONRAD HEINDORFF and OTTO AURICH: O^6-Methylguanine, but not N7-methylguanine or N3-methyladenine, induces gene mutations, sister-chromatid exchanges and chromosomal aberrations in Chinese hamster cells. Mutat. Res. 108 (1983) S. 279–292.
KNÜPFFER, HELMUT: Nutzung der EDV zur Dokumentation in Genbanken – Aufbau eines Kleinrechner-Datenspeichersystems für die Gaterslebener Genbank der Kulturpflanzen. Diss. A, Akad. Wiss. DDR, Berlin (1983).
– Computer in Genbanken – eine Übersicht. Kulturpflanze 31 (1983) S. 77–143.
KOBLITZ, HELMUT and DOROTHEA KOBLITZ: Tissue and protoplast culture studies in *Lycopersicon esculentum* Miller var. *flammatum* Lehm. cv. 'Bonner Beste' and its mutant *chloronerva*. Plant Cell Rep. 2 (1983) S. 194–197.
–, DOROTHEA KOBLITZ, H.-P. SCHMAUDER and D. GRÖGER: Studies on tissue cultures of the genus *Cinchona* L. *In vitro* mass propagation through meristem-derived plants. Plant Cell Rep. 2 (1983) S. 95–97.
–, DOROTHEA KOBLITZ, H.-P. SCHMAUDER and D. GRÖBER: Studies on tissue cultures of the genus *Cinchona* L. Alkaloid production in cell suspension cultures. Plant Cell Rep. 2 (1983) S. 122–125.
–, DOROTHEA KOBLITZ, H.-P. SCHMAUDER und D. GRÖGER: *In vitro* Verfahren zur Bewurzelung von Pflanzen der Gattung *Cinchona*. DDR-Wirtschaftspatent A 01 N/249 091/7.

LEHMANN, CHRISTIAN O. und KARL HAMMER: Botanische Ergebnisse einer Reise nach Spanien 1978 zur Sammlung kultivierter und wildwachsender Leguminosen. Kulturpflanze 31 (1983) S. 185–206.

– und KARL HAMMER: Bericht über eine Reise in die Sozialistische Libysche Arabische Volksjamahiriya 1982 zur Sammlung autochthoner Sippen von Kulturpflanzen. Kulturpflanze 31 (1983) S. 207–217.

– and KARL HAMMER: Cereal and vegetable collection in Libya. Plant Genetic Resources Newsletter (1983) Nr. 55, S. 26–27.

LINDENHAHN, MONIKA and INGO SCHUBERT: On the origin of hydroxyurea–induced chromatid aberrations in G2 chromosomes with BrdUrd in only one of the sister chromatids. Mutat. Res. 108 (1983) S. 301–316.

MACHOLD, OTTO: Translational sites of thylakoid membrane proteins. Studies based on selective inhibition of cytoplasmic and chloroplast ribosomes by cycloheximide and chloramphenicol. Biochim. et Biophys. Acta 740 (1983) S. 57–63.

MANSFELD, JOHANNA: Verfahren zur Beschleunigung von Diffusionsvorgängen. DDR-Wirtschaftspatent C 12 N/253 352/7.

MANTEUFFEL, RENATE and ERNST WEBER: Fluorografic detection of tritium-labelled proteins in immunoelectropherograms with the water-soluble fluor, sodium salicylate. J. biochem. & biophys. Methods 7 (1983) S. 293–297.

MATZK, FRITZ: Interspecific and intergeneric crosses in grasses. Otdalečena chibridizacija pri rastenijata. Bulg. Akad. Naukite, Sofija (1983) S. 184–192.

MENDEL, RALF R.: Release of molybdenum co-factor from nitrate reductase and xanthine oxidase by heat treatment. Phytochemistry 22 (1983) S. 817–819.

– and ZEREKBAI A. ALIKULOV: Reversible immobilization of molybdenum cofactor on a gel matrix via sulphydryl groups. J. Chromatography 267 (1983) S. 409–413.

– and ANDREAS J. MÜLLER: Molybdenum metabolism is connected to nitrate assimilation in higher plants – Biochemical and genetic aspects. Abstracts 2nd internat. Symposium Mineral Nutrition of Plants, Varna/VR Bulgarien (1983) S. 19–20.

MÜLLER, ANDREAS J.: Genetic analysis of nitrate reductase-deficient tobacco plants regenerated from mutant cells. Evidence for duplicate structural genes. Molecular & gen. Genetics 192 (1983) S. 275–281.

MÜNTZ, KLAUS und UWE SCHLÜTER: Platten-Elektrophorese System. Spectrum 14 (1983) Nr. 3, S. 21.

MOSKALENKO, ANDREJ A., OTTO MACHOLD and YURIN, E. EROCHIN: Lokalisazia polipeptidov is pigment-belkowych-kompleksow *Chromatium minutissimum*. Doklady Akad. Nauk SSSR, Ser. Biochimia 270 (1983) S. 251–254.

NÀTR, LUBOMIR and PETER APEL: The effect of phosphorus and nitrogen deficiency on growth of seedlings of spring barley in dependence on irradiance: growth analysis. Biol. Plantarum 25 (1983) S. 425–432.

–, PETER APEL and SWATAVA FIALOVA: The effect of phosphorus and nitrogen deficiency on growth of seedlings of spring barley in dependence on irradiance: Content of chlorophyll, nitrogen and phosphorus. Biol. Plantarum 25 (1983) S. 433–439.

PEISKER, MARTIN, INGRID TICHÁ, JIRI ČATSKÝ, MIROSLAV KAŠE and HANS-WOLFGANG JANK: Dependence of carbon dioxide compensation concentration on photon fluence rate in french bean leaves and its relation to quantum yield and dark respiration. Photosynthetica 17 (1983) S. 344–351.

PERRINO, PIETRO and KARL HAMMER: Collection of land-races of cultivated plants in South Italy 1982. Kulturpflanze 31 (1983) S. 219–226.

– and KARL HAMMER: Sicilian wheat varieties. Kulturpflanze 31 (1983) 227–279.

– and KARL HAMMER: Collecting in southern Italy. Plant Genetic Resources Newsletter (1983) Nr. 54, S. 26–27.

PISTRICK, KLAUS: Beitrag zur Moosflora der Dübener Heide. Mitt. florist. Kartier., Halle 9 (1983) S. 63–88.

PLIETZ, PAUL, GREGOR DAMASCHUN, JÜRGEN JOACHIM MÜLLER and BERNHARD SCHLESIER: Comparison of the structure of the 7S globulin from *Phaseolus vulgaris* in solution with the crystal structure of 7S globulin from *Canavalia ensiformis* by small angle X-ray scattering. FEBS Letters 162 (1983) S. 43–46.

PLIETZ, PAUL, GREGOR DAMASCHUN, DIETRICH ZIRWER, KLAUS GAST and BERNHARD SCHLESIER: Structure of 7S seed globulin from *Phaseolus vulgaris* L. in solution. Internat. J. biol. Macromolecules 5 (1983) S. 356–360.

POPOVA, OLGA, RUDOLF JUNG and YURI MITIN: Some properties of amino acid and peptide phenacyl esters. In: BLÁHA, K. and P. MALOŇ (Eds.): Peptides 1982. Berlin, New York (1983) S. 137–140.

RIETH, ALFRED: Eine *Spirogyra* von der Ostsee bei Zingst. Kulturpflanze 31 (1983) S. 317–326.

SCHÖNEICH, JÖRG: Gefährdung des menschlichen Erbgutes durch mutagene Chemikalien. Diss. B, Akad. Wiss. DDR, Berlin (1983).

–, KLAUS BECKER and ROLF BRAUN: Investigations on the mutagenic activity of STS 557. Exper. Clin. Endocrinol. 81 (1983) S. 210–216.

– and ROLF BRAUN: Effect of pharmacokinetics on mutagenicity of drugs in mammals. Abstracts Symposium "Current problems of environmental mutagenesis", Poznań/VR Polen (1983) S. 11.

SCHOLZ, GÜNTER: A report on iron uptake by higher plants. Biol. Zbl. 102 (1983) S. 65–75.

– Nicotianamine, an effector of root elongation for the tomato mutant *chloronerva* (*Lycopersicon esculentum* MILL.). Plant Sci. Letters 32 (1983) S. 327–332.

–, RENATE MANTEUFFEL, KLAUS MÜNTZ and ARMIN RUDOLPH: Low-molecular–weight polypeptides of vicilin from *Vicia faba* L. are products of proteolytic breakdown. Eur. J. Biochem. 132 (1983) S. 103–107.

SCHUBERT, INGO, HARALD OHLE and PETER HANELT: Phylogenetic conclusions from Giemsa banding and NOR staining in top onions (Liliaceae). Plant Syst. & Evolut. 143 (1983) S. 245–256.

–, RIGOMAR RIEGER and ARND MICHAELIS: A method for directed production of definite aneuploids of *Vicia faba* L. FABIS Newsletter **7** (1983) S. 13–18.

SCHULTZE-MOTEL, JÜRGEN: Literatur über archäologische Kulturpflanzenreste (1981/1982). Kulturpflanze 31 (1983) S. 281–297.

–, REINHARD FRITSCH, KARL HAMMER, PETER HANELT, JOACHIM KRUSE, HARALD OHLE and KLAUS PISTRICK: Taxonomy and evolution of cultivated plants: Literature review 1981/1982. Kulturpflanze 31 (1983) S. 299–316.

SCHUMANN, URSULA and HELMUT KOBLITZ: Studies on plant recovery from mesophyll protoplasts of *Solanum tuberosum* L. and *Solanum phureja* JUZ. et BUK. Biol. Plantarum 25 (1983) S. 180–186.

SERFLING, EDGAR, LUTZ MEYER, ARMIN RUDOLPH and KERSTIN STEINER: Secretory proteins and Balbiani ring gene activities in salivary glands of *Chironomus thummi* larvae. Chromosoma 88 (1983) S. 16–23.

SIEBER, M., ST. KÖNIG, G. HÜBNER and A. SCHELLENBERGER: A rapid procedure for the preparation of highly purified pyruvate decarboxylase from brewer's yeast. Biomed. Biochim. Acta, Berlin 42 (1983) S. 343–349.

STEPHAN, UDO W.: The influence of nicotianamine, an unusual amino acid, on the iron uptake by cell cultures of *Lycopersicon esculentum*, mutant '*chloronerva*'. Abstracts 2nd internat. Symposium Mineral Nutrition of Plants, Varna/VR Bulgarien (1983) S. 8.

SÜSS, KARL-HEINZ: A new isolation method and properties of the light-harvesting chlorophyll a/b-protein complex of higher plants. Photobiochem. & Photobiophys. 5 (1983) S. 317–324.

– and RENATE MANTEUFFEL: The chloroplast ATP-synthetase consists of the subunits α, β, γ, δ, ε and proteolipid only. Evidence that the components termed I and II are proteolytically altered light–harvesting proteins. FEBS Letters 153 (1983) S. 134–140.

THUST, RUDOLF, JÜRGEN MENDEL, BERND KAINA and ROLF BRAUN: Genetic and cytogenetic effects of 1,3-dimethyl-3-phenyl-1-nitrosourea in *Salmonella typhimurium* and Chinese hamster V79 cells. Carcinogenesis 4 (1983) S. 409–413.

WOBUS, ULRICH: Gentechnik, Neurobiologie und Verhaltensforschung – Molekulare Analyse eines angeborenen Verhaltensmusters bei der Meeresschnecke *Aplysia*. Biol. Rdsch. 21 (1983) S. 309–310.

– Ribozym – eine Ribonukleinsäure mit enzymatischen Eigenschaften. Biochemie Informat. (1983) Nr. 3, S. 114–115.

Nachtrag zum Verzeichnis der Veröffentlichungen aus dem Institut 1982

APEL, PETER und REINHARD FRANK: Korrelationen zwischen Jungpflanzenmerkmalen und Ertragskomponenten bei Sommergerste und ihre Eignung zur Frühselektion. Ber. Arbeitstag. 1982 „Arbeitsgemeinschaft der Saatzuchtleiter", Gumpenstein (1982) S. 61–69.

BOGAJEWSKI, JERZY and JÖRG SCHÖNEICH: W sprawie konieczności badania mutagennych właściwości nowych leków. Pol. Tyg. lék. 37 (1982) Nr. 2, S. 63–67.

BRÄUTIGAM, E., J. SCHIEMANN and R. WOLLGIEHN: Transcription activity of a DNA-protein complex isolated from chloroplasts of *Euglena gracilis*. In: ZELINKA, J. and J. BALAN (Eds.): Metabolism and enzymology of nucleic acids. Bratislava (1982) S. 425–437.

HANELT, PETER: Die Erhaltung der genetischen Variabilität bei Kulturpflanzen. In: SCHLOSSER, S. (Hrsg.): Genressourcen für Forschung und Nutzung. Naturschutzarbeit in den Bezirken Halle und Magdeburg 19 (1982) Beiheft, S. 22–25.

KÜNZEL, GOTTFRIED: Differences between genetic and physical centromere distances in the case of two genes for male sterility in barley. Theoret. & appl. Genetics 64 (1982) S. 25–29.

PERRINO, PIETRO and KARL HAMMER: *Triticum monococcum* L. and *T. dicoccum* SCHÜBLER (Syn. of *T. dicoccon* SCHRANK) are still cultivated in Italy. Genetica Agraria 36 (1982) S. 343–352.

POPOVA, OLGA JU., RUDOLF JUNG and JURI V. MITIN: Issledovanie fenacilovych ėfirov aminokislot i peptidov. Ž. org. chimii 18 (1982) S. 1716–1720.

SCHÖNEICH, JÖRG, RADIM J. ŠRÁM and JERZY BOGAJEWSKI: Recommendations for the mutagenicity testing of new pharmaceuticals. Pol. J. Pharmacol. & Pharmacy 34 (1982) S. 201–205.

SCHOLZ, FRIEDRICH and GOTTFRIED KÜNZEL: Progress and problems with hybrid barley. Barley Genetics IV, Proc. 4th internat. Barley Genetics Symposium, Edinburgh (1981) S. 758–765.

SÜSS, KARL-HEINZ: Evidence for the presence of phosphorylated electron transport proteins and the absence of phosphoryl-enzyme intermediates in the ATPase complex (CF_1–CF_0) of chloroplasts. Proc. 2nd internat. Youth Symposium "Plant Metabolism Regulation", Sofia/VR Bulgarien (1982) S. 243–246.

D. Vorträge von Mitarbeitern bei wissenschaftlichen Veranstaltungen anderer Institutionen

Dr. ADLER, BARBARA: Reaktionen der Bakterienzelle auf DNA-Schädigung: Die SOS-Antwort. – Arbeitstagung des Fachverbands Physiologie und Biochemie der Mikroorganismen der Gesellschaft für Allgemeine und Technische Mikrobiologie der Biologischen Gesellschaft der DDR „Strategien der Mikroorganismen zur Auseinandersetzung mit Streßfaktoren der zellulären Umwelt", Kloster/Hiddensee, 13. bis 17. September 1983 (15. September 1983).

Dr. APEL, PETER: Kritische Entwicklungsstadien und Stoffwechselprozesse bei der Ertragsbildung von Getreide. – Kolloquium am Institut für Züchtungsforschung der Akademie der Landwirtschaftswissenschaften der DDR, Quedlinburg (26. Mai 1983).

– Möglichkeiten einer gezielten Ertragszüchtung aus ertragsphysiologischer Sicht. – Tagung der Züchtergemeinschaft Sommergerste der DDR, Biesdorf (22. Juni 1983).

– Photosynthetic properties in the genus *Moricandia*. – International Symposium "Kinetics of photosynthetic carbon metabolism in C_3-plants", Tallinn/UdSSR, 10. bis 13. August 1983 (11. August 1983).

Dr. BÄUMLEIN, HELMUT: Gentechnische Analyse von Samenproteingenen von *Vicia faba* als Voraussetzung für ihre gezielte Modifikation. – Weiterbildungsveranstaltung der

Sektion Genetik der Biologischen Gesellschaft der DDR und des Instituts für Getreideforschung der Akademie der Landwirtschaftswissenschaften der DDR, Bernburg, 2. bis 3. März 1983 (3. März 1983).

– Identification of seed protein specific cDNA clones and selection of recombinant phages from a genomic library. – Koordinierungsberatung und wissenschaftliches Symposium zum Thema IV–1.3 „Regulation of genom expression in plants", Moskva/UdSSR, 12. bis 17. September 1983 (14. September 1983).

– Konstruktion von Genbanken. – Kolloquium am Lehrstuhl für Virologie der Humboldt-Universität zu Berlin, Berlin (24. November 1983).

–, Dr. Ronald Bassüner, Dipl.-Chem. Ute Heim, Dr. Rudolf Jung, Prof. Dr. Klaus Müntz, Dr. Reinhard Panitz, Dr. Winfriede Weschke and Dr. Ulrich Wobus (vorgetragen von Dr. H. Bäumlein): Structure and function of seed protein genes. – 2nd Symposium of Young Scientists in Physiology and Biochemistry of Plants, Reinhardsbrunn, 17. bis 20. Oktober 1983 (17. Oktober 1983).

Dr. Bauwe, Hermann: Different biochemical bases of C_3–C_4 intermediate photosynthesis in genera *Flaveria* and *Moricandia*. – International Symposium "Kinetics of photosynthetic carbon metabolism in C_3-plants", Tallinn/UdSSR, 10. bis 13. August 1983 (10. August 1983).

– Evolution and carbon metabolism of C_3–C_4 intermediate photosynthetic species. – Kolloquium of the A. R. C. Research Group on Photosynthesis, University of Sheffield, Sheffield/Großbritannien (19. September 1983).

Dr. Borriss, Rainer: β-Glucanase-Mutanten von *B. subtilis*. – 1. Gentechnik-Symposium DDR – UdSSR „Mechanismen der Genexpression", Minsk/UdSSR, 19. bis 24. Juni 1983 (22. Juni 1983).

– Klonierung von β-Glucanase aus *Bacillus amyloliquefaciens*. – Poster anläßlich der 6. Vortragstagung und Hauptversammlung der Gesellschaft für Allgemeine und Technische Mikrobiologie der Biologischen Gesellschaft der DDR, Reinhardsbrunn, 9. bis 13. Oktober 1983 (12. Oktober 1983).

–, Dr. J. Zemek and Dr. L. Kuniak: Affinity of beta-1,3-1,4-glucanase to crosslinked lichenan. – Poster presented at the 3rd Symposium of Socialist Countries on Biotechnology, Bratislava/ČSSR, 25. bis 29. April 1983.

Dr. Braun, Rolf: Aussagefähigkeit mikrobieller Mutagenitätsteste. – Weiterbildungskurs des VEB Pharmazeutisches Kombinat GERMED „Moderne Methoden in der Toxikologie", Unterweißbach, 17. bis 21. Januar 1983 (21. Januar 1983).

– Verfahren zum Nachweis mutagener Wirkungen toxischer Stoffe. – Lehrgang des ZAM Berlin „Industrietoxikologie II", Berlin, 16. bis 20. Mai 1983 (16. Mai 1983).

–, Dr. Jörg Schöneich, Dr. Heinrich V. Malling and Dr. Aftab A. Ansari (vorgetragen von Dr. R. Braun): Sensitivity of somatic tissues and spermatogonia of the mouse for induction of point mutations by alkylating agents. – Symposium of the Institute of Human Genetics, Polish Academy of Sciences, "Current problems of environmental mutagenesis", Poznań/VR Polen, 8. bis 11. November 1983 (10. November 1983).

Dr. Brecht, Eberhard: Stationäre Fluoreszenz photosynthetischer Objekte. – 2. Frühjahrsschule „Optische Spektroskopie" der Sektion Chemie der Friedrich-Schiller-Universität Jena „Fluoreszenzspektroskopie – Methodische Entwicklungen und Anwendungen in Chemie, Biologie und Medizin", Georgenthal, 17. bis 19. März 1983 (18. März 1983).

–, Dr. Klaus Adler und Prof. Dr. Agnes Faludi-Daniel: Spektrale Eigenschaften des mit Galaktolipiden rekombinierten Light-Harvesting-Complex (LHC) aus *Vicia faba*. – Poster anläßlich des III. Symposiums „Biophysik pflanzlicher Systeme" der Sektion Biophysik der Gesellschaft für Physikalische und Mathematische Biologie der DDR, Reinhardsbrunn, 6. bis 11. März 1983 (7. März 1983).

–, Dr. Klaus Adler and Prof. Dr. Agnes Faludi-Daniel: Spectral properties of the light-harvesting chlorophyll a/b-protein complex from *Vicia faba* reconstituted with galactolipids. – Poster presented at the RGW-Symposium "Biological conversion of solar energy", Visegrad/Ungarische VR, 3. bis 8. Oktober 1983 (5. Oktober 1983).

Dr. El-Tarras, A., Dr. Rolf Braun, Dr. E. Stenz und Dr. G. Schuster: Prüfung antiphytoviraler Verbindungen in Bakterien-Mutationstests vor und nach Biokonversion durch Fraktionen an pflanzlichen oder tierischen Zellhomogenaten. – Poster anläßlich der 6. Vortragstagung und Hauptversammlung der Gesellschaft für Allgemeine und Technische Mikrobiologie der Biologischen Gesellschaft der DDR, Reinhardsbrunn, 9. bis 13. Oktober 1983.

Dr. FAUST, JÜRGEN, Dr. ALFRED PREISS, Dr. HELMUT RIPPERGER, Dr. GÜNTER SCHOLZ und Prof. Dr. KLAUS SCHREIBER: Synthesen von Strukturanaloga des Phytosiderophors Nicotianamin. – Poster anläßlich der Chemiedozenten-Tagung der Chemischen Gesellschaft der DDR, Jena 6. bis 8. April 1983.

–, Dr. ALFRED PREISS, Dr. HELMUT RIPPERGER, Dr. GÜNTER SCHOLZ and Prof. Dr. KLAUS SCHREIBER: Synthesis of structure analogues of the phytosiderophore nicotianamine. – Poster presented at the 2nd International Conference on Chemistry and Biotechnology of Biologically Active Natural Products, Budapest/Ungarische VR, 15. bis 19. August 1983.

Dipl.-Biol. FRANK, PETRA und Dr. PETER APEL (vorgetragen von Dr. P. APEL): Vergleichende Untersuchungen des Ontogeneseverlaufs von Sommergerstensorten mit unterschiedlichem Ertragsniveau. – Wissenschaftliche Vortragstagung des Instituts für Getreideforschung der Akademie der Landwirtschaftswissenschaften der DDR „Steigerung und Stabilisierung der Getreideerträge mit erhöhter Effektivität", Bernburg, 22. bis 23. Februar 1983 (23. Februar 1983).

Dr. FRANK, REINHARD and Dipl.-Biol. PETRA FRANK (vorgetragen von Dr. R. FRANK): Pre-anthesis distribution of assimilates in relation to ear development of spring barley. – 2nd Symposium of Young Scientists in Physiology and Biochemistry of Plants, Reinhardsbrunn, 17. bis 20. Oktober 1983 (18. Oktober 1983).

Dr. GRAFE, REINHARD: Nitrate reductase deficient mutants in higher plants. – Seminar am INRA Institut Physiologie Végétale, Montpellier/Frankreich (14. Februar 1983).

– and Dr. ANDREAS J. MÜLLER (vorgetragen von Dr. R. GRAFE): Genetic characterization of nitrate reductase deficient mutants in *N. tabacum*. – Seminar am INRA Institut Physiologie Végétale, Versailles/Frankreich (17. Januar 1983).

Dipl.-Ing. agr. GREBENŠČIKOV, IGOR: Bemerkungen zur Verbreitung und Systematik der laparosticten Scarabaeoidea (Coleoptera), die aus der Mongolischen Volksrepublik bekannt sind. – Symposium der Sektion Biowissenschaften der Martin-Luther-Universität Halle – Wittenberg und der Biologischen Gesellschaft der DDR „Erforschung biologischer Ressourcen der Mongolischen Volksrepublik", Halle, 29. August bis 2. September 1983 (1. September 1983).

Dr. GRÖBER, KURT: Entwicklung und Nutzung der apomiktischen Samenausbildung bei Kulturpflanzen. – Beratung der Arbeitsgruppe Grundlagen der Züchtungsforschung, Sektion Züchtungsforschung bei Pflanzen, Akademie der Landwirtschaftswissenschaften der DDR, Berlin (27. April 1983).

Dr. HAMMER, KARL: Italien – Bilder einer botanischen Sammelreise. – Kulturbund der DDR, Quedlinburg (8. Februar 1983).

– und Dr. CHRISTIAN O. LEHMANN (vorgetragen von Dr. K. HAMMER): Sammlung und Erhaltung genetischer Kulturpflanzenressourcen im Zentralinstitut für Genetik und Kulturpflanzenforschung der Akademie der Wissenschaften der DDR. – 15. Arbeitstagung Botanischer Gärten der DDR, Dresden und Tharandt, 8. bis 11. Juni 1983 (9. Juni 1983).

Dipl.-Landwirt HELLMUND, DETLEF: Investigations on the transcription of *Xenopus* 5S RNA genes after injection into *Xenopus* oocyte nuclei. – Seminar am Institute of Biochemistry of the Biological Research Center, Szeged/Ungarische VR (25. März 1983).

Dr. HOFEMEISTER, JÜRGEN: Plasmidintegration und Insertionsmutagenese bei *Bacillus subtilis*. – 1. Gentechnik-Symposium DDR – UdSSR „Mechanismen der Genexpression", Minsk/UdSSR, 19. bis 24. Juni 1983 (22. Juni 1983).

– *Bacillus subtilis* in der Gentechnik. – 1. Gentechnik-Symposium DDR – UdSSR „Mechanismen der Genexpression", Minsk/UdSSR, 19. bis 24. Juni 1983 (22. Juni 1983).

– Plasmid pE194-Integration im Chromosom von *Bacillus subtilis*. – Poster anläßlich der 6. Vortragstagung und Hauptversammlung der Gesellschaft für Allgemeine und Technische Mikrobiologie der Biologischen Gesellschaft der DDR, Reinhardsbrunn, 9. bis 13. Oktober 1983 (12. Oktober 1983).

Dr. GERHARD HÜBNER und Dr. JÖRG BÄR: Kinetik der Produkt-Aktivierung der Phosphofruktokinase. – Poster anläßlich der Gemeinschaftstagung der Biochemischen Gesellschaften der UdSSR und der DDR, Leipzig, 29. Mai bis 3. Juni 1983.

– und Dr. JÖRG BÄR: Funktion und Reaktivität der SH-Gruppen der Phosphofruktokinase. – Poster anläßlich der Jahrestagung der Biochemischen Gesellschaft der DDR, Gera, 7. bis 9. November 1983.

Dr. JÄKEL, HANS-PETER: Pränataldiagnostik der Mukoviszidose – Gegenwärtiger Stand, Probleme, Sicherheit. – Tagung der Forschungsrichtung „Genetische Defekte" der HFR M 30 in Zusammenarbeit mit der Gesellschaft für Humangenetik der DDR, Neubrandenburg, 14. bis 16. November 1983 (14. November 1983).

Dipl.-Biol. JUNG, RUDOLF: Chemische Synthese der hypothetischen ancestralen Sequenz Ca-modulierter Eiweiße und analoger langkettiger Peptide mit potentieller Ca-Bindungsfähigkeit. – Promotion A, Martin-Luther-Universität Halle – Wittenberg, Sektion Biowissenschaften, Vortrag anläßlich der Verteidigung, Halle (26. April 1983).

Dipl.-Math. KNÜPFFER, HELMUT: Kleinrechner-Datenspeicher für die Genbank am Zentralinstitut für Genetik und Kulturpflanzenforschung Gatersleben. – Vortragstagung der Kommission Datenverarbeitung des Forschungszentrums für Molekularbiologie und Medizin der Akademie der Wissenschaften der DDR, Berlin-Buch (12. Oktober 1983).

Prof. Dr. KOBLITZ, HELMUT: Ursachen und Mechanismen der Habituierung in pflanzlichen Zell- und Gewebekulturen. – 2. Arbeitstagung der Arbeitsgemeinschaft Pflanzliche Zell- und Gewebekultur der Biologischen Gesellschaft der DDR, Elbingerode, 2. bis 4. Mai 1983 (3. Mai 1983).

Dr. KÜNZEL, GOTTFRIED: Stand und Entwicklungstendenzen auf dem Gebiet der Cytogenetik. – Beratung der Arbeitsgruppe Grundlagen der Züchtungsforschung, Sektion Züchtungsforschung bei Pflanzen, Akademie der Landwirtschaftswissenschaften der DDR, Berlin (27. April 1983).

Dr. LEHMANN, CHRISTIAN O.: Die genetischen Ressourcen der Kulturpflanzen und die Genbank im Zentralinstitut für Genetik und Kulturpflanzenforschung. – Weiterbildungslehrgang der VVB Saat- und Pflanzgut, Neugattersleben (16. November 1983).

Prof. Dr. MACHOLD, OTTO: Aufbau des Photosyntheseapparates unter besonderer Berücksichtigung der Pigmentsysteme. – Problemseminar der Sektion Chemie der Friedrich-Schiller-Universität Jena „Photoprozesse in Natur und Technik", Jena (22. März 1983).

– Chlorophyll a/b-proteins in their relation to the light-harvesting complex. – 6th International Congress on Photosynthesis, Brüssel/Belgien, 1. bis 6. August 1983 (1. August 1983).

– The present state of chlorophyll-protein research under special consideration of PS II chlorophyll a/b-proteins. – RGW-Symposium "Biological conversion of solar energy", Visegrad/Ungarische VR, 3. bis 8. Oktober 1983 (5. Oktober 1983).

– Chlorophyll a/b-proteins and light-harvesting complex characteristics of different plant species. – Poster presented at the Workshop of the Institut für Molekularbiologie und Biophysik der Eidgenössischen Technischen Hochschule Zürich „Molecular structure and function of light-harvesting pigment-protein complexes and reaction centers", Zürich/Schweiz, 25. bis 28. Juli 1983.

Dr. MEISTER, ARMIN: Nachweis der spektralen Chlorophyllformen *in vivo*. – III. Symposium „Biophysik pflanzlicher Systeme" der Sektion Biophysik der Gesellschaft für Physikalische und Mathematische Biologie der DDR, Reinhardsbrunn, 6. bis 11. März 1983 (7. März 1983).

– and Dr. EBERHARD BRECHT: Principal component analysis of fluorescence spectra from chloroplast and LHC. – Poster presented at the RGW-Symposium "Biological conversion of solar energy", Visegrad/Ungarische VR, 3. bis 8. Oktober 1983 (5. Oktober 1983).

Dr. MENDEL, RALF R. and ZEREKBAI A. ALIKULOV (vorgetragen von Dr. R. R. MENDEL): At least four genes are involved in molybdenum cofactor biosynthesis and nitrate reduction in higher plants. – 4th International Symposium on Genetic of Nitrogen Fixation, Varna/VR Bulgarien, 9. bis 13. Mai 1983 (11. Mai 1983).

– and Dr. ANDREAS J. MÜLLER (vorgetragen von Dr. R. R. MENDEL): Molybdenum metabolism is connected to nitrate assimilation in higher plants – Biochemical and genetic aspects. – 2nd International Symposium on Mineral Nutrition of Plants, Varna/VR Bulgarien, 18. bis 24. September 1983 (20. September 1983).

Dr. MENDEL, RALF R. and Dr. ANDREAS J. MÜLLER (vorgetragen von Dr. R. R. MENDEL): At least four genes are involved in nitrate reduction in higher plants – Biochemical and genetic aspects. – 3rd International Youth Symposium "Plant Metabolism Regulation", Varna/VR Bulgarien, 24. bis 29. September 1983 (26. September 1983).
– and Dr. ANDREAS J. MÜLLER: At least four genes are involved in nitrate reduction in a higher plant. – Poster presented at the International Symposium on Genetic Manipulation of Plants and its Application to Agriculture, University of London, London/Großbritannien, 19. bis 20. Dezember 1983 (19. Dezember 1983).
Dipl.-Biol. MEYER, LUTZ: Charakterisierung larvaler Sekretproteine von *Chironomus thummi* zur Identifizierung der Balbianiring-Genprodukte. – Tag des Jungen Wissenschaftlers, Forschungszentrum für Molekularbiologie und Medizin der Akademie der Wissenschaften der DDR, Berlin (16. Dezember 1983).
Dr. MÜLLER, ANDREAS J.: Utilization of cell culture methods for studying the genetic control of nitrate assimilation in higher plants. – Kolloquium an der University of California, Davis/USA (6. Oktober 1983).
– Genetic regulation of nitrate assimilation in tobacco. – Kolloquium an der University of California, Berkeley/USA (14. Oktober 1983).
– Selection and characterization of nitrate reductase mutants in higher plants. – Kolloquium an der Cornell University, Ithaca/USA (8. November 1983).
Prof. Dr. MÜNTZ, KLAUS: Struktur und Biosynthese pflanzlicher Reserveproteine. – 5. Reinhardsbrunner Weiterbildungsveranstaltung der Biochemischen Gesellschaft der DDR, Reinhardsbrunn, 3. bis 7. Januar 1983 (5. Januar 1983).
– Investigación de las proteinas de reserva y la ingenería genetica. – Kolloquium am Nationalen Forschungszentrum (CNIC) der Republik Kuba, La Habana (3. März 1983).
– Investigación de las proteinas de reserva y la ingenería genetica. – Kolloquium am Institut für die Erforschung der Grundlagen der tropischen Landwirtschaft (INIFAT) der Akademie der Wissenschaften der Republik Kuba, Santiago de las Vegas (1. März 1983).
– Hat die genetische Manipulation von pflanzlichen Reserveproteinen eine Zukunft? – Sitzung der Klasse Biowissenschaften der Akademie der Wissenschaften der DDR, Berlin (12. Mai 1983).
– Storage protein gene expression in legumes. – Koordinierungsberatung und wissenschaftliches Symposium zum Thema IV-1.3 "Regulation of genom expression in plants", Moskva/UdSSR, 12. bis 17. September 1983 (14. September 1983).
– Möglichkeiten zur Erhöhung und Verbesserung der Proteinzusammensetzung bei großkörnigen Leguminosen. – ERWIN-BAUR-Symposium der Akademie der Landwirtschaftswissenschaften der DDR, Eberswalde, 6. bis 7. Dezember 1983 (6. Dezember 1983).
– Reserveglobuline und Gentechnik. – Kolloquium am Institut für Biochemie der Pflanzen der Akademie der Wissenschaften der DDR, Halle (6. Oktober 1983).
– und Dipl.-Chem. BÄRBEL OSTERMANN: Enzymkomplementation als Heterosisfrühtest an Mais und Roggen. – Kolloquium am Institut für Getreideforschung der Akademie der Landwirtschaftswissenschaften der DDR, Bernburg (28. September 1983).
Dr. NIEDEN, UTA ZUR, Dr. DIETER NEUMANN und Dr. ERNST WEBER (vorgetragen von Dr. D. NEUMANN): Doppelmarkierung mit der Protein A-Gold-Technik an pflanzlichen Objekten. – 12. Arbeitstagung „Markierte Antikörper" des Staatlichen Instituts für Immunpräparate und Nährmedien Berlin, Berlin (26. Mai 1983).
Dr. PEISKER, MARTIN: CO_2-Kompensationskonzentration und Kohlenstoff-Isotopendiskriminierung bei C_3-C_4-Intermediärformen. – III. Symposium „Biophysik pflanzlicher Systeme" der Sektion Biophysik der Gesellschaft für Physikalische und Mathematische Biologie der DDR, Reinhardsbrunn, 6. bis 11. März 1983 (8. März 1983).
– Modelle des Kohlenstoffmetabolismus bei C_3-C_4-Intermediärformen. – Poster anläßlich des III. Symposiums „Biophysik pflanzlicher Systeme" der Sektion Biophysik der Gesellschaft für Physikalische und Mathematische Biologie der DDR, Reinhardsbrunn, 6. bis 11. März 1983.
– Models of CO_2 compensation concentration and carbon isotope discrimination in C_3–C_4 intermediate species. – International Symposium "Kinetics of photosynthetic carbon metabolism in C_3-plants", Tallinn/UdSSR, 10. bis 13. August 1983 (12. August 1983).

Dr. Peisker, Martin, Dr. Ingrid Tichá and Dr. Peter Apel (vorgetragen von Dr. P. Apel): CO_2 compensation concentration and quantum yield in C_3 and C_3-C_4 intermediate species of *Moricandia*. – Internationales Symposium „Experimentelle Grundlagen für die Entwicklung von Modellen der Photosynthese", Praha/ČSSR, 5. bis 9. September 1983 (7. September 1983).

Prof. Dr. Rieger, Rigomar: *Vicia faba* as a cytogenetical model. – International *Vicia faba* Cytogenetics Review Meeting, Centre for European Agricultural Studies, Wye/Großbritannien, 7. bis 9. Februar 1983 (8. Februar 1983).

– Inducible DNA repair in higher plants. – Kolloquium am John Innes Institute, Norwich/Großbritannien (11. Februar 1983).

– Reconstructed karyotypes as a tool in mutation research. – Kolloquium am Queen Mary College, School of Biology, University of London, London/Großbritannien (19. Februar 1983).

Dr. Schiemann, Joachim: Frontiers in the application of genetic engineering methods to plant systems. – 2nd Symposium of Young Scientists in Physiology and Biochemistry of Plants, Reinhardsbrunn, 17. bis 20. Oktober 1983 (17. Oktober 1983).

– Attempts to isolate NR genes from eukaryotes. – Seminar der Gruppe Kondorosi, Biologisches Forschungszentrum der Ungarischen Akademie der Wissenschaften, Szeged/Ungarische VR (17. Oktober 1983).

–, Dr. Ralf R. Mendel and Dr. Andreas J. Müller (vorgetragen von Dr. J. Schiemann): Nitrate reductase deficient mutants of *N. tabacum*. – Kolloquium am Biologischen Institut der Medizinischen Universität Debrecen, Debrecen/Ungarische VR (20. Mai 1983).

Dr. Schmauder, Hans-Peter, Dr. Kranz, Prof. Dr. Detlef Gröger und Prof. Dr. Helmut Koblitz: Biotechnologische Aspekte der Kultur pflanzlicher Zellen. – Poster anläßlich der IV. Allunionskonferenz über pflanzliche Zellkultur und Biotechnologie, Kišinev/UdSSR, 3. bis 6. Oktober 1983.

Dr. Schöneich, Jörg: Genetic hazards to man from chemical mutagens. – 10. Jahrestagung der Ungarischen Gesellschaft für Humangenetik, Budapest/Ungarische VR, 25. Januar bis 1. Februar 1983 (26. Januar 1983).

– Kann die Embryonalentwicklung der Säugetiere genetisch manipuliert werden? – Kulturbund der DDR, Quedlinburg (1. März 1983).

– Prüfung von Arzneimitteln auf Mutagenität. – Lehrgang der Akademie für Gesundheits- und Sozialwesen des Bezirks Gera, Gera (17. März 1983).

– Genetic manipulation of the mammalian embryo. – Tagung der Ungarischen Gesellschaft für Humangenetik "Perspectives in medical genetics", Szeged/Ungarische VR, 4. bis 7. Juli 1983 (4. Juli 1983).

– Möglichkeiten zur Übertragung genetischer Information in die Keimbahn von Säugetieren. – Sitzung der Klasse Biowissenschaften der Akademie der Wissenschaften der DDR, Berlin (20. Oktober 1983).

– Genetische Manipulation der Säuger. – Tagung der Forschungsrichtung „Genetische Defekte" der HFR M 30 in Zusammenarbeit mit der Gesellschaft für Humangenetik der DDR, Neubrandenburg, 14. bis 16. November 1983 (14. November 1983).

– Möglichkeiten der genetischen Manipulation der Säuger. – Kolloquium der Interdisziplinären Arbeitsgemeinschaft Genetik der Martin-Luther-Universität Halle – Wittenberg, Halle (30. November 1983).

– and Dr. Rolf Braun (vorgetragen von Dr. J. Schöneich): Effect of pharmacokinetics on mutagenicity of drugs in mammals. – Symposium of the Institute of Human Genetics, Polish Academy of Sciences, "Current problems of environmental mutagenesis", Poznań/VR Polen, 8. bis 11. November 1983 (9. November 1983).

Dr. Scholz, Günter and Dr. Gerda Schlesier: The effect of nicotianamine upon the iron absorption by plants. – Poster presented at the International Symposium on Membrane Transport in Plants, Praha/ČSSR, 15. bis 21. August 1983 (20. August 1983).

Dr. Schubert, Ingo: Eine Methode zur gezielten Gewinnung aneuploider *Vicia faba* Karyotypen. – Weiterbildungsveranstaltung der Sektion Genetik der Biologischen Gesellschaft der DDR und des Instituts für Getreideforschung der Akademie der Landwirtschaftswissenschaften der DDR, Bernburg, 2. bis 3. März 1983 (2. März 1983).

Dr. SCHUBERT, INGO: Specific reactions of plant NORs on mutagen treatment. – 16th Annual Meeting of the Cytogenetic Section, Czechoslovak Biological Society, Bratislava/ČSSR, 23. bis 25. August 1983 (24. August 1983).
– and Dr. PETER DÖBEL: BrdU labelling method for measuring replication cycles of plant chromosomes. – Poster presented at the 6th European Cell Cycle Workshop "Progress in cell cycle control", Praha/ČSSR, 4. bis 8. April 1983 (5. April 1983).
Dr. SERFLING, EDGAR: Manipulation der Wachstumshormon-cDNA des Rindes zur effektiven Expression *in vivo*. – 1. Symposium der Arbeitsgemeinschaft Molekularbiologie der Biochemischen Gesellschaft der DDR, Reinhardsbrunn, 4. bis 8. Oktober 1983 (4. Oktober 1983).
– Genomstruktur und Genexpression. – 11. Weiterbildungskurs des Zentralinstituts für Molekularbiologie der Akademie der Wissenschaften der DDR „Entwicklungsbiologie", Schönfeld, 25. Oktober bis 1. November 1983 (1. November 1983).
Dr. STEINBORN, GERHARD: Klonierung von *Bacillus licheniformis* alpha–Amylase. – Poster anläßlich der 6. Vortragstagung und Hauptversammlung der Gesellschaft für Allgemeine und Technische Mikrobiologie der Biologischen Gesellschaft der DDR, Reinhardsbrunn, 9. bis 13. Oktober 1983 (12. Oktober 1983).
Dr. STEPHAN, UDO W.: The influence of nicotianamine, an unusual amino acid, on the iron uptake by cell cultures of *Lycopersicon esculentum*, mutant '*chloronerva*'. – Poster presented at the 2nd International Symposium on Mineral Nutrition of Plants, Varna/VR Bulgarien, 18. bis 24. September 1983 (19. September 1983).
Dr. SÜSS, KARL-HEINZ: Organisation und Funktion der Untereinheiten H^+-transportierender ATP Synthetase energieumwandelnder Membranen. – Kolloquium an der Sektion Biowissenschaften der Humboldt-Universität zu Berlin, Berlin (15. Februar 1983).
– Molecular action of heat on the biosynthesis and incorporation of chloroplast components. – Kolloquium am Institute of Plant Physiology, Sofia/VR Bulgarien (18. Mai 1983).
– On the molecular composition and arrangement of subunits in the chloroplast H^+-translocating ATP synthetase (CF_1–CF_0). – 3rd International Youth Symposium "Plant Metabolism Regulation", Varna/VR Bulgarien, 24. bis 29. September 1983 (25. September 1983).
Dipl.-Biol. SÜSS, MARINA and Dr. KARL-HEINZ SÜSS: On the molecular mechanism of the action of heavy metals on plants. – Poster presented at the 3rd International Youth Symposium "Plant Metabolism Regulation", Varna/VR Bulgarien, 24. bis 29. September 1983 (26. September 1983).
Dr. TICHÁ, INGRID, Dr. JIŘI ČATSKÝ, Dr. MARTIN PEISKER and MIROSLAV KAŠE: The ontogenetic pattern of leaf photosynthesis as affected by irradiance, carbon dioxide concentration and temperature. – Poster presented at the 6th International Congress on Photosynthesis, Brüssel/Belgien, 1. bis 6. August 1983 (2. August 1983).
Dr. ULBRICH, RENATE, Dr. JOHANNA MANSFELD and Prof. Dr. ALFRED SCHELLENBERGER: Thermostability of enzyme-matrix-systems. – Poster presented at the 3rd Symposium of Socialist Countries on Biotechnology, Bratislava/ČSSR, 25. bis 29. April 1983.
– and Prof. Dr. ALFRED SCHELLENBERGER: Thermoinactivation of immobilized enzymes. – Poster anläßlich des 2. Sommerkurses der Ingenieur-Hochschule Köthen ,,Enzyme engineering", Köthen, 17. bis 24. Juli 1983.
Dipl.-Biol. WALTER, GERALD: H-Y-Antigen und Geschlechtsdetermination. – Tagung der Arbeitsgemeinschaft Pädiatrische Endokrinologie der Gesellschaft für Pädiatrie und der Gesellschaft für Endokrinologie und Stoffwechselkrankheiten der DDR „Intersexualität im Kindesalter", Magdeburg, 1. bis 3. März 1983 (1. März 1983).
– Genetische Manipulation von Säugerembryonen. – 11. Weiterbildungskurs des Zentralinstituts für Molekularbiologie der Akademie der Wissenschaften der DDR „Entwicklungsbiologie", Schönfeld, 25. Oktober bis 1. November 1983 (30. Oktober 1983).
Dr. WOBUS, ANNA M.: *In vivo* and *in vitro* differentiation of cultivated embryonal cells from mouse blastocysts. – Kongress der Europäischen Gesellschaft für Gewebekultur (ETCS) und der Europäischen Gesellschaft des Retikuloendothelialen Systems (EURES), Budapest/Ungarische VR, 9. bis 13. Mai 1983 (12. Mai 1983).
– Entwicklungsbiologische Möglichkeiten der genetischen Manipulation am Säuger. –

Kolloquium der Gesellschaft für Zell- und Gewebezüchtung der DDR „Genmutationen somatischer Zellen", Greifswald (30. Mai 1983).

– Teratokarzinome-Modell für Differenzierung und Embryogenese. – 11. Weiterbildungskurs des Zentralinstituts für Molekularbiologie der Akademie der Wissenschaften der DDR „Entwicklungsbiologie", Schönfeld, 25. Oktober bis 1. November 1983 (30. Oktober 1983).

Dr. Wobus, Ulrich: Klonierung und Charakterisierung von Samenprotein-Genen der Ackerbohne. – Kolloquium an der Sektion Biologie der Ernst-Moritz-Arndt-Universität Greifswald, Greifswald (16. September 1983).

–, Dr. Helmut Bäumlein, Dr. Ronald Bassüner, Dipl.-Chem. Ute Heim und Dr. Wilfriede Weschke (vorgetragen von Dr. U. Wobus): Molekulare Charakterisierung von Samenprotein-Genen der Ackerbohne *Vicia faba*. – 1. Symposium der Arbeitsgemeinschaft Molekularbiologie der Biochemischen Gesellschaft der DDR, Reinhardsbrunn, 4. bis 8. Oktober 1983 (4. Oktober 1983).

Dr. Zacharias, Martin und Dr. Friedrich Scholz (vorgetragen von Dr. M. Zacharias): Nutzung von Mutationen in der Pflanzenzüchtung. – Beratung der Arbeitsgruppe Grundlagen der Züchtungsforschung, Sektion Züchtungsforschung bei Pflanzen, Akademie der Landwirtschaftswissenschaften der DDR, Berlin (27. April 1983).

E. Tagungen im Institut

7. Arbeitstagung „Mikrobengenetik 1983" des Zentralinstituts für Genetik und Kulturpflanzenforschung Gatersleben, der Interessengemeinschaft Mikrobengenetik der Gesellschaft für Allgemeine und Technische Mikrobiologie und der Sektion Genetik der Biologischen Gesellschaft der DDR in Elbingerode („Haus Büchenberg") vom 21. bis 25. März 1983

Vorträge

Adler, B. und J. Hofemeister, Gatersleben/DDR: Plasmid-R46 abhängige Lückenreparatur ist fehlerfrei.

Bäumlein, H., Gatersleben/DDR: Anlage und Nutzung von Genbanken.

Baldauf, F. und A. Kiss (vorgetragen von F. Baldauf), Halle/DDR und Szeged/Ungarische VR: Molekulare Klonierung und Expression eines Modifikationsmethylase-Gens von *Bacillus subtilis*.

Banfalvi, G., Budapest/Ungarische VR: RNA-linked DNA-fragments synthesized in permeable cells of *Bacillus subtilis*.

Barth, G., Jena/DDR: Stand der rekombinanten DNA-Forschung an Hefen.

Becher, D., Greifswald/DDR: Kerntransfer bei Hefen.

Borris, R., Gatersleben/DDR: Genetik der β-Glucanasebildung bei *Bacillus subtilis*.

Brandt, R., H. J. Schelenz, E. Günther und H. Herrmann (vorgetragen von E. Günther), Greifswald/DDR: Mapping in *Pseudomonas aeruginosa*.

Breznovits, A., Budapest/Ungarische VR: Genetic mapping of 16–3 phage of *R. meliloti*.

Cullum, J., Manchester/Großbritannien: Transposable elements.

– *Streptomyces* research.

Dünnhaupt, K. und W. Witte (vorgetragen von K. Dünnhaupt), Wernigerode/DDR; Plasmidmuster mehrfach resistenter *Staphylococcus aureus* Hospitalstämme.

Eitner, G., Gatersleben/DDR: Das recA-Gen aus *Proteus mirabilis* in *Bacillus subtilis* rec-Mutanten.

Herrmann, H. und M. Roy (vorgetragen von H. Herrmann), Greifswald/DDR: Characterization of the plasmid pHR 1 involved in the degradation of lower N-alkanes.

Hofemeister, J., Gatersleben/DDR: Interaktion von Plasmid pE 194 mit dem Chromosom in *Bacillus subtilis*.

Jacob, H. E., X. X. Shivarova und W. Förster (vorgetragen von H. E. Jacob), Jena/DDR und Praha/ČSSR: Transformation von *Bacillus* mit Plasmid-DNA unter Wirkung des elektrischen Feldes.

Klaus, S., J. Meyer und J. Störl (vorgetragen von S. Klaus), Jena/DDR: DNA des *Streptomyces*-Phagen SH-10: Bindungsorte für *Streptomyces* RNA-Polymerase und *in vitro* Transkriptionsrate.

Kruegel, H., Jena/DDR: Klonierungsstrategien für Streptokokken.

Krüger, D., M. Reuter, H. A. Rosenthal, T. A. Bickle und W. Arber (vorgetragen von D. Krüger), Berlin/DDR und Basel/Schweiz: Sequenzspezifische DNA-Methylierung und Endonukleolyse in *Escherichia coli* Zellen und deren Beeinflussung durch ein viruscodiertes Protein.

Lanzov, S. V., Leningrad/UdSSR: Enzymes of general recombination in *E. coli*.

Linde, K., Leipzig/DDR: Chromosomale Antibiotika-Resistenz: Zugang zur Target-spezifischen Stoffwechseldrift.

Piechocki, R. und G. Langhammer (vorgetragen von R. Piechocki), Halle/DDR: Mechanismus der Entstehung von Deletionen und Duplikationen.

Prager, R., Wernigerode/DDR: Charakterisierung von kryptischen Transferplasmiden und ihre Bedeutung für uropathogene *E. coli*-Stämme.

Quinones, A. und R. Piechocki (vorgetragen von A. Quinones), Halle/DDR: Genetische Charakterisierung von *E. coli* Antimutatoren.

Rosenthal, S., Berlin/DDR: Sequenz des Karpfen-Präproinsulingens und Entwicklung der Insulingene.

Tietze, G., Wernigerode/DDR: Charakterisierung von Bakterienisolaten durch ihre Plasmidmuster.

Tschäpe, H., Wernigerode/DDR: Charakterisierung der Inkompatibilität als Plasmid-spezies anzeigendes Merkmal.

Walter, F., M. Hartmann und S. Klaus (vorgetragen von F. Walter), Jena/DDR: Untersuchungen zur Struktur der DNA des *Streptomyces*-phagen SH-3 bzw. des SH-3 Prophagen.

Witte, W. und K. Dünnhaupt (vorgetragen von W. Witte), Wernigerode/DDR: Unterschiedliche Funktion von *Staphylococcus aureus*-Phagen bei der Übertragung von Resistenzdeterminanten.

Posterdemonstration

Barth, G., Jena/DDR: Isozitratlyase defekte Mutanten der Hefe *Saccharomycopsis lipolytica*.

Beer, J. und M. Schulze, Leipzig/DDR: Untersuchungen zum Mechanismus der Veränderung von Leistungsparametern bei ausgewählten Stoffwechseldrift-Mutanten.

Brantl, S. und J. Hofemeister, Gatersleben/DDR: Streptokokken-Plasmide in *Bacillus subtilis*.

Herrmann, H., Greifswald/DDR: Hfr in *Pseudomonas aeruginosa* PAO.

Hofemeister, J., Gatersleben/DDR: Rekombinationsorte im Chromosom von *B. subtilis* und Plasmid pE 194.

Jantschak, J., U. Nöthel und G. Nyakatura, Berlin/DDR: Isolierung integrierter Provirus-DNA des Bovine leucaemie-Virus.

Krüger, D., S. Hansen and L. S. Chernin, Berlin/DDR and Moskva/UdSSR: The abortive infection of F-plasmid-containing *Escherichia coli* cells by bacterial virus T 7 is determined by the right end of T 7 gene 1.

Kurischko, C., Jena/DDR: Genetic analysis of several markers in new breeding stocks of *Saccharomycopsis lipolytica*.

Reuter, M., J. G. Bogdarina, C. Schroeder, S. Hansen und D. H. Krüger, Berlin/DDR: Blockierung intrazellulärer DNA-Restriktionsendonukleasen durch die ocr$^+$-Genfunktion des Virus T 7 und Reinigung des ocr$^+$-Proteins.

Schmidt, G., J. Berrhold und D. H. Krüger, Berlin/DDR: Plaquebildung durch Bakterienviren auf nicht-permissivem Zellrasen.

SPANIER, E. und R. PIECHOCKI, Halle/DDR: Neue Methoden zur Isolierung von *E. coli*-Mutatoren.
STÖRL, K. and C. MUND, Jena/DDR: Effects of undermethylation of DNA in *E. coli*.
TIETZE, E., R. PRAGER and H. TSCHÄPE, Wernigerode/DDR: Characterization of the transposons Tn 1822 (Tc) and Tn 1824 (Tp Sm).
WONITZKI, C. und J. BEER, Leipzig/DDR: Virulenz-beeinflussende pleiotrope Effekte bei Rif^R Mutanten von *Salmonella typhimurium*.

Tag des Jungen Wissenschaftlers (25. Mai 1983)

Dipl.-Biol. U. CONRAD und Dipl.-Biochem. D. HELBING (vorgetragen von Dipl.-Biol. U. CONRAD), Gatersleben: Untersuchungen zur serologischen Kreuzreaktivität von Glucose-Phosphat-Isomerase-Alloenzymen der Maus.
Dr. R. JUNG, Gatersleben: Chemische Synthese der hypothetischen ancestralen Sequenz Ca-modulierter Eiweiße und analoger langkettiger Peptide mit potentieller Ca-Bindungsfähigkeit.
Dipl.-Biol. L. MEYER, Gatersleben: Charakterisierung larvaler Sekretproteine von *Chironomus thummi* zur Identifizierung der Balbianiring-Genprodukte.

3rd Symposium on Seed Proteins des Zentralinstiuts für Genetik und Kulturpflanzenforschung Gatersleben und der Sektion Physiologie und Biochemie der Pflanzen der Biologischen Gesellschaft der DDR ,,Genetics of Seed Proteins" in Gatersleben vom 31. August bis 2. September 1983.

Die Vorträge und Poster des Symposiums sind in dem vorliegenden Band der Schriftenreihe „Die Kulturpflanze", S. S 1–S 268, enthalten.

Workshop des Zentralinstituts für Genetik und Kulturpflanzenforschung Gatersleben ,,Structure and Evolution of Balbiani Ring Genes" in Altenbrak (Ferienheim „Windenhütte") vom 24. bis 28. Oktober 1983

Themenkomplex 1

,,Aspects of Balbiani ring gene fine structure and evolution at the sequence level": S. CASE, R. GALLER, C. HÖÖG, F. C. KAFATOS, U. LENDAHL, J. PUSTELL, L. WIESLANDER, U. WOBUS.

Themenkomplex 2

,,Aspects of overall structure and cytological organization of Balbiani rings": H. BÄUMLEIN, I. I. KIKNADSE, N. KOLEŠNIKOV, R. PANITZ, G. ZAINIEV.

Themenkomplex 3

,,Chromatin and RNP-particles": B. DANEHOLT, M. LEZZI.

Themenkomplex 4

,,Protein products of Balbiani ring genes": S. CASE, J.-E. EDSTRÖM, S. HAMODRAKAS, U. LENDAHL, E. SERFLING.

Sachregister

FSC
www.fsc.org
MIX
Papier | Fördert gute Waldnutzung
FSC® C083411

Kolibri 360 GmbH
Ferdinand-Jühlke-Straße 7
99095 Erfurt, Deutschland
produktsicherheit@kolibri360.de